MOBILE RADIO CHANNELS

MOBILE RADIO CHANNELS

Second Edition

Matthias Pätzold
University of Agder, Norway

WILEY
A John Wiley & Sons, Ltd., Publication

This edition first published 2012

Registered office
John Wiley & Sons Ltd, The Atrium, Southern Gate, Chichester, West Sussex, PO19 8SQ, United Kingdom

For details of our global editorial offices, for customer services and for information about how to apply for permission to reuse the copyright material in this book please see our website at www.wiley.com.

Library of Congress Cataloging-in-Publication Data

Pätzold, Matthias, professor of mobile communications.
Mobile radio channels / Matthias Pätzold. – 2nd ed.
p. cm.
Includes bibliographical references and index.
ISBN 978-0-470-51747-5 (cloth)
1. Mobile communication systems. 2. Radio wave propagation. 3. Radio resource management (Wireless communications). I. Title.
TK5103.2.P385 2012
621.3845–dc23

2011015882

A catalogue record for this book is available from the British Library.

Print ISBN: 978-0-470-51747-5
ePDF ISBN: 978-1-119-97412-3
oBook ISBN: 978-1-119-97411-6
ePub ISBN: 978-1-119-97525-0
Mobi ISBN: 978-1-119-97526-7

Typeset in 10/12pt Times by Aptara Inc., New Delhi, India
Printed and bound in Singapore by Markono Print Media Pte Ltd

Contents

Preface to the Second Edition

With this book at your fingertips, you, the reader, and I have something in common. We share the same interest in mobile radio channels. This area attracted my interest first in autumn 1992 when I moved from industry to academia to find a challenge in my life and to pursue a scientific career. Since then, I consider myself as a student of the mobile radio channel who lives for modelling, analyzing, and simulating them. While the first edition of this book resulted from my teaching and research activities at the Technical University of Hamburg-Harburg (TUHH), Germany, the present second edition is entirely an outcome of my work at the University of Agder, Norway.

To share my passion with the reader, my objective was to write a comprehensive book that begins with the basics and moves gradually to advanced research topics. This makes the present edition of interest to beginners and experts alike. It is addressed especially to mobile radio engineers, telecommunication engineers, and physicists working in industry or in research institutes in the rapidly growing wireless and mobile communications market. In addition to that, it is also suitable for experts who have a professional interest in subjects dealing with mobile radio channel issues. Last, but not least, this book is also addressed to Master's students specializing in mobile radio communications.

The study of this book assumes prior basic knowledge of statistics and systems theory. Master's students in general have sufficient background knowledge in these areas. To simplify comprehension, all fundamental mathematical tools, which are relevant for the objectives of this book, are recapitulated at the beginning. Starting from this basic knowledge, nearly all statements made in this book are derived in detail, so that a high degree of mathematical transparency and explicitness can be achieved. Thanks to sufficient guidance and help, it is guaranteed that the interested reader can verify the results with reasonable effort. Longer derivations interrupting the flow of the text have been relegated to the appendices. Much emphasis has been placed on the confirmation of the theoretical results by numerical simulations. To illustrate the theoretical and experimental results, a large number of figures have been included. Their multiple interpretations and meanings enrich the reader's understanding of the text. The use of abbreviations has in general been avoided, which in my experience eases readability considerably. Furthermore, a large number of references are provided leading the reader to further sources for the almost inexhaustible topic of mobile radio channel modelling. A huge set of selected MATLAB®-programs for simulating and analyzing the mobile radio channel models has been made available for free downloading at the companion website www.wiley.com/go/paetzold. They provide valuable support in simulating the channel models and give practical guidelines for applying the powerful analysis tools described in the book.

My aim was to introduce the reader to the fundamentals of modelling, analysis, and simulation of mobile fading channels. One of the main focusses of this book is the treatment of stochastic and deterministic sum-of-sinusoids processes. They establish the basis for the development of efficient channel simulators. For the design of sum-of-sinusoids processes with given correlation properties, nearly all methods known in the literature up to now are presented, analyzed, and assessed on their performance. The focus is also on the derivation and analysis of stochastic geometrical-based channel models as well as on the development of highly precise channel simulators for many classes of frequency-selective and -nonselective mobile radio channels for single-input single-output (SISO) as well as for multiple-input multiple-output (MIMO) systems. Another important topic covered in this book is the fitting of the statistical properties of the developed channel models to the statistics of real-world channels.

Chapter 1 begins with an overview of the evolution of mobile radio systems, continues with the basics of mobile radio channels, and ends with a description of the organization of the book.

Chapter 2 reviews the fundamentals of random variables, stochastic processes, and systems theory. This chapter introduces the most important definitions, terms, and formulas, which are often used throughout the book.

Chapter 3 builds on the terms introduced in the previous chapter and introduces Rayleigh and Rice processes as reference models for characterizing frequency-nonselective mobile radio channels.

Chapter 4 presents an introduction to sum-of-sinusoids processes from which high-performance channel simulators with low realization complexity will be derived. Another substantial part of this chapter deals with processes comprised of a sum of complex-valued sinusoids (cisoids). A sum-of-cisoids process allows a simple physical interpretation as a plane wave model. This makes such processes very attractive for the development of mobile radio channel models under real-world propagation conditions, where non-isotropic scattering becomes the norm.

Chapter 5 treats the parametrization of sum-of-sinusoids processes. It provides a comprehensive description and analysis of the most important procedures presently known for computing the model parameters of sum-of-sinusoids processes. The performance of each parameter computation method is assessed and their individual advantages and disadvantages are highlighted. This chapter also provides solutions to the parametrization of the plane wave models.

Chapter 6 is concerned with the development of frequency-nonselective channel models. A variety of sophisticated combined stochastic processes is introduced enabling the modelling of frequency-nonselective mobile radio channels. The usefulness of the presented channel models is demonstrated by fitting their statistical properties to measurement data. The final part of this chapter delves into the modelling of nonstationary land mobile satellite channels.

Chapter 7 is dedicated to the modelling, analysis, and simulation of frequency-selective channel models. The core of this chapter is devoted to Bello's theory of linear time-variant stochastic systems. Special attention will be paid to the so-called wide-sense stationary uncorrelated scattering (WSSUS) model. Another substantial part deals with the simulation of wideband channels. Furthermore, methods are introduced for the modelling of given power delay profiles. The last part of this chapter presents a general method for the modelling and simulation of measured wideband mobile radio channels.

Chapter 8 focusses on the modelling, analysis, and simulation of MIMO channels. Starting from specific geometrical scattering models, a universal technique is presented for the derivation of stochastic reference MIMO channel models under the assumption of isotropic and non-isotropic scattering. The proposed procedure provides an important framework for designers of advanced mobile communication systems to verify new transmission concepts employing MIMO techniques under realistic propagation conditions.

Chapter 9 deals with the derivation, analysis, and realization of high-speed channel simulators. It is shown how simulation models can be developed by just using adders, storage elements, and simple address generators. The proposed techniques for the design of high-speed channel simulators are suitable for all types of channel models presented in the previous chapters.

Chapter 10 concludes the book with three selected topics in mobile radio channel modelling. The first topic addresses the problem of designing multiple uncorrelated Rayleigh fading channels. The second is devoted to the modelling of shadow fading, and the last elaborates on the development of frequency hopping mobile radio channel models.

In the course of writing this book, I have had the pleasure to work with a number of PhD students on a variety of subjects that influenced the contents of the second edition in one way or the other. I am especially indebted to Dr. B. O. Hogstad, Dr. C. A. Gutierrez, Dr. B. Talha, Dr. G. Rafiq, and Ms. Y. Ma. I will never forget the great and fruitful time that we spent together pulling on the same end of the scientific rope. I am also grateful to Mark Hammond, the Editorial Director at John Wiley & Sons, as well as to Sarah Tilley and Susan Barclay, my Project Editors, for their constant support, ultimate patience, and for giving me all the freedom that I needed to publish this edition in its present form. Finally, my special thanks go to my wife Katharina. This book truly would not have been written without her steady encouragement and professional assistance. Of course, all errors are entirely mine.

Matthias Pätzold
Grimstad
March 2011

List of Acronyms

ACeS	Asian Cellular System
ACF	autocorrelation function
ADF	average duration of fades
AMPS	Advanced Mobile Phone System
AOA	angle-of-arrival
AOD	angle-of-departure
ARIB	Association of Radio Industries and Businesses
ATDMA	Advanced Time Division Multiple Access
AWGN	additive white Gaussian noise
B-ISDN	Broadband Integrated Services Digital Network
BMFT	Bundesministerium für Forschung und Technologie
BPSK	binary phase shift keying
BRAN	Broadband Radio Access Networks
BS	base station
BU	Bad Urban
CCF	cross-correlation function
CDF	cumulative distribution function
CDMA	code division multiple access
CEPT	Conference of European Postal and Telecommunications Administrations
CF	correlation function
CMOS	complementary metal oxide semiconductor
COST	European Cooperation in the Field of Scientific and Technical Research
D-AMPS	Digital Advanced Mobile Phone Service
DC	direct current
DCS	Digital Cellular System
DECT	Digital European Cordless Telephone
DLR	German Aerospace Center (German: Deutsches Zentrum für Luft- und Raumfahrt e.V.)
DPSK	differential phase shift keying
DS-CDMA	direct sequence code division multiple access
DSP	digital signal processor
EA	elevation angle
EDGE	Enhanced Data Rates for GSM Evolution
EMEDS	extended method of exact Doppler spread

ESA	European Space Agency
ETSI	European Telecommunications Standards Institute
FCF	frequency correlation function
FDD	frequency division duplex
FDMA	frequency division multiple access
FH	frequency hopping
FIR	finite impulse response
FM	frequency modulation
FPGA	field programmable gate array
FPLMTS	Future Public Land Mobile Telecommunications System
GaAs	gallium arsenide
GEO	geostationary earth orbit
GMEA	generalized method of equal areas
GMEDS_q	generalized method of exact Doppler spread
GSM	Global System for Mobile Communications (formerly: Groupe Spécial Mobile)
GWSSUS	Gaussian wide-sense stationary uncorrelated scattering
HEO	highly elliptical orbit
HF	high frequency
HIPERLAN	High Performance Radio Local Area Network
HIPERLAN/2	High Performance Radio Local Area Network Type 2
HT	Hilly Terrain
IEEE	Institute of Electrical and Electronics Engineers
IIR	infinite impulse response
IMT-2000	International Mobile Telecommunications 2000
INMARSAT	International Maritime Satellite Organization
IS-95	Interim Standard 95
ISI	intersymbol interference
ITU	International Telecommunications Union
JM	Jakes method
LAN	local area network
LCR	level-crossing rate
LDPC	low-density parity-check
LEO	low earth orbit
LMS	land mobile satellite
LOS	line-of-sight
LPNM	L_p-norm method
LTE	Long Term Evolution
MBS	Mobile Broadband System
MC-CDMA	multi-carrier code division multiple access
MCM	Monte Carlo method
MEA	method of equal areas
MED	method of equal distances
MEDS	method of exact Doppler spread
MEDS-SP	method of exact Doppler spread with set partitioning
MEO	medium earth orbit

MIMO	multiple-input multiple-output
MISO	multiple-input single-output
MMEA	modified method of equal areas
MMEDS	modified method of exact Doppler spread
MS	mobile station
MSEM	mean-square-error method
NGEO	non-geostationary earth orbit
NLOS	non-line-of-sight
NMT	Nordic Mobile Telephone
NTT	Nippon Telephone and Telegraph
NTTPC	Nippon Telephone and Telegraph Public Corporation
OFDM	orthogonal frequency division multiplexing
OFDMA	orthogonal frequency division multiple access
PCN	Personal Communications Network
PDC	Personal Digital Cellular
PDF	probability density function
PDP	power delay profile
PSD	power spectral density
QPSK	quadrature phase shift keying
RA	Rural Area
RACE	Research and Development in Advanced Communications Technologies in Europe
RMEDS	randomized method of exact Doppler spread
RMS	root mean square
RSM	Riemann sum method
RTM	Radio Telephone Mobile
RV	random variable
SAW	surface acoustic wave
SCM	spatial channel model
SIMO	single-input multiple-output
SISO	single-input single-output
SMS	short message services
SNR	signal-to-noise ratio
SOC	sum-of-cisoids
SOCUS	sum-of-cisoids uncorrelated scattering
SOS	sum-of-sinusoids
SOSUS	sum-of-sinusoids uncorrelated scattering
SUI	Stanford University Interim
TACS	Total Access Communication System
TDMA	time division multiple access
TU	Typical Urban
ULA	uniform linear array
UMTS	Universal Mobile Telecommunications System
US	uncorrelated scattering
UTRA	UMTS Terrestrial Radio Access
UWC-136	Universal Wireless Communications 136

VLSI	very large scale integration
WARC	World Administration Radio Conference
WCDMA	wideband code division multiple access
WGN	white Gaussian noise
WiMAX	Worldwide Interoperability for Microwave Access
WLAN	Wireless Local Area Network
WSS	wide-sense stationary
WSSUS	wide-sense stationary uncorrelated scattering
2D, 3D	two-, three-dimensional
1G, 2G	1st, 2nd generation
3G, 4G	3rd, 4th generation
3GPP	3rd Generation Partnership Project
3GPP2	3rd Generation Partnership Project 2

List of Symbols

Set Theory

$\mathbb{C}$	set of complex numbers
$\mathbb{N}$	set of natural numbers
$\mathbb{R}$	set of real numbers
$\mathbb{Z}$	set of integer numbers
$\in$	is an element of
$\notin$	is not an element of
$\forall$	for all
$\subset$	subset
$\cup$	union
$\cap$	intersection
$A \backslash B$	difference of set A and set B
$\emptyset$	empty set or null set
$[a, b]$	set of real numbers within the closed interval from a to b, i.e., $[a, b] = \{x \in \mathbb{R} \mid a \leq x \leq b\}$
$[a, b)$	set of real numbers within the right-hand side open interval from a to b, i.e., $[a, b) = \{x \in \mathbb{R} \mid a \leq x < b\}$
$(a, b]$	set of real numbers within the left-hand side open interval from a to b, i.e., $(a, b] = \{x \in \mathbb{R} \mid a < x \leq b\}$
$\{x_n\}_{n=1}^N$	set of elements $x_1, x_2, \ldots, x_N$

Operators and Miscellaneous Symbols

$\arg\{x\}$	argument of $x = x_1 + jx_2$
$\text{Cov}\{x_1, x_2\}$	covariance of x_1 and x_2
e^x	exponential function
$\exp\{x\}$	exponential function
$E\{x\}$	(statistical) mean value or expected value of x
$\mathcal{F}\{x(t)\}$	Fourier transform of $x(t)$
$\mathcal{F}^{-1}\{X(f)\}$	inverse Fourier transform of $X(f)$
$\gcd\{x_n\}_{n=1}^N$	greatest common divisor (also known as highest common factor) of $x_1, x_2, \ldots, x_N$
$\mathcal{H}\{x(t)\}$	Hilbert transform of $x(t)$

$\mathcal{H}^{-1}\{X(f)\}$	inverse Hilbert transform of $X(f)$
$\mathrm{Im}\{x\}$	imaginary part of $x = x_1 + jx_2$
$\mathrm{lcm}\{x_n\}_{n=1}^{N}$	least common multiple of $x_1, x_2, \ldots, x_N$
lim	limit
$\ln x$	natural logarithm of x
$\log_a x$	logarithm of x to base a
$\max\{x_n\}_{n=1}^{N}$	largest element of the set $\{x_1, x_2, \ldots, x_N\}$
$\min\{x_n\}_{n=1}^{N}$	smallest element of the set $\{x_1, x_2, \ldots, x_N\}$
mod	modulo operation
$n!$	factorial function
$P\{\mu \leq x\}$	probability that the event μ is less than or equal to x
$\mathrm{Re}\{x\}$	real part of $x = x_1 + jx_2$
$\mathrm{round}\{x\}$	nearest integer to the real-valued number x
$\mathrm{sgn}\,(x)$	sign of the number x: 1 if $x > 0$, 0 if $x = 0$, -1 if $x < 0$
$\mathrm{Var}\{x\}$	variance of x
$x_1(t) * x_2(t)$	convolution of $x_1(t)$ and $x_2(t)$
x^*	complex conjugate of the complex number $x = x_1 + jx_2$
$\lvert x \rvert$	absolute value of x
$\sqrt{x}$	principal value of the square root of x, i.e., $\sqrt{x} \geq 0$ for $x \geq 0$
$\prod_{n=1}^{N}$	multiple product
$\sum_{n=1}^{N}$	multiple sum
$\int_a^b x(t)dt$	integral of the function $x(t)$ over the interval $[a, b]$
$< x(t) >$	time average operator, i.e., $< x(t) > := \lim_{T\to\infty} \frac{1}{2T} \int_{-T}^{T} x(t)\, dt$
$\dot{x}(t)$	derivative of the function $x(t)$ with respect to time t
$\check{x}(t)$	Hilbert transform of $x(t)$
$x \to a$	x tends to a or x approaches a
$\lceil x \rceil$	ceiling function, the smallest integer greater than or equal to x
$\lfloor x \rfloor$	floor function, the greatest integer less than or equal to x
$\approx$	approximately equal
$\sim$	distributed according to (statistics) or asymptotically equal (analysis)
$\leq$	less than or equal to
$\ll$	much less than
$=$	equal
$\neq$	unequal
$\circ\!\!-\!\!\bullet$	Fourier transform

Matrices and Vectors

$(a_{m,n})$	matrix with $a_{m,n}$ as the entry of the mth row and the nth column
$\boldsymbol{A}^H$	Hermitian transpose (or conjugate transpose) of the matrix $\boldsymbol{A}$
$\boldsymbol{A}^T$	transpose matrix of the matrix $\boldsymbol{A}$
$\boldsymbol{A}^{-1}$	inverse matrix of the matrix $\boldsymbol{A}$
$\boldsymbol{C}_{\mu_\rho}$	covariance matrix of the vector process $\boldsymbol{\mu}_\rho(t) = (\mu_{\rho_1}(t), \mu_{\rho_2}(t), \dot{\mu}_{\rho_1}(t), \dot{\mu}_{\rho_2}(t))^T$
$\det \boldsymbol{A}$	determinant of the matrix $\boldsymbol{A}$

J	Jacobian determinant
$\boldsymbol{m}$	column vector of $m_1, m_2, \dot{m}_1$, and $\dot{m}_2$, i.e., $\boldsymbol{m} = (m_1, m_2, \dot{m}_1, \dot{m}_2)^T$
$\boldsymbol{R}_\mu$	autocorrelation matrix of the vector process $\boldsymbol{\mu}(t) = (\mu_1(t), \mu_2(t), \dot{\mu}_1(t), \dot{\mu}_2(t))^T$
$\mathrm{tr}(\boldsymbol{A})$	trace of the matrix $\boldsymbol{A} = (a_{m,n}) \in \mathbb{R}^{N \times N}$, i.e., $\mathrm{tr}(\boldsymbol{A}) = \sum_{n=1}^{N} a_{n,n}$
$\boldsymbol{x}$	column vector of $x_1, x_2, \dot{x}_1$, and $\dot{x}_2$, i.e., $\boldsymbol{x} = (x_1, x_2, \dot{x}_1, \dot{x}_2)^T$
$\boldsymbol{\Omega}$	parameter vector

Special Functions

$\mathrm{erf}\,(\cdot)$	error function
$\mathrm{erf}^{-1}(\cdot)$	inverse error function
$\mathrm{erfc}\,(\cdot)$	complementary error function
$E(\cdot, \cdot)$	elliptic integral of the second kind
$\boldsymbol{E}(\cdot)$	complete elliptic integral of the second kind
$F(\cdot, \cdot; \cdot; \cdot)$	hypergeometric function (series)
${}_1F_1(\cdot; \cdot; \cdot)$	generalized hypergeometric function (series)
$H_0(\cdot)$	Struve function of order zero
$I_\nu(\cdot)$	modified Bessel function of the first kind of order ν
$J_\nu(\cdot)$	Bessel function of the first kind of order ν
$\mathbf{K}(\cdot)$	complete elliptic integral of the first kind
$K_0(\cdot)$	modified Bessel function
$P_\nu(\cdot)$	Legendre function of the first kind of order ν
$Q(\cdot, \cdot)$	Marcum Q-function
$Q_m(\cdot, \cdot)$	generalized Marcum Q-function
$\mathrm{rect}\,(\cdot)$	rectangular function
$\mathrm{sgn}\,(\cdot)$	sign function
$\mathrm{sinc}\,(\cdot)$	sinc function
$\delta(\cdot)$	Dirac delta function
$\gamma(\cdot, \cdot)$	incomplete gamma function
$\Gamma(\cdot)$	gamma function

Stochastic Processes

a	semi-major axis length of an ellipse
A_k^R	kth receiver antenna element
A_l^T	lth transmitter antenna element
A_1	time share factor
b	semi-minor axis length of an ellipse
B	signal bandwidth or system bandwidth
B_C	coherence bandwidth
$B_{\mu_i\mu_i}^{(1)}$	average Doppler shift of $\mu_i(t)$
$B_{\mu_i\mu_i}^{(2)}$	Doppler spread of $\mu_i(t)$
$B_{\tau'}^{(1)}$	average delay
$B_{\tau'}^{(2)}$	delay spread

c_R	Rice factor
c_0	speed of light
$\mathcal{C}$	number of clusters of scatterers
D	distance from the transmitter to the receiver
D_c	decorrelation distance
D_n	total length of the nth path
E_n	path gain introduced by the nth scatterer S_n
$E_2(\mathbf{\Omega})$	mean-square-error norm
f	Doppler frequency
f_c	cut-off frequency
$f_{\max}$	maximum Doppler frequency
$f_{\max}^R$	maximum Doppler frequency caused by the movement of the receiver
$f_{\max}^T$	maximum Doppler frequency caused by the movement of the transmitter
$f_{\min}$	lower cut-off frequency of the left-sided restricted Jakes power spectral density
f_n	Doppler frequency of the nth plane wave
f_n^R	Doppler frequency of the nth plane wave caused by the movement of the receiver
f_n^T	Doppler frequency of the nth plane wave caused by the movement of the transmitter
f_s	sampling rate (or sampling frequency)
f_{sym}	symbol rate
f_ρ	Doppler frequency of the line-of-sight component $m(t)$
f_0	carrier frequency
Δf	Doppler frequency resolution
$F_w(r)$	cumulative distribution function of Weibull processes w (t)
$F_{\zeta_-}(r)$	cumulative distribution function of Rayleigh processes $\zeta(t)$
$F_{\eta_-}(r)$	cumulative distribution function of Suzuki processes $\eta(t)$
$F_{\eta_+}(r)$	complementary cumulative distribution function of Suzuki processes $\eta(t)$
$F_\vartheta(\varphi)$	cumulative distribution function of the phase $\vartheta(t)$ of $\mu(t) = \mu_1(t) + j\mu_2(t)$
$F_{\lambda_-}(r)$	cumulative distribution function of spatial lognormal processes $\lambda(x)$
$\hat{F}_{\lambda_-}(r)$	cumulative distribution function of spatial stochastic SOS lognormal processes $\hat{\lambda}(x)$
$F_{\mu_i}(r)$	cumulative distribution function of Gaussian random processes $\mu_i(t)$
$F_{\xi_-}(r)$	cumulative distribution function of Rice processes $\xi(t)$
$F_{\xi_+}(r)$	complementary cumulative distribution function of Rice processes $\xi(t)$
$F_{\varrho_-}(r)$	cumulative distribution function of Loo processes $\varrho(t)$
$F_{\varrho_+}(r)$	complementary cumulative distribution function of Loo processes $\varrho(t)$
$F_\omega(r)$	cumulative distribution function of Nakagami-m processes $\omega(t)$
$h(\tau')$	time-invariant impulse response
$h(\tau', t)$	time-variant impulse response
$h_{kl}(t)$	complex channel gain (reference model) of the link from A_l^T to A_k^R
$\hat{h}_{kl}(t)$	complex channel gain (stochastic simulation model) of the link from A_l^T to A_k^R
$h_{kl}(\tau', t)$	time-variant impulse response of the link from A_l^T to A_k^R

$H(f)$	transfer function of linear time-invariant systems
$H(f', t)$	time-variant transfer function
$H_{kl}(f', t)$	time-variant transfer function of the link from A_l^T to A_k^R
$\check{H}(f)$	Hilbert transformer
$\boldsymbol{H}(t)$	channel matrix (reference model)
$\hat{\boldsymbol{H}}(t)$	channel matrix (stochastic simulation model)
$\mathcal{J}$	number of observed transitions from state S_1 to S_2 and vice versa
$\vec{k}_m^T$	wave vector pointing in the direction of the mth transmitted plane wave
$\vec{k}_n^R$	wave vector pointing in the direction of the nth received plane wave
k_0	free-space wave number
K	number of measured signal samples
$\mathcal{L}$	number of discrete paths
$\mathcal{L}^{\star}$	number of measured discrete paths
$m(t)$	(time-variant) line-of-sight component
m_L	area mean
m_s'	sampling rate ratio, i.e., $m_s' = f_s'/f_s = T_s/T_s'$
m_α	mean direction of the angle-of-arrival
m_{μ_i}	mean value of $\mu_i(t)$
M	number of channel states characterizing the M-state Markov model
M_r	number of signal levels at which measurements were taken
M_R	number of receiver antenna elements
M_T	number of transmitter antenna elements
$n(t)$	additive white Gaussian noise
$\boldsymbol{n}(t)$	noise vector process
N	number of propagation paths (number of local scatterers)
$N_\zeta(r)$	level-crossing rate of Rayleigh processes $\zeta(t)$
$N_\eta(r)$	level-crossing rate of Suzuki processes $\eta(t)$
$N_\lambda(r)$	level-crossing rate of spatial lognormal processes $\lambda(x)$
$\hat{N}_\lambda(r)$	level-crossing rate of spatial stochastic SOS lognormal processes $\hat{\lambda}(x)$
$N_\xi(r)$	level-crossing rate of Rice processes $\xi(t)$
$N_\varrho(r)$	level-crossing rate of Loo processes $\varrho(t)$
N_0	noise power
$p_{0_-}(\tau_-; r)$	probability density function of the fading intervals τ_- of Rayleigh processes $\zeta(t)$
$p_{1_-}(\tau_-; r)$	approximate solution for $p_{0_-}(\tau_-; r)$
p_{ij}	transition probabilities of Markov chains
$p_{Lutz}(z)$	probability density function of the envelope of Lutz's model
$p_w(x)$	Weibull distribution
$p_\alpha(\alpha)$	distribution of the angle-of-arrival α
$p_\zeta(z)$	Rayleigh distribution
$p_\eta(z)$	Suzuki distribution
$p_\vartheta(\theta)$	probability density function of the phase $\vartheta(t)$
$p_\lambda(z)$	lognormal distribution
$\hat{p}_\lambda(z)$	probability density function of spatial stochastic SOS lognormal processes $\hat{\lambda}(x)$

$p_{\mu_i}(x)$	Gaussian distribution
$p_{\mu_{\rho_1}\mu_{\rho_2}\dot{\mu}_{\rho_1}\dot{\mu}_{\rho_2}}$	joint probability density function of $\mu_{\rho_1}(t)$, $\mu_{\rho_2}(t)$, $\dot{\mu}_{\rho_1}(t)$, and $\dot{\mu}_{\rho_2}(t)$
$\hat{p}_\nu(x)$	probability density function of spatial stochastic SOS processes $\hat{\nu}(x)$
$p_\xi(z)$	Rice distribution
$p_\varrho(z)$	probability density function of Loo processes $\varrho(t)$
$p_{\tau'}(\tau')$	probability density function of the propagation delays τ'
$p_\omega(z)$	Nakagami$-m$ distribution
$p_{\xi\dot{\xi}}(z,\theta)$	joint probability density function of $\xi(t)$ and $\dot{\xi}(t)$
$p_{\xi\dot{\xi}\vartheta\dot{\vartheta}}(z,\dot{z},\theta,\dot{\theta})$	joint probability density function of $\xi(t)$, $\dot{\xi}(t)$, $\vartheta(t)$, and $\dot{\vartheta}(t)$
$P_S(f)$	cumulative power function
Q	sample space or certain event
$Q(\cdot,\cdot)$	Marcum Q-function
$Q_m(\cdot,\cdot)$	generalized Marcum Q-function
r	signal level
$\boldsymbol{r}(t)$	receive signal vector
$r_{h_{kl}}(\tau)$	time autocorrelation function of $h_{kl}(t)$
$\hat{r}_{h_{kl}}(\tau)$	time autocorrelation function of $\hat{h}_{kl}(t)$
$r_k(t)$	complex envelope of the signal received at the antenna element A_k^R $(k = 1, 2, \ldots, M_R)$
$\vec{r}_R$	spatial translation vector at the receiver side
$\vec{r}_T$	spatial translation vector at the transmitter side
$r_{hh}(\cdot,\cdot;\cdot,\cdot)$	autocorrelation function of $h(\tau',t)$
$r_{HH}(\upsilon',\tau)$	time-frequency correlation function of WSSUS models
$r_{HH}(\cdot,\cdot;\cdot,\cdot)$	autocorrelation function of $H(f',t)$
$r_{ss}(\cdot,\cdot;\cdot,\cdot)$	autocorrelation function of $s(\tau',f)$
$r_{TT}(\cdot,\cdot;\cdot,\cdot)$	autocorrelation function of $T(f',f)$
$r_{xx}(t_1,t_2)$	autocorrelation function of $x(t)$, i.e., $r_{xx}(t_1,t_2) = E\{x^*(t_1)\,x(t_2)\}$
$r_{yy}(t_1,t_2)$	autocorrelation function of $y(t)$, i.e., $r_{yy}(t_1,t_2) = E\{y^*(t_1)\,y(t_2)\}$
$r^c_{\zeta\zeta}(\cdot)$	autocovariance function of $\zeta(t)$
$r_{\lambda\lambda}(\Delta x)$	spatial autocorrelation function of $\lambda(x)$
$\hat{r}_{\lambda\lambda}(\Delta x)$	spatial autocorrelation function of $\hat{\lambda}(x)$
$r_{\mu\mu}(\tau)$	autocorrelation function of $\mu(t) = \mu_1(t) + j\mu_2(t)$
$r_{\mu_i\mu_i}(\tau)$	autocorrelation function of $\mu_i(t)$
$\hat{r}_{\mu_i\mu_i}(\tau)$	autocorrelation function of $\hat{\mu}_i(t)$
$r_{\mu_1\mu_2}(\tau)$	cross-correlation function of $\mu_1(t)$ and $\mu_2(t)$
$r_{\mu\mu'}(\tau,\chi)$	time-frequency cross-correlation function of $\mu(t)$ and $\mu'(t)$
$r_{\mu_i\mu_i'}(\tau,\chi)$	time-frequency cross-correlation function of $\mu_i(t)$ and $\mu_i'(t)$
$r_{\mu_1\mu_2'}(\tau,\chi)$	time-frequency cross-correlation function of $\mu_1(t)$ and $\mu_2'(t)$
$r^c_{\mu_1\mu_2}(\tau)$	cross-covariance function of $\mu_1(t)$ and $\mu_2(t)$
$r_{\nu\nu}(\Delta x)$	spatial autocorrelation function of $\nu(x)$
$\hat{r}_{\nu\nu}(\Delta x)$	spatial autocorrelation function of $\hat{\nu}(x)$
$r^\star_{\nu\nu}(\Delta x)$	measured spatial autocorrelation function
$r_{\tau'}(\upsilon')$	frequency correlation function
R	ring radius
R_R	radius of the ring of scatterers around the receiver

R_T	radius of the ring of scatterers around the transmitter
$s(\tau',f)$	Doppler-variant impulse response
$\mathbf{s}(t)$	transmit signal vector
$s_l(t)$	complex envelope of the signal transmitted from the antenna element A_l^T $(l=1,2,\ldots,M_T)$
s_n	spatial frequencies
$\mathbf{S}^{\star}$	measured scattering matrix
$S(\tau',f)$	scattering function of WSSUS models
$\mathbf{S}^{\star}(\tau',f)$	measured scattering function
S_n	nth local scatterer
S_n^R	nth local scatterer around the receiver
S_n^T	nth local scatterer around the transmitter
$S_{hh}(\tau',\tau)$	delay cross-power spectral density of WSSUS models
$S_{TT}(\upsilon',f)$	Doppler cross-power spectral density of WSSUS models
$S_{\zeta\zeta}^c(f)$	autocovariance power spectral density of $\zeta(t)$
$S_{\mu\mu}(f)$	power spectral density of $\mu(t)=\mu_1(t)+j\mu_2(t)$
$S_{\mu_i\mu_i}(f)$	power spectral density of $\mu_i(t)$
$S_{\mu_1\mu_2}(f)$	cross-power spectral density of $\mu_1(t)$ and $\mu_2(t)$
$S_{\mu_1\mu_2}^c(f)$	cross-covariance power spectral density of $\mu_1(t)$ and $\mu_2(t)$
$S_{\nu\nu}(s)$	power spectral density of spatial Gaussian processes $\nu(x)$
$S_{\tau'}(\tau')$	delay power spectral density (or power delay profile)
t	time variable
$T(f',f)$	Doppler-variant transfer function
T_C	coherence time
T_s	sampling interval
$T_s^{\star}$	sampling interval of measured signals
T_{sym}	symbol interval (or symbol duration)
$T_{\zeta_-}(r)$	average duration of fades of Rayleigh processes $\zeta(t)$
$T_{\eta_-}(r)$	average duration of fades of Suzuki processes $\eta(t)$
$T_{\lambda_-}(r)$	average duration of fades of spatial lognormal processes $\lambda(x)$
$T_{\xi_-}(r)$	average duration of fades of Rice processes $\xi(t)$
$T_{\varrho_-}(r)$	average duration of fades of Loo processes $\varrho(t)$
$u_{i,n}$	uniformly distributed random variable
v	speed of the mobile unit
v_R	speed of the mobile receiver
v_T	speed of the mobile transmitter
$\vec{\mathrm{v}}$	mobile speed vector
$\vec{\mathrm{v}}_R$	mobile speed vector of the receiver
$\vec{\mathrm{v}}_T$	mobile speed vector of the transmitter
$\mathrm{w}(t)$	Weibull process
$W_i(\cdot)$	weighting function
$x(t)$	input signal
$x_{BP}(t)$	bandpass signal
$x_{LP}(t)$	equivalent lowpass signal
$y(t)$	output signal
$\alpha_{\max}^T$	maximum of the angle-of-departure

α_n	angle-of-arrival of the nth plane wave (path)
α_n^R	angle-of-arrival of the nth plane wave seen from the receiver
α_n^T	angle-of-departure of the nth plane wave seen from the transmitter
α_v	angle-of-motion
α_v^R	angle-of-motion of the receiver
α_v^T	angle-of-motion of the transmitter
β	negative curvature of the autocorrelation function $r_{\mu_i\mu_i}(\tau)$ at the origin, i.e., $\beta = \beta_i = -\ddot{r}_{\mu_i\mu_i}(0) \quad (i = 1, 2)$
β_R	tilt angle of the receiver antenna array
β_T	tilt angle of the transmitter antenna array
γ	negative curvature of the autocorrelation function $r_{\nu_3\nu_3}(\tau)$ at the origin, i.e., $\gamma = -\ddot{r}_{\nu_3\nu_3}(0)$
$\hat{\gamma}$	negative curvature of the spatial autocorrelation function $\hat{r}_{\nu\nu}(\Delta x)$ at the origin, i.e., $\hat{\gamma} = -\ddot{\hat{r}}_{\nu\nu}(0)$
δ_R	antenna element spacing at the receiver
δ_T	antenna element spacing at the transmitter
$\zeta(t)$	Rayleigh process
$\eta(t)$	Suzuki process
θ_n	phase shift due to the interaction with the nth scatterer S_n
θ_ρ	phase of the line-of-sight component $m(t)$
θ_0	constant phase shift
$\vartheta(t)$	phase of $\mu_\rho(t)$, i.e., $\vartheta(t) = \arg\{\mu_\rho(t)\}$
κ	angular spread parameter of the von Mises distribution
κ_c	frequency ratio $f_{\max}$ over f_c
κ_0	frequency ratio $f_{\min}$ over $f_{\max}$
$\lambda(t)$	lognormal process
$\lambda(x)$	spatial lognormal process
$\hat{\lambda}(x)$	spatial stochastic SOS lognormal process
λ_0	wavelength
$\mu(t)$	zero-mean complex Gaussian random process
$\mu_i(t)$	real Gaussian random process (stochastic reference model)
$\hat{\mu}_i(t)$	real stochastic process (stochastic simulation model)
$\mu_\rho(t)$	complex Gaussian random process with mean $m(t)$
$\nu(x)$	real spatial Gaussian process
$\hat{\nu}(x)$	real spatial stochastic SOS process
$\nu_i(t)$	white Gaussian noise process
$\xi(t)$	Rice process
ρ	amplitude of the line-of-sight component $m(t)$
$\rho(\cdot,\cdot)$	2D space cross-correlation function of $h_{kl}(t)$ and $h_{k'l'}(t)$
$\hat{\rho}(\cdot,\cdot)$	2D space cross-correlation function of $\hat{h}_{kl}(t)$ and $\hat{h}_{k'l'}(t)$
$\rho_{kl,k'l'}(\cdot,\cdot,\cdot)$	3D space-time cross-correlation function of $h_{kl}(t)$ and $h_{k'l'}(t)$
$\hat{\rho}_{kl,k'l'}(\cdot,\cdot,\cdot)$	3D space-time cross-correlation function of $\hat{h}_{kl}(t)$ and $\hat{h}_{k'l'}(t)$
$\rho_R(\delta_R, \tau)$	receiver correlation function (stochastic reference model)
$\rho_T(\delta_T, \tau)$	transmitter correlation function (stochastic reference model)
$\varrho(t)$	Loo process

σ_L	shadow standard deviation
σ_0^2	mean power of $\mu_i(t)$
τ	time difference between t_2 and t_1, i.e., $\tau = t_2 - t_1$
τ_-	fading interval
τ_+	connecting interval
$\tau_q(r)$	length of the time interval that comprises q per cent of all fading intervals of the process $\zeta(t)$ at the signal level r
τ'	continuous propagation delay
τ'_ℓ	discrete propagation delay of the ℓth path
$\tau'_{\max}$	maximum propagation delay
$\Delta_{\max}$	maximum transition time interval
$\Delta_{\min}$	minimum transition time interval
Δ_0	transition time interval
$\Delta\tau'$	delay resolution
$\Delta\tau'_\ell$	propagation delay difference between τ'_ℓ and $\tau'_{\ell-1}$, i.e., $\Delta\tau'_\ell = \tau'_\ell - \tau'_{\ell-1}$
υ'	frequency difference between f'_2 and f'_1, i.e., $\upsilon' = f'_2 - f'_1$
ϕ_0	symbol denoting the cross-correlation function $r_{\mu_1\mu_2}(\tau)$ at $\tau = 0$
χ	frequency separation variable (measures the size of a frequency hop from f_0 to f'_0, i.e., $\chi = f'_0 - f_0$)
ψ_0	symbol for the autocorrelation function $r_{\mu_i\mu_i}(\tau)$ at $\tau = 0$
$\Psi_{\mu_i}(\nu)$	characteristic function of $\mu_i(t)$
$\Psi_{\mu_1\mu_2}(\nu_1, \nu_2)$	joint characteristic function of $\mu_1(t)$ and $\mu_2(t)$
$\boldsymbol{\Omega}$	parameter vector of the reference model

Continuous-Time Deterministic Processes

$\tilde{a}_\ell$	path gain of the ℓth path
$\tilde{B}_C$	coherence bandwidth of SOSUS (SOCUS) models
$\tilde{B}^{(1)}_{\mu_i\mu_i}$	average Doppler shift of $\tilde{\mu}_i(t)$
$\tilde{B}^{(2)}_{\mu_i\mu_i}$	Doppler spread of $\tilde{\mu}_i(t)$
$\tilde{B}^{(1)}_{\tau'}$	average delay of SOSUS (SOCUS) models
$\tilde{B}^{(2)}_{\tau'}$	delay spread of SOSUS (SOCUS) models
$c_{i,n}$	path gain of the nth component of $\tilde{\mu}_i(t)$
$c_{i,n,\ell}$	path gain of the nth component of $\tilde{\mu}_{i,\ell}(t)$
$\tilde{C}(t)$	channel capacity of deterministic SOC MIMO channel models
$e_2(\tau)$	square error of the sample mean autocorrelation function $\bar{r}_{\mu_i\mu_i}(\tau)$
$e_T(\delta_T, \tau)$	absolute error of $\tilde{\rho}_T(\delta_T, \tau)$
$E_{p_{\mu_i}}$	mean-square error of $\tilde{p}_{\mu_i}(x)$
$E_{r_{\mu_i\mu_i}}$	mean-square error of $\tilde{r}_{\mu_i\mu_i}(\tau)$
$E^{(p)}$	L_p-norm
$f_{i,n}$	discrete Doppler frequency of the nth component of $\tilde{\mu}_i(t)$
$f_{i,n,\ell}$	discrete Doppler frequency of the nth component of $\tilde{\mu}_{i,\ell}(t)$
f_n	discrete Doppler frequency of the nth component of $\tilde{\mu}(t)$ (or $\tilde{h}_{kl}(t)$)
F_i	greatest common divisor $f_{i,1}, f_{i,2}, \ldots, f_{i,N_i}$, i.e., $F_i = \gcd\{f_{i,n}\}_{n=1}^{N_i}$

$\tilde{F}_{\zeta_-}(r)$	cumulative distribution function of deterministic Rayleigh processes $\tilde{\zeta}(t)$
$\tilde{F}_{\eta_-}(r)$	cumulative distribution function of deterministic Suzuki processes $\tilde{\eta}(t)$
$\tilde{F}_{\vartheta}(\varphi)$	cumulative distribution function of the phase $\tilde{\vartheta}(t)$ of $\tilde{\mu}(t) = \tilde{\mu}_1(t) + j\tilde{\mu}_2(t)$
$\tilde{F}_{\mu_i}(r)$	cumulative distribution function of deterministic Gaussian processes $\tilde{\mu}_i(t)$
$\tilde{F}_{\xi_-}(r)$	cumulative distribution function of deterministic Rice processes $\tilde{\xi}(t)$
$\tilde{F}_{\varrho_-}(r)$	cumulative distribution function of deterministic Loo processes $\tilde{\varrho}(t)$
$\tilde{h}(\tau')$	time-invariant impulse response of SOSUS (SOCUS) models
$\tilde{h}(\tau', t)$	time-variant impulse response of SOSUS (SOCUS) models
$\tilde{h}_{kl}(t)$	complex channel gain of deterministic SOC MIMO models of the link from A_l^T to A_k^R
$\tilde{H}(f', t)$	time-variant transfer function of SOSUS (SOCUS) models
$\tilde{\boldsymbol{H}}(t)$	channel matrix of deterministic SOC MIMO models
$\tilde{m}_{\mu_i}$	mean value of $\tilde{\mu}_i(t)$
N	number of cisoids (number of paths or number of scatterers)
N_i	number of sinusoids of $\tilde{\mu}_i(t)$
$N_{i,\ell}$	number of sinusoids of $\tilde{\mu}_{i,\ell}(t)$
N_i'	virtual number of sinusoids of $\tilde{\mu}_i(t)$
$N_{\max}$	largest number of N_1 and N_2, i.e., $N_{\max} = \max\{N_1, N_2\}$
N_s	number of sampling values
$\tilde{N}_{\zeta}(r)$	level-crossing rate of deterministic Rayleigh processes $\tilde{\zeta}(t)$
$\tilde{N}_{\eta}(r)$	level-crossing rate of deterministic Suzuki processes $\tilde{\eta}(t)$
$\tilde{N}_{\xi}(r)$	level-crossing rate of deterministic Rice processes $\tilde{\xi}(t)$
$\tilde{N}_{\varrho}(r)$	level-crossing rate of deterministic Loo processes $\tilde{\varrho}(t)$
$\tilde{p}_{0_-}(\tau_-; r)$	probability density function of the fading intervals τ_- of $\tilde{\zeta}(t)$
$\tilde{p}_{0_{-+}}(\tau_-, \tau_+; r)$	joint probability density function of fading and connecting intervals of $\tilde{\zeta}(t)$
$\tilde{p}_{1_-}(\tau_-; r)$	approximate solution for $\tilde{p}_{0_-}(\tau_-; r)$
$\tilde{p}_{\zeta}(z)$	probability density function of deterministic Rayleigh processes $\tilde{\zeta}(t)$
$\tilde{p}_{\eta}(z)$	probability density function of deterministic Suzuki processes $\tilde{\eta}(t)$
$\tilde{p}_{\vartheta}(\theta)$	probability density function of the phase $\tilde{\vartheta}(t)$ of $\tilde{\mu}(t) = \tilde{\mu}_1(t) + j\tilde{\mu}_2(t)$
$\tilde{p}_{\mu_i}(x)$	probability density function of deterministic Gaussian processes $\tilde{\mu}_i(t)$
$\tilde{p}_{\xi}(z)$	probability density function of deterministic Rice processes $\tilde{\xi}(t)$
$\tilde{p}_{\varrho}(z)$	probability density function of deterministic Loo processes $\tilde{\varrho}(t)$
$\tilde{p}_{\xi\dot{\xi}}(z, \theta)$	joint probability density function of $\tilde{\xi}(t)$ and $\dot{\tilde{\xi}}(t)$
$\tilde{r}_{hh}(\cdot, \cdot; \cdot, \cdot)$	autocorrelation function $\tilde{h}(\tau', t)$
$\tilde{r}_{h_{kl}}(\tau)$	time autocorrelation function of $\tilde{h}_{kl}(t)$
$\tilde{r}_{HH}(\upsilon', \tau)$	time-frequency correlation function of SOSUS (SOCUS) models
$\tilde{r}_{ss}(\cdot, \cdot; \cdot, \cdot)$	autocorrelation function of $\tilde{s}(\tau', f)$
$\tilde{r}_{TT}(\cdot, \cdot; \cdot, \cdot)$	autocorrelation function of $\tilde{T}(f', f)$
$\tilde{r}_{\mu\mu}(\tau)$	autocorrelation function of $\tilde{\mu}(t) = \tilde{\mu}_1(t) + j\tilde{\mu}_2(t)$
$\tilde{r}_{\mu_i\mu_i}(\tau)$	autocorrelation function of $\tilde{\mu}_i(t)$
$\bar{r}_{\mu_i\mu_i}(\tau)$	sample mean autocorrelation function of $\tilde{r}^{(k)}_{\mu_i\mu_i}(\tau)$

$\tilde{r}_{\mu_{i,\ell}\mu_{i,\ell}}(\tau)$	autocorrelation function of $\tilde{\mu}_{i,\ell}(t)$
$\tilde{r}_{\mu_\ell\mu_\ell}(\tau)$	autocorrelation function of $\tilde{\mu}_\ell(t)$
$\tilde{r}_{\mu_1\mu_2}(\tau)$	cross-correlation function of $\tilde{\mu}_1(t)$ and $\tilde{\mu}_2(t)$
$\tilde{r}_{\mu\mu'}(\tau, \chi)$	time-frequency cross-correlation function of $\tilde{\mu}(t)$ and $\tilde{\mu}'(t)$
$\tilde{r}_{\mu_i\mu_i'}(\tau, \chi)$	time-frequency cross-correlation function of $\tilde{\mu}_i(t)$ and $\tilde{\mu}_i'(t)$
$\tilde{r}_{\mu_1\mu_2'}(\tau, \chi)$	time-frequency cross-correlation function of $\tilde{\mu}_1(t)$ and $\tilde{\mu}_2'(t)$
$\tilde{r}_{\tau'}(\upsilon')$	frequency correlation function of SOSUS (SOCUS) models
$\tilde{s}(\tau', f)$	Doppler-variant impulse response of SOSUS (SOCUS) models
$\tilde{S}(\tau', f)$	scattering function of SOSUS (SOCUS) models
$\tilde{S}_{hh}(\tau', \tau)$	delay cross-power spectral density of SOSUS (SOCUS) models
$\tilde{S}_{TT}(\upsilon', f)$	Doppler cross-power spectral density of SOSUS (SOCUS) models
$\tilde{S}_{\mu_i\mu_i}(f)$	power spectral density of $\tilde{\mu}_i(t)$
$\tilde{S}_{\mu_\ell\mu_\ell}(f)$	power spectral density of $\tilde{\mu}_\ell(t)$
$\tilde{S}_{\mu_1\mu_2}(f)$	cross-power spectral density of $\tilde{\mu}_1(t)$ and $\tilde{\mu}_2(t)$
$\tilde{S}_{\tau'}(\tau')$	delay power spectral density of SOSUS (SOCUS) models
$\tilde{T}(f', f)$	Doppler-variant transfer function of SOSUS (SOCUS) models
$\tilde{T}_C$	coherence time of SOSUS (SOCUS) models
T_i	period of $\tilde{\mu}_i(t)$
T_s, T_s'	sampling intervals
T_{sim}	simulation time
$\tilde{T}_{\zeta_-}(r)$	average duration of fades of deterministic Rayleigh processes $\tilde{\zeta}(t)$
$\tilde{T}_{\eta_-}(r)$	average duration of fades of deterministic Suzuki processes $\tilde{\eta}(t)$
$\tilde{T}_{\xi_-}(r)$	average duration of fades of deterministic Rice processes $\tilde{\xi}(t)$
$\tilde{T}_{\varrho_-}(r)$	average duration of fades of deterministic Loo processes $\tilde{\varrho}(t)$
$\alpha_{i,0}^{(k)}$	angle-of-rotation
α_n^R	discrete angle-of-arrival of the nth component of $\tilde{h}_{kl}(t)$
α_n^T	discrete angle-of-departure of the nth component of $\tilde{h}_{kl}(t)$
$\tilde{\beta}_i$	negative curvature of the autocorrelation function $\tilde{r}_{\mu_i\mu_i}(\tau)$ at the origin, i.e., $\tilde{\beta}_i = -\ddot{\tilde{r}}_{\mu_i\mu_i}(0) \quad (i = 1, 2)$
$\Delta\beta_i$	model error of the simulation model, i.e., $\Delta\beta_i = \tilde{\beta}_i - \beta$
$\tilde{\gamma}$	negative curvature of the autocorrelation function of $\tilde{r}_{\nu_3\nu_3}(\tau)$ at the origin, i.e., $\tilde{\gamma} = -\ddot{\tilde{r}}_{\nu_3\nu_3}(0)$
ε_{N_ξ}	relative error of the level-crossing rate $\tilde{N}_\xi(r)$
$\varepsilon_{T_{\xi_-}}$	relative error of the average duration of fades $\tilde{T}_{\xi_-}(r)$
$\tilde{\zeta}(t)$	continuous-time deterministic Rayleigh process
$\tilde{\eta}(t)$	continuous-time deterministic Suzuki process
$\tilde{\eta}^{(0)}(t)$	handover process (deterministic Suzuki process with time-variant parameters)
$\theta_{i,n}$	phase of the nth component of $\tilde{\mu}_i(t)$
$\theta_{i,n,\ell}$	phase of the nth component of $\tilde{\mu}_{i,\ell}(t)$
$\vec{\theta}_i$	phase vector
θ_0	phase difference between $\tilde{\mu}_{1,n}(t)$ and $\tilde{\mu}_{2,n}(t)$
$\vec{\Theta}_i$	standard phase vector

$\tilde{\vartheta}(t)$	phase of $\tilde{\mu}_\rho(t)$, i.e., $\tilde{\vartheta}(t) = \arg\{\tilde{\mu}_\rho(t)\}$
$\tilde{\lambda}(t)$	continuous-time deterministic SOS lognormal process
$\tilde{\lambda}(x)$	spatial deterministic SOS lognormal process
$\tilde{\mu}(t)$	zero-mean complex continuous-time deterministic Gaussian process
$\tilde{\mu}_i(t)$	zero-mean real continuous-time deterministic Gaussian process
$\tilde{\mu}_{i,\ell}(t)$	real deterministic Gaussian process of the ℓth path of SOSUS (SOCUS) models
$\tilde{\mu}_{i,n}(t)$	nth elementary sinusoidal function of $\tilde{\mu}_i(t)$
$\tilde{\mu}_\ell(t)$	complex deterministic Gaussian process of the ℓth path of SOSUS (SOCUS) models
$\tilde{\mu}_\rho(t)$	complex deterministic Gaussian process with mean value $m(t)$
$\tilde{\xi}(t)$	continuous-time deterministic Rice process
$\tilde{\rho}(\cdot,\cdot)$	2D space cross-correlation function of $\tilde{h}_{kl}(t)$ and $\tilde{h}_{k'l'}(t)$
$\tilde{\rho}_{kl,k'l'}(\cdot,\cdot,\cdot)$	3D space-time cross-correlation function of $\tilde{h}_{kl}(t)$ and $\tilde{h}_{k'l'}(t)$
$\tilde{\rho}_R(\delta_R,\tau)$	receiver correlation function (deterministic simulation model)
$\tilde{\rho}_T(\delta_T,\tau)$	transmitter correlation function (deterministic simulation model)
$\tilde{\rho}^{(k,l)}_{\mu_i\mu_\lambda}$	correlation coefficient of the deterministic processes $\tilde{\mu}_i^{(k)}(t)$ and $\tilde{\mu}_\lambda^{(l)}(t)$
$\hat{\rho}^{(k,l)}_{\mu_i\mu_\lambda}$	upper limit of the correlation coefficient $\tilde{\rho}^{(k,l)}_{\mu_i\mu_\lambda}$
$\tilde{\varrho}(t)$	continuous-time deterministic Loo process
$\tilde{\sigma}^2_\mu$	mean power of $\tilde{\mu}(t)$
$\tilde{\sigma}^2_{\mu_i}$	mean power of $\tilde{\mu}_i(t)$
$\tilde{\tau}'_\ell$	discrete propagation delay of the ℓth path
$\Delta\tilde{\tau}'_\ell$	propagation delay difference between $\tilde{\tau}'_\ell$ and $\tilde{\tau}'_{\ell-1}$, i.e., $\Delta\tilde{\tau}'_\ell = \tilde{\tau}'_\ell - \tilde{\tau}'_{\ell-1}$
$\tilde{\tau}_q(r)$	length of the time interval that comprises q per cent of all fading intervals of the process $\tilde{\zeta}(t)$ at the signal level r
$\tilde{\phi}_0$	symbol for the cross-correlation function $\tilde{r}_{\mu_1\mu_2}(\tau)$ at $\tau = 0$
$\tilde{\psi}_0$	symbol for the autocorrelation function $\tilde{r}_{\mu_i\mu_i}(\tau)$ at $\tau = 0$
$\tilde{\Xi}_\ell(f)$	Fourier transform of $\tilde{\mu}_\ell(t)$
$\tilde{\Psi}_{\mu_i}(\nu)$	characteristic function of $\tilde{\mu}_i(t)$
$\tilde{\boldsymbol{\Omega}}$	parameter vector of the simulation model
$\Omega_{i,n}$	normalized discrete Doppler frequency, i.e., $\Omega_{i,n} = 2\pi f_{i,n} T_s$

Discrete-Time Deterministic Processes

$a_{i,n}[k]$	address of the look-up table $\mathrm{Tab}_{i,n}$ at the discrete time k
$\bar{B}^{(1)}_{\mu_i\mu_i}$	average Doppler shift of $\bar{\mu}_i[k]$
$\bar{B}^{(2)}_{\mu_i\mu_i}$	Doppler spread of $\bar{\mu}_i[k]$
$c_{i,n}$	path gain of the nth component of $\bar{\mu}_i[k]$
$\bar{f}_{i,n}$	quantized Doppler frequency of the nth component of $\bar{\mu}_i[k]$
f_s	sampling frequency (or sampling rate)
$f_{s,\min}$	minimum sampling frequency
$\bar{F}_{\zeta_-}(r)$	cumulative distribution function of discrete-time deterministic Rayleigh processes $\bar{\zeta}[k]$
$\bar{F}_\vartheta(\varphi)$	cumulative distribution function of the phase $\bar{\vartheta}[k]$ of $\bar{\mu}[k] = \bar{\mu}_1[k] + j\bar{\mu}_2[k]$

$\bar{F}_{\mu_i}(r)$	cumulative distribution function of discrete-time deterministic Gaussian processes $\bar{\mu}_i[k]$
k	discrete time variable ($t = kT_s$)
K	number of simulated samples of a discrete-time deterministic process
L	period of $\bar{\zeta}[k]$
$\hat{L}$	upper limit of the period of $\bar{\zeta}[k]$
L_i	period of $\bar{\mu}_i[k]$
$\hat{L}_i$	upper limit of the period of $\bar{\mu}_i[k]$
$L_{i,n}$	period of the nth component of $\bar{\mu}_i[k]$
$\bar{m}_{\mu_i}$	mean value of the sequence $\bar{\mu}_i[k]$
$\boldsymbol{M}_i$	channel matrix; contains the complete information required for the reconstruction of $\bar{\mu}_i[k]$
$\bar{N}_\zeta(r)$	level-crossing rate of discrete-time deterministic Rayleigh processes $\bar{\zeta}[k]$
$\bar{p}_\zeta(z)$	probability density function of discrete-time deterministic Rayleigh processes $\bar{\zeta}[k]$
$\bar{p}_\vartheta(\theta)$	probability density function of the phase $\bar{\vartheta}[k]$ of $\bar{\mu}[k] = \bar{\mu}_1[k] + j\bar{\mu}_2[k]$
$\bar{p}_{\mu_i}(x)$	probability density function of discrete-time deterministic Gaussian processes $\bar{\mu}_i[k]$
$\text{Reg}_{i,n}$	register; contains one period of the elementary sinusoidal sequence $\bar{\mu}_{i,n}[k]$
$\bar{r}_{\mu_i\mu_i}[\kappa]$	autocorrelation sequence of $\bar{\mu}_i[k]$
$\bar{r}_{\mu_1\mu_2}[\kappa]$	cross-correlation sequence of $\bar{\mu}_1[k]$ and $\bar{\mu}_2[k]$
$\boldsymbol{S}_i$	selection matrix
$\bar{S}_{\mu_i\mu_i}(f)$	power spectral density of $\bar{\mu}_i[k]$
$\bar{S}_{\mu_1\mu_2}(f)$	cross-power spectral density of $\bar{\mu}_1[k]$ and $\bar{\mu}_2[k]$
$\text{Tab}_{i,n}$	look-up table; contains one period of the elementary sinusoidal sequence $\bar{\mu}_{i,n}[k]$
T_s	sampling interval
$T_s^\star$	sampling interval of measured signals
T_{sim}	simulation time
ΔT_{sim}	iteration time
$\bar{T}_{\zeta_-}(r)$	average duration of fades of discrete-time deterministic Rayleigh processes $\bar{\zeta}[k]$
$\bar{\beta}_i$	negative curvature of the autocorrelation sequence $\bar{r}_{\mu_i\mu_i}[\kappa]$ at the origin, i.e., $\bar{\beta}_i = -\ddot{\bar{r}}_{\mu_i\mu_i}[0] \quad (i = 1, 2)$
$\Delta\bar{\beta}_i$	model error of $\ddot{\bar{r}}_{\mu_i\mu_i}[0]$, i.e., $\Delta\bar{\beta}_i = \bar{\beta}_i - \beta$
$\Delta_{n,m}^{(i,j)}$	auxiliary function for the determination of the minimum sampling frequency $f_{s,\min}$
$\varepsilon_{\bar{f}_{i,n}}$	relative error of the quantized Doppler frequencies $\bar{f}_{i,n}$
$\bar{\zeta}[k]$	discrete-time deterministic Rayleigh process
$\bar{\theta}_{i,n}$	quantized phase of the nth component of $\bar{\mu}_i[k]$
$\bar{\vartheta}[k]$	phase of $\bar{\mu}[k] = \bar{\mu}_1[k] + j\bar{\mu}_2[k]$, i.e., $\bar{\vartheta}[k] = \arg\{\bar{\mu}[k]\}$
κ	time difference between the discrete time instants k_2 and k_1, i.e., $\kappa = k_2 - k_1$
$\bar{\mu}[k]$	complex discrete-time deterministic Gaussian process
$\bar{\mu}_i[k]$	real discrete-time deterministic Gaussian process
$\bar{\mu}_{i,n}[k]$	nth elementary sinusoidal function of $\bar{\mu}_i[k]$
$\bar{\sigma}_{\mu_i}^2$	mean power of $\bar{\mu}_i[k]$

1

Introduction

1.1 The Evolution of Mobile Radio Systems

For several decades now, the mobile communications sector has been the fastest growing market segment in telecommunications. At present, we are still in the early stages of a global growth trend in mobile communications, which will most likely continue for many years to come. In trying to define the reasons for this development, one can readily identify a broad range of factors. We have seen the international liberalization of telecommunications services, the opening and deregulation of major world markets, the extension of the frequency range around and beyond 1 GHz, improved modulation and coding techniques, as well as impressive progress in semiconductor technology (e.g., by using VLSI[1] circuits based on FPGA[2]-, CMOS[3]-, and GaAs[4]-technologies), and, last but not least, greater knowledge of the propagation characteristics of electromagnetic waves in an extraordinarily complex environment have undoubtedly contributed to the stellar success of the telecommunications sector worldwide. The beginning of the success story of mobile communications can be traced back by more than 50 years — a period of half a century that spans over four generations of mobile communication systems.

First generation (1G) mobile communication systems were introduced in the early 1980s. They were based entirely on analog transmission techniques. The objective of 1G mobile communication systems was to offer voice services over mobile radio channels. The technology employed was based on analog frequency modulation (FM) and frequency division multiple access (FDMA) schemes. 1G systems were strictly limited in their subscriber capacity and their accessibility. Moreover, they suffered from an inherently inefficient use of the frequency spectrum.

A variety of 1G analog cellular mobile radio standards has been developed in Europe, the United States, and Japan. In Europe, the first 1G standard was the Nordic Mobile Telephone (NMT) standard, which was developed jointly in Sweden, Norway, Denmark, Finland, and

[1] VLSI: very large scale integration.
[2] FPGA: field programmable gate arrays.
[3] CMOS: complementary metal oxide semiconductor.
[4] GaAs: gallium arsenide.

Mobile Radio Channels, Second Edition. Matthias Pätzold.

Iceland. The first fully operational NMT systems were inaugurated in Sweden and Norway in 1981, in Denmark and Finland in 1982, and in Iceland in 1986. This system operated originally in the 450 MHz frequency band (NMT-450) and later, from 1986, also in the 900 MHz frequency band (NMT-900). In Germany, the first cellular mobile radio system was called randomly A-Net. It was in service between 1958 and 1977. The A-Net was based on manual switching techniques, so that human operators were required to connect calls. Direct dialing first became possible with the B-Net, which was in service from 1972 until 1994. The capacity limit of 27 000 subscribers was reached fairly quickly. In order to reach a subscriber, the calling party had to know the location of the called party, because the handset required knowledge of the local area code of the base station serving it. Handover was not possible, but roaming calls could be made between neighbouring countries (Austria, The Netherlands, and Luxembourg) that had also implemented the B-Net standard. The B-Net was taken out of service on 31 December 1994. Automatic localization of the mobile subscriber and handover to the next cell was first made possible with the technically superior cellular C-Net, which was officially put into operation on May 1, 1985. It operated in the 450 MHz frequency band and had a Germany-wide accessibility. The C-Net service reached a peak of around 800 000 subscribers in the early 1990s. The C-Net service was shut down in most parts of Germany on 31 December 2000. Other important 1G analog systems developed in Europe include the Total Access Communication System (TACS), which was largely used in the United Kingdom and Ireland, as well as the NMT-F and RadioCom 2000 systems used in France, and the Radio Telephone Mobile (RTM) system that operated in Italy in the 450 MHz frequency band. In the United States, the Advanced Mobile Phone System (AMPS) standard developed by the Bell labs was officially introduced in 1983. The AMPS systems operated in the 800 MHz frequency band. In Japan, the first commercial 1G service was provided by the Nippon Telephone and Telegraph (NTT) Public Corporation (NTTPC) in 1979.

Today, 1G analogue mobile systems are not in use anymore. Many countries have reallocated the frequency resources to other mobile system standards. The mobile market in the 1G era was fragmented in the sense that an efficient harmonization and interoperability/roaming was either a non-issue or at best a very complicated process. This was especially seen as a huge problem in Europe. Hence, one of the requirements for the next generation mobile was the use of common standards and the creation of a single market for mobile services. Another main requirement for the new standards was an improved utilization of the frequency resources. This requirement has been fulfilled by selecting digital technology as the foundation for the next standards.

Second generation (2G) mobile communication systems were developed in the early 1990s. These systems differ from the previous generation in their use of digital transmission techniques instead of analog techniques. The primary objectives of 2G mobile communication systems were to facilitate pan-European roaming, to improve the transmission quality, and to offer both voice services and data services over mobile radio channels. The new system uses digital modulation techniques and provides higher voice quality and improved spectral efficiency at lower costs to consumers.

The Global System for Mobile Communications (GSM[5]) standard is generally recognized as the most elaborate 2G standard worldwide. In 1982, the Conference of European Postal

[5] Formerly, the acronym GSM stood for "Groupe Spécial Mobile". As the original pan-European GSM standard became more global, the meaning of the acronym GSM was changed to its present meaning.

and Telecommunications Administrations (CEPT) established a working group called Groupe Spécial Mobile with the mandate to define standards for future pan-European cellular radio systems. Later, in 1989, GSM was taken over by the European Telecommunications Standards Institute (ETSI), which finalized the GSM standard in 1990. GSM uses a combination of time division multiple access (TDMA) and FDMA techniques. It supports voice calls and data services with possible data rates of 2.4, 4.8, and 9.6 kbit/s, together with the transmission of short message services (SMS) [1]. In Germany, the so-called D-Net, which is based on the GSM standard, was brought into service in 1992. It operates in the 900 MHz frequency band and offers all subscribers Europe-wide coverage. In addition, the E-Net (Digital Cellular System, DCS 1800) operating in the 1800 MHz frequency band has been running in parallel to the D-Net since 1994. These two GSM networks differ mainly in their respective frequency range. In Great Britain, the DCS 1800 is known as the Personal Communications Network (PCN). In the United States and Canada, GSM operates in the 850 MHz and 1900 MHz frequency bands. The original European GSM standard has become in the meantime a worldwide mobile communication standard that had been adopted by 222 (210) countries by the end of 2009 (2005). In 2009, the network operators altogether ran worldwide 1050 GSM networks with over 3.8 billion GSM subscribers. This means that approximately 55 per cent of the world's population use GSM services.

In addition to the GSM standard, a new standard for cordless telephones, named the Digital European Cordless Telephone (DECT) standard, was introduced by ETSI. The DECT standard allows subscribers moving at a fair pace by using cordless telephones within a maximum range of about 300 m. Other important 2G standards include the Interim Standard 95 (IS-95), IS-54, IS-136, as well as the Personal Digital Cellular (PDC) standard. The brand name for IS-95 is cdmaOne, which was the first digital cellular standard developed by Qualcomm. IS-95 systems are based on code division multiple access (CDMA) techniques. They are widely used in America, particulary in the United States and Canada, and parts of Asia. IS-54 and IS-136 are also known as Digital Advanced Mobile Phone Service (D-AMPS), which is the digital version of the 1G analog cellular phone standard AMPS. D-AMPS uses digital TDMA techniques and operates in the 800 and 1900 MHz frequency bands. D-AMPS systems were once widely used in the United States and Canada, but today they are considered end-of-life, and existing networks have mostly been replaced by GSM or CDMA2000 networks. The PDC standard was defined in Japan in April 1991 and launched by NTT DoCoMo in March 1993. PDC systems are TDMA-based and used exclusively in Japan. Although 2G mobile communication systems are still widely in use in many parts of the world, their underlying technology has been superseded by newer technologies, such as 2.5G, 2.75G, 3G, and 4G.

Third generation (3G) mobile communication systems were developed in the early 2000s. The prime objective of 3G mobile systems is to achieve a fully integrated digital mobile terrestrial (satellite) communication network that offers voice, data, and multimedia services (mobile Internet) at anytime and anywhere in the world with seamless global roaming. The key factors of 3G systems include worldwide usage, global coverage by integration of satellite and terrestrial systems, and high spectrum efficiency. 3G systems provide a wide range of telecommunications services (voice, data, multimedia, Internet), and they are able to operate in all radio environments (cellular, satellite, cordless, and LAN[6]). In addition, they support both packet-switched and circuit-switched data transmissions. Depending on the environment,

[6] LAN: local area network.

3G wireless systems offer a wide range of data rates, ranging from 9.6 kbit/s for satellite users over 144 kbit/s for vehicular users (high mobility) and 384 kbit/s for pedestrian users (restricted mobility) up to a maximum data rate of 2.048 Mbit/s for users in stationary indoor office environments. The first commercial 3G system was launched by NTT DoCoMo in Japan in October 2001. Its technology was based on wideband CDMA (WCDMA).

In Europe, 3G mobile communication systems are usually referred to as Universal Mobile Telecommunications System (UMTS). With UMTS, one is aiming at integrating the various services offered by 2G systems into one universal system. An individual subscriber can be called at anytime, from any place (car, train, aircraft, etc.) and is able to use mobile Internet services via a universal terminal. Apart from that, UMTS also provides a variety of application services that were not previously available to 2G mobile phone users, such as mobile TV, video on demand, video conferencing, telemedicine, and location-based services.

Originally it was the intention to have only one common global standard for 3G systems. For the first time, this would enable worldwide roaming with a single handset. But unfortunately, during the standardization process led by the International Telecommunications Union (ITU[7]), it became clear that for both technical and political reasons, the ITU was not in a position to enforce a single unified worldwide standard. Instead, a set of globally harmonized standards fulfilling the specifications set by the ITU has been specified under the umbrella known as International Mobile Telecommunications 2000 (IMT-2000[8]). IMT-2000 operates in the frequency bands 1885–2025 MHz and 2110–2200 MHz, which were assigned to 3G systems by the World Administration Radio Conference (WARC) in March 1992 for worldwide use. At the 18th ITU Task Group 8/1 meeting held in Helsinki from 25 October to 5 November 1999, a set of five terrestrial and satellite radio interface standards was approved for IMT-2000. The IMT-2000 family of standards accommodates the following five terrestrial radio interface standards (see Figure 1.1):

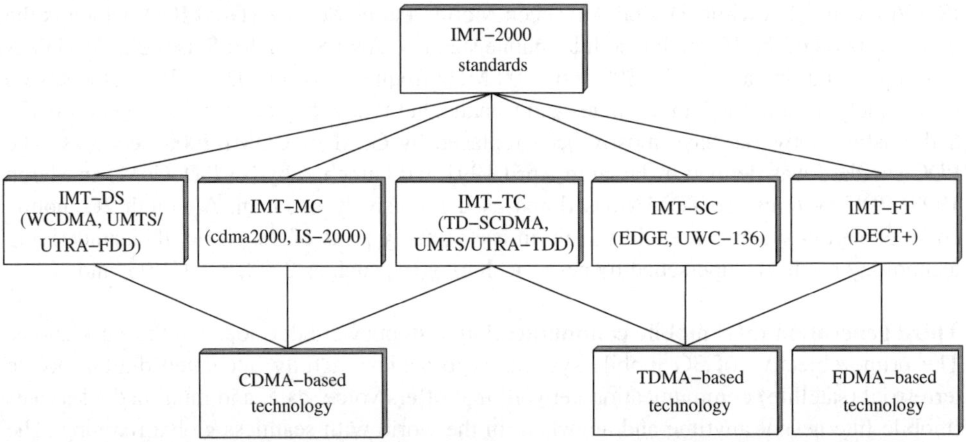

Figure 1.1 The IMT-2000 family of standards for terrestrial radio interfaces.

[7] The ITU is the leading United Nations agency for information and communication technology issues.

[8] IMT-2000 was formally known as Future Public Land Mobile Telecommunications System (FPLMTS).

- IMT-DS: This terrestrial radio interface standard is based on direct sequence CDMA (DS-CDMA) technology. The frequency division duplex (FDD) mode is used for symmetrical applications requiring the same amount of radio resources in the uplink as in the downlink. IMT-DS is also known as wideband CDMA (WCDMA), WCDMA-FDD, and UMTS/UTRA[9]-FDD. This standard is suitable for applications in public macrocell and microcell environments. IMT-DS is supported by the GSM network operators and vendors as well as by Japan's Association of Radio Industries and Businesses (ARIB).
- IMT-MC: This terrestrial radio interface standard falls under the multi-carrier CDMA (MC-CDMA) category, which is based mainly on FDD frameworks. IMT-MC is also known as cdma2000 and IS-2000. The cdma2000 standard is an evolutionary outgrowth of cdmaOne, which is supported by US cellular network operators and vendors.
- IMT-TC: This standard is based on a combination of TDMA and WCDMA technologies. IMT-TC is also known as UMTS/UTRA-TDD, TD-CDMA, and TD-SCDMA. UMTS/UTRA-TDD is an evolutionary outgrowth of the TDMA-based GSM standard. TD-SCDMA is proposed by China. The IMT-TC standard is optimized for symmetrical and asymmetrical applications with high data rates. It aims to provide 3G services in public microcell and picocell environments.
- IMT-SC: This standard falls under the TDMA single-carrier category. IMT-SC is also known as EDGE (Enhanced Data Rates for GSM Evolution) and UWC-136 (Universal Wireless Communications 136). IMT-SC is an evolutionary outgrowth of GSM and TDMA-136, which is achieved by building upon enhanced versions of GSM and TDMA-136. Many EDGE physical layer parameters are identical to those of GSM, including GSM's TDMA frame structure and carrier spacing. EDGE was developed to enable operators to offer multimedia and other IP-based services at speeds of up to 472 kbits/s in wide area networks when all eight time slots are used.
- IMT-FT: This standard falls into both the FDMA and the TDMA category. IMT-FT is also well known in Europe as DECT+, which is an evolution of the DECT standard. IMT-FT is used mainly to provide 3G services in indoor environments.

IMT-2000 provides smooth evolution paths from the various widely deployed existing 2G to 3G mobile networks. The trend is currently that people are moving rapidly from 2G to 3G networks, in both developed and developing countries [2]. The ITU estimates that at the end of 2010 there will be 940 million mobile subscriptions to 3G services worldwide, which corresponds to 18 per cent of the total number of subscriptions. In 2010, 143 countries were offering 3G services commercially, compared to 95 in 2007 [2]. Some countries, including Sweden, Norway, Ukraine, and the United States, are already moving to 4G.

Mobile satellites systems are an integral part of UMTS/IMT-2000. The advantages of mobile satellite systems are that they provide global coverage and offer cost-effective services in large areas with low user density and limited traffic density. Their role is not to compete with terrestrial mobile communication systems, but to complement them in a geographical sense (cost-effective coverage of remote areas) and in a service sense (cost effective for broadcast/multicast services). Satellite communication systems can be classified into geostationary earth orbit (GEO) and non-geostationary earth orbit (NGEO) satellite systems. GEO satellites are placed in an equatorial orbit approximately 36 000 km above the earth such that the

[9] UTRA: UMTS Terrestrial Radio Access.

satellites orbit synchronously with the rotating earth and seem to be fixed in the sky [3]. Global coverage can be reached by just three geostationary satellites, but their high altitude results in large propagation delays and causes a very high signal attenuation. NGEO satellite systems include low earth orbit (LEO), medium earth orbit (MEO), and highly elliptical orbit (HEO) satellite systems. LEO satellites (700–1000 km altitude) have relatively short propagation delays and low signal attenuations, but require a large number of satellites to cover the earth's surface. A compromise is provided by MEO satellites (6000–20 000 km altitude), which avoid both the large propagation delays and the high signal attenuation of geostationary satellites, while still providing global coverage with a comparatively small number of about 10 satellites. A thorough survey on mobile satellite systems is provided in [4].

Typical representatives of GEO satellite systems are Inmarsat, Thuraya, and the Asian Cellular System (ACeS). Inmarsat operates currently three global constellations of 11 GEO telecommunications satellites. They provide seamless mobile voice and data communications around the world, enabling users to make phone calls or connect to the Internet on land, at sea or in the air. In addition, Inmarsat offers global maritime distress and safety services to ships and aircraft for free. Thuraya runs two active communications satellites (Thuraya-2 and Thuraya-3) and provides GSM-compatible mobile telephone services to over 140 countries around the world. Its coverage area encompasses the Middle East, North and Central Africa, Europe, Central Asia, and the Indian subcontinent. ACeS is a regional satellite telecommunications company that operates one GEO satellite (Garuda 1), which was launched in 2000. It offers GSM-like satellite telephony services to the Asian market. The coverage area includes South East Asia, Japan, China, and some parts of India.

The first global LEO satellite system was Iridium, which was launched on 1 November 1998 to provide handheld telephone and paging satellite services. The Iridium system consists of 66 satellites covering 100 per cent of the globe and circulating the earth in six polar LEO planes at a height of 781 km. Iridium plans to replace its current satellite constellation by Iridium NEXT, the world's largest LEO satellite system, which is expected to begin launching in 2015. Iridium NEXT will offer truly global mobile communication services on land, at sea, and in the sky. Other representatives of LEO satellite systems include Globalstar (48 satellites, 1414 km altitude)[10] and Teledesic (288 satellites, 1400 km altitude)[11] [5].

Satellite phones are no longer big and expensive. Around the millennium, the price of a satellite phone was about $3000, and the cost of making voice calls was about $7 per minute. Ten years later, satellite phones are available for around $500 to $1200, calling plans can fall under $1 per minute, and the handset weight came down from 400 g to only 130 g [6].

The Mobile Broadband System (MBS) has been considered as a necessary step towards the next generation of mobile communication systems [7]. The research on MBS was initiated by the Research and Development in Advanced Communications Technologies in Europe (RACE II) program. MBS plans mobile broadband services up to a data rate of 155 Mbit/s in the 40 and 60 GHz frequency bands. Services of MBS include voice, video, and high demanding data applications, such as the wireless transmission of high-quality digital TV and video

[10] Globalstar's second generation satellite constellation will consist of 32 LEO satellites.

[11] Teledesic had originally planned in 1995 to operate 924 satellites (840 active satellites plus 84 on-orbit spares) circling around the earth in 21 orbits at an altitude between 695 and 705 km. In 2002, Teledesic has officially suspended its work on satellite construction.

conferencing signals. MBS can be considered as a wireless extension of the wired B-ISDN[12] system. It provides radio coverage restricted to small indoor and outdoor areas (e.g., sports arenas, factory halls, television studios) and supports the wireless communication between MBS terminals and terminals directly connected to the B-ISDN network. The underlying technology of MBS is IP-based.

Fourth generation (4G) mobile communication systems are currently under development. They are often referred to as IMT-Advanced (International Mobile Telecommunications Advanced) systems, whose requirements have been stated in the ITU-R report [8]. The prime objective of 4G mobile systems is to achieve a fully integrated digital mobile terrestrial (satellite) communication network that offers voice, data, and next generation multimedia services (mobile Internet) at anytime and anywhere in the world with seamless global roaming. 4G systems will offer enhanced peak data rates of 100 Mbit/s for high mobility devices and 1 Gbit/s for low mobility devices. Some other requirements and features that have been identified for 4G systems are increased spectral efficiency, interworking with other radio access systems, compatibility of services, smooth handovers across heterogeneous networks, and the ability to offer high quality of service for multimedia support. The principal technologies used in 4G systems include multiple-input multiple-output (MIMO) techniques, turbo coding techniques, adaptive modulation and error-correcting coding schemes, orthogonal frequency division multiple access (OFDMA) techniques, as well as fixed relaying and cooperative relaying networks. Proper candidates for 4G standards are LTE-Advanced (Long Term Evolution Advanced) and IEEE 802.16m. The present LTE [9] and WiMAX[13] [10] systems are widely considered as pre-4G systems, as they do not fully comply with the LTE-Advanced requirements regarding the peak data rates of 100 Mbit/s for high mobility devices and 1 Gbit/s for low mobility devices.

Before the introduction of newly developed mobile communication systems, a large number of theoretical and experimental investigations have to be made. These help to answer open questions, such as how existing resources (energy, frequency range, labour, ground, capital) can be used economically with a growing number of subscribers and how reliable secure data transmission can be provided for the user as cheap and as simple to handle as possible. Also included are estimates of environmental and health risks that almost inevitably exist when mass-market technologies are introduced and that are only to a certain extent tolerated by a public that is becoming more and more critical. Another boundary condition growing in importance during the development of new transmission techniques is often the demand for compatibility with existing systems. To solve the technical problems related to these boundary conditions, it is necessary to have a firm knowledge of the specific characteristics of the mobile radio channel. The term mobile radio channel in this context is the physical medium that is used to send the signal from the transmitter to the receiver [11]. However, when the channel is modelled, the characteristics of the transmitting and the receiving antenna are in general included in the channel model. The basic characteristics of mobile radio channels are explained subsequently. The thermal noise is not taken into consideration in the following and has to be added separately to the output signal of the mobile radio channel, if necessary.

[12] B-ISDN: Broadband Integrated Services Digital Network.
[13] WiMAX: Worldwide Interoperability for Microwave Access.

1.2 Basic Knowledge of Mobile Radio Channels

The three basic propagation phenomena are known as *reflection*, *diffraction*, and *scattering*. Reflection occurs when a plane wave encounters an object with size A that is very large compared to the wavelength λ_0, i.e., $A \gg \lambda_0$. According to the law of reflection, the direction of the incident plane wave and the direction of the reflected plane wave make the same angle α with respect to the surface normal. Diffraction arises when a plane wave strikes an object with size A that is in the order of the wavelength λ_0, i.e., $A \approx \lambda_0$. According to Huygens' principle, the interaction of a plane wave with a diffracting object generates secondary waves behind the object. Scattering occurs when a plane wave incidents on an object with size A that is very small compared to the wavelength λ_0, i.e., $A \ll \lambda_0$. A scattering object redirects the energy of the incident plane wave in many directions. The three basic propagation phenomena are illustrated in Figure 1.2.

In mobile radio communications, the emitted electromagnetic waves often do not reach the receiving antenna directly because of obstacles blocking the line-of-sight path. In fact, the received waves are a superposition of waves coming from many different directions due to reflection, diffraction, and scattering caused by buildings, trees, and other obstacles. This effect is known as *multipath propagation*. A typical scenario for the terrestrial mobile radio channel is shown in Figure 1.3. Without loss of generality, we assume in the following that the base station acts as the transmitter, while the mobile station is the receiver. Due to multipath propagation, the received signal is composed of an infinite sum of attenuated, delayed, and phase-shifted replicas of the transmitted signal, each influencing each other. Depending on the phase constellations of the received plane waves, the superposition can be constructive or destructive. A constructive (destructive) superposition of the received wave components corresponds to a high (low) received signal level. Apart from that, when transmitting digital signals, the form of the transmitted impulse can be distorted during transmission and often several individually distinguishable impulses occur at the receiver due to multipath propagation. This effect is known as the *impulse dispersion*. The size of the impulse dispersion depends on the propagation delay differences and the amplitude relations of the plane waves. We will see later on that

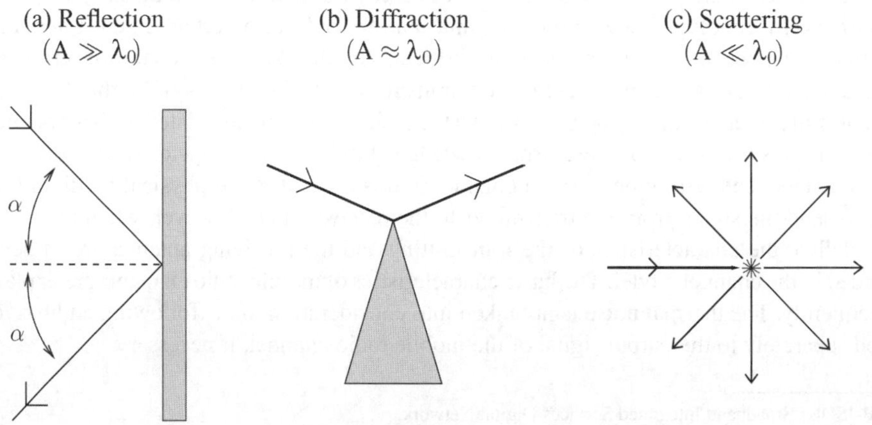

Figure 1.2 The three basic propagation phenomena: (a) reflection, (b) diffraction, and (c) scattering.

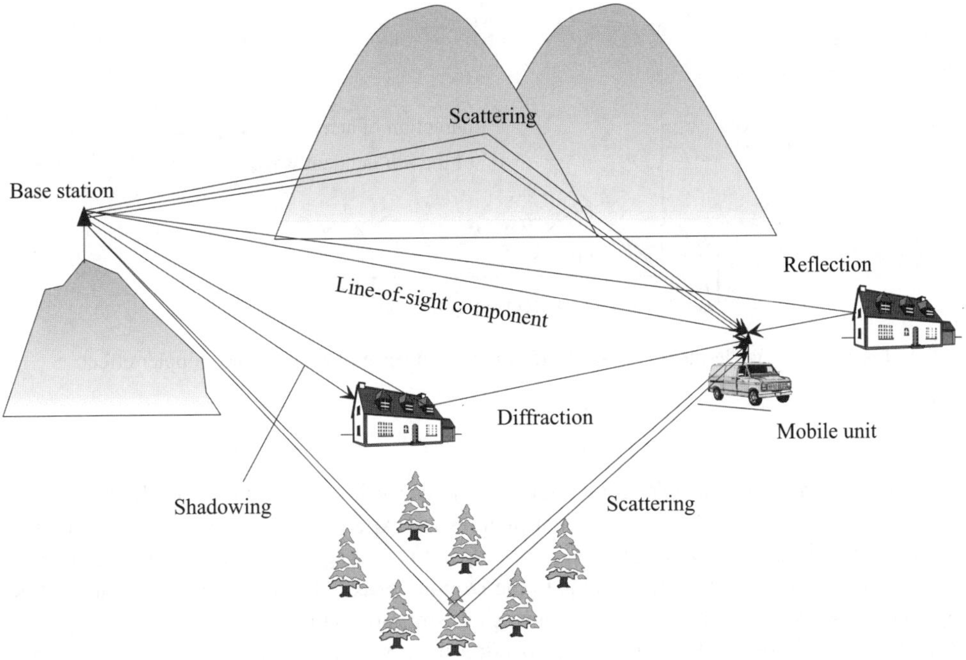

Figure 1.3 A typical mobile radio scenario illustrating the effect of multipath propagation due to reflection, diffraction, and scattering in a terrestrial mobile radio environment.

multipath propagation manifests itself in the frequency domain in a non-ideal frequency response of the transfer function of the mobile radio channel. As a consequence, the channel distorts the frequency response characteristic of the transmitted signal. This effect can generally be neglected in narrowband wireless systems, but not in wideband wireless systems, where the impulse dispersion of multipath channels results in *intersymbol interferences*. The distortions caused by multipath propagation are linear and have to be compensated in wideband wireless systems in the receiver, for example, by using equalization techniques.

Besides the multipath propagation, the *Doppler effect* also has a negative impact on the performance of mobile radio communication systems. Due to the movement of the mobile station, the Doppler effect causes a frequency shift of each of the incident plane waves. The *angle-of-arrival* α_n, which is defined by the direction of arrival of the nth incident wave and the direction of motion of the mobile station, as shown in Figure 1.4, determines the *Doppler frequency* (or *Doppler shift*) of the nth incident plane wave according to the relation

$$f_n := f_{\max} \cos \alpha_n \,, \tag{1.1}$$

where $f_{\max}$ denotes the *maximum Doppler frequency*. The maximum Doppler frequency $f_{\max}$ is related to the speed v of the mobile station, the speed of light c_0, and the carrier frequency f_0 through the equation

$$f_{\max} = \frac{\mathrm{v}}{c_0} f_0 \,. \tag{1.2}$$

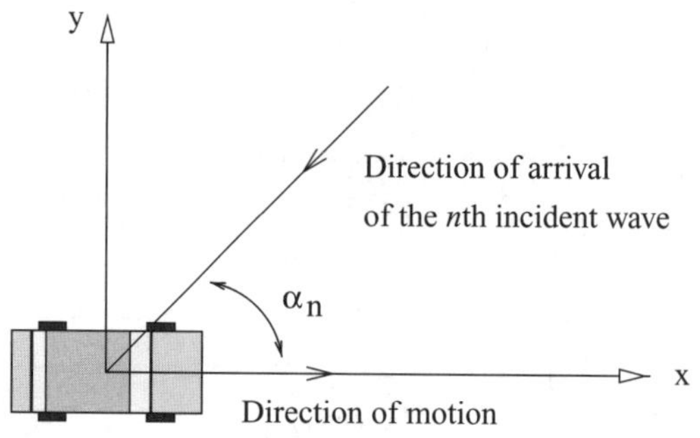

Figure 1.4 Angle-of-arrival α_n of the nth incident wave illustrating the Doppler effect.

Note that the maximum Doppler frequency $f_{\max}$ increases linearly with the mobile speed v and the carrier frequency f_0. The nth incident plane wave experiences the maximum (minimum) Doppler shift if $\alpha_n = 0$ ($\alpha_n = \pm\pi$), i.e., $f_n = f_{\max}$ ($f_n = -f_{\max}$). The Doppler shift is zero ($f_n = 0$) if $\alpha_n = \pi/2$ or $\alpha_n = 3\pi/2$. Due to the Doppler effect, the spectrum of the transmitted signal undergoes a frequency expansion during transmission. This effect is called the *frequency dispersion*. The size of the frequency dispersion mainly depends on the maximum Doppler frequency and the amplitudes of the received plane waves. In the time domain, the Doppler effect implicates that the impulse response of the channel becomes time-variant. One can easily show that mobile radio channels fulfil the principle of superposition [12], and therefore they are linear systems. Due to the time-variant behaviour of the impulse response, mobile radio channels fall generally into the class of linear time-variant systems.

Multipath propagation in connection with the movement of the receiver and/or the transmitter leads to drastic and random fluctuations of the received signal. Fades of 30 to 40 dB and more below the mean value of the received signal level can occur several times per second, depending on the speed of the mobile station and the carrier frequency [13]. A typical example of the behaviour of the received signal in mobile communications is shown in Figure 1.5. In this case, the speed of the mobile unit is v = 110 km/h and the carrier frequency is $f_0 = 900$ MHz. According to (1.2), this corresponds to a maximum Doppler frequency of $f_{\max} = 91$ Hz. In the present example, the distance covered by the mobile station during the chosen period of time from 0 to 0.327 s is equal to 10 m.

In digital data transmission, the fading of the received signal causes *burst errors* or *error bursts*. A burst error of length t_e is a sequence of t_e symbols, the first and the last of which are in error [14]. A fading interval produces burst errors, where the burst length t_e is closely related to the duration of the fading interval for which the term *duration of fades* has been coined [15]. Corresponding to this, a *connecting interval* produces a symbol sequence almost free of errors. Its length depends on the duration of the connecting interval, which is known as the *connecting time interval* [15]. As suitable measures for error protection and error correction, high-performance channel coding schemes with burst-error-correcting capabilities are called in to help. The development of error correction schemes requires detailed knowledge of the statistical distribution of the duration of fades as well as of the connecting time intervals. The tasks of channel modelling now are to identify, to analyze, and to model the main characteristic

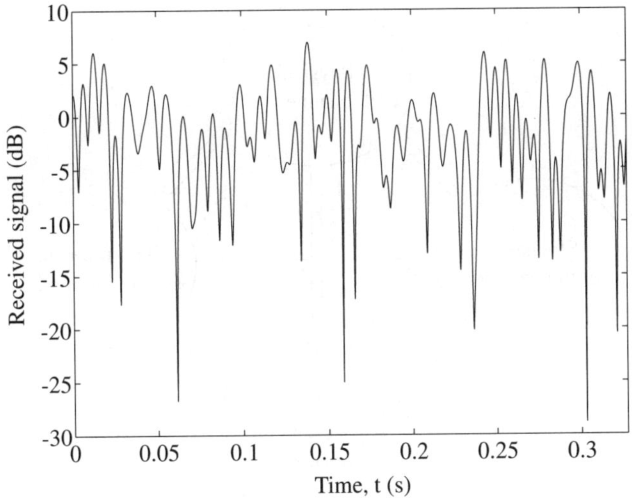

Figure 1.5 Typical behaviour of the received signal in mobile communications.

properties of the channel and to thus create a basis for the development, optimization, and test of digital transmission systems.

Classical methods of modelling the fading behaviour of mobile radio channels are characterized by the modelling of the transmission link between a base station and a mobile station. In the early stage of channel modelling, the aim was to characterize the statistical properties of real-world channels mainly with respect to the probability density function (first order statistics) of the channel's envelope. The time characteristics, and later the frequency characteristics of the mobile radio channel, have been included in the design procedure only to a limited degree. Modern methods of channel modelling aim to characterize the envelope fading regarding the first order statistics and the second order statistics, which includes the level-crossing rate and the average duration of fades. They also try to capture accurately the space-time-frequency characteristic of the mobile radio channel in a variety of environments. Questions connected to this theme will be treated in detail in this book. Mainly, two goals are aimed at. The first is to find proper stochastic processes, which are suitable for the modelling of the temporal, frequency, and spatial characteristics of mobile radio channels. In this context, we will refer to channel models described by ideal (non-realizable) stochastic processes as *reference models* or as *analytical models*. The second goal is to provide fundamental methods for the design of efficient simulation models enabling the simulation of a huge class of mobile radio channels on a software or hardware platform. The simulation model is usually derived from the underlying reference model or directly from measurements of a physical (real-world) channel. The usefulness and the importance of a reference model and the corresponding simulation model are ultimately judged on how well their statistical properties can be matched to the statistical properties of specified or measured channels. Following these primary goals, Figure 1.6 illustrates the relationships between the physical channel, the stochastic reference model, and the simulation model derived therefrom. These relationships will accompany us throughout the book.

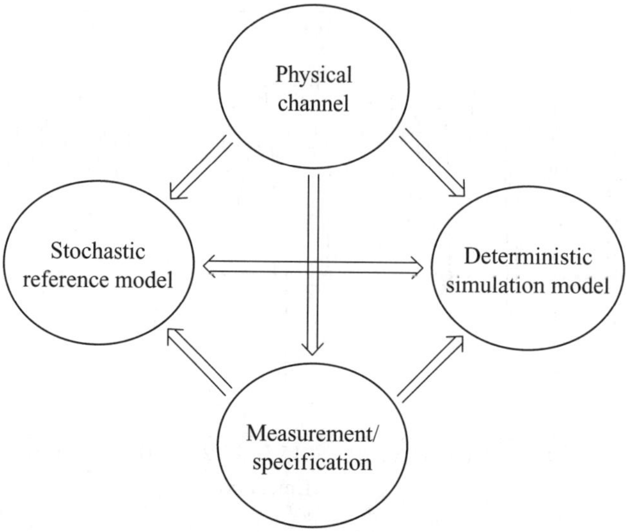

Figure 1.6 Relationships between the physical channel, the stochastic reference model, the deterministic simulation model, and the measurement or specification.

1.3 Structure of this Book

This book provides both fundamental and advanced topics in the area of mobile radio channel modelling. It serves as a basic introduction to concepts indispensable in the fascinating world of channel modelling and guides the reader step by step to the forefront of research. To this end, the book is split into ten chapters. A brief synopsis of each chapter is given in the following.

Chapter 2 outlines the basics of statistics and systems theory, which provide powerful tools for active research scientists as well as for engineers in practice. The main objective of this chapter is to provide a sound platform upon which a deeper understanding of mobile radio channels can be developed. Therefore, the most important definitions, terms, and formulas often referred to in the following chapters will be introduced. As a sideline, Chapter 2 makes the reader familiar with the nomenclature used consistently throughout the book.

Chapter 3 builds on the terms introduced in the previous chapter and introduces Rayleigh and Rice processes as reference models for characterizing frequency-nonselective mobile radio channels. This chapter opens with a system theoretical analysis of multipath fading channels. It continues with a formal description of Rayleigh and Rice channels. Then, their correlation properties and spectral characteristics are studied in detail. The most frequently used Doppler power spectral densities, known as the Jakes (or Clarke) power spectral density and the Gaussian power spectral density, are discussed and their characteristic quantities, such as the mean Doppler shift and the Doppler spread, are presented. This chapter elaborates further on the statistical analysis of Rayleigh and Rice processes by deriving their first order statistical properties (probability density function of the envelope and phase) as well as their second order statistical properties (level-crossing rate and average duration of fades). The last topic of this chapter is devoted to the analysis of the distribution of the fading intervals of Rayleigh channels.

Chapter 4 presents an introduction to sum-of-sinusoids channel models. From the systems developer's point of view, Rayleigh and Rice channels represent to a certain extent, like many other analytical channel models, non-realizable reference models. An important task in channel modelling is to find a flexible simulation model with low realization complexity that has approximately the same statistical properties as a given reference model. To solve this problem, various stochastic and deterministic methods have been proposed in the literature. The core of many methods is based on the well-known fact that filtered Gaussian random processes can be approximated by a finite sum of weighted sinusoids. This procedure can be traced back to the seminal work of S. O. Rice [16,17]. Starting from the original Rice method, the principle of deterministic channel modelling is developed at the beginning of Chapter 4. It follows a study of the elementary properties of sum-of-sinusoids processes, including the autocorrelation function, power spectral density, and Doppler spread. The analysis of the elementary properties of sum-of-sinusoids processes is performed by applying concepts of systems theory and signal theory. This is in contrast to the analysis of their statistical properties for which we will refer to concepts of probability theory and statistics. Of interest are both the first and the second order statistical properties. An overview of the various classes of sum-of-sinusoids processes is given and their stationary and ergodic properties are briefly described. Another substantial part of this chapter deals with processes comprising a sum of complex-valued sinusoids (cisoids). A sum-of-cisoids process allows a simple physical interpretation as a wave propagation model, which makes such processes very attractive for the development of mobile radio channel models. In this regard, the relationship and the main differences between sum-of-sinusoids and sum-of-cisoids processes are highlighted. Finally, the most important quality criteria for the performance evaluation of fading channel simulators are presented. The application of these criteria will turn out to be useful in subsequent chapters for the development of high-performance channel simulators.

Chapter 5 treats the parametrization of sum-of-sinusoids processes. It provides a comprehensive description and analysis of the most important procedures presently known for computing the model parameters of sum-of-sinusoids processes. The model parameters of sum-of-sinusoids processes are the gains, frequencies, and phases. Depending on the underlying philosophy of the parameter computation methods, they can be classified in deterministic and stochastic methods. Deterministic methods provide constant values for all model parameters, while stochastic methods result in random variables for at least one type of model parameters (gains, frequencies, phases). Deterministic (stochastic) parameter computation methods result in deterministic (stochastic) sum-of-sinusoids processes. The performance of each parameter computation method will be assessed with the help of the quality criteria introduced in the previous chapter. This chapter strives to fairly compare the performance of the proposed methods and highlights their individual advantages and disadvantages. This chapter also analyzes the duration of fades of Rayleigh fading channel simulators designed by using deterministic and stochastic sum-of-sinusoids processes. Chapter 5 ends with solutions to the parametrization problem of sum-of-cisoids processes, which unfold their full potential when modelling and simulating temporally and spatially correlated mobile radio channels in non-isotropic scattering environments.

Chapter 6 is concerned with the development of frequency-nonselective channel models. It is well known that the first and second order statistics of Rayleigh and Rice channels can only be controlled by a small number of parameters. On the one hand, this simplifies considerably the mathematical description of these models, but on the other, it restricts severely their flexibility in the sense that their main statistical properties (probability density function, level-crossing rate, and average duration of fades) can only be varied in a very limited range. A consequence of the small number of available parameters is that the statistical properties of real-world channels can only roughly be modelled by Rayleigh and Rice processes. To achieve a better fit to real-world channels, one therefore calls for more flexible stochastic model processes. Chapter 6 presents sophisticated combined stochastic processes for the modelling of frequency-nonselective mobile radio channels. The so-called extended Suzuki processes of Type I and Type II as well as the generalized Rice and Suzuki processes are derived and their statistical properties are analyzed. Apart from that, a modified version of the Loo model is introduced, containing the classical Loo model as a special case. To demonstrate the usefulness of all channel models suggested in this chapter, the statistical properties of each channel model in terms of the probability density function of the channel envelope, level-crossing rate, and average duration of fades are fitted to measurement results available in the literature. Starting from each underlying reference channel model, the corresponding simulation model is derived by using the concept of deterministic channel modelling, which provides us with the ability to confirm all theoretical results by simulations. The final part of this chapter delves into the modelling of nonstationary land mobile satellite channels. It includes an approach for the modelling and simulation of nonstationary real-world land mobile satellite channels.

Chapter 7 is dedicated to the development, analysis, and simulation of frequency-selective channel models. This chapter begins with a review of the ellipse model introduced originally by Parsons and Bajwa for describing the path geometry of multipath fading channels. It follows a system theoretical analysis of linear time-variant systems. With the help of systems theory, four important system functions are introduced allowing for a description of the input-output relationship of linear time-variant systems in alternative forms. The core of Chapter 7 is devoted to Bello's theory of linear time-variant stochastic systems going back to 1963 [18]. In this connection, all relevant stochastic system functions and the related characteristic quantities of frequency-selective stochastic channel models will be derived. Special attention will be paid to the so-called wide-sense stationary uncorrelated scattering (WSSUS) model. For typical propagation environments, the COST 207[14] channel models, specified by the European working group COST 207 [19], and the HIPERLAN/2[15] channel models according to ETSI[16] BRAN[17] [20] are presented. Another substantial part of Chapter 7 is devoted to the design and analysis of frequency-selective sum-of-sinusoids channel models enabling the simulation of wideband channels. In addition, this chapter includes an overview of methods for the modelling of given power delay profiles. The last part of Chapter 7

[14] COST: European Cooperation in the Field of Scientific and Technical Research.
[15] HIPERLAN/2: High Performance Radio Local Area Network Type 2.
[16] ETSI: European Telecommunications Standards Institute.
[17] BRAN: Broadband Radio Access Networks.

presents a general method for the modelling and simulation of measured wideband mobile radio channels.

Chapter 8 focusses on the modelling, analysis, and simulation of multiple-input multiple-output (MIMO) fading channels. A MIMO mobile system employs multiple antennas at the transmitter side and the receiver side, and a MIMO channel is the wireless link between the multiple transmitter and the multiple receiver antennas. MIMO channel models are important for the optimization, test, and performance evaluation of space-time coding schemes [21] as well as space-time processing techniques [22]. The main focus in this chapter is on geometry-based MIMO channel models. Starting from specific geometrical scattering models, a universal technique is presented for the derivation of stochastic reference MIMO channel models under the assumption of isotropic and non-isotropic scattering. By way of example, we apply the technique to the most important geometrical models, which are known as the one-ring model, the two-ring model, and the elliptical model. For all presented geometry-based MIMO channel models, the complex channel gains of the reference models are derived starting from a wave propagation model. The statistical properties of the derived MIMO channel models are studied in detail. General analytical solutions are provided for the three-dimensional (3D) space-time cross-correlation function from which other important correlation functions, such as the 2D space cross-correlation function and the time autocorrelation function, can easily be derived. Furthermore, from the non-realizable reference models, stochastic and deterministic simulation models are derived using a limited number of complex-valued sinusoids (cisoids). It is shown how the parameters of the simulation models can be determined for any given distribution of the angle-of-departure and the angle-of-arrival. In the case of isotropic scattering, closed-form solutions are presented for the parameter computation problem. The principal theoretical results for the designed reference and simulation models are illustrated and validated by simulations. The proposed procedure provides an important framework for designers of advanced mobile communication systems to verify new transmission concepts employing MIMO techniques under realistic propagation conditions.

Chapter 9 deals with the derivation, analysis, and realization of high-speed channel simulators. For the derivation of high-speed channel simulators, the periodicity of sinusoidal functions is exploited. It is shown how alternative structures for the simulation of sum-of-sinusoids processes can be derived. In particular, for complex Gaussian random processes, it is extraordinarily easy to develop simulation models by just using adders, storage elements, and simple address generators. During the actual simulation of the complex-valued channel envelope, not only time-consuming trigonometric operations but even multiplications can be avoided. This results in high-speed channel simulators, which are suitable for all kinds of channel models presented in previous chapters. Since the proposed principle can be generalized easily, we will restrict our attention to the derivation of high-speed channel simulators for Rayleigh channels. Therefore, we employ exclusively the discrete-time representation. At the beginning of Chapter 9, we introduce so-called discrete-time deterministic processes. These processes open up new possibilities for indirect realization forms. The three most important of them will be introduced in the second part of Chapter 9. In the third part, the elementary and statistical properties of discrete-time deterministic processes are examined. The second to last part deals with the analysis of the required realization complexity and with the measurement of the simulation speed of the designed high-speed channel simulators. Finally, Chapter 9 ends with a comparison of the Rice method and the filter method.

Chapter 10 concludes the book with an outline of three selected topics in mobile radio channel modelling. The first topic addresses the problem of designing multiple uncorrelated Rayleigh fading waveforms. After a short problem description, the reader will find a class of parameter computation methods, which enables him to design theoretically an infinite number of uncorrelated Rayleigh fading waveforms under the assumption of isotropic scattering conditions. The second topic is devoted to spatial channel models for shadow fading. Several correlation models for shadow fading are described, including the Gudmundson correlation model, the Gaussian correlation model, the Butterworth correlation model, and a measurement-based correlation model. Finally, the third and last topic elaborates on the modelling of frequency hopping mobile radio channels with applications to typical frequency hopping scenarios in GSM.

2

Random Variables, Stochastic Processes, and Deterministic Signals

Besides clarifying the terminology employed, this chapter introduces some important terms, which will later often be used in the context of modelling and simulation of mobile radio channels. The primary aim, however, is to familiarize the reader with the basic principles and definitions of probability, random signals, and systems theory, as far as it is important for the understanding of this book. It is anticipated that the reader has already accumulated a certain level of understanding in these areas. Therefore, a complete and detailed description of random variables, stochastic processes, and deterministic signals will not be presented here. Instead, links to key references are provided for further studies.

This chapter is split into four sections. Section 2.1 is devoted to a review of probability theory and random variables. Section 2.2 introduces the fundamentals of stochastic processes. Section 2.3 summarizes the main characteristics of deterministic signals. Finally, Section 2.4 provides a list of selected references for further reading.

2.1 Random Variables

In the context of this book, random variables are of central importance, not only to the statistical but also to the deterministic modelling of mobile radio channels. For this reason, we will at first review some basic definitions and terms which are frequently used in connection with random variables.

2.1.1 Basic Definitions of Probability Theory

An experiment whose outcome is not known in advance is called a *random experiment*. We will call points representing the outcomes of a random experiment *sample points* s. A collection of possible outcomes of a random experiment is an *event* A. The event $A = \{s\}$ consisting of

Mobile Radio Channels, Second Edition. Matthias Pätzold.

a single element s is an *elementary event*. The set of all possible outcomes of a given random experiment is called the *sample space* Q of the experiment. Hence, a sample point is an element of the event, i.e., $s \in A$, and the event itself is a subset of the sample space, i.e., $A \subset Q$. The sample space Q is called the *certain event*, and the *empty set* or *null set*, denoted by $\emptyset$, is the *impossible event*. Let $\mathcal{A}$ be a class (collection) of subsets of a sample space Q. In probability theory, $\mathcal{A}$ is often called *σ-field* (or *σ-algebra*), if and only if the following conditions are fulfilled:

(i) The empty set $\emptyset \in \mathcal{A}$.
(ii) If $A \in \mathcal{A}$, then also $Q - A \in \mathcal{A}$, i.e., if the event A is an element of the class $\mathcal{A}$, then so is its complement.
(iii) If $A_n \in \mathcal{A}$ $(n = 1, 2, \ldots)$, then also $\cup_{n=1}^{\infty} A_n \in \mathcal{A}$, i.e., if the events A_n are all elements of the class $\mathcal{A}$, then so is their countable union.

A pair $(Q, \mathcal{A})$ consisting of a sample space Q and a σ-field $\mathcal{A}$ is called a *measurable space*.

A mapping $P : \mathcal{A} \to \mathbb{R}$ is called the *probability measure* or in short *probability*, if the following conditions are fulfilled:

(i) If $A \in \mathcal{A}$, then $0 \leq P(A) \leq 1$.
(ii) $P(Q) = 1$.
(iii) If $A_n \in \mathcal{A}$ $(n = 1, 2, \ldots)$ with $\cup_{n=1}^{\infty} A_n \in \mathcal{A}$ and $A_n \cap A_k = \emptyset$ for any $n \neq k$, then also $P(\cup_{n=1}^{\infty} A_n) = \sum_{n=1}^{\infty} P(A_n)$.

A *probability space* is the triple $(Q, \mathcal{A}, P)$.

A *random variable* $\mu \in Q$ is a mapping which assigns to every outcome s of a random experiment a number $\mu(s)$, i.e.,

$$\mu : Q \to \mathbb{R}, \quad s \longmapsto \mu(s) . \tag{2.1}$$

This mapping has the property that the set $\{s \mid \mu(s) \leq x\}$ is an event of the considered σ-algebra for all $x \in \mathbb{R}$, i.e., $\{s \mid \mu(s) \leq x\} \in \mathcal{A}$. Hence, a random variable is a function of the elements of a sample space Q.

For the probability that the random variable μ is less or equal to x, we use the simplified notation

$$P\{\mu \leq x\} := P(\{s \mid \mu(s) \leq x\}) \tag{2.2}$$

in the sequel.

Cumulative distribution function (CDF): The function $F_\mu(x)$, defined by

$$F_\mu : \mathbb{R} \to [0, 1], \quad x \longmapsto F_\mu(x) = P\{\mu \leq x\} , \tag{2.3}$$

is called the *cumulative distribution function* of the random variable μ. The cumulative distribution function $F_\mu(x)$ satisfies the following properties:

(i) $F_\mu(x)$ is bounded between $F_\mu(-\infty) = 0$ and $F_\mu(+\infty) = 1$.
(ii) $F_\mu(x)$ is a non-decreasing function of x, i.e., if $x_1 \leq x_2$ then $F_\mu(x_1) \leq F_\mu(x_2)$.

(iii) $F_\mu(x)$ is continuous from the right, i.e., $\lim_{\varepsilon\to 0} F_\mu(x+\varepsilon) = F_\mu(x)$.
(iv) The probability of a number x equals $P\{\mu = x\} = F_\mu(x) - \lim_{\varepsilon\to 0} F_\mu(x-\varepsilon)$.
(v) The probability that μ is between x_1 and x_2 is given by $P\{x_1 < \mu \leq x_2\} = F_\mu(x_2) - F_\mu(x_1)$.

Probability density function (PDF): The function $p_\mu(x)$, defined by

$$p_\mu : \mathbb{R} \to \mathbb{R}, \quad x \longmapsto p_\mu(x) = \frac{dF_\mu(x)}{dx}, \tag{2.4}$$

is called the *probability density function* (or *probability density* or simply *density*) of the random variable μ, where it is assumed that the cumulative distribution function $F_\mu(x)$ is differentiable with respect to x. The probability density function $p_\mu(x)$ satisfies the following properties:

(i) Since $F_\mu(x)$ is a non-decreasing function of x, it follows that $p_\mu(x)$ is non-negative, because $p_\mu(x) = \frac{dF_\mu(x)}{dx} = \lim_{\Delta x\to 0} \frac{F_\mu(x+\Delta x)-F_\mu(x)}{\Delta x} \geq 0$ holds for all x.
(ii) From (2.4), we obtain $F_\mu(x) = \int_{-\infty}^{x} p_\mu(y)\,dy$.
(iii) Since $F_\mu(+\infty) = 1$, it follows from the previous property that $\int_{-\infty}^{\infty} p_\mu(x)dx = 1$ holds, i.e., the area under the probability density function $p_\mu(x)$ equals one.
(iv) Furthermore, from Property (ii), we also get $P\{x_1 < \mu \leq x_2\} = F_\mu(x_2) - F_\mu(x_1) = \int_{x_1}^{x_2} p_\mu(x)\,dx$.

Joint cumulative distribution function: The function $F_{\mu_1\mu_2}(x_1, x_2)$, defined by

$$F_{\mu_1\mu_2} : \mathbb{R}^2 \to [0, 1], \quad (x_1, x_2) \longmapsto F_{\mu_1\mu_2}(x_1, x_2) = P\{\mu_1 \leq x_1, \mu_2 \leq x_2\}, \tag{2.5}$$

is called the *joint cumulative distribution function* (or *bivariate cumulative distribution function*) of the random variables μ_1 and μ_2.

Joint probability density function: The function $p_{\mu_1\mu_2}(x_1, x_2)$, defined by

$$p_{\mu_1\mu_2} : \mathbb{R}^2 \to \mathbb{R}, \quad (x_1, x_2) \longmapsto p_{\mu_1\mu_2}(x_1, x_2) = \frac{\partial^2 F_{\mu_1\mu_2}(x_1, x_2)}{\partial x_1 \partial x_2}, \tag{2.6}$$

is called the *joint probability density function* (or *bivariate density function* or simply *bivariate density*) of the random variables μ_1 and μ_2, where it is assumed that the joint cumulative distribution function $F_{\mu_1\mu_2}(x_1, x_2)$ is partially differentiable with respect to both x_1 and x_2.

Independence: The random variables μ_1 and μ_2 are said to be *statistically independent* if the events $\{s \mid \mu_1(s) \leq x_1\}$ and $\{s \mid \mu_2(s) \leq x_2\}$ are independent for all $x_1, x_2 \in \mathbb{R}$. In this case, we can write $F_{\mu_1\mu_2}(x_1, x_2) = F_{\mu_1}(x_1) \cdot F_{\mu_2}(x_2)$ and $p_{\mu_1\mu_2}(x_1, x_2) = p_{\mu_1}(x_1) \cdot p_{\mu_2}(x_2)$.

Marginal probability density function: The *marginal probability density functions* (or *marginal densities*) of the joint probability density function $p_{\mu_1\mu_2}(x_1, x_2)$ are obtained by

$$p_{\mu_1}(x_1) = \int_{-\infty}^{\infty} p_{\mu_1\mu_2}(x_1, x_2)\, dx_2, \tag{2.7a}$$

$$p_{\mu_2}(x_2) = \int_{-\infty}^{\infty} p_{\mu_1\mu_2}(x_1, x_2)\, dx_1. \tag{2.7b}$$

Expected value (mean value): The quantity

$$m_\mu = E\{\mu\} = \int_{-\infty}^{\infty} x\, p_\mu(x)\, dx \tag{2.8}$$

is called the *expected value* (or *mean value* or *statistical average*) of the random variable μ, where $E\{\cdot\}$ denotes the *expected value operator*. The expected value of a constant $a \in \mathbb{R}$ equals $E\{a\} = a$. The expected value operator $E\{\cdot\}$ is linear, i.e., the relations $E\{a\mu\} = aE\{\mu\}$ ($a \in \mathbb{R}$) and $E\{\mu_1 + \mu_2\} = E\{\mu_1\} + E\{\mu_2\}$ hold. If μ_1 and μ_2 are statistically independent, then $E\{\mu_1 \cdot \mu_2\} = E\{\mu_1\} \cdot E\{\mu_2\}$. Let $f(\mu)$ be a function of the random variable μ. Then, the expected value of $f(\mu)$ can be determined by applying the fundamental relationship

$$E\{f(\mu)\} = \int_{-\infty}^{\infty} f(x) p_\mu(x)\, dx. \tag{2.9}$$

The extension to two random variables μ_1 and μ_2 leads to

$$E\{f(\mu_1, \mu_2)\} = \int_{-\infty}^{\infty}\int_{-\infty}^{\infty} f(x_1, x_2) p_{\mu_1\mu_2}(x_1, x_2)\, dx_1\, dx_2. \tag{2.10}$$

Variance: The value

$$\begin{aligned}\sigma_\mu^2 &= \mathrm{Var}\,\{\mu\} = E\left\{(\mu - E\{\mu\})^2\right\} \\ &= E\{\mu^2\} - \left(E\{\mu\}\right)^2 \\ &= E\{\mu^2\} - m_\mu^2\end{aligned} \tag{2.11}$$

is called the *variance* of the random variable μ, where $\mathrm{Var}\,\{\cdot\}$ denotes the *variance operator*. The variance of a random variable μ is a measure of the concentration of μ around its expected value. The variance of a constant $a \in \mathbb{R}$ is zero, i.e., $\mathrm{Var}\,\{a\} = 0$. The variance operator is nonlinear, as the relation $\mathrm{Var}\,\{a\mu + b\} = \alpha^2\,\mathrm{Var}\,\{\mu\}$ $(a, b \in \mathbb{R})$ holds. In the case of statistically independent random variables $\mu_1, \mu_2, \ldots, \mu_N$, we may write $\mathrm{Var}\,\{\sum_{n=1}^{N} \mu_n\} = \sum_{n=1}^{N} \mathrm{Var}\,\{\mu_n\}$. The positive constant $\sigma_\mu = \sqrt{\mathrm{Var}\,\{\mu\}} = \sqrt{E\{(\mu - E\{\mu\})^2\}}$ is known as the *standard deviation*, which represents the root-mean-square value of the random variable μ around its expected value.

Covariance: The *covariance* of two random variables μ_1 and μ_2 is defined by

$$\begin{aligned} C_{\mu_1\mu_2} = \operatorname{Cov}\{\mu_1, \mu_2\} &= E\big\{(\mu_1 - E\{\mu_1\})(\mu_2 - E\{\mu_2\})\big\} \\ &= E\{\mu_1\mu_2\} - E\{\mu_1\}E\{\mu_2\} \\ &= E\{\mu_1\mu_2\} - m_{\mu_1}m_{\mu_2} \,. \end{aligned} \tag{2.12}$$

Uncorrelatedness: Two random variables μ_1 and μ_2 are called *uncorrelated* if their covariance $C_{\mu_1\mu_2}$ is zero. This can be phrased in the following two equivalent forms:

$$C_{\mu_1\mu_2} = 0 \quad \text{and} \quad E\{\mu_1\mu_2\} = E\{\mu_1\}E\{\mu_2\} \,. \tag{2.13a,b}$$

If two random variables μ_1 and μ_2 are statistically independent, then they are also uncorrelated. The inverse statement is generally not true, i.e., if two random variables μ_1 and μ_2 are uncorrelated, then they are not necessarily independent. For normal random variables, however, uncorrelatedness is equivalent to independence.

Moments: The *kth moment* and the *kth central moment* of the random variable μ are defined by

$$E\{\mu^k\} = \int_{-\infty}^{\infty} x^k \, p_\mu(x) \, dx, \quad k = 0, 1, \ldots, \tag{2.14}$$

and

$$E\big\{(\mu - E\{\mu\})^k\big\} = \int_{-\infty}^{\infty} (\mu - E\{\mu\})^k \, p_\mu(x) \, dx, \quad k = 0, 1, \ldots, \tag{2.15}$$

respectively. Note that the mean and variance can be seen to be defined in terms of the first two moments $E\{\mu\}$ and $E\{\mu^2\}$.

Characteristic function: The characteristic function of a random variable μ is defined as

$$\Psi_\mu(\nu) = E\left\{e^{j2\pi\nu\mu}\right\} \tag{2.16a}$$

$$= \int_{-\infty}^{\infty} p_\mu(x) \, e^{j2\pi\nu x} \, dx \,, \tag{2.16b}$$

where ν is a real-valued variable. The two expressions given above motivate the following two interpretations of the characteristic function. In the first expression (2.16a), we can view $\Psi_\mu(\nu)$ as the expected value of the function $e^{j2\pi\nu\mu}$. In the second expression (2.16b), $\Psi_\mu(\nu)$ can be identified as (the complex conjugate of) the Fourier transform of the probability density function $p_\mu(x)$. From this fact it follows that the properties of the characteristic function are essentially the same as the properties of the Fourier transform. In particular, from the Fourier transform inversion formula, it follows that $p_\mu(x)$ can be expressed in terms of $\Psi_\mu(\nu)$

$$p_\mu(x) = \int_{-\infty}^{\infty} \Psi_\mu(\nu)\, e^{-j2\pi \nu x}\, d\nu\,. \tag{2.17}$$

The concept of the characteristic function often provides a simple technique for determining the probability density function of a sum of statistically independent random variables.

Convolution theorem: Let μ_1 and μ_2 be two statistically independent random variables. Then, the characteristic function of the sum $\mu = \mu_1 + \mu_2$ equals

$$\Psi_\mu(\nu) = \Psi_{\mu_1}(\nu) \cdot \Psi_{\mu_2}(\nu)\,. \tag{2.18}$$

In Subsection 2.1.3, we will see that the density $p_\mu(y)$ of the sum $\mu = \mu_1 + \mu_2$ of two independent random variables μ_1 and μ_2 is equal to the convolution of their probability density functions. From this and (2.18) it follows that the characteristic function of the convolution of two densities corresponds to the product of their characteristic functions. This result is often used when dealing with sums of random variables.

Moment theorem: The moment theorem states that the kth moment of a random variable μ is given by

$$E\{\mu^k\} = \frac{1}{(j2\pi)^k} \frac{d^k}{d\nu^k} \Psi_\mu(\nu)\bigg|_{\nu=0}\,, \quad k = 0, 1, \ldots \tag{2.19}$$

We mention that under certain conditions, the characteristic function $\Psi_\mu(\nu)$ and hence the probability density function $p_\mu(x)$ of a random variable μ are uniquely determined by the moments of μ.

Joint characteristic function: The *joint characteristic function* of the random variables μ_1 and μ_2 is defined as

$$\Psi_{\mu_1\mu_2}(\nu_1, \nu_2) = E\left\{e^{\,j2\pi(\nu_1\mu_1+\nu_2\mu_2)}\right\} \tag{2.20a}$$

$$= \int_{-\infty}^{\infty}\int_{-\infty}^{\infty} p_{\mu_1\mu_2}(x_1, x_2)\, e^{\,j2\pi(\nu_1\mu_1+\nu_2\mu_2)}\, dx_1\, dx_2\,. \tag{2.20b}$$

By analogy to the one-dimensional case, the two expressions given above can be interpreted as follows. In the first expression (2.20a), we can regard $\Psi_{\mu_1\mu_2}(\nu_1, \nu_2)$ as the expected value of the function $e^{\,j2\pi(\nu_1\mu_1+\nu_2\mu_2)}$. In the second expression (2.20b), $\Psi_{\mu_1\mu_2}(\nu_1, \nu_2)$ can be identified as (the complex conjugate of) the two-dimensional Fourier transform of the joint probability density function $p_{\mu_1\mu_2}(x_1, x_2)$.

The inversion formula for the two-dimensional Fourier transform implies that the joint probability density function is given by

$$p_{\mu_1\mu_2}(x_1, x_2) = \int_{-\infty}^{\infty}\int_{-\infty}^{\infty} \Psi_{\mu_1\mu_2}(\nu_1, \nu_2)\, e^{-j2\pi(\nu_1 x_1+\nu_2 x_2)}\, d\nu_1\, d\nu_2\,. \tag{2.21}$$

Marginal characteristic function: The *marginal characteristic functions* $\Psi_{\mu_1}(\nu)$ and $\Psi_{\mu_2}(\nu)$ of the joint characteristic function $\Psi_{\mu_1\mu_2}(\nu_1, \nu_2)$ can be obtained by

$$\Psi_{\mu_1}(\nu) = \Psi_{\mu_1\mu_2}(\nu, 0) \quad \text{and} \quad \Psi_{\mu_2}(\nu) = \Psi_{\mu_1\mu_2}(0, \nu)\,. \tag{2.22a,b}$$

If the two random variables μ_1 and μ_2 are statistically independent, then the joint characteristic function is the product of the marginal characteristic functions, i.e.,

$$\Psi_{\mu_1\mu_2}(\nu_1, \nu_2) = \Psi_{\mu_1}(\nu_1) \cdot \Psi_{\mu_2}(\nu_2)\,. \tag{2.23}$$

Chebyshev inequality: Let μ be an arbitrary random variable with a finite expected value and a finite variance. The *Chebyshev inequality* states that

$$P\{|\mu - E\{\mu\}| \geq \epsilon\} \leq \frac{\mathrm{Var}\,\{\mu\}}{\epsilon^2} \tag{2.24}$$

holds for any $\epsilon > 0$. The Chebyshev inequality is often used to obtain bounds on the probability of finding μ outside of the interval $E\{\mu\} \pm \epsilon\sqrt{\mathrm{Var}\,\{\mu\}}$.

Markov inequality: Suppose that μ is a non-negative random variable, i.e., $p_\mu(x) = 0$ for $x < 0$. Then, the Markov inequality states that

$$P\{\mu \geq \varepsilon\} \leq \frac{E\{\mu\}}{\varepsilon}\,, \tag{2.25}$$

where $\varepsilon > 0$.

Central limit theorem of Lindeberg-Lévy: Let μ_n $(n = 1, 2, \ldots, N)$ be N statistically independent, identically distributed (i.i.d.) random variables, each having a finite expected value $E\{\mu_n\} = m$ and a finite variance $\mathrm{Var}\,\{\mu_n\} = \sigma_0^2 > 0$. Then, the central limit theorem of Lindeberg-Lévy states that the normalized random variable

$$\mu = \lim_{N\to\infty} \frac{1}{\sqrt{N}} \sum_{n=1}^{N} (\mu_n - m_{\mu_n}) \tag{2.26}$$

is asymptotically normally distributed with the expected value $E\{\mu\} = 0$ and the variance $\mathrm{Var}\,\{\mu\} = \sigma_0^2$.

The central limit theorem plays an important role in stochastic limit theory. Notice that the random variables μ_n can have any distribution as long as they have a finite expected value and a finite variance. The density of the sum in (2.26) of merely 7 to 20 i.i.d. random variables (with almost identical variance) often results in a good approximation to the normal distribution. In fact, if the probability density function $p_{\mu_n}(x)$ is smooth, then values of N as low as five can be used. It should be emphasized, however, that the approximation to the tails of the normal distribution is generally poor if N is small.

In the following variation of the central limit theorem, the assumption of identically distributed random variables is abandoned in favour of a condition on the third absolute central moment of the random variables μ_n.

Central limit theorem of Lyapunov: Let μ_n $(n = 1, 2, \ldots, N)$ be N statistically independent random variables, each having a finite expected value $E\{\mu_n\} = m_{\mu_n}$ and a finite variance $\text{Var}\,\{\mu_n\} = \sigma^2_{\mu_n} > 0$. Furthermore, let

$$r_N = \sqrt[3]{\sum_{n=1}^{N} E\{|\mu_n - m_{\mu_n}|^3\}} \quad \text{and} \quad s_N = \sqrt{\sum_{n=1}^{N} \sigma^2_{\mu_n}}\,. \tag{2.27a,b}$$

The central limit theorem of Lyapunov states that if the condition $\lim_{N\to\infty} r_N/s_N = 0$ is fulfilled, then the normalized random variable

$$\mu = \lim_{N\to\infty} \frac{1}{\sqrt{N}} \sum_{n=1}^{N} (\mu_n - m_{\mu_n}) \tag{2.28}$$

is asymptotically normally distributed with the expected value $E\{\mu\} = 0$ and the variance $\text{Var}\,\{\mu\} = \sigma^2_\mu = \lim_{N\to\infty} \frac{1}{N} \sum_{n=1}^{N} \sigma^2_{\mu_n}$.

2.1.2 *Important Probability Density Functions*

In the following, a summary of some important probability density functions often used in connection with mobile radio channel modelling will be presented. The corresponding statistical properties such as the expected value and the variance will be dealt with as well.

Uniform distribution: Let θ be a real-valued random variable with the probability density function

$$p_\theta(x) = \begin{cases} \dfrac{1}{2\pi}, & x \in [-\pi, \pi)\,, \\ 0\,, & \text{else}\,. \end{cases} \tag{2.29}$$

Then, $p_\theta(x)$ is called the *uniform distribution* and θ is said to be *uniformly distributed* in the interval $[-\pi, \pi)$. To indicate that θ is uniformly distributed in the interval $[-\pi, \pi)$, we will often write $\theta \sim U[-\pi, \pi)$. The expected value and the variance of a uniformly distributed random variable θ are $E\{\theta\} = 0$ and $\text{Var}\,\{\theta\} = \pi^2/3$, respectively.

Gaussian distribution (normal distribution): Let μ be a real-valued random variable with the probability density function

$$p_\mu(x) = \frac{1}{\sqrt{2\pi}\sigma_\mu} e^{-\frac{(x-m_\mu)^2}{2\sigma^2_\mu}}\,, \quad x \in \mathbb{R}\,. \tag{2.30}$$

Then, $p_\mu(x)$ is called the *Gaussian distribution* (or *normal distribution*) and μ is said to be *Gaussian distributed* (or *normally distributed*). In the equation above, the quantity $m_\mu \in \mathbb{R}$ denotes the expected value and $\sigma^2_\mu \in (0, \infty)$ is the variance of μ, i.e.,

$$E\{\mu\} = m_\mu \tag{2.31}$$

and

$$\mathrm{Var}\,\{\mu\} = E\{\mu^2\} - m_\mu^2 = \sigma_\mu^2\,. \tag{2.32}$$

To describe the distribution properties of Gaussian distributed random variables μ, we often use the short notation $\mu \sim N(m_\mu, \sigma_\mu^2)$ instead of giving the complete expression (2.30). In particular, for $m_\mu = 0$ and $\sigma_\mu^2 = 1$, $N(0, 1)$ is called the *standard normal distribution*. The cumulative distribution function $F_\mu(x)$ of μ can be expressed as

$$F_\mu(x) = \frac{1}{2}\left[1 + \mathrm{erf}\left(\frac{x - m_\mu}{\sqrt{2}\sigma_\mu}\right)\right], \tag{2.33}$$

where erf$(\cdot)$ denotes the *error function*, which is defined as

$$\mathrm{erf}\,(x) = \frac{2}{\sqrt{\pi}}\int_0^x e^{-z^2}dz. \tag{2.34}$$

The error function is an odd function, i.e., $\mathrm{erf}(-x) = -\mathrm{erf}(x)$, and can be related to the *complementary error function* $\mathrm{erfc}\,(x)$ by

$$\mathrm{erfc}\,(x) = 1 - \mathrm{erf}\,(x) = \frac{2}{\sqrt{\pi}}\int_x^\infty e^{-z^2}dz\,. \tag{2.35}$$

We note that if μ is Gaussian (normally) distributed with zero mean, then its absolute value $|\mu|$ is *one-sided Gaussian (half-normally) distributed.*

Multivariate Gaussian distribution: Let us consider N real-valued Gaussian distributed random variables $\mu_1, \mu_2, \ldots, \mu_N$ with the expected values m_{μ_n} $(n = 1, 2, \ldots, N)$ and the variances $\sigma_{\mu_n}^2$ $(n = 1, 2, \ldots, N)$. The *multivariate Gaussian distribution* (or *multivariate normal distribution*) of the Gaussian random variables $\mu_1, \mu_2, \ldots, \mu_N$ is defined by

$$p_{\mu_1\mu_2\ldots\mu_N}(x_1, x_2, \ldots, x_N) = \frac{1}{\left(\sqrt{2\pi}\right)^N\sqrt{\det \boldsymbol{C}_\mu}}\, e^{-\frac{1}{2}(\boldsymbol{x}-\boldsymbol{m}_\mu)^T\boldsymbol{C}_\mu^{-1}(\boldsymbol{x}-\boldsymbol{m}_\mu)}, \tag{2.36}$$

where T denotes the transpose of a vector (or a matrix). In the above expression, $\boldsymbol{x}$ and $\boldsymbol{m}_\mu$ are column vectors, which are given by

$$\boldsymbol{x} = \begin{pmatrix} x_1 \\ x_2 \\ \vdots \\ x_N \end{pmatrix} \in \mathbb{R}^{N\times 1} \tag{2.37a}$$

and

$$\boldsymbol{m}_\mu = \begin{pmatrix} E\{\mu_1\} \\ E\{\mu_2\} \\ \vdots \\ E\{\mu_N\} \end{pmatrix} = \begin{pmatrix} m_{\mu_1} \\ m_{\mu_2} \\ \vdots \\ m_{\mu_N} \end{pmatrix} \in \mathbb{R}^{N\times 1}, \tag{2.37b}$$

respectively, and $\det \boldsymbol{C}_\mu$ ($\boldsymbol{C}_\mu^{-1}$) denotes the determinant (inverse) of the *covariance matrix*

$$\boldsymbol{C}_\mu = \begin{pmatrix} C_{\mu_1\mu_1} & C_{\mu_1\mu_2} & \cdots & C_{\mu_1\mu_N} \\ C_{\mu_2\mu_1} & C_{\mu_2\mu_2} & \cdots & C_{\mu_2\mu_N} \\ \vdots & \vdots & \ddots & \vdots \\ C_{\mu_N\mu_1} & C_{\mu_N\mu_2} & \cdots & C_{\mu_N\mu_N} \end{pmatrix} \in \mathbb{R}^{N\times N}. \quad (2.38)$$

The elements of the covariance matrix $\boldsymbol{C}_\mu$ are given by

$$C_{\mu_i\mu_j} = \text{Cov}\{\mu_i, \mu_j\} = E\{(\mu_i - m_{\mu_i})(\mu_j - m_{\mu_j})\}, \quad \forall i, j = 1, 2, \ldots, N. \quad (2.39)$$

If the N random variables μ_n are normally distributed and uncorrelated in pairs, then the covariance matrix $\boldsymbol{C}_\mu$ results in a diagonal matrix with diagonal entries $\sigma_{\mu_n}^2$. In this case, the joint probability density function in (2.36) decomposes into a product of N Gaussian distributions of the normally distributed random variables $\mu_n \sim N(m_{\mu_n}, \sigma_{\mu_n}^2)$. This implies that the random variables μ_n are statistically independent for all $n = 1, 2, \ldots, N$.

Rayleigh distribution: Let us consider two zero-mean statistically independent normally distributed random variables μ_1 and μ_2, each having the variance σ_0^2, i.e., $\mu_1, \mu_2 \sim N(0, \sigma_0^2)$. Furthermore, let us derive a new random variable from μ_1 and μ_2 according to $\zeta = \sqrt{\mu_1^2 + \mu_2^2}$. Then, ζ represents a *Rayleigh distributed* random variable. The probability density function $p_\zeta(x)$ of Rayleigh distributed random variables ζ is given by

$$p_\zeta(x) = \begin{cases} \dfrac{x}{\sigma_0^2} e^{-\frac{x^2}{2\sigma_0^2}}, & x \geq 0, \\ 0, & x < 0. \end{cases} \quad (2.40)$$

The corresponding cumulative distribution function $F_\zeta(x)$ of ζ is

$$F_\zeta(x) = \begin{cases} 1 - e^{-\frac{x^2}{2\sigma_0^2}}, & x \geq 0, \\ 0, & x < 0. \end{cases} \quad (2.41)$$

Rayleigh distributed random variables ζ have the expected value

$$E\{\zeta\} = \sigma_0 \sqrt{\frac{\pi}{2}} \quad (2.42)$$

and the variance

$$\text{Var}\{\zeta\} = \sigma_0^2 \left(2 - \frac{\pi}{2}\right). \quad (2.43)$$

The Rayleigh distribution and the corresponding cumulative distribution function are illustrated in Figures 2.1(a) and 2.1(b), respectively.

In mobile radio channel modelling, the Rayleigh distribution is often used to characterize the distribution of the fading envelope if no line-of-sight path exists between the transmitter and the receiver.

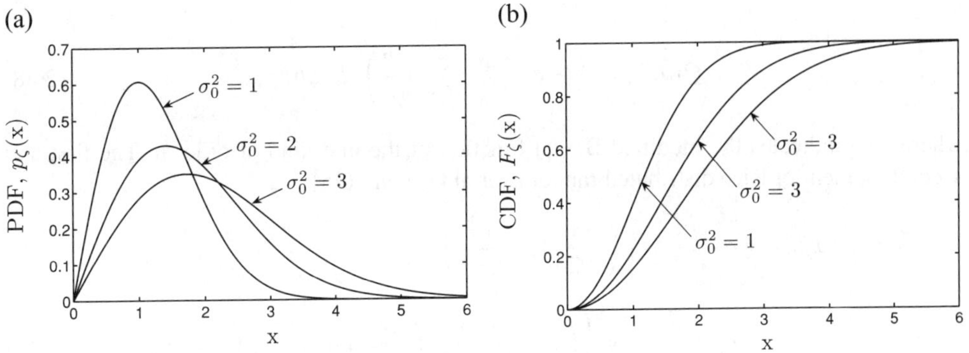

Figure 2.1 (a) The Rayleigh distribution $p_\zeta(x)$ and (b) the corresponding cumulative distribution function $F_\zeta(x)$ for various values of σ_0^2.

Rice distribution: Let $\mu_1,\ \mu_2 \sim N(0,\ \sigma_0^2)$ and $\rho \in \mathbb{R}$. Then, the random variable $\xi = \sqrt{(\mu_1 + \rho)^2 + \mu_2^2}$ is the so-called *Rice distributed* random variable. The probability density function $p_\xi(x)$ of Rice distributed random variables ξ is

$$p_\xi(x) = \begin{cases} \dfrac{x}{\sigma_0^2}\, e^{-\frac{x^2+\rho^2}{2\sigma_0^2}}\, I_0\left(\dfrac{x\rho}{\sigma_0^2}\right), & x \geq 0\,, \\ 0\,, & x < 0\,, \end{cases} \tag{2.44}$$

where $I_0(\cdot)$ denotes the modified Bessel function of the first kind of order zero [23, Equation (8.431.5)]. An important parameter characterizing the Rice distribution $p_\xi(x)$ is the *Rice factor* c_R, which is defined as $c_R = \rho^2/(2\sigma_0^2)$. For $\rho = 0$ ($c_R = 0$), the Rice distribution $p_\xi(x)$ results in the Rayleigh distribution $p_\zeta(x)$ described above. The cumulative distribution function $F_\xi(x)$ of ξ can be expressed as

$$F_\xi(x) = \begin{cases} 1 - Q\left(\dfrac{\rho}{\sigma_0}, \dfrac{x}{\sigma_0}\right), & x \geq 0\,, \\ 0\,, & x < 0\,, \end{cases} \tag{2.45}$$

where Q (·, ·) denotes the *Marcum Q-function* [24]

$$Q(a,b) = \int_b^\infty z\, e^{-\frac{z^2+a^2}{2}} I_0(az)\, dz\,. \tag{2.46}$$

The Marcum Q-function may be expanded in an infinite series giving

$$Q(a,b) = e^{-\frac{a^2+b^2}{2}} \sum_{n=0}^{\infty} \left(\frac{a}{b}\right)^n I_n(ab) \tag{2.47}$$

or

$$Q(a,b) = 1 - e^{-\frac{a^2+b^2}{2}} \sum_{n=1}^{\infty} \left(\frac{b}{a}\right)^n I_n(ab)\,, \tag{2.48}$$

where $I_n(\cdot)$ indicates the modified Bessel function of the first kind of order n. The first and second moment of Rice distributed random variables ξ are [25]

$$\begin{aligned} E\{\xi\} &= \sigma_0\sqrt{\frac{\pi}{2}}\,{}_1F_1\left(-\frac{1}{2};1;-\frac{\rho^2}{2\sigma_0^2}\right) \\ &= \sigma_0\sqrt{\frac{\pi}{2}}\,e^{-\frac{\rho^2}{4\sigma_0^2}}\left\{\left(1+\frac{\rho^2}{2\sigma_0^2}\right)I_0\left(\frac{\rho^2}{4\sigma_0^2}\right)+\frac{\rho^2}{2\sigma_0^2}I_1\left(\frac{\rho^2}{4\sigma_0^2}\right)\right\} \end{aligned} \tag{2.49}$$

and

$$E\{\xi^2\} = 2\sigma_0^2 + \rho^2\,, \tag{2.50}$$

respectively, where ${}_1F_1(\cdot;\cdot;\cdot)$ is the generalized hypergeometric function. From (2.49), (2.50), and by using (2.11), the variance of ξ can easily be calculated. For various values of ρ, the Rice distribution and the corresponding cumulative distribution function are shown in Figures 2.2(a) and 2.2(b), respectively.

The Rice distribution is frequently used in mobile communications to model the statistics of the fading envelope in cases where a line-of-sight component is present.

Lognormal distribution: Let μ be a Gaussian distributed random variable with the expected value m_μ and the variance σ_μ^2, i.e., $\mu \sim N(m_\mu, \sigma_\mu^2)$. Then, the random variable $\lambda = e^\mu$ is said to be *lognormally distributed*. The probability density function $p_\lambda(x)$ of lognormally distributed random variables λ is given by

$$p_\lambda(x) = \begin{cases} \dfrac{1}{\sqrt{2\pi}\sigma_\mu x}\,e^{-\frac{(\ln(x)-m_\mu)^2}{2\sigma_\mu^2}}\,, & x \geq 0\,, \\ 0\,, & x < 0\,. \end{cases} \tag{2.51}$$

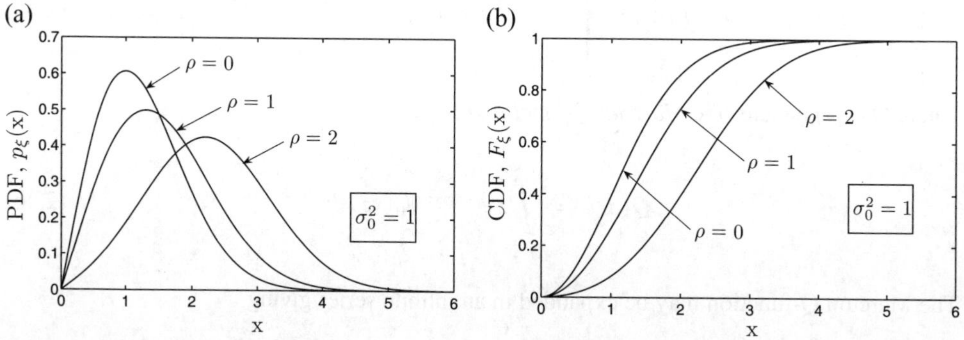

Figure 2.2 (a) The Rice distribution $p_\xi(x)$ and (b) the corresponding cumulative distribution function $F_\xi(x)$ for various values of ρ ($\sigma_0^2 = 1$).

The cumulative distribution function $F_\lambda(x)$ of λ can be expressed in terms of the error function [see (2.34)] as

$$F_\lambda(x) = \begin{cases} \frac{1}{2} + \frac{1}{2}\,\mathrm{erf}\left[\frac{\ln(x)-m_\mu}{\sqrt{2}\sigma_\mu}\right], & x \geq 0\,, \\ 0\,, & x < 0\,. \end{cases} \tag{2.52}$$

Lognormally distributed random variables λ have the expected value

$$E\{\lambda\} = e^{m_\mu + \frac{\sigma_\mu^2}{2}} \tag{2.53}$$

and the variance

$$\mathrm{Var}\,\{\lambda\} = e^{2m_\mu + \sigma_\mu^2}\left(e^{\sigma_\mu^2} - 1\right)\,. \tag{2.54}$$

The lognormal distribution and the corresponding cumulative distribution function are illustrated in Figures 2.3(a) and 2.3(b), respectively, for various values of σ_μ^2 ($m_\mu = 0$).

In mobile radio channel modelling, the lognormal distribution is an appropriate stochastic model to describe the envelope changes caused by shadowing. Shadowing effects occur when propagation paths between the transmitter and the receiver are obscured by large obstructions such as tall buildings and hills.

Suzuki distribution: Consider a Rayleigh distributed random variable ζ with the probability density function $p_\zeta(x)$ according to (2.40) and a lognormally distributed random variable λ with the probability density function $p_\lambda(x)$ given by (2.51). Let us assume that ζ and λ are statistically independent. Furthermore, let η be a random variable defined by the product $\eta = \zeta \cdot \lambda$. Then, the probability density function $p_\eta(x)$ of η, that is

$$p_\eta(x) = \begin{cases} \dfrac{x}{\sqrt{2\pi}\sigma_0^2\sigma_\mu} \displaystyle\int_0^\infty \frac{1}{y^3}\, e^{-\frac{x^2}{2y^2\sigma_0^2}}\, e^{-\frac{(\ln y - m_\mu)^2}{2\sigma_\mu^2}}\, dy\,, & x \geq 0\,, \\ 0\,, & x < 0\,, \end{cases} \tag{2.55}$$

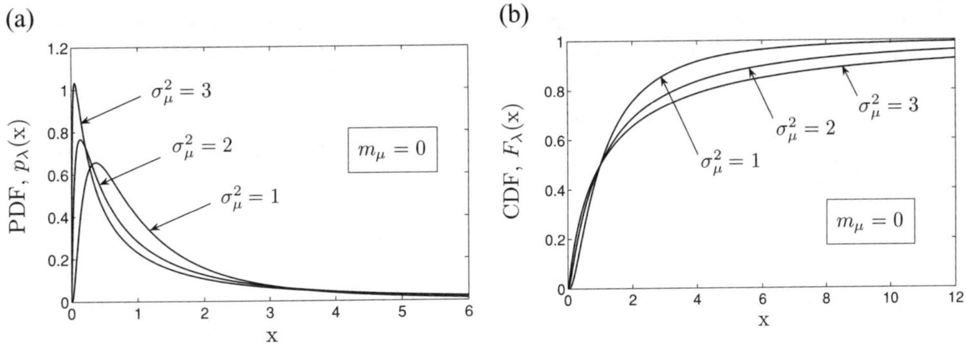

Figure 2.3 (a) The lognormal distribution $p_\lambda(x)$ and (b) the corresponding cumulative distribution function $F_\lambda(x)$ for various values of σ_μ^2 ($m_\mu = 0$).

is called the *Suzuki distribution* [26]. Suzuki distributed random variables η have the expected value

$$E\{\eta\} = \sigma_0 \sqrt{\frac{\pi}{2}} e^{m_\mu + \frac{\sigma_\mu^2}{2}} \tag{2.56}$$

and the variance

$$\mathrm{Var}\{\eta\} = \sigma_0^2 \, e^{2m_\mu + \sigma_\mu^2} \left(2e^{\sigma_\mu^2} - \frac{\pi}{2} \right). \tag{2.57}$$

Figure 2.4(a) illustrates the Suzuki distribution for various values of σ_μ^2 and σ_0^2 ($m_\mu = 0$). The graphs of the corresponding cumulative distribution function, obtained by using $F_\eta(x) = \int_0^x p_\eta(z)\, dz$, are shown in Figure 2.4(b).

The Suzuki distribution is often used in mobile communications to describe the envelope fading caused by the combined effects of multipath propagation and shadowing.

Nakagami-*m* distribution: Consider a random variable ω distributed according to the probability density function

$$p_\omega(x) = \begin{cases} \dfrac{2m^m x^{2m-1} e^{-(m/\Omega)x^2}}{\Gamma(m)\, \Omega^m}, & m \geq 1/2, \quad x \geq 0, \\ 0, & x < 0. \end{cases} \tag{2.58}$$

Then, ω denotes a *Nakagami-m distributed* (or simply *Nakagami distributed*) random variable and the corresponding probability density function $p_\omega(x)$ is called the *Nakagami-m distribution* (or *Nakagami distribution*) [27]. In (2.58), the symbol $\Gamma(\cdot)$ represents the gamma function,

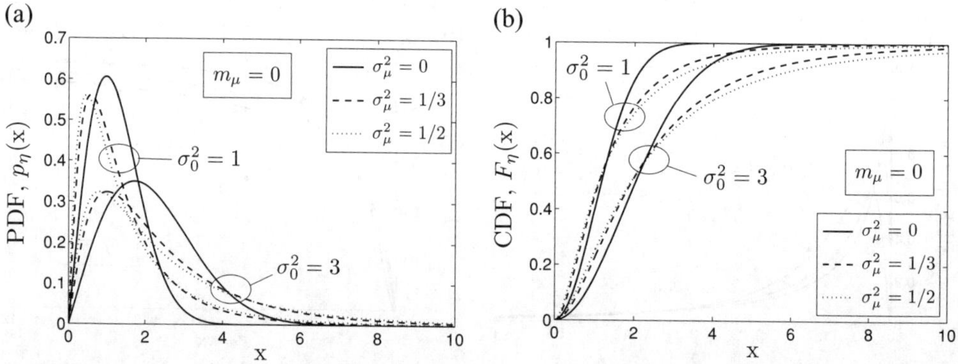

Figure 2.4 (a) The Suzuki distribution $p_\eta(x)$ and (b) the corresponding cumulative distribution function $F_\eta(x)$ for various values of σ_μ^2 and σ_0^2 ($m_\mu = 0$).

which is according to Euler defined as[1]

$$\Gamma(x) = \int_0^\infty z^{x-1}\, e^{-z}\, dz\,, \tag{2.59}$$

where x is a positive real number. The parameter Ω denotes the second moment of the random variable ω, i.e., $\Omega = E\{\omega^2\}$, and m is called the *Nakagami shape factor*, which is defined as the reciprocal value of the variance of ω^2 normalized to Ω^2, i.e., $m = \Omega^2/E\{(\omega_1^2 - \Omega)^2\}$. The cumulative distribution function of Nakagami-m distributed random variables ω can be expressed as

$$F_\omega(x) = \begin{cases} \dfrac{\gamma\left(m, \frac{m}{\Omega}x^2\right)}{\Gamma(m)}\,, & m \geq 1/2\,,\ x \geq 0\,, \\ 0\,, & x < 0\,, \end{cases} \tag{2.60}$$

where $\gamma(\,\cdot, \cdot)$ denotes the *incomplete gamma function* [23, Equation (8.350.1)]

$$\gamma(\alpha, x) = \int_0^x z^{\alpha-1} e^{-z}\, dz\,, \qquad \mathrm{Re}\,\{\alpha\} > 0\,. \tag{2.61}$$

The kth moment and the variance of ω are given by

$$E\{\omega^k\} = \frac{\Gamma\left(m + \frac{k}{2}\right)}{\Gamma(m)} \left(\frac{\Omega}{m}\right)^{\frac{k}{2}} \tag{2.62}$$

and

$$\mathrm{Var}\,\{\omega\} = \Omega \left[1 - \frac{1}{m}\left(\frac{\Gamma\left(m + \frac{1}{2}\right)}{\Gamma(m)}\right)^2\right], \tag{2.63}$$

respectively. From the Nakagami-m distribution, we obtain the one-sided Gaussian distribution if $m = 1/2$ and the Rayleigh distribution if $m = 1$. If the Nakagami shape factor m is in the range $1/2 \leq m < 1$, then the tail of the Nakagami-m distribution is larger than that of the Rayleigh distribution. On the other hand, if $m > 1$, then the tail of the Nakagami-m distribution decays faster compared to the Rayleigh distribution. In certain limits, the Nakagami-m distribution appears similar to the Rice distribution if the parameters of the Nakagami-m distribution and the parameters of the Rice distribution are related by [27, 28]

$$m = \frac{(c_R + 1)^2}{2c_R + 1} \tag{2.64}$$

$$\Omega = \frac{2\sigma_0^2}{1 - \sqrt{1 - m^{-1}}}\,. \tag{2.65}$$

[1] The gamma function $\Gamma(x)$ satisfies the recurrence equation $\Gamma(x+1) = x\,\Gamma(x) = x!$ for $x > 0$, which allows the interpretation of the gamma function as an extension of the factorial function to real numbers. If x is a non-negative integer $m = 0, 1, \ldots$, then $\Gamma(m+1) = m!$ holds.

Nakagami random variables can be expressed in terms of Gaussian random variables as follows. Let μ_i $(i = 1, 2, \ldots, n)$ be $n = 2m$ real-valued uncorrelated Gaussian random variables, each with zero mean and variance σ_0^2, then the transformed random variable

$$\omega = \sqrt{\sum_{i=1}^{n} \mu_i^2} \tag{2.66}$$

follows the Nakagami-m distribution with parameters m and $\Omega = n\sigma_0^2$ for any integer $n = 2m \geq 1$ [29]. We note that if $n = 1$ $(m = 1/2)$, then $\omega = |\mu_1|$ is one-sided Gaussian distributed. We note also that this model includes the Rayleigh case, which is obtained when $n = 2$ $(m = 1)$. Figure 2.5 illustrates the Nakagami-m distribution and the corresponding cumulative distribution function for various values of m $(\Omega = 1)$.

Originally, the Nakagami-m distribution was introduced by Minoru Nakagami [27] to characterize the rapid fading effects in long-distance high frequency (HF) channels. In mobile communications, the Nakagami-m distribution was selected because it often provides a closer fit to measured data than the Rayleigh or the Rice distribution [28, 26]. The formula (2.58) is rather elegant and has proven useful as it often results in closed-form solutions in system performance studies.

Nakagami-*q* distribution: Let μ_1 and μ_2 be two zero-mean statistically independent normally distributed random variables with variances σ_1^2 and σ_2^2, respectively, i.e., $\mu_i \sim N(0, \sigma_i^2)$, $i = 1, 2$, where the variances σ_1^2 and σ_2^2 can be different. Then, the random variable $X = |\mu_1 + j\mu_2| = \sqrt{\mu_1^2 + \mu_2^2}$ is said to be *Nakagami-q distributed*.[2] The probability density function $p_X(x)$ of X is known as the *Nakagami-q distribution*, which is given by [27, 30]

$$p_X(x) = \frac{1+q^2}{q\Omega_X} x\, e^{-\frac{(1+q^2)^2}{4q^2\Omega_X}x^2} I_0\left(\frac{1-q^4}{4q^2\Omega_X}x^2\right). \tag{2.67}$$

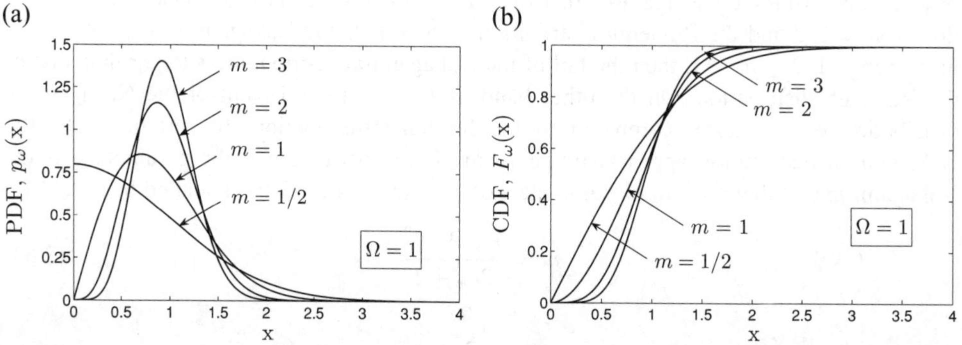

Figure 2.5 (a) The Nakagami-m distribution $p_\omega(x)$ and (b) the corresponding cumulative distribution function $F_\omega(x)$ for various values of m $(\Omega = 1)$.

[2] The Nakagami-q distribution is sometimes also referred to as the *Hoyt distribution* or the *Nakagami-Hoyt distribution.*

The parameter Ω_X denotes the second moment of the random variable X, i.e., $\Omega_X = E\{X^2\} = \sigma_1^2 + \sigma_2^2$, and q is called the *Nakagami-q shape factor*, which is defined as the ratio of σ_2 and σ_1, i.e., $q = \sigma_2/\sigma_1$ ($\sigma_1 \neq 0, 0 \leq q \leq 1$).

The cumulative distribution function of Nakagami-q distributed random variables can be expressed in closed form as the difference of two Marcum Q-functions [31]

$$F_X(x) = Q\left(a(q)\frac{x}{\sqrt{\Omega_X}}, b(q)\frac{x}{\sqrt{\Omega_X}}\right) - Q\left(b(q)\frac{x}{\sqrt{\Omega_X}}, a(q)\frac{x}{\sqrt{\Omega_X}}\right), \tag{2.68}$$

where

$$a(q) = \sqrt{\frac{1+q^4}{2q}}\sqrt{\frac{1+q}{1-q}}, \tag{2.69}$$

$$b(q) = \sqrt{\frac{1-q^4}{2q}}\sqrt{\frac{1-q}{1+q}} = a(q)\sqrt{\frac{1-q}{1+q}}. \tag{2.70}$$

An alternative expression for the Nakagami-q cumulative distribution function has been derived in [32].

Nakagami-q distributed random variables have the expected value

$$E\{X\} = 2\sqrt{\pi\Omega_X}\frac{q^2}{(1+q^2)^2}\,F\left(\frac{3}{4}, \frac{5}{4}; 1; \left(\frac{1-q^2}{1+q^2}\right)^2\right) \tag{2.71}$$

and the variance

$$\mathrm{Var}\,\{X\} = \Omega_X\left[1 - 4\pi\frac{q^4}{(1+q^2)^4}\,F^2\left(\frac{3}{4}, \frac{5}{4}; 1; \left(\frac{1-q^2}{1+q^2}\right)^2\right)\right], \tag{2.72}$$

where the symbol $F(\cdot, \cdot; \cdot; \cdot)$ stands for the hypergeometric function [23, Equation (9.100)].

From the Nakagami-q distribution, we obtain the one-sided Gaussian distribution if $q = 0$ and the Rayleigh distribution if $q = 1$. For $1/2 \leq m \leq 1$, the Nakagami-m distribution is very close to the Nakagami-q distribution if the parameters of the Nakagami-m distribution and the parameters of the Nakagami-q distribution are related by [27]

$$\Omega = \frac{\Omega_X}{2}, \tag{2.73}$$

$$m = \frac{(1+q^2)^2}{(1+q^2)^2 + (1-q^2)^2}. \tag{2.74}$$

Figure 2.6 illustrates the Nakagami-q distribution and the corresponding cumulative distribution function for various values of the Nakagami-q shape factor.

The Nakagami-q (Hoyt) distribution was first introduced by Hoyt [30] in 1947 and later revisited by Nakagami [27] in 1960. In mobile fading channel modelling, the Nakagami-q distribution is usually used to study the performance of mobile communication systems under fading conditions which are more severe than Rayleigh fading. The Nakagami-q channel model is an empirical model, which lacks a clear physical interpretation. Nevertheless, Nakagami-q

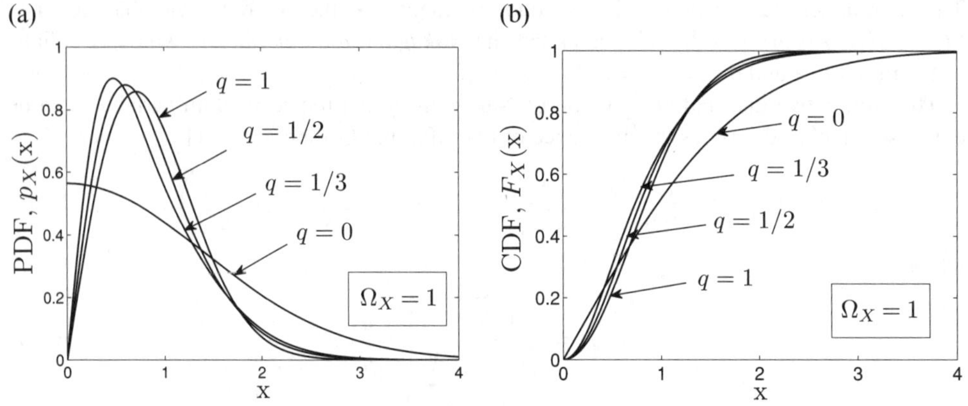

Figure 2.6 (a) The Nakagami-q distribution $p_X(x)$ and (b) the corresponding cumulative distribution function $F_X(x)$ for various values of q ($\Omega_X = 1$).

channel models are quite useful, as it was shown in [33] that they allow in some cases a better match to measured channels than Rayleigh channel models. The Nakagami-q model allows one to simulate Nakagami-m channels efficiently if the Nakagami shape factor m is between 0.5 and 1.

Weibull distribution: Let μ_1 and μ_2 be two zero-mean statistically independent normally distributed random variables, each having the variance σ_0^2, i.e., $\mu_1, \mu_2 \sim N(0, \sigma_0^2)$. Then, a *Weibull distributed* random variable w is obtained through the transformation $\mathrm{w} = |\mu_1 + j\mu_2|^{2/\beta_\mathrm{w}} = (\mu_1^2 + \mu_2^2)^{1/\beta_\mathrm{w}}$, where β_w is a positive real constant value. The probability density function $p_\mathrm{w}(x)$ of w is given by the *Weibull distribution*

$$p_\mathrm{w}(x) = \begin{cases} \frac{\beta_\mathrm{w}}{\Omega}\, x^{\beta_\mathrm{w}-1}\, e^{-x^{\beta_\mathrm{w}}/\Omega}, & x \geq 0, \\ 0, & x < 0, \end{cases} \tag{2.75}$$

where $\Omega = E\{\mathrm{w}^{\beta_\mathrm{w}}\} = 2\sigma_0^2$. The parameter β_w is referred to as the *Weibull shape factor*. The cumulative distribution function $F_\mathrm{w}(x)$ of w is given by

$$F_\mathrm{w}(x) = \begin{cases} 1 - e^{-x^{\beta_\mathrm{w}}/\Omega}, & x \geq 0, \\ 0, & x < 0. \end{cases} \tag{2.76}$$

Weibull distributed random variables have the expected value

$$E\{\mathrm{w}\} = \Omega^{1/\beta_\mathrm{w}}\,\Gamma\left(1 + \frac{1}{\beta_\mathrm{w}}\right) \tag{2.77}$$

and the variance

$$\mathrm{Var}\,\{\mathrm{w}\} = \Omega^{2/\beta_\mathrm{w}}\left[\Gamma\left(1 + \frac{2}{\beta_\mathrm{w}}\right) - \Gamma^2\left(1 + \frac{1}{\beta_\mathrm{w}}\right)\right], \tag{2.78}$$

where $\Gamma(\cdot)$ is the gamma function in (2.59). From the Weibull distribution, we obtain the negative exponential distribution if $\beta_\mathrm{w} = 1$ and the Rayleigh distribution if $\beta_\mathrm{w} = 2$. For

$\beta_W = 3.4$, the Weibull distribution is close to the Gaussian distribution. The Weibull distribution and the corresponding cumulative distribution function are illustrated in Figure 2.7 for various Weibull shape factors β_W $(\Omega = 1)$.

The Weibull distribution was first introduced by Waloddi Weibull in 1937 for estimating machinery lifetime and became widely known in 1951 through his famous publication [34]. At the present day, the Weibull distribution is one of the most popular probability density functions in reliability engineering and failure data analysis [35]. In mobile communications, the Weibull distribution became attractive because it exhibits a good fit to experimental measurements of fading channels in both indoor [36, 37] and outdoor environments [38–40].

2.1.3 Functions of Random Variables

In some parts of this book, we will deal with functions of two and more random variables. In particular, we will often make use of fundamental rules in connection with the addition, multiplication, and transformation of random variables. In the following, the mathematical principles necessary for this will be reviewed briefly.

Addition of two random variables: Let μ_1 and μ_2 be two random variables, which are statistically characterized by the joint probability density function $p_{\mu_1\mu_2}(x_1, x_2)$. Then, the probability density function of the sum $\mu = \mu_1 + \mu_2$ can be obtained as follows

$$p_\mu(y) = \int_{-\infty}^{\infty} p_{\mu_1\mu_2}(x_1, y - x_1)\, dx_1 = \int_{-\infty}^{\infty} p_{\mu_1\mu_2}(y - x_2, x_2)\, dx_2 \,. \tag{2.79}$$

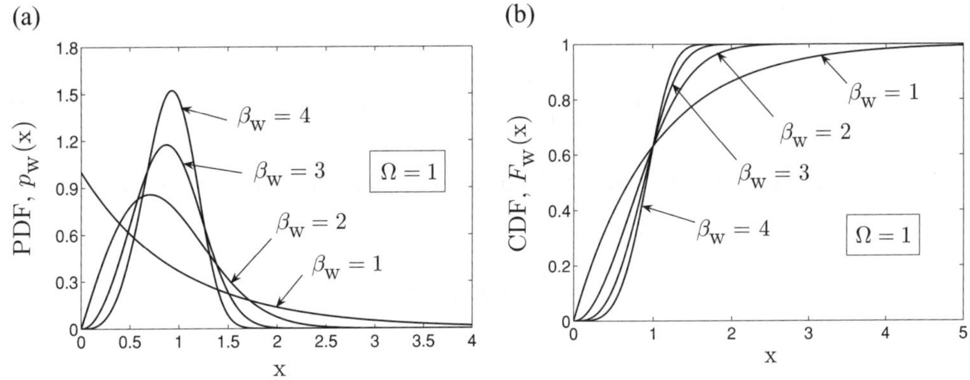

Figure 2.7 (a) The Weibull distribution $p_W(x)$ and (b) the corresponding cumulative distribution function $F_W(x)$ for various values of β_W $(\Omega = 1)$.

If the two random variables μ_1 and μ_2 are statistically independent, then it follows that the probability density function of μ is given by the convolution of the probability densities of μ_1 and μ_2. Thus,

$$\begin{aligned} p_\mu(y) &= p_{\mu_1}(y) * p_{\mu_2}(y) \\ &= \int_{-\infty}^{\infty} p_{\mu_1}(x_1)\, p_{\mu_2}(y - x_1)\, dx_1 \\ &= \int_{-\infty}^{\infty} p_{\mu_1}(y - x_2)\, p_{\mu_2}(x_2)\, dx_2 , \end{aligned} \tag{2.80}$$

where $*$ denotes the convolution operator.

Multiplication of two random variables: Let ζ and λ be two random variables, which are statistically described by the joint probability density function $p_{\zeta\lambda}(x, y)$. Then, the probability density function of the random variable $\eta = \zeta \cdot \lambda$ is equal to

$$p_\eta(z) = \int_{-\infty}^{\infty} \frac{1}{|y|}\, p_{\zeta\lambda}\left(\frac{z}{y}, y\right) dy . \tag{2.81}$$

From this relation, we obtain the expression

$$p_\eta(z) = \int_{-\infty}^{\infty} \frac{1}{|y|}\, p_\zeta\left(\frac{z}{y}\right) p_\lambda(y)\, dy \tag{2.82}$$

for statistically independent random variables ζ, λ.

Quotient of two random variables: Let ζ and λ be two random variables with joint probability density function $p_{\zeta\lambda}(x, y)$. Then, the probability density function of the random variable $\eta = \zeta/\lambda$ is given by

$$p_\eta(z) = \int_{-\infty}^{\infty} |y|\, p_{\zeta\lambda}(yz, y)\, dy . \tag{2.83}$$

If the two variables ζ and λ are statistically independent, then the above expression simplifies to

$$p_\eta(z) = \int_{-\infty}^{\infty} |y|\, p_\zeta(yz)\, p_\lambda(y)\, dy . \tag{2.84}$$

Transformation of one random variable: Suppose that μ is a random variable with probability density function $p_\mu(x)$, and let $f(\cdot)$ be a real-valued function. Furthermore, let y be a particular value of the transformed random variable $\xi = f(\mu)$. If the equation $y = f(x)$ has

$m \geq 1$ real-valued roots $x_1, x_2, \ldots, x_m$, i.e.,

$$y = f(x_1) = f(x_2) = \cdots = f(x_m), \tag{2.85}$$

then the probability density function $p_\xi(y)$ of $\xi = f(\mu)$ can be expressed in terms of the function $f(\cdot)$ and the probability density function $p_\mu(x)$ of μ as

$$p_\xi(y) = \sum_{\nu=1}^{m} \frac{p_\mu(x_\nu)}{|f'(x)|_{x=x_\nu}}, \tag{2.86}$$

where $f'(x)$ denotes the derivative of $f(x)$.

Transformation of *n* random variables: Suppose that $\mu_1, \mu_2, \ldots, \mu_n$ are random variables, which are statistically described by the joint probability density function $p_{\mu_1\mu_2\ldots\mu_n}(x_1, x_2, \ldots, x_n)$. Furthermore, let us assume that the functions $f_1, f_2, \ldots, f_n$ are given. If the system of equations $f_i(x_1, x_2, \ldots, x_n) = y_i$ $(i = 1, 2, \ldots, n)$ has real-valued solutions $x_{1\nu}, x_{2\nu}, \ldots, x_{n\nu}$ $(\nu = 1, 2, \ldots, m)$, then the joint probability density function of the random variables $\xi_1 = f_1(\mu_1, \mu_2, \ldots, \mu_n)$, $\xi_2 = f_2(\mu_1, \mu_2, \ldots, \mu_n), \ldots, \xi_n = f_n(\mu_1, \mu_2, \ldots, \mu_n)$ can be expressed by

$$p_{\xi_1\xi_2\ldots\xi_n}(y_1, y_2, \ldots, y_n) = \sum_{\nu=1}^{m} \frac{p_{\mu_1\mu_2\ldots\mu_n}(x_{1\nu}, x_{2\nu}, \ldots, x_{n\nu})}{|J(x_{1\nu}, x_{2\nu}, \ldots, x_{n\nu})|}, \tag{2.87}$$

where

$$J(x_1, x_2, \ldots, x_n) = \begin{vmatrix} \frac{\partial f_1}{\partial x_1} & \frac{\partial f_1}{\partial x_2} & \cdots & \frac{\partial f_1}{\partial x_n} \\ \frac{\partial f_2}{\partial x_1} & \frac{\partial f_2}{\partial x_2} & \cdots & \frac{\partial f_2}{\partial x_n} \\ \vdots & \vdots & \ddots & \vdots \\ \frac{\partial f_n}{\partial x_1} & \frac{\partial f_n}{\partial x_2} & \cdots & \frac{\partial f_n}{\partial x_n} \end{vmatrix} \tag{2.88}$$

denotes the *Jacobian determinant*.

One can compute the joint probability density function of the random variables $\xi_1, \xi_2, \ldots, \xi_k$ for $k < n$ by using (2.87) as follows

$$p_{\xi_1\xi_2\ldots\xi_k}(y_1, y_2, \ldots, y_k) = \int_{-\infty}^{\infty}\int_{-\infty}^{\infty}\cdots\int_{-\infty}^{\infty} p_{\xi_1\xi_2\ldots\xi_n}(y_1, y_2, \ldots, y_n)\, dy_{k+1}\, dy_{k+2} \cdots dy_n\,. \tag{2.89}$$

2.2 Stochastic Processes

Let $(Q, \mathcal{A}, P)$ be a probability space. Now let us assign to every particular outcome $s = s_i \in Q$ of a random experiment a particular function of time $\mu(t, s_i)$ according to a rule. Hence, for a particular $s_i \in Q$, the function $\mu(t, s_i)$ denotes a mapping from $\mathbb{R}$ to $\mathbb{R}$ (or $\mathbb{C}$) according to

$$\mu(\cdot, s_i) : \mathbb{R} \to \mathbb{R} \text{ (or } \mathbb{C}), \quad t \mapsto \mu(t, s_i) . \tag{2.90}$$

The individual functions $\mu(t, s_i)$ of time are called *realizations* or *sample functions*. A *stochastic process* $\mu(t, s)$ is a family (or an ensemble) of sample functions $\mu(t, s_i)$, i.e., $\mu(t, s) = \{\mu(t, s_i) \,|\, s_i \in Q\} = \{\mu(t, s_1), \mu(t, s_2), \ldots\}$.

On the other hand, at a particular time instant $t = t_0 \in \mathbb{R}$, the stochastic process $\mu(t_0, s)$ only depends on the outcome s and thus equals a random variable. Hence, for a particular $t_0 \in \mathbb{R}$, $\mu(t_0, s)$ denotes a mapping from Q to $\mathbb{R}$ (or $\mathbb{C}$) according to

$$\mu(t_0, \cdot) : Q \to \mathbb{R} \text{ (or } \mathbb{C}), \quad s \mapsto \mu(t_0, s) . \tag{2.91}$$

The probability density function of the random variable $\mu(t_0, s)$ is determined by the occurrence of the outcomes.

Therefore, a stochastic process is a function of two variables $t \in \mathbb{R}$ and $s \in Q$, so that the correct notation is $\mu(t, s)$. Henceforth, however, we will drop the second argument and simply write $\mu(t)$ as in common practice.

From the statements above, we can conclude that a stochastic process $\mu(t)$ can be interpreted as follows [41]:

(i) If t is a variable and s is a random variable, then $\mu(t)$ represents a family or an ensemble of sample functions $\mu(t, s)$.
(ii) If t is a variable and $s = s_0$ is a constant, then $\mu(t) = \mu(t, s_0)$ is a realization or a sample function of the stochastic process.
(iii) If $t = t_0$ is a constant and s is a random variable, then $\mu(t_0)$ is a random variable as well.
(iv) If both $t = t_0$ and $s = s_0$ are constants, then $\mu(t_0)$ is a real-valued (complex-valued) number.

The relationships following from the Statements (i)–(iv) made above are illustrated in Figure 2.8.

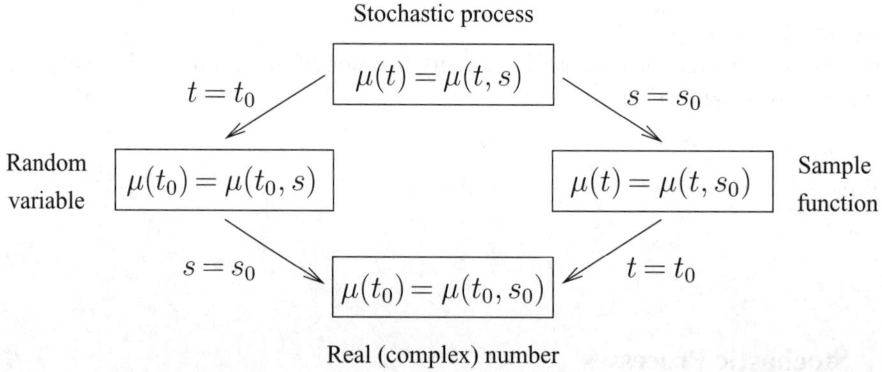

Figure 2.8 Relationships between stochastic processes, random variables, sample functions, and real-valued (complex-valued) numbers.

Let $\mu_1, \mu_2, \ldots, \mu_n$ be a sequence of n random variables obtained by sampling the stochastic process $\mu(t)$ at the time instants $t_1, t_2, \ldots, t_n$, i.e.,

$$\mu_1 = \mu(t_1), \quad \mu_2 = \mu(t_2), \ldots, \quad \mu_n = \mu(t_n) . \tag{2.92}$$

Then, the statistical properties of the stochastic process $\mu(t)$ are completely determined by the nth-order distribution of the random variables $\mu_1, \mu_2, \ldots, \mu_n$.

***n*th-order distribution:** The *nth-order distribution* of a stochastic process $\mu(t)$ is the joint distribution

$$F(x_1, x_2, \ldots, x_n; t_1, t_2, \ldots, t_n) = P\{\mu(t_1) \leq x_1, \mu(t_2) \leq x_2, \ldots, \mu(t_n) \leq x_n\} \tag{2.93}$$

of the random variables $\mu_1, \mu_2, \ldots, \mu_n$.

***n*th-order density:** The function, defined as

$$p(x_1, x_2, \ldots, x_n; t_1, t_2, \ldots, t_n) = \frac{\partial^n F(x_1, x_2, \ldots, x_n; t_1, t_2, \ldots, t_n)}{\partial x_1 \, \partial x_2 \cdots \partial x_n} \tag{2.94}$$

is called the *nth-order density* of the process $\mu(t)$.

Complex-valued stochastic processes: Let $\mu_1(t)$ and $\mu_2(t)$ be two real-valued stochastic processes, then a *(complex-valued) stochastic process* is defined by the sum

$$\mu(t) = \mu_1(t) + j\mu_2(t) . \tag{2.95}$$

Vector processes: A *vector process* $\boldsymbol{\mu}(t)$ (or an *n-dimensional process*) is a family of n stochastic processes $\mu_1(t), \mu_2(t), \ldots, \mu_n(t)$

$$\boldsymbol{\mu}(t) = (\mu_1(t), \mu_2(t), \ldots, \mu_n(t)) . \tag{2.96}$$

We have stated above that a stochastic process $\mu(t)$ can be interpreted as a random variable for fixed values of $t \in \mathbb{R}$. This random variable can again be described by a distribution function $F_\mu(x; t) = P\{\mu(t) \leq x\}$ or a probability density function $p_\mu(x; t) = dF_\mu(x; t)/dx$. The extension of the expected value concept, which was introduced for random variables, to stochastic processes leads to the *expected value function*

$$m_\mu(t) = E\{\mu(t)\} . \tag{2.97}$$

Let us consider the random variables $\mu(t_1)$ and $\mu(t_2)$, which are assigned to a stochastic process $\mu(t)$ at the time instants t_1 and t_2, then

$$r_{\mu\mu}(t_1, t_2) = E\{\mu^*(t_1)\, \mu(t_2)\} \tag{2.98}$$

is called the *autocorrelation function* of $\mu(t)$, where the superscripted asterisk * denotes the complex conjugation operator. Here, the complex conjugation is associated with the first independent variable in $r_{\mu\mu}(t_1, t_2)$.[3] Note in particular that the autocorrelation function $r_{\mu\mu}(t_1, t_2)$

[3] It should be mentioned that in the literature, the complex conjugation is often also associated with the second independent variable of the autocorrelation function $r_{\mu\mu}(t_1, t_2)$, i.e., $r_{\mu\mu}(t_1, t_2) = E\{\mu(t_1)\, \mu^*(t_2)\}$.

on the diagonal $t = t_1 = t_2$ equals the mean power of $\mu(t)$, i.e., $r_{\mu\mu}(t, t) = E\{|\mu(t)|^2\} \geq 0$. The so-called *variance function* of a complex-valued stochastic process $\mu(t)$ is defined as

$$\begin{aligned}\sigma_\mu^2(t) &= \text{Var}\{\mu(t)\} = E\{|\mu(t) - E\{\mu(t)\}|^2\} \\ &= E\{\mu^*(t)\mu(t)\} - E\{\mu^*(t)\}E\{\mu(t)\} \\ &= r_{\mu\mu}(t,t) - |m_\mu(t)|^2 ,\end{aligned} \tag{2.99}$$

where $r_{\mu\mu}(t, t)$ denotes the autocorrelation function in (2.98) at the time instant $t_1 = t_2 = t$, and $m_\mu(t)$ represents the expected value function according to (2.97). The expression

$$r_{\mu_1\mu_2}(t_1, t_2) = E\{\mu_1^*(t_1)\,\mu_2(t_2)\} \tag{2.100}$$

introduces the *cross-correlation function* of the stochastic processes $\mu_1(t)$ and $\mu_2(t)$ at the time instants t_1 and t_2, respectively. Finally,

$$r_{\mu_1\mu_2}^c(t_1, t_2) = r_{\mu_1\mu_2}(t_1, t_2) - m_{\mu_1}^*(t_1)\, m_{\mu_2}(t_2) \tag{2.101}$$

defines the *cross-covariance function* of $\mu_1(t)$ and $\mu_2(t)$. The stochastic processes $\mu_1(t)$ and $\mu_2(t)$ are called *uncorrelated* if $r_{\mu_1\mu_2}^c(t_1, t_2) = 0$ for every t_1 and t_2.

2.2.1 *Stationary Processes*

Stationary processes are of crucial importance to the modelling of mobile radio channels and will therefore be dealt with briefly here. One often distinguishes between strict-sense stationary processes and wide-sense stationary processes.

Strict-sense stationary processes: A stochastic process $\mu(t)$ is said to be *strict-sense stationary*, if its statistical properties are invariant to a shift of the origin, i.e., $\mu(t)$ and $\mu(t + c)$ have the same statistics for all values of $c \in \mathbb{R}$. As a consequence, the nth-order density of a strict-sense stationary process must also be invariant to a shift of the time instants, i.e.,

$$p(x_1, x_2, \ldots, x_n; t_1, t_2, \ldots, t_n) = p(x_1, x_2, \ldots, x_n; t_1 + c, t_2 + c, \ldots, t_n + c) \tag{2.102}$$

for all values of $c \in \mathbb{R}$. Strict-sense stationary processes have the following properties:

(i) The 1st-order density is independent of time, i.e.,

$$p_\mu(x; t) = p_\mu(x; t + c) = p_\mu(x) \tag{2.103}$$

for all t and c.

(ii) The 2nd-order density depends only on the time difference $\tau = t_1 - t_2$ between two samples, i.e.,

$$p_{\mu_1\mu_2}(x_1, x_2; t_1, t_2) = p_{\mu_1\mu_2}(x_1, x_2;\ \tau) \tag{2.104}$$

for all t_1 and t_2.

(iii) The first property implies that the expected value of $\mu(t)$ is constant, i.e.,

$$E\{\mu(t)\} = m_\mu = \text{const.} \tag{2.105}$$

for all t.

(iv) The second property implies that the autocorrelation function of $\mu(t)$ depends only on the time difference $\tau = t_1 - t_2$, i.e.,

$$r_{\mu\mu}(t_1, t_2) = r_{\mu\mu}(|t_1 - t_2|) \tag{2.106}$$

for all t_1 and t_2.

If (2.102) holds not for every n, but only for $k \leq n$, then the stochastic process $\mu(t)$ is said to be *kth-order stationary*. In many technical problems, we cannot determine whether a stochastic process is strict-sense stationary, since (2.102) is difficult to examine. But we can often determine whether the expected value is constant and whether the autocorrelation function is a function of $\tau = t_1 - t_2$ only. This motivates us to introduce a further important class of stochastic processes.

Wide-sense stationary processes: A stochastic process $\mu(t)$ is said to be *wide-sense stationary* if only the last two conditions (2.105) and (2.106) are fulfilled. In this case, the expected value function $E\{\mu(t)\}$ is independent of t and thus simplifies to the expected value m_μ introduced for random variables. Furthermore, the autocorrelation function $r_{\mu\mu}(t_1, t_2)$ merely depends on the time difference $t_1 - t_2$. A strict-sense stationary process is always wide-sense stationary, but a wide-sense stationary process needs not to be strict-sense stationary. Let $\mu(t)$ be a wide-sense stationary process, then it follows from (2.98) and (2.106) with $t_1 = t$ and $t_2 = t + \tau$ for $\tau > 0$

$$r_{\mu\mu}(\tau) = r_{\mu\mu}(t, t+\tau) = E\{\mu^*(t)\,\mu(t+\tau)\}\,. \tag{2.107}$$

Note that in particular $r_{\mu\mu}(0) = E\{|\mu(t)|^2\}$ represents the *mean power* of $\mu(t)$. Analogously, for the cross-correlation function (2.100) of two wide-sense stationary processes $\mu_1(t)$ and $\mu_2(t)$, we obtain

$$r_{\mu_1\mu_2}(\tau) = E\{\mu_1^*(t)\,\mu_2(t+\tau)\} = r^*_{\mu_2\mu_1}(-\tau)\,. \tag{2.108}$$

Similarly, the *cross-covariance function* of two wide-sense stationary processes $\mu_1(t)$ and $\mu_2(t)$ equals

$$r^c_{\mu_1\mu_2}(\tau) = r_{\mu_1\mu_2}(\tau) - m^*_{\mu_1}\, m_{\mu_2}\,. \tag{2.109}$$

If $r^c_{\mu_1\mu_2}(\tau) = 0$ for every value of τ, then the stochastic processes $\mu_1(t)$ and $\mu_2(t)$ are called *uncorrelated*. Suppose that $\mu_1(t)$ and $\mu_2(t)$ are two real-valued and wide-sense stationary processes. Then, the autocorrelation function of the complex process $\mu(t) = \mu_1(t) + j\mu_2(t)$ can be obtained as follows

$$\begin{aligned} r_{\mu\mu}(\tau) &= E\{\mu^*(t)\mu(t+\tau)\} \\ &= E\{(\mu_1(t) - j\mu_2(t))(\mu_1(t+\tau) + j\mu_2(t+\tau))\} \\ &= E\{\mu_1(t)\,\mu_1(t+\tau)\} + E\{\mu_2(t)\,\mu_2(t+\tau)\} \\ &\quad + jE\{\mu_1(t)\,\mu_2(t+\tau)\} - jE\{\mu_2(t)\,\mu_1(t+\tau)\} \\ &= r_{\mu_1\mu_1}(\tau) + r_{\mu_2\mu_2}(\tau) + j(r_{\mu_1\mu_2}(\tau) - r_{\mu_2\mu_1}(\tau))\,. \end{aligned} \tag{2.110}$$

Thus, the autocorrelation function of a complex process depends in general on the autocorrelation function of the inphase and quadrature component as well as on the cross-correlation functions.

Let $\mu(t)$ be a wide-sense stationary process. The Fourier transform of the autocorrelation function $r_{\mu\mu}(\tau)$ of $\mu(t)$, defined by

$$S_{\mu\mu}(f) := \int_{-\infty}^{\infty} r_{\mu\mu}(\tau)\, e^{-j2\pi f\tau}\, d\tau\,, \quad f \in \mathbb{R}\,, \tag{2.111}$$

is called the *power spectral density* (or *power density spectrum*). This general relation between the power spectral density and the autocorrelation function is known as the *Wiener-Khinchin theorem* (also known as the *Wiener-Khintchine theorem* or the *Wiener-Khinchin-Einstein theorem*). From the *inverse Fourier transform* of the power spectral density $S_{\mu\mu}(f)$, we obtain the autocorrelation function $r_{\mu\mu}(\tau)$ of $\mu(t)$, i.e.,

$$r_{\mu\mu}(\tau) := \int_{-\infty}^{\infty} S_{\mu\mu}(f)\, e^{j2\pi f\tau}\, df\,, \quad \tau \in \mathbb{R}\,. \tag{2.112}$$

If $\mu_1(t)$ and $\mu_2(t)$ are two wide-sense stationary processes with cross-correlation function $r_{\mu_1\mu_2}(\tau)$, then the Fourier transform of $r_{\mu_1\mu_2}(\tau)$, defined by

$$S_{\mu_1\mu_2}(f) := \int_{-\infty}^{\infty} r_{\mu_1\mu_2}(\tau)\, e^{-j2\pi f\tau}\, d\tau\,, \quad f \in \mathbb{R}\,, \tag{2.113}$$

is called the *cross-power spectral density* (or *cross-power spectrum*). Taking (2.108) into account, we immediately realize that $S_{\mu_1\mu_2}(f) = S^*_{\mu_2\mu_1}(f)$ holds. Analogously, the Fourier transform of the cross-covariance function $r^c_{\mu_1\mu_2}(\tau)$ is called the *cross-covariance power spectral density* (or *cross-covariance spectrum*) $S^c_{\mu_1\mu_2}(f)$.

It should be noted that, strictly speaking, a stationary processes cannot exist in the real world. Stationary processes are merely used as mathematical models for processes, which hold their statistical properties over a relatively long time. From now on, a stochastic process will be assumed as a strict-sense stationary stochastic process, as long as nothing else is said.

2.2.2 Ergodic Processes

The description of the statistical properties of stochastic processes, like the expected value or the autocorrelation function, is based on ensemble averages (statistical averages) by taking all possible sample functions of the stochastic process into account. In practice, however, one almost always observes and records only a finite number of sample functions (mostly even only one single sample function). Nevertheless, in order to make statements on the statistical properties of stochastic process, one refers to the ergodicity hypothesis.

The ergodicity hypothesis deals with the question, whether it is possible to evaluate only a single sample function of a stationary stochastic process instead of averaging over the whole ensemble of sample functions at one or more specific time instants. Of particular importance is the question whether the expected value and the autocorrelation function of a stochastic process $\mu(t)$ equal the temporal means taken over any arbitrarily sample function $\mu(t, s_i)$. According to the *ergodic theorem*, a stochastic process $\mu(t)$ is called *mean ergodic* if its expected value $E\{\mu(t)\} = m_\mu$ equals the temporal average of $\mu(t, s_i)$, i.e.,

$$m_\mu = \tilde{m}_\mu := \lim_{T\to\infty} \frac{1}{2T} \int_{-T}^{+T} \mu(t, s_i)\, dt\,, \quad \forall\, i = 1, 2, \ldots \tag{2.114}$$

Analogously, a stochastic process $\mu(t)$ is called *autocorrelation ergodic* if its autocorrelation function $r_{\mu\mu}(\tau) = E\{\mu^*(t)\,\mu(t+\tau)\}$ equals the temporal autocorrelation function of $\mu(t, s_i)$, i.e.,

$$r_{\mu\mu}(\tau) = \tilde{r}_{\mu\mu}(\tau) := \lim_{T\to\infty} \frac{1}{2T} \int_{-T}^{+T} \mu^*(t, s_i)\,\mu(t+\tau, s_i)\, dt\,, \quad \forall\, i = 1, 2, \ldots \tag{2.115}$$

A stationary stochastic process $\mu(t)$ is said to be *strict-sense ergodic*, if all expected values, which take all possible sample functions into account, are identical to the respective temporal averages taken over an arbitrary sample function. If this condition is only fulfilled for the expected value and the autocorrelation function, i.e., if only (2.114) and (2.115) are fulfilled, then the stochastic process $\mu(t)$ is said to be *wide-sense ergodic*. A strict-sense ergodic process is always stationary. The inverse statement is not always true, although commonly assumed.

2.2.3 Level-Crossing Rate and Average Duration of Fades

Apart from the probability density function and the autocorrelation function, other characteristic quantities describing the statistics of mobile fading channels are of importance, such as the *level-crossing rate* and the *average duration of fades*.

As we know, the received signal in mobile radio communications often undergoes heavy statistical fluctuations, which can span a range of up to 30 dB and more. In digital communications, a heavy decline of the received signal leads directly to a drastic increase in the bit error rate. For the optimization of coding schemes, which are required for error correction, it is not only important to know how often the received signal crosses a given threshold level per time unit, but also for how long on average the signal remains below a certain level. Suitable measures for the occurrence of these events are the level-crossing rate and the average duration of fades.

Level-crossing rate: The *level-crossing rate*, denoted by $N_\zeta(r)$, describes how often a stochastic process $\zeta(t)$ crosses in average a given signal level r from down to up (or from up to down) within a time interval of one second. According to [16, 17], the level-crossing rate $N_\zeta(r)$ can be calculated by

$$N_\zeta(r) = \int_0^\infty \dot{x}\, p_{\zeta\dot{\zeta}}(r, \dot{x})\, d\dot{x}\,, \quad r \geq 0\,, \tag{2.116}$$

where $p_{\zeta\dot{\zeta}}(x, \dot{x})$ denotes the joint probability density function of the process $\zeta(t)$ and its time derivative $\dot{\zeta}(t) = d\zeta(t)/dt$ at the same time instant t. The proof of (2.116) can be found in Appendix 2.A.

Analytical expressions for the level-crossing rate of Rayleigh and Rice processes can easily be calculated. Let us consider two uncorrelated real-valued zero-mean Gaussian random processes $\mu_1(t)$ and $\mu_2(t)$ with identical autocorrelation functions, i.e., $r_{\mu_1\mu_1}(\tau) = r_{\mu_2\mu_2}(\tau)$. Then, for the level-crossing rate of the resulting Rayleigh process $\zeta(t) = \sqrt{\mu_1^2(t) + \mu_2^2(t)}$, we obtain the following expression [13, 42]

$$\begin{aligned} N_\zeta(r) &= \sqrt{\frac{\beta}{2\pi}} \cdot \frac{r}{\sigma_0^2} e^{-\frac{r^2}{2\sigma_0^2}} \\ &= \sqrt{\frac{\beta}{2\pi}} \cdot p_\zeta(r), \quad r \geq 0, \end{aligned} \tag{2.117}$$

where $\sigma_0^2 = r_{\mu_i\mu_i}(0)$ denotes the mean power of the underlying Gaussian random processes $\mu_i(t)$ $(i = 1, 2)$. Here, β is a short notation for the negative curvature of the autocorrelation functions $r_{\mu_1\mu_1}(\tau)$ and $r_{\mu_2\mu_2}(\tau)$ at the origin $\tau = 0$, i.e.,

$$\beta = -\left. \frac{d^2}{d\tau^2} r_{\mu_i\mu_i}(\tau) \right|_{\tau=0} = -\ddot{r}_{\mu_i\mu_i}(0), \quad i = 1, 2. \tag{2.118}$$

For Rice processes $\xi(t) = \sqrt{(\mu_1(t) + \rho)^2 + \mu_2^2(t)}$, the level-crossing rate can be expressed as [42]

$$\begin{aligned} N_\xi(r) &= \sqrt{\frac{\beta}{2\pi}} \cdot \frac{r}{\sigma_0^2} e^{-\frac{r^2+\rho^2}{2\sigma_0^2}} I_0\left(\frac{r\rho}{\sigma_0^2}\right) \\ &= \sqrt{\frac{\beta}{2\pi}} \cdot p_\xi(r), \quad r \geq 0. \end{aligned} \tag{2.119}$$

Average duration of fades: The *average duration of fades*, denoted by $T_{\zeta_-}(r)$, is the expected value of the length of the time intervals in which the stochastic process $\zeta(t)$ is below a given signal level r. The average duration of fades $T_{\zeta_-}(r)$ can be calculated by means of [13, 42]

$$T_{\zeta_-}(r) = \frac{F_{\zeta_-}(r)}{N_\zeta(r)}, \tag{2.120}$$

where $F_{\zeta_-}(r)$ denotes the cumulative distribution function of the stochastic process $\zeta(t)$ being the probability that $\zeta(t)$ is less than or equal to the signal level r, i.e.,

$$F_{\zeta_-}(r) = P\{\zeta(t) \leq r\} = \int_0^r p_\zeta(x)\, dx. \tag{2.121}$$

For Rayleigh processes $\zeta(t)$, the average duration of fades is given by

$$T_{\zeta_-}(r) = \sqrt{\frac{2\pi}{\beta}} \cdot \frac{\sigma_0^2}{r} \left(e^{\frac{r^2}{2\sigma_0^2}} - 1 \right), \quad r \geq 0, \tag{2.122}$$

where the quantity β is given by (2.118).

For Rice processes $\xi(t)$, however, we find by using (2.44) and substituting (2.121) and (2.119) in (2.120) the following integral expression

$$T_{\xi_-}(r) = \sqrt{\frac{2\pi}{\beta}} \cdot \frac{e^{\frac{r^2}{2\sigma_0^2}}}{r I_0(\frac{r\rho}{\sigma_0^2})} \int_0^r x e^{-\frac{x^2}{2\sigma_0^2}} I_0\left(\frac{x\rho}{\sigma_0^2}\right) dx, \quad r \geq 0, \tag{2.123}$$

which has to be evaluated numerically.

Analogously, the *average connecting time interval*$T_{\zeta_+}(r)$ can be introduced. This quantity describes the expected value of the length of the time intervals in which the stochastic process $\zeta(t)$ is above a given signal level r. Thus,

$$T_{\zeta_+}(r) = \frac{F_{\zeta_+}(r)}{N_\zeta(r)}, \tag{2.124}$$

where $F_{\zeta_+}(r)$ is called the *complementary cumulative distribution function* of $\zeta(t)$. This function describes the probability that $\zeta(t)$ is larger than r, i.e., $F_{\zeta_+}(r) = P\{\zeta(t) > r\}$. The complementary cumulative distribution function F_{ζ_+} and the cumulative distribution function $F_{\zeta_-}(r)$ are related by $F_{\zeta_+}(r) = 1 - F_{\zeta_-}(r)$.

2.2.4 Linear Systems with Stochastic Inputs

Linear systems can be classified into two main categories: linear time-invariant systems and linear time-variant systems. In many cases, physical systems are modelled as linear time-invariant systems. In mobile radio channel modelling, however, it is often more reasonable to model the channel as a linear time-variant system. In this subsection, we introduce both linear time-invariant and linear time-variant systems and we develop techniques for computing the response of these systems driven by stochastic input signals.

2.2.4.1 Linear Time-Invariant Systems

Let us consider a system in which a stochastic input process $x(t)$ is mapped into a stochastic output process $y(t)$ by the transformation

$$y(t) = L\{x(t)\}. \tag{2.125}$$

The system is said to be *linear time-invariant* if it satisfies the following properties:

(i) The system is *linear*, i.e.,

$$L\{a_1 x_1(t) + a_2 x_2(t)\} = a_1 L\{x_1(t)\} + a_2 L\{x_2(t)\} \tag{2.126}$$

for any constants a_1, a_2 and stochastic input processes $x_1(t)$, $x_2(t)$.

(ii) The system is *time-invariant*, i.e., if $y(t) = L\{x(t)\}$, then

$$y(t - t_0) = L\{x(t - t_0)\} \tag{2.127}$$

for any $t_0 \in \mathbb{R}$.

A linear time-invariant system is completely determined by its *impulse response*. The impulse response, denoted by $h(t)$, is defined by

$$h(t) := L\{\delta(t)\} \tag{2.128}$$

where $\delta(t)$ is the *Dirac delta function*. The Fourier transform of the impulse response $h(t)$ is known as the *transfer function*

$$H(f) = \int_{-\infty}^{\infty} h(t)\, e^{-j2\pi f t} dt\,. \tag{2.129}$$

The output process $y(t)$ of a linear time-invariant system is determined by the convolution of the input process $x(t)$ and the impulse response $h(t)$ of the system, i.e.,

$$y(t) = x(t) * h(t) = \int_{-\infty}^{\infty} x(t - \tau)\, h(\tau)\, d\tau\,. \tag{2.130}$$

Moreover, the following relations hold:

$$r_{xy}(\tau) = r_{xx}(\tau) * h(\tau) \quad \circ\!\!-\!\!\bullet \quad S_{xy}(f) = S_{xx}(f) \cdot H(f), \tag{2.131a, b}$$

$$r_{yx}(\tau) = r_{xx}(\tau) * h^*(-\tau) \quad \circ\!\!-\!\!\bullet \quad S_{yx}(f) = S_{xx}(f) \cdot H^*(f), \tag{2.131c, d}$$

$$r_{yy}(\tau) = r_{xx}(\tau) * h(\tau) * h^*(-\tau) \quad \circ\!\!-\!\!\bullet \quad S_{yy}(f) = S_{xx}(f) \cdot |H(f)|^2, \tag{2.131e, f}$$

where the symbol $\circ\!\!-\!\!\bullet$ denotes the Fourier transform.

Hilbert transformer: A system with the transfer function

$$\check{H}(f) = -j\,\mathrm{sgn}\,(f) = \begin{cases} -j, & f > 0 \\ 0, & f = 0 \\ j, & f < 0 \end{cases} \tag{2.132}$$

is called the *Hilbert transformer*. We observe that this system causes a phase shift of $-\pi/2$ for $f > 0$ and a phase shift of $+\pi/2$ for $f < 0$. Thus, the Hilbert transformer converts sine functions into cosine functions and vice versa. The inverse Fourier transform of the transfer function $\check{H}(f)$ results in the impulse response

$$\check{h}(t) = \frac{1}{\pi t}\,. \tag{2.133}$$

Since $\check{h}(t) \neq 0$ holds for $t < 0$, it follows that the Hilbert transformer is a non-causal system. Let $x(t)$ with $E\{x(t)\} = 0$ be a real-valued input process of the Hilbert transformer, then the output process

$$\check{x}(t) = x(t) * \check{h}(t) = \frac{1}{\pi} \int_{-\infty}^{\infty} \frac{x(t')}{t - t'} dt' \tag{2.134}$$

is said to be the *Hilbert transform* of $x(t)$. One should note that the computation of the integral in (2.134) must be performed according to Cauchy's principal value [43]. To denote that $\check{x}(t)$ is the Hilbert transform of $x(t)$, we also use the notation $\check{x}(t) = \mathcal{H}\{x(t)\}$, where $\mathcal{H}\{\cdot\}$ is called the *Hilbert transform operator*. The Hilbert transform operator applied twice to a real function $x(t)$ gives the same function, but with an altered sign, i.e., $\mathcal{H}\{\mathcal{H}\{x(t)\}\} = -x(t)$.

With (2.131) and (2.133), the following relations hold:

$$r_{x\check{x}}(\tau) = \check{r}_{xx}(\tau) \quad \circ\!\!-\!\!\bullet \quad S_{x\check{x}}(f) = -j\,\mathrm{sgn}\,(f) \cdot S_{xx}(f), \tag{2.135a, b}$$

$$r_{x\check{x}}(\tau) = -r_{\check{x}x}(\tau) \quad \circ\!\!-\!\!\bullet \quad S_{x\check{x}}(f) = -S_{\check{x}x}(f), \tag{2.135c, d}$$

$$r_{\check{x}\check{x}}(\tau) = r_{xx}(\tau) \quad \circ\!\!-\!\!\bullet \quad S_{\check{x}\check{x}}(f) = S_{xx}(f). \tag{2.135e, f}$$

2.2.4.2 Linear Time-Variant Systems

If only the first condition (2.126) is fulfilled, then the system is said to be *linear time-variant*. A linear time-variant system is completely specified by its time-variant impulse response. The time-variant impulse response, denoted by $h(t_0, t)$, is the response of the system observed at time t to a delta impulse that stimulated the system at time t_0, i.e.,

$$h_0(t_0, t) := L\{\delta(t - t_0)\}\,. \tag{2.136}$$

In Section 7.2, we will show that the output process $y(t)$ of a linear time-variant system can be expresses in terms of the input process $x(t)$ and the impulse response $h_0(t_0, t)$ as

$$y(t) = \int_{-\infty}^{\infty} x(t_0)\, h_0(t_0, t)\, dt_0\,. \tag{2.137}$$

For convenience, we substitute t_0 by $t_0 = t - \tau'$ and obtain thus the equivalent relation

$$y(t) = \int_{-\infty}^{\infty} x(t - \tau')\, h(\tau', t)\, d\tau'\,, \tag{2.138}$$

where $h(\tau', t) := h_0(t - \tau', t)$. The time-variant impulse response $h(\tau', t)$ is the response of the system observed at time t to a delta impulse that stimulated the system at time $t - \tau'$. The Fourier transform of the time-variant impulse response $h(\tau', t)$ with respect to τ', i.e.,

$$H(f', t) := \int_{-\infty}^{\infty} h(\tau', t)\, e^{-j2\pi f'\tau'} d\tau' \tag{2.139}$$

is known as the *time-variant transfer function* of the system. The fact that $h(\tau', t)$ and $H(f, t)$ are forming a Fourier pair is also expressed symbolically by $h(\tau', t) \overset{\tau'\, f'}{\circ\!\!-\!\!\bullet} H(f', t)$. All real physical systems must obey the law of causality: $h(\tau', t) = 0$ if $\tau' < 0$, which means that the system cannot respond before it was excited. For causal linear-time variant systems, the lower limits of the integrals in (2.138) and (2.139) can be set to zero.

2.3 Deterministic Signals

In principle, one distinguishes between continuous-time and discrete-time signals. For deterministic signals, we will in what follows use the continuous-time representation wherever it is reasonable. Only in those sections, where the numerical simulations of channel models play a significant role, we prefer the discrete-time representation of signals.

2.3.1 Deterministic Continuous-Time Signals

A *deterministic* (*continuous-time*) signal is usually defined over $\mathbb{R}$. The set $\mathbb{R}$ is considered as the time space in which the variable t takes its values, i.e., $t \in \mathbb{R}$. A deterministic signal is described by a function (mapping) such that each value of t is definitely assigned to a real-valued (or complex-valued) number. Furthermore, in order to clearly distinguish between deterministic signals and stochastic processes, we will place a tilde sign above the symbols chosen for deterministic signals. Thus, under a deterministic signal $\tilde{\mu}(t)$, we will understand a mapping of the kind

$$\tilde{\mu} : \mathbb{R} \to \mathbb{R} \quad (\text{or } \mathbb{C}), \quad t \longmapsto \tilde{\mu}(t). \tag{2.140}$$

In connection with deterministic signals, the following terms are of importance.

Mean value: The *mean value* of a deterministic signal $\tilde{\mu}(t)$ is defined by[4]

$$\tilde{m}_\mu := \lim_{T\to\infty} \frac{1}{2T} \int_{-T}^{T} \tilde{\mu}(t)\, dt. \tag{2.141}$$

Mean power: The *mean power* of a deterministic signal $\tilde{\mu}(t)$ is defined by

$$\tilde{\sigma}_\mu^2 := \lim_{T\to\infty} \frac{1}{2T} \int_{-T}^{T} |\tilde{\mu}(t)|^2\, dt. \tag{2.142}$$

Henceforth, it is assumed that the power of a deterministic signal is finite.

[4] To denote the time average of $\tilde{\mu}(t)$, we alternatively use the notation $< \tilde{\mu}(t) > := \lim_{T\to\infty} \frac{1}{2T} \int_{-T}^{T} \tilde{\mu}(t)\, dt$, where $<\cdot>$ is called the *time average operator*.

Autocorrelation function: Let $\tilde{\mu}(t)$ be a deterministic signal. Then, the *autocorrelation function* of $\tilde{\mu}(t)$ is defined by

$$\tilde{r}_{\mu\mu}(\tau) := \lim_{T\to\infty} \frac{1}{2T} \int_{-T}^{T} \tilde{\mu}^*(t)\, \tilde{\mu}(t+\tau)\, dt\,, \quad \tau \in \mathbb{R}\,. \tag{2.143}$$

Comparing (2.142) and (2.143), we realize that the value of $\tilde{r}_{\mu\mu}(\tau)$ at $\tau = 0$ is identical to the mean power of $\tilde{\mu}(t)$, i.e., the relation $\tilde{r}_{\mu\mu}(0) = \tilde{\sigma}_\mu^2$ holds.

Cross-correlation function: Let $\tilde{\mu}_1(t)$ and $\tilde{\mu}_2(t)$ be two deterministic signals. Then, the *cross-correlation function* of $\tilde{\mu}_1(t)$ and $\tilde{\mu}_2(t)$ is defined by

$$\tilde{r}_{\mu_1\mu_2}(\tau) := \lim_{T\to\infty} \frac{1}{2T} \int_{-T}^{T} \tilde{\mu}_1^*(t)\, \tilde{\mu}_2(t+\tau)\, dt\,, \quad \tau \in \mathbb{R}\,. \tag{2.144}$$

Here, $\tilde{r}_{\mu_1\mu_2}(\tau) = \tilde{r}^*_{\mu_2\mu_1}(-\tau)$ holds.

Power spectral density: Let $\tilde{\mu}(t)$ be a deterministic signal. Then, the Fourier transform of the autocorrelation function $\tilde{r}_{\mu\mu}(\tau)$, defined by

$$\tilde{S}_{\mu\mu}(f) := \int_{-\infty}^{\infty} \tilde{r}_{\mu\mu}(\tau)\, e^{-j2\pi f\tau}\, d\tau\,, \quad f \in \mathbb{R}\,, \tag{2.145}$$

is called the *power spectral density* (or *power density spectrum*) of $\tilde{\mu}(t)$. From the power spectral density $\tilde{S}_{\mu\mu}(f)$, we obtain the autocorrelation function $\tilde{r}_{\mu\mu}(\tau)$ via the *inverse Fourier transform*

$$\tilde{r}_{\mu\mu}(\tau) := \int_{-\infty}^{\infty} \tilde{S}_{\mu\mu}(f)\, e^{j2\pi f\tau}\, df\,, \quad \tau \in \mathbb{R}\,. \tag{2.146}$$

Cross-power spectral density: Let $\tilde{\mu}_1(t)$ and $\tilde{\mu}_2(t)$ be two deterministic signals. Then, the Fourier transform of the cross-correlation function $\tilde{r}_{\mu_1\mu_2}(\tau)$

$$\tilde{S}_{\mu_1\mu_2}(f) := \int_{-\infty}^{\infty} \tilde{r}_{\mu_1\mu_2}(\tau)\, e^{-j2\pi f\tau}\, d\tau\,, \quad f \in \mathbb{R}\,, \tag{2.147}$$

is called the *cross-power spectral density* (or *cross-power density spectrum*). From (2.147) and the relation $\tilde{r}_{\mu_1\mu_2}(\tau) = \tilde{r}^*_{\mu_2\mu_1}(-\tau)$ it follows that $\tilde{S}_{\mu_1\mu_2}(f) = \tilde{S}^*_{\mu_2\mu_1}(f)$ holds.

Let $\tilde{\nu}(t)$ and $\tilde{\mu}(t)$ be the deterministic input signal and the deterministic output signal, respectively, of a linear time-invariant stable system with the transfer function $H(f)$. Then, the power spectral density $\tilde{S}_{\mu\mu}(f)$ of the output signal $\tilde{\mu}(t)$ can be expressed in terms of the power spectral density $\tilde{S}_{\nu\nu}(f)$ of the input signal $\tilde{\nu}(t)$ via the relationship

$$\tilde{S}_{\mu\mu}(f) = |H(f)|^2\, \tilde{S}_{\nu\nu}(f)\,. \tag{2.148}$$

2.3.2 Deterministic Discrete-Time Signals

By equidistant sampling of a continuous-time signal $\tilde{\mu}(t)$ at the discrete time instants $t = t_k = kT_s$, where $k \in \mathbb{Z}$ and T_s symbolizes the *sampling interval*, we obtain a sequence of numbers $\{\tilde{\mu}(kT_s)\} = \{\ldots, \tilde{\mu}(-T_s), \tilde{\mu}(0), \tilde{\mu}(T_s), \ldots\}$. In specific questions of many engineering disciplines, one occasionally distinguishes strictly between the sequence $\{\tilde{\mu}(kT_s)\}$ itself, which is then called a *discrete-time signal*, and the kth element $\tilde{\mu}(kT_s)$ of it. For our purposes, however, this differentiation does not provide any advantage worth mentioning. In what follows, we will therefore simply write $\tilde{\mu}(kT_s)$ for discrete-time signals or sequences, and we will make use of the notation $\bar{\mu}[k] := \tilde{\mu}(kT_s) = \tilde{\mu}(t)|_{t=kT_s}$.

It is obvious that by sampling a deterministic continuous-time signal $\tilde{\mu}(t)$, we obtain a discrete-time signal $\bar{\mu}[k]$, which is deterministic as well. Under a deterministic discrete-time signal $\bar{\mu}[k]$, we will understand a mapping of the kind

$$\bar{\mu} : \mathbb{Z} \to \mathbb{R} \quad (\text{or } \mathbb{C}), \quad k \longmapsto \bar{\mu}[k]. \tag{2.149}$$

The terms such as the mean value, autocorrelation function, and power spectral density, which were previously introduced for deterministic continuous-time signals, can similarly be defined for deterministic discrete-time signals. Here, only the most important definitions and relationships will be reviewed briefly, as far as they are actually used, especially in Chapter 9. The reader can find an in-depth representation of the relationships, e.g., in [12, 44].

Mean value: The *mean value* of a deterministic sequence $\bar{\mu}[k]$ is defined by

$$\bar{m}_\mu := \lim_{K\to\infty} \frac{1}{2K+1} \sum_{k=-K}^{K} \bar{\mu}[k]. \tag{2.150}$$

Mean power: The *mean power* of a deterministic sequence $\bar{\mu}[k]$ is defined by

$$\bar{\sigma}_\mu^2 := \lim_{K\to\infty} \frac{1}{2K+1} \sum_{k=-K}^{K} |\bar{\mu}[k]|^2. \tag{2.151}$$

Autocorrelation sequence: Let $\bar{\mu}[k]$ be a deterministic sequence, then the corresponding *autocorrelation sequence* is defined by

$$\bar{r}_{\mu\mu}[\kappa] := \lim_{K\to\infty} \frac{1}{2K+1} \sum_{k=-K}^{K} \bar{\mu}^*[k]\,\bar{\mu}[k+\kappa], \quad \kappa \in \mathbb{Z}. \tag{2.152}$$

Thus, in connection with (2.151), it follows $\bar{\sigma}_\mu^2 = \bar{r}_{\mu\mu}[0]$.

Cross-correlation sequence: Let $\bar{\mu}_1[k]$ and $\bar{\mu}_2[k]$ be two deterministic sequences, then the *cross-correlation sequence* is defined by

$$\bar{r}_{\mu_1\mu_2}[\kappa] := \lim_{K\to\infty} \frac{1}{2K+1} \sum_{k=-K}^{K} \bar{\mu}_1^*[k]\,\bar{\mu}_2[k+\kappa], \quad \kappa \in \mathbb{Z}. \tag{2.153}$$

Here, the relationship $\bar{r}_{\mu_1\mu_2}[\kappa] = \bar{r}^*_{\mu_2\mu_1}[-\kappa]$ holds.

Power spectral density: Let $\bar{\mu}[k]$ be a deterministic sequence, then the *discrete Fourier transform* of the autocorrelation sequence $\bar{r}_{\mu\mu}[\kappa]$, defined by

$$\bar{S}_{\mu\mu}(f) := \sum_{\kappa=-\infty}^{\infty} \bar{r}_{\mu\mu}[\kappa]\, e^{-j2\pi f T_s \kappa}, \quad f \in \mathbb{R}, \tag{2.154}$$

is called the *power spectral density* or *power density spectrum* of $\bar{\mu}[k]$.

Between (2.154) and (2.145), the relation

$$\bar{S}_{\mu\mu}(f) := \frac{1}{T_s} \sum_{m=-\infty}^{\infty} \tilde{S}_{\mu\mu}(f - mf_s) \tag{2.155}$$

holds, where $f_s = 1/T_s$ is called the *sampling frequency* or the *sampling rate*. Obviously, the power spectral density $\bar{S}_{\mu\mu}(f)$ is periodic with period f_s, since $\bar{S}_{\mu\mu}(f) = \bar{S}_{\mu\mu}(f - mf_s)$ holds for all $m \in \mathbb{Z}$. The relation (2.155) states that the power spectral density $\bar{S}_{\mu\mu}(f)$ of $\bar{\mu}[k]$ follows from the power spectral density $\tilde{S}_{\mu\mu}(f)$ of $\tilde{\mu}(t)$, if the latter is weighted by $1/T_s$ and periodically continued at instants mf_s, where $m \in \mathbb{Z}$.

The *inverse discrete Fourier transform* of the power spectral density $\bar{S}_{\mu\mu}(f)$ results in the autocorrelation sequence $\bar{r}_{\mu\mu}[\kappa]$ of $\bar{\mu}[k]$, i.e.,

$$\bar{r}_{\mu\mu}[\kappa] := \frac{1}{f_s} \int_{-f_s/2}^{f_s/2} \bar{S}_{\mu\mu}(f)\, e^{j2\pi f T_s \kappa}\, df, \quad \kappa \in \mathbb{Z}. \tag{2.156}$$

Cross-power spectral density: Let $\bar{\mu}_1[k]$ and $\bar{\mu}_2[k]$ be two deterministic sequences. Then, the discrete Fourier transform of the cross-correlation sequence $\bar{r}_{\mu_1\mu_2}[\kappa]$, defined by

$$\bar{S}_{\mu_1\mu_2}(f) := \sum_{\kappa=-\infty}^{\infty} \bar{r}_{\mu_1\mu_2}[\kappa]\, e^{-j2\pi f T_s \kappa}, \quad f \in \mathbb{R}, \tag{2.157}$$

is called the *cross-power spectral density* or the *cross-power density spectrum*. From the above equation and $\bar{r}_{\mu_1\mu_2}[\kappa] = \bar{r}^*_{\mu_2\mu_1}[-\kappa]$ it follows that $\bar{S}_{\mu_1\mu_2}(f) = \bar{S}^*_{\mu_2\mu_1}(f)$ holds.

Sampling theorem: Let $\tilde{\mu}(t)$ be a band-limited continuous-time signal with the cut-off frequency f_c. If the signal $\tilde{\mu}(t)$ is sampled with a sampling frequency f_s that exceeds twice the cut-off frequency f_c, i.e.,

$$f_s > 2f_c, \tag{2.158}$$

then $\tilde{\mu}(t)$ is completely determined by the resulting sampling values $\bar{\mu}[k] = \tilde{\mu}(kT_s)$. In particular, the continuous-time signal $\tilde{\mu}(t)$ can be reconstructed from the sequence $\bar{\mu}[k]$ by means of the *Whittaker-Shannon interpolation formula*

$$\tilde{\mu}(t) = \sum_{k=-\infty}^{\infty} \bar{\mu}[k]\, \mathrm{sinc}\left(\pi \frac{t - kT_s}{T_s}\right), \tag{2.159}$$

where $\mathrm{sinc}\,(\cdot)$ denotes the *sinc function*, which is defined by $\mathrm{sinc}\,(x) = \sin(x)/x$.

It should be added that the sampling condition in (2.158) can be replaced by the less restrictive condition $f_s \geq 2f_c$, if the power spectral density $\tilde{S}_{\mu\mu}(f)$ has no δ-components at the limits $f = \pm f_c$ [45]. In this case, even on condition that $f_s \geq 2f_c$ holds, the validity of the sampling theorem is absolutely guaranteed.

2.4 Further Reading

As technical literature for the subject of probability theory, random variables, and stochastic processes, the books by Papoulis and Pillai [41], Peebles [46], Therrien [47], Leon-Garcia [48], Gubner [49], Dupraz [50], as well as Shanmugan and Breipohl [51] are recommended, especially for electrical and computer engineers. Also the classical works of Middleton [52], Davenport [53], and the book by Davenport and Root [54] are even nowadays still worth reading. A short introduction into the principles of probability and stochastic processes can be found in Chapter 2 of [11]. A more mathematical treatment of probability and statistics is provided by Rohatgi and Saleh [55]. Finally, the excellent textbooks by Allen and Mills [44], Oppenheim and Schafer [12], Papoulis [56], Rabiner and Gold [57], Kailath [58], Ingle and Proakis [59], and Stearns [60] provide a deep insight into systems theory as well as into the principles of digital signal processing.

Appendix 2.A Derivation of Rice's General Formula for the Level-Crossing Rate

Let A be the elementary event that the stochastic process $\zeta(t)$ crosses the signal level r in the interval $[t, t + \Delta t)$ from down to up. Such an up-crossing event is illustrated in Figure 2.A.1 for a specific sample function $\zeta(t, s_0)$ of the process $\zeta(t)$.

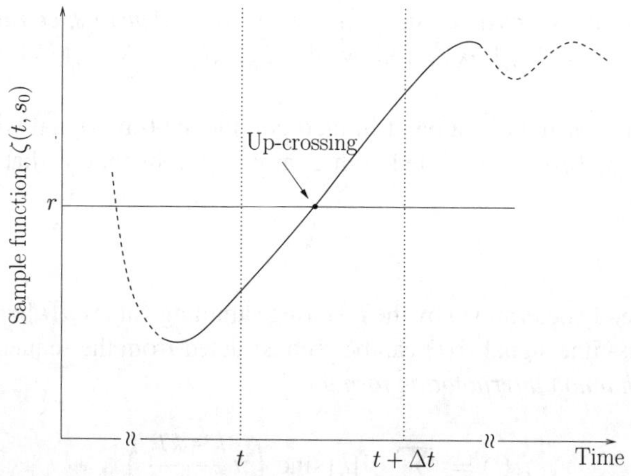

Figure 2.A.1 Illustration of an up-crossing of a sample function $\zeta(t, s_0)$ through the signal level r in the interval $[t, t + \Delta t)$.

The event A implies that

$$\text{(i)} \qquad \zeta(t) \leq r\,, \tag{2.A.1}$$

$$\text{(ii)} \qquad \zeta(t + \Delta t) > r\,. \tag{2.A.2}$$

From $\zeta(t + \Delta t) \approx \zeta(t) + \dot{\zeta}(t)\Delta t > r$ and (2.A.1), we may conclude that $r - \dot{\zeta}(t)\Delta t < \zeta(t) \leq r$ holds. The probability of the event A can then be expressed as

$$\begin{aligned} P\{A\} &= P\left\{r - \dot{x}\,\Delta t < \zeta(t) \leq r,\ \dot{\zeta}(t) \geq 0\right\} \\ &= \int_0^\infty \int_{r-\dot{x}\Delta t}^{r} p_{\zeta\dot{\zeta}}(x, \dot{x})\, dx\, d\dot{x}\,, \end{aligned} \tag{2.A.3}$$

where $p_{\zeta\dot{\zeta}}(x, \dot{x})$ is the joint probability density function of the process $\zeta(t)$ and its time derivative $\dot{\zeta}(t)$ at the same point in time. Using $dx \approx \dot{x}\Delta t$, the equation above results in

$$P\{A\} = \int_0^\infty \dot{x}\, p_{\zeta\dot{\zeta}}(r; \dot{x})\, d\dot{x}\, \Delta t \quad \text{as} \quad \Delta t \to 0\,, \tag{2.A.4}$$

which is the expected number of crossings of the process $\zeta(t)$ with a positive slope through the signal level r within an infinitesimal time interval of duration Δt. Hence, the expected number of up-crossings of $\zeta(t)$ through the signal level r within a time interval of one second equals the quantity

$$N_\zeta(r) = \lim_{\Delta t \to 0} \frac{P\{A\}}{\Delta t} = \int_0^\infty \dot{x}\, p_{\zeta\dot{\zeta}}(r, \dot{x})\, d\dot{x}\,, \tag{2.A.5}$$

which is known as the *level-crossing rate*. Analogously, one can derive the expected number of down-crossings of $\zeta(t)$ through r within a time interval of one second. The result is the same as in (2.A.5).

3

Rayleigh and Rice Channels

Rayleigh and Rice channels are the most important channel models in mobile communications for several reasons. They can be interpreted physically, they are comparatively easy to describe, and they have been confirmed by measurements. Moreover, Rayleigh and Rice channel models with given correlation properties can be implemented in software and hardware very efficiently and with a high degree of precision. Because of this, they are widely used in system performance studies. This chapter provides a thorough analysis of Rayleigh and Rice channels starting from a geometrical multipath propagation model for terrestrial mobile radio channels.

In order to better assess the performance of deterministic and stochastic simulation models, we will often refer to stochastic reference models. Depending on the objective, a reference model can be a Rayleigh, a Rice, or just a Gaussian process. The aim of this chapter is not only to describe such reference models, but also to provide the foundations for the modelling and simulation of more elaborated channel models that have been proved to be useful for capturing the characteristics of specific propagation environments.

In densely built-up areas, such as cities, the transmitted electromagnetic waves mostly do not arrive at the antenna of the receiver over the direct path. Owing to reflections from buildings, from the ground, and from other obstacles with vast surfaces, as well as scatterers, such as trees, vehicles, and other large objects, a multitude of partial waves arrive at the receiver antenna from different directions. This effect is known as *multipath propagation*. A system theoretical description of multipath channels is the topic of Section 3.1.

An often made simplifying assumption is that the propagation delay differences among the scattered signal components at the receiver are negligible compared to the symbol interval. We will understand that this assumption allows us to model the channel as a *frequency-nonselective channel*, which is also often called *narrowband channel* or *flat fading channel*. In this case, the fluctuations of the received signal can be modelled by multiplying the transmitted signal by an appropriate stochastic process. After extensive measurements [61, 62, 63] of the received envelope in urban and suburban areas, i.e., in regions where the line-of-sight component is often blocked by obstacles, the Rayleigh process was suggested as a suitable stochastic process. In rural areas, however, where the line-of-sight component is often a part of the received signal, it turned out that the Rice process is the more suitable stochastic model. A formal description of Rayleigh and Rice channels is presented in Section 3.2.

Mobile Radio Channels, Second Edition. Matthias Pätzold.

Owing to multipath propagation, the received partial waves superimpose each other at the receiver antenna. The superposition can be constructive or destructive, depending on the phase relations of the received waves. Consequently, the received electromagnetic field strength, and thus also the received signal are both strongly fluctuating functions of the receiver's position [64] or, in case of a moving receiver, strongly fluctuating functions of time. Besides, as a result of the Doppler effect, the motion of the receiver leads to a *frequency shift* (*Doppler shift*)[1] of the partial waves impinging on the antenna. Depending on the angle-of-arrival of the partial waves, different Doppler shifts occur, so that for the sum of all scattered components, we finally obtain a continuous spectrum of Doppler frequencies, which is called the *Doppler power spectral density*. A close examination of typical Doppler power spectral densities and related quantities is the purpose of Section 3.3.

The validity of Rayleigh and Rice models is limited to relatively small areas with dimensions in the order of about a few tens of wavelengths, where the local mean of the envelope is approximately constant [13]. In larger areas, however, the local mean fluctuates due to shadowing effects and is approximately lognormally distributed [63, 65]. The knowledge of the statistical properties of the received signal envelope is necessary for the development of digital communication systems and for the planning of mobile radio networks. Usually, Rayleigh and Rice processes are preferred for modelling *fast-term fading*, whereas *slow-term fading* is modelled by a lognormal process [65]. Slow-term fading not only has a strong influence on the channel availability, selection of the carrier frequency, handover, etc., but is also important in the planning of mobile radio networks. For the choice of the transmission technique and the design of digital receivers, however, the properties of the fast-term statistics are of vital importance [66]. In Section 3.4, we study the most important statistical properties of Rayleigh and Rice channels, such as the probability density function of the envelope, the squared envelope, and the phase. Moreover, we analyze the level-crossing rate, the average duration of fades, and the distribution of the fading intervals.

Finally, Chapter 3 closes with a further reading section to direct readers to sources of additional information that are beyond the scope of this chapter.

3.1 System Theoretical Description of Multipath Channels

In this section, we derive Rice and Rayleigh processes starting from the geometrical model shown in Figure 3.1. This figure illustrates a typical propagation scenario in terrestrial mobile radio environments. Without loss of generalization, we assume that the base station (BS) serves as transmitter and the mobile station (MS) is the receiver. Both terminals are equipped with single omnidirectional antennas. The base station is fixed and the mobile station is moving with speed $\mathrm{v} = |\vec{\mathrm{v}}|$ in the direction indicated by the mobile speed vector $\vec{\mathrm{v}}$. It is further assumed that the mobile station is surrounded by N local scatterers, denoted by S_n ($n = 1, 2, \ldots, N$). The distances between the terminals and the scattering objects are supposed to be large compared to the wavelength, so that the far-field condition is met, and thus the wave propagation can be described by using the plane wave model. When an incident plane wave impinges on a scatterer, then the energy of the wave is redistributed in all directions, meaning that some

[1] In the two-dimensional horizontal plane, the Doppler shift (Doppler frequency) of an elementary wave is equal to $f = f_{\max}\cos\alpha$, where α is the angle-of-arrival as illustrated in Figure 1.2 and $f_{\max} = \mathrm{v}f_0/c_0$ denotes the maximum Doppler frequency (v: velocity of the vehicle, f_0: carrier frequency, c_0: speed of light) [13].

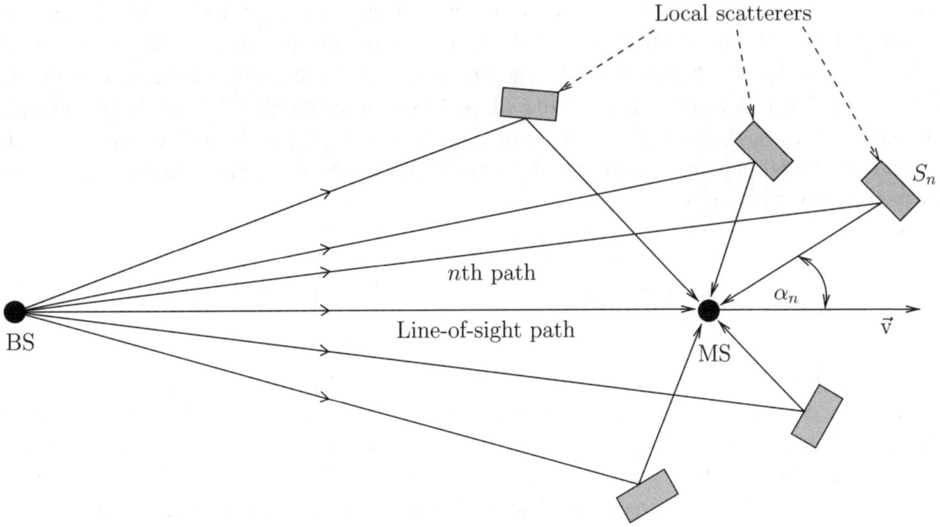

Figure 3.1 Typical propagation scenario illustrating the effect of multipath propagation in terrestrial mobile radio environments (BS: base station, MS: mobile station).

energy is scattered towards the receiver. Depending on the material of the scatterer, the incident plane wave experiences an attenuation and a phase shift. Since the path loss increases rapidly with the propagation distance, we can neglect the contributions of remote scatterers to the total energy of the received signal. All local scatterers are assumed to be fixed, so that the mobile radio environment can be regarded as (quasi-) stationary for a short period of time.

The geometrical model in Figure 3.1 is an appropriate model to describe typical propagation scenarios in urban and suburban areas, where the base station is usually elevated, so that only the mobile station is surrounded by a large number of scatterers, but not the base station. The straight path from the transmitter antenna to the receiver antenna is called the *line-of-sight path*. Depending on the propagation environment, the line-of-sight path is either present or blocked by obstructions such as buildings and mountains. The analysis in the remainder of this section is restricted to the two-dimensional horizontal plane, and it is supposed that the line-of-sight path is not present.

To examine the effects of multipath propagation on a transmitted signal, we start from the transmitted bandpass signal that can generally be represented as

$$x_{\mathrm{BP}}(t) = \mathrm{Re}\left\{x_{\mathrm{LP}}(t)\, e^{j2\pi f_0 t}\right\}, \tag{3.1}$$

where $x_{LP}(t)$ denotes the transmitted equivalent lowpass signal and f_0 is the carrier frequency. From Figure 3.1, we observe that the transmitted signal travels over local scatterers along multiple paths before impinging on the receiver antenna from various directions. Each scatterer causes an attenuation and a phase shift. Furthermore, by taking into account that each path is associated with a delay, the received bandpass signal might be expressed as

$$y_{\mathrm{BP}}(t) = \mathrm{Re}\left\{\sum_{n=1}^{N(t)} c_n(t)\, e^{j\phi_n(t)} x_{\mathrm{LP}}(t-\tau_n'(t))\, e^{j2\pi f_0 (t-\tau_n'(t))}\right\}, \tag{3.2}$$

where $N(t)$ is the number of propagation paths that equals the number of local scatterers at time t. The real-valued quantities $c_n(t)$ and $\phi_n(t)$ represent the attenuation and the phase shift, respectively, which are caused by the interaction of the transmitted signal with the nth scatterer S_n. The propagation delay of the nth path is denoted by $\tau'_n(t)$. Notice that the channel parameters $N(t), c_n(t), \phi_n(t)$, and $\tau'_n(t)$ are in general time-variant as a result of changes in the structure of the propagation environment. From (3.2), it is obvious that the received equivalent lowpass signal is given by

$$y_{\mathrm{LP}}(t) = \sum_{n=1}^{N(t)} c_n(t)\, e^{j(\phi_n(t) - 2\pi f_0 \tau'_n(t))}\, x_{\mathrm{LP}}(t - \tau'_n(t)). \tag{3.3}$$

This result provides an insight into the effects caused by multipath propagation. If a signal is transmitted over a time-varying multipath channel, then the received signal appears in the equivalent baseband domain as a sum of weighted, delayed copies of the transmitted signal, where the weighing factors are complex-valued and time-variant. Hence, one major characteristic of a multipath channel is that the channel stretches the transmitted signal in time, so that the duration of the received signal is greater than that of the transmitted signal. This effect is known as *time dispersion*.

From the input-output relationship in (3.3), it follows that the time-variant impulse response of the multipath channel can be expressed in the equivalent baseband by

$$h(\tau', t) = \sum_{n=1}^{N(t)} c_n(t)\, e^{j(\phi_n(t) - 2\pi f_0 \tau'_n(t))}\, \delta(\tau' - \tau'_n(t)). \tag{3.4}$$

The time-variant impulse response $h(\tau', t)$ describes the response of the channel at time t due to an impulse that stimulated the channel at time $t - \tau'$. Since the multipath channel is causal, it follows that the time-variant impulse response is zero for negative propagation delays, i.e., $h(\tau', t) = 0$ if $\tau' < 0$. Owing to the countless number of propagation environments combined with the lack of detailed knowledge about the channel parameters, it is reasonable to describe the time-variant impulse response $h(\tau', t)$ statistically. To this end, we simplify the channel model in (3.4) by restricting its validity to a small observation time interval of duration T_0, i.e., $t \in [t_0, t_0 + T_0]$, where t_0 denotes a reference point in time, which can be set to zero without loss of generality. If T_0 is sufficiently small, then we may consider the number of propagation paths $N(t)$, the attenuation factors $c_n(t)$, the phases $\phi_n(t)$, the angle-of-arrival $\alpha_n(t)$, and the mobile speed $\mathrm{v}(t)$ as time independent, which allows us to write $N = N(t)$, $c_n = c_n(t)$, $\phi_n = \phi_n(t)$, $\alpha_n = \alpha_n(t)$, and $\mathrm{v} = \mathrm{v}(t)$. In the case under consideration, we can represent the propagation delays $\tau'_n(t)$ as a linear function of time according to

$$\begin{aligned} \tau'_n(t) &= \tau'_n(t_0) - t\frac{\mathrm{v}}{c_0}\cos(\alpha_n) \\ &= \tau'_n(t_0) - t \cdot \frac{f_{\max}}{f_0}\cos(\alpha_n) \\ &= \tau'_n(t_0) - t \cdot \frac{f_n}{f_0}, \end{aligned} \tag{3.5}$$

where c_0 is the speed of light, f_{max} denotes the maximum Doppler frequency [see (1.2)], and f_n describes the Doppler frequency of the nth path as introduced in (1.1). Now, substituting (3.5) in (3.4), using the approximation $\delta(\tau' - \tau_n'(t_0) + tf_n/f_0) \approx \delta(\tau' - \tau_n'(t_0))$, and simplifying the notation by writing τ_n' instead of $\tau_n'(t_0)$ results in

$$h(\tau', t) = \sum_{n=1}^{N} c_n \, e^{\, j(2\pi f_n t + \phi_n - 2\pi f_0 \tau_n')} \, \delta(\tau' - \tau_n'). \tag{3.6}$$

The third phase term in the expression above can be rewritten as

$$2\pi f_0 \tau_n' = \frac{2\pi}{\lambda_0} c_0 \, \tau_n' = k_0 \, D_n, \tag{3.7}$$

where λ_0 is the wavelength, $k_0 = 2\pi/\lambda_0$ denotes the free-space wave number, and $D_n = c_0 \, \tau_n'$ is the length of the total distance which the emitted signal travels from the transmitting antenna via the local scatterer S_n to the receiving antenna. Thus, we obtain

$$h(\tau', t) = \sum_{n=1}^{N} c_n \, e^{\, j(2\pi f_n t + \phi_n - k_0 D_n)} \, \delta(\tau' - \tau_n'). \tag{3.8}$$

The phase change in $h(\tau', t)$ is determined by three terms. The first term $2\pi f_n t$ describes the phase change due to the Doppler effect caused by the motion of the receiver. The second term ϕ_n captures the phase change that is caused — along with the attenuation c_n — by the interaction of the transmitted signal with the scatterer S_n. Finally, the phase change of the third term $k_0 D_n$ in (3.8) results from the total distance travelled. Note that the phase term $k_0 D_n = 2\pi D_n/\lambda_0$ changes by 2π when D_n changes by λ_0. In GSM, for example, the wavelength λ_0 is about 30 cm, and hence the phase $k_0 D_n$ can change considerably with relatively small motions of the mobile.

To proceed further with a stochastic description of the channel model in (3.8), we introduce various random variables for the model parameters. This can be done in a number of ways. A reasonable assumption is that the phases ϕ_n and $\phi_n' = k_0 D_n$ are independent random variables, each having a uniform distribution in the interval from 0 to 2π, i.e., $\phi_n, \phi_n' \sim U[0, 2\pi)$. The differential phase $\beta_n = \phi_n - \phi_n'$ of the nth path describes then a new random variable, which is also uniformly distributed in the interval $[0, 2\pi)$.[2] Hence, the channel impulse response takes the form

$$h(\tau', t) = \sum_{n=1}^{N} c_n \, e^{\, j(2\pi f_n t + \theta_n)} \, \delta(\tau' - \tau_n'). \tag{3.9}$$

According to (2.139), the time-variant channel transfer function $H(f', t)$ is obtained by taking the Fourier transform of $h(\tau', t)$ with respect to τ', which gives

$$H(f', t) = \sum_{n=1}^{N} c_n \, e^{\, j[\theta_n + 2\pi(f_n t - f' \tau_n')]}. \tag{3.10}$$

[2] We mention that if $\phi_n, \phi_n' \sim U[0, 2\pi)$, then the distribution of $\beta_n = \phi_n - \phi_n'$ has a triangular shape limited on the interval $(-\pi, 3\pi)$. However, if we consider θ_n mod 2π, then the resulting random variable is uniformly distributed over $[0, 2\pi)$.

Note that the absolute value of the time-variant channel transfer function, $|H(f', t)|$, is in general dependent on the frequency variable f'. Such channels are said to be *frequency-selective channels*, which will be studied in Chapter 7.

An important special case is when the differential propagation delays $\tau'_n - \tau'_m$ are small in comparison to the symbol duration T_{sym}, i.e., $\max|\tau'_n - \tau'_m| \ll T_{sym}$. If this condition is fulfilled, then the propagation delays τ'_n can be approximated by $\tau'_0 \approx \tau'_n \ \forall \ n = 1, 2, \ldots, N$. As a consequence, the channel impulse response simplifies further to

$$h(\tau', t) = \mu(t) \cdot \delta(\tau' - \tau'_0), \tag{3.11}$$

where

$$\mu(t) = \sum_{n=1}^{N} c_n \, e^{\,j(2\pi f_n t + \theta_n)} \tag{3.12}$$

describes a complex stochastic process that represents the sum of all scattered components. Taking the Fourier transform of $h(\tau', t)$ in (3.11) with respect to τ' results in the time-variant channel transfer function $H(f', t)$

$$H(f', t) = \mu(t) \cdot e^{-j2\pi f' \tau'_0}, \tag{3.13}$$

which provides further insight into the characteristics of multipath channels. Since the absolute value of the time-variant channel transfer function, $|H(f', t)| = |\mu(t)|$, is independent of the frequency variable f', we call such channels *frequency-nonselective channels* or *flat fading channels*. They form an important class of channels for which the input-output relationship in (3.3) can be simplified by using (2.138) as follows

$$\begin{aligned} y_{\mathrm{LP}}(t) &= \int_0^\infty h(\tau', t)\, x_{\mathrm{LP}}(t - \tau')\, d\tau' \\ &= \mu(t) \int_0^\infty \delta(\tau' - \tau'_0)\, x_{\mathrm{LP}}(t - \tau')\, d\tau' \\ &= \mu(t) \cdot x_{\mathrm{LP}}(t - \tau'_0). \end{aligned} \tag{3.14}$$

This result shows clearly that the received signal of a flat fading channel is given by the delayed version of the transmitted signal multiplied by a complex stochastic process $\mu(t)$. The distribution of this process depends on the number of paths N and the assumptions made on the model parameters c_n, f_n, and β_n. If the number of paths N tends to infinity, then we may invoke the central limit theorem, which states that

$$\mu(t) = \lim_{N \to \infty} \sum_{n=1}^{N} c_n \, e^{\,j(2\pi f_n t + \theta_n)} \tag{3.15}$$

equals a complex-valued Gaussian random process with zero mean and variance $2\sigma_0^2 = \mathrm{Var}\{\mu(t)\} = \lim_{N\to\infty} \sum_{n=1}^{N} E\{c_n^2\}$.

When the mobile station moves from one place to another, then the phase term $2\pi f_n t$ in (3.15) changes fast with time. Now it becomes apparent that the superposition of all scattered signal components results in a multipath channel which is characterized by rapid changes in the received signal strength as a function of time or distance. This effect is called *fading*. The analysis of the fading effects of Rayleigh and Rice channels is the topic of the following sections.

3.2 Formal Description of Rayleigh and Rice Channels

In the previous section, we have seen that the transmission of a narrowband signal over a frequency-nonselective mobile radio channel results in a sum of scattered components, which can be modelled in the equivalent complex baseband by a zero-mean complex-valued Gaussian random process

$$\mu(t) = \mu_1(t) + j\mu_2(t). \tag{3.16}$$

The real-valued Gaussian random processes $\mu_1(t)$ and $\mu_2(t)$ are also called the *inphase component* and the *quadrature component*, respectively. Usually, it is assumed that $\mu_1(t)$ and $\mu_2(t)$ are statistically uncorrelated. Let the variance of the processes $\mu_i(t)$ be equal to $\text{Var}\{\mu_i(t)\} = \sigma_0^2$ for $i = 1, 2$, then the variance of $\mu(t)$ is given by $\text{Var}\{\mu(t)\} = 2\sigma_0^2$.

In the following, the line-of-sight component of the received signal will be described by a complex sinusoid (cisoid) of the form

$$m(t) = m_1(t) + jm_2(t) = \rho\, e^{j(2\pi f_\rho t + \theta_\rho)}, \tag{3.17}$$

where ρ, f_ρ, and θ_ρ denote the amplitude, the Doppler frequency, and the phase of the line-of-sight component, respectively. If not otherwise stated, then it is supposed that the parameters ρ, f_ρ, and θ_ρ are constant; meaning that the line-of-sight component $m(t)$ is deterministic. One should note that due to the Doppler effect, the relation $f_\rho = 0$ only holds if the direction of arrival of the dominant wave is orthogonal to the direction of motion of the mobile station. Consequently, (3.17) then becomes a time-invariant component, i.e.,

$$m = m_1 + jm_2 = \rho\, e^{j\theta_\rho}. \tag{3.18}$$

At the receiver antenna, we receive the superposition of the sum of the scattered components and the line-of-sight component. In the model chosen here, the superposition is equal to the addition of the components in (3.16) and (3.17). This provides the rational for introducing a further complex Gaussian random process

$$\mu_\rho(t) = \mu_{\rho_1}(t) + j\mu_{\rho_2}(t) = \mu(t) + m(t) \tag{3.19}$$

with time-variant mean value $m(t) = E\{\mu_\rho(t)\}$.

As we know, taking the absolute values of (3.16) and (3.19) leads to Rayleigh and Rice processes [42], respectively. In order to distinguish these processes clearly from each other, we will henceforth denote Rayleigh processes by

$$\zeta(t) = |\mu(t)| = |\mu_1(t) + j\mu_2(t)| \tag{3.20}$$

and Rice processes by

$$\xi(t) = |\mu_\rho(t)| = |\mu(t) + m(t)|. \tag{3.21}$$

3.3 Elementary Properties of Rayleigh and Rice Channels

3.3.1 Autocorrelation Function and Spectrum of the Complex Envelope

The shape of the power spectral density (PSD) of the complex Gaussian random process (3.19) is identical to the Doppler power spectral density, which is determined by both the power of the electromagnetic waves impinging on the receiver antenna and the distribution of the angles-of-arrival. In addition to that, the antenna radiation pattern of the receiving antenna has also a decisive influence on the shape of the Doppler power spectral density.

By modelling mobile radio channels, one frequently simplifies matters by assuming that the propagation of the electromagnetic waves occurs in the horizonal two-dimensional plane. Furthermore, the idealized assumption is often made that the angles-of-arrival α_n [see Figure 3.1] are uniformly distributed over the interval from 0 to 2π. For omnidirectional antennas, we can then easily calculate the (Doppler) power spectral density $S_{\mu\mu}(f)$ of the scattered components $\mu(t) = \mu_1(t) + j\mu_2(t)$. For $S_{\mu\mu}(f)$, one finds the following expression [13, 67]

$$S_{\mu\mu}(f) = S_{\mu_1\mu_1}(f) + S_{\mu_2\mu_2}(f), \tag{3.22}$$

where

$$S_{\mu_i\mu_i}(f) = \begin{cases} \dfrac{\sigma_0^2}{\pi f_{\max}\sqrt{1-(f/f_{\max})^2}}, & |f| \leq f_{\max}, \\ 0, & |f| > f_{\max}, \end{cases} \tag{3.23}$$

holds for $i = 1, 2$ and $f_{\max}$ denotes the *maximum Doppler frequency*. In the literature, (3.23) is often called *Jakes power spectral density* (*Jakes PSD*), although it was originally derived by Clarke [67]. For a detailed derivation of the Jakes power spectral density, the reader is referred to Appendix 3.A.

In principle, the electromagnetic waves arriving at the receiver antenna have besides the horizontal (azimuth) also a vertical (elevation) component. Taking both components into account leads to the three-dimensional propagation model derived in [68]. The main difference between the resulting power spectral density and (3.23) is that there are no singularities at $f = \pm f_{\max}$. Apart from that, the graph of the power spectral density is very similar to that obtained from (3.23).

Considering (3.22) and (3.23), we see that the function $S_{\mu\mu}(f)$ is positive, real, and even. The even property is no longer fulfilled, if either buildings or other obstructions prevent a uniform distribution of the angles-of-arrival or a sector antenna with a directional antenna pattern is used at the receiver [67, 69]. The electromagnetic characteristics of the environment can be such that waves from certain directions are scattered with different intensities. In these cases, the Doppler power spectral density $S_{\mu\mu}(f)$ of the complex Gaussian random process in (3.16) is also asymmetrical [70]. We will return to this subject in Chapter 6.

It is shown in Appendix 3.A that the inverse Fourier transform of the Jakes power spectral density $S_{\mu\mu}(f)$ results in the following autocorrelation function (ACF)

$$r_{\mu\mu}(\tau) = r_{\mu_1\mu_1}(\tau) + r_{\mu_2\mu_2}(\tau), \tag{3.24}$$

where

$$r_{\mu_i\mu_i}(\tau) = \sigma_0^2 J_0(2\pi f_{\max}\tau), \quad i = 1, 2, \tag{3.25}$$

holds, and $J_0(\cdot)$ denotes the zeroth-order Bessel function of the first kind.

By way of illustration, the Jakes power spectral density introduced in (3.23) is presented together with the corresponding autocorrelation function introduced in (3.25) in Figures 3.2(a) and 3.2(b), respectively.

Besides the Jakes power spectral density, the so-called *Gaussian power spectral density* (*Gaussian PSD*)

$$S_{\mu_i\mu_i}(f) = \frac{\sigma_0^2}{f_c}\sqrt{\frac{\ln 2}{\pi}}\, e^{-\ln 2\left(\frac{f}{f_c}\right)^2}, \quad i = 1, 2, \tag{3.26}$$

will also play an important role in the following, where f_c denotes the 3-dB-cut-off frequency.

Theoretical investigations in [71] have shown that the Doppler power spectral density of aeronautical channels has a Gaussian shape. Further information on measurements concerning the propagation characteristics of aeronautical satellite channels can be found, for example, in [72]. Although no absolute correspondence to the obtained measurements could be proved, (3.26) can in most cases very well be used as a sufficiently good approximation [73]. For signal bandwidths up to some 10 kHz, the aeronautical satellite channel belongs to the class of frequency-nonselective mobile radio channels [73].

Especially for frequency-selective mobile radio channels, it has been shown [74] that the Doppler power spectral density of far echoes deviates strongly from the shape of the Jakes power spectral density. In some cases, the Doppler power spectral density is approximately Gaussian shaped and is generally shifted from the origin of the frequency plane, because the far echoes mostly dominate from a certain direction of preference. Specifications of frequency-shifted Gaussian power spectral densities for the pan-European, terrestrial, cellular GSM system can be found in [19].

The inverse Fourier transform of the Gaussian power spectral density in (3.26) results in the autocorrelation function

$$r_{\mu_i\mu_i}(\tau) = \sigma_0^2\, e^{-\left(\pi \frac{f_c}{\sqrt{\ln 2}}\tau\right)^2}. \tag{3.27}$$

In Figure 3.3, the Gaussian power spectral density in (3.26) is illustrated together with the corresponding autocorrelation function according to (3.27).

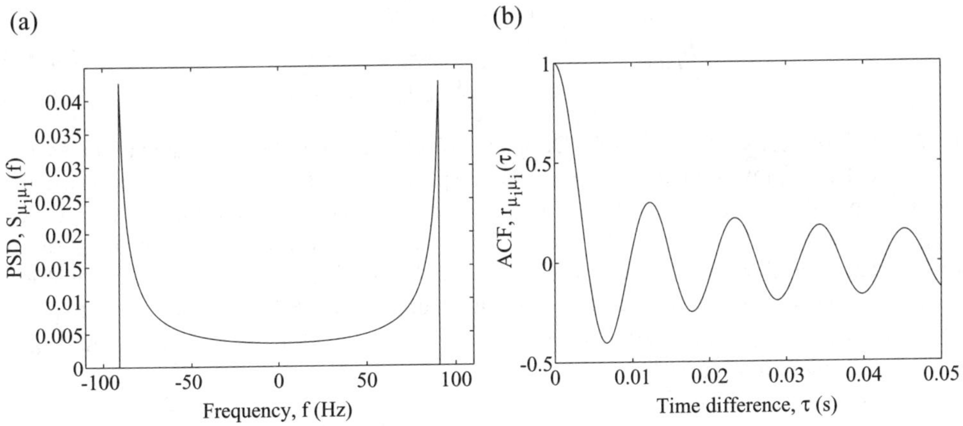

Figure 3.2 (a) Jakes power spectral density $S_{\mu_i\mu_i}(f)$ and (b) the corresponding autocorrelation function $r_{\mu_i\mu_i}(\tau)$ ($f_{\max} = 91$ Hz, $\sigma_0^2 = 1$).

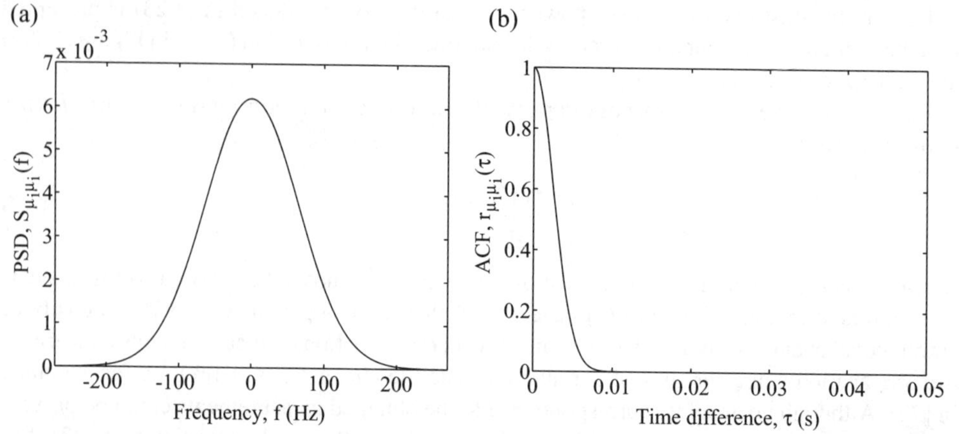

Figure 3.3 (a) Gaussian power spectral density $S_{\mu_i\mu_i}(f)$ and (b) corresponding autocorrelation function $r_{\mu_i\mu_i}(\tau)$ ($f_c = \sqrt{\ln 2} f_{\max}$, $f_{\max} = 91$ Hz, $\sigma_0^2 = 1$).

Important characteristic quantities of the Doppler power spectral density $S_{\mu_i\mu_i}(f)$ are the *average Doppler shift* $B^{(1)}_{\mu_i\mu_i}$ and the *Doppler spread* $B^{(2)}_{\mu_i\mu_i}$ [18]. The average Doppler shift (Doppler spread) describes the average frequency shift (frequency spread) that a carrier signal experiences during the transmission over a multipath mobile radio channel. The average Doppler shift $B^{(1)}_{\mu_i\mu_i}$ is defined as the first moment of $S_{\mu_i\mu_i}(f)$ and the Doppler spread $B^{(2)}_{\mu_i\mu_i}$ as the square root of the second central moment of $S_{\mu_i\mu_i}(f)$. Mathematically, the two quantities $B^{(1)}_{\mu_i\mu_i}$ and $B^{(2)}_{\mu_i\mu_i}$ can be defined in terms of the Doppler power spectral density $S_{\mu_i\mu_i}(f)$ by

$$B^{(1)}_{\mu_i\mu_i} := \frac{\int_{-\infty}^{\infty} f S_{\mu_i\mu_i}(f)\, df}{\int_{-\infty}^{\infty} S_{\mu_i\mu_i}(f)\, df} \tag{3.28a}$$

and

$$B^{(2)}_{\mu_i\mu_i} := \sqrt{\frac{\int_{-\infty}^{\infty} \left(f - B^{(1)}_{\mu_i\mu_i}\right)^2 S_{\mu_i\mu_i}(f)\, df}{\int_{-\infty}^{\infty} S_{\mu_i\mu_i}(f)\, df}}, \tag{3.28b}$$

for $i = 1, 2$, respectively. The average Doppler shift $B^{(1)}_{\mu_i\mu_i}$ can thus be considered as a measure of the centre of gravity of the Doppler power spectral density $S_{\mu_i\mu_i}(f)$. Analogously, the Doppler spread $B^{(2)}_{\mu_i\mu_i}$ can be interpreted as a measure of the range of frequencies f over which $S_{\mu_i\mu_i}(f)$ is essentially different from zero. Equivalent expressions, which are often easier to calculate than those in (3.28a) and (3.28b), can be obtained by using Fourier transform techniques. This allows us to express the average Doppler shift $B^{(1)}_{\mu_i\mu_i}$ and the Doppler spread $B^{(2)}_{\mu_i\mu_i}$ in terms of the autocorrelation function $r_{\mu_i\mu_i}(\tau)$ as well as its first and second time derivative at the origin as follows

$$B^{(1)}_{\mu_i\mu_i} := \frac{1}{2\pi j} \cdot \frac{\dot{r}_{\mu_i\mu_i}(0)}{r_{\mu_i\mu_i}(0)} \quad \text{and} \quad B^{(2)}_{\mu_i\mu_i} = \frac{1}{2\pi}\sqrt{\left(\frac{\dot{r}_{\mu_i\mu_i}(0)}{r_{\mu_i\mu_i}(0)}\right)^2 - \frac{\ddot{r}_{\mu_i\mu_i}(0)}{r_{\mu_i\mu_i}(0)}}, \tag{3.29a,b}$$

for $i = 1, 2$, respectively.

For the important special case where the Doppler power spectral densities $S_{\mu_1\mu_1}(f)$ and $S_{\mu_2\mu_2}(f)$ are identical and symmetrical, we have $\dot{r}_{\mu_i\mu_i}(0) = 0$ $(i = 1, 2)$. Hence, by using (3.22), we obtain the following expressions

$$B_{\mu\mu}^{(1)} = B_{\mu_i\mu_i}^{(1)} = 0 \quad \text{and} \quad B_{\mu\mu}^{(2)} = B_{\mu_i\mu_i}^{(2)} = \frac{\sqrt{\beta}}{2\pi\sigma_0}, \tag{3.30a,b}$$

where $\sigma_0^2 = r_{\mu_i\mu_i}(0) \geq 0$ and $\beta = -\ddot{r}_{\mu_i\mu_i}(0) \geq 0$.

Especially for the Jakes power spectral density [see (3.23)] and the Gaussian power spectral density [see (3.26)], the expressions given above can further be simplified to

$$B_{\mu_i\mu_i}^{(1)} = B_{\mu\mu}^{(1)} = 0 \quad \text{and} \quad B_{\mu_i\mu_i}^{(2)} = B_{\mu\mu}^{(2)} = \begin{cases} \dfrac{f_{\max}}{\sqrt{2}}, & \text{Jakes PSD,} \\ \dfrac{f_c}{\sqrt{2\ln 2}}, & \text{Gaussian PSD,} \end{cases} \tag{3.31a,b}$$

for $i = 1, 2$. From this result, it follows that the Doppler spread of the Jakes power spectral density is identical to that of the Gaussian power spectral density if the 3-dB-cut-off frequency f_c and the maximum Doppler frequency $f_{\max}$ are related by $f_c = \sqrt{\ln 2} f_{\max}$.

The Doppler spread allows to classify mobile radio channels in fast fading channels and slow fading channels. If the Doppler spread $B_{\mu\mu}^{(2)}$ is much larger than the symbol rate f_{sym}, i.e., $B_{\mu\mu}^{(2)} \gg f_{sym}$, then the channel envelope changes rapidly within each symbol period. Such channels are called *fast fading channels*. Otherwise, if the Doppler spread $B_{\mu\mu}^{(2)}$ is much smaller than the symbol rate f_{sym}, i.e., $B_{\mu\mu}^{(2)} \ll f_{sym}$, then the channel envelope remains approximately constant during the duration of the transmission of one symbol, and we call the channel a *slow fading channel.*

3.3.2 *Autocorrelation Function and Spectrum of the Envelope*

The derivation of the autocorrelation function of the envelope $\zeta(t) = |\mu(t)|$ of a zero-mean complex Gaussian random process $\mu(t) = \mu_1(t) + j\mu_2(t)$ is a cumbersome problem. To simplify the mathematics, it is supposed that the underlying real Gaussian random processes $\mu_1(t)$ and $\mu_2(t)$ are uncorrelated and have identical autocorrelation functions. This means that the relation $r_{\mu\mu}(\tau) = r_{\mu_1\mu_1}(\tau) + r_{\mu_2\mu_2}(\tau) = 2r_{\mu_i\mu_i}(\tau)$ holds, as $r_{\mu_1\mu_1}(\tau) = r_{\mu_2\mu_2}(\tau)$ and $r_{\mu_1\mu_2}(\tau) = r_{\mu_2\mu_1}(\tau) = 0$. The power spectral density $S_{\mu\mu}(f)$ of $\mu(t)$ is then positive, real, and even, i.e., $S_{\mu\mu}^*(f) = S_{\mu\mu}(f) = S_{\mu\mu}(-f) \geq 0$. For this case, it is shown in Appendix 3.B that the autocorrelation function of the envelope $\zeta(t)$ can be expressed as

$$\begin{aligned} r_{\zeta\zeta}(\tau) &= E\{\zeta(t)\zeta(t+\tau)\} \\ &= \sigma_0^2 \frac{\pi}{2} F\left(-\frac{1}{2}, -\frac{1}{2}; 1; \frac{r_{\mu_i\mu_i}^2(\tau)}{\sigma_0^4}\right), \end{aligned} \tag{3.32}$$

where $F(\cdot, \cdot; \cdot; \cdot)$ denotes the hypergeometric function and $\sigma_0^2 = r_{\mu_i\mu_i}(0)$ $(i = 1, 2)$. A useful approximation of the above result can be obtained by using the following series expansion of the hypergeometric function [23, Equation (9.100)]

$$F(\alpha, \beta; \gamma; x) = 1 + \frac{\alpha\beta}{\gamma \cdot 1} x + \frac{\alpha(\alpha+1)\beta(\beta+1)}{\gamma(\gamma+1) \cdot 1 \cdot 2} x^2 + \cdots . \tag{3.33}$$

Fortunately, for $\alpha = -1/2$, $\beta = -1/2$, and $\gamma = 1$, the series converges very fast, which allows us to neglect all terms beyond second order. Hence, the autocorrelation function $r_{\zeta\zeta}(\tau)$ can be approximated by

$$r_{\zeta\zeta}(\tau) \approx \sigma_0^2 \frac{\pi}{2} \left(1 + \frac{r^2_{\mu_i\mu_i}(\tau)}{4\sigma_0^4} \right) . \tag{3.34}$$

From this expression, the mean power of the envelope is obtained as $r_{\zeta\zeta}(0) \approx \sigma_0^2 5\pi/8$, whereas the exact result equals $r_{\zeta\zeta}(0) = \sigma_0^2 \pi F(-1/2, -1/2; 1; 1)/2 = 2\sigma_0^2$. Since the relative error of the envelope power is only -1.83 per cent, we can consider (3.34) as a good approximation to the exact autocorrelation function $r_{\zeta\zeta}(\tau)$ presented in (3.32) [13, p. 27].

The exact solution of the envelope power spectral density $S_{\zeta\zeta}(f)$ cannot be obtained in closed form. However, $S_{\zeta\zeta}(f)$ can be approximated by computing the Fourier transform of the autocorrelation function $r_{\zeta\zeta}(\tau)$ in (3.34) as follows

$$\begin{aligned} S_{\zeta\zeta}(f) &\approx \sigma_0^2 \frac{\pi}{2} \delta(f) + \frac{\pi}{8\sigma_0^2} S_{\mu_i\mu_i}(f) * S_{\mu_i\mu_i}(f) \\ &= \sigma_0^2 \frac{\pi}{2} \delta(f) + \frac{\pi}{8\sigma_0^2} \int_{-\infty}^{\infty} S_{\mu_i\mu_i}(f - \upsilon) \, S_{\mu_i\mu_i}(\upsilon) \, d\upsilon . \end{aligned} \tag{3.35}$$

From this result and the fact that $S_{\mu_i\mu_i}(f)$ is positive, real, and even, we may conclude that $S_{\zeta\zeta}(f)$ is also positive, real, and even. Using $S_{\mu_i\mu_i}(f) = 0$ if $|f| > f_{\max}$, we can finally write

$$S_{\zeta\zeta}(f) \approx \sigma_0^2 \frac{\pi}{2} \delta(f) + \frac{\pi}{8\sigma_0^2} \int_{-f_{\max}}^{f_{\max}-|f|} S_{\mu_i\mu_i}(|f| + \upsilon) \, S_{\mu_i\mu_i}(\upsilon) \, d\upsilon, \quad |f| \leq 2 f_{\max} . \tag{3.36}$$

Note that the bandwidth of the envelope spectra $S_{\zeta\zeta}(f)$ is $4 f_{\max}$. The presence of the Dirac delta function in $S_{\zeta\zeta}(f)$ is a consequence of the mean value of $\zeta(t)$ [cf. (2.42)]. Actually, the factor in front of the Dirac delta function equals the squared mean value of $\zeta(t)$, i.e., $m_\zeta^2 = E^2\{\zeta(t)\} = \sigma_0^2 \pi/2$. To remove the influence of the mean m_ζ, one prefers to study the *autocovariance function* of the envelope

$$r^c_{\zeta\zeta}(\tau) = r_{\zeta\zeta}(\tau) - m_\zeta^2 \tag{3.37a}$$

$$= \sigma_0^2 \frac{\pi}{2} \left[F\left(-\frac{1}{2}, -\frac{1}{2}; 1; \frac{r^2_{\mu_i\mu_i}(\tau)}{\sigma_0^4} \right) - 1 \right] \tag{3.37b}$$

$$\approx \frac{\pi}{8\sigma_0^2} r^2_{\mu_i\mu_i}(\tau), \quad i = 1, 2. \tag{3.37c}$$

Its Fourier transform is known as the *autocovariance spectrum* $S^c_{\zeta\zeta}(f)$ of the envelope, which can be identified as the second term of (3.36), i.e.,

$$S^c_{\zeta\zeta}(f) \approx \frac{\pi}{8\sigma_0^2} \int\limits_{-f_{\max}}^{f_{\max}-|f|} S_{\mu_i\mu_i}(|f|+\upsilon)\, S_{\mu_i\mu_i}(\upsilon)\, d\upsilon, \quad |f| \leq 2f_{\max}. \tag{3.38}$$

As an illustrative example, we consider the two-dimensional isotropic scattering case, for which the autocorrelation function $r_{\mu_i\mu_i}(\tau)$ is given by (3.25). Substituting (3.25) in (3.37c) gives the autocovariance function in the form

$$r^c_{\zeta\zeta}(\tau) \approx \sigma_0^2 \frac{\pi}{8} J_0^2(2\pi f_{\max}\tau). \tag{3.39}$$

Solving (3.38) or, alternatively, computing the Fourier transform of $r^c_{\zeta\zeta}(\tau)$ in (3.39) results in the autocovariance spectrum (see Appendix 3.C)

$$S^c_{\zeta\zeta}(f) \approx \begin{cases} \dfrac{\sigma_0^2}{8\pi f_{\max}} \boldsymbol{K}\left(\sqrt{1-\left(\dfrac{f}{2f_{\max}}\right)^2}\right), & |f| < 2f_{\max}, \\ 0, & |f| > 2f_{\max}, \end{cases} \tag{3.40}$$

where $\boldsymbol{K}(\cdot)$ denotes the *complete elliptic integral of the first kind*, which is defined as [23, Equation (8.110.3)]

$$\boldsymbol{K}(k) = \int\limits_0^1 \frac{1}{\sqrt{(1-x^2)(1-k^2x^2)}}\, dx \tag{3.41a}$$

$$= \int\limits_0^{\pi/2} \frac{1}{\sqrt{1-k^2\sin^2\varphi}}\, d\varphi. \tag{3.41b}$$

Since the complete elliptic integral $\boldsymbol{K}(k)$ has a singularity at $k = 1$, it follows that $S^c_{\zeta\zeta}(f) \to \infty$ as $f \to 0$.

The results for the approximation of the autocovariance function $r^c_{\zeta\zeta}(\tau)$ in (3.39) and the corresponding autocovariance spectrum $S^c_{\zeta\zeta}(f)$ in (3.40) are illustrated in Figures 3.4(a) and 3.4(b), respectively. For comparison, the exact solutions are also presented. In the case of the autocovariance function $r^c_{\zeta\zeta}(\tau)$, the exact solution follows from the evaluation of (3.37b) using (3.25). Finally, the exact solution of the autocovariance spectrum $S^c_{\zeta\zeta}(f)$ has been obtained by computing numerically the Fourier transform of $r^c_{\zeta\zeta}(\tau)$ in (3.37b).

3.3.3 Autocorrelation Function and Spectrum of the Squared Envelope

The signal-to-noise ratio (SNR) in mobile communication systems is proportional to the squared envelope of the received signal. A characterization of the squared envelope is therefore also of interest. In this subsection, we derive the autocorrelation function and the spectrum of the squared envelope $\zeta^2(t) = |\mu(t)|^2$ for the case where $\mu(t)$ is a zero-mean complex Gaussian

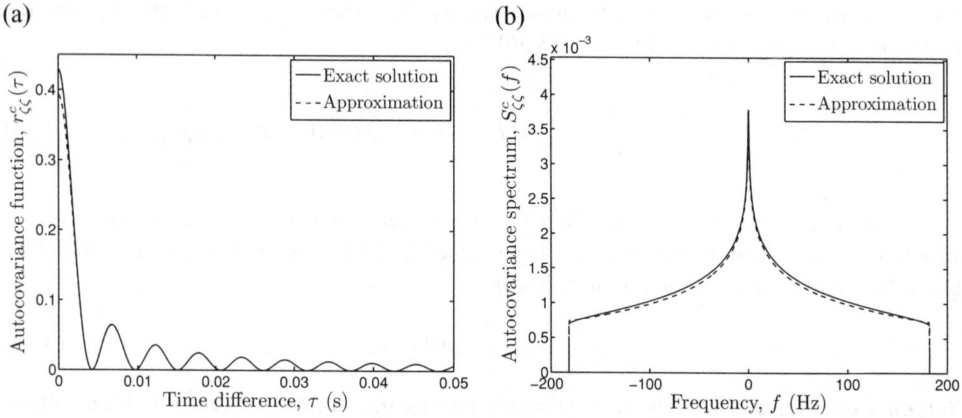

Figure 3.4 (a) Autocovariance function $r^c_{\zeta\zeta}(\tau)$ and (b) corresponding autocovariance spectrum $S^c_{\zeta\zeta}(f)$ of the envelope ($f_{\max} = 91\,\text{Hz}$, $\sigma_0^2 = 1$).

random process. Our starting point is the definition of the autocorrelation function of the squared envelope

$$r_{\zeta^2\zeta^2}(\tau) = E\{\zeta^2(t)\zeta^2(t+\tau)\}. \tag{3.42}$$

In this expression, we substitute $\zeta^2(t) = |\mu(t)|^2 = \mu_1^2(t) + \mu_2^2(t)$ and obtain

$$\begin{aligned} r_{\zeta^2\zeta^2}(\tau) &= E\{\mu_1^2(t)\,\mu_1^2(t+\tau) + \mu_2^2(t)\,\mu_2^2(t+\tau) \\ &\quad + \mu_1^2(t)\mu_2^2(t+\tau) + \mu_2^2(t)\mu_1^2(t+\tau)\} \\ &= r_{\mu_1^2\mu_1^2}(\tau) + r_{\mu_2^2\mu_2^2}(\tau) + r_{\mu_1^2\mu_2^2}(\tau) + r_{\mu_2^2\mu_1^2}(\tau). \end{aligned} \tag{3.43}$$

By using, e.g., the mathematical model of a zero-mean complex Gaussian random process introduced in (3.15), one can show that the four terms in the expression above can be written as

$$r_{\mu_1^2\mu_1^2}(\tau) = r_{\mu_2^2\mu_2^2}(\tau) = r^2_{\mu_1\mu_1}(0) + 2\,r^2_{\mu_1\mu_1}(\tau) \tag{3.44a}$$

$$r_{\mu_1^2\mu_2^2}(\tau) = r_{\mu_2^2\mu_1^2}(\tau) = r^2_{\mu_1\mu_1}(0) + 2\,r^2_{\mu_1\mu_2}(\tau). \tag{3.44b}$$

With these relations and $\sigma_0^4 = r^2_{\mu_1\mu_1}(0)$, the squared-envelope autocorrelation function can be expressed as

$$\begin{aligned} r_{\zeta^2\zeta^2}(\tau) &= 4\,\sigma_0^4 + 4\,r^2_{\mu_1\mu_1}(\tau) + 4\,r^2_{\mu_1\mu_2}(\tau) \\ &= 4\,\sigma_0^4 + |r_{\mu\mu}(\tau)|^2. \end{aligned} \tag{3.45}$$

Taking the Fourier transform of $r_{\zeta^2\zeta^2}(\tau)$ and applying the convolution theorem

$$r_{\mu\mu}(\tau)\cdot r^*_{\mu\mu}(\tau) \quad \circ\!\!-\!\!\bullet \quad S_{\mu\mu}(f) * S^*_{\mu\mu}(-f) \tag{3.46}$$

with $S^*_{\mu\mu}(-f) = S_{\mu\mu}(-f)$, enables us to present the spectrum of the squared envelope in the form

$$S_{\zeta^2\zeta^2}(f) = 4\sigma_0^4\,\delta(f) + S_{\mu\mu}(f) * S_{\mu\mu}(-f). \tag{3.47}$$

To remove the influence of the mean $m_{\zeta^2} = E\{\zeta^2(t)\} = 2\sigma_0^2$, we proceed by studying the squared-envelope autocovariance function

$$\begin{aligned} r^c_{\zeta^2\zeta^2}(\tau) &= r_{\zeta^2\zeta^2}(\tau) - m^2_{\zeta^2} \\ &= |r_{\mu\mu}(\tau)|^2 \end{aligned} \tag{3.48}$$

and the squared-envelope autocovariance spectrum

$$S^c_{\zeta^2\zeta^2}(f) = S_{\mu\mu}(f) * S_{\mu\mu}(-f). \tag{3.49}$$

As an example, we consider again the two-dimensional isotropic scattering case for which the autocorrelation function of $\mu(t)$ equals $r_{\mu\mu}(\tau) = 2\sigma_0^2 J_0(2\pi f_{\max}\tau)$. Hence, the squared-envelope autocovariance function simplifies to

$$r^c_{\zeta^2\zeta^2}(\tau) = 4\sigma_0^4\,J_0^2(2\pi f_{\max}\tau), \tag{3.50}$$

and the squared-envelope autocovariance spectrum can be written as

$$S^c_{\zeta^2\zeta^2}(f) = \begin{cases} \dfrac{4\sigma_0^4}{\pi^2 f_{\max}} \boldsymbol{K}\left(\sqrt{1-\left(\dfrac{f}{2f_{\max}}\right)^2}\right), & |f| < 2f_{\max}, \\ 0, & |f| > 2f_{\max}. \end{cases} \tag{3.51}$$

By comparing (3.51) and (3.40), it turns out that the exact solution of the squared-envelope autocovariance spectrum $S^c_{\zeta^2\zeta^2}(f)$ differs from the approximate solution of the envelope autocovariance spectrum $S^c_{\zeta\zeta}(f)$ only by a multiplicative factor.

3.4 Statistical Properties of Rayleigh and Rice Channels

Besides the probability density function of the envelope, the squared envelope, and the phase, we will also analyze in this section the level-crossing rate as well as the average duration of fades of Rice processes $\xi(t) = |\mu(t) + m(t)|$ [see (3.21)] with time-variant line-of-sight components $m(t)$. When analyzing the influence of the power spectral density $S_{\mu\mu}(f)$ of the complex Gaussian random process $\mu(t)$ on the statistical properties of $\xi(t)$, we will restrict ourselves to the Jakes and Gaussian power spectral densities introduced in Subsection 3.3.1.

3.4.1 Probability Density Function of the Envelope and the Phase

The probability density function of Rice processes $\xi(t)$, denoted by $p_\xi(x)$, is described by the so-called Rice distribution [42]

$$p_\xi(x) = \begin{cases} \dfrac{x}{\sigma_0^2} e^{-\frac{x^2+\rho^2}{2\sigma_0^2}} I_0\left(\dfrac{x\rho}{\sigma_0^2}\right), & x \geq 0, \\ 0, & x < 0, \end{cases} \tag{3.52}$$

where $I_0(\cdot)$ is the zeroth-order modified Bessel function of the first kind and $\sigma_0^2 = r_{\mu_i\mu_i}(0) = r_{\mu\mu}(0)/2$ again denotes the mean power of the real-valued Gaussian random process $\mu_i(t)$ $(i = 1, 2)$. Obviously, neither the time variation of the mean (3.17) caused by the Doppler frequency of the line-of-sight component nor the exact shape of the Doppler power spectral density $S_{\mu\mu}(f)$ influences the probability density function $p_\xi(x)$. This means that the distribution of the envelope $\xi(t)$ does not provide any insight into the rate of fading. Merely the amplitude of the line-of-sight component ρ and the mean power σ_0^2 of the real part or the imaginary part of the scattered components determine the behaviour of $p_\xi(x)$.

Of particular interest is in this context the *Rice factor*, denoted by c_R, which describes the ratio of the power of the line-of-sight component to the sum of the power of all scattered components. Thus, the Rice factor is defined by

$$c_R := \frac{\rho^2}{2\sigma_0^2}. \tag{3.53}$$

From the limit $\rho \to 0$, i.e., $c_R \to 0$, the Rice process $\xi(t)$ results in the Rayleigh process $\zeta(t)$, whose statistical signal variations are described by the Rayleigh distribution [41]

$$p_\zeta(x) = \begin{cases} \dfrac{x}{\sigma_0^2} e^{-\frac{x^2}{2\sigma_0^2}}, & x \geq 0, \\ 0, & x < 0. \end{cases} \tag{3.54}$$

The probability density functions $p_\xi(x)$ and $p_\zeta(x)$ according to (3.52) and (3.54) are shown in Figures 3.5(a) and 3.5(b), respectively.

As mentioned above, the exact shape of the Doppler power spectral density $S_{\mu\mu}(f)$ has no effect on the probability density function of the absolute value of a complex Gaussian random process, i.e., $\xi(t) = |\mu_\rho(t)|$. Analogously, this statement is also valid for the probability density function of the phase $\vartheta(t) = \arg\{\mu_\rho(t)\}$, where $\vartheta(t)$ can be expressed by means of (3.16), (3.17), and (3.19) as follows

$$\vartheta(t) = \arctan\left\{\frac{\mu_2(t) + \rho\sin(2\pi f_\rho t + \theta_\rho)}{\mu_1(t) + \rho\cos(2\pi f_\rho t + \theta_\rho)}\right\}. \tag{3.55}$$

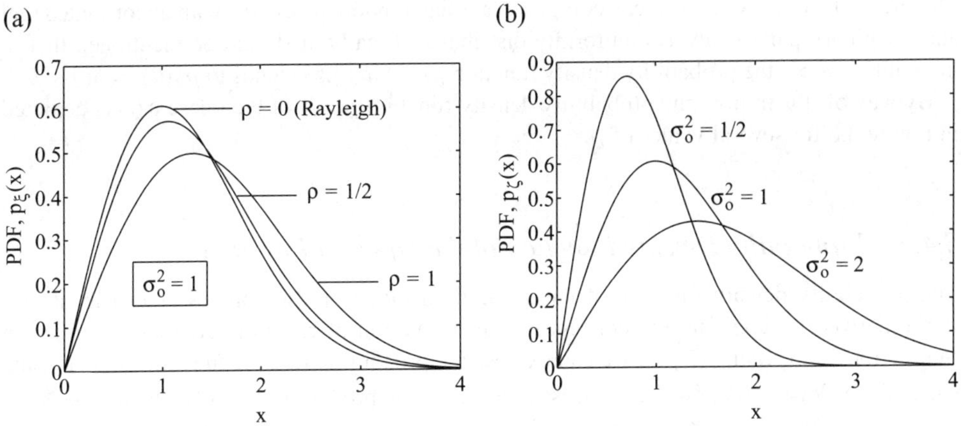

Figure 3.5 The probability density function of (a) Rice and (b) Rayleigh processes.

In order to confirm this statement, we study the probability density function $p_\vartheta(\theta; t)$ of the phase $\vartheta(t)$ given by the following relation [75]

$$p_\vartheta(\theta; t) = \frac{e^{-\frac{\rho^2}{2\sigma_0^2}}}{2\pi} \left\{ 1 + \frac{\rho}{\sigma_0}\sqrt{\frac{\pi}{2}} \cos(\theta - 2\pi f_\rho t - \theta_\rho)\, e^{\frac{\rho^2 \cos^2(\theta - 2\pi f_\rho t - \theta_\rho)}{2\sigma_0^2}} \left[1 + \operatorname{erf}\left(\frac{\rho \cos(\theta - 2\pi f_\rho t - \theta_\rho)}{\sigma_0\sqrt{2}} \right) \right] \right\}, \quad -\pi < \theta \le \pi, \tag{3.56}$$

where erf (·) is called the *error function*, the definition of which is already known from (2.34). The dependency of the probability density function $p_\vartheta(\theta; t)$ on time t is due to the Doppler frequency f_ρ of the line-of-sight component $m(t)$. According to the discussions in Subsection 2.2.1, the stochastic process $\vartheta(t)$ is not stationary in the strict sense, because condition (2.103) is violated. Only in the special case where $f_\rho = 0$ ($\rho \neq 0$), the phase $\vartheta(t)$ is a strict-sense stationary process, which is then described by the probability density function [65]

$$p_\vartheta(\theta) = \frac{e^{-\frac{\rho^2}{2\sigma_0^2}}}{2\pi} \left\{ 1 + \frac{\rho}{\sigma_0}\sqrt{\frac{\pi}{2}} \cos(\theta - \theta_\rho)\, e^{\frac{\rho^2 \cos^2(\theta - \theta_\rho)}{2\sigma_0^2}} \left[1 + \operatorname{erf}\left(\frac{\rho \cos(\theta - \theta_\rho)}{\sigma_0\sqrt{2}} \right) \right] \right\}, \quad -\pi < \theta \le \pi. \tag{3.57}$$

As $\rho \to 0$, it follows $\mu_\rho(t) \to \mu(t)$ and thus $\xi(t) \to \zeta(t)$, and from (3.57), we obtain the uniform distribution

$$p_\vartheta(\theta) = \frac{1}{2\pi}, \quad -\pi < \theta \le \pi. \tag{3.58}$$

Therefore, the phase of zero-mean complex Gaussian random processes with uncorrelated real and imaginary parts is always uniformly distributed. Finally, it should be mentioned that in the limit $\rho \to \infty$, the probability density function $p_\vartheta(\theta)$ in (3.57) tends to $p_\vartheta(\theta) = \delta(\theta - \theta_\rho)$.

By way of illustration, the probability density function $p_\vartheta(\theta)$ of the phase $\vartheta(t)$ is depicted in Figure 3.6 for several values of ρ.

3.4.2 Probability Density Function of the Squared Envelope

The probability density function of the squared envelope of Rice processes can most easily be derived by applying the concept of transformation of random variables described in Subsection 2.1.3. Using (2.86), we can express the probability density function $p_{\xi^2}(x)$ of the squared envelope $\xi^2(t)$ of a Rice process in terms of the probability density function $p_\xi(x)$ of the envelope $\xi(t)$ as $p_{\xi^2}(x) = p_\xi(\sqrt{x})/(2\sqrt{x})$. Applying this relationship in connection with (3.52) leads to the density

$$p_{\xi^2}(x) = \begin{cases} \dfrac{1}{2\sigma_0^2} e^{-\frac{x+\rho^2}{2\sigma_0^2}} I_0\left(\dfrac{\sqrt{x}\rho}{\sigma_0^2}\right), & x \geq 0, \\ 0, & x < 0. \end{cases} \tag{3.59}$$

The squared envelope $\xi^2(t)$ of a Rice process has the expected value

$$\begin{aligned} E\left\{\xi^2(t)\right\} &= E\left\{|\mu_\rho(t)|^2\right\} \\ &= E\left\{(\mu_1(t) + m_1(t))^2 + (\mu_2(t) + m_2(t))^2\right\} \\ &= 2\,\sigma_0^2 + \rho^2 \end{aligned} \tag{3.60}$$

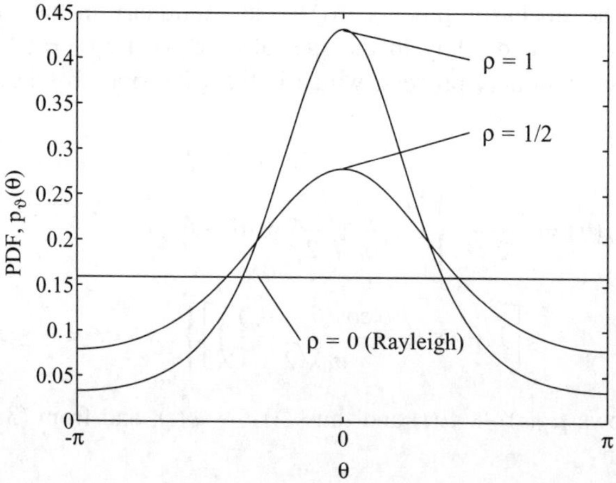

Figure 3.6 The probability density function $p_\vartheta(\theta)$ of the phase $\vartheta(t)$ ($f_\rho = 0$, $\theta_\rho = 0$, $\sigma_0^2 = 1$).

and the variance

$$\begin{aligned}\text{Var}\left\{\xi^2(t)\right\} &= E\left\{\xi^4(t)\right\} - \left(E\left\{\xi^2(t)\right\}\right)^2 \\ &= E\left\{\left|\mu_\rho(t)\right|^4\right\} - \left(E\left\{\left|\mu_\rho(t)\right|^2\right\}\right)^2 \\ &= 4\sigma_0^2\left(\sigma_0^2 + \rho^2\right). \end{aligned} \tag{3.61}$$

In non-line-of-sight situations, where $\rho \to 0$, it follows $\xi^2(t) \to \zeta^2(t)$, and thus from (3.59), we obtain immediately the distribution of the squared envelope of a Rayleigh process in the form of the one-sided negative exponential distribution

$$p_{\zeta^2}(x) = \begin{cases} \dfrac{e^{-\frac{x}{2\sigma_0^2}}}{2\sigma_0^2}, & x \geq 0, \\ 0, & x < 0. \end{cases} \tag{3.62}$$

Hence, the squared envelope $\zeta^2(t)$ of a Rayleigh process follows the negative exponential distribution. From (3.60) and (3.61), we realize that the squared envelope $\zeta^2(t)$ of a Rayleigh process has the mean $E\{\zeta^2(t)\} = 2\sigma_0^2$ and the variance $\text{Var}\{\zeta^2(t)\} = 4\sigma_0^4$.

The probability density functions $p_{\xi^2}(x)$ and $p_{\zeta^2}(x)$ according to (3.59) and (3.62) are illustrated in Figures 3.7(a) and 3.7(b), respectively. A comparison of Figures 3.7 and 3.5 shows that the squared envelope reaches low and high signal levels with a much higher probability than the envelope does, whereas the inverse statement is true for medium signal levels.

3.4.3 Level-Crossing Rate and Average Duration of Fades

As further statistical quantities, we will in this subsection study the level-crossing rate (LCR) and the average duration of fades (ADF). Therefore, we turn our interest at first to Rice

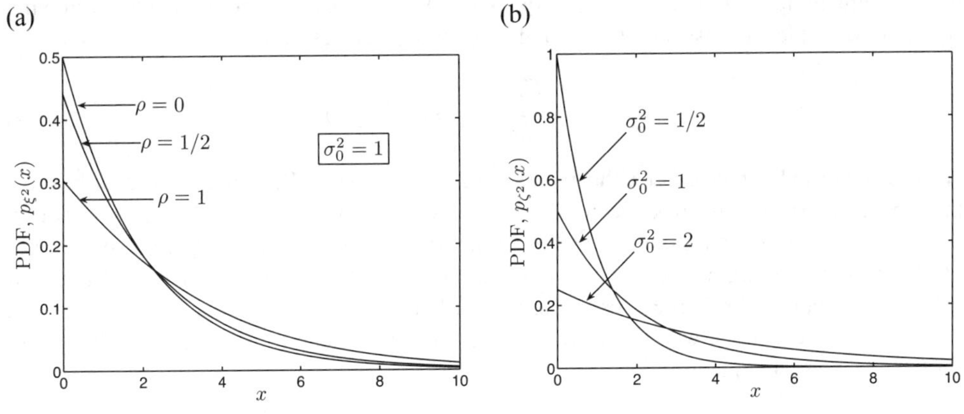

Figure 3.7 The probability density function of the squared envelope of (a) Rice and (b) Rayleigh processes.

processes $\xi(t)$ introduced by (3.21), where we impose on the reference model that the real-valued zero-mean Gaussian random processes $\mu_1(t)$ and $\mu_2(t)$ are uncorrelated and that these processes have identical autocorrelation functions, i.e., $r_{\mu_1\mu_2}(\tau) = 0$ and $r_{\mu_1\mu_1}(\tau) = r_{\mu_2\mu_2}(\tau)$ $\forall\,\tau$. When calculating the level-crossing rate $N_\xi(r)$ of Rice processes $\xi(t) = |\mu_\rho(t)|$, however, it must be taken into consideration that a correlation exists between the real and imaginary part of the complex Gaussian random process $\mu_\rho(t)$ [see (3.19)] due to the time-variant line-of-sight component $m(t)$ introduced in (3.17).

For the level-crossing rate $N_\xi(r)$ it then holds [75]

$$N_\xi(r) = \frac{r\sqrt{2\beta}}{\pi^{3/2}\sigma_0^2}\, e^{-\frac{r^2+\rho^2}{2\sigma_0^2}} \int_0^{\pi/2} \cosh\left(\frac{r\rho}{\sigma_0^2}\cos\theta\right) \cdot \left\{ e^{-(\alpha\rho\sin\theta)^2} + \sqrt{\pi}\alpha\rho\sin(\theta)\ \mathrm{erf}\,(\alpha\rho\sin\theta)\right\} d\theta, \quad r \geq 0, \tag{3.63}$$

where the quantities α and β are given by

$$\alpha = 2\pi f_\rho \big/ \sqrt{2\beta} \tag{3.64}$$

and

$$\beta = \beta_i = -\ddot{r}_{\mu_i\mu_i}(0), \quad i = 1, 2, \tag{3.65}$$

respectively. Considering (3.64), we notice that the Doppler frequency f_ρ of the line-of-sight component $m(t)$ has an influence on the level-crossing rate $N_\xi(r)$. However, if $f_\rho = 0$, and thus $\alpha = 0$, it follows from (3.63) the relation known from (2.119), which will be presented here again for completeness, i.e.,

$$N_\xi(r) = \sqrt{\frac{\beta}{2\pi}} \cdot p_\xi(r), \quad r \geq 0. \tag{3.66}$$

Therefore, (3.66) describes the level-crossing rate of Rice processes with a time-invariant line-of-sight component. For $\rho \to 0$, it follows $p_\xi(r) \to p_\zeta(r)$, so that we obtain the following equation for the level-crossing rate $N_\zeta(r)$ of Rayleigh processes $\zeta(t)$

$$N_\zeta(r) = \sqrt{\frac{\beta}{2\pi}} \cdot p_\zeta(r), \quad r \geq 0. \tag{3.67}$$

For Rice and Rayleigh processes, the expressions (3.66) and (3.67), respectively, clearly show the proportional relationship between the level-crossing rate and the corresponding probability density function of the envelope. The value of the proportional factor $\sqrt{\beta/(2\pi)}$ is, because of (3.65), only depending on the negative curvature of the autocorrelation function of the real-valued Gaussian random processes at the origin. Especially for the Jakes and the Gaussian power spectral densities, we obtain, by using (3.25), (3.27), and (3.65), the following result for the characteristic quantity β:

$$\beta = \begin{cases} 2(\pi f_{\max}\sigma_0)^2, & \text{Jakes PSD}, \\ 2(\pi f_c\sigma_0)^2/\ln 2, & \text{Gaussian PSD}. \end{cases} \tag{3.68}$$

Despite the large differences existing between the shapes of the Jakes and the Gaussian power spectral densities, both Doppler power spectral densities enable the modelling of Rice or Rayleigh processes with identical level-crossing rates, as long as $f_{\max}$ and f_c are related by $f_c = \sqrt{\ln 2}\, f_{\max}$.

The influence of the parameters f_ρ and ρ on the normalized level-crossing rate $N_\xi(r)/f_{\max}$ is illustrated in Figures 3.8(a) and 3.8(b), respectively. Thereby, Figure 3.8(a) points out that an increase in $|f_\rho|$ leads to an increase in the level-crossing rate $N_\xi(r)$.

In some sections of this book, the case $r_{\mu_1\mu_1}(0) = r_{\mu_2\mu_2}(0)$ with $\beta_1 = -\ddot{r}_{\mu_1\mu_1}(0) \neq -\ddot{r}_{\mu_2\mu_2}(0) = \beta_2$ will be relevant for us. Based on this condition, it is shown in Appendix 3.D that the level-crossing rate $N_\xi(r)$ of the Rice processes $\xi(t)$ can be expressed as

$$N_\xi(r) = \sqrt{\frac{\beta_1}{2\pi}}\frac{r}{\sigma_0^2}e^{-\frac{r^2+\rho^2}{2\sigma_0^2}}\frac{1}{\pi}\int_0^\pi \cosh\left[\frac{r\rho}{\sigma_0^2}\cos(\theta-\theta_\rho)\right]\sqrt{1-k^2\sin^2\theta}\,d\theta, \quad r \geq 0, \tag{3.69}$$

where $k = \sqrt{(\beta_1 - \beta_2)/\beta_1}$, $\beta_1 \geq \beta_2$. In this case, the level-crossing rate is in general no longer proportional to the probability density function of the Rice process.

On the other hand, we again obtain the expected proportional relationship for Rayleigh processes $\zeta(t)$, whose level-crossing rate $N_\zeta(r)$ is obtained from (3.69) by taking the limit $\rho \to 0$, which results in

$$N_\zeta(r) = \sqrt{\frac{\beta_1}{2\pi}} \cdot \frac{r}{\sigma_0^2}e^{-\frac{r^2}{2\sigma_0^2}} \cdot \frac{1}{\pi}\int_0^\pi \sqrt{1-k^2\sin^2\theta}\,d\theta, \quad r \geq 0. \tag{3.70}$$

In the literature (see [23, Equation (8.111.3)]), the integral above having the form

$$E(\varphi, k) = \int_0^\varphi \sqrt{1-k^2\sin^2\theta}\,d\theta \tag{3.71}$$

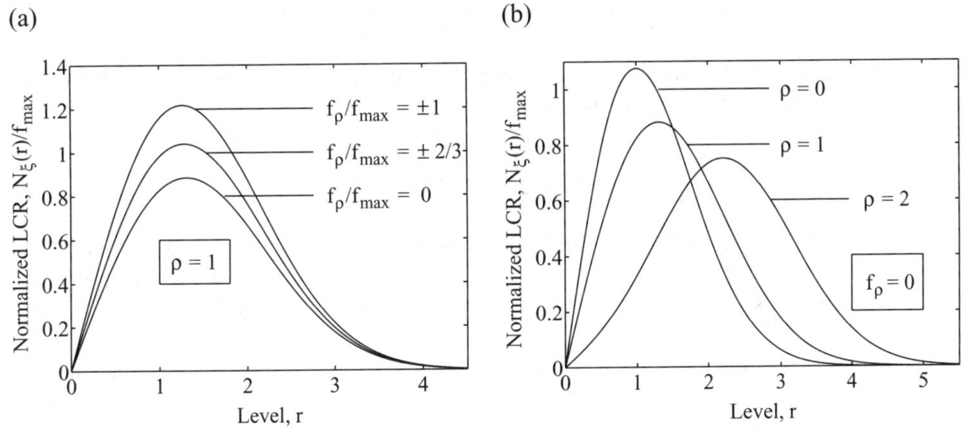

Figure 3.8 Normalized level-crossing rate $N_\xi(r)/f_{\max}$ of Rice processes illustrating the dependency on the parameters (a) f_ρ and (b) ρ (Jakes PSD, $f_{\max} = 91$ Hz, $\sigma_0^2 = 1$).

is known as the *elliptic integral of the second kind*. The parameter k denotes the modulus of the integral. For $\varphi = \pi/2$, this integral is also called the *complete elliptic integral of the second kind* for which we write $\boldsymbol{E}(k) = E(\frac{\pi}{2}, k)$.

Now, by using (3.54), the level-crossing rate of Rayleigh processes can be presented in the following form

$$N_\zeta(r) = \sqrt{\frac{\beta_1}{2\pi}} p_\zeta(r) \cdot \frac{2}{\pi} \boldsymbol{E}(k), \quad r \geq 0, \tag{3.72}$$

where the modulus k is again given by $k = \sqrt{(\beta_1 - \beta_2)/\beta_1}$ for $\beta_1 \geq \beta_2$. Thus, it is shown that the level-crossing rate of Rayleigh processes is proportional to the probability density function of the envelope, even for the case $\beta_1 \neq \beta_2$. The proportional factor is here not only determined by β_1, but also by the difference $\beta_1 - \beta_2$.

Furthermore, we are interested in the level-crossing rate $N_\zeta(r)$ for the case that the relative deviation between β_1 and β_2 is very small. To investigate this case, let us assume that a positive number $\varepsilon = \beta_1 - \beta_2 \geq 0$ with $\varepsilon/\beta_1 \ll 1$ exists, so that

$$k = \sqrt{\frac{\beta_1 - \beta_2}{\beta_1}} = \sqrt{\frac{\varepsilon}{\beta_1}} \ll 1 \tag{3.73}$$

holds. Next, we make use of the relation (see [23, Equation (8.114.1)])

$$\begin{aligned} \boldsymbol{E}(k) &= \frac{\pi}{2} F\left(-\frac{1}{2}, \frac{1}{2}; 1; k^2\right) \\ &= \frac{\pi}{2} \left\{ 1 - \sum_{n=1}^{\infty} \left[\frac{1 \cdot 3 \cdot 5 \cdot \ldots \cdot (2n-1)}{2^n \quad n!} \right]^2 \frac{k^{2n}}{2n-1} \right\}, \end{aligned} \tag{3.74}$$

where $F(.,.;.;.)$ denotes the *hypergeometric function*. By using the first two terms of the above series expansion for $\boldsymbol{E}(k)$, we obtain the following approximation formula

$$\boldsymbol{E}(k) \approx \frac{\pi}{2}\left(1 - \frac{k^2}{4}\right) \approx \frac{\pi}{2}\sqrt{1 - \frac{k^2}{2}}, \quad k \ll 1. \tag{3.75}$$

Now, substituting (3.73) into (3.75) and taking (3.72) into account, allows us to approximate the level-crossing rate $N_\zeta(r)$ by the formula

$$N_\zeta(r) \approx \sqrt{\frac{\beta}{2\pi}} \cdot p_\zeta(r), \quad r \geq 0, \tag{3.76}$$

which is valid if the condition $(\beta_1 - \beta_2)/\beta_1 \ll 1$ is met, where in (3.76) the quantity β stands for $\beta = (\beta_1 + \beta_2)/2$. Hence, (3.67) keeps its validity approximately if the relative deviation between β_1 and β_2 is small, and if $\beta = \beta_1 = \beta_2$ is replaced in (3.67) by the arithmetical mean $\beta = (\beta_1 + \beta_2)/2$.

The average duration of fades is a measure for the average length of the time intervals within which the channel envelope is below a given signal level r. According to (2.120), the average duration of fades is defined as the quotient of the distribution function of the channel envelope over the level-crossing rate. The probability density function and the level-crossing rate of Rice and Rayleigh processes have already been studied in detail, so that the analysis of the corresponding average duration of fades can easily be carried out. For completeness, however, the resulting relations will again be given here. For Rice processes with $f_\rho = 0$ and Rayleigh processes, we obtain the following expressions for the average duration of fades [see also (2.123) and (2.122)]

$$T_{\xi_-}(r) = \frac{F_{\xi_-}(r)}{N_\xi(r)} = \sqrt{\frac{2\pi}{\beta}} \cdot \frac{e^{\frac{r^2}{2\sigma_0^2}}}{r\, I_0\left(\frac{r\rho}{\sigma_0^2}\right)} \int_0^r x\, e^{-\frac{x^2}{2\sigma_0^2}} I_0\left(\frac{x\rho}{\sigma_0^2}\right) dx, \quad r \geq 0, \tag{3.77a}$$

and

$$T_{\zeta_-}(r) = \frac{F_{\zeta_-}(r)}{N_\zeta(r)} = \sqrt{\frac{2\pi}{\beta}} \cdot \frac{\sigma_0^2}{r} \left(e^{\frac{r^2}{2\sigma_0^2}} - 1\right), \quad r \geq 0, \tag{3.77b}$$

respectively, where $F_{\xi_-}(r) = P\{\xi(t) \leq r\}$ and $F_{\zeta_-}(r) = P\{\zeta(t) \leq r\}$ denote the respective cumulative distribution function of the Rice and Rayleigh processes.

In channel modelling, we are especially interested in the behaviour of the average duration of fades at low signal levels r. We therefore wish to analyze this case separately. For this purpose, let $r \ll 1$, so that for moderate Rice factors, we may write $r\rho/\sigma_0^2 \ll 1$ and, consequently, both $I_0(r\rho/\sigma_0^2)$ and $I_0(x\rho/\sigma_0^2)$ can be approximated by 1 in (3.77a), since the independent variable x is within the relevant interval $[0, r]$. After a series expansion of the integrand in (3.77a), $T_{\xi_-}(r)$ can be presented in closed form. By this means, it quickly turns out that at low signal levels r, the average duration of fades $T_{\xi_-}(r)$ converges to $T_{\zeta_-}(r)$ given by (3.77b). The relation (3.77b) can furthermore be simplified by using $e^x \approx 1 + x$ $(x \ll 1)$, so that we finally obtain the approximations

$$T_{\xi_-}(r) \approx T_{\zeta_-}(r) \approx r\sqrt{\frac{\pi}{2\beta}}, \quad r \ll 1, \tag{3.78}$$

which are valid if $r\rho/\sigma_0^2 \ll 1$ holds. The result above shows that at low signal levels r, the average duration of fades of Rice and Rayleigh processes is approximately proportional to r.

An illustration of the results is shown in Figure 3.9. In Figure 3.9(a) it can be seen that an increase in $|f_\rho|$ leads to a decrease in the average duration of fades $T_{\xi_-}(r)$. Figure 3.9(b) shows impressively that the approximations in (3.78) are very good if the signal level r is low. It can also be observed that Rice and Rayleigh processes have almost the same average duration of fades at low signal levels.

3.4.4 *The Statistics of the Fading Intervals of Rayleigh Channels*

The statistical properties of Rayleigh and Rice processes analyzed so far are independent of the behaviour of the autocorrelation function $r_{\mu_i\mu_i}(\tau)$ $(i = 1, 2)$ of the underlying Gaussian random processes for values $\tau > 0$. For example, we have seen that the probability density function of the envelope $\zeta(t) = |\mu(t)|$ is totally determined by the behaviour of the autocorrelation function $r_{\mu_i\mu_i}(\tau)$ at the origin, i.e., by the variance $\sigma_0^2 = r_{\mu_i\mu_i}(0)$. The behaviour of $r_{\mu_i\mu_i}(\tau)$ at the origin determines the level-crossing rate $N_\zeta(r)$ and the average duration of fades $T_{\zeta_-}(r)$. These quantities are, besides on the variance $\sigma_0^2 = r_{\mu_i\mu_i}(0)$, also dependent on the

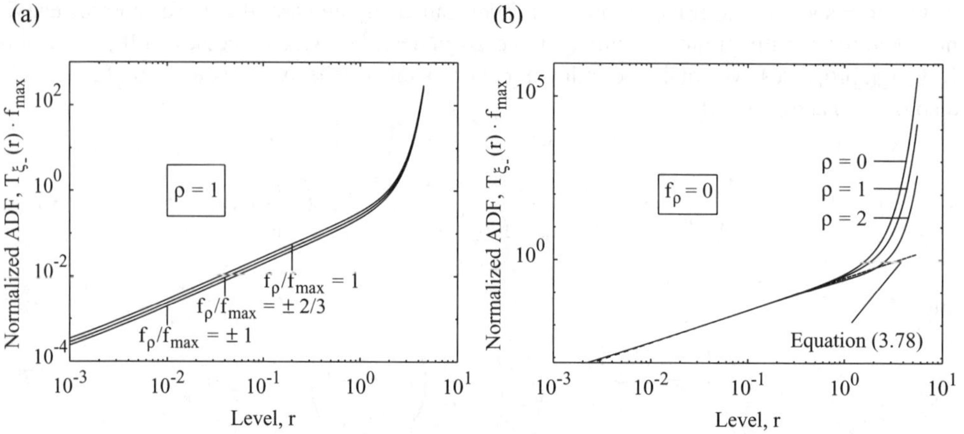

Figure 3.9 Normalized average duration of fades $T_{\xi\_}(r) \cdot f_{\max}$ of Rice processes illustrating the influence of the parameters (a) f_ρ and (b) ρ (Jakes PSD, $f_{\max} = 91$ Hz, $\sigma_0^2 = 1$).

negative curvature of the autocorrelation function at the origin described by $\beta = -\ddot{r}_{\mu_i\mu_i}(0)$. If we now ask ourselves which relevant statistical properties are at all affected by the behaviour of the autocorrelation function $r_{\mu_i\mu_i}(\tau)$ $(i = 1, 2)$ for $\tau > 0$, then this leads to the statistical distribution of the fading intervals.

The conditional probability density function for the event that a Rayleigh process $\zeta(t)$ crosses a given signal level r in an infinitesimal time interval $(t + \tau_-, t + \tau_- + d\tau_-)$ upwards for the first time on condition that the last down-crossing occurred within the time interval $(t, t + dt)$ is denoted as $p_{0\_}(\tau_-; r)$. An exact theoretical derivation for $p_{0\_}(\tau_-; r)$ is still today even for Rayleigh processes an unsolved problem. In [76], Rice, however, managed to derive the probability density $p_{1\_}(\tau_-; r)$ for the case that the Rayleigh process $\zeta(t)$ crosses the signal level r in the order mentioned, where no information on the behaviour of $\zeta(t)$ between t and $t + \tau_-$ is available. For small τ_--values at which the probability that further level crossings occur between t and $t + \tau_-$ is very low, $p_{1\_}(\tau_-; r)$ can be considered as a reasonable approximation for the desired probability density function $p_{0\_}(\tau_-; r)$. On the other hand, for large τ_--values, $p_{1\_}(\tau_-; r)$ cannot be used any further as a suitable approximation for $p_{0\_}(\tau_-; r)$.

The determination of $p_{1\_}(\tau_-; r)$ requires the numerical calculation of the threefold integral [76]

$$p_{1\_}(\tau_-; r) = \frac{r\, M_{22}\, e^{\frac{r^2}{2}}}{\sqrt{2\pi\beta}(1 - r^2_{\mu_i\mu_i}(\tau_-))^2} \int_0^{2\pi} J(a, b)\, e^{-r^2 \frac{1 - r_{\mu_i\mu_i}(\tau_-)\cdot\cos\varphi}{1 - r^2_{\mu_i\mu_i}(\tau_-)}}\, d\varphi, \tag{3.79}$$

where

$$J(a, b) = \frac{1}{2\pi\sqrt{1 - a^2}} \int_b^\infty \int_b^\infty (x - b)(y - b)\, e^{-\frac{x^2 + y^2 - 2axy}{2(1 - a^2)}}\, dx\, dy, \tag{3.80}$$

$$a = \cos\varphi \cdot \frac{M_{23}}{M_{22}}, \tag{3.81}$$

$$b = \frac{r\dot{r}_{\mu_i\mu_i}(\tau_-) \cdot (r_{\mu_i\mu_i}(\tau_-) - \cos\varphi)}{1 - r^2_{\mu_i\mu_i}(\tau_-)} \cdot \sqrt{\frac{1 - r^2_{\mu_i\mu_i}(\tau_-)}{M_{22}}}, \tag{3.82}$$

$$M_{22} = \beta(1 - r^2_{\mu_i\mu_i}(\tau_-)) - \dot{r}^2_{\mu_i\mu_i}(\tau_-), \tag{3.83}$$

$$M_{23} = \ddot{r}_{\mu_i\mu_i}(\tau_-)(1 - r^2_{\mu_i\mu_i}(\tau_-)) + r_{\mu_i\mu_i}(\tau_-)\dot{r}^2_{\mu_i\mu_i}(\tau_-), \tag{3.84}$$

and β is again the characteristic quantity defined by (3.65).

Figures 3.10 and 3.11 show the evaluation results of the probability density function $p_{1_-}(\tau_-; r)$ by using the Jakes and Gaussian power spectral densities, respectively. For the 3-dB-cut-off frequency of the Gaussian power spectral density, the value $f_c = \sqrt{\ln 2} f_{\max}$ was chosen. For the quantity β, we hereby obtain identical values for the Jakes and Gaussian power spectral density due to (3.68). Observing Figures 3.10(a) and 3.11(a), we see that at low signal levels ($r = 0.1$) the graphs of the probability density function $p_{1_-}(\tau_-; r)$ are identical. With increasing values of r, however, the shapes of the graphs differ more and more from each other (cf. Figures 3.10(b) and 3.11(b) for medium signal levels ($r = 1$) as well as Figures 3.10(c) and 3.11(c) for high signal levels ($r = 2.5$)). In these figures, it can be observed that $p_{1_-}(\tau_-; r)$ does not converge to zero at medium and large values of the signal level r. Obviously, $p_{1_-}(\tau_-; r) \geq 0$ holds if $\tau_- \to \infty$, which extremely jeopardizes the accuracy of (3.79) — at least for the range of medium and high signal levels of r in connection with long fading intervals τ_-.

The validity of the approximate solution (3.79) can ultimately only be determined by simulating the level-crossing behaviour. Therefore, simulation models are needed, which reproduce the Gaussian random processes $\mu_i(t)$ of the reference model extremely accurately with respect to the probability density function $p_{\mu_i}(x)$ and the autocorrelation function $r_{\mu_i\mu_i}(\tau)$. We will return to this subject in Section 5.3. For our purposes, at first only the awareness that the probability density function of the fading intervals of Rayleigh channels at medium and high signal levels r depends decisively on the behaviour of the autocorrelation function $r_{\mu_i\mu_i}(\tau)$ for $\tau \geq 0$ is of significance.

In the following, we will analyze the statistics of the deep fades. The knowledge of the statistics of the deep fades is of great importance in mobile radio communications, because the occurrence of bit or symbol errors is closely related to the occurrence of deep fades. Hence, let $r \ll 1$. In this case, the durations of fades τ_- are short. Thus, the probability that further level-crossings occur between t and $t + \tau_-$ is very low, and consequently the approximation $p_{0_-}(\tau_-; r) \approx p_{1_-}(\tau_-; r)$ is of high accuracy. In [76], it has been shown that the probability density function in (3.79) converges to

$$p_{1_-}(\tau_-; r) = -\frac{1}{T_{\zeta_-}(r)} \frac{d}{du}\left[\frac{2}{u} I_1(z) e^{-z}\right] \tag{3.85}$$

as $r \to 0$, where $z = 2/(\pi u^2)$ and $u = \tau_-/T_{\zeta_-}(r)$ hold. After some algebraic manipulations, we find the following expression

$$p_{1_-}(\tau_-; r) = \frac{2\pi z^2 e^{-z}}{T_{\zeta_-}(r)} \left[I_0(z) - \left(1 + \frac{1}{2z}\right) I_1(z)\right], \quad r \to 0, \tag{3.86}$$

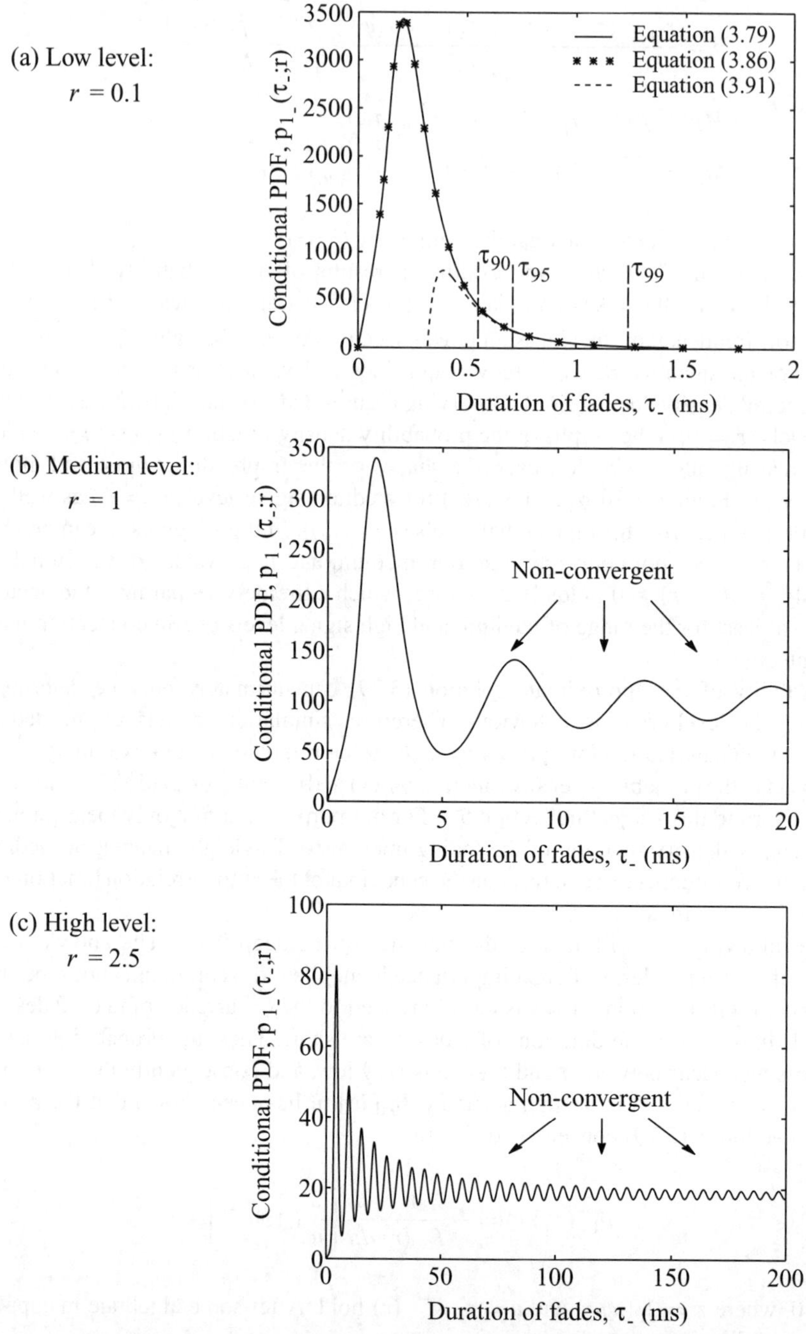

Figure 3.10 The probability density function $p_{1-}(\tau_-; r)$ of the fading intervals of Rayleigh processes when using the Jakes power spectral density ($f_{max} = 91$ Hz, $\sigma_0^2 = 1$).

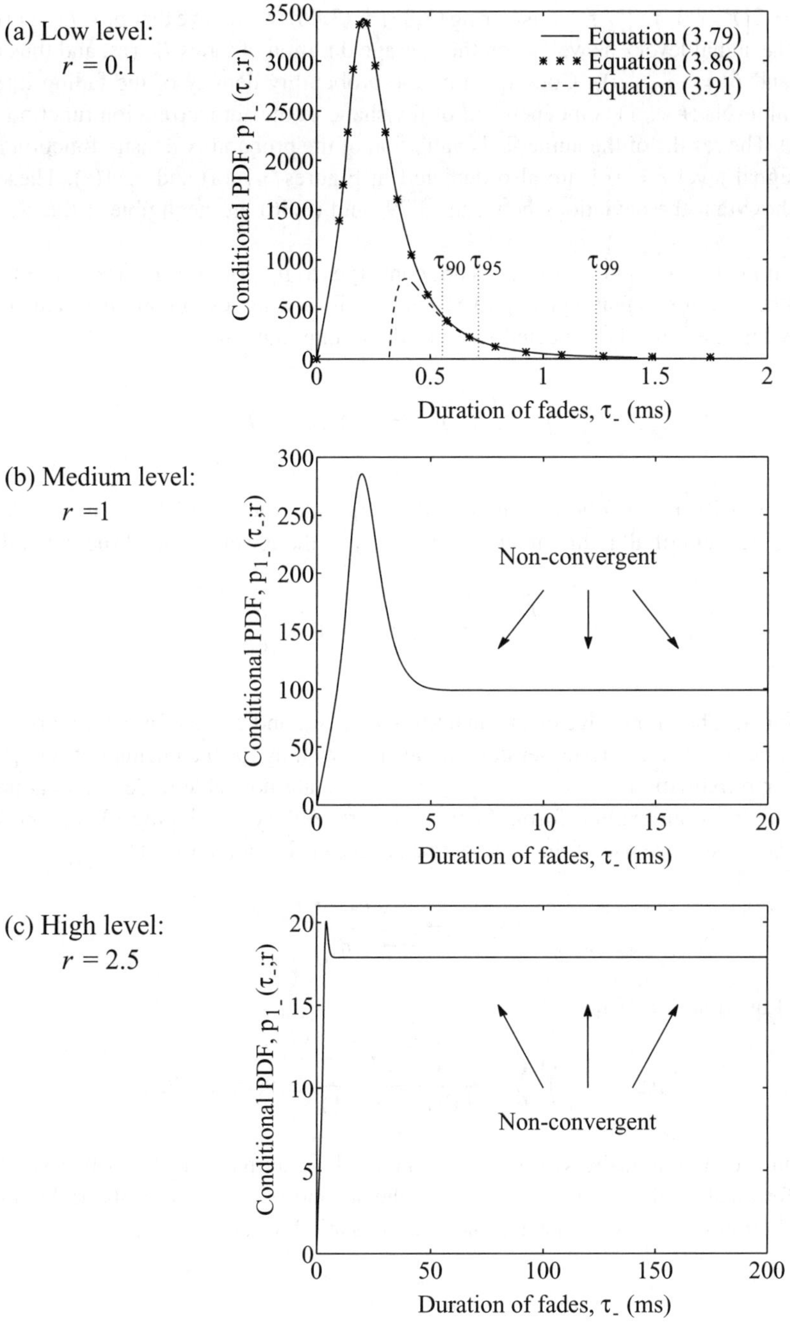

Figure 3.11 The probability density function $p_{1\_}(\tau_-; r)$ of the fading intervals of Rayleigh processes when using the Gaussian power spectral density ($f_c = \sqrt{\ln 2} f_{\max}$, $f_{\max} = 91\,\text{Hz}$, $\sigma_0^2 = 1$).

where $z = 2\left[T_{\zeta_-}(r)/\tau_-\right]^2/\pi$. Considering (3.85) or (3.86), we realize that $p_1-(\tau_-;r)$ depends only on the signal level r as well as on the average duration of fades $T_{\zeta_-}(r)$, and thus on $\sigma_0^2 = r_{\mu_i\mu_i}(0)$ and $\beta = -\ddot{r}_{\mu_i\mu_i}(0)$. Consequently, the probability density of the fading intervals at low signal levels ($r \ll 1$) is independent of the shape of the autocorrelation function $r_{\mu_i\mu_i}(\tau)$ for $\tau > 0$. The results of the numerical evaluation of the probability density function in (3.86) for the signal level $r = 0.1$ are also depicted in Figures 3.10(a) and 3.11(a). These figures clearly show that the deviations between (3.79) and (3.86) are negligible if the signal level r is low.

In the limits $\tau_- \to 0$ and $\tau_- \to \infty$, (3.86) converges to $p_{1_-}(0; r) = p_{1_-}(\infty; r) = 0$. Finally, it should be mentioned that by using (3.86) one finds — after a short auxiliary calculation — the following result for the expected value of the fading intervals τ_-

$$E\{\tau_-\} = \int_0^\infty \tau_-\, p_{1_-}(\tau_-; r)\, d\tau_- = T_{\zeta_-}(r). \tag{3.87}$$

With τ_q, we will in the following denote the time interval of the duration of fades which includes q per cent of all fading intervals. Thus, by τ_q the lower bound of the integral

$$\int_{\tau_q}^\infty p_{0_-}(\tau_-; r)\, d\tau_- = 1 - \frac{q}{100} \tag{3.88}$$

is determined. The knowledge of the quantities τ_{90}, τ_{95}, and τ_{99} is of great importance to the (optimal) design of the interleaver/deinterleaver as well as for the channel encoder/decoder. With the approximation $p_{0_-}(\tau_-; r) \approx p_{1_-}(\tau_-; r)$, we are now able to derive an approximate solution for τ_q in an explicit form. At first, we proceed by developing (3.86) into a power series, where we make use of the series expansions [77, Equation (4.2.1)]

$$e^{-z} = \sum_{n=0}^\infty \frac{(-z)^n}{n!} \tag{3.89}$$

and [77, Equation (9.6.10)]

$$I_\nu(z) = \left(\frac{z}{2}\right)^\nu \sum_{n=0}^\infty \frac{(z^2/4)^n}{n!\,\Gamma(\nu+n+1)}, \quad \nu = 0, 1, 2, \ldots \tag{3.90}$$

In the latter expression, the symbol $\Gamma(\cdot)$ denotes the gamma function, which is defined in (2.59). Terminating the resulting series after the second term, leads for the right-side tail of the distribution $p_{1_-}(\tau_-; r)$ to the approximation usable for our purposes

$$p_{1_-}(\tau_-; r) \approx \frac{\pi z^2}{2}(3 - 5z)/T_{\zeta_-}(r), \tag{3.91}$$

where z again represents $z = 2\left[T_{\zeta_-}(r)/\tau_-\right]^2/\pi$. If we now replace the probability density $p_{0_-}(\tau_-; r)$ in (3.88) by (3.91), then an explicit expression for the quantity $\tau_q = \tau_q(r)$ can be derived from the result of the integration.

Finally, we obtain the approximation

$$\tau_q(r) \approx \frac{T_{\zeta_-}(r)}{\left\{\frac{\pi}{4}\left[1-\sqrt{1-4(1-\frac{q}{100})}\right]\right\}^{\frac{1}{3}}}, \quad r \ll 1, \tag{3.92}$$

which is valid for $75 \leq q \leq 100$ [78]. This equation clearly shows that the quantity $\tau_q(r)$ is at deep fades proportional to the average duration of fades. Especially for the quantities $\tau_{90}(r)$, $\tau_{95}(r)$, and $\tau_{99}(r)$, we obtain from (3.92) the following expressions:

$$\tau_{90}(r) \approx 1.78 \cdot T_{\zeta_-}(r), \tag{3.93}$$

$$\tau_{95}(r) \approx 2.29 \cdot T_{\zeta_-}(r), \tag{3.94}$$

$$\tau_{99}(r) \approx 3.98 \cdot T_{\zeta_-}(r). \tag{3.95}$$

Further simplifications are possible if we approximate the average duration of fades $T_{\zeta_-}(r)$ for $r \ll 1$ by $T_{\zeta_-}(r) \approx r\sqrt{\pi/(2\beta)}$ [cf. (3.78)]. If we replace β in this relation by the formula (3.68) found for the Jakes and Gaussian power spectral density, then we obtain, e.g., for the quantity $\tau_{90}(r)$ the approximation

$$\tau_{90}(r) \approx \begin{cases} \dfrac{r}{2\sigma_0 f_{\max}}, & \text{Jakes PSD,} \\[2ex] \dfrac{r\sqrt{\ln 2}}{2\sigma_0 f_c}, & \text{Gaussian PSD,} \end{cases} \tag{3.96}$$

which is valid for $r \ll 1$. By means of this result, we see that the quantity $\tau_{90}(r)$ and thus the general quantity $\tau_q(r)$ $(75 \leq q \leq 100)$ are proportional to r and reciprocally proportional to $f_{\max}$ (or f_c) for low signal levels r. Hereby, it is of major importance that the exact shape of the power spectral density of the complex Gaussian random process, which generates the Rayleigh process, does not have any influence on the behaviour of $\tau_q(r)$. Hence, for the Jakes and the Gaussian power spectral densities, we again obtain identical values for $\tau_q(r)$ by choosing $f_c = \sqrt{\ln 2}\, f_{\max}$. Therefore, one may also compare Figures 3.10(a) and 3.11(a), where the approximation (3.91) and the quantities $\tau_{90}(r)$, $\tau_{95}(r)$, and $\tau_{99}(r)$ [see (3.93)–(3.95)] derived therefrom are illustrated. It should be noted that the relative deviations between the approximations (3.93)–(3.95) and the quantities $\tau_q(r)$, obtained by solving (3.88) numerically, are less than one per thousand if the signal level r equals 0.1. The validity of all these approximate solutions for $\tau_q(r)$ can again ultimately only be judged by simulating the level-crossing behaviour. In Section 5.3, we will see that the approximations introduced in this section match the simulation results obtained there quite well.

In [25], computer simulations of probability density functions $p_{0_-}(\tau_-; r)$ were also performed for Rice processes. Thereby, it turned out that a Rice process has practically the same probability density function of the fading intervals as the corresponding Rayleigh process. These results are at least for low signal levels not surprising anymore, because from (3.86) it follows that $p_{1_-}(\tau_-; r)$ merely depends on $T_{\zeta_-}(r)$, and on the other hand, we have seen in Subsection 3.4.3 that $T_{\xi_-}(r) \approx T_{\zeta_-}(r)$ holds if $r \ll 1$ and $r\rho/\sigma_0^2 \ll 1$. The analytical approximations obtained for Rayleigh processes for $p_{0_-}(\tau_-; r)$ and $\tau_q(r)$ can in such cases be adopted directly for Rice processes.

3.5 Further Reading

A stochastic model for a land mobile radio channel with communication between two moving vehicles (mobile-to-mobile communication) was introduced in [79]. There it was shown that the channel can again be represented by a narrow-band complex Gaussian random process with symmetrical Doppler power spectral density, which has now singularities at the frequencies $f = \pm(f_{\max_1} - f_{\max_2})$. Here, $f_{\max_1}$ ($f_{\max_2}$) denotes the maximum Doppler frequency due to the motion of the receiver (transmitter). The shape of the Doppler spectrum differs considerably from the Jakes power spectral density in (3.23), but contains it as a special case for $f_{\max_1} = 0$ or $f_{\max_2} = 0$. The statistical properties (of second order) for this mobile-to-mobile channel model were analyzed in another paper [80].

The analysis of the probability density function of fading intervals carried out by Rice [76] has triggered further research in this field (e.g., [81–85]). These had the aim of deriving new and more precise approximations than the approximate solution proposed by Rice in (3.79). The mathematical treatment of the so-called level-crossing problem is even for Rayleigh channels fraught with considerable difficulties and an exact general solution is still to be found. Special attention in this field should be paid to the work in [25, 86–91]. In [87], results have been reported about a 4-state model which gives a good approximation to the probability density function $p_{0_-}(\tau_-; r)$ over a much greater region than that over which (3.79) is valid [25]. The obtained approximate solutions could again be noticeably improved by extending the 4-state model to 6- and 8-state models [88–91]. However, investigations of generalized Gaussian random processes, the so-called spherical invariant stochastic processes [86], have shown [92] that the 4- and 6-state models of this process class often do not provide satisfying results, whereas the approximation suggested in [85] does quite well. Approximate solutions to the distribution of the fading intervals of lognormal shadow fading channels can be found in [93]. An attempt to tackle the problem of finding the distribution of the fading intervals of Nakagami processes by using an orthogonal series expansion has been made in [94]. It is shown that the distribution of the fading intervals can be approximated in closed form by the gamma distribution for asymptotic and non-asymptotic threshold levels.

In spite of all the progress that has been made to tackle the level-crossing problem, the expenditure of mathematical and numerical calculations is considerable. Moreover, the reliability of all theoretically obtained approximations is not guaranteed from the start, so that we cannot get by without an experimental verification of the theoretical results. From this perspective, it seems to be more sensible to give up the lavish analytical calculations aiming to solve the level-crossing problem and instead to carry out simulations of accurately generated sample functions [92]. This background will be taken into consideration in the following two chapters, where we will discuss various methods for the efficient design of highly precise simulation models enabling the generation of fading waveforms with desired statistics.

Appendix 3.A Derivation of the Jakes Power Spectral Density and the Corresponding Autocorrelation Function

The derivation of the Jakes power spectral density is based on the assumption of two-dimensional isotropic scattering for which the following three conditions must be fulfilled:

(i) The propagation of the electromagnetic waves takes place in the two-dimensional (horizontal) plane, and the receiver is located in the centre of an isotropic scattering area.

(ii) The angles-of-arrival α of the waves impinging on the receiving antenna are uniformly distributed in the interval $[-\pi, \pi)$.

(iii) The antenna radiation pattern of the receiving antenna is circular symmetric (omnidirectional antenna).

Owing to the assumption that the angles-of-arrival α are random variables with the probability density function

$$p_\alpha(\alpha) = \begin{cases} \dfrac{1}{2\pi}, & \alpha \in [-\pi, \pi), \\ 0, & \text{else}, \end{cases} \tag{3.A.1}$$

it follows that the Doppler frequencies, defined by

$$f = f(\alpha) := f_{\max} \cos(\alpha), \tag{3.A.2}$$

are also random variables. The probability density function of the Doppler frequencies f, denoted by $p_f(f)$, can easily be computed by using (2.86). Applying (2.86) to the present problem enables us to write the probability density function $p_f(f)$ in the following form

$$p_f(f) = \sum_{\nu=1}^{m} \frac{p_\alpha(\alpha_\nu)}{\left|\frac{d}{d\alpha} f(\alpha)\right|_{\alpha=\alpha_\nu}}, \tag{3.A.3}$$

where m is the number of solutions of (3.A.2) within the interval $[-\pi, \pi)$. For $|f| > f_{\max}$, the equation $f = f_{\max} \cos(\alpha)$ has no real-valued solution and, consequently, $p_f(f) = 0$ for $|f| > f_{\max}$. However, due to the ambiguity of the inverse function of the cosine function within the interval $[-\pi, \pi)$, two solutions exist for $|f| < f_{\max}$, namely,

$$\alpha_1 = -\alpha_2 = \arccos(f/f_{\max}), \tag{3.A.4}$$

so that $m = 2$. After elementary computations and by using (3.A.1)–(3.A.4), we find the following result for the probability density function $p_f(f)$ of the Doppler frequencies

$$p_f(f) = \begin{cases} \dfrac{1}{\pi f_{\max}\sqrt{1-(f/f_{\max})^2}}, & |f| < f_{\max}, \\ 0, & |f| > f_{\max}. \end{cases} \tag{3.A.5}$$

It can easily be verified that the probability density function $p_f(f)$ of the Doppler frequencies is directly proportional to the power spectral density $S_{\mu\mu}(f)$ of the scattered components $\mu(t) = \mu_1(t) + j\mu_2(t)$ received at the receiving antenna. To show this, we imagine that $\mu(t)$ can be represented by a superposition of an infinite number of complex sinusoids according to (3.15), i.e.,

$$\mu(t) = \lim_{N\to\infty} \sum_{n=1}^{N} c_n \, e^{j(2\pi f_n t + \theta_n)}. \tag{3.A.6}$$

As a consequence of the idealized assumption of isotropic scattering, it follows that all amplitudes $c_n = \sigma_0\sqrt{2/N}$ have the same size. The Doppler frequencies f_n in (3.A.6) are independent and identically distributed (i.i.d.) random variables whose probability density functions are determined by (3.A.5). Likewise, the phases β_n are i.i.d. random variables, but they are uniformly distributed in the interval $[0, 2\pi)$. It should be noted that the power spectral density $S_{\mu\mu}(f)$ of $\mu(t)$ in (3.A.6) is composed of an infinite number of discrete spectral lines and that the average power within an infinitesimal frequency interval df is determined by $S_{\mu\mu}(f)\,df$. This infinitesimal amount of power is obviously proportional to the number of spectral lines contained in df. On the other hand, it follows from (3.A.5) that the number of spectral lines contained in the frequency interval df can also be represented by $p_f(f)\,df$. Hence, the following relation can be established

$$S_{\mu\mu}(f)\,df \sim p_f(f)\,df, \tag{3.A.7}$$

and thus

$$S_{\mu\mu}(f) \sim p_f(f). \tag{3.A.8}$$

Consequently, due to $\int_{-\infty}^{\infty} S_{\mu\mu}(f)\,df = 2\sigma_0^2$ and $\int_{-\infty}^{\infty} p_f(f)\,df = 1$, it follows the relation

$$S_{\mu\mu}(f) = 2\sigma_0^2\, p_f(f). \tag{3.A.9}$$

Thus, by taking (3.A.5) into account, we find the power spectral density

$$S_{\mu\mu}(f) = \begin{cases} \dfrac{2\sigma_0^2}{\pi f_{\max}\sqrt{1-(f/f_{\max})^2}}, & |f| \leq f_{\max}, \\ 0, & |f| > f_{\max}, \end{cases} \tag{3.A.10}$$

which in the literature is often called *Jakes power spectral density* or *Clarke power spectral density*. Strictly speaking, in the equation above, we should have used the less strict inequality $|f| < f_{\max}$ instead of $|f| \leq f_{\max}$. In other publications, however, the poles at $f = \pm f_{\max}$ are commonly assigned to the range of the Jakes power spectral density. Without wanting to go into a detailed analysis of $S_{\mu\mu}(f)$ at $f = \pm f_{\max}$, we will follow the conventional notation, particularly since this small modification does not have any effect on the subsequent computations anyway.

For the power spectral density of the real part and the imaginary part of $\mu(t) = \mu_1(t) + j\mu_2(t)$, the relation

$$S_{\mu_i\mu_i}(f) = \frac{S_{\mu\mu}(f)}{2} = \begin{cases} \dfrac{\sigma_0^2}{\pi f_{\max}\sqrt{1-(f/f_{\max})^2}}, & |f| \leq f_{\max}, \\ 0, & |f| > f_{\max}, \end{cases} \tag{3.A.11}$$

holds for $i = 1$ and $i = 2$, respectively.

Finally, we also compute the autocorrelation function $r_{\mu\mu}(\tau)$ of the scattered components $\mu(t) = \mu_1(t) + j\mu_2(t)$ by using two different approaches. In the first approach, we determine $r_{\mu\mu}(\tau)$ via the inverse Fourier transform of the Jakes power spectral density in (3.A.10). This leads, by taking into account that $S_{\mu\mu}(f)$ is an even function, to the expression

$$\begin{aligned} r_{\mu\mu}(\tau) &= \int_{-\infty}^{\infty} S_{\mu\mu}(f)\, e^{j2\pi f\tau}\, df \\ &= \frac{4\sigma_0^2}{\pi f_{\max}} \int_0^{f_{\max}} \frac{\cos(2\pi f\tau)}{\sqrt{1-(f/f_{\max})^2}}\, df. \end{aligned} \tag{3.A.12}$$

The substitution of f by $f_{\max}\cos(\alpha)$ first of all leads to

$$r_{\mu\mu}(\tau) = \sigma_0^2 \frac{4}{\pi} \int_0^{\pi/2} \cos(2\pi f_{\max}\tau \cos\alpha)\, d\alpha, \tag{3.A.13}$$

from which, by using the integral representation of the zeroth-order Bessel function of the first kind [23, Equation (3.715.19)]

$$J_0(z) = \frac{2}{\pi} \int_0^{\pi/2} \cos(z\cos\alpha)\, d\alpha, \tag{3.A.14}$$

the result

$$r_{\mu\mu}(\tau) = 2\sigma_0^2\, J_0(2\pi f_{\max}\tau) \tag{3.A.15}$$

follows immediately.

An alternative approach is the following one. Starting with the definition of the autocorrelation function

$$r_{\mu\mu}(\tau) := E\{\mu^*(t)\,\mu(t+\tau)\}, \tag{3.A.16}$$

introduced by (2.107) and using (3.A.6), we find

$$r_{\mu\mu}(\tau) = \lim_{N\to\infty} \lim_{M\to\infty} \sum_{n=1}^{N} \sum_{m=1}^{M} c_n c_m\, E\left\{e^{j[2\pi(f_m - f_n)t + 2\pi f_m \tau + \theta_m - \theta_n]}\right\}. \tag{3.A.17}$$

The calculation of the expected value has to be performed with respect to the uniformly distributed phases as well as with respect to the Doppler frequencies distributed according to (3.A.5). Determining the expected value with respect to θ_m and θ_n, and noticing that $E\{e^{\,j(\theta_m-\theta_n)}\}$ equals 1 for $m = n$ and 0 for $m \neq n$, results in

$$r_{\mu\mu}(\tau) = \lim_{N\to\infty} \sum_{n=1}^{N} c_n^2\, E\left\{e^{j2\pi f_n \tau}\right\}. \tag{3.A.18}$$

With the probability density function (3.A.5), we can — after a short intermediate computation similar to that of the first procedure — represent the expected value appearing at the right-hand side of (3.A.18) by

$$\begin{aligned} E\left\{e^{j2\pi f_n \tau}\right\} &= \int_{-\infty}^{\infty} p_f(f)\, e^{j2\pi f\tau}\, df \\ &= J_0(2\pi f_{\max}\tau). \end{aligned} \tag{3.A.19}$$

Finally, we recall that the amplitudes c_n are determined, according to the assumptions made before, by $c_n = \sigma_0\sqrt{2/N}$. Thus, from (3.A.18) and under consideration of (3.A.19), it follows the expression

$$r_{\mu\mu}(\tau) = 2\sigma_0^2\, J_0(2\pi f_{\max}\tau), \tag{3.A.20}$$

which is identical to the result in (3.A.15) obtained by computing the inverse Fourier transform of the Jakes power spectral density.

Appendix 3.B Derivation of the Autocorrelation Function of the Envelope

The starting point for proving the correctness of (3.32) is the definition of the autocorrelation function of the envelope

$$r_{\zeta\zeta}(\tau) = E\{\zeta(t)\,\zeta(t+\tau)\}. \tag{3.B.1}$$

Since $\zeta(t) = |\mu(t)| = \sqrt{\mu_1^2(t) + \mu_2^2(t)}$, the autocorrelation function $r_{\zeta\zeta}(\tau)$ can be expressed for fixed values of $t = t_0$ as

$$r_{\zeta\zeta}(\tau) = E\left\{\sqrt{\left(\mu_1^2 + \mu_2^2\right)\left(\mu_1'^2 + \mu_2'^2\right)}\right\}, \tag{3.B.2}$$

where $\mu_i = \mu_i(t_0)$ and $\mu_i' = \mu_i(t_0 + \tau)$ are zero-mean Gaussian random variables described by the autocorrelation function $r_{\mu_i\mu_i'}(\tau) = E\{\mu_i\mu_i'\}$ $(i = 1, 2)$. The extension of the fundamental relationship in (2.10) to four random variables leads to

$$r_{\zeta\zeta}(\tau) = \int_{-\infty}^{\infty}\int_{-\infty}^{\infty}\int_{-\infty}^{\infty}\int_{-\infty}^{\infty} \sqrt{\left(x_1^2 + x_2^2\right)\left(x_1'^2 + x_2'^2\right)}\, p_{\mu_1\mu_2\mu_1'\mu_2'}\left(x_1, x_2, x_1', x_2'\right) dx_1\, dx_2\, dx_1'\, dx_2', \tag{3.B.3}$$

where $p_{\mu_1\mu_2\mu_1'\mu_2'}(\cdot,\cdot,\cdot,\cdot)$ denotes the joint probability density function of μ_1, μ_2, μ_1', and μ_2'. Notice that μ_i and μ_i' are correlated Gaussian random variables, whereas μ_i and μ_j are uncorrelated for $i, j = 1, 2$ $(i \neq j)$. By recalling that uncorrelatedness is equivalent to independence for Gaussian random variables, we can thus write

$$p_{\mu_1\mu_2\mu_1'\mu_2'}(x_1, x_2, x_1', x_2') = p_{\mu_1\mu_1'}(x_1, x_1') \cdot p_{\mu_2\mu_2'}(x_2, x_2'), \tag{3.B.4}$$

where the joint probability density function $p_{\mu_i\mu_i'}(x_i, x_i')$ of μ_i and μ_i' is obtained from the multivariate Gaussian distribution in (2.36) as

$$p_{\mu_i\mu_i'}(x_i, x_i') = \frac{e^{-\frac{x_i^2 - 2\rho(\tau)x_i x_i' + x_i'^2}{2\sigma_0^2\left(1-\rho^2(\tau)\right)}}}{2\pi\sigma_0^2\sqrt{1-\rho^2(\tau)}} \tag{3.B.5}$$

with $\rho(\tau) = r_{\mu_i\mu_i'}(\tau)/\sigma_0^2$ for $i = 1, 2$. Substituting (3.B.4) with (3.B.5) in (3.B.3) and using the transformations of variables

$$x_1 = z\cos\theta, \qquad x_1' = z'\cos\theta' \tag{3.B.6}$$

$$x_2 = z\sin\theta, \qquad x_2' = z'\sin\theta' \tag{3.B.7}$$

enables us to reduce the four-fold integral in (3.B.3) to the double integral

$$r_{\zeta\zeta}(\tau) = \frac{2\alpha(\tau)}{\sigma_0^2} \int_0^\infty \int_0^\infty (zz')^2 \, e^{-\alpha(\tau)(z^2+z'^2)} I_0\left(2\alpha(\tau)\,\rho(\tau)\,zz'\right) dz\,dz', \tag{3.B.8}$$

where

$$\alpha(\tau) = \frac{1}{2\sigma_0^2\left[1-\rho^2(\tau)\right]}. \tag{3.B.9}$$

According to [77, Equation (9.6.3)], we can replace $I_0(x)$ by $J_0(jx)$ in (3.B.8) to obtain, by using [23, Equation (6.631.1)]

$$\int_0^\infty x^2 \, e^{-\alpha x^2} J_0(\beta x)\, dx = \frac{\sqrt{\pi}}{4\alpha^{3/2}} \, {}_1F_1\left(\frac{3}{2}; 1; -\frac{\beta^2}{4\alpha}\right), \tag{3.B.10}$$

the following expression

$$\begin{aligned} r_{\zeta\zeta}(\tau) &= \frac{\sqrt{\pi}}{2\sigma_0^2\sqrt{\alpha(\tau)}} \int_0^\infty z^2 \, e^{-\alpha(\tau)z^2} \, {}_1F_1\left(\frac{3}{2}; 1; \alpha(\tau)\,\rho^2(\tau)\,z^2\right) dz \\ &= \frac{\sqrt{\pi}}{4\sigma_0^2\sqrt{\alpha(\tau)}} \int_0^\infty \sqrt{x}\, e^{-\alpha(\tau)x} \, {}_1F_1\left(\frac{3}{2}; 1; \alpha(\tau)\,\rho^2(\tau)\,x\right) dx, \end{aligned} \tag{3.B.11}$$

where ${}_1F_1(\cdot\,;\cdot;\cdot)$ denotes the *generalized hypergeometric series*. Next, by using [23, Equation (7.525.1)]

$$\int_0^\infty \sqrt{x}\, e^{-\alpha x} \, {}_1F_1(a; b; \lambda x)\, dx = \Gamma\left(\frac{3}{2}\right) \alpha^{-\frac{3}{2}} \, {}_2F_1\left(a, \frac{3}{2}; b; \frac{\lambda}{\alpha}\right) \tag{3.B.12}$$

and the relation [23, Equation (9.14.2)]

$${}_2F_1(a, b; c; z) = F(a, b; c; z), \tag{3.B.13}$$

we can write the autocorrelation function $r_{\zeta\zeta}(\tau)$ in terms of the hypergeometric function $F(\cdot, \cdot; \cdot; \cdot)$ as follows

$$r_{\zeta\zeta}(\tau) = \frac{\pi}{8\sigma_0^2\alpha^2(\tau)} F\left(\frac{3}{2}, \frac{3}{2}; 1; \rho^2(\tau)\right). \tag{3.B.14}$$

Finally, by using the transformation formula [23, Equation (9.131.1)]

$$F(a, b; c; z) = (1-z)^{c-a-b} F(c-a, c-b; c; z) \tag{3.B.15}$$

and (3.B.9), we have shown that $r_{\zeta\zeta}(\tau)$ can be brought into the form

$$r_{\zeta\zeta}(\tau) = \sigma_0^2 \, \frac{\pi}{2} F\left(-\frac{1}{2}, -\frac{1}{2}; 1; \rho^2(\tau)\right). \tag{3.B.16}$$

From this result, it follows immediately (3.32), as $\rho(\tau) = r_{\mu_i\mu_i'}(\tau)/\sigma_0^2$ and

$$\begin{aligned} r_{\mu_i\mu_i'}(\tau) &= E\{\mu_i\mu_i'\} \\ &= E\{\mu_i(t_0)\,\mu_i(t_0+\tau)\} \\ &= E\{\mu_i(t)\,\mu_i(t+\tau)\} \\ &= r_{\mu_i\mu_i}(\tau), \quad i = 1, 2. \end{aligned} \tag{3.B.17}$$

An abridged version of this proof can also be found in [95]. Another source for the key result in (3.B.16) is [54, p. 170], where the proof is left as an exercise for the reader.

Appendix 3.C Derivation of the Autocovariance Spectrum of the Envelope Under Isotropic Scattering Conditions

Under the assumption of isotropic scattering, we prove in this appendix that the autocovariance spectrum $S_{\zeta\zeta}^c(f)$ of the envelope $\zeta(t)$ can be approximated as

$$S_{\zeta\zeta}^c(f) \approx \begin{cases} \dfrac{\sigma_0^2}{8\pi f_{\max}} \boldsymbol{K}\left(\sqrt{1-\left(\dfrac{f}{2f_{\max}}\right)^2}\right), & |f| < 2f_{\max}, \\ 0, & |f| > 2f_{\max}, \end{cases} \tag{3.C.1}$$

where $\boldsymbol{K}(\cdot)$ is the *complete elliptic integral of the first kind*. The proof can be carried out in a number of ways. One possibility is to solve the convolution integral in (3.38) by using the Jakes power spectral density introduced in (3.23). Here, we proceed in an alternative way by computing the Fourier transform of the autocovariance function [see (3.39)]

$$r_{\zeta\zeta}^c(\tau) \approx \frac{\pi}{8}\sigma_0^2 J_0^2\,(2\pi f_{\max}\tau) \tag{3.C.2}$$

which gives

$$\begin{aligned} S_{\zeta\zeta}^c(\tau) &\approx \frac{\pi}{8}\sigma_0^2 \int_{-\infty}^{\infty} J_0^2\,(2\pi f_{\max}\tau)\, e^{-j2\pi f\tau}\, d\tau \\ &= \frac{\pi}{4}\sigma_0^2 \int_{0}^{\infty} J_0^2\,(2\pi f_{\max}\tau) \cos(2\pi f\tau)\, d\tau \\ &= \frac{\sigma_0^2}{8f_{\max}} \int_{0}^{\infty} J_0^2(x) \cos\left(\frac{f}{f_{\max}}x\right) dx. \end{aligned} \tag{3.C.3}$$

Using [23, Equation (6.672.2)]

$$\int_0^\infty J_0^2(x)\cos(bx)\,dx = \begin{cases} \dfrac{1}{2}P_{-\frac{1}{2}}\left(\dfrac{b^2}{2}-1\right), & |b| < 2, \\ 0, & |b| > 2, \end{cases} \tag{3.C.4}$$

where $P_\nu(\cdot)$ denotes the *Legendre function of the first kind*, and identifying b as $f/f_{\max}$ gives

$$S^c_{\zeta\zeta}(f) \approx \begin{cases} \dfrac{\sigma_0^2}{16 f_{\max}} P_{-\frac{1}{2}}\left(\dfrac{f^2}{2f_{\max}^2} - 1\right), & |f| < 2f_{\max}, \\ 0, & |f| > 2f_{\max}. \end{cases} \tag{3.C.5}$$

Furthermore, with the relations [23, Equation (8.820.1)]

$$P_\nu(x) = F\left(-\nu, \nu+1; 1; \frac{1-x}{2}\right) \tag{3.C.6}$$

and [23, Equation (8.113.1)]

$$\boldsymbol{K}(k) = \frac{\pi}{2} F\left(\frac{1}{2}, \frac{1}{2}; 1; k^2\right), \tag{3.C.7}$$

we can establish the identity

$$P_{-\frac{1}{2}}(x) = \frac{2}{\pi} \boldsymbol{K}\left(\sqrt{\frac{1-x}{2}}\right). \tag{3.C.8}$$

Finally, by using the above result in (3.C.5), we immediately obtain (3.C.1).

Appendix 3.D Derivation of the Level-Crossing Rate of Rice Processes with Different Spectral Shapes of the Underlying Gaussian Random Processes

Let $\mu_1(t)$ and $\mu_2(t)$ be two uncorrelated zero-mean Gaussian random processes with identical variances but different spectral shapes, i.e., the corresponding autocorrelation functions are subject to the following conditions:

$$\text{(i)} \quad r_{\mu_1\mu_1}(0) = r_{\mu_2\mu_2}(0) = \sigma_0^2, \tag{3.D.1}$$

$$\text{(ii)} \quad r_{\mu_1\mu_1}(\tau) \neq r_{\mu_2\mu_2}(\tau), \quad \text{if } \tau > 0, \tag{3.D.2}$$

$$\text{(iii)} \quad \frac{d^n}{d\tau^n} r_{\mu_1\mu_1}(\tau) \neq \frac{d^n}{d\tau^n} r_{\mu_2\mu_2}(\tau), \quad \text{if } \tau \geq 0, \quad n = 1, 2, \ldots \tag{3.D.3}$$

For further simplification of the problem, we assume that $f_\rho = 0$, i.e., the line-of-sight component m is supposed to be time invariant, and is thus determined by (3.18).

The starting point for the computation of the level-crossing rate of the resulting Rice process is the joint probability density function of the stationary processes $\mu_{\rho_1}(t)$, $\mu_{\rho_2}(t)$, $\dot{\mu}_{\rho_1}(t)$, and $\dot{\mu}_{\rho_2}(t)$ [see (3.19)] at the same time t. Here, we have to take the following fact into account: If $\mu_{\rho_i}(t)$ is a real-valued (stationary) Gaussian random process with mean value $E\{\mu_{\rho_i}(t)\} = m_i \neq 0$ and variance $\text{Var}\{\mu_{\rho_i}(t)\} = \text{Var}\{\mu_i(t)\} = r_{\mu_i\mu_i}(0) = \sigma_0^2$, then its derivative with respect to time, denoted by $\dot{\mu}_{\rho_i}(t)$, is also a real-valued (stationary) Gaussian random process but with mean value $E\{\dot{\mu}_{\rho_i}(t)\} = \dot{m}_i = 0$ and variance $\text{Var}\{\dot{\mu}_{\rho_i}(t)\} = \text{Var}\{\dot{\mu}_i(t)\} = r_{\dot{\mu}_i\dot{\mu}_i}(0) = -\ddot{r}_{\mu_i\mu_i}(0) = \beta_i$ ($i = 1, 2$). Due to (3.D.3), the inequality $\beta_1 \neq \beta_2$ holds for the characteristic quantity β_i. Furthermore, the processes $\mu_{\rho_i}(t)$ and $\dot{\mu}_{\rho_i}(t)$ are in pairs uncorrelated at the same time t. From this fact, it follows that the joint probability density function $p_{\mu_{\rho_1}\mu_{\rho_2}\dot{\mu}_{\rho_1}\dot{\mu}_{\rho_2}}(x_1, x_2, \dot{x}_1, \dot{x}_2)$ is given by the multivariate Gaussian distribution, which can be represented by using (2.36) as

$$p_{\mu_{\rho_1}\mu_{\rho_2}\dot{\mu}_{\rho_1}\dot{\mu}_{\rho_2}}(x_1, x_2, \dot{x}_1, \dot{x}_2) = \frac{e^{-\frac{(x_1-m_1)^2}{2\sigma_0^2}}}{\sqrt{2\pi}\sigma_0} \cdot \frac{e^{-\frac{(x_2-m_2)^2}{2\sigma_0^2}}}{\sqrt{2\pi}\sigma_0} \cdot \frac{e^{-\frac{\dot{x}_1^2}{2\beta_1}}}{\sqrt{2\pi\beta_1}} \cdot \frac{e^{-\frac{\dot{x}_2^2}{2\beta_2}}}{\sqrt{2\pi\beta_2}}. \tag{3.D.4}$$

The transformation of the Cartesian coordinates (x_1, x_2) into polar coordinates (z, θ) by means of $z = \sqrt{x_1^2 + x_2^2}$ and $\theta = \arctan(x_2/x_1)$ leads to the following system of equations:

$$\begin{aligned} x_1 &= z\cos\theta, \quad & \dot{x}_1 &= \dot{z}\cos\theta - \dot{\theta}z\sin\theta \\ x_2 &= z\sin\theta, \quad & \dot{x}_2 &= \dot{z}\sin\theta + \dot{\theta}z\cos\theta \end{aligned} \tag{3.D.5}$$

for $z \geq 0$ and $|\theta| \leq \pi$. The application of the transformation rule (2.87) then results in the joint probability density function

$$p_{\xi\dot{\xi}\vartheta\dot{\vartheta}}(z, \dot{z}, \theta, \dot{\theta}) = |J|^{-1} p_{\mu_{\rho_1}\mu_{\rho_2}\dot{\mu}_{\rho_1}\dot{\mu}_{\rho_2}}(z\cos\theta, z\sin\theta, \dot{z}\cos\theta - \dot{\theta}z\sin\theta, \dot{z}\sin\theta + \dot{\theta}z\cos\theta), \tag{3.D.6}$$

where

$$J = J(z) = \begin{vmatrix} \frac{\partial x_1}{\partial z} & \frac{\partial x_1}{\partial \dot{z}} & \frac{\partial x_1}{\partial \theta} & \frac{\partial x_1}{\partial \dot{\theta}} \\ \frac{\partial x_2}{\partial z} & \frac{\partial x_2}{\partial \dot{z}} & \frac{\partial x_2}{\partial \theta} & \frac{\partial x_2}{\partial \dot{\theta}} \\ \frac{\partial \dot{x}_1}{\partial z} & \frac{\partial \dot{x}_1}{\partial \dot{z}} & \frac{\partial \dot{x}_1}{\partial \theta} & \frac{\partial \dot{x}_1}{\partial \dot{\theta}} \\ \frac{\partial \dot{x}_2}{\partial z} & \frac{\partial \dot{x}_2}{\partial \dot{z}} & \frac{\partial \dot{x}_2}{\partial \theta} & \frac{\partial \dot{x}_2}{\partial \dot{\theta}} \end{vmatrix}^{-1} = -\frac{1}{z^2} \tag{3.D.7}$$

denotes the Jacobian determinant [see (2.88)]. Inserting (3.D.5) and (3.D.7) into (3.D.6) results, after some algebraic calculations, in the following expression for the joint probability density function $p_{\xi\dot{\xi}\vartheta\dot{\vartheta}}(z, \dot{z}, \theta, \dot{\theta})$

$$\begin{aligned} p_{\xi\dot{\xi}\vartheta\dot{\vartheta}}(z, \dot{z}, \theta, \dot{\theta}) = {} & \frac{z^2}{(2\pi\sigma_0)^2\sqrt{\beta_1\beta_2}} e^{-\frac{1}{2\sigma_0^2}[z^2+\rho^2-2z\rho\cos(\theta-\theta_\rho)]} \\ & \cdot e^{-\frac{\dot{z}^2}{2}\left(\frac{\cos^2\theta}{\beta_1}+\frac{\sin^2\theta}{\beta_2}\right)-\frac{z^2\dot{\theta}^2}{2}\left(\frac{\cos^2\theta}{\beta_2}+\frac{\sin^2\theta}{\beta_1}\right)-z\dot{z}\dot{\theta}\left(\frac{\beta_1-\beta_2}{\beta_1\beta_2}\right)\cos\theta\sin\theta} \end{aligned} \tag{3.D.8}$$

for $z \geq 0$, $|\dot{z}| < \infty$, $|\theta| \leq \pi$, and $|\dot{\theta}| < \infty$. According to (2.89), we can now compute the joint probability density function of the processes $\xi(t)$ and $\dot{\xi}(t)$ at the same time t by using the relation

$$p_{\xi\dot{\xi}}(z, \dot{z}) = \int_{-\infty}^{\infty}\int_{-\pi}^{\pi} p_{\xi\dot{\xi}\vartheta\dot{\vartheta}}(z, \dot{z}, \theta, \dot{\theta})\, d\theta\, d\dot{\theta}, \quad z \geq 0, \quad |\dot{z}| < \infty. \tag{3.D.9}$$

Inserting (3.D.8) into (3.D.9) finally results in

$$p_{\xi\dot{\xi}}(z,\dot{z}) = \frac{z}{(2\pi)^{3/2}\sigma_0^2} e^{-\frac{z^2+\rho^2}{2\sigma_0^2}} \int_{-\pi}^{\pi} e^{\frac{z\rho}{\sigma_0^2}\cos(\theta-\theta_\rho)} \cdot \frac{e^{-\frac{\dot{z}^2}{2(\beta_1\cos^2\theta+\beta_2\sin^2\theta)}}}{\sqrt{\beta_1\cos^2\theta+\beta_2\sin^2\theta}} \, d\theta. \qquad (3.D.10)$$

Since the level-crossing rate $N_\xi(r)$ of Rice process $\xi(t)$ is generally defined by

$$N_\xi(r) := \int_0^\infty \dot{z} p_{\xi\dot{\xi}}(r,\dot{z}) \, d\dot{z}, \quad r \geq 0, \qquad (3.D.11)$$

we obtain, by using the above expression (3.D.10), the following result

$$N_\xi(r) = \frac{r e^{-\frac{r^2+\rho^2}{2\sigma_0^2}}}{(2\pi)^{3/2}\sigma_0^2} \cdot \int_{-\pi}^{\pi} e^{\frac{r\rho}{\sigma_0^2}\cos(\theta-\theta_\rho)} \sqrt{\beta_1\cos^2\theta+\beta_2\sin^2\theta} \, d\theta, \qquad (3.D.12)$$

which holds for $\beta_1 \neq \beta_2$. Without restriction of generality, we may assume that $\beta_1 \geq \beta_2$ holds. On this condition, we can also express (3.D.12) by

$$N_\xi(r) = \sqrt{\frac{\beta_1}{2\pi}} \cdot \frac{r}{\sigma_0^2} e^{-\frac{r^2+\rho^2}{2\sigma_0^2}} \cdot \frac{1}{\pi} \int_0^{\pi} \cosh\left[\frac{r\rho}{\sigma_0^2}\cos(\theta-\theta_\rho)\right] \sqrt{1-k^2\sin^2\theta} \, d\theta, \quad r \geq 0, \qquad (3.D.13)$$

where $k = \sqrt{(\beta_1-\beta_2)/\beta_1}$. It should be mentioned that for $\beta = \beta_1 = \beta_2 \neq 0$, i.e., $k = 0$, and by using the relation [77, Equation (9.6.16)]

$$I_0(z) = \frac{1}{\pi} \int_0^{\pi} \cosh(z\cos\theta) \, d\theta, \qquad (3.D.14)$$

the above expression for the level-crossing rate $N_\xi(r)$ can be reduced to the form (3.66), as it was to be expected.

At the end of this appendix, we consider an approximation for the case that the relative deviation between β_1 and β_2 is very small. Thus, for a positive number ε for which $\varepsilon/\beta_1 \ll 1$ holds, we can write

$$\beta_1 = \beta_2 + \varepsilon. \qquad (3.D.15)$$

Due to $k = \sqrt{(\beta_1-\beta_2)/\beta_1} = \sqrt{\varepsilon/\beta_1} \ll 1$, we may use the approximation

$$\begin{aligned} \sqrt{1-k^2\sin^2\theta} &\approx 1 - \frac{k^2}{2}\sin^2\theta \\ &= 1 - \frac{\varepsilon}{2\beta_1}\sin^2\theta, \end{aligned} \qquad (3.D.16)$$

so that the relation in (3.D.13) can be simplified for $\theta_\rho = 0$ to the following expressions

$$N_\xi(r)|_{\beta_1\approx\beta_2} \approx \sqrt{\frac{\beta_1}{2\pi}} \cdot \frac{r}{\sigma_0^2} e^{-\frac{r^2+\rho^2}{2\sigma_0^2}} \left[I_0\left(\frac{r\rho}{\sigma_0^2}\right) - \frac{\varepsilon}{2\beta_1} I_1\left(\frac{r\rho}{\sigma_0^2}\right) \Big/ \left(\frac{r\rho}{\sigma_0^2}\right) \right]$$
$$\approx \sqrt{\frac{\beta_1}{2\pi}} \cdot \frac{r}{\sigma_0^2} e^{-\frac{r^2+\rho^2}{2\sigma_0^2}} I_0\left(\frac{r\rho}{\sigma_0^2}\right)$$
$$\approx \sqrt{\frac{\beta_1}{2\pi}} \cdot p_\xi(r). \tag{3.D.17}$$

For the derivation of these relations, we have made use of the integral representation of the first-order modified Bessel function of the first kind [77, Equation (9.6.18)]

$$I_1(z) = \frac{z}{\pi} \int_0^\pi e^{\pm z\cos\theta} \sin^2\theta \, d\theta. \tag{3.D.18}$$

Hence, (3.D.17) shows that in cases where $\beta_1 \approx \beta_2$ holds, the expression (3.66) is still approximately valid for the level-crossing rate of Rice processes $\xi(t)$, if the quantity β is replaced there by β_1.

4

Introduction to Sum-of-Sinusoids Channel Models

The majority of channel models studied in the sequel of this book are based on the utilization of coloured Gaussian random processes. In the previous chapter, for example, we have seen that the modelling of classical Rayleigh or Rice processes requires the realization of two real-valued coloured Gaussian random processes. Whereas for a Suzuki process [26], which is defined as the product process of a Rayleigh process and a lognormal process, three real-valued coloured Gaussian random processes are needed. Regarding digital data transmission over land mobile radio channels, we often refer to such processes (Rayleigh, Rice, Suzuki) as appropriate stochastic models in order to describe the random envelope fluctuations of the received signal in the equivalent complex baseband. Mobile radio channels, whose statistical envelope behaviour can be described by Rayleigh, Rice or Suzuki processes, will therefore be denoted as Rayleigh, Rice or Suzuki channels. These models can be classified into the group of frequency-nonselective channels [11]. A further example is given by the modelling of frequency-selective channels [11] using finite impulse response (FIR) filters with $\mathcal{L}$ time-variant complex-valued random coefficients. This requires the realization of $2\mathcal{L}$ real-valued coloured Gaussian random processes. Last but not least, we mention that for the modelling of multiple-input multiple-output channels with M_T transmitter and M_R receiver antennas, we strive for the efficient design of $M_T \cdot M_R$ complex-valued coloured Gaussian processes, which are in general correlated in space, time, and frequency. With the help of these few examples, it becomes strikingly clear that the development of efficient methods for the realization of coloured Gaussian random processes is of utmost importance for the modelling of both frequency-nonselective and frequency-selective mobile radio channels for single and multiple antenna systems.

To solve the problem of modelling efficiently Gaussian processes with given correlation properties, we describe in this chapter a fundamental method, which is based on a superposition of a finite number of sinusoids. The basic idea behind this method can be traced back to Rice's sum-of-sinusoids [16, 17]. This chapter presents an introduction to the theory of sum-of-sinusoids processes, which serve as a further important pillar for the concepts on which this book is built.

Mobile Radio Channels, Second Edition. Matthias Pätzold.

At the present day, the application of the sum-of-sinusoids principle ranges from the development of channel simulators for relatively simple time-variant Rayleigh fading channels [13, 96, 97] and Nakagami channels [33, 98] over frequency-selective channels [99–103] up to elaborated space-time narrowband [104, 105] and wideband [106–109] channels. Further applications can be found in research areas dealing with the design of multiple cross-correlated [110] and multiple uncorrelated [111–114] Rayleigh fading channels. Such channel models are of special interest, e.g., in system performance studies of multiple-input multiple-output systems [115–117] and diversity schemes [118, Chap. 6], [119]. Moreover, it has been shown that the sum-of-sinusoids principle enables the design of fast fading channel simulators [120] and facilitates the development of perfect channel models [121, 122]. A perfect channel model is a model, whose scattering function can be fitted perfectly to any given measured scattering function obtained from snap-shot measurements carried out in real-world environments. Finally, it should be mentioned that the sum-of-sinusoids principle has been applied successfully to the modelling and simulation of mobile-to-mobile radio channels in cooperative networks [123, 124] and to the development of channel capacity simulators [125, 126].

The organization of this chapter is as follows. Section 4.1 introduces the principle of deterministic channel modelling. Section 4.2 studies the elementary properties of deterministic sum-of-sinusoids processes, such as the autocorrelation function, power spectral density, and Doppler spread. The analysis of the statistical properties of these processes is the subject of study in Section 4.3. Section 4.4 provides an overview of the various classes of sum-of-sinusoids processes and discusses briefly their stationary and ergodic properties. The basics of sum-of-cisoids processes will be introduced in Section 4.5. In this connection, we will also highlight the main differences between sum-of-sinusoids and sum-of-cisoids processes. Section 4.6 presents suitable quality criteria on the basis of which a fair assessment of the performance of the parameter computation methods discussed in Chapter 5 can be carried out. The application of these criteria enables us to establish guidelines for the development of high-performance parameter computation methods. Moreover, the problems caused by employing less suitable methods will thereby become more comprehensible. Finally, Section 4.7 points to some additional references complementary to the topics covered in this chapter.

4.1 Principle of Deterministic Channel Modelling

In the literature, one essentially finds two fundamental methods for the modelling of coloured Gaussian random processes: the *filter method* and the *Rice method*.

When using the filter method, a white Gaussian noise (WGN) process $\nu_i(t)$ as shown in Figure 4.1(a) is given to the input of a linear time-invariant filter, whose transfer function is denoted by $H_i(f)$. In the following, we assume that the filter is ideal in the sense that the transfer function $H_i(f)$ can be fitted to any given frequency response with arbitrary precision. If $\nu_i(t) \sim N(0, 1)$, then we obtain a zero-mean stochastic Gaussian random process $\mu_i(t)$ at the filter output, where according to (2.131f) the power spectral density $S_{\mu_i\mu_i}(f)$ of $\mu_i(t)$ matches the squared absolute value of the transfer function, i.e., $S_{\mu_i\mu_i}(f) = |H_i(f)|^2$. Hence, a coloured Gaussian random process $\mu_i(t)$ can be considered as the result of filtering white Gaussian noise $\nu_i(t)$.

The principle of the Rice method [16, 17] is illustrated in Figure 4.1(b). It is based on a superposition of an infinite number of sinusoids with constant gains, equidistant frequencies,

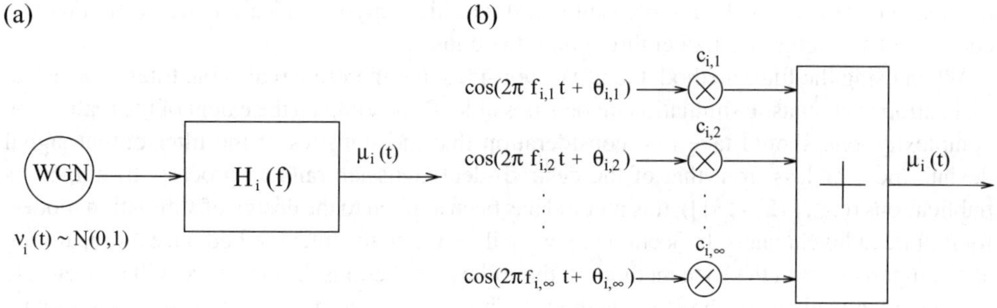

Figure 4.1 Stochastic reference models for coloured Gaussian random processes $\mu_i(t)$: (a) filter method and (b) Rice method.

and random phases. According to this principle, a stochastic Gaussian process $\mu_i(t)$ can be described mathematically by an infinite sum-of-sinusoids (SOS)[1]

$$\mu_i(t) = \lim_{N_i \to \infty} \sum_{n=1}^{N_i} c_{i,n} \cos\left(2\pi f_{i,n} t + \theta_{i,n}\right), \tag{4.1}$$

where the gains $c_{i,n}$ and frequencies $f_{i,n}$ are constant quantities given by

$$c_{i,n} = 2\sqrt{\Delta f_i S_{\mu_i\mu_i}(f_{i,n})}, \tag{4.2a}$$

$$f_{i,n} = n \cdot \Delta f_i. \tag{4.2b}$$

The phases $\theta_{i,n}$ ($n = 1, 2, \ldots, N_i$) are independent, identically distributed random variables, each having a uniform distribution over the interval $(0, 2\pi]$, and the quantity Δf_i is here chosen in such a way that (4.2b) covers the whole relevant frequency range, where it is assumed that $\Delta f_i \to 0$ as $N_i \to \infty$.

As we know, a Gaussian random process is completely characterized by its mean value and its autocorrelation function or, alternatively, by its power spectral density. According to Rice [16, 17], the expression (4.1) represents a zero-mean Gaussian random process with the power spectral density $S_{\mu_i\mu_i}(f)$. Consequently, the analytical models shown in Figures 4.1(a) and 4.1(b) are equivalent, i.e., the two introduced methods — the filter method and the Rice method — result in identical stochastic processes. For both methods, however, one should take into account that the resulting processes are not exactly realizable. When using the filter method, an exact realization is prevented by the assumption that the filter should be ideal. Strictly speaking, the input signal of the filter — the white Gaussian noise — can also not be realized exactly. When using the Rice method, a realization is impossible because an infinite number of sinusoids N_i cannot be implemented on a computer or on a hardware platform. Hence, concerning the modelling of coloured Gaussian random processes, the filter

[1] A *sinusoid* or *plane wave* is a mathematical function of the form $x(t) = A \sin(2\pi ft + \theta)$, where A, f, and θ are the amplitude, frequency, and phase, respectively. A cosine wave is also said to be a sinusoid, because $A \cos(2\pi ft + \theta) = A \sin(2\pi ft + \theta + \pi/2)$ equals a sinusoid with a phase shift of $\pi/2$. This relationship justifies the term *sum-of-sinusoids* for the mathematical model in (4.1).

method and the Rice method only result in a stochastic analytical (ideal) model, which will be considered as a reference model throughout the book.

When using the filter method, the use of non-ideal but therefore realizable filters makes the realization of stochastic simulation models possible. Depending on the extent of the realization complexity, one should take into consideration that the statistics of the filter output signal deviate more or less from that of the desired ideal Gaussian random process. In numerous publications (e.g., [127–131]), this method has been applied to the design of simulation models for mobile radio channels. In Section 9.5, we will return to the filter method once more, aiming at a comparison with the Rice method. In the following sections, however, we will concentrate on a detailed analysis of the Rice method. It should be noted that many of the results found for the Rice method can be applied directly to the filter method.

If the Rice method is used with only a finite number of sinusoids $N_i < \infty$, then we obtain another stochastic sum-of-sinusoids process denoted by

$$\hat{\mu}_i(t) = \sum_{n=1}^{N_i} c_{i,n} \cos(2\pi f_{i,n} t + \theta_{i,n}). \tag{4.3}$$

Let us assume for the moment that the parameters $c_{i,n}$ and $f_{i,n}$ are still given by (4.2a) and (4.2b), respectively, and that the phases $\theta_{i,n}$ are again uniformly distributed random variables. Then we can now apply this method to the realization of a simulation model, whose general structure is shown in Figure 4.2(a). It is obvious that $\hat{\mu}_i(t) \to \mu_i(t)$ holds as $N_i \to \infty$. At this point, it should be emphasized that the simulation model is still of stochastic nature, since the phases $\theta_{i,n}$ are uniformly distributed random variables for all $n = 1, 2, \ldots, N_i$.

Only after the phases $\theta_{i,n}$ $(n = 1, 2, \ldots, N_i)$ are taken out of a random generator with a uniform distribution in the interval $(0, 2\pi]$, the phases $\theta_{i,n}$ no longer represent random variables but constant quantities, since they are now realizations (outcomes) of a random variable. Thus, in connection with (4.2a), (4.2b), and (4.3), it becomes obvious that

$$\tilde{\mu}_i(t) = \sum_{n=1}^{N_i} c_{i,n} \cos(2\pi f_{i,n} t + \theta_{i,n}) \tag{4.4}$$

is a *deterministic sum-of-sinusoids process*, which is also often called just a *deterministic process* or *waveform* for brevity. Hence, from the stochastic simulation model shown in

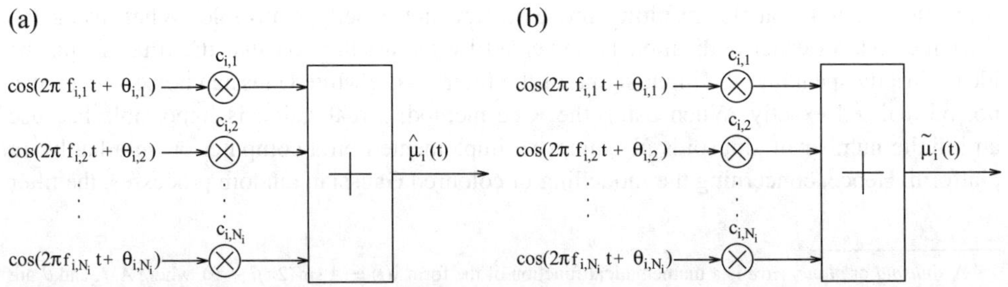

Figure 4.2 Simulation models for coloured Gaussian random processes: (a) stochastic simulation model (random phases $\theta_{i,n}$) and (b) deterministic simulation model (constant phases $\theta_{i,n}$).

Figure 4.2(a), a deterministic simulation model follows, whose structure is presented in Figure 4.2(b) in its continuous-time form of representation. Note that in the limit $N_i \to \infty$, the deterministic process $\tilde{\mu}_i(t)$ tends to a sample function of the stochastic process $\mu_i(t)$.

The statements above allow us to interpret stochastic and deterministic simulation models for Gaussian random processes $\mu_i(t)$ as follows:

(i) If t is a variable, N_i is infinite, and $\theta_{i,n}$ are random variables, then $\mu_i(t)$ is a Gaussian process, which is a family or an ensemble of sample functions $\mu_i(t, \theta_{i,n}^{(k)})$ $(k = 1, 2, \ldots)$, i.e., $\mu_i(t) = \mu_i(t, \theta_{i,n}) = \{\mu_i(t, \theta_{i,n}^{(k)}) \,|\, \theta_{i,n}^{(k)} \in (0, 2\pi]\} = \{\mu_i(t, \theta_{i,n}^{(1)}), \mu_i(t, \theta_{i,n}^{(2)}), \ldots\}$. The Gaussian process $\mu_i(t)$ represents the reference model.

(ii) If t is a variable, N_i is finite, and $\theta_{i,n}$ are random variables, then $\hat{\mu}_i(t)$ is a stochastic process, which can be interpreted as a family or an ensemble of sample functions $\hat{\mu}_i(t, \theta_{i,n}^{(k)})$, i.e., $\hat{\mu}_i(t) = \hat{\mu}_i(t, \theta_{i,n}) = \{\hat{\mu}_i(t, \theta_{i,n}^{(k)}) \,|\, \theta_{i,n}^{(k)} \in (0, 2\pi]\} = \{\hat{\mu}_i(t, \theta_{i,n}^{(1)}), \hat{\mu}_i(t, \theta_{i,n}^{(2)}), \ldots\}$. The realization of the stochastic process $\hat{\mu}_i(t)$ is called the stochastic simulation model.

(iii) If t is a variable, N_i is finite, and $\theta_{i,n} = \theta_{i,n}^{(k)}$ are constants, then $\tilde{\mu}_i(t) = \hat{\mu}_i(t, \theta_{i,n}^{(k)})$ is a realization or a sample function of the stochastic process $\hat{\mu}_i(t)$. The sample function $\tilde{\mu}_i(t)$ is deterministic and its realization is called the deterministic simulation model.

(iv) If $t = t_0$ is a constant, N_i is finite, and $\theta_{i,n}$ are random variables, then $\hat{\mu}_i(t_0)$ is a random variable.

(v) If $t = t_0$ is a constant, N_i is finite, and $\theta_{i,n}^{(k)}$ are constants, then $\tilde{\mu}_i(t_0)$ is a real-valued number.

The relationships following from the Statements (i)–(v) are illustrated in Figure 4.3.

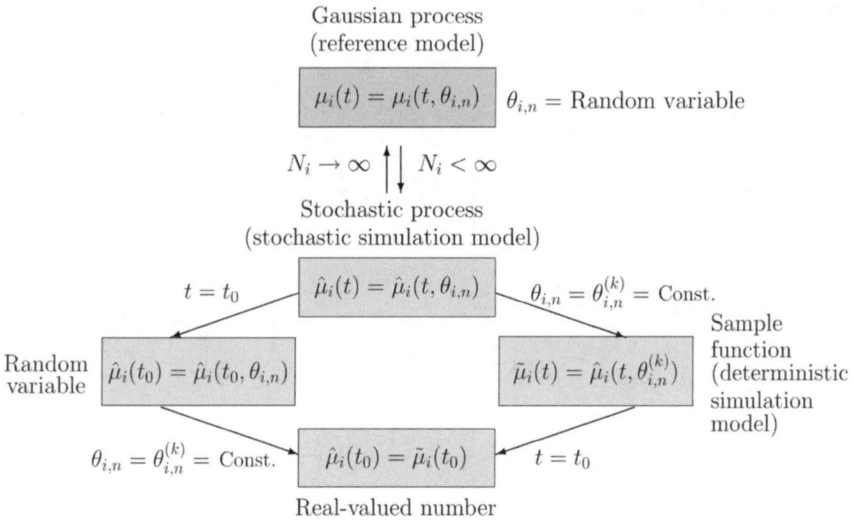

Figure 4.3 Relationships between Gaussian random processes $\mu_i(t)$ (reference model), stochastic processes $\hat{\mu}_i(t)$ (stochastic simulation model), sample functions $\tilde{\mu}_i(t)$ (deterministic simulation model), random variables, and real-valued numbers.

In Section 4.3 and Chapter 5, it will be shown that by choosing the parameters describing the deterministic process (4.4) appropriately, a very good approximation can be achieved such that the statistical properties of $\tilde{\mu}_i(t)$ are very close to those of the underlying zero-mean coloured Gaussian random process $\mu_i(t)$. For this reason, $\tilde{\mu}_i(t)$ will be called *real-valued deterministic (SOS) Gaussian process* and analogously

$$\tilde{\mu}(t) = \tilde{\mu}_1(t) + j\tilde{\mu}_2(t) \tag{4.5}$$

will be referred to as the *complex-valued deterministic (SOS) Gaussian process*. With reference to (3.20), a so-called *deterministic (SOS) Rayleigh process*

$$\tilde{\zeta}(t) = |\tilde{\mu}(t)| = |\tilde{\mu}_1(t) + j\tilde{\mu}_2(t)| \tag{4.6}$$

follows by taking the absolute value of (4.5). Logically, by taking the absolute value of $\tilde{\mu}_\rho(t) = \tilde{\mu}(t) + m(t)$, a *deterministic (SOS) Rice process*

$$\tilde{\xi}(t) = |\tilde{\mu}_\rho(t)| = |\tilde{\mu}(t) + m(t)| \tag{4.7}$$

can be introduced, where $m(t)$ again describes the line-of-sight component of the received signal, as defined by (3.17).

The resulting structure of the simulation model for deterministic Rice processes is shown in Figure 4.4. The corresponding discrete-time simulation model, which is required for computer simulations, can be obtained directly from the continuous-time simulation model by replacing the time variable t by $t = kT_s$, where T_s denotes the sampling interval and k is an integer. To carry out computer simulations, one generally proceeds by determining the parameters of the simulation model $c_{i,n}$, $f_{i,n}$, and $\theta_{i,n}$ for $n = 1, 2, \ldots, N_i$ during the simulation set-up phase. During the simulation run phase following this, all parameters are kept constant for the whole duration of the simulation.

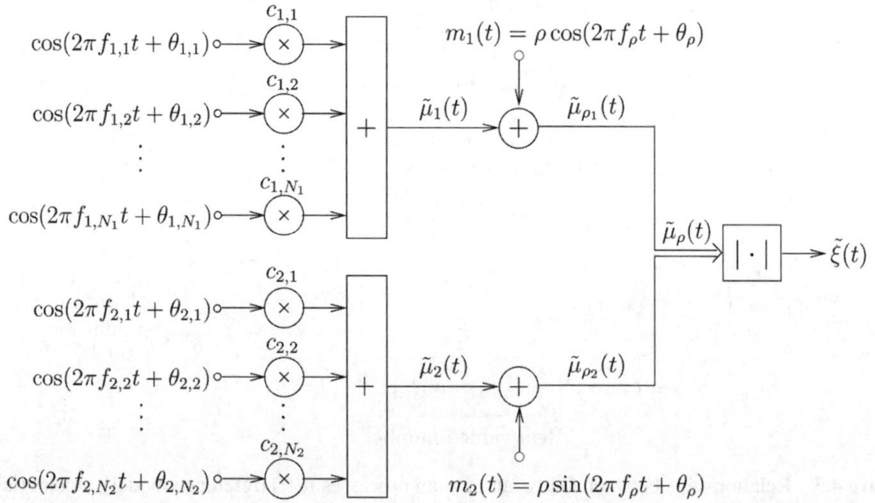

Figure 4.4 Deterministic sum-of-sinusoids simulation model for Rice fading channels.

The procedure explained above for the design of sum-of-sinusoids channel simulators is known as the *principle of deterministic channel modelling*, which was first introduced in [96]. This principle is very general, as it allows the efficient design of simulation models for a huge number of classes of mobile fading channels, ranging from frequency-nonselective and frequency-selective channel models over mobile-to-mobile and amplify-and-forward channel models to more elaborate space-time-frequency channels models for wideband mobile communication systems with multiple transmitter and receiver antennas. The principle of deterministic channel modelling can be summarized in the following five design steps:

Step 1: Starting point is a (non-realizable) reference model based on one or several Gaussian processes, each with prescribed autocorrelation function.
Step 2: Derive a stochastic simulation model from the reference model by replacing each Gaussian process by a finite sum-of-sinusoids with fixed gains, fixed frequencies, and random phases.
Step 3: Determine a deterministic simulation model by fixing all model parameters of the stochastic simulation model, including the phases.
Step 4: Compute the model parameters of the simulation model by fitting the relevant statistical properties of the deterministic (or stochastic) simulation model to those of the reference model (see Chapter 5).
Step 5: Perform the simulation of one or some few sample functions (deterministic processes or waveforms).

The five steps of the principle of deterministic channel modelling are illustrated in Figure 4.5.

For our purpose, the deterministic processes are exclusively used for the modelling of the time-variant fading behaviour caused by the Doppler effect. Therefore, the parameters $c_{i,n}$,

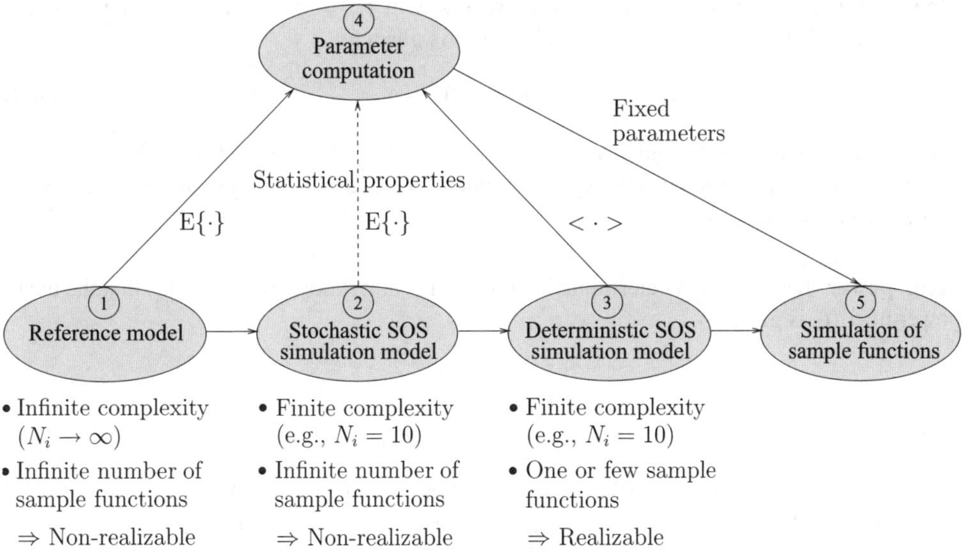

Figure 4.5 Illustration of the principle of deterministic channel modelling.

$f_{i,n}$, and $\theta_{i,n}$ describing the deterministic process in (4.4) will be called in the following the *path gains, discrete Doppler frequencies*, and *phases*, respectively. One aim of this book is to present methods by which the model parameters $(c_{i,n}, f_{i,n}, \theta_{i,n})$ can be determined such that the statistical properties of the deterministic process $\tilde{\mu}_i(t)$ or $\tilde{\mu}_i(kT)$ match those of the (ideal) stochastic process $\mu_i(t)$ as closely as possible. Of course, this aim is pursued under the boundary condition that the realization complexity should be kept as low as possible. Since the realization complexity is mainly determined by the number of sinusoids N_i, we have to solve the fitting problem under the constraint that N_i should be as small as possible. However, before tuning our attention to this topic, we will first of all investigate the fundamental properties of deterministic processes.

4.2 Elementary Properties of Deterministic Sum-of-Sinusoids Processes

The interpretation of $\tilde{\mu}_i(t)$ as a deterministic process, i.e., as a mapping of the form

$$\tilde{\mu}_i : \mathbb{R} \to \mathbb{R}, \quad t \mapsto \tilde{\mu}_i(t), \tag{4.8}$$

enables us to derive for this kind of processes simple analytical closed-form solutions for most of the fundamental characteristic quantities, such as the autocorrelation function, power spectral density, average Doppler shift, and Doppler spread.

First, a discussion of the terms introduced in Section 2.3 follows, where the definitions introduced there are now applied to the deterministic processes $\tilde{\mu}_i(t)$ $(i = 1, 2)$ defined by (4.4). If not stated otherwise, then it is assumed in the following that the model parameters $(c_{i,n}, f_{i,n}, \theta_{i,n})$ are real-valued and constant.

Mean value: Let $\tilde{\mu}_i(t)$ be a deterministic process with $f_{i,n} \neq 0\ (n = 1, 2, \ldots, N_i)$. Then, it follows from (2.141) that the mean value of $\tilde{\mu}_i(t)$ is given by

$$\tilde{m}_{\mu_i} = 0. \tag{4.9}$$

It is assumed that $f_{i,n} \neq 0$ holds for all $n = 1, 2, \ldots, N_i$ and $i = 1, 2$. Under this condition, the mean value of $\tilde{\mu}_i(t)$ equals the expected value of the underlying Gaussian random process $\mu_i(t)$, i.e., $\tilde{m}_{\mu_i} = m_{\mu_i} = 0$.

Mean power: Let $\tilde{\mu}_i(t)$ be a deterministic process. Then, it follows from (2.142) that the mean power of $\tilde{\mu}_i(t)$ is given by

$$\tilde{\sigma}^2_{\mu_i} = \sum_{n=1}^{N_i} \frac{c^2_{i,n}}{2}. \tag{4.10}$$

Obviously, the mean power $\tilde{\sigma}^2_{\mu_i}$ depends on the number of sinusoids N_i and the gains $c_{i,n}$, but not on the discrete Doppler frequencies $f_{i,n}$ and the phases $\theta_{i,n}$. By choosing, e.g., $c_{i,n} = \sigma_0\sqrt{2/N_i}$, we can guarantee that the mean power of $\tilde{\mu}_i(t)$ equals the variance of the underlying Gaussian random process $\mu_i(t)$, i.e., $\tilde{\sigma}^2_{\mu_i} = \sigma_0^2$.

Autocorrelation function: Let $\tilde{\mu}_i(t)$ be a deterministic process. Then, it follows from (2.143) that the autocorrelation function of $\tilde{\mu}_i(t)$ can be expressed in closed form as

$$\tilde{r}_{\mu_i\mu_i}(\tau) = \sum_{n=1}^{N_i} \frac{c_{i,n}^2}{2} \cos(2\pi f_{i,n}\tau). \tag{4.11}$$

One should note that $\tilde{r}_{\mu_i\mu_i}(\tau)$ depends on the number of sinusoids N_i, the gains $c_{i,n}$, and the discrete Doppler frequencies $f_{i,n}$, but not on the phases $\theta_{i,n}$. Note also that the mean power $\tilde{\sigma}_{\mu_i}^2$ is identical to the autocorrelation function $\tilde{r}_{\mu_i\mu_i}(\tau)$ at $\tau = 0$, i.e., $\tilde{\sigma}_{\mu_i}^2 = \tilde{r}_{\mu_i\mu_i}(0)$.

Cross-correlation function: Let $\tilde{\mu}_1(t)$ and $\tilde{\mu}_2(t)$ be two deterministic processes. Then, it follows from (2.144) that the cross-correlation function of $\tilde{\mu}_1(t)$ and $\tilde{\mu}_2(t)$ can be written as

$$\tilde{r}_{\mu_1\mu_2}(\tau) = 0, \quad \text{if } f_{1,n} \neq \pm f_{2,m}, \tag{4.12}$$

holds for all $n = 1, 2, \ldots, N_1$ and $m = 1, 2, \ldots, N_2$. This result shows that the deterministic processes $\tilde{\mu}_1(t)$ and $\tilde{\mu}_2(t)$ are uncorrelated if the absolute values of the respective discrete Doppler frequencies are different from each other. We can also say that $\tilde{r}_{\mu_1\mu_2}(\tau) = 0$ holds if the sets $\{f_{1,n}\}$ and $\{\pm f_{2,m}\}$ are disjoint, i.e., $\{f_{1,n}\} \cap \{\pm f_{2,m}\} = \emptyset$. However, if $f_{1,n} = \pm f_{2,m}$ holds for some or all pairs of (n, m), then $\tilde{\mu}_1(t)$ and $\tilde{\mu}_2(t)$ are correlated. Under this condition, we obtain the following expression for the cross-correlation function

$$\tilde{r}_{\mu_1\mu_2}(\tau) = \sum_{\substack{n=1 \\ f_{1,n}=\pm f_{2,m}}}^{N_{\max}} \frac{c_{1,n}c_{2,m}}{2} \cos(2\pi f_{1,n}\tau - \theta_{1,n} \pm \theta_{2,m}), \tag{4.13}$$

where $N_{\max}$ denotes the largest number of N_1 and N_2, i.e., $N_{\max} = \max\{N_1, N_2\}$. One should note that $\tilde{r}_{\mu_1\mu_2}(\tau)$ depends in this case on all model parameters, including the phases $\theta_{i,n}$. The cross-correlation function $\tilde{r}_{\mu_2\mu_1}(\tau)$ of $\tilde{\mu}_2(t)$ and $\tilde{\mu}_1(t)$ can easily be obtained from the relation $\tilde{r}_{\mu_2\mu_1}(\tau) = \tilde{r}^*_{\mu_1\mu_2}(-\tau) = \tilde{r}_{\mu_1\mu_2}(-\tau)$.

Autocorrelation function of the envelope: Let $\tilde{\mu}(t) = \tilde{\mu}_1(t) + j\tilde{\mu}_2(t)$ be a complex deterministic Gaussian process with uncorrelated real and imaginary parts or equivalently disjoint sets $\{f_{1,n}\}$ and $\{\pm f_{2,n}\}$, i.e., $\{f_{1,n}\} \cap \{\pm f_{2,n}\} = \emptyset$. If the distribution of $\tilde{\mu}_i(t)$ $(i = 1, 2)$ is close to the Gaussian distribution with zero mean and variance $\tilde{\sigma}_0^2$ and if $\tilde{r}_{\mu_1\mu_1}(\tau) \approx \tilde{r}_{\mu_2\mu_2}(\tau)$, then the autocorrelation function $\tilde{r}_{\zeta\zeta}(\tau)$ of the envelope $\tilde{\zeta}(t) = |\tilde{\mu}(t)|$, defined by $\tilde{r}_{\zeta\zeta}(\tau) := \langle \tilde{\zeta}(t)\,\tilde{\zeta}(t+\tau) \rangle$, can be approximated by

$$\tilde{r}_{\zeta\zeta}(\tau) \approx \tilde{\sigma}_0^2 \frac{\pi}{2} F\left(-\frac{1}{2}, -\frac{1}{2}; 1; \frac{\tilde{r}_{\mu_i\mu_i}^2(\tau)}{\tilde{\sigma}_0^4}\right) \tag{4.14}$$

$$\approx \tilde{\sigma}_0^2 \frac{\pi}{2} \left(1 + \frac{\tilde{r}_{\mu_i\mu_i}^2(\tau)}{4\tilde{\sigma}_0^4}\right), \tag{4.15}$$

where $F(\cdot, \cdot; \cdot; \cdot)$ denotes the hypergeometric function and $\tilde{\sigma}_0^2 = \tilde{r}_{\mu_i\mu_i}(0)$ $(i = 1, 2)$. The approximate results above follow readily from the results obtained for Rayleigh processes [see (3.32) and (3.34)] after replacing $r_{\mu_i\mu_i}(\tau)$ by $\tilde{r}_{\mu_i\mu_i}(\tau)$.

Autocorrelation function of the squared envelope: Let $\tilde{\mu}(t) = \tilde{\mu}_1(t) + j\tilde{\mu}_2(t)$ be a complex deterministic Gaussian process. Then, it can be shown (see Appendix 4.A) that the autocorrelation function $\tilde{r}_{\zeta^2\zeta^2}(\tau)$ of the squared envelope $\tilde{\zeta}^2(t) = |\mu(t)|^2$, defined by $\tilde{r}_{\zeta^2\zeta^2}(\tau) := < \tilde{\zeta}^2(t)\,\tilde{\zeta}^2(t+\tau) >$, can be expressed in terms of the autocorrelation functions of $\tilde{\mu}_1(t)$ and $\tilde{\mu}_2(t)$ as

$$\tilde{r}_{\zeta^2\zeta^2}(\tau) = \left(\tilde{r}_{\mu_1\mu_1}(0) + \tilde{r}_{\mu_2\mu_2}(0)\right)^2 + 2\left(\tilde{r}^2_{\mu_1\mu_1}(\tau) + \tilde{r}^2_{\mu_2\mu_2}(\tau)\right). \tag{4.16}$$

With reference to Appendix 4.A, it should be mentioned that the result above only holds if conditions (4.A.1) and (4.A.2) are fulfilled. Let $c_{i,n} = \sigma_0\sqrt{2/N_i}$ for all $n = 1, 2, \ldots, N_i$ $(i = 1, 2)$, then $\tilde{r}_{\mu_1\mu_1}(0) = \tilde{r}_{\mu_2\mu_2}(0) = \sigma_0^2$ holds, and (4.16) can be written as

$$\tilde{r}_{\zeta^2\zeta^2}(\tau) = 4\sigma_0^4 + \tilde{r}^2_{\mu\mu}(\tau) + \left(\tilde{r}_{\mu_1\mu_1}(\tau) - \tilde{r}_{\mu_2\mu_2}(\tau)\right)^2, \tag{4.17}$$

where $\tilde{r}_{\mu\mu}(\tau) = \tilde{r}_{\mu_1\mu_1}(\tau) + \tilde{r}_{\mu_2\mu_2}(\tau)$. As a consequence of the condition in (4.A.1), we have taken into account here that the deterministic processes $\tilde{\mu}_1(t)$ and $\tilde{\mu}_2(t)$ are uncorrelated, i.e., $\tilde{r}_{\mu_1\mu_2}(\tau) = 0$.

Power spectral density: Let $\tilde{\mu}_i(t)$ be a deterministic process. Then, it follows by substituting (4.11) in (2.145) that the power spectral density of $\tilde{\mu}_i(t)$ can be expressed as

$$\tilde{S}_{\mu_i\mu_i}(f) = \sum_{n=1}^{N_i} \frac{c_{i,n}^2}{4}\left[\delta(f - f_{i,n}) + \delta(f + f_{i,n})\right]. \tag{4.18}$$

Hence, the power spectral density of $\tilde{\mu}_i(t)$ is a symmetrical line spectrum, i.e., $\tilde{S}_{\mu_i\mu_i}(f) = \tilde{S}_{\mu_i\mu_i}(-f)$. The spectral lines are located at the discrete frequencies $f = \pm f_{i,n}$ and are weighted by the factor $c_{i,n}^2/4$.

Cross-power spectral density: Let $\tilde{\mu}_1(t)$ and $\tilde{\mu}_2(t)$ be two deterministic processes. Then, it follows from (2.147) by using (4.12) and (4.13) that the cross-power spectral density of $\tilde{\mu}_1(t)$ and $\tilde{\mu}_2(t)$ is given by

$$\tilde{S}_{\mu_1\mu_2}(f) = 0, \quad \text{if } f_{1,n} \neq \pm f_{2,m}, \tag{4.19}$$

and

$$\tilde{S}_{\mu_1\mu_2}(f) = \sum_{\substack{n=1 \\ f_{1,n}=\pm f_{2,m}}}^{N} \frac{c_{1,n}c_{2,m}}{4}\left[\delta(f - f_{1,n})\,e^{-j(\theta_{1,n}\mp\theta_{2,m})} + \delta(f + f_{1,n})\,e^{j(\theta_{1,n}\mp\theta_{2,m})}\right] \tag{4.20}$$

for all $n = 1, 2, \ldots, N_1$ and $m = 1, 2, \ldots, N_2$, where $N = \max\{N_1, N_2\}$. The cross-power spectral density $\tilde{S}_{\mu_2\mu_1}(f)$ of $\tilde{\mu}_2(t)$ and $\tilde{\mu}_1(t)$ can directly be obtained from the relation $\tilde{S}_{\mu_2\mu_1}(f) = \tilde{S}^*_{\mu_1\mu_2}(f)$.

Power spectral density of the envelope: Let $\tilde{\mu}(t) = \tilde{\mu}_1(t) + j\tilde{\mu}_2(t)$ be a complex deterministic Gaussian process with uncorrelated inphase and quadrature components and approximately identical autocorrelation functions, i.e., $\tilde{r}_{\mu_1\mu_2}(\tau) = 0$ and $\tilde{r}_{\mu_1\mu_1}(\tau) \approx \tilde{r}_{\mu_2\mu_2}(\tau)$. Then, it follows by substituting (4.15) in (2.145) that the power spectral density of the envelope $\tilde{\zeta}(t) = |\mu(t)|$ can be approximated by

$$\tilde{S}_{\zeta\zeta}(f) \approx \tilde{\sigma}_0^2 \frac{\pi}{2} \delta(f) + \frac{\pi}{8\tilde{\sigma}_0^2} \tilde{S}_{\mu_i\mu_i}(f) * \tilde{S}_{\mu_i\mu_i}(f), \tag{4.21}$$

where the convolution term is given by

$$\tilde{S}_{\mu_i\mu_i}(f) * \tilde{S}_{\mu_i\mu_i}(f) = \sum_{n=1}^{N_1} \sum_{m=1}^{N_2} \frac{c_{i,n}^2 c_{i,m}^2}{16} \big[\delta(f - f_{i,n} - f_{i,m}) + \delta(f - f_{i,n} + f_{i,m}) + \delta(f + f_{i,n} - f_{i,m}) + \delta(f + f_{i,n} + f_{i,m})\big]. \tag{4.22}$$

Hence, the power spectral density of the envelope $\tilde{\zeta}(t)$ is a symmetrical line spectrum, i.e., $\tilde{S}_{\zeta\zeta}(f) = \tilde{S}_{\zeta\zeta}(-f)$. The spectral lines are located at the discrete frequencies $f = 0$ and $f = \pm f_{i,n} \pm f_{i,m}$ and are weighted by the factors $\tilde{\sigma}_0^2 \pi/2$ and $\pi c_{i,n}^2 c_{i,m}^2/(128\tilde{\sigma}_0^2)$, respectively. The first term of $\tilde{S}_{\zeta\zeta}(f)$ on the right-hand side of (4.21) is due to the direct current (DC) component of $\tilde{\zeta}(t)$.

Power spectral density of the squared envelope: Let $\tilde{\mu}(t) = \tilde{\mu}_1(t) + j\tilde{\mu}_2(t)$ be a complex deterministic Gaussian process, which satisfies the conditions in (4.A.1) and (4.A.2). Then, it follows by substituting (4.16) in (2.145) that the power spectral density of the squared envelope $\tilde{\zeta}^2(t) = |\mu(t)|^2$ can be expressed as

$$\tilde{S}_{\zeta^2\zeta^2}(f) = \tilde{\sigma}_\mu^4 \cdot \delta(f) + 2 \sum_{i=1}^{2} \tilde{S}_{\mu_i\mu_i}(f) * \tilde{S}_{\mu_i\mu_i}(f), \tag{4.23}$$

where $\tilde{\sigma}_\mu^2 = \tilde{\sigma}_{\mu_1}^2 + \tilde{\sigma}_{\mu_2}^2$ and $\tilde{S}_{\mu_i\mu_i}(f) * \tilde{S}_{\mu_i\mu_i}(f)$ is the line spectrum given by (4.22). A comparison of (4.23) and (4.21) shows that the exact solution of the squared envelope spectrum $\tilde{S}_{\zeta^2\zeta^2}(f)$ differs from the approximate solution of the envelope spectrum $\tilde{S}_{\zeta\zeta}(f)$ basically only by the weighting factors of the spectral lines.

Average Doppler shift: Let $\tilde{\mu}_i(t)$ be a deterministic process described by the power spectral density $\tilde{S}_{\mu_i\mu_i}(f)$. Then, by analogy to (3.28a) and (3.29a), the average Doppler shift $\tilde{B}_{\mu_i\mu_i}^{(1)}$ of $\tilde{\mu}_i(t)$ can be defined by

$$\tilde{B}_{\mu_i\mu_i}^{(1)} := \frac{\int_{-\infty}^{\infty} f\,\tilde{S}_{\mu_i\mu_i}(f)\,df}{\int_{-\infty}^{\infty} \tilde{S}_{\mu_i\mu_i}(f)\,df} = \frac{1}{2\pi j} \cdot \frac{\dot{\tilde{r}}_{\mu_i\mu_i}(0)}{\tilde{r}_{\mu_i\mu_i}(0)}. \tag{4.24}$$

Owing to the symmetry property $\tilde{S}_{\mu_i\mu_i}(f) = \tilde{S}_{\mu_i\mu_i}(-f)$, it follows that

$$\tilde{B}_{\mu_i\mu_i}^{(1)} = 0. \tag{4.25}$$

On condition that the real part $\tilde{\mu}_1(t)$ and the imaginary part $\tilde{\mu}_2(t)$ are uncorrelated, one obtains analogously for complex deterministic processes $\tilde{\mu}(t) = \tilde{\mu}_1(t) + j\tilde{\mu}_2(t)$, the following relation between the average Doppler shifts of $\tilde{\mu}(t)$ and $\tilde{\mu}_i(t)$

$$\tilde{B}^{(1)}_{\mu\mu} = \tilde{B}^{(1)}_{\mu_i\mu_i} = 0, \quad i = 1, 2. \tag{4.26}$$

By comparing the relations (4.26) and (3.30a), it turns out that the average Doppler shift of the simulation model is identical to that of the reference model.

Doppler spread: Let $\tilde{\mu}_i(t)$ be a deterministic process with the power spectral density $\tilde{S}_{\mu_i\mu_i}(f)$. Then, by analogy to (3.28b) and (3.29b), the Doppler spread $\tilde{B}^{(2)}_{\mu_i\mu_i}$ of $\tilde{\mu}_i(t)$ can be defined by

$$\begin{aligned}\tilde{B}^{(2)}_{\mu_i\mu_i} &:= \sqrt{\frac{\int_{-\infty}^{\infty} (f - \tilde{B}^{(1)}_{\mu_i\mu_i})^2 \tilde{S}_{\mu_i\mu_i}(f)\, df}{\int_{-\infty}^{\infty} \tilde{S}_{\mu_i\mu_i}(f)\, df}} \\ &= \frac{1}{2\pi}\sqrt{\left(\frac{\dot{\tilde{r}}_{\mu_i\mu_i}(0)}{\tilde{r}_{\mu_i\mu_i}(0)}\right)^2 - \frac{\ddot{\tilde{r}}_{\mu_i\mu_i}(0)}{\tilde{r}_{\mu_i\mu_i}(0)}}. \end{aligned} \tag{4.27}$$

Using (4.10) and (4.11), the last equation can be written as

$$\tilde{B}^{(2)}_{\mu_i\mu_i} = \frac{\sqrt{\tilde{\beta}_i}}{2\pi\tilde{\sigma}_{\mu_i}}, \tag{4.28}$$

where

$$\tilde{\beta}_i = -\ddot{\tilde{r}}_{\mu_i\mu_i}(0) = 2\pi^2 \sum_{n=1}^{N_i} (c_{i,n} f_{i,n})^2. \tag{4.29}$$

The comparison of the Equation (3.30b) with (4.28) shows that the Doppler spreads $B^{(2)}_{\mu_i\mu_i}$ and $\tilde{B}^{(2)}_{\mu_i\mu_i}$ are identical, if the gains $c_{i,n}$ and the discrete Doppler frequencies $f_{i,n}$ are determined such that $\tilde{\sigma}^2_{\mu_i} = \sigma_0^2$ and $\tilde{\beta}_i = \beta$ hold. (In particular, it is sufficient that the condition $\tilde{\beta}_i/\tilde{\sigma}_0^2 = \beta/\sigma_0^2$ is fulfilled.)

In an analogous manner, the Doppler spread $\tilde{B}^{(2)}_{\mu\mu}$ corresponding to the power spectral density $\tilde{S}_{\mu\mu}(f)$ of the complex deterministic process $\tilde{\mu}(t) = \tilde{\mu}_1(t) + j\tilde{\mu}_2(t)$ can be determined. On condition that $\tilde{\mu}_1(t)$ and $\tilde{\mu}_2(t)$ are uncorrelated, the Doppler spread $\tilde{B}^{(2)}_{\mu\mu}$ can be expressed as

$$\tilde{B}^{(2)}_{\mu\mu} = \frac{\sqrt{\tilde{\beta}}}{2\pi\tilde{\sigma}_{\mu}}, \tag{4.30}$$

where $\tilde{\sigma}^2_\mu = \tilde{\sigma}^2_{\mu_1} + \tilde{\sigma}^2_{\mu_2} > 0$ and $\tilde{\beta} = \tilde{\beta}_1 + \tilde{\beta}_2 > 0$ hold. In Chapter 5, we will get acquainted with methods for the design of deterministic processes $\tilde{\mu}_1(t)$ and $\tilde{\mu}_2(t)$ having the properties

$\tilde{\sigma}_{\mu_1}^2 = \tilde{\sigma}_{\mu_2}^2$ and $\tilde{\beta}_1 \neq \tilde{\beta}_2$. Especially for such cases, the Doppler spread $\tilde{B}_{\mu\mu}^{(2)}$ can be calculated from the quadratic mean of $\tilde{B}_{\mu_1\mu_1}^{(2)}$ and $\tilde{B}_{\mu_2\mu_2}^{(2)}$, i.e.,

$$\tilde{B}_{\mu\mu}^{(2)} = \sqrt{\frac{\left(\tilde{B}_{\mu_1\mu_1}^{(2)}\right)^2 + \left(\tilde{B}_{\mu_2\mu_2}^{(2)}\right)^2}{2}}. \tag{4.31}$$

Finally, it should be mentioned that if $\tilde{\sigma}_0^2 = \tilde{\sigma}_{\mu_1}^2 = \tilde{\sigma}_{\mu_2}^2$ and $\tilde{\beta} = \tilde{\beta}_1 = \tilde{\beta}_2$ hold, then we obtain the result

$$\tilde{B}_{\mu\mu}^{(2)} = \tilde{B}_{\mu_i\mu_i}^{(2)} = \frac{\sqrt{\tilde{\beta}}}{2\pi\tilde{\sigma}_0}, \quad i = 1, 2, \tag{4.32}$$

which is closely related to (3.30b). However, if the deviations between $\tilde{\beta}_1$ and $\tilde{\beta}_2$ are small, which is often the case, then the above expression might be considered as a very good approximation for $\tilde{B}_{\mu\mu}^{(2)}$ if $\tilde{\beta}$ is replaced there by $\tilde{\beta} = \tilde{\beta}_1 \approx \tilde{\beta}_2$.

Periodicity: Let $\tilde{\mu}_i(t)$ be a deterministic process with arbitrary but non-zero parameters $c_{i,n}$, $f_{i,n}$ (and $\theta_{i,n}$). If the greatest common divisor[2] of the discrete Doppler frequencies

$$F_i = \gcd\{f_{i,1}, f_{i,2}, \ldots, f_{i,N_i}\} \neq 0 \tag{4.33}$$

exists, then $\tilde{\mu}_i(t)$ is periodic with period $T_i = 1/F_i$, i.e., it holds $\tilde{\mu}_i(t + T_i) = \tilde{\mu}_i(t)$ and $\tilde{r}_{\mu_i\mu_i}(\tau + T_i) = \tilde{r}_{\mu_i\mu_i}(\tau)$.

The proof of this theorem is relatively simple and will therefore be presented here only briefly. Since F_i is the greatest common divisor of $f_{i,1}, f_{i,2}, \ldots, f_{i,N_i}$, there are integers $q_{i,n} \in \mathbb{Z}$, so that $f_{i,n} = q_{i,n} \cdot F_i$ holds for all $n = 1, 2, \ldots, N_i$ and $i = 1, 2$. By putting $f_{i,n} = q_{i,n} \cdot F_i = q_{i,n}/T_i$ into (4.4) and (4.11), the validity of $\tilde{\mu}_i(t + T_i) = \tilde{\mu}_i(t)$ and $\tilde{r}_{\mu_i\mu_i}(\tau + T_i) = \tilde{r}_{\mu_i\mu_i}(\tau)$ can be proved directly.

4.3 Statistical Properties of Deterministic Sum-of-Sinusoids Processes

Even though a discussion of the elementary properties of deterministic processes could be performed in the previous section without any problems, an analysis of their statistical properties at first seems to be absurd, since statistical methods can only be applied meaningfully to random variables and stochastic processes. From the statistical point of view, however, it is obvious that the application of statistical concepts to deterministic processes seems to make no sense. In order to still gain access to the most important statistical quantities like the probability density function, the level-crossing rate, and the average duration of fades, we will study the behaviour of deterministic processes $\tilde{\mu}_i(t)$ at random time instants t. If nothing else is explicitly mentioned, we will assume throughout this section that the time variable t is a random variable uniformly distributed over the interval $\mathbb{R}$. It should also be noted that both the time variable t and the phases $\theta_{i,n}$ are in the argument of the cosine functions of (4.4).

[2] Here, the greatest common divisor of $f_{i,1}, f_{i,2}, \ldots, f_{i,N_i}$, denoted by $\gcd\{f_{i,1}, f_{i,2}, \ldots, f_{i,N_i}\}$, is defined as follows. Let $f_{i,n} = q_{i,n} \cdot F_i$, where $q_{i,n}$ are integers for $n = 1, 2, \ldots, N_i$ $(i = 1, 2)$ and F_i signifies a real-valued constant, then $\gcd\{f_{i,1}, f_{i,2}, \ldots, f_{i,N_i}\} = F_i$.

Therefore, we could assume alternatively that the time $t = t_0$ is a constant and the phases $\theta_{i,n}$ are uniformly distributed random variables. As the following analysis will show, we obtain in both cases approximately the same results.

4.3.1 Probability Density Function of the Envelope and the Phase

In this subsection, we will analyze the probability density function of the envelope and phase of complex deterministic processes $\tilde{\mu}(t) = \tilde{\mu}_1(t) + j\tilde{\mu}_2(t)$. It will be shown that these probability density functions are completely determined by the number of sinusoids N_i and the choice of the gains $c_{i,n}$.

The starting point is a single sinusoidal function of the form

$$\tilde{\mu}_{i,n}(t) = c_{i,n}\cos(2\pi f_{i,n}t + \theta_{i,n}), \tag{4.34}$$

where $c_{i,n}$, $f_{i,n}$, and $\theta_{i,n}$ are arbitrary but constant parameters different from zero and t is a uniformly distributed random variable. Due to the periodicity of $\tilde{\mu}_{i,n}(t)$, it is sufficient to restrict t on the open interval $(0, f_{i,n}^{-1})$ with $f_{i,n} \neq 0$. Since t was assumed to be a uniformly distributed random variable, $\tilde{\mu}_{i,n}(t)$ is no longer a deterministic process but a random variable as well, whose probability density function is given by [41, p. 135]

$$\tilde{p}_{\mu_{i,n}}(x) = \begin{cases} \dfrac{1}{\pi|c_{i,n}|\sqrt{1-(x/c_{i,n})^2}}, & |x| < c_{i,n}, \\ 0, & |x| \geq c_{i,n}. \end{cases} \tag{4.35}$$

The expected value and the variance of $\tilde{\mu}_{i,n}(t)$ are equal to 0 and $c_{i,n}^2/2$, respectively. Assuming that the random variables $\tilde{\mu}_{i,n}(t)$ are statistically independent, then the probability density function $\tilde{p}_{\mu_i}(x)$ of the sum

$$\tilde{\mu}_i(t) = \tilde{\mu}_{i,1}(t) + \tilde{\mu}_{i,2}(t) + \cdots + \tilde{\mu}_{i,N_i}(t) \tag{4.36}$$

can be obtained from the convolution of the individual probability density functions $\tilde{p}_{\mu_{i,n}}(x)$, i.e.,

$$\tilde{p}_{\mu_i}(x) = \tilde{p}_{\mu_{i,1}}(x) * \tilde{p}_{\mu_{i,2}}(x) * \cdots * \tilde{p}_{\mu_{i,N_i}}(x). \tag{4.37}$$

The expected value $\tilde{m}_{\mu_i}$ and the variance $\tilde{\sigma}_{\mu_i}^2$ of $\tilde{\mu}_i(t)$ are then given by

$$\tilde{m}_{\mu_i} = 0 \tag{4.38a}$$

and

$$\tilde{\sigma}_{\mu_i}^2 = \sum_{n=1}^{N_i} \frac{c_{i,n}^2}{2}, \tag{4.38b}$$

respectively. In principle, a rule for the computation of $\tilde{p}_{\mu_i}(x)$ is given by (4.37). But with regard to the following procedure, it is more advantageous to apply the concept of the characteristic function [see (2.16b)]. After substituting (4.35) into (2.16b), we find the following expression for the characteristic function $\Psi_{\mu_{i,n}}(\nu)$ of the random variables $\tilde{\mu}_{i,n}(t)$

$$\Psi_{\mu_{i,n}}(\nu) = J_0(2\pi c_{i,n}\nu). \tag{4.39}$$

The N_i-fold convolution of the probability density functions $\tilde{p}_{\mu_{i,n}}(x)$ in (4.37) can now be formulated as an N_i-fold product of the corresponding characteristic functions $\tilde{\Psi}_{\mu_{i,n}}(\nu)$, i.e.,

$$\begin{aligned}\tilde{\Psi}_{\mu_i}(\nu) &= \tilde{\Psi}_{\mu_{i,1}}(\nu)\cdot\tilde{\Psi}_{\mu_{i,2}}(\nu)\cdots\tilde{\Psi}_{\mu_{i,N_i}}(\nu)\\ &= \prod_{n=1}^{N_i} J_0(2\pi c_{i,n}\nu), \quad i=1,2.\end{aligned} \tag{4.40}$$

Concerning (4.37), an alternative expression for the probability density function $\tilde{p}_{\mu_i}(x)$ is then given by the inverse Fourier transform of $\tilde{\Psi}_{\mu_i}(-\nu) = \tilde{\Psi}_{\mu_i}(\nu)$ [132]

$$\begin{aligned}\tilde{p}_{\mu_i}(x) &= \int_{-\infty}^{\infty} \tilde{\Psi}_{\mu_i}(\nu)\, e^{j2\pi\nu x}\, d\nu\\ &= 2\int_0^{\infty}\left[\prod_{n=1}^{N_i} J_0(2\pi c_{i,n}\nu)\right]\cos(2\pi\nu x)\, d\nu, \quad i=1,2.\end{aligned} \tag{4.41}$$

It is important to realize that the probability density function $\tilde{p}_{\mu_i}(x)$ of $\tilde{\mu}_i(t)$ is completely determined by the number of sinusoids N_i and the gains $c_{i,n}$, whereas the discrete Doppler frequencies $f_{i,n}$ and the phases $\theta_{i,n}$ have no influence on $\tilde{p}_{\mu_i}(x)$.

In the following, let $c_{i,n} = \sigma_0\sqrt{2/N_i}$ and $f_{i,n} \neq 0$ for all $n = 1, 2, \ldots, N_i$ and $i = 1, 2$. Then, due to (4.38a) and (4.38b), the sum $\tilde{\mu}_i(t)$ introduced by (4.36) is a random variable with the expected value 0 and the variance

$$\tilde{\sigma}_0^2 = \tilde{\sigma}_{\mu_1}^2 = \tilde{\sigma}_{\mu_2}^2 = \sigma_0^2. \tag{4.42}$$

By invoking the central limit theorem [see (2.26)], it turns out that in the limit $N_i \to \infty$, the sum $\tilde{\mu}_i(t)$ of the random variables in (4.36) tends to a normally distributed random variable having the expected value 0 and the variance σ_0^2, i.e.,

$$\lim_{N_i\to\infty} \tilde{p}_{\mu_i}(x) = p_{\mu_i}(x) = \frac{1}{\sqrt{2\pi}\sigma_0} e^{-\frac{x^2}{2\sigma_0^2}}. \tag{4.43}$$

Hence, after computing the Fourier transform of this equation, one obtains the following relation for the corresponding characteristic functions

$$\lim_{N_i\to\infty} \tilde{\Psi}_{\mu_i}(\nu) = \Psi_{\mu_i}(\nu) = e^{-2(\pi\sigma_0\nu)^2}, \tag{4.44}$$

from which — by using (4.40) — the remarkable property

$$\lim_{N_i\to\infty}\left[J_0\left(2\pi\sigma_0\sqrt{\frac{2}{N_i}}\nu\right)\right]^{N_i} = e^{-2(\pi\sigma_0\nu)^2} \tag{4.45}$$

finally follows.

Of course, for a finite number of sinusoids N_i, we have to write: $\tilde{p}_{\mu_i}(x) \approx p_{\mu_i}(x)$ and $\tilde{\Psi}_{\mu_i}(\nu) \approx \Psi_{\mu_i}(\nu)$. From Figure 4.6(a), illustrating $\tilde{p}_{\mu_i}(x)$ according to (4.41) with $c_{i,n} = \sigma_0\sqrt{2/N_i}$ for $N_i \in \{3, 5, 7, \infty\}$, it follows that in fact for $N_i \geq 7$, the approximation $\tilde{p}_{\mu_i}(x) \approx$

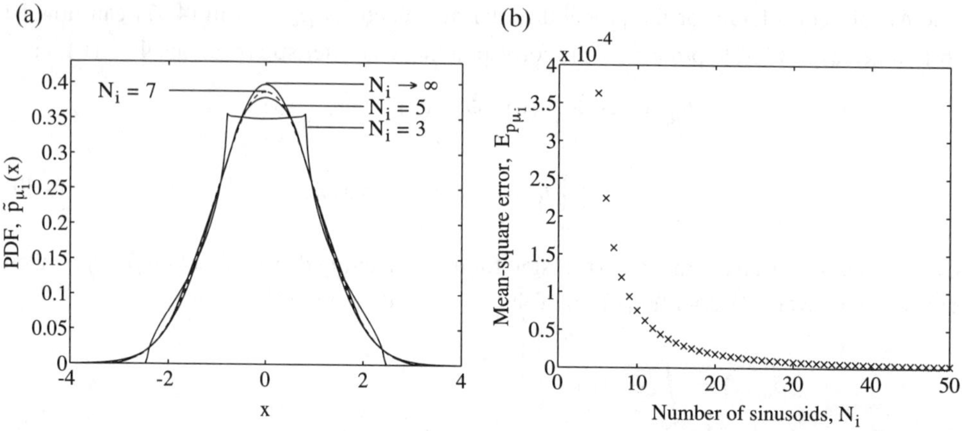

Figure 4.6 (a) Probability density function $\tilde{p}_{\mu_i}(x)$ for $N_i \in \{3, 5, 7, \infty\}$, (b) mean-square error $E_{p_{\mu_i}}$ in terms of the number of sinusoids N_i. (Theoretical results obtained by using $c_{i,n} = \sigma_0\sqrt{2/N_i}$, $\sigma_0^2 = 1$.)

$p_{\mu_i}(x)$ is astonishingly good. An appropriate measure of the approximation error is the mean-square error of the probability density function $\tilde{p}_{\mu_i}(x)$ defined by

$$E_{p_{\mu_i}} := \int_{-\infty}^{\infty} \left(p_{\mu_i}(x) - \tilde{p}_{\mu_i}(x)\right)^2 dx. \tag{4.46}$$

The behaviour of the mean-square error $E_{p_{\mu_i}}$ in terms of the number of sinusoids N_i is shown in Figure 4.6(b). This figure gives us an impression of how fast $\tilde{p}_{\mu_i}(x)$ converges to $p_{\mu_i}(x)$ if N_i increases.

Owing to the good convergence behaviour that $\tilde{p}_{\mu_i}(x)$ exhibits in conjunction with $c_{i,n} = \sigma_0\sqrt{2/N_i}$, we will occasionally assume (without causing a large error) that the identity

$$\tilde{p}_{\mu_i}(x) = p_{\mu_i}(x), \quad \text{if } N_i \geq 7, \tag{4.47}$$

holds. In this case, many analytical problems, which are otherwise difficult to overcome, can then be solved in a relatively easy way.

Next, we will derive the probability density function of the absolute value and the phase of the complex-valued random variable

$$\tilde{\mu}_\rho(t) = \tilde{\mu}_{\rho_1}(t) + j\tilde{\mu}_{\rho_2}(t), \tag{4.48}$$

where $\tilde{\mu}_{\rho_i}(t) = \tilde{\mu}_i(t) + m_i(t)$ $(i = 1, 2)$. In regard to the line-of-sight component $m_i(t)$ [see (3.17)], we have to discuss the two cases $f_\rho = 0$ and $f_\rho \neq 0$ separately.

At first, we consider the case $f_\rho = 0$. By doing this, $m_i(t)$ becomes independent of the random variable t. Consequently, the line-of-sight $m_i(t)$ is a constant, i.e., $m_i(t) = m_i$, whose probability density function is described by $p_{m_i}(x) = \delta(x - m_i)$. From (2.80), it follows that the probability density function $\tilde{p}_{\mu_{\rho_i}}(x)$ of $\tilde{\mu}_{\rho_i}(t)$ can now be expressed in terms of $\tilde{p}_{\mu_i}(x)$ as

$$\begin{aligned} \tilde{p}_{\mu_{\rho_i}}(x) &= \tilde{p}_{\mu_i}(x) * p_{m_i}(x) \\ &= \tilde{p}_{\mu_i}(x - m_i). \end{aligned} \tag{4.49}$$

On the assumption that $\tilde{\mu}_{\rho_1}(t)$ and $\tilde{\mu}_{\rho_2}(t)$ are statistically independent, and thus $f_{1,n} \neq \pm f_{2,m}$ holds for all $n = 1, 2, \ldots, N_1$ and $m = 1, 2, \ldots, N_2$, the joint probability density function of the random variables $\tilde{\mu}_{\rho_1}(t)$ and $\tilde{\mu}_{\rho_2}(t)$, denoted by $\tilde{p}_{\mu_{\rho_1}\mu_{\rho_2}}(x_1, x_2)$, can be expressed by

$$\tilde{p}_{\mu_{\rho_1}\mu_{\rho_2}}(x_1, x_2) = \tilde{p}_{\mu_{\rho_1}}(x_1) \cdot \tilde{p}_{\mu_{\rho_2}}(x_2). \tag{4.50}$$

The transformation of the Cartesian coordinates (x_1, x_2) into polar coordinates (z, θ) by means of

$$x_1 = z\cos\theta, \quad x_2 = z\sin\theta \tag{4.51a,b}$$

allows us to derive the joint probability density function $\tilde{p}_{\xi\vartheta}(z, \theta)$ of the envelope $\tilde{\xi}(t) = |\tilde{\mu}_\rho(t)|$ and the phase $\tilde{\vartheta}(t) = \arg\{\tilde{\mu}_\rho(t)\}$ with the help of (2.87) as follows:

$$\begin{aligned}
\tilde{p}_{\xi\vartheta}(z, \theta) &= z\,\tilde{p}_{\mu_{\rho_1}\mu_{\rho_2}}(z\cos\theta, z\sin\theta) \\
&= z\,\tilde{p}_{\mu_{\rho_1}}(z\cos\theta) \cdot \tilde{p}_{\mu_{\rho_2}}(z\sin\theta) \\
&= z\,\tilde{p}_{\mu_1}(z\cos\theta - \rho\cos\theta_\rho) \cdot \tilde{p}_{\mu_2}(z\sin\theta - \rho\sin\theta_\rho).
\end{aligned} \tag{4.52}$$

By using the last equation in conjunction with (2.89), the probability density functions of the envelope $\tilde{\xi}(t)$ and phase $\tilde{\vartheta}(t)$ can now be brought into the forms:

$$\tilde{p}_\xi(z) = z \int_{-\pi}^{\pi} \tilde{p}_{\mu_1}(z\cos\theta - \rho\cos\theta_\rho) \cdot \tilde{p}_{\mu_2}(z\sin\theta - \rho\sin\theta_\rho)\, d\theta \tag{4.53a}$$

$$\tilde{p}_\vartheta(\theta) = \int_0^\infty z\tilde{p}_{\mu_1}(z\cos\theta - \rho\cos\theta_\rho) \cdot \tilde{p}_{\mu_2}(z\sin\theta - \rho\sin\theta_\rho)\, dz. \tag{4.53b}$$

Putting (4.41) into the last two expressions gives us the following threefold integrals for the desired probability density functions:

$$\begin{aligned}
\tilde{p}_\xi(z) = 4z \int_{-\pi}^{\pi} &\left\{ \int_0^\infty \left[\prod_{n=1}^{N_1} J_0(2\pi c_{1,n}\nu_1) \right] g_1(z, \theta, \nu_1)\, d\nu_1 \right\} \\
&\cdot \left\{ \int_0^\infty \left[\prod_{m=1}^{N_2} J_0(2\pi c_{2,m}\nu_2) \right] g_2(z, \theta, \nu_2)\, d\nu_2 \right\} d\theta,
\end{aligned} \tag{4.54a}$$

$$\begin{aligned}
\tilde{p}_\vartheta(\theta) = 4 \int_0^\infty z &\left\{ \int_0^\infty \left[\prod_{n=1}^{N_1} J_0(2\pi c_{1,n}\nu_1) \right] g_1(z, \theta, \nu_1)\, d\nu_1 \right\} \\
&\cdot \left\{ \int_0^\infty \left[\prod_{m=1}^{N_2} J_0(2\pi c_{2,m}\nu_2) \right] g_2(z, \theta, \nu_2)\, d\nu_2 \right\} dz,
\end{aligned} \tag{4.54b}$$

where

$$g_1(z, \theta, \nu_1) = \cos[2\pi\nu_1(z\cos\theta - \rho\cos\theta_\rho)], \tag{4.55a}$$

$$g_2(z, \theta, \nu_2) = \cos[2\pi\nu_2(z\sin\theta - \rho\sin\theta_\rho)]. \tag{4.55b}$$

Up to now, there are no further simplifications known for (4.54b), and hence, the remaining three integrals must be solved numerically. In contrast to that, it is possible to reduce the threefold integral on the right-hand side of (4.54a) to a double integral by making use of the expression [23, Equation (3.876.7)]

$$\int_0^1 \frac{\cos\left(2\pi\nu_2 z\sqrt{1-x^2}\right)}{\sqrt{1-x^2}}\cos(2\pi\nu_1 z x)\,dx = \frac{\pi}{2}J_0\left(2\pi z\sqrt{\nu_1^2+\nu_2^2}\right). \tag{4.56}$$

After some algebraic manipulations, we finally come to the result that the density of $\tilde{\xi}(t)$ can be expressed as

$$\tilde{p}_\xi(z) = 4\pi z\int_0^\pi\int_0^\infty\left[\prod_{n=1}^{N_1}J_0(2\pi c_{1,n}y\cos\theta)\right]\left[\prod_{m=1}^{N_2}J_0(2\pi c_{2,m}y\sin\theta)\right]$$
$$\cdot J_0(2\pi zy)\cos\left[2\pi\rho y\cos(\theta-\theta_\rho)\right]y\,dy\,d\theta. \tag{4.57}$$

The results of the numerical evaluations of $\tilde{p}_\xi(z)$ and $\tilde{p}_\vartheta(\theta)$ for the special case $c_{i,n} = \sigma_0\sqrt{2/N_i}$ ($\sigma_0^2 = 1$) are illustrated in Figures 4.7(a) and 4.7(b), respectively. These illustrations again make clear that the approximation error caused by a limited number of sinusoids N_i can in general be ignored if $N_i \geq 7$.

We also want to show that if the gains $c_{i,n}$ are given by $c_{i,n} = \sigma_0\sqrt{2/N_i}$, then from (4.54a) and (4.54b) it follows in the limit $N_i \to \infty$ the expected results: $\tilde{p}_\xi(z) \to p_\xi(z)$ and $\tilde{p}_\vartheta(\theta) \to$

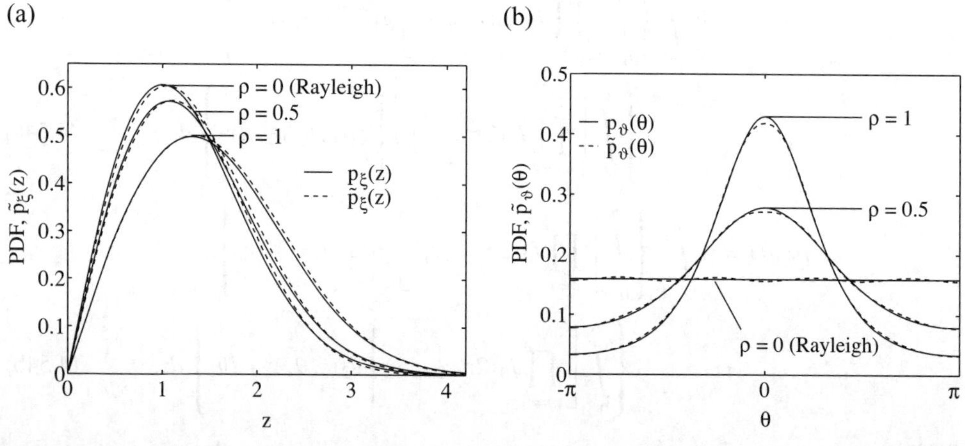

Figure 4.7 (a) Probability density function $\tilde{p}_\xi(z)$ of the envelope $\tilde{\xi}(t) = |\tilde{\mu}_\rho(t)|$ and (b) probability density function $\tilde{p}_\vartheta(\theta)$ of the phase $\tilde{\vartheta}(t) = \arg\{\tilde{\mu}_\rho(t)\}$ for $N_1 = N_2 = 7$. (Analytical results obtained by using $c_{i,n} = \sigma_0\sqrt{2/N_i}$ and $\sigma_0^2 = 1$.)

$p_\vartheta(\theta)$, respectively. To show this, we apply the property (4.45), which enables us to express (4.54a) and (4.54b) as

$$\lim_{N_i\to\infty} \tilde{p}_\xi(z) = 4z \int_{-\pi}^{\pi} \left[\int_0^\infty e^{-2(\pi\sigma_0\nu_1)^2} g_1(z,\theta,\nu_1)\, d\nu_1 \right] \cdot \left[\int_0^\infty e^{-2(\pi\sigma_0\nu_2)^2} g_2(z,\theta,\nu_2)\, d\nu_2 \right] d\theta \tag{4.58}$$

and

$$\lim_{N_i\to\infty} \tilde{p}_\vartheta(\theta) = 4 \int_0^z z \left[\int_0^\infty e^{-2(\pi\sigma_0\nu_1)^2} g_1(z,\theta,\nu_1)\, d\nu_1 \right] \cdot \left[\int_0^\infty e^{-2(\pi\sigma_0\nu_2)^2} g_2(z,\theta,\nu_2)\, d\nu_2 \right] dz, \tag{4.59}$$

respectively. The use of the integral [23, Equation (3.896.2)]

$$\int_{-\infty}^{\infty} e^{-q^2x^2} \cos(px) dx = \frac{\sqrt{\pi}}{q} e^{-\frac{p^2}{4q^2}} \tag{4.60}$$

allows us to present the expressions (4.58) and (4.59) as

$$\lim_{N_i\to\infty} \tilde{p}_\xi(z) = \frac{z}{\sigma_0^2} e^{-\frac{z^2+\rho^2}{2\sigma_0^2}} \cdot \frac{1}{\pi} \int_0^\pi e^{\frac{z\rho}{\sigma_0^2}\cos(\theta-\theta_\rho)} d\theta \tag{4.61}$$

and

$$\lim_{N_i\to\infty} \tilde{p}_\vartheta(\theta) = \frac{1}{2\pi\sigma_0^2} e^{-\frac{\rho^2}{2\sigma_0^2}} \int_0^\infty z e^{-\frac{z^2}{2\sigma_0^2} + \frac{z\rho}{\sigma_0^2}\cos(\theta-\theta_\rho)} dz, \tag{4.62}$$

respectively. With the integral representation of the modified Bessel function of the first kind of order zero [77, Equation (9.6.16)]

$$I_0(z) = \frac{1}{\pi} \int_0^\pi e^{\pm z\cos\theta}\, d\theta, \tag{4.63}$$

we can immediately identify (4.61) with the Rice distribution $p_\xi(z)$ defined by (3.52), and from (4.62), using [23, Equation (3.462.5)]

$$\int_0^\infty z e^{-qz^2-2pz} dz = \frac{1}{2q} - \frac{p}{2q}\sqrt{\frac{\pi}{q}}\, e^{\frac{p^2}{q}} \left[1 - \operatorname{erf}\left(\frac{p}{\sqrt{q}}\right) \right] \tag{4.64}$$

for $|\arg\{p\}| < \pi/2$ and $\text{Re}\{q\} > 0$, we obtain, after elementary calculations, the probability density function $p_\vartheta(\theta)$ given by (3.57).

Furthermore, we will pay attention to the general case, where $f_\rho \neq 0$. The line-of-sight component $m(t) = m_1(t) + jm_2(t)$ [see (3.17)] will now be considered as time-variant mean value, whose real and imaginary parts can consequently be described by the probability density functions

$$p_{m_1}(x_1; t) = \delta(x_1 - m_1(t)) = \delta(x_1 - \rho\cos(2\pi f_\rho t + \theta_\rho)) \tag{4.65}$$

and

$$p_{m_2}(x_2; t) = \delta(x_2 - m_2(t)) = \delta(x_2 - \rho\sin(2\pi f_\rho t + \theta_\rho)), \tag{4.66}$$

respectively. The derivation of the probability density functions $\tilde{p}_\xi(z; t)$ and $\tilde{p}_\vartheta(\theta; t)$ can be performed analogously to the case $f_\rho = 0$. For these densities, one obtains expressions, which coincide with the right-hand side of (4.54a) and (4.54b), if one replaces there the functions $g_i(z, \theta, \nu_i)$ $(i = 1, 2)$ by

$$g_1(z, \theta, \nu_1) = \cos\left\{2\pi\nu_1[z\cos\theta - \rho\cos(2\pi f_\rho t + \theta_\rho)]\right\} \tag{4.67a}$$

and

$$g_2(z, \theta, \nu_2) = \cos\left\{2\pi\nu_2[z\sin\theta - \rho\sin(2\pi f_\rho t + \theta_\rho)]\right\}. \tag{4.67b}$$

Concerning the convergence behaviour, it can be shown that for $N_i \to \infty$ with $c_{i,n} = \sigma_0\sqrt{2/N_i}$, it follows $\tilde{p}_\xi(z; t) \to p_\xi(z)$ and $\tilde{p}_\vartheta(\theta; t) \to p_\vartheta(\theta; t)$ as expected, where $p_\xi(z)$ and $p_\vartheta(\theta; t)$ are the probability density functions described by (3.52) and (3.56), respectively.

In order to complete this topic, we will verify the correctness of the derived analytical expressions obtained for the probability density functions $\tilde{p}_\xi(z)$ and $\tilde{p}_\vartheta(\theta)$ by simulations. In principle, we can proceed here by employing the simulation model shown in Figure 4.4, where we have to substitute the time variable t by a uniformly distributed random variable, which already helped us to achieve our aim in deriving the analytical expressions. In the following, however, we will keep the conventional approach, i.e., we replace the continuous-time variable t by $t = kT_s$, where T_s denotes the sampling interval and $k = 1, 2, \ldots, K$. One should note here that the value chosen for the sampling interval T_s must be sufficiently small to ensure that the statistical analysis and the evaluation of the deterministic sequences $\tilde{\xi}(kT_s) = |\tilde{\mu}_\rho(kT_s)|$ and $\tilde{\vartheta}(kT_s) = \arg\left\{\tilde{\mu}_\rho(kT_s)\right\}$ can be performed as precisely as possible. It is not sufficient in this context if the sampling frequency $f_s = 1/T_s$ is merely given by the value $f_s = 2 \cdot \max\left\{f_{i,n}\right\}_{n=1}^{N_i}$; although this value would suffice to fulfil the sampling theorem according to (2.158). For our purposes, the inequality $f_s \gg \max\left\{f_{i,n}\right\}_{n=1}^{N_i}$ should rather hold. Experimental investigations have shown that a good compromise between computational complexity and attainable precision can be achieved in most applications, if f_s is in the range from $f_s \approx 20 \cdot \max\{f_{i,n}\}_{n=1}^{N_i}$ up to $f_s \approx 100 \cdot \max\{f_{i,n}\}_{n=1}^{N_i}$, and if the number of iterations K is in order of 10^6. From the simulation of the discrete-time signals $\tilde{\xi}(kT_s)$ and $\tilde{\vartheta}(kT_s)$, the probability density functions $\tilde{p}_\xi(z)$ and $\tilde{p}_\vartheta(\theta)$ can then be determined by means of the histograms of the generated samples. Here, the choice of the discrete Doppler frequencies $f_{i,n}$ is not decisive. On these parameters, we only impose that they should all be unequal and different from zero. Moreover, due to the periodic behaviour of $\tilde{\mu}_i(t)$, the discrete Doppler frequencies $f_{i,n}$ have to be determined such that the period $T_i = 1/\gcd\{f_{i,n}\}_{n=1}^{N_i}$ is greater than or equal to the simulation time T_{sim}, i.e., $T_i \geq T_{\text{sim}} = KT_s$.

By way of example, some illustrative simulation results for the probability density functions $\tilde{p}_{\xi}(z)$ and $\tilde{p}_{\vartheta}(\theta)$ are depicted in Figures 4.8(a) and 4.8(b) for the case $c_{i,n} = \sigma_0\sqrt{2/N_i}$, where $N_i = 7$ $(i = 1, 2)$ and $\sigma_0^2 = 1$. These figures confirm that the probability density functions, which have been obtained from the simulation of deterministic processes, are identical to the analytical expressions describing the statistics of the underlying random variables. In the remainder of this book, we will therefore designate $\tilde{p}_{\mu_i}(x)$ [see (4.41)] as the probability density function of the deterministic process $\tilde{\mu}_i(t)$. Analogously, $\tilde{p}_{\xi}(z)$ [see (4.54a)] and $\tilde{p}_{\vartheta}(\theta)$ [see (4.54b)] will be referred to as the probability density functions of the envelope $\tilde{\xi}(t)$ and the phase $\tilde{\vartheta}(t)$ of the complex deterministic process $\tilde{\mu}_{\rho}(t) = \tilde{\mu}_{\rho_1}(t) + j\tilde{\mu}_{\rho_2}(t)$, respectively.

4.3.2 *Level-Crossing Rate and Average Duration of Fades*

In this subsection, we will derive general analytical expressions for the level-crossing rate $\tilde{N}_{\xi}(r)$ and the average duration of fades $\tilde{T}_{\xi_-}(r)$ of the deterministic simulation model for Rice processes shown in Figure 4.4. The knowledge of analytical solutions makes the determination of $\tilde{N}_{\xi}(r)$ and $\tilde{T}_{\xi_-}(r)$ by means of time-consuming simulations superfluous. Moreover, they enable a deeper insight into the cause and the effect of statistical degradations, which can be attributed to the finite number of used sinusoids N_i or to the method used for the computation of the model parameters.

In Subsection 4.3.1, we have seen that the probability density function $\tilde{p}_{\mu_i}(x)$ of the deterministic process $\tilde{\mu}_i(t)$ is almost identical to the probability density function $p_{\mu_i}(x)$ of the (ideal) stochastic process $\mu_i(t)$, provided that the number of sinusoids N_i is sufficiently large, meaning $N_i \geq 7$. On condition that the relations

$$\text{(i)} \quad \tilde{p}_{\mu_i}(x) = p_{\mu_i}(x), \tag{4.68a}$$

$$\text{(ii)} \quad \tilde{\beta} = \tilde{\beta}_1 = \tilde{\beta}_2 \tag{4.68b}$$

hold, the level-crossing rate $\tilde{N}_{\xi}(r)$ is still given by (3.63), if there the quantities α and β of the stochastic reference model are replaced by the corresponding quantities $\tilde{\alpha}$ and $\tilde{\beta}$ of the

(a) (b)

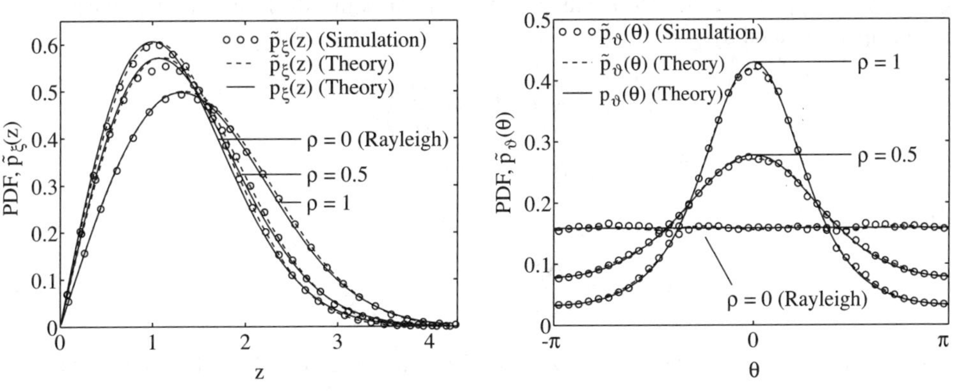

Figure 4.8 (a) Probability density function $\tilde{p}_{\xi}(z)$ of the envelope $\tilde{\xi}(t)$ and (b) probability density function $\tilde{p}_{\vartheta}(\theta)$ of the phase $\tilde{\vartheta}(t)$ for $N_1 = N_2 = 7 (c_{i,n} = \sigma_0\sqrt{2/N_i}, \sigma_0^2 = 1, f_{\rho} = 0, \theta_{\rho} = 0)$.

deterministic simulation model. Thus, for the case $f_\rho \neq 0$, one obtains the following expression for the level-crossing rate $\tilde{N}_\xi(r)$ of deterministic Rice processes $\tilde{\xi}(t)$

$$\tilde{N}_\xi(r) = \frac{r\sqrt{2\tilde{\beta}}}{\pi^{3/2}\sigma_0^2} e^{-\frac{r^2+\rho^2}{2\sigma_0^2}} \int_0^{\pi/2} \cosh\left(\frac{r\rho}{\sigma_0^2}\cos\theta\right) \cdot \left\{e^{-(\tilde{\alpha}\rho\sin\theta)^2} + \sqrt{\pi}\tilde{\alpha}\rho\sin(\theta)\operatorname{erf}(\tilde{\alpha}\rho\sin\theta)\right\} d\theta, \quad r \geq 0, \tag{4.69}$$

where

$$\tilde{\alpha} = 2\pi f_\rho \Big/ \sqrt{2\tilde{\beta}}, \tag{4.70a}$$

$$\tilde{\beta} = \tilde{\beta}_i = -\ddot{\tilde{r}}_{\mu_i\mu_i}(0) = 2\pi^2 \sum_{n=1}^{N_i} (c_{i,n} f_{i,n})^2. \tag{4.70b}$$

For the special case $f_\rho = 0$, it follows from (4.70a) that $\tilde{\alpha} = 0$ holds, so that (4.69) simplifies to the following expression

$$\tilde{N}_\xi(r) = \sqrt{\frac{\tilde{\beta}}{2\pi}} \cdot p_\xi(r), \quad r \geq 0, \tag{4.71}$$

which is identical to (3.66) after replacing β by $\tilde{\beta}$ there. Obviously, the quality of the approximation $\tilde{\beta} \approx \beta$ decisively determines the deviation of the level-crossing rate of the deterministic simulation model from that of the underlying stochastic reference model.

For further analysis, we write

$$\tilde{\beta} = \beta + \Delta\beta, \tag{4.72}$$

where $\Delta\beta$ describes the *true error of* $\tilde{\beta}$ caused by the method chosen for the computation of the model parameters $c_{i,n}$ and $f_{i,n}$. In the following, we will call $\Delta\beta$ the *model error* for short. Let us assume that the *relative model error* $\Delta\beta/\beta$ is small, then we can use the approximation $\sqrt{\beta + \Delta\beta} \approx \sqrt{\beta}(1 + \Delta\beta/(2\beta))$, which allows us to show that the level-crossing rate in (4.71) can be approximated by

$$\begin{aligned}\tilde{N}_\xi(r) &\approx N_\xi(r)\left(1 + \frac{\Delta\beta}{2\beta}\right)\\ &= N_\xi(r) + \Delta N_\xi(r), \quad r \geq 0,\end{aligned} \tag{4.73}$$

where

$$\Delta N_\xi(r) = \frac{\Delta\beta}{2\beta} N_\xi(r) = \frac{\Delta\beta}{2\sqrt{2\pi\beta}} p_\xi(r). \tag{4.74}$$

Notice that $\Delta N_\xi(r)$ describes the *true error of* $\tilde{N}_\xi(r)$. In this case, $\Delta N_\xi(r)$ behaves proportionally to $\Delta\beta$, or in other words: For any given signal level r, the ratio $\Delta N_\xi(r)/\Delta\beta$ is constant, and thus independent of the model error $\Delta\beta$ characterizing the simulation model.

For $\rho \to 0$, it follows $\tilde{\xi}(t) \to \tilde{\zeta}(t)$, and thus $\tilde{p}_\xi(r) \to \tilde{p}_\zeta(r)$. Consequently, with reference to the assumptions in (4.68a) and (4.68b), we obtain the following expression for the level-crossing rate $\tilde{N}_\zeta(r)$ of deterministic Rayleigh processes $\tilde{\zeta}(t)$

$$\tilde{N}_\zeta(r) = \sqrt{\frac{\tilde{\beta}}{2\pi}} \cdot p_\zeta(r), \quad r \geq 0. \tag{4.75}$$

Now, it is obvious that the approximation (4.73) also holds for $\tilde{N}_\zeta(r)$ in connection with (4.74) if the index ξ is replaced by ζ in both equations.

For illustration purposes, the analytical expression for $\tilde{N}_\xi(r)$ given by (4.71) is shown in Figure 4.9 assuming a relative model error $\Delta\beta/\beta$ in the range of ± 10 per cent. This figure also shows the ideal conditions, where $\tilde{\beta}$ equals β, which we have already inspected in Figure 3.8(b).

We now want to concentrate on the analysis of the average duration of fades of deterministic Rice processes, where we again hold on to the assumptions made in (4.68a) and (4.68b). In particular, it follows from (4.68a) that the cumulative distribution function $\tilde{F}_{\xi_-}(r)$ of deterministic Rice processes $\tilde{\xi}(t)$ is identical to that of stochastic Rice processes $\xi(t)$, i.e., $\tilde{F}_{\xi_-}(r) = F_{\xi_-}(r)$. Hence, by taking the definition (2.120) into consideration, it turns out that in this case the average duration of fades $\tilde{T}_{\xi_-}(r)$ of deterministic Rice processes $\tilde{\xi}(t)$ can be expressed by

$$\tilde{T}_{\xi_-}(r) = \frac{F_{\xi_-}(r)}{\tilde{N}_\xi(r)}, \quad r \geq 0, \tag{4.76}$$

where $F_{\xi_-}(r)$ can be found in (2.45), and $\tilde{N}_\xi(r)$ is given by (4.69).

For the special case $f_\rho = 0$, simple approximate solutions can again be obtained if the relative model error $\Delta\beta/\beta$ is small. Hence, after substituting (4.73) into (4.76) and using the

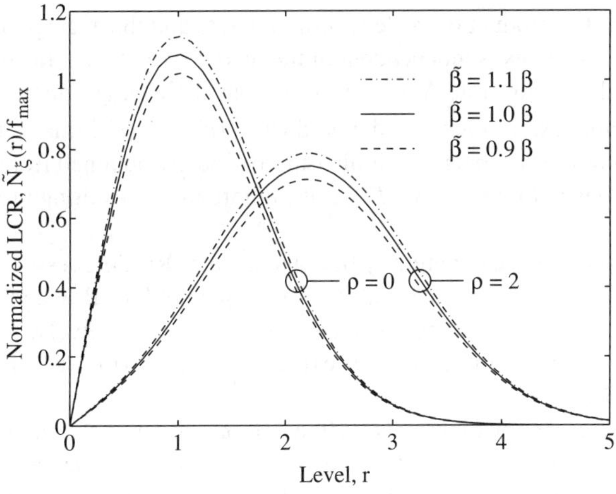

Figure 4.9 Normalized level-crossing rate $\tilde{N}_\xi(r)/f_{\max}$ of deterministic Rice processes $\tilde{\xi}(t)$ for various values of $\tilde{\beta} = \beta + \Delta\beta$ (Jakes PSD, $f_{\max} = 91$ Hz, $f_\rho = 0$).

approximation formula $1/[1+\Delta\beta/(2\beta)] \approx 1-\Delta\beta/(2\beta)$, we obtain the expression

$$\begin{aligned}\tilde{T}_{\xi_-}(r) &\approx T_{\xi_-}(r)\left(1-\frac{\Delta\beta}{2\beta}\right)\\ &= T_{\xi_-}(r)+\Delta T_{\xi_-}(r), \quad r \geq 0,\end{aligned} \tag{4.77}$$

where

$$\Delta T_{\xi_-}(r) = -\frac{\Delta\beta}{2\beta}T_{\xi_-}(r) \tag{4.78}$$

denotes the *true error of* $\tilde{T}_{\xi_-}(r)$. Most notably, the approximation (4.77) states that the average duration of fades of the deterministic Rice process decreases (increases) approximately linearly with an increasing (decreasing) model error $\Delta\beta$.

For low signal levels r and moderate Rice factors $c_R = \rho^2/(2\sigma_0^2)$, we obtain immediately the following approximation after substituting (3.78) in the first expression of (4.77)

$$\tilde{T}_{\xi_-}(r) \approx \tilde{T}_{\zeta_-}(r) \approx r\sqrt{\frac{\pi}{2\beta}}\left(1-\frac{\Delta\beta}{2\beta}\right). \tag{4.79}$$

Notice that this result is only valid for low signal levels $r \ll 1$ and if $r\rho/\sigma_0^2 \ll 1$. From the relations above, we may conclude that the mean value of the fading intervals of deterministic Rice and Rayleigh processes are approximately identical at low signal levels, provided that the signal strength of the line-of-sight component is moderate. By comparing (3.78) with (4.79) it also becomes obvious that the relative model error again determines the deviation of the simulation model's average duration of fades from the average duration of fades of the corresponding reference model.

An interesting statement can also be made on the product $\tilde{N}_\xi(r)\cdot\tilde{T}_{\xi_-}(r)$. Namely, from (2.120) and (4.76), the *model error law* of deterministic channel modelling

$$\tilde{N}_\xi(r)\cdot\tilde{T}_{\xi_-}(r) = N_\xi(r)\cdot T_{\xi_-}(r) \tag{4.80}$$

follows, stating that the product of the level-crossing rate and the average duration of fades of deterministic Rice processes is independent of the model error $\Delta\beta$. With an increasing model error $\Delta\beta$, the level-crossing rate $\tilde{N}_\xi(r)$ may rise, but the average duration of fades $\tilde{T}_{\xi_-}(r)$ decreases by the same extent such that the product $\tilde{N}_\xi(r)\cdot\tilde{T}_{\xi_-}(r)$ remains unchanged for any given signal level $r = \text{const}$. This result is also approximately obtained from the multiplication of the approximations in (4.73) and (4.77), if we ignore in the resulting product the quadratic term $[\Delta\beta/(2\beta)]^2$.

Since Rayleigh processes can naturally be considered as Rice processes for the special case $\rho = 0$, the relations (4.76)–(4.78) and (4.80) in principle hold for Rayleigh processes as well. Only the indices ξ and ξ_- have to be replaced by ζ and ζ_-, respectively.

The evaluation results of the analytical expression for $\tilde{T}_{\xi_-}(r)$ [see (4.76)] are shown in Figure 4.10 for $\Delta\beta/\beta \in \{-0.1, 0, +0.1\}$.

In Chapter 5, where various methods for the determination of the model parameters ($c_{i,n}$, $f_{i,n}$, $\theta_{i,n}$) are analyzed, we will see that the condition (4.68b) can often not be fulfilled exactly. In most cases, however, the relative deviation between $\tilde{\beta}_1$ and $\tilde{\beta}_2$ is very small. From the analysis of the level-crossing rate and the average duration of fades of Rice and Rayleigh processes, which was the topic of Subsection 3.4.3, we already know that for small relative deviations between β_1 and β_2, the ideal relations, derived on condition that $\beta = \beta_1 = \beta_2$ holds,

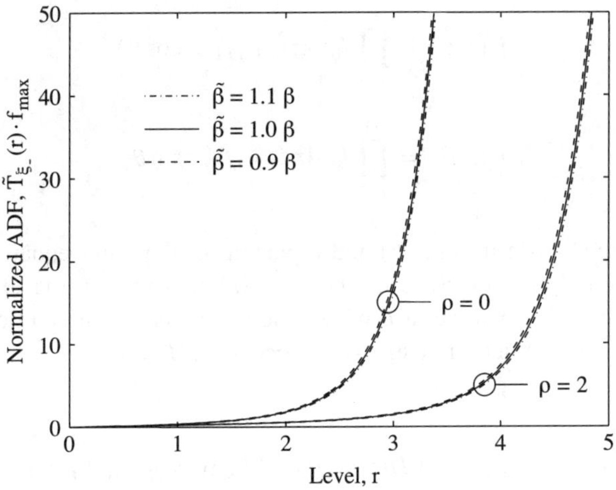

Figure 4.10 Normalized average duration of fades $\tilde{T}_{\xi_-}(r) \cdot f_{\max}$ of deterministic Rice processes $\tilde{\xi}(t)$ for various values of $\tilde{\beta} = \beta + \Delta\beta$ (Jakes PSD, $f_{\max} = 91$ Hz, $f_\rho = 0$).

will still keep their validity in a very good approximation, if we replace the quantity $\beta = \beta_1 = \beta_2$ by the arithmetical mean $\beta = (\beta_1 + \beta_2)/2$ [see (3.76)] in the corresponding expressions or if we directly identify β with β_1, i.e., $\beta = \beta_1 \approx \beta_2$ [see also Appendix 3.D, Equation (3.D.17)]. Similar results can be found for the deterministic model as well. For further simplification, we will therefore set $\tilde{\beta} = \tilde{\beta}_1 \approx \tilde{\beta}_2$ in the following, in case the relative deviation between $\tilde{\beta}_1$ and $\tilde{\beta}_2$ is small.

In consideration of this subject, it should finally be noted that even without the stated conditions (4.68a) and (4.68b), the level-crossing rate and thus also the average duration of fades of deterministic Rice processes can be calculated exactly. However, the numerical complexity for solving the obtained integral equations is considerably high. Apart from that the achievable improvements are often only small, even if a small number of sinusoids N_i is used, so that the comparatively high numerical complexity does not seem to be justified. Not only against this background, it turns out that especially the condition (4.68a) is meaningful, even though — strictly speaking — this condition is only fulfilled exactly as $N_i \to \infty$.

For completeness, the exact solution of both the level-crossing rate and the average duration of fades of deterministic Rice processes for any number of sinusoids N_i is presented in Appendix 4.B, where both conditions (4.68a) and (4.68b) have been dropped. In Appendix 4.B, one finds the following analytical expression for the level-crossing rate $\tilde{N}_\xi(r)$

$$\tilde{N}_\xi(r) = 2r \int_0^\infty \int_{-\pi}^{\pi} w_1(r,\theta)\, w_2(r,\theta) \int_0^\infty j_1(z,\theta)\, j_2(z,\theta)\, \dot{z} \cos(2\pi z \dot{z})\, dz\, d\theta\, d\dot{z}, \tag{4.81}$$

where

$$w_1(r,\theta) = \tilde{p}_{\mu_1}(r\cos\theta - \rho\cos\theta_\rho), \tag{4.82a}$$

$$w_2(r,\theta) = \tilde{p}_{\mu_2}(r\sin\theta - \rho\sin\theta_\rho), \tag{4.82b}$$

$$\tilde{j}_1(z,\theta) = \prod_{n=1}^{N_1} J_0(4\pi^2 c_{1,n} f_{1,n} z \cos\theta), \tag{4.82c}$$

$$\tilde{j}_2(z,\theta) = \prod_{n=1}^{N_2} J_0(4\pi^2 c_{2,n} f_{2,n} z \sin\theta). \tag{4.82d}$$

Now, by substituting (4.81) into (4.76) and replacing in the latter equation the cumulative distribution function $F_{\xi_-}(r)$ by $\tilde{F}_{\xi_-}(r)$, where $\tilde{F}_{\xi_-}(r)$ is given by the expression (4.B.40) derived in Appendix 4.B, then we also obtain the exact analytical solution for the average duration of fades $\tilde{T}_{\xi_-}(r)$ of deterministic Rice processes $\tilde{\xi}(t)$.

4.3.3 *Statistics of the Fading Intervals at Low Signal Levels*

In this subsection, we will discuss the statistical properties of the fading intervals of deterministic Rayleigh processes. Here, we will restrict ourselves to low signal levels, as in this case very precise approximate solutions can be derived by analytical means. At medium and high signal levels, however, we have to rely on simulations to which we will turn our attention subsequently in Section 5.3.

At first, we will study the probability density function $\tilde{p}_{0_-}(\tau_-; r)$ of the fading intervals of deterministic Rayleigh processes. This density characterizes the conditional probability density function for the case that a deterministic Rayleigh process $\tilde{\zeta}(t)$ crosses a signal level r for the first time at the time instant $t_2 = t_1 + \tau_-$, provided that the last down-crossing occurred at the time instant t_1. If no further statements are made about the level-crossing behaviour of $\tilde{\zeta}(t)$ between t_1 and t_2, then the corresponding probability density function is denoted by $\tilde{p}_{1_-}(\tau_-; r)$. During the analysis carried out for the stochastic reference model in Subsection 3.4.4, it was pointed out that $p_{1_-}(\tau_-; r)$ according to (3.86) can be considered as a very good approximation for $p_{0_-}(\tau_-; r)$ if the signal level r is low. The same statement holds for the deterministic model, which allows us to write: $\tilde{p}_{0_-}(\tau_-; r) \to \tilde{p}_{1_-}(\tau_-; r)$ if $r \to 0$, where $\tilde{p}_{1_-}(\tau_-; r)$ follows directly from (3.86) if we replace there $T_{\zeta_-}(r)$ by $\tilde{T}_{\zeta_-}(r)$ [78], i.e.,

$$\tilde{p}_{1_-}(\tau_-; r) = \frac{2\pi \tilde{z}^2 e^{-\tilde{z}}}{\tilde{T}_{\zeta_-}(r)} \left[I_0(\tilde{z}) - \left(1 + \frac{1}{2\tilde{z}}\right) I_1(\tilde{z}) \right], \quad 0 \le r \ll 1, \tag{4.83}$$

where $\tilde{z} = 2[\tilde{T}_{\zeta_-}(r)/\tau_-]^2/\pi$. The use of the result (4.77) now gives us the opportunity to investigate analytically the influence of the model error $\Delta\beta$ on the probability density function of the fading intervals of deterministic Rayleigh processes at deep signal levels. The evaluation of $\tilde{p}_{1_-}(\tau_-; r)$ according to (4.83) for various values of $\tilde{\beta} = \beta + \Delta\beta$ is shown in Figure 4.11 for a signal level r of $r = 0.1$. In this figure, one recognizes that a positive model error $\Delta\beta > 0$ causes a distinct decrease (increase) in the probability density function $\tilde{p}_{1_-}(\tau_-; r)$ in the range of relatively large (small) fading intervals τ_-. For negative model errors $\Delta\beta < 0$, the inverse behaviour can be observed.

In a similar way, we can obtain an expression for $\tilde{\tau}_q = \tilde{\tau}_q(r)$ describing the length of the time interval of those fading intervals of deterministic Rayleigh processes $\tilde{\zeta}(t)$ which include

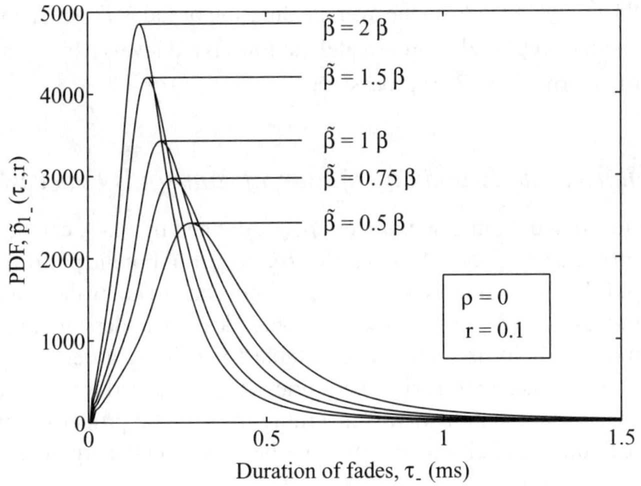

Figure 4.11 The influence of $\tilde{\beta} = \beta + \Delta\beta$ on the probability density function $\tilde{p}_{1\_}(\tau_-; r)$ of the fading intervals τ_- at low signal levels (Jakes PSD, $f_{\max} = 91$ Hz, $\sigma_0^2 = 1$).

q per cent of all fading intervals at the signal level r. The quantity $\tilde{\tau}_q(r)$ follows directly from (3.92) if we replace there $T_{\zeta_-}(r)$ by $\tilde{T}_{\zeta_-}(r)$. Hence, for $75 \leq q \leq 100$ it follows

$$\tilde{\tau}_q(r) \approx \frac{\tilde{T}_{\zeta_-}(r)}{\left\{\frac{\pi}{4}\left[1 - \sqrt{1 - 4\left(1 - \frac{q}{100}\right)}\right]\right\}^{1/3}}, \quad r \ll 1. \tag{4.84}$$

In order to make the influence of the model error $\Delta\beta$ transparent, we substitute $\tilde{T}_{\zeta_-}(r)$ by $\tilde{T}_{\zeta_-}(r) = T_{\zeta_-}(r)[1 - \Delta\beta/(2\beta)]$ in (4.84). In connection with (3.92), we can now rewrite the relation above in terms of $\tau_q(r)$ as

$$\tilde{\tau}_q(r) \approx \tau_q(r)\left(1 - \frac{\Delta\beta}{2\beta}\right), \quad r \ll 1. \tag{4.85}$$

This relation makes clear that a relative error of $\tilde{\beta}$ in the order of $\pm\varepsilon$ approximately causes a relative error of $\tilde{\tau}_q(r)$ in the order of $\mp\varepsilon/2$.

In particular, after substituting (3.93)–(3.95) into (4.85), the quantities $\tilde{\tau}_{90}(r)$, $\tilde{\tau}_{95}(r)$, and $\tilde{\tau}_{99}(\tau)$ can now approximately be expressed by:

$$\tilde{\tau}_{90}(r) \approx 1.78 \cdot T_{\zeta_-}(r)[1 - \Delta\beta/(2\beta)], \tag{4.86a}$$

$$\tilde{\tau}_{95}(r) \approx 2.29 \cdot T_{\zeta_-}(r)[1 - \Delta\beta/(2\beta)], \tag{4.86b}$$

$$\tilde{\tau}_{99}(r) \approx 3.98 \cdot T_{\zeta_-}(r)[1 - \Delta\beta/(2\beta)]. \tag{4.86c}$$

At this point, it should be explicitly emphasized that all approximations for $\tilde{p}_{0\_}(\tau_-; r)$ and $\tilde{\tau}_q(r)$, which were specially derived in this subsection for deterministic Rayleigh processes, are also valid for deterministic Rice processes with moderate Rice factors. This statement becomes clear immediately if we take into consideration that $\tilde{p}_{1\_}(\tau_-; r)$ [see (4.83)] as well as $\tilde{\tau}_q(r)$ [see (4.84)] only depends on the average duration of fades $\tilde{T}_{\zeta_-}(r)$, and from (4.79) we know that $\tilde{T}_{\zeta_-}(r)$ can be replaced approximately at low signal levels r by the average duration of fades $\tilde{T}_{\xi_-}(r)$ of deterministic Rice processes.

4.3.4 Stationarity and Ergodicity of Sum-of-Sinusoids Processes

As already mentioned, a deterministic process $\tilde{\mu}_i(t)$, defined by (4.4), can be interpreted as a sample function of the corresponding stochastic process $\hat{\mu}_i(t)$. For the performance evaluation of sum-of-sinusoids fading channel simulators, it is important to know the conditions on which the stochastic process $\hat{\mu}_i(t)$ is stationary and ergodic. A clear understanding of the concepts of stationarity and ergodicity is therefore inevitable for gaining a deeper understanding of the subject. It is thus advisable to review here briefly some related terms, such as first-order stationarity and wide-sense stationarity. Concerning the ergodic properties of $\hat{\mu}_i(t)$, we will distinguish between the ergodicity with respect to the mean and the ergodicity with respect to the autocorrelation function.

First-order stationarity: A stochastic process $\hat{\mu}_i(t)$ is said to be *first-order stationary* [41, p. 392] if the first-order density of $\hat{\mu}_i(t)$ is independent of time, i.e.,

$$\hat{p}_{\mu_i}(x; t) = \hat{p}_{\mu_i}(x; t + c) = \hat{p}_{\mu_i}(x) \tag{4.87}$$

holds for all values of t and $c \in \mathbb{R}$. This implies that the mean and the variance of $\hat{\mu}_i(t)$ are independent of time as well.

Wide-sense stationarity: A stochastic process $\hat{\mu}_i(t)$ is said to be *wide-sense stationary* [41, p. 388] if $\hat{\mu}_i(t)$ satisfies the following two conditions:

(i) The expected value of $\hat{\mu}_i(t)$ is constant, i.e.,

$$E\{\hat{\mu}_i(t)\} = \hat{m}_{\mu_i} = \text{const.} \tag{4.88}$$

holds for all t.

(ii) The autocorrelation function of $\hat{\mu}_i(t)$ depends only on the time difference $\tau = t_1 - t_2$, i.e.,

$$\hat{r}_{\mu_i\mu_i}(t_1, t_2) = \hat{r}_{\mu_i\mu_i}(\tau) = E\{\hat{\mu}_i(t)\,\hat{\mu}_i(t+\tau)\} \tag{4.89}$$

holds for all t_1 and t_2.

Mean ergodicity: A stochastic process $\hat{\mu}_i(t)$ is said to be *mean ergodic* if the temporal mean value of $\tilde{\mu}_i(t)$, computed over the interval $[-T, T]$, converges in the limit $T \to \infty$ to the statistical average $\hat{m}_{\mu_i} := E\{\hat{\mu}_i(t)\}$, i.e.,

$$\hat{m}_{\mu_i} = \tilde{m}_{\mu_i} := \lim_{T\to\infty} \frac{1}{2T} \int_{-T}^{T} \tilde{\mu}_i(t)\, dt. \tag{4.90}$$

Since the phases $\theta_{i,n}$ of the stochastic process $\hat{\mu}_i(t)$ are random variables, which are uniformly distributed in the interval $(0, 2\pi]$, the left-hand side of the equation above is equal to zero. The right-hand side equals also zero if all discrete Doppler frequencies $f_{i,n}$ are unequal to zero, i.e., if $f_{i,n} \neq 0$ for all $n = 1, 2, \ldots, N_i$ and $i = 1, 2$. This requirement ($f_{i,n} \neq 0$) can be fulfilled without any difficulty by all parameter computation methods introduced in the next chapter. Hence, $\hat{m}_{\mu_i} = \tilde{m}_{\mu_i} = 0$ holds, and thus the stochastic process $\hat{\mu}_i(t)$ is ergodic with respect to the mean value.

Autocorrelation ergodicity: A stochastic process $\hat{\mu}_i(t)$ is said to be *autocorrelation ergodic* if the temporal mean of $\tilde{\mu}_i(t)\tilde{\mu}_i(t+\tau)$, computed over the interval $[-T, T]$, converges in the limit $T \to \infty$ to the statistical average $\hat{r}_{\mu_i\mu_i}(\tau) := E\{\hat{\mu}_i(t)\,\hat{\mu}_i(t+\tau)\}$, i.e.,

$$\hat{r}_{\mu_i\mu_i}(\tau) = \tilde{r}_{\mu_i\mu_i}(\tau) := \lim_{T\to\infty} \frac{1}{2T} \int_{-T}^{T} \tilde{\mu}_i(t)\,\tilde{\mu}_i(t+\tau)\,dt. \tag{4.91}$$

The application of the concepts of stationarity and ergodicity on various classes of sum-of-sinusoids processes will be the topic of the next subsection.

4.4 Classes of Sum-of-Sinusoids Processes

For any given number $N_i > 0$, the sum-of-sinusoids depends on three types of parameters (gains, discrete Doppler frequencies, and phases), each of which can be a collection of random variables or constants. However, at least one random variable is required to obtain a stochastic process $\hat{\mu}_i(t)$ — otherwise we get a deterministic process $\tilde{\mu}_i(t)$, as was pointed out in Figure 4.3. Therefore, altogether $2^3 = 8$ classes of sum-of-sinusoids-based simulation models for Rayleigh fading channels can be defined, seven of which are stochastic simulation models and one is completely deterministic. For example, one class of stochastic channel simulators is defined by postulating random values for the gains $c_{i,n}$, discrete Doppler frequencies $f_{i,n}$, and phases $\theta_{i,n}$. The definition and analysis of the various classes of simulation models with respect to their stationary and ergodic properties was the topic of the study in [133, 134]. The obtained results are summarized in Table 4.1. Among the eight classes of sum-of-sinusoids channel simulators, Classes I, II, and IV are the most important classes, which will be studied in detail below.

Whenever the gains $c_{i,1}, c_{i,2}, \ldots, c_{i,N_i}$ are random variables, it is assumed that they are statistically independent and identically distributed. The same shall hold for the sequences of random frequencies $f_{i,1}, f_{i,2}, \ldots, f_{i,N_i}$ and phases $\theta_{i,1}, \theta_{i,2}, \ldots, \theta_{i,N_i}$. From the physical point of view, it is reasonable to assume that the gains $c_{i,n}$, frequencies $f_{i,n}$, and phases $\theta_{i,n}$ are mutually independent. We will therefore impose in addition the independence of the random variables $c_{i,n}$, $f_{i,n}$, and $\theta_{i,n}$ on our model. In cases, where the gains $c_{i,n}$ and discrete Doppler frequencies $f_{i,n}$ are constant quantities, it is assumed that they are different from zero, so that $c_{i,n} \neq 0$ and $f_{i,n} \neq 0$ hold for all values of $n = 1, 2, \ldots, N_i$ and $i = 1, 2$. We might also impose further constraints on the sum-of-sinusoids model. For example, we require that the absolute values of all frequencies, $|f_{i,n}|$, are different, i.e., (i) $|f_{i,1}| \neq |f_{i,2}| \neq \cdots \neq |f_{i,N_i}|$ for $i = 1, 2$ and (ii) $\{|f_{1,n}|\}_{n=1}^{N_1} \cap \{|f_{2,n}|\}_{n=1}^{N_2} = \emptyset$. The former condition (i) is introduced as a measure to avoid intra-correlations, i.e., correlations within $\tilde{\mu}_i(t)$ $(i = 1, 2)$, and the latter condition (ii) ensures that the cross-correlation (inter-correlation) of $\tilde{\mu}_1(t)$ and $\tilde{\mu}_2(t)$ is zero.

Table 4.1 Classes of deterministic and stochastic sum-of-sinusoids channel simulators and their stationary and ergodic properties [133, 134].

Class	Gains $c_{i,n}$	Freq. $f_{i,n}$	Phases $\theta_{i,n}$	First-order stat.	Wide-sense stat.	Mean ergodic	Auto-correlation ergodic
I	Const.	Const.	Const.	–	–	–	–
II	Const.	Const.	RV	Yes	Yes	Yes	Yes
III	Const.	RV	Const.	No/Yes^{a,b,c,d}	No/Yes^{a,b,c}	No/Yes^{a,b}	No
IV	Const.	RV	RV	Yes	Yes	Yes	No
V	RV	Const.	Const.	No	No	No/Yes^{e}	No
VI	RV	Const.	RV	Yes	Yes	Yes	No
VII	RV	RV	Const.	No/Yes^{a,b,c,d}	No/Yes^{a,b,c}	No/Yes$^{e \text{ or } a,b}$	No
VIII	RV	RV	RV	Yes	Yes	Yes	No

[a] If the density of $f_{i,n}$ is an even function.
[b] If the boundary condition $\sum_{n=1}^{N_i} \cos(\theta_{i,n}) = 0$ is fulfilled.
[c] If the boundary condition $\sum_{n=1}^{N_i} \cos(2\theta_{i,n}) = 0$ is fulfilled.
[d] Only in the limit $t \to \pm\infty$.
[e] If the gains $c_{i,n}$ have zero mean, i.e., $E\{c_{i,n}\} = 0$.

Class I sum-of-sinusoids channel simulators: The channel simulators of Class I are defined by the set of deterministic processes $\tilde{\mu}_i(t)$ [see (4.4)] with constant gains $c_{i,n}$, constant frequencies $f_{i,n}$, and constant phases $\theta_{i,n}$. Since all model parameters are constants, there is no meaning in examining the stationary and ergodic properties of this class of channel simulators. The reason is that the concept of stationarity and ergodicity only makes sense for stochastic processes and not for deterministic processes. However, we will benefit from the investigation of the mean and the autocorrelation function of $\tilde{\mu}_i(t)$.

Class II sum-of-sinusoids channel simulators: The channel simulators of Class II are defined by the set of stochastic processes $\hat{\mu}_i(t)$ with constant gains $c_{i,n}$, constant frequencies $f_{i,n}$, and random phases $\theta_{i,n}$, which are uniformly distributed in the interval $(0, 2\pi]$. In this case, the stochastic process $\hat{\mu}_i(t)$ has exactly the same form as in (4.3).

The distribution of a sum-of-sinusoids with random phases has first been studied in [132], where it was shown that the first-order density $\hat{p}_{\mu_i}(x)$ of $\hat{\mu}_i(t)$ is given by

$$\hat{p}_{\mu_i}(x) = 2 \int_0^\infty \left[\prod_{n=1}^{N_i} J_0(2\pi c_{i,n} \nu) \right] \cos(2\pi x \nu)\, d\nu. \tag{4.92}$$

Note that $\hat{p}_{\mu_i}(x)$ is independent of time and depends only on the number of sinusoids N_i and the gains $c_{i,n}$. If all gains are equal to $c_{i,n} = \sigma_0\sqrt{2/N_i}$, then it follows from the central limit theorem of Lindeberg-Lévy [see (2.26)] that the density $\hat{p}_{\mu_i}(x)$ in (4.92) approaches the Gaussian density $p_{\mu_i}(x)$ with zero mean and variance σ_0^2 if N_i tends to infinity, i.e., $\hat{p}_{\mu_i}(x) \to p_{\mu_i}(x)$ as $N_i \to \infty$. However, it is widely accepted that the approximation $\hat{p}_{\mu_i}(x) \approx p_{\mu_i}(x)$ is sufficiently good if $N_i \geq 7$.

From (4.3), it follows that the mean $\hat{m}_{\mu_i}$ of $\hat{\mu}_i(t)$ is constant and equal to zero, because

$$\begin{aligned}\hat{m}_{\mu_i} &= E\{\hat{\mu}_i(t)\}\\ &= \sum_{n=1}^{N_i} c_{i,n} E\{\cos(2\pi f_{i,n} t + \theta_{i,n})\}\\ &= 0.\end{aligned} \tag{4.93}$$

Substituting (4.3) in (4.89) results in the autocorrelation function

$$\hat{r}_{\mu_i\mu_i}(\tau) = \sum_{n=1}^{N_i} \frac{c_{i,n}^2}{2} \cos(2\pi f_{i,n}\tau), \tag{4.94}$$

which is a function of the time difference $\tau = t_1 - t_2$.

From (4.92)–(4.94), we may conclude that $\hat{\mu}_i(t)$ is both first-order stationary and wide-sense stationary, because the density $\hat{p}_{\mu_i}(x)$ is independent of time and conditions (i)–(ii) [see (4.88) and (4.89)] are fulfilled. For a specific realization of the random phases $\theta_{i,n}$, it follows from Figure 4.3 that the stochastic process $\hat{\mu}_i(t)$ results in a deterministic process (sample function) $\tilde{\mu}_i(t)$. In other words, Class I is a subset of Class II. We realize that the identity $\hat{m}_{\mu_i} = \tilde{m}_{\mu_i}$ holds, which states with reference to (4.90) that $\hat{\mu}_i(t)$ is mean ergodic. A comparison of (4.94) and (4.11) shows that $\hat{\mu}_i(t)$ is also autocorrelation ergodic, since the criterion $\hat{r}_{\mu_i\mu_i}(\tau) = \tilde{r}_{\mu_i\mu_i}(\tau)$ is fulfilled. Class II is the most important class among the classes of stochastic sum-of-sinusoids channel simulators.

Class IV sum-of-sinusoids channel simulators: The channel simulators of Class IV are defined by the set of stochastic processes $\hat{\mu}_i(t)$ with constant gains $c_{i,n}$, random frequencies $f_{i,n}$, and random phases $\theta_{i,n}$, which are uniformly distributed in the interval $(0, 2\pi]$. Thus, the stochastic process $\hat{\mu}_i(t)$ has the form of (4.3), but with random frequencies $f_{i,n}$.

The assumption that the frequencies $f_{i,n}$ are random variables has no effect on the density $\hat{p}_{\mu_i}(x)$ of $\hat{\mu}_i(t)$. Hence, the density $p_{\tilde{\mu}_i}(x)$ of $\hat{\mu}_i(t)$ is still given by (4.92). Also the mean $\hat{m}_{\mu_i}$ is equal to zero, and thus identical to (4.93). But the autocorrelation function $\hat{r}_{\mu_i\mu_i}(\tau)$ in (4.94) has to be averaged with respect to the distribution of the frequencies $f_{i,n}$, i.e.,

$$\hat{r}_{\mu_i\mu_i}(\tau) = \sum_{n=1}^{N_i} \frac{c_{i,n}^2}{2} E\{\cos(2\pi f_{i,n}\tau)\}. \tag{4.95}$$

Alternatively, the expression in (4.95) can be derived by using the result of Example 9-14 in [41, pp. 391–392]. Independent of a specific distribution of $f_{i,n}$, we can say that $\hat{\mu}_i(t)$ is first-order stationary and wide-sense stationary, since the density $\hat{p}_{\mu_i}(x)$ in (4.92) is independent of time and conditions (i)–(ii) [see (4.88) and (4.89)] are fulfilled. Furthermore, the condition $\hat{m}_{\mu_i} = \tilde{m}_{\mu_i}$ is fulfilled, which implies that $\hat{\mu}_i(t)$ is mean ergodic.

When applying the Monte Carlo method [99, 100], the gains $c_{i,n}$ and frequencies $f_{i,n}$ are given by $c_{i,n} = \sigma_0\sqrt{2/N_i}$ and $f_{i,n} = f_{\max}\sin(\pi u_{i,n})$, respectively, where $u_{i,n}$ is a random variable which is uniformly distributed in the interval (0, 1]. Using these equations in (4.95) results in

$$\hat{r}_{\mu_i\mu_i}(\tau) = \sigma_0^2 J_0(2\pi f_{\max}\tau). \tag{4.96}$$

Obviously, the autocorrelation function $\hat{r}_{\mu_i\mu_i}(\tau)$ of the stochastic simulation model is identical to the autocorrelation function $r_{\mu_i\mu_i}(\tau)$ of the reference model described by (3.25). However, a comparison of (4.96) and (4.11) shows that $\hat{r}_{\mu_i\mu_i}(\tau) \neq \tilde{r}_{\mu_i\mu_i}(\tau)$. Thus, the stochastic processes $\hat{\mu}_i(t)$ of the Class IV channel simulators are non-autocorrelation ergodic.

4.5 Basics of Sum-of-Cisoids Channel Models

A special case of the sum-of-sinusoids model is obtained if $c_n = c_{1,n} = c_{2,n}$, $f_n = f_{1,n} = f_{2,n}$, and $\theta_n = \theta_{1,n} = \theta_{2,n} + \pi/2$ for all $n = 1, 2, \ldots, N$, where $N = N_1 = N_2$. In this case, the complex process $\hat{\mu}(t) = \hat{\mu}_1(t) + j\hat{\mu}_2(t)$ with inphase and quadrature components according to (4.3) can be written as

$$\hat{\mu}(t) = \sum_{n=1}^{N} c_n \, e^{j(2\pi f_n t + \theta_n)} . \tag{4.97}$$

Hence, $\hat{\mu}(t)$ represents a sum of complex-valued sinusoids (cisoids), which constitutes the sum-of-cisoids (SOC) model. This model is quite useful for the modelling and simulation of mobile radio channels under realistic non-isotropic scattering conditions, while the SOS model is the better choice when isotropic scattering conditions are assumed [135]. It is worth noting that the SOC model in (4.97) has the same form as the multipath fading model described by (3.12), which has been derived under the assumption of flat fading. This relationship allows a simple physical interpretation of the SOC model in form of a plane wave model. The number of cisoids N equals the number of scatterers, each of which introduces a gain c_n and a phase shift θ_n, while the movement of the receiver causes a Doppler shift f_n of the nth received plane wave. Thus, the sum of plane waves (cisoids) models the diffuse component of a multipath channel under flat fading channel conditions.

The absolute value and the phase of $\hat{\mu}(t)$ will be denoted by $\hat{\zeta}(t) = |\hat{\mu}(t)|$ and $\hat{\vartheta}(t) = \arg\{\hat{\mu}(t)\}$, respectively. The software or hardware implementation of $\hat{\zeta}(t)$ is called the *SOC Rayleigh fading channel simulator*. Note that $\hat{\zeta}(t)$ is lower bounded by $\hat{\zeta}_{\min} = \min|\hat{\mu}(t)| = 0$ and upper bounded by $\hat{\zeta}_{\max} = \max|\hat{\mu}(t)| = \sum_{n=1}^{N} |c_n|$.

The SOC model in (4.97) is an appropriate model for the scattered component of flat fading channels. In microcellular environments and rural areas, however, there exists often a line-of-sight component. In the following, we assume that this component is given by

$$m(t) = \rho \, e^{j(2\pi f_\rho t + \theta_\rho)} , \tag{4.98}$$

where the amplitude ρ, the frequency f_ρ, and the phase θ_ρ are constant quantities.

The sum of the scattered component $\hat{\mu}(t)$ and the line-of-sight component $m(t)$ defines the non-zero-mean complex random process

$$\hat{\mu}_\rho(t) = \hat{\mu}(t) + m(t) . \tag{4.99}$$

By taking the absolute value of $\hat{\mu}_\rho(t)$, we obtain the envelope process

$$\hat{\xi}(t) = |\hat{\mu}_\rho(t)| \tag{4.100}$$

and the argument of $\hat{\mu}_\rho(t)$ results in the phase process

$$\hat{\vartheta}(t) = \arg\{\hat{\mu}_\rho(t)\}. \tag{4.101}$$

The implementation of $\hat{\xi}(t)$ on a hardware or software platform is called the *SOC Rice fading channel simulator*. The structure of the resulting simulation model is shown in the normally used time-continuous form in Figure 4.12.

4.5.1 Elementary Properties of Stochastic Sum-of-Cisoids Processes

In this section, we analyze the most important statistical properties of SOC models, such as the mean value, variance, correlation functions, and the power spectral density. Another topic in this section is to point out the differences between SOC and SOS models. The following presentation is guided by the results in [135], which are valid for the class of SOC models with fixed gains c_n, fixed frequencies f_n, and random phases θ_n.

Mean value: Let $\hat{\mu}(t) = \hat{\mu}_1(t) + j\hat{\mu}_2(t)$ be an SOC process with $f_n \neq 0$ for all $n = 1, 2, \ldots, N$, then the mean value of $\hat{\mu}_i(t)$ equals

$$\hat{m}_{\mu_i} = E\{\hat{\mu}_i(t)\} = 0. \tag{4.102}$$

Note that the same result holds for SOS processes [see (4.9)].

Variance: Let $\hat{\mu}(t) = \hat{\mu}_1(t) + j\hat{\mu}_2(t)$ be an SOC process, then the variance of $\hat{\mu}_i(t)$ can be obtained as

$$\hat{\sigma}^2_{\mu_i} = \text{Var}\{\hat{\mu}_i(t)\} = \sum_{n=1}^{N} \frac{c_n^2}{2}. \tag{4.103}$$

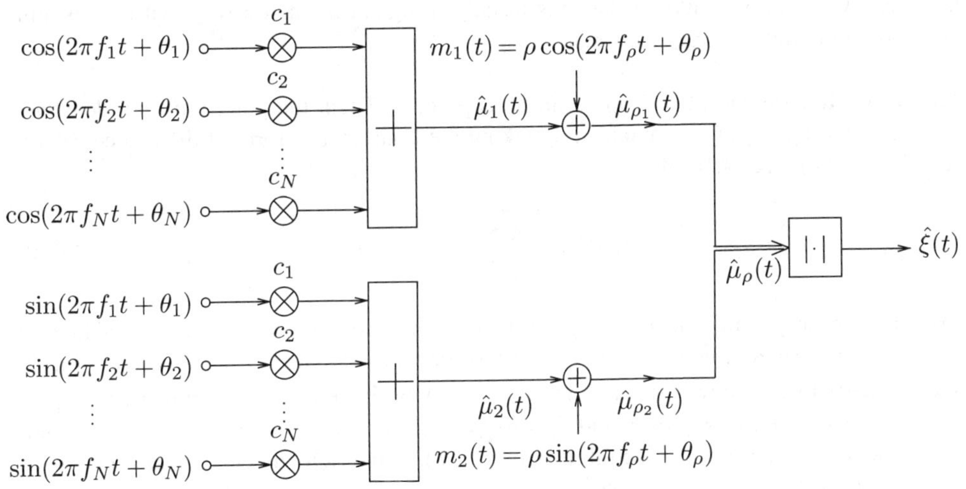

Figure 4.12 Stochastic sum-of-cisoids simulation model for Rice fading channels.

Obviously, the variance $\hat{\sigma}^2_{\mu_i}$ depends only on the number of cisoids N and the gains c_n, but not on the frequencies f_n and phases θ_n. The same result is known for SOS processes [see (4.10)].

Correlation functions: Let $\hat{\mu}(t)$ be an SOC process, then its autocorrelation function is given by

$$\begin{aligned}\hat{r}_{\mu\mu}(\tau) &= E\{\hat{\mu}^*(t)\,\hat{\mu}(t+\tau)\}\\ &= \sum_{n=1}^{N} c_n^2\, e^{j2\pi f_n \tau}.\end{aligned} \tag{4.104}$$

From (4.104), the autocorrelation function $\hat{r}_{\mu_i\mu_i}(\tau)$ of $\hat{\mu}_i(t)$ ($i = 1, 2$) as well as the cross-correlation function $\hat{r}_{\mu_1\mu_2}(\tau)$ of $\hat{\mu}_1(t)$ and $\hat{\mu}_2(t)$ can be obtained directly as

$$\hat{r}_{\mu_i\mu_i}(\tau) = \frac{1}{2}\,\mathrm{Re}\{\hat{r}_{\mu\mu}(\tau)\} = \sum_{n=1}^{N} \frac{c_n^2}{2}\cos(2\pi f_n \tau) \tag{4.105}$$

and

$$\hat{r}_{\mu_1\mu_2}(\tau) = \frac{1}{2}\,\mathrm{Im}\{\hat{r}_{\mu\mu}(\tau)\} = \sum_{n=1}^{N} \frac{c_n^2}{2}\sin(2\pi f_n \tau), \tag{4.106}$$

respectively. The cross-correlation function $\hat{r}_{\mu_2\mu_1}(\tau)$ of $\hat{\mu}_2(t)$ and $\hat{\mu}_1(t)$ can be obtained from the relation $\hat{r}_{\mu_2\mu_1}(\tau) = \hat{r}^*_{\mu_1\mu_2}(-\tau) = \hat{r}_{\mu_1\mu_2}(-\tau)$. The result in (4.106) shows that the cross-correlation function $\hat{r}_{\mu_1\mu_2}(\tau)$ is generally different from zero, and thus it follows that the inphase component $\hat{\mu}_1(t)$ and the quadrature component $\hat{\mu}_2(t)$ of the complex process $\hat{\mu}(t) = \hat{\mu}_1(t) + j\hat{\mu}_2(t)$ are correlated. This is different from SOS models, where the correlation between $\hat{\mu}_1(t)$ and $\hat{\mu}_2(t)$ can easily be avoided by designing the processes $\hat{\mu}_1(t)$ and $\hat{\mu}_2(t)$ using disjoint sets of discrete Doppler frequencies [see (4.12)]. One of the main differences between SOC and SOS models is that the autocorrelation function $\hat{r}_{\mu\mu}(\tau)$ of the former model is generally complex, while the latter is usually designed in such a way that the resulting autocorrelation function $\hat{r}_{\mu\mu}(\tau)$ is real.

Power spectral density: Let $\hat{\mu}(t)$ be an SOC process. Then, the power spectral density of $\hat{\mu}(t)$, denoted by $\hat{S}_{\mu\mu}(f)$, is obtained by taking the Fourier transform of the autocorrelation function $\hat{r}_{\mu\mu}(\tau)$ [see (4.104)] as

$$\hat{S}_{\mu\mu}(f) = \sum_{n=1}^{N} c_n^2\,\delta(f - f_n). \tag{4.107}$$

Thus, the power spectral density $\hat{S}_{\mu\mu}(f)$ of $\hat{\mu}(t)$ is a line spectrum, where the spectral lines are located at the discrete frequencies $f = f_n$ and weighted by the factor c_n^2. The result in (4.107) shows clearly that the power spectral density $\hat{S}_{\mu\mu}(f)$ has in general an asymmetrical shape. A special case is given if N is even and $f_n = -f_{N-n}$, $c_n = \pm c_{N-n}$ $\forall\ n = 1, 2, \ldots, N/2$. Under these conditions, the power spectral density $\hat{S}_{\mu\mu}(f)$ of the SOC model is also symmetrical, i.e., $\hat{S}_{\mu\mu}(f) = \hat{S}_{\mu\mu}(-f)$. Hence, the SOC model can be used to model both symmetrical as well as asymmetrical power spectral densities. This is in contrast to SOS models, which are used primarily to model symmetrical, especially Jakes-type, power spectral densities.

Average Doppler shift: Let $\hat{\mu}(t)$ be an SOC process. Then, the average Doppler shift $\hat{B}^{(1)}_{\mu\mu}$ of $\hat{\mu}(t)$ can be obtained as

$$\hat{B}^{(1)}_{\mu\mu} := \frac{\int\limits_{-\infty}^{\infty} f \hat{S}_{\mu\mu}(f)\, df}{\int\limits_{-\infty}^{\infty} \hat{S}_{\mu\mu}(f)\, df} = \frac{1}{2\pi j} \cdot \frac{\dot{\hat{r}}_{\mu\mu}(0)}{\hat{r}_{\mu\mu}(0)}$$
$$= \frac{\sum\limits_{n=1}^{N} f_n c_n^2}{\sum\limits_{n=1}^{N} c_n^2}, \tag{4.108}$$

where $\dot{\hat{r}}_{\mu\mu}(\tau)$ denotes the derivative of $\hat{r}_{\mu\mu}(\tau)$ with regard to the variable τ.

Doppler spread: Let $\hat{\mu}(t)$ be an SOC process. Then, the Doppler spread $\hat{B}^{(2)}_{\mu\mu}$ of $\hat{\mu}(t)$ can be expressed as follows

$$\hat{B}^{(2)}_{\mu\mu} := \sqrt{\frac{\int\limits_{-\infty}^{\infty} \left(f - \hat{B}^{(1)}_{\mu\mu}\right)^2 \hat{S}_{\mu\mu}(f)\, df}{\int\limits_{-\infty}^{\infty} \hat{S}_{\mu\mu}(f)\, df}} = \frac{1}{2\pi} \cdot \sqrt{\left(\frac{\dot{\hat{r}}_{\mu\mu}(0)}{\hat{r}_{\mu\mu}(0)}\right)^2 - \frac{\ddot{\hat{r}}_{\mu\mu}(0)}{\hat{r}_{\mu\mu}(0)}}$$
$$= \sqrt{\frac{\hat{\beta}}{(2\pi\hat{\sigma}_\mu)^2} - \left(\hat{B}^{(1)}_{\mu\mu}\right)^2}, \tag{4.109}$$

where $\hat{\beta} = -\ddot{\hat{r}}_{\mu\mu}(0) = (2\pi)^2 \sum_{n=1}^{N} (c_n f_n)^2$, and $\hat{\sigma}_\mu^2$ is the variance of $\hat{\mu}(t)$ as given by (4.103). Note that in the symmetrical case, i.e., $f_n = f_{N-n}$ and $c_n = \pm c_{N-n} \forall\, n = 1, 2, \ldots, N/2$ (N even), it follows $\hat{B}^{(1)}_{\mu\mu} = 0$ and $\hat{B}^{(2)}_{\mu\mu} = \sqrt{\hat{\beta}}/(2\pi\hat{\sigma}_\mu)$. These results are identical to those of SOS models [see (4.30)]. Thus, the average Doppler shift and the Doppler spread of the SOC model include the corresponding quantities of the SOS model as a special case.

4.5.2 Probability Density Function of the Envelope and Phase

In the following, we derive the probability density function of the envelope and phase of SOC processes $\hat{\mu}(t) = \hat{\mu}_1(t) + j\hat{\mu}_2(t)$ by taking into account that the inphase component $\hat{\mu}_1(t)$ and the quadrature component $\hat{\mu}_2(t)$ are correlated.

The starting point is a weighted elementary complex sinusoid of the form

$$\hat{\mu}_n(t) = c_n e^{j(2\pi f_n t + \theta_n)}, \tag{4.110}$$

where c_n and f_n are non-zero constants and θ_n is a random variable having a uniform distribution over the interval $(-\pi, \pi)$. For fixed values of $t = t_0$, $\hat{\mu}_n(t_0) = \hat{\mu}_{1,n}(t_0) + j\hat{\mu}_{2,n}(t_0)$ represents a complex random variable whose real part $\hat{\mu}_{1,n}(t_0) = c_n \cos(2\pi f_n t_0 + \theta_n)$ and imaginary part

$\hat{\mu}_{2,n}(t_0) = c_n \sin(2\pi f_n t_0 + \theta_n)$ obey the distribution [41, p. 135]

$$\hat{p}_{\mu_{i,n}}(x_i) = \begin{cases} \dfrac{1}{\pi |c_n| \sqrt{1 - (x_i/c_n)^2}}, & |x_i| < c_n \\ 0, & |x_i| \geq c_n \end{cases} \tag{4.111}$$

for $i = 1, 2$. We notice that $\hat{\mu}_{1,n}(t_0)$ and $\hat{\mu}_{2,n}(t_0)$ are statistically dependent. To find the joint distribution $\hat{p}_{\mu_{1,n}\mu_{2,n}}(x_1, x_2)$ of the two dependent random variables $\hat{\mu}_{1,n}(t_0)$ and $\hat{\mu}_{2,n}(t_0)$, we make use of [136, Equation (3.15)], which provides us with the key for the solution of our problem in the form of

$$\hat{p}_{\mu_{1,n}\mu_{2,n}}(x_1, x_2) = \hat{p}_{\mu_{1,n}}(x_1) \cdot \delta(x_2 - g(x_1)), \tag{4.112}$$

where

$$g(x_1) = \begin{cases} c_n \sqrt{1 - (x_1/c_n)^2}, & |x_1| \leq c_n \\ 0, & |x_1| > c_n. \end{cases} \tag{4.113}$$

Recall that the joint characteristic function of the random variables $\hat{\mu}_{1,n}(t_0)$ and $\hat{\mu}_{2,n}(t_0)$ is by definition the double integral [see (2.20b)]

$$\hat{\Psi}_{\mu_{1,n}\mu_{2,n}}(\nu_1, \nu_2) = \int_{-\infty}^{\infty} \int_{-\infty}^{\infty} \hat{p}_{\mu_{1,n}\mu_{2,n}}(x_1, x_2)\, e^{j2\pi(\nu_1 x_1 + \nu_2 x_2)}\, dx_1\, dx_2. \tag{4.114}$$

Substituting (4.112) in (4.114) and using [23, Equation (3.937-1)] results after some algebraic manipulations in

$$\hat{\Psi}_{\mu_{1,n}\mu_{2,n}}(\nu_1, \nu_2) = J_0(2\pi |c_n| \sqrt{\nu_1^2 + \nu_2^2}). \tag{4.115}$$

From the assumption that θ_n are i.i.d. random variables, it follows that the terms of $\hat{\mu}_i(t_0) = \hat{\mu}_{i,1}(t_0) + \hat{\mu}_{i,2}(t_0) + \cdots + \hat{\mu}_{i,n}(t_0)$ are also i.i.d. random variables. This allows us to express the joint characteristic function $\hat{\Psi}_{\mu_1\mu_2}(\nu_1, \nu_2)$ of $\hat{\mu}_1(t_0)$ and $\hat{\mu}_2(t_0)$ as the N-fold product of the joint characteristic functions $\hat{\Psi}_{\mu_{1,n}\mu_{2,n}}(\nu_1, \nu_2)$ of $\hat{\mu}_{1,n}(t_0)$ and $\hat{\mu}_{2,n}(t_0)$, i.e.,

$$\hat{\Psi}_{\mu_1\mu_2}(\nu_1, \nu_2) = \prod_{n=1}^{N} J_0(2\pi |c_n| \sqrt{\nu_1^2 + \nu_2^2}). \tag{4.116}$$

From this and the two-dimensional inversion formula for Fourier transforms, it follows that the joint probability density function $\hat{p}_{\mu_1\mu_2}(x_1, x_2)$ of $\hat{\mu}_1(t_0)$ and $\hat{\mu}_2(t_0)$ can be expressed by using [23, Equation (3.937-1)] as

$$\hat{p}_{\mu_1\mu_2}(x_1, x_2) = 2\pi \int_0^{\infty} \left[\prod_{n=1}^{N} J_0(2\pi |c_n| y) \right] J_0(2\pi y \sqrt{x_1^2 + x_2^2})\, y\, dy. \tag{4.117}$$

Notice that the joint probability density function $\hat{p}_{\mu_1\mu_2}(x_1, x_2)$ cannot be written as the product of the marginal probability density functions $\hat{p}_{\mu_1}(x_1)$ and $\hat{p}_{\mu_2}(x_2)$, i.e., $\hat{p}_{\mu_1\mu_2}(x_1, x_2) \neq \hat{p}_{\mu_1}(x_1) \cdot \hat{p}_{\mu_2}(x_2)$, which is a consequence of the fact that $\hat{\mu}_1(t)$ and $\hat{\mu}_2(t)$ are statistically dependent. Next we compute the joint probability density function $\hat{p}_{\mu_{\rho_1}\mu_{\rho_2}}(x_1, x_2)$ of

$\hat{\mu}_{\rho_1}(t) = \hat{\mu}_1(t) + m_1$ and $\hat{\mu}_{\rho_2}(t) = \hat{\mu}_2(t) + m_2$. From our assumption that $m = m_1 + jm_2$ has constant parameters, it follows that the joint probability density function $p_{m_1m_2}(x_1, x_2)$ of m_1 and m_2 equals $p_{m_1m_2}(x_1, x_2) = \delta(x_1 - m_1, x_2 - m_2)$. Hence, the joint probability density function $\hat{p}_{\mu_{\rho_1}\mu_{\rho_2}}(x_1, x_2)$ can be expressed as

$$\begin{aligned}\hat{p}_{\mu_{\rho_1}\mu_{\rho_2}}(x_1, x_2) &= \hat{p}_{\mu_1\mu_2}(x_1, x_2) * \delta(x_1 - m_1, x_2 - m_2)\\ &= \hat{p}_{\mu_1\mu_2}(x_1 - m_1, x_2 - m_2). \end{aligned} \tag{4.118}$$

The transformation of the Cartesian coordinates (x_1, x_2) into polar coordinates (z, θ) by means of $x_1 = z\cos\theta$ and $x_2 = z\sin\theta$ enables us to derive the joint probability density function $\hat{p}_{\xi\vartheta}(z, \theta)$ of the envelope $\hat{\xi}(t) = |\hat{\mu}_\rho(t)|$ and the phase $\hat{\vartheta}(t) = \arg\{\hat{\mu}_\rho(t)\}$ as follows:

$$\begin{aligned}\hat{p}_{\xi\vartheta}(z, \theta) &= z\hat{p}_{\mu_{\rho_1}\mu_{\rho_2}}(z\cos\theta, z\sin\theta)\\ &= z\hat{p}_{\mu_{\rho_1}\mu_{\rho_2}}(z\cos\theta - \rho\cos\theta_\rho, z\sin\theta - \rho\sin\theta_\rho)\\ &= 2\pi z\int_0^\infty\left[\prod_{n=1}^N J_0(2\pi|c_n|x)\right]J_0(2\pi x\sqrt{z^2 + \rho^2 - 2z\rho\cos(\theta - \theta_\rho)})\,x\,dx\end{aligned} \tag{4.119}$$

for $z \geq 0$ and $\theta \in [-\pi, \pi)$. Now, the probability density function $\hat{p}_\xi(z)$ of the envelope $\hat{\xi}(t)$ can be obtained from the joint probability density function $\hat{p}_{\xi\vartheta}(z, \theta)$ by integrating over θ, which results finally in

$$\hat{p}_\xi(z) = (2\pi)^2 z\int_0^\infty\left[\prod_{n=1}^N J_0(2\pi|c_n|x)\right]J_0(2\pi zx)\,J_0(2\pi\rho x)\,x\,dx \tag{4.120}$$

for $z \geq 0$. It is important to note that the probability density function $\hat{p}_\xi(z)$ is completely determined by the number of cisoids N, the gains c_n, and the line-of-sight amplitude ρ, whereas the other model parameters $(f_n, \theta_n, \theta_\rho)$ have no influence. From the result in (4.120), we can easily obtain the distribution of the envelope $\hat{\zeta}(t) = |\hat{\mu}(t)|$ of the SOC model without a line-of-sight component ($\rho = 0$) as

$$\hat{p}_\zeta(z) = (2\pi)^2 z\int_0^\infty\left[\prod_{n=1}^N J_0(2\pi|c_n|x)\right]J_0(2\pi zx)\,x\,dx \tag{4.121}$$

for $z \geq 0$. Subsequently, we prove that the probability density function $\hat{p}_\xi(z)$ in (4.120) approaches the Rice density if $N \to \infty$ and $c_n = \sigma_0\sqrt{2/N}$.

Finally, the probability density function $\hat{p}_\vartheta(\theta)$ of the phase $\hat{\vartheta}(t)$ can be derived from the joint probability density function $\hat{p}_{\xi\vartheta}(z, \theta)$ in (4.119) by integrating over z, i.e.,

$$\hat{p}_\vartheta(\theta) = 2\pi\int_0^\infty\int_0^\infty\left[\prod_{n=1}^N J_0(2\pi|c_n|x)\right]J_0\left(2\pi x\sqrt{z^2 + \rho^2 - 2z\rho\cos(\theta - \theta_\rho)}\right)xz\,dx\,dz \tag{4.122}$$

for $-\pi < \theta \leq \pi$. It is interesting to note that if $\rho \to 0$, then the probability density function $\hat{p}_{\vartheta}(\theta)$ reduces to the uniform distribution

$$\hat{p}_{\vartheta}(\theta) = \frac{1}{2\pi}, \quad -\pi < \theta \leq \pi \tag{4.123}$$

for all $N \geq 1$. Without proof, we mention that if $c_n = \sigma_0\sqrt{2/N}$ and $N \to \infty$, then the probability density function $\hat{p}_{\vartheta}(\theta)$ of the phase of the SOC model approaches the probability density function $p_{\vartheta}(\theta)$ of the phase of the reference model, which is given by (3.56) for $f_\rho \neq 0$ and by (3.57) for $f_\rho = 0$.

Choosing $c_n = \sigma_0\sqrt{2/N}$ ($\sigma_0^2 = 1$) for all $n = 1, 2, \ldots, N$ and substituting these quantities in (4.121), allows us to study the distribution of the probability density function $\hat{p}_{\zeta}(z)$ of the envelope $\hat{\zeta}(t) = |\hat{\mu}(t)|$ of the SOC model as a function of the number of cisoids N. The obtained results are presented in Figure 4.13. This figure shows also the simulation results of the distribution of $\hat{\zeta}(t) = |\hat{\mu}(t)|$ obtained from the simulation of $\hat{\mu}(t)$ given by (4.97) and averaging over 100 trials. It can be observed that the simulation results confirm the correctness of the analytical solution presented in (4.121). From Figure 4.13, we can also conclude that the density $\hat{p}_{\zeta}(z)$ is sufficiently close to the Rayleigh distribution if $N \geq 10$.

Choosing $N = 10$ and evaluating the probability density function $\hat{p}_{\xi}(z)$ of the envelope $\hat{\xi}(t) = |\hat{\mu}_\rho(t)|$ according to (4.120) gives the results presented in Figure 4.14 for various values of the amplitude ρ of the line-of-sight component. For the purpose of comparison, the Rice density $p_{\xi}(z)$ [see (3.54)] is also shown. Figure 4.14 demonstrates impressively that an SOC consisting of $N = 10$ terms is sufficient to obtain an excellent approximation $\hat{p}_{\xi}(z) \approx p_{\xi}(z)$. The corresponding graphs for the probability density function of the phase are plotted in Figure 4.15. The graphs referring to the simulation model (theory) and the reference model (theory) have been obtained by evaluating the probability density functions in (4.122) and (3.57), respectively. Observe that our theoretical results shown in Figures 4.14 and 4.15 for the simulation model line up exactly with the corresponding simulation results.

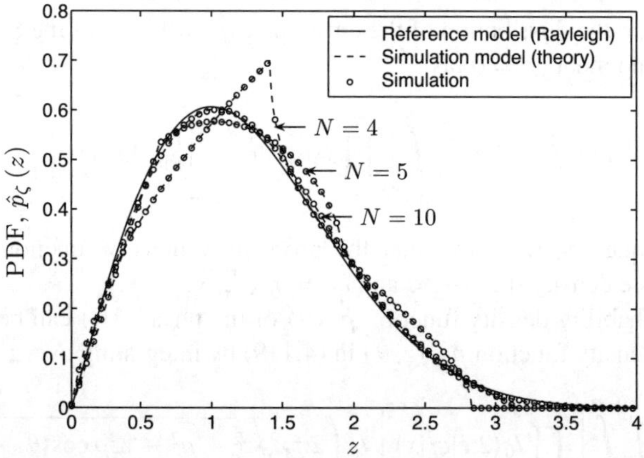

Figure 4.13 Probability density function $\hat{p}_{\zeta}(z)$ of the envelope $\hat{\zeta}(t) = |\hat{\mu}(t)|$ of the SOC model for $N \in \{4, 5, 10, \infty\}$ ($c_n = \sigma_0\sqrt{2/N}$, $\sigma_0^2 = 1$).

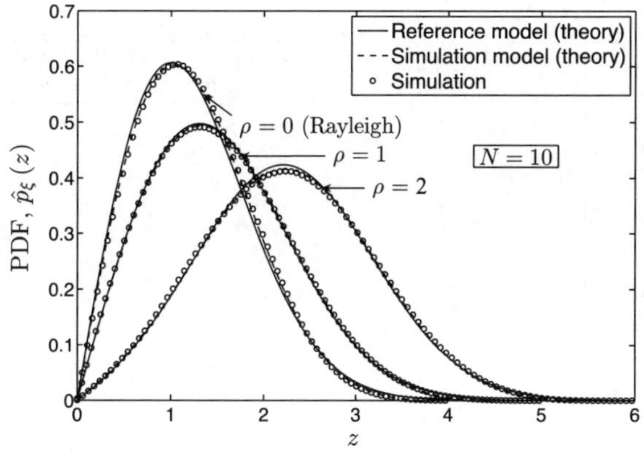

Figure 4.14 Probability density function $\hat{p}_\xi(z)$ of the envelope $\hat{\xi}(t) = |\hat{\mu}_\rho(t)|$ of the SOC model including a line-of-sight component $m = \rho e^{j\theta_\rho}$ for various values of the amplitude ρ ($c_n = \sigma_0\sqrt{2/N}$, $\sigma_0^2 = 1$).

A measure for the quality of the approximation $p_\zeta(z) \approx \hat{p}_\zeta(z)$ is the mean-square error E_{p_ζ} of $\hat{p}_\zeta(z)$, which is defined by

$$E_{p_\zeta} = \int_0^\infty (p_\zeta(z) - \hat{p}_\zeta(z))^2 \, dz, \tag{4.124}$$

where $p_\zeta(z)$ denotes the Rayleigh distribution and $\hat{p}_\zeta(z)$ is the density of the SOC model, as presented in (4.121). The evaluation of E_{p_ξ} is shown in Figure 4.16 in terms of N ranging from 5 to 30. For comparative purposes, we have also evaluated the mean-square error E_{p_ζ} in

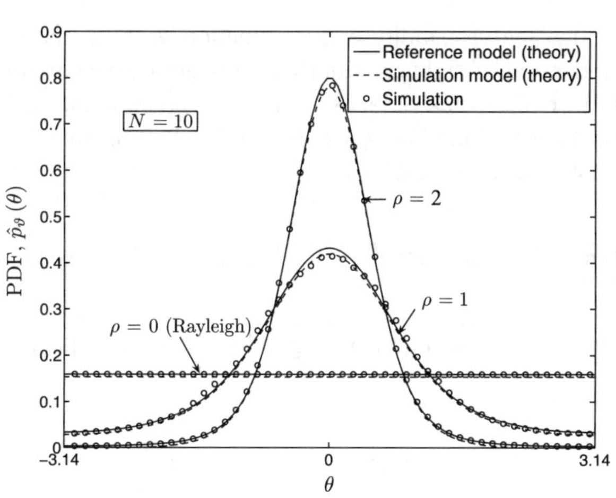

Figure 4.15 Probability density function $\hat{p}_\vartheta(\theta)$ of the phase $\hat{\vartheta}(t) = \arg\{\hat{\mu}_\rho(t)\}$ of the SOC model including a line-of-sight component $m = \rho e^{j\theta_\rho}$ for various values of the amplitude ρ ($c_n = \sigma_0\sqrt{2/N}$, $\sigma_0^2 = 1, \theta_\rho = 0$).

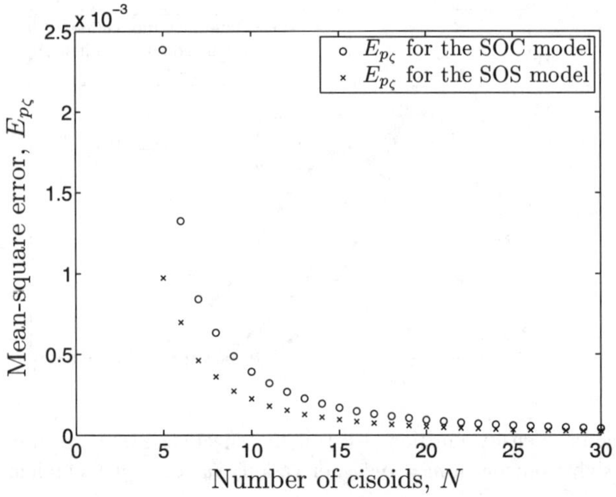

Figure 4.16 Mean-square error E_{p_ς} of the probability density function $\hat{p}_\varsigma(z)$ evaluated in terms of N for the SOC model ($c_n = \sigma_0\sqrt{2/N}$, $\sigma_0^2 = 1$) and the SOS model ($c_{i,n} = \sigma_0\sqrt{2/N_i}$ ($i = 1, 2$), $\sigma_0^2 = 1$, $N = N_1 = N_2$).

(4.124) when using the SOS model, the density of which can be found in (4.57). The results in Figure 4.16 show that the envelope distribution of the SOS model converges slightly faster to the Rayleigh distribution than that of the SOC model. Obviously, the cross-correlation inherent to the SOC model causes a small performance loss, which can only be compensated by increasing N. For the SOS model, one often considers $N = 7$ (or $N = 8$) as a sufficient number for the generation of quasi Rayleigh fading waveforms. From Figure 4.16, we conclude that the SOC model requires at least $N = 10$ terms to obtain nearly the same value for the mean-square error E_{p_ς}.

Finally, we prove that the probability density function $\hat{p}_\xi(z)$ of the envelope of an SOC model approaches under line-of-sight conditions the probability density function $p_\xi(z)$ of Rice processes if the number of cisoids N tends to infinity and $c_n = \sigma_0\sqrt{2/N}$. Our proof starts from the probability density function $\hat{p}_\xi(z)$ in (4.120). Putting in this equation the gains $c_n = \sigma_0\sqrt{2/N}$ and making use of (4.45) gives

$$\lim_{N\to\infty} \hat{p}_\xi(z) = (2\pi)^2 z \int_0^\infty e^{-2(\pi\sigma_0 x)^2} J_0(2\pi z x)\, J_0(2\pi x\rho)\, x\, dx. \tag{4.125}$$

Finally, the solution of the aforementioned integral by using [23, Equation (6.333-2)], results in the density

$$\lim_{N\to\infty} \hat{p}_\xi(z) = p_\xi(z) = \frac{z}{\sigma_0^2} e^{-\frac{z^2+\rho}{2\sigma_0^2}} I_0\left(\frac{z\rho}{\sigma_0^2}\right), \tag{4.126}$$

which can be identified as the Rice density. It is worth mentioning that in the Rayleigh case ($\rho = 0$), the probability density function $\hat{p}_\xi(z)$ of $\hat{\xi}(t) = |\hat{\mu}(t)|$ in (4.120) approaches the Rayleigh density if $c_n = \sigma_0\sqrt{2/N}$ and $N \to \infty$.

4.6 Criteria for the Performance Evaluation

In this subsection, we propose some criteria for the assessment of the performance, which will play a significant role in the following chapter. In some aspects of channel modelling it is not especially crucial whether the stochastic process $\hat{\mu}_i(t)$ is ergodic with respect to the mean value or the autocorrelation function. The deviations of the statistical properties of the deterministic process $\tilde{\mu}_i(t)$ from the statistical properties of the underlying ideal stochastic process $\mu_i(t)$ are decisive. From these deviations, criteria can be established for the performance evaluation of the parameter computation methods presented in the next chapter.

Mean-square error of the probability density function: Since the process $\mu_i(t)$ was introduced here as a zero-mean Gaussian random process, i.e., $\mu_i(t) \sim N(0, \sigma_0^2)$, the mean-square error of the probability density function $\tilde{p}_{\mu_i}(x)$ [cf. (4.46)]

$$E_{p_{\mu_i}} := \int_{-\infty}^{\infty} \left(p_{\mu_i}(x) - \tilde{p}_{\mu_i}(x)\right)^2 dx \tag{4.127}$$

defines the first important criterion for the performance evaluation [96].

Mean-square error of the autocorrelation function: As is well known, a real-valued Gaussian random process is completely described by its probability density function and its autocorrelation function. A further important criterion for the performance evaluation is therefore the mean-square error of the autocorrelation function $\tilde{r}_{\mu_i\mu_i}(\tau)$ defined by

$$E_{r_{\mu_i\mu_i}} := \frac{1}{\tau_{max}} \int_{0}^{\tau_{max}} \left(r_{\mu_i\mu_i}(\tau) - \tilde{r}_{\mu_i\mu_i}(\tau)\right)^2 d\tau. \tag{4.128}$$

For the parameter τ_{max}, the value $\tau_{max} = N_i/(2f_{max})$ has turned out to be suitable, especially for the Jakes power spectral density, as we will see in the next chapter.

Model error: Meanwhile, it has already been mentioned several times that the statistical properties of deterministic simulation models can deviate considerably from those of the underlying ideal stochastic reference model. We have seen that for many important statistical quantities, the model error $\Delta\beta$ could be held responsible for this. Therefore, a good parameter computation procedure should only cause a small relative model error $\Delta\beta/\beta$, even when the realization complexity is low, i.e., for a small number of sinusoids N_i. Hence, the model error $\Delta\beta$ and its convergence property $\Delta\beta \to 0$ or $\tilde{\beta} \to \beta$ for $N_i \to \infty$ will receive attention in the next chapter as well.

4.7 Further Reading

The fundamental properties of sum-of-cisoids channel simulators have been treated here comparatively briefly. A more comprehensive study of the statistics of sum-of-cisoids processes and their applications in fading channel modelling for beyond-3G wireless communication systems can be found in [137]. An analytical investigation of the level-crossing rate and the average duration of fades of the envelope of sum-of-cisoids processes is presented in [138].

The stationary and ergodic properties of seven classes of stochastic sum-of-cisoids processes have been explored in [139] and [140], respectively. Finally, the effect of the number of scatterers on the capacity of multipath fading channels has been studied in [141] by using the sum-of-cisoids model.

Test procedures and performance assessment strategies for fading channel simulators have been proposed in [142]. Information on the performance of deterministic and stochastic sum-of-sinusoids Rayleigh fading channel simulators with respect to the bit error probability of digital modulation schemes can be found in [143]. There, the influence of the number of sinusoids on the bit error probability of quadrature phase shift keying (QPSK) and differential phase shift keying (DPSK) with coherent demodulation has been analyzed. In addition, the impact of the model error on the bit error probability of noncoherent DPSK has been studied. The concept was later extended in [144], where the performance of wideband sum-of-sinusoid channel simulators with respect to the bit error probability of orthogonal frequency division multiplexing (OFDM) systems has been analyzed. In this paper, analytical expressions have been derived for the bit error probability of binary phase-shift keying (BPSK) OFDM modulation schemes in the presence of perfect and imperfect channel state information. A performance analysis of wideband sum-of-cisoids channel simulators with regard to the bit error probability of DPSK-OFDM systems can be found in [145]. Performance studies of this kind are important mainly for two reasons. First, the derived analytical expressions are useful for studying the degradation effects on the bit error probability introduced by an imperfect channel simulator. This enables system designers to minimize the channel simulator's realization complexity without introducing significant performance degradations. Second, the analytical results provide us with a powerful tool for assessing the performance of parameter computation methods for the design of sum-of-sinusoids and sum-of-cisoids channel simulators.

An important problem is the computation of the model parameters of the sum-of-sinusoids and sum-of-cisoids processes. The most important solutions to this problem will be described in detail in the next chapter.

Appendix 4.A Derivation of the Autocorrelation Function of the Squared Envelope of Complex Deterministic Gaussian Processes

In this appendix, we derive the autocorrelation function of the squared envelope of a complex deterministic Gaussian process $\tilde{\mu}(t) = \tilde{\mu}_1(t) + j\tilde{\mu}_2(t)$, where it is supposed that the model parameters $f_{i,n}$ fulfill the following conditions:

$$(i) \quad \{f_{1,n}\}_{n=1}^{N_1} \cap \{\pm f_{2,n}\}_{n=1}^{N_2} = \emptyset \tag{4.A.1}$$

$$(ii) \quad f_{i,n} + p_1 f_{i,m} + p_2 f_{i,n'} + p_3 f_{i,m'} = 0 \quad \text{only if} \tag{4.A.2}$$

$$\text{a)} \quad n = m \quad \text{and} \quad n' = m' \quad (p_1 = -1, p_2 = \mp 1, p_3 = \pm 1)$$

$$\text{b)} \quad n = n' \quad \text{and} \quad m = m' \quad (p_1 = \mp 1, p_2 = -1, p_3 = \pm 1)$$

$$\text{c)} \quad n = m' \quad \text{and} \quad m = n' \quad (p_1 = \mp 1, p_2 = \pm 1, p_3 = \mp 1)$$

for $n, n', m, m' = 1, 2, \ldots, N_i$ $(i = 1, 2)$. The first condition [see (4.A.1)] assures the uncorrelatedness of $\tilde{\mu}_1(t)$ and $\tilde{\mu}_2(t)$, i.e., $\tilde{r}_{\mu_1\mu_2}(\tau) = \tilde{r}_{\mu_2\mu_1}(\tau) = 0$. The second condition

(4.A.2) imposes on the model parameters $f_{i,n}$ that each element of the set $\{f_{i,n}\}_{n=1}^{N_i}$ cannot be represented as a combination (sum or difference) of three other elements of the same set.

Now, the substitution of the squared envelope $\tilde{\zeta}^2(t) = |\tilde{\mu}(t)|^2 = \tilde{\mu}_1^2(t) + \tilde{\mu}_2^2(t)$ in the definition of the autocorrelation function for deterministic signals [see (2.144)]

$$\tilde{r}_{\zeta^2\zeta^2}(\tau) = < \tilde{\zeta}^2(t)\,\tilde{\zeta}^2(t+\tau) > \tag{4.A.3}$$

gives

$$\begin{aligned}\tilde{r}_{\zeta^2\zeta^2}(\tau) &= < (\tilde{\mu}_1^2(t) + \tilde{\mu}_2^2(t))\,(\tilde{\mu}_1^2(t+\tau) + \tilde{\mu}_2^2(t+\tau)) > \\ &= < \tilde{\mu}_1^2(t)\tilde{\mu}_1^2(t+\tau) > + < \tilde{\mu}_2^2(t)\tilde{\mu}_2^2(t+\tau) > \\ &\quad + < \tilde{\mu}_1^2(t)\tilde{\mu}_2^2(t+\tau) > + < \tilde{\mu}_2^2(t)\tilde{\mu}_1^2(t+\tau) > \\ &= \tilde{r}_{\mu_1^2\mu_1^2}(\tau) + \tilde{r}_{\mu_2^2\mu_2^2}(\tau) + \tilde{r}_{\mu_1^2\mu_2^2}(\tau) + \tilde{r}_{\mu_2^2\mu_1^2}(\tau).\end{aligned} \tag{4.A.4}$$

Using the elementary trigonometric identity [77, Equation (4.3.32)] and taking the second boundary condition (4.A.2) into account, provides us after some algebraic manipulations with the following formula for the first and second term of the preceding equation

$$\tilde{r}_{\mu_i^2\mu_i^2}(\tau) = \tilde{r}^2_{\mu_i\mu_i}(0) + 2\tilde{r}^2_{\mu_i\mu_i}(\tau), \tag{4.A.5}$$

where $i = 1, 2$. On condition that the first boundary condition (4.A.1) is fulfilled, we can show in a similar way that the third and fourth term in (4.A.4) can be expressed as

$$\tilde{r}_{\mu_1^2\mu_2^2}(\tau) = \tilde{r}_{\mu_2^2\mu_1^2}(\tau) = \tilde{r}_{\mu_1\mu_1}(0) \cdot \tilde{r}_{\mu_2\mu_2}(0). \tag{4.A.6}$$

Finally, by substituting (4.A.5) and (4.A.6) in (4.A.4), it follows that the autocorrelation function of the squared envelope $\tilde{\zeta}^2(t)$ can be expressed in terms of the autocorrelation functions of $\tilde{\mu}_1(t)$ and $\tilde{\mu}_2(t)$ by

$$\tilde{r}_{\zeta^2\zeta^2}(\tau) = (\tilde{r}_{\mu_1\mu_1}(0) + \tilde{r}_{\mu_2\mu_2}(0))^2 + 2(\tilde{r}^2_{\mu_1\mu_1}(\tau) + \tilde{r}^2_{\mu_2\mu_2}(\tau)). \tag{4.A.7}$$

Since $\tilde{r}_{\mu\mu}(\tau) = \tilde{r}_{\mu_1\mu_1}(\tau) + \tilde{r}_{\mu_2\mu_2}(\tau)$, we can also write

$$\tilde{r}_{\zeta^2\zeta^2}(\tau) = \tilde{\sigma}_\mu^4 + \tilde{r}^2_{\mu\mu}(\tau) + (\tilde{r}_{\mu_1\mu_1}(\tau) - \tilde{r}_{\mu_2\mu_2}(\tau))^2, \tag{4.A.8}$$

where $\tilde{\sigma}_\mu^2$ is the average power of $\tilde{\mu}(t)$ given by $\tilde{\sigma}_\mu^2 = \tilde{r}_{\mu\mu}(0) = \tilde{r}_{\mu_1\mu_1}(0) + \tilde{r}_{\mu_2\mu_2}(0)$.

Appendix 4.B Derivation of the Exact Solution of the Level-Crossing Rate and the Average Duration of Fades of Deterministic Rice Processes

We start with the derivation of the exact solution of the level-crossing rate of deterministic Rice processes using a finite number of sinusoids. The assumptions (4.68a) and (4.68b), which were made in Subsection 4.3.2 for the purpose of simplification, will be abandoned here. This appendix closes with the computation of the corresponding average duration of fades.

Let us consider two uncorrelated zero-mean deterministic Gaussian processes

$$\tilde{\mu}_i(t) = \sum_{n=1}^{N_i} c_{i,n}\cos(2\pi f_{i,n} t + \theta_{i,n}), \quad i = 1, 2, \tag{4.B.1}$$

with identical variances equal to $\text{Var}\{\tilde{\mu}_i(t)\} = \tilde{\sigma}^2_{\mu_i} = \sum_{n=1}^{N_i} c^2_{i,n}/2$, where the parameters $c_{i,n}$, $f_{i,n}$, and $\theta_{i,n}$ are non-zero real-valued constants. We demand that the discrete Doppler frequencies have to be different from each other for all $n = 1, 2, \ldots, N_i$ and $i = 1, 2$, so that in particular the sets $\{f_{1,n}\}_{1n=1}^{N}$ and $\{f_{2,n}\}_{2n=1}^{N}$ are disjoint, guaranteeing that the deterministic Gaussian processes $\tilde{\mu}_1(t)$ and $\tilde{\mu}_2(t)$ are uncorrelated. According to (4.41), the probability density function of $\tilde{\mu}_i(t)$ reads as follows

$$\tilde{p}_{\mu_i}(x) = 2\int_0^{\infty}\left[\prod_{n=1}^{N_i} J_0(2\pi c_{i,n}\nu)\right]\cos(2\pi\nu x)\,d\nu, \quad i = 1, 2. \tag{4.B.2}$$

Since the differentiation with respect to time is a linear operation, it follows from (4.B.1) that

$$\dot{\tilde{\mu}}_i(t) = -2\pi\sum_{n=1}^{N_i} c_{i,n}f_{i,n}\sin(2\pi f_{i,n}t + \theta_{i,n}), \quad i = 1, 2, \tag{4.B.3}$$

also describes two uncorrelated zero-mean deterministic Gaussian processes, where the variance of these processes is equal to $\text{Var}\{\dot{\tilde{\mu}}_i(t)\} = \tilde{\beta}_i = 2\pi^2\sum_{n=1}^{N_i}(c_{i,n}f_{i,n})^2$. For the corresponding probability density function $\tilde{p}_{\dot{\mu}_i}(\dot{x})$ of $\dot{\tilde{\mu}}_i(t)$, the expression

$$\tilde{p}_{\dot{\mu}_i}(\dot{x}) = 2\int_0^{\infty}\left[\prod_{n=1}^{N_i} J_0\left[(2\pi)^2 c_{i,n}f_{i,n}\nu\right]\right]\cos(2\pi\nu\dot{x})\,d\nu, \quad i = 1, 2, \tag{4.B.4}$$

holds.

In this connection it has to be taken into account that with (4.13), the cross-correlation function of $\tilde{\mu}_i(t)$ and $\dot{\tilde{\mu}}_i(t)$ can be expressed by

$$\tilde{r}_{\mu_i\dot{\mu}_i}(\tau) = \dot{\tilde{r}}_{\mu_i\mu_i}(\tau) = -\pi\sum_{n=1}^{N_i} c^2_{i,n}f_{i,n}\sin(2\pi f_{i,n}\tau), \tag{4.B.5}$$

and thus it becomes clear that $\tilde{\mu}_i(t)$ and $\dot{\tilde{\mu}}_i(t)$ are in general correlated. For the computation of the level-crossing rate, however, we are only interested in the behaviour of $\tilde{\mu}_i(t_1)$ and $\dot{\tilde{\mu}}_i(t_2)$ at the same time instant $t = t_1 = t_2$, which is equivalent to $\tau = t_2 - t_1 = 0$. Observe that from (4.B.5) it follows $\tilde{r}_{\mu_i\dot{\mu}_i}(\tau) = 0$ for $\tau = 0$, i.e., the deterministic Gaussian processes $\tilde{\mu}_i(t)$ and $\dot{\tilde{\mu}}_i(t)$ are uncorrelated at the same time t. Consequently, also the deterministic processes $\tilde{\mu}_1(t)$, $\tilde{\mu}_2(t)$, $\dot{\tilde{\mu}}_1(t)$, and $\dot{\tilde{\mu}}_2(t)$, are uncorrelated in pairs at the same time t. We know that if two random variables are uncorrelated, then they are not necessarily statistically independent. However, for Gaussian distributed random variables, uncorrelatedness is equivalent to independence [41, p. 212]. In the present case, the probability density functions $\tilde{p}_{\mu_i}(x)$ and $\tilde{p}_{\dot{\mu}_i}(\dot{x})$ [see (4.B.2) and (4.B.4), respectively] are both almost identical to the Gaussian distribution if $N_i \geq 7$. Therefore, we may assume that $\tilde{\mu}_1(t)$, $\tilde{\mu}_2(t)$, $\dot{\tilde{\mu}}_1(t)$, and $\dot{\tilde{\mu}}_2(t)$ are mutually statistically independent at the same time t. As a consequence, the joint probability density function of these processes can be expressed as the product of the individual probability density functions, i.e.,

$$\tilde{p}_{\mu_1\mu_2\dot{\mu}_1\dot{\mu}_2}(x_1, x_2, \dot{x}_1, \dot{x}_2) = \tilde{p}_{\mu_1}(x_1)\cdot\tilde{p}_{\mu_2}(x_2)\cdot\tilde{p}_{\dot{\mu}_1}(\dot{x}_1)\cdot\tilde{p}_{\dot{\mu}_2}(\dot{x}_2). \tag{4.B.6}$$

Considering the line-of-sight component (3.2), we assume — in order to simplify matters — that $f_\rho = 0$ holds, so that $m = m_1 + jm_2$ is a complex-valued constant, whose real and imaginary part is characterized by the discrete probability density function $p_{m_i}(x_i) = \delta(x_i - m_i)$, $i = 1$, 2. For the probability density functions of the complex deterministic processes $\tilde{\mu}_{\rho_i}(t) = \tilde{\mu}_i(t) + m_i$ and $\dot{\tilde{\mu}}_{\rho_i}(t) = \dot{\tilde{\mu}}_i(t) + \dot{m}_i = \dot{\tilde{\mu}}_i(t)$, the following relations hold for $i = 1, 2$:

$$\tilde{p}_{\mu_{\rho_i}}(x_i) = \tilde{p}_{\mu_i}(x_i) * p_{m_i}(x_i) = \tilde{p}_{\mu_i}(x_i - m_i), \tag{4.B.7a}$$

$$\tilde{p}_{\dot{\mu}_{\rho_i}}(\dot{x}_i) = \tilde{p}_{\dot{\mu}_i}(\dot{x}_i) * p_{\dot{m}_i}(\dot{x}_i) = \tilde{p}_{\dot{\mu}_i}(\dot{x}_i). \tag{4.B.7b}$$

Thus, for the joint probability density function of the deterministic processes $\tilde{\mu}_{\rho_1}(t)$, $\tilde{\mu}_{\rho_2}(t)$, $\dot{\tilde{\mu}}_{\rho_1}(t)$, and $\dot{\tilde{\mu}}_{\rho_2}(t)$, we may write

$$\tilde{p}_{\mu_{\rho_1}\mu_{\rho_2}\dot{\mu}_{\rho_1}\dot{\mu}_{\rho_2}}(x_1, x_2, \dot{x}_1, \dot{x}_2) = \tilde{p}_{\mu_1}(x_1 - m_1) \cdot \tilde{p}_{\mu_2}(x_2 - m_2) \cdot \tilde{p}_{\dot{\mu}_1}(\dot{x}_1) \cdot \tilde{p}_{\dot{\mu}_2}(\dot{x}_2). \tag{4.B.8}$$

The transformation of the Cartesian coordinates $(x_1, x_2, \dot{x}_1, \dot{x}_2)$ to polar coordinates $(z, \dot{z}, \theta, \dot{\theta})$ [cf. Appendix 3.D, Equation (3.D.5)] results in the joint probability density function of the processes $\tilde{\xi}(t)$, $\dot{\tilde{\xi}}(t)$, $\tilde{\vartheta}(t)$, and $\dot{\tilde{\vartheta}}(t)$ at the same time t according to

$$\tilde{p}_{\xi\dot{\xi}\vartheta\dot{\vartheta}}(z, \dot{z}, \theta, \dot{\theta}) = z^2 \cdot \tilde{p}_{\mu_1}(z\cos\theta - \rho\cos\theta_\rho) \cdot \tilde{p}_{\mu_2}(z\sin\theta - \rho\sin\theta_\rho) \cdot \tilde{p}_{\dot{\mu}_1}(\dot{z}\cos\theta - \dot{\theta}z\sin\theta) \cdot \tilde{p}_{\dot{\mu}_2}(\dot{z}\sin\theta + \dot{\theta}z\cos\theta), \tag{4.B.9}$$

for $0 \le z < \infty$, $|\dot{z}| < \infty$, $|\theta| \le \pi$, and $|\dot{\theta}| < \infty$. From this expression, the joint probability density function $\tilde{p}_{\xi\dot{\xi}}(z, \dot{z})$ of the deterministic process $\tilde{\xi}(t)$ and its time derivative $\dot{\tilde{\xi}}(t)$ can be obtained after applying the relation (2.89). Hence,

$$\tilde{p}_{\xi\dot{\xi}}(z, \dot{z}) = z^2 \int_{-\infty}^{\infty}\int_{-\pi}^{\pi} \tilde{p}_{\mu_1}(z\cos\theta - \rho\cos\theta_\rho) \cdot \tilde{p}_{\mu_2}(z\sin\theta - \rho\sin\theta_\rho) \cdot \tilde{p}_{\dot{\mu}_1}(\dot{z}\cos\theta - \dot{\theta}z\sin\theta) \cdot \tilde{p}_{\dot{\mu}_2}(\dot{z}\sin\theta + \dot{\theta}z\cos\theta)\, d\theta\, d\dot{\theta}, \tag{4.B.10}$$

where $0 \le z < \infty$ and $|\dot{z}| < \infty$. If we substitute the equation above into the definition of the level-crossing rate $\tilde{N}_\xi(r)$ of deterministic Rice processes $\tilde{\xi}(t)$

$$\tilde{N}_\xi(r) := \int_0^{\infty} \dot{z}\tilde{p}_{\xi\dot{\xi}}(r, \dot{z})\, d\dot{z}, \quad r \ge 0, \tag{4.B.11}$$

then we obtain the expression

$$\tilde{N}_\xi(r) = r^2 \int_{-\pi}^{\pi} \tilde{p}_{\mu_1}(r\cos\theta - \rho\cos\theta_\rho) \cdot \tilde{p}_{\mu_2}(r\sin\theta - \rho\sin\theta_\rho) \cdot \int_0^{\infty} \dot{z} \int_{-\infty}^{\infty} \tilde{p}_{\dot{\mu}_1}(\dot{z}\cos\theta - \dot{\theta}r\sin\theta) \cdot \tilde{p}_{\dot{\mu}_2}(\dot{z}\sin\theta + \dot{\theta}r\cos\theta)\, d\dot{\theta}\, d\dot{z}\, d\theta. \tag{4.B.12}$$

It is mathematically convenient to express (4.B.12) as

$$\tilde{N}_\xi(r) = r^2 \int_{-\pi}^{\pi} w_1(r,\theta)\, w_2(r,\theta) \int_0^\infty \dot{z} f(r,\dot{z},\theta)\, d\dot{z}\, d\theta, \tag{4.B.13}$$

where $w_1(r,\theta)$, $w_2(r,\theta)$, and $f(r,\dot{z},\theta)$ are auxiliary functions defined by

$$w_1(r,\theta) := \tilde{p}_{\mu_1}(r\cos\theta - \rho\cos\theta_\rho), \tag{4.B.14a}$$

$$w_2(r,\theta) := \tilde{p}_{\mu_2}(r\sin\theta - \rho\sin\theta_\rho), \tag{4.B.14b}$$

and

$$f(r,\dot{z},\theta) := 2\int_0^\infty \left[\prod_{n=1}^{N_1} J_0(4\pi^2 c_{1,n} f_{1,n} \nu_1)\right] \int_0^\infty \left[\prod_{m=1}^{N_2} J_0(4\pi^2 c_{2,m} f_{2,m} \nu_2)\right]$$
$$\cdot \int_{-\infty}^{\infty} \left\{\cos\left[2\pi\dot{z}(\nu_1\cos\theta - \nu_2\sin\theta) - 2\pi\dot{\theta}r(\nu_1\sin\theta + \nu_2\cos\theta)\right]\right.$$
$$\left. + \cos\left[2\pi\dot{z}(\nu_1\cos\theta + \nu_2\sin\theta) - 2\pi\dot{\theta}r(\nu_1\sin\theta - \nu_2\cos\theta)\right]\right\} d\dot{\theta}\, d\nu_1\, d\nu_2, \tag{4.B.15}$$

respectively. The integration over $\dot{\theta}$ in (4.B.15) results in

$$\int_{-\infty}^{\infty} \cos\left[2\pi\dot{z}(\nu_1\cos\theta \mp \nu_2\sin\theta) - 2\pi\dot{\theta}r(\nu_1\sin\theta \pm \nu_2\cos\theta)\right] d\dot{\theta}$$
$$= \cos[2\pi\dot{z}(\nu_1\cos\theta \mp \nu_2\sin\theta)] \cdot \delta[r(\nu_1\sin\theta \pm \nu_2\cos\theta)]. \tag{4.B.16}$$

Putting the relation

$$\delta(r(\nu_1\sin\theta \pm \nu_2\cos\theta)) = \frac{\delta(\tan\theta \pm \nu_2/\nu_1)}{|r\nu_1\cos\theta|} \tag{4.B.17}$$

into (4.B.16) and using the transformation of the variables $\varphi = \tan\theta$, then (4.B.13) can be represented by

$$\tilde{N}_\xi(r) = 2r^2 \int_{-\infty}^{\infty} w_1(r,\arctan\varphi)\, w_2(r,\arctan\varphi)$$
$$\cdot \int_0^\infty \dot{z} f(r,\dot{z},\arctan\varphi) \cos^2(\arctan\varphi)\, d\dot{z}\, d\varphi, \tag{4.B.18}$$

where

$$f(r, \dot{z}, \arctan\varphi) = 2\int_0^\infty\int_0^\infty \frac{\left[\prod_{n=1}^{N_1} J_0(4\pi^2 c_{1,n} f_{1,n} \nu_1)\right]\left[\prod_{m=1}^{N_2} J_0(4\pi^2 c_{2,m} f_{2,m} \nu_2)\right]}{|r\nu_1 \cos(\arctan\varphi)|}$$
$$\cdot \left\{\cos\left[2\pi\dot{z}\nu_2\cos(\arctan\varphi)\left(\frac{\nu_1}{\nu_2}-\varphi\right)\right]\cdot\delta\left(\varphi+\frac{\nu_2}{\nu_1}\right)\right.$$
$$\left. + \cos\left[2\pi\dot{z}\nu_2\cos(\arctan\varphi)\left(\frac{\nu_1}{\nu_2}+\varphi\right)\right]\cdot\delta\left(\varphi-\frac{\nu_2}{\nu_1}\right)\right\} d\nu_1\, d\nu_2. \tag{4.B.19}$$

If we now substitute (4.B.19) into (4.B.18) and subsequently transform the Cartesian coordinates (ν_1, ν_2) to polar coordinates (z, θ) by means of $(\nu_1, \nu_2) \rightarrow (z\cos\theta, z\sin\theta)$, then we obtain

$$\tilde{N}_\xi(r) = 2r\int_0^\infty\int_0^\pi w_1(r,\theta)[w_2(r,\theta)+w_2(r,-\theta)]$$
$$\cdot\int_0^\infty j_1(z,\theta)\, j_2(z,\theta)\,\dot{z}\cos(2\pi z\dot{z})\, dz\, d\theta\, d\dot{z}, \tag{4.B.20}$$

where

$$j_1(z,\theta) = \prod_{n=1}^{N_1} J_0(4\pi^2 c_{1,n} f_{1,n} z\cos\theta), \tag{4.B.21a}$$

$$j_2(z,\theta) = \prod_{n=1}^{N_2} J_0(4\pi^2 c_{2,n} f_{2,n} z\sin\theta), \tag{4.B.21b}$$

and $w_1(r, \theta)$, $w_2(r, \theta)$ are the auxiliary functions introduced by (4.B.14a) and (4.B.14b), respectively. For the derivation of (4.B.20), we exploited the fact that $w_1(r, \theta)$ is an even function in θ, i.e., $w_1(r,\theta) = w_1(r, -\theta)$. Since $w_2(r, \theta)$ is neither even nor odd in θ if $\rho \neq 0$ (or $\theta_\rho \neq k\pi$, $k = 0, \pm 1, \pm 2, \ldots$), we may also write for the level-crossing rate of deterministic Rice processes [146]

$$\tilde{N}_\xi(r) = 2r\int_0^\infty\int_{-\pi}^\pi w_1(r,\theta)\, w_2(r,\theta)\int_0^\infty j_1(z,\theta)\, j_2(z,\theta)\dot{z}\cos(2\pi z\dot{z})\, dz\, d\theta\, d\dot{z}. \tag{4.B.22}$$

Further slight simplifications are possible for the level-crossing rate $\tilde{N}_\zeta(r)$ of deterministic Rayleigh processes $\tilde{\zeta}(t)$. Since $\rho = 0$ holds in this case, it follows that $w_2(r, \theta)$ is an even function in θ as well, so that from (4.B.22), the expression

$$\tilde{N}_{\zeta}(r) = 4r \int_0^{\infty} \int_0^{\pi} w_1(r,\theta)\, w_2(r,\theta) \int_0^{\infty} j_1(z,\theta) j_2(z,\theta) \dot{z} \cos(2\pi z\dot{z})\, dz\, d\theta\, d\dot{z} \tag{4.B.23}$$

can be obtained, where $w_1(r, \theta)$ and $w_2(r, \theta)$ have to be computed according to (4.B.14a) and (4.B.14b), respectively, by taking into account that $\rho = 0$ holds. In (4.B.23), $j_1(z, \theta)$ and $j_2(z, \theta)$ again denote the functions introduced in (4.B.21a) and (4.B.21b), respectively.

By means of the exact solution of the level-crossing rate $\tilde{N}_{\xi}(r)$ of deterministic Rice processes $\tilde{\xi}(t)$ it now becomes obvious that apart from ρ and the number of sinusoids N_i, $\tilde{N}_{\xi}(r)$ also depends on the quantities $c_{i,n}$ and $f_{i,n}$. In contrast to that, the phases $\theta_{i,n}$ have no influence on $\tilde{N}_{\xi}(r)$. Thus, for a given number of sinusoids N_i, the deviations between the level-crossing rate of the simulation model and that of the reference model are essentially determined by the method applied for the computation of the model parameters $c_{i,n}$ and $f_{i,n}$. For the purpose of illustration and verification of the obtained theoretical results, the normalized level-crossing rates, computed according to (4.B.22) and (4.B.23), are depicted in Figure 4.B.1 together with the pertinent simulation results. The method of exact Doppler spread (MEDS), method of equal area (MEA), and the Monte Carlo method (MCM), which have been applied for the determination of the model parameters $c_{i,n}$ and $f_{i,n}$, are described in detail in Chapter 5. For the MCM it should in addition be noted that the results shown in Figure 4.B.1 are only valid for a specific realization of the set of discrete Doppler frequencies $\{f_{i,n}\}_{n=1}^{N_i}$. Another realization for the set of parameters $\{f_{i,n}\}_{n=1}^{N_i}$ may give better or worse results for $\tilde{N}_{\xi}(r)$. The reason for this lies in the nature of the MCM, according to which the discrete Doppler frequencies $f_{i,n}$ are random variables, so that the deviations between $\tilde{N}_{\xi}(r)$ and $N_{\xi}(r)$ can only be described statistically. Further details on this subject are described in Subsection 5.1.4.

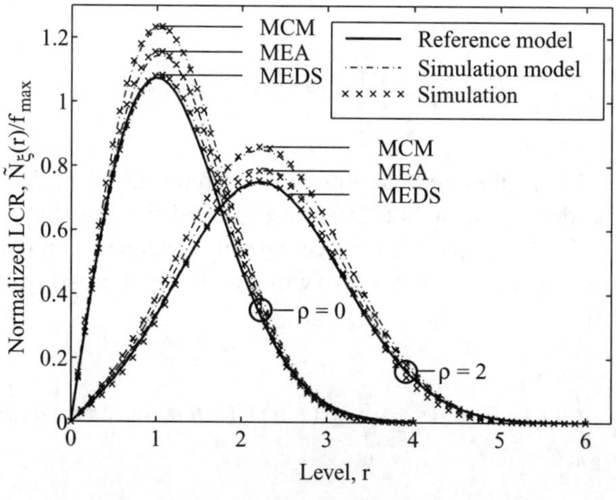

Figure 4.B.1 Normalized level-crossing rate of deterministic Rice and Rayleigh processes, where both have been realized by using $N_1 = 7$ and $N_2 = 8$ (Jakes PSD, $f_{\max} = 91$ Hz, $\sigma_0^2 = 1$, $\theta_\rho = \pi/4$).

Next, we want to show that the level-crossing rate of deterministic Rice processes indeed converges to the level-crossing rate of the reference model as N_i tends to infinity, i.e.,

$$\tilde{N}_\xi(r) \to N_\xi(r), \quad N_i \to \infty. \tag{4.B.24}$$

Therefore, we merely assume that the autocorrelation function $\tilde{r}_{\mu_i\mu_i}(\tau)$ of $\tilde{\mu}_i(t)$ fulfils the following two conditions:

$$\text{(i)} \quad \tilde{r}_{\mu_i\mu_i}(0) = r_{\mu_i\mu_i}(0) \quad \Longleftrightarrow \quad \tilde{\sigma}^2_{\mu_i} = \tilde{\sigma}^2_0 = \sigma^2_0, \tag{4.B.25a}$$

$$\text{(ii)} \quad \ddot{\tilde{r}}_{\mu_i\mu_i}(0) = \ddot{r}_{\mu_i\mu_i}(0) \quad \Longleftrightarrow \quad \tilde{\beta}_i = \tilde{\beta} = \beta. \tag{4.B.25b}$$

The first condition (i) imposes the so-called *power constraint* on the simulation model. If the power constraint is fulfilled, then the mean power of the deterministic process $\tilde{\mu}_i(t)$ is identical to the variance of the stochastic process $\mu_i(t)$. By analogy to the power constraint (4.B.25a), we will in the following denote (4.B.25b) as *curvature constraint*. The curvature constraint imposes on the simulation model that the curvature of the autocorrelation function $\tilde{r}_{\mu_i\mu_i}(\tau)$ of $\tilde{\mu}_i(t)$ at $\tau = 0$ is identical to the curvature of the autocorrelation function $r_{\mu_i\mu_i}(\tau)$ of $\mu_i(t)$ at $\tau = 0$. It should be mentioned that the power constraint is a necessary condition and the curvature constraint is a sufficient condition for the validity of the relation (4.B.24).

In order to prove (4.B.24), we once again consider the time t as a uniformly distributed random variable. Recall that if $N_i \to \infty$, then it follows from the central limit theorem (2.26) that the probability density function in (4.B.1) converges to a Gaussian distribution with mean 0 and variance $\tilde{\sigma}^2_0$, i.e.,

$$\lim_{N_i\to\infty} \tilde{p}_{\mu_i}(x_i) = \frac{1}{\sqrt{2\pi}\,\tilde{\sigma}_0}\, e^{-\frac{x_i^2}{2\tilde{\sigma}_0^2}}, \quad i = 1, 2, \tag{4.B.26}$$

where

$$\tilde{\sigma}^2_0 = \lim_{N_i\to\infty} \tilde{r}_{\mu_i\mu_i}(0) = \lim_{N_i\to\infty} \sum_{n=1}^{N_i} \frac{c_{i,n}^2}{2}. \tag{4.B.27}$$

If we now substitute the result (4.B.26) into (4.B.14a) and (4.B.14b), then it follows

$$w_1(r,\theta) = \frac{1}{\sqrt{2\pi}\,\tilde{\sigma}_0}\, e^{-\frac{(r\cos\theta - \rho\cos\theta_\rho)^2}{2\tilde{\sigma}_0^2}}, \quad \text{as} \quad N_1 \to \infty, \tag{4.B.28a}$$

$$w_2(r,\theta) = \frac{1}{\sqrt{2\pi}\,\tilde{\sigma}_0}\, e^{-\frac{(r\sin\theta - \rho\sin\theta_\rho)^2}{2\tilde{\sigma}_0^2}}, \quad \text{as} \quad N_2 \to \infty. \tag{4.B.28b}$$

Applying the Fourier transform on the right-hand side of (4.B.2) and (4.B.26), we realize that (4.45) can be expressed more generally by

$$\lim_{N_i\to\infty} \prod_{n=1}^{N_i} J_0(2\pi c_{i,n}\nu) = e^{-2(\pi\tilde{\sigma}_0\nu)^2}, \tag{4.B.29}$$

where $\tilde{\sigma}_0^2$ is given by (4.B.27). Furthermore, by replacing the quantities $c_{i,n}$ by $2\pi c_{i,n} f_{i,n}$ in (4.B.29), the relation

$$\lim_{N_i \to \infty} \prod_{n=1}^{N_i} J_0(4\pi^2 f_{i,n} c_{i,n} \nu) = e^{-2\tilde{\beta}_i (\pi \nu)^2} \tag{4.B.30}$$

can easily be derived, where $\tilde{\beta}_i$ denotes the quantity introduced in (4.29). Thus, it becomes obvious that in the limit $N_i \to \infty$, the functions $j_1(z, \theta)$ [see (4.B.21a)] and $j_2(z, \theta)$ [see (4.B.21b)] converge to

$$j_1(z, \theta) = e^{-2\tilde{\beta}_1 (\pi z \cos\theta)^2}, \quad \text{as} \quad N_1 \to \infty, \tag{4.B.31a}$$

$$j_2(z, \theta) = e^{-2\tilde{\beta}_2 (\pi z \sin\theta)^2}, \quad \text{as} \quad N_2 \to \infty, \tag{4.B.31b}$$

respectively. If we now substitute the obtained results (4.B.28a), (4.B.28b), (4.B.31a), and (4.B.31b) into $\tilde{N}_\xi(r)$ given by (4.B.22), and if we furthermore take into account that the condition $\tilde{\beta} = \tilde{\beta}_1 = \tilde{\beta}_2$ holds, then it follows

$$\lim_{N_i \to \infty} \tilde{N}_\xi(r) = \frac{r}{\pi \tilde{\sigma}_0^2} e^{-\frac{r^2+\rho^2}{2\tilde{\sigma}_0^2}} \int_0^\infty \int_{-\pi}^{\pi} \dot{z}\, e^{\frac{r\rho}{\tilde{\sigma}_0^2}\cos(\theta-\theta_\rho)} \int_0^\infty e^{-2\tilde{\beta}(\pi z)^2} \cos(2\pi z \dot{z})\, dz\, d\theta\, d\dot{z}. \tag{4.B.32}$$

Using the integral [23, Equation (3.896.4)]

$$\int_0^\infty e^{-ux^2} \cos(bx)\, dx = \frac{1}{2}\sqrt{\frac{\pi}{u}}\, e^{-\frac{b^2}{4u}}, \quad \text{Re}\{u\} > 0, \tag{4.B.33}$$

(4.B.32) can be simplified to

$$\lim_{N_i \to \infty} \tilde{N}_\xi(r) = \frac{r}{\sqrt{2\pi \tilde{\beta} \tilde{\sigma}_0^2}} e^{-\frac{r^2+\rho^2}{2\tilde{\sigma}_0^2}} \cdot \frac{1}{2\pi} \int_{-\pi}^{\pi} e^{\frac{r\rho}{\tilde{\sigma}_0^2}\cos(\theta-\theta_\rho)}\, d\theta \cdot \int_0^\infty \dot{z} e^{-\frac{\dot{z}^2}{2\tilde{\beta}}}\, d\dot{z}. \tag{4.B.34}$$

The remaining two integrals over θ and $\dot{z}$ can be solved without great expenditure by using the integral representation of the zeroth-order modified Bessel function of the first kind [77, Equation (9.6.16)]

$$I_0(z) = \frac{1}{\pi} \int_0^\pi e^{\pm z \cos\theta}\, d\theta \tag{4.B.35}$$

and the integral [23, Equation (3.461.3)]

$$\int_0^\infty x^{2n+1} e^{-px^2}\, dx = \frac{n!}{2p^{n+1}}, \quad p > 0. \tag{4.B.36}$$

Finally, we obtain

$$\lim_{N_i \to \infty} \tilde{N}_\xi(r) = \sqrt{\frac{\tilde{\beta}}{2\pi}} \cdot \frac{r}{\tilde{\sigma}_0^2} e^{-\frac{r^2+\rho^2}{2\tilde{\sigma}_0^2}} I_0\left(\frac{r\rho}{\tilde{\sigma}_0^2}\right). \tag{4.B.37}$$

Taking the power constraint (4.B.25a) and the curvature constraint (4.B.25b) into account, the right-hand side of the equation above can now directly be identified with (2.119), which proves the correctness of (4.B.24).

For completeness, we will also give the exact solution for the average duration of fades $\tilde{T}_{\xi_-}(r)$ of deterministic Rice processes $\tilde{\xi}(t)$. For this purpose, we need an expression for the cumulative distribution function $\tilde{F}_{\xi_-}(r)$ of $\tilde{\xi}(t)$, which can be derived by substituting the probability density function $\tilde{p}_\xi(z)$ introduced in (4.57) into

$$\tilde{F}_{\xi_-}(r) = \int_0^r \tilde{p}_\xi(z)\, dz. \tag{4.B.38}$$

The integration over z can be carried out by using the indefinite integral [23, Equation (5.56.2)]

$$\int z J_0(z)\, dz = z J_1(z), \tag{4.B.39}$$

so that after some algebraic manipulations, the following result is obtained

$$\tilde{F}_{\xi_-}(r) = 2r \int_0^\infty J_1(2\pi r y) \int_0^\pi h_1(y,\theta) h_2(y,\theta) \cos[2\pi \rho y \cos(\theta - \theta_\rho)]\, d\theta\, dy, \tag{4.B.40}$$

where

$$h_1(y,\theta) = \prod_{n=1}^{N_1} J_0(2\pi c_{1,n} y \cos\theta), \tag{4.B.41a}$$

$$h_2(y,\theta) = \prod_{n=1}^{N_2} J_0(2\pi c_{2,n} y \sin\theta). \tag{4.B.41b}$$

With the cumulative distribution function $\tilde{F}_{\xi_-}(r)$ presented above and the solution for the level-crossing rate $\tilde{N}_\xi(r)$ presented before in (4.B.22), the average duration of fades $\tilde{T}_{\xi_-}(r)$ of deterministic Rice processes $\tilde{\xi}(t)$ can now be studied analytically using

$$\tilde{T}_{\xi_-}(r) = \frac{\tilde{F}_{\xi_-}(r)}{\tilde{N}_\xi(r)}. \tag{4.B.42}$$

The results obtained are illustrated in Figure 4.B.2, which shows the normalized average duration of fades of deterministic Rice and Rayleigh processes, according to the theoretical results (4.B.2), in comparison with the corresponding simulation results.

Subsequently, we want to prove that in the conditions (4.B.25a) and (4.B.25b), the average duration of fades $\tilde{T}_{\xi_-}(r)$ of deterministic Rice processes $\tilde{\xi}(t)$ converges to the average duration

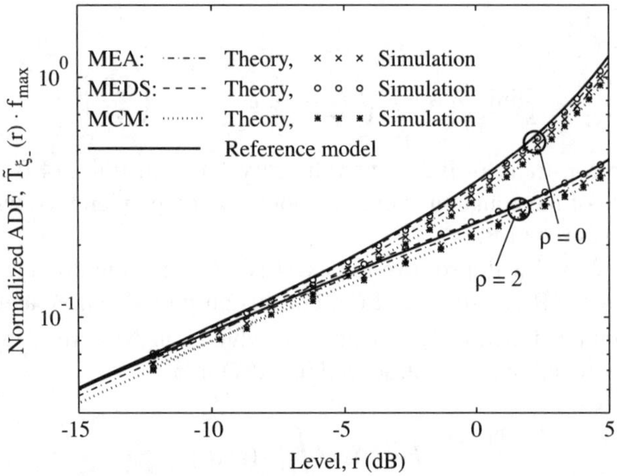

Figure 4.B.2 Normalized average duration of fades of deterministic Rice and Rayleigh processes, where both have been realized by using $N_1 = 7$ and $N_2 = 8$ (Jakes PSD, $f_{\max} = 91\,\text{Hz}$, $\sigma_0^2 = 1$, $\theta_\rho = \pi/4$).

of fades $T_{\xi_-}(r)$ of stochastic Rice processes $\xi(t)$ as the number of sinusoids N_i tends to infinity, i.e.,

$$\tilde{T}_{\xi_-}(r) = T_{\xi_-}(r), \quad \text{as } N_i \to \infty. \tag{4.B.43}$$

Owing to (4.B.24) and the general relation (4.B.42), it is sufficient here to show that

$$\tilde{F}_{\xi_-}(r) = F_{\xi_-}(r), \quad \text{as } N_i \to \infty, \tag{4.B.44a}$$

or, equivalently, that

$$\tilde{p}_{\xi}(r) = p_{\xi}(r), \quad \text{as } N_i \to \infty, \tag{4.B.44b}$$

holds. Due to (4.B.29), we first realize that the functions $h_1(y,\theta)$ [see (4.B.41a)] and $h_2(y,\theta)$ [see (4.B.41b)] tend to

$$h_1(y,\theta) = e^{-2(\pi\tilde{\sigma}_0 y\cos\theta)^2}, \quad \text{as } N_1 \to \infty, \tag{4.B.45a}$$

and

$$h_2(y,\theta) = e^{-2(\pi\tilde{\sigma}_0 y\sin\theta)^2}, \quad \text{as } N_2 \to \infty, \tag{4.B.45b}$$

respectively. With this result it follows from (4.57)

$$\lim_{N_i\to\infty} \tilde{p}_{\xi}(z) = (2\pi)^2 z \int_0^{\infty} e^{-2(\pi\tilde{\sigma}_0 y)^2} J_0(2\pi z y)\, \frac{1}{\pi} \int_0^{\pi} \cos[2\pi\rho y\cos(\theta-\theta_\rho)]\, d\theta\, y\, dy. \tag{4.B.46}$$

The integral representation of the zeroth-order Bessel function [77, Equation (9.1.18)]

$$J_0(z) = \frac{1}{\pi} \int_0^{\pi} \cos(z \cos\theta)\, d\theta \tag{4.B.47}$$

enables us to write the expression in (4.B.46) in the form

$$\lim_{N_i \to \infty} \tilde{p}_{\xi}(z) = (2\pi)^2 z \int_0^{\infty} e^{-2(\pi\tilde{\sigma}_0 y)^2} J_0(2\pi z y)\, J_0(2\pi \rho y)\, y\, dy. \tag{4.B.48}$$

The remaining integral can be solved by using [23, Equation (6.633.2)]

$$\int_0^{\infty} e^{-(ax)^2} J_0(\alpha x)\, J_0(\beta x)\, x\, dx = \frac{1}{2a^2}\, e^{-\frac{\alpha^2+\beta^2}{4a^2}}\, I_0\left(\frac{\alpha\beta}{2a^2}\right). \tag{4.B.49}$$

Thus, we finally obtain

$$\tilde{p}_{\xi}(z) = \frac{z}{\tilde{\sigma}_0^2}\, e^{-\frac{z^2+\rho^2}{2\tilde{\sigma}_0^2}}\, I_0\left(\frac{z\rho}{\tilde{\sigma}_0^2}\right), \quad N_i \to \infty. \tag{4.B.50}$$

With the power constraint (4.B.25a), i.e., $\tilde{\sigma}_0^2 = \sigma_0^2$, the Rice distribution (2.44) follows from the right-hand side of (4.B.50) proving the validity of (4.B.44b) and, consequently, also the validity of (4.B.43).

5

Parametrization of Sum-of-Sinusoids Channel Models

The modelling and analysis of mobile fading channels has been a topic of research since the earliest beginnings of mobile radio communications. Once a new mathematical channel model is developed, its statistical properties need to be confirmed by measurements. If the usefulness of the mathematical channel model is widely accepted by the wireless community, then it might serve as a reference model for the design of a channel simulator. The task of any channel simulator is to reproduce the statistical properties of the reference model with sufficient accuracy while keeping the simulation model's complexity to a minimum. To accomplish the task of designing accurate and efficient channel simulators, one often employs the sum-of-sinusoids processes described in the previous chapter. However, to fully exploit the advantages inherent in sum-of-sinusoids-based simulation models, one must carefully determine the model's parameters. The objective of this chapter is to present solutions to the parametrization problem of sum-of-sinusoids channel simulators.

Parametrization is the process of deciding and defining the parameters necessary for the specification of a model. In the present case, the model is a sum-of-sinusoids channel simulator implying that the parameters to be determined are the discrete Doppler frequencies, the path gains, and the phases. Since the phases are of no importance with regard to long-time simulation aspects, we will focus mainly on the computation of the principal model parameters, namely discrete Doppler frequencies and path gains.

The classification of parameter computation methods can be done in a number of ways. One is to classify them into deterministic and stochastic methods. The philosophy behind deterministic methods is to compute constant values for all principal model parameters, while stochastic methods assume that the discrete Doppler frequencies and/or the gains are random variables. A typical representative of the class of deterministic methods is the Jakes method (JM) [13], which is the parametrization scheme behind the well-known Jakes simulator. Other deterministic methods include the original Rice method [16, 17], the method of equal distances (MED) [147, 148], the mean-square-error method (MSEM) [148], the method of equal areas (MEA) [147, 148], the method of exact Doppler spread (MEDS) [96, 149], the MEDS with

Mobile Radio Channels, Second Edition. Matthias Pätzold.

set partitioning (MEDS-SP) [150], the L_p-norm method (LPNM) [96], the generalized MEA (GMEA) [151], and the Riemann sum method [152]. The latter three methods can be used to model arbitrarily (symmetrical and asymmetrical) Doppler power spectral densities, while the other procedures assume that the Doppler power spectral density has a symmetrical shape. Typical representatives of stochastic methods include the Monte Carlo method (MCM) [99, 100], the harmonic decomposition technique [102], and the randomized MEDS (RMEDS) [111, 153].

Each methodology — deterministic or stochastic — has its own advantages and disadvantages, which have been discussed in numerous papers, e.g., [78, 111, 112]. One advantage of the deterministic channel modelling approach is that the resulting simulation model is ergodic, while stochastic methods result always in non-ergodic fading emulators [133, 134, 154]. By using non-ergodic simulation models, one has to perform several simulation runs with different sets of model parameters. However, the statistics of non-ergodic simulation models can be improved by averaging over the obtained results. On the other hand, if ergodic simulation models are employed, then a single run (or some few runs) are sufficient in most cases. It lies in the nature of deterministic models that their statistical properties cannot be improved in general by averaging over several waveforms. As a consequence, the performance of the majority of deterministic channel simulators can only be improved by increasing the number of sinusoids and thus the complexity, but not by averaging over the simulation results achieved by multiple simulation runs. The only exception is the MEDS-SP [150]. A study of the performance of this method has shown that the approach outperforms existing MCMs by far with respect to single trials as well as averaging over multiple trials [150].

For an infinite number of sinusoids, all these methods with the exception of the Jakes method result in sum-of-sinusoids processes having identical statistical properties, which even match those of the reference model exactly. However, as soon as only a finite number of sinusoids are used, we obtain sum-of-sinusoids processes with completely different statistical properties, which can deviate considerably from those of the reference model. The discussion of these properties will be one objective of this chapter. Thus, in order to compute the principal model parameters, we will proceed by describing and analyzing nine selected design procedures in such a way that they are generally applicable. Afterwards, the methods will be applied to often used power spectral densities. For the most important practical applications, simple closed-form solutions could be found in many cases that ease considerably the computation of the model parameters. The characteristic properties as well as the advantages and disadvantages of each method will be discussed. To facilitate a fair performance comparison, we will apply the assessment criteria and metrics introduced previously in Section 4.6. A further objective is to point out the relationships between the various methods, whenever they exist.

This chapter is organized as follows. Section 5.1 introduces altogether nine different parameter computation methods for sum-of-sinusoids processes. For each method, the characteristic properties as well as the advantages and disadvantages will be discussed. Section 5.2 proposes a stochastic and a deterministic method for the computation of the phases of sum-of-sinusoids models. At that stage, the relevance of the phases with regard to the statistical properties of sum-of-sinusoids processes will be analyzed in more detail. Section 5.3 provides examples of the results achieved for the probability density function of the fading intervals of deterministic Rayleigh processes by using some of the presented parameter computation methods. Section 5.4 introduces three solutions to the parametrization problem of sum-of-cisoids processes. Two of the solutions enable the modelling of non-isotropic scattering scenarios, while the third is restricted to isotropic scattering. Finally, Section 5.5 contains some concluding remarks and presents suggestions for further reading.

5.1 Methods for Computing the Doppler Frequencies and Gains

5.1.1 Method of Equal Distances (MED)

One of the main characteristics of the method of equal distances (MED) [147, 148] is that discrete Doppler frequencies, which are found in neighbouring pairs, have the same distance. This property is achieved by defining the discrete Doppler frequencies $f_{i,n}$ as

$$f_{i,n} := \frac{\Delta f_i}{2}(2n-1), \quad n = 1, 2, \ldots, N_i, \tag{5.1}$$

where

$$\Delta f_i = f_{i,n} - f_{i,n-1}, \quad n = 2, 3, \ldots, N_i, \tag{5.2}$$

denotes the distance between two neighbouring discrete Doppler frequencies of the ith deterministic process $\tilde{\mu}_i(t)$ $(i = 1, 2)$.

In order to compute the path gains $c_{i,n}$, we take a look at the frequency interval

$$I_{i,n} := \left[f_{i,n} - \frac{\Delta f_i}{2}, f_{i,n} + \frac{\Delta f_i}{2} \right), \quad n = 1, 2, \ldots, N_i, \tag{5.3}$$

and demand that within this interval, the mean power of the power spectral density $S_{\mu_i\mu_i}(f)$ of the stochastic reference model is identical to that of the power spectral density $\tilde{S}_{\mu_i\mu_i}(f)$ of the deterministic simulation model, i.e.,

$$\int\limits_{f \in I_{i,n}} S_{\mu_i\mu_i}(f)\, df = \int\limits_{f \in I_{i,n}} \tilde{S}_{\mu_i\mu_i}(f)\, df \tag{5.4}$$

for all $n = 1, 2, \ldots, N_i$ and $i = 1, 2$. Thus, after substituting (4.18) in the equation above, the path gains $c_{i,n}$ are determined by the expression

$$c_{i,n} = 2\sqrt{\int\limits_{f \in I_{i,n}} S_{\mu_i\mu_i}(f)\, df}. \tag{5.5}$$

By substituting (5.5) in (4.11), one can easily prove that $\tilde{r}_{\mu_i\mu_i}(\tau) \to r_{\mu_i\mu_i}(\tau)$ holds as $N_i \to \infty$. Referring to the central limit theorem, it can furthermore be shown that the convergence property $\tilde{p}_{\mu_i}(x) \to p_{\mu_i}(x)$ holds as $N_i \to \infty$. Thus, for an infinite number of sinusoids, the deterministic processes designed by using the method of equal distances can be interpreted as sample functions of the underlying ideal Gaussian random process.

The major disadvantage of this method is the resulting poor periodicity property of $\tilde{\mu}_i(t)$. To make this clear, we start from (4.33), and in connection with (5.1) it follows that the greatest common divisor of the discrete Doppler frequencies equals

$$F_i = \gcd\{f_{i,n}\}_{n=1}^{N_i} = \frac{\Delta f_i}{2}. \tag{5.6}$$

Consequently, $\tilde{\mu}_i(t)$ is periodic with period $T_i = 1/F_i = 2/\Delta f_i$. To obtain a large value for T_i means that Δf_i must be small, which results in the case of the method of equal distances in a high realization complexity, as we will see below.

Jakes power spectral density: The frequency range of the Jakes power spectral density [see (3.23)] is limited to the range $|f| \leq f_{\max}$, so that for a given number of sinusoids N_i, a reasonable value for the difference between two neighbouring discrete Doppler frequencies Δf_i is given by $\Delta f_i = f_{\max}/N_i$. Thus, from (5.1), we obtain the following relation for the discrete Doppler frequencies $f_{i,n}$

$$f_{i,n} = \frac{f_{\max}}{2N_i}(2n-1) \tag{5.7}$$

for all $n = 1, 2, \ldots, N_i$ and $i = 1, 2$. The corresponding path gains $c_{i,n}$ can now easily be computed with (3.23), (5.3), (5.5), and (5.7). After an elementary computation, we find the expression

$$c_{i,n} = \frac{2\sigma_0}{\sqrt{\pi}}\left[\arcsin\left(\frac{n}{N_i}\right) - \arcsin\left(\frac{n-1}{N_i}\right)\right]^{1/2} \tag{5.8}$$

for all $n = 1, 2, \ldots, N_i$ and $i = 1, 2$.

The deterministic processes $\tilde{\mu}_i(t)$ designed with (5.7) and (5.8) obviously have the mean value $\tilde{m}_{\mu_i} = 0$ and the mean power

$$\tilde{\sigma}^2_{\mu_i} = \tilde{r}_{\mu_i\mu_i}(0) = \sum_{n=1}^{N_i} \frac{c^2_{i,n}}{2} = \sigma_0^2. \tag{5.9}$$

Hence, both the mean value and the mean power of the deterministic process $\tilde{\mu}_i(t)$ match exactly the corresponding quantities of the stochastic process $\mu_i(t)$, i.e., the expected value and the variance.

Designing the complex deterministic processes $\tilde{\mu}(t) = \tilde{\mu}_1(t) + j\tilde{\mu}_2(t)$, the uncorrelatedness of $\tilde{\mu}_1(t)$ and $\tilde{\mu}_2(t)$ must be guaranteed. This can be ensured without difficulty by choosing N_2 in accordance with $N_2 = N_1 + 1$, so that due to (5.7) it follows: $f_{1,n} \neq f_{2,m}$ for $n = 1, 2, \ldots, N_1$ and $m = 1, 2, \ldots, N_2$. This again leads to the desired property that $\tilde{\mu}_1(t)$ and $\tilde{\mu}_2(t)$ are uncorrelated [cf. (4.12)].

As an example, the power spectral density $\tilde{S}_{\mu_i\mu_i}(f)$ and the corresponding autocorrelation function $\tilde{r}_{\mu_i\mu_i}(\tau)$ are depicted in Figure 5.1, where the value 25 has been chosen for the number of sinusoids N_i.

For comparison, the autocorrelation function $r_{\mu_i\mu_i}(\tau)$ of the reference model is also presented in Figure 5.1(b). The shape of $\tilde{r}_{\mu_i\mu_i}(\tau)$ shown in this figure makes the periodic behaviour clearly recognizable. In general, the following relation holds

$$\tilde{r}_{\mu_i\mu_i}(\tau + mT_i/2) = \begin{cases} \tilde{r}_{\mu_i\mu_i}(\tau), & m \text{ even}, \\ -\tilde{r}_{\mu_i\mu_i}(\tau), & m \text{ odd}, \end{cases} \tag{5.10}$$

where $T_i = 1/F_i = 2/\Delta f_i = 2N_i/f_{\max}$. If we now choose the value $\tau_{\max} = T_i/4 = N_i/(2f_{\max})$ for the upper limit of the integral in (4.128), then the mean-square error $E_{r_{\mu_i\mu_i}}$ [see (4.128)] can

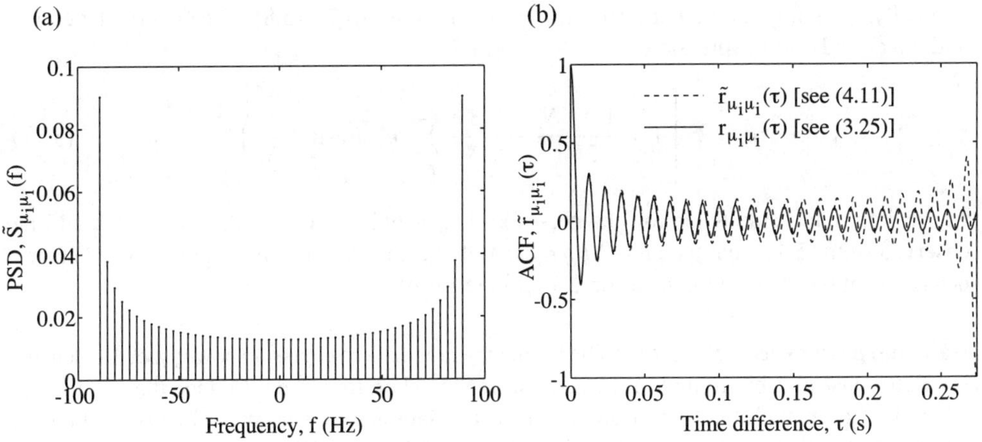

Figure 5.1 (a) Power spectral density $\tilde{S}_{\mu_i\mu_i}(f)$ and (b) autocorrelation function $\tilde{r}_{\mu_i\mu_i}(\tau)$ for $N_i = 25$ (MED, Jakes PSD, $f_{\max} = 91$ Hz, $\sigma_0^2 = 1$).

be evaluated in terms of the number of sinusoids N_i determining the realization complexity of $\tilde{\mu}_i(t)$. The evaluation of the performance criteria $E_{r_{\mu_i\mu_i}}$ and $E_{p_{\mu_i}}$ according to (4.128) and (4.127), respectively, was performed on the basis of the method of equal distances. The obtained results are presented in Figures 5.2(a) and 5.2(b) showing the influence of the used number of sinusoids N_i. For a better assessment of the performance of this method, the results obtained for $c_{i,n} = \sigma_0\sqrt{2/N_i}$ are also shown in Figure 5.2(b). Hence, one realizes that the approximation of the Gaussian distribution using the path gains $c_{i,n}$ according to (5.8) is worse in comparison with the results obtained by using $c_{i,n} = \sigma_0\sqrt{2/N_i}$.

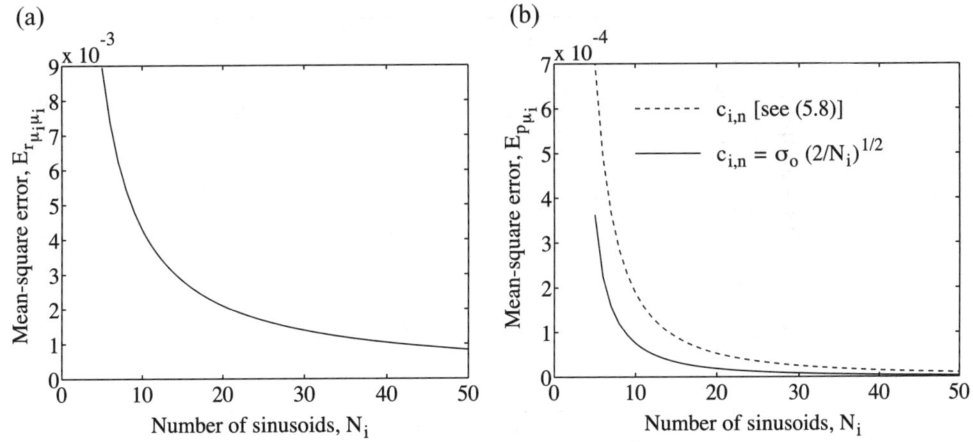

Figure 5.2 Mean-square errors: (a) $E_{r_{\mu_i\mu_i}}$ and (b) $E_{p_{\mu_i}}$ (MED, Jakes PSD, $f_{\max} = 91$ Hz, $\sigma_0^2 = 1$, $\tau_{\max} = N_i/(2f_{\max})$).

Finally, we study the model error $\Delta\beta_i = \tilde{\beta}_i - \beta$. With (5.7), (5.8), (3.68), and (4.29), we find the closed-form expression

$$\Delta\beta_i = \beta \left[1 + \frac{1-4N_i}{2N_i^2} - \frac{8}{\pi N_i^2} \sum_{n=1}^{N_i-1} n \cdot \arcsin\left(\frac{n}{N_i}\right) \right], \tag{5.11}$$

whose right-hand side tends to 0 as $N_i \to \infty$, i.e., it holds $\lim_{N_i\to\infty} \Delta\beta_i = 0$. It should be observed from (5.11) that the ratio $\Delta\beta_i/\beta$ can only be controlled by N_i. Figure 5.3 depicts the behaviour of the relative model error $\Delta\beta_i/\beta$ in terms of N_i.

Gaussian power spectral density: The frequency range of the Gaussian power spectral density in (3.26) must first be limited to the relevant range. Therefore, we introduce the quantity κ_c which is chosen such that the mean power of the Gaussian power spectral density obtained within the frequency range $|f| \leq \kappa_c f_c$ makes up at least 99.99 per cent of its total mean power. This demand is fulfilled with $\kappa_c = 2\sqrt{2/\ln 2}$. Depending on the number of sinusoids N_i, the difference between two neighbouring discrete Doppler frequencies Δf_i can then be described by $\Delta f_i = \kappa_c f_c / N_i$. Thus, with (5.1), we obtain the following expression for the discrete Doppler frequencies $f_{i,n}$

$$f_{i,n} = \frac{\kappa_c f_c}{2N_i}(2n-1) \tag{5.12}$$

for all $n = 1, 2, \ldots, N_i$ and $i = 1, 2$. Now, by using (3.26), (5.3), and (5.5), we can compute the path gains $c_{i,n}$, which results in the closed-form solution

$$c_{i,n} = \sigma_0\sqrt{2}\left[\operatorname{erf}\left(\frac{n\kappa_c\sqrt{\ln 2}}{N_i}\right) - \operatorname{erf}\left(\frac{(n-1)\kappa_c\sqrt{\ln 2}}{N_i}\right)\right]^{\frac{1}{2}} \tag{5.13}$$

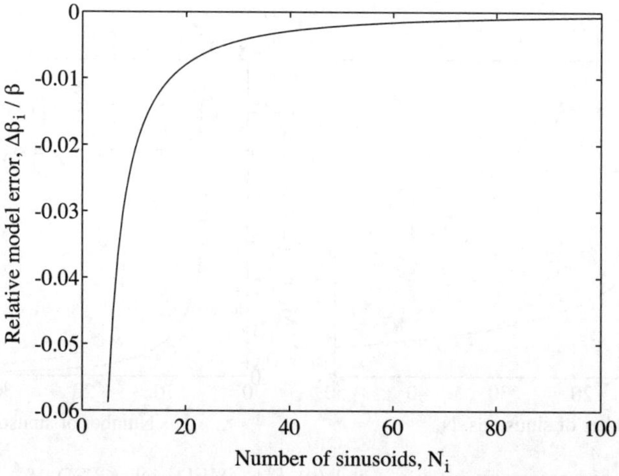

Figure 5.3 Relative model error $\Delta\beta_i/\beta$ (MED, Jakes PSD).

for all $n = 1, 2, \ldots, N_i$ and $i = 1, 2$. Deterministic processes $\tilde{\mu}_i(t)$ designed with (5.12) and (5.13) have a mean value of zero and a mean power of

$$\begin{aligned}\tilde{\sigma}_{\mu_i}^2 = \tilde{r}_{\mu_i\mu_i}(0) &= \sum_{n=1}^{N_i} \frac{c_{i,n}^2}{2} \\ &= \sigma_0^2 \operatorname{erf}\left(\kappa_c\sqrt{\ln 2}\right) \\ &= 0.9999366 \cdot \sigma_0^2 \approx \sigma_0^2,\end{aligned} \tag{5.14}$$

provided that κ_c is chosen as suggested, i.e., $\kappa_c = 2\sqrt{2/\ln 2}$. In the present case, the period of $\tilde{\mu}_i(t)$ is given by $T_i = 2/\Delta f_i = 2N_i/(\kappa_c f_c)$.

Figure 5.4(a) shows the power spectral density $\tilde{S}_{\mu_i\mu_i}(f)$ for $N_i = 25$, and Figure 5.4(b) illustrates the corresponding behaviour of the autocorrelation function $\tilde{r}_{\mu_i\mu_i}(\tau)$ in comparison with the autocorrelation function $r_{\mu_i\mu_i}(\tau)$ of the reference model in the range $0 \leq \tau \leq T_i/2$.

A suitable value for the upper limit of the integral in (4.128) is also in this case a quarter of the period T_i, i.e., $\tau_{\max} = T_i/4 = N_i/(2\kappa_c f_c)$. Evaluating the mean-square error $E_{r_{\mu_i\mu_i}}$ [see (4.128)] by using this value for $\tau_{\max}$ leads to the graph presented in Figure 5.5(a), which illustrates the influence of the number of sinusoids N_i. Figure 5.5(b) presents the results of the evaluation of the performance criterion $E_{p_{\mu_i}}$ according to (4.127). For comparison, the corresponding results obtained by using $c_{i,n} = \sigma_0\sqrt{2/N_i}$ are shown in this figure as well.

Finally, we will analyze the model error $\Delta\beta_i$. Using (4.29), (5.12), (5.13), and (3.68), we find the following closed-form solution for $\Delta\beta_i = \tilde{\beta}_i - \beta$

$$\Delta\beta_i = \beta\left\{2\ln 2\kappa_c^2\left[\left(1 - \frac{1}{2N_i}\right)^2 \operatorname{erf}\left(\kappa_c\sqrt{\ln 2}\right) - \frac{2}{N_i^2}\sum_{n=1}^{N_i-1} n \operatorname{erf}\left(\frac{n\kappa_c\sqrt{\ln 2}}{N_i}\right)\right] - 1\right\}. \tag{5.15}$$

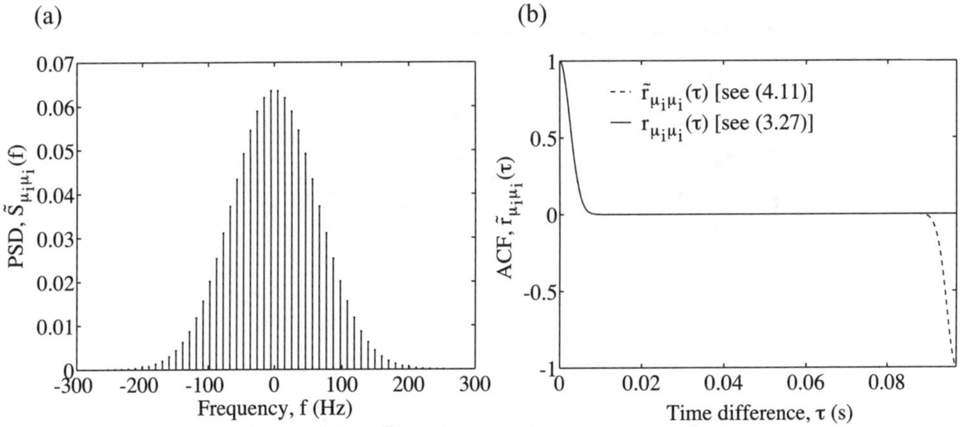

Figure 5.4 (a) Power spectral density $\tilde{S}_{\mu_i\mu_i}(f)$ and (b) autocorrelation function $\tilde{r}_{\mu_i\mu_i}(\tau)$ for $N_i = 25$ (MED, Gaussian PSD, $f_c = \sqrt{\ln 2} f_{\max}$, $f_{\max} = 91$ Hz, $\sigma_0^2 = 1$, $\kappa_c = 2\sqrt{2/\ln 2}$).

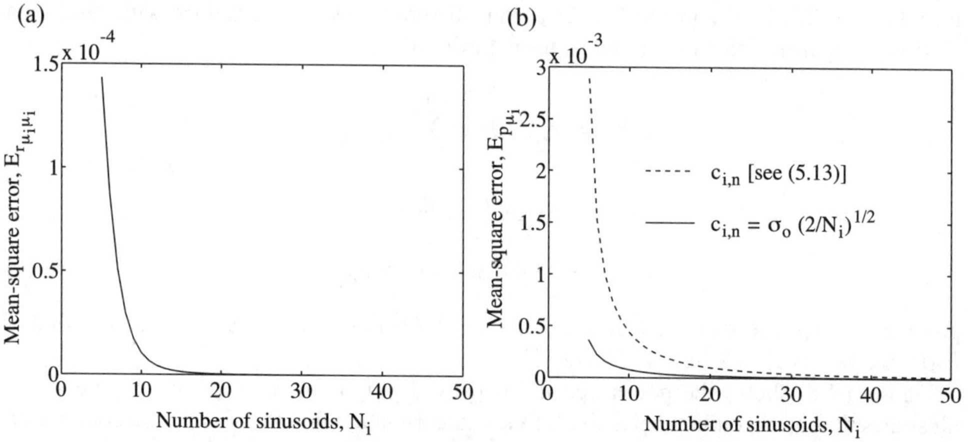

Figure 5.5 Mean-square errors: (a) $E_{r_{\mu_i\mu_i}}$ and (b) $E_{p_{\mu_i}}$ (MED, Gaussian PSD, $f_c = \sqrt{\ln 2}\, f_{\max}$, $f_{\max} = 91$ Hz, $\sigma_0^2 = 1$, $\tau_{\max} = N_i/(2\kappa_c f_c)$, $\kappa_c = 2\sqrt{2/\ln 2}$).

Let us choose $\kappa_c = 2\sqrt{2/\ln 2}$ again. From the equation above, we then obtain the following expression for the relative model error $\Delta\beta_i/\beta$

$$\frac{\Delta\beta_i}{\beta} = 16\left[\left(1 - \frac{1}{2N_i}\right)^2 \operatorname{erf}\left(2\sqrt{2}\right) - \frac{2}{N_i^2}\sum_{n=1}^{N_i-1} n \cdot \operatorname{erf}\left(\frac{n2\sqrt{2}}{N_i}\right)\right] - 1, \tag{5.16}$$

whose behaviour is depicted in Figure 5.6 in terms of N_i. In addition to the rather small values for $\Delta\beta_i/\beta$, the fast convergence behaviour is to be assessed positively. When considering the limit $N_i \to \infty$, it turns out that the model error $\Delta\beta_i$ is very small but still larger than 0, because,

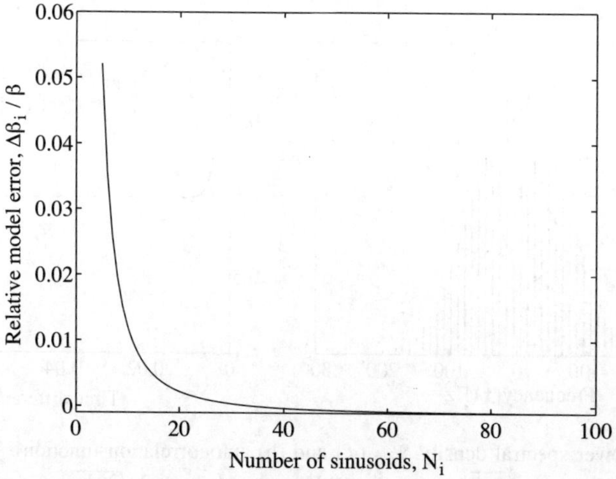

Figure 5.6 Relative model error $\Delta\beta_i/\beta$ (MED, Gaussian PSD, $\kappa_c = 2\sqrt{2/\ln 2}$).

due to the finite value for κ_c, the frequency range of the Gaussian power spectral density in (3.26) is not covered completely by the discrete Doppler frequencies.

In order to avoid correlations between $\tilde{\mu}_1(t)$ and $\tilde{\mu}_2(t)$, N_2 is again defined by $N_2 := N_1 + 1$. Hence, $\Delta\beta = \Delta\beta_1 \approx \Delta\beta_2$ holds, and we can easily analyze the characteristic quantities $\tilde{N}_\xi(r)$, $\tilde{T}_{\xi_-}(r)$, and $\tilde{\tau}_q(r)$ of deterministic Rice processes $\tilde{\xi}(t)$ by making use of (4.73), (4.77), and (4.85), respectively. Concerning the simulation of $\tilde{\xi}(t)$, it must be taken into account that the simulation time T_{sim} does not exceed the period T_i, i.e., $T_{sim} \leq T_i = 2N_i/f_{max}$ (Jakes PSD). As an example, we consider $N_i = 25$ and $f_{max} = 91$ Hz (v $= 110$ km/h, $f_0 = 900$ MHz). This results in a maximum simulation time of $T_{sim} = 0.549$ s. Within this time, the vehicle covers a distance of 16.775 m, so that the model of the underlying mobile radio channel can be regarded as wide-sense stationary.[1] Nevertheless, this simulation time is far from being sufficient to determine typical characteristic quantities such as $\tilde{N}_\xi(r)$, $\tilde{T}_{\xi_-}(r)$, and $\tilde{\tau}_q(r)$ with acceptable precision. A more exact measurement of these quantities for the same parameter sets $\{f_{i,n}\}$ and $\{c_{i,n}\}$ can be achieved by ensemble averaging (statistical averaging). Therefore, various realizations of $\tilde{\xi}(t)$ are required, which can be generated by means of various sets for the phases $\{\theta_{i,n}\}$. Due to the relatively small period T_i, which only increases linearly with N_i, the method of equal distances is not recommended for long-time simulations. For this reason, the properties of this method will not be investigated here in any detail. Further results for this approach can be found in [148].

5.1.2 *Mean-Square-Error Method (MSEM)*

The mean-square-error method (MSEM) is based on the idea that the model parameter sets $\{c_{i,n}\}$ and $\{f_{i,n}\}$ are computed in such a way that the mean-square error (4.128)

$$E_{r_{\mu_i\mu_i}} = \frac{1}{\tau_{max}} \int_0^{\tau_{max}} (r_{\mu_i\mu_i}(\tau) - \tilde{r}_{\mu_i\mu_i}(\tau))^2 \, d\tau \tag{5.17}$$

becomes minimal [148]. Here, $r_{\mu_i\mu_i}(\tau)$ can be any autocorrelation function of the process $\mu_i(t)$ describing a theoretical reference model. Alternatively, $r_{\mu_i\mu_i}(\tau)$ can also be obtained from the measurement data of a real-world channel. The autocorrelation function $\tilde{r}_{\mu_i\mu_i}(\tau)$ of the deterministic model is again given by (4.11). In the equation above, τ_{max} describes an appropriate time-lag interval over which the approximation of the autocorrelation function $r_{\mu_i\mu_i}(\tau)$ is of interest. Unfortunately, a simple and closed-form solution for this problem only exists, if the discrete Doppler frequencies $f_{i,n}$ are defined by (5.1) and, consequently, they are equidistant.

After substituting (4.11) in (5.17) and setting the partial derivatives of $E_{r_{\mu_i\mu_i}}$ with respect to the path gains $c_{i,n}$ equal to zero, i.e., $\partial E_{r_{\mu_i\mu_i}}/\partial c_{i,n} = 0$, we obtain, in connection with (5.1), the following formula [148]

$$c_{i,n} = 2\sqrt{\frac{1}{\tau_{max}} \int_0^{\tau_{max}} r_{\mu_i\mu_i}(\tau) \cos(2\pi f_{i,n}\tau) \, d\tau} \tag{5.18}$$

[1] Measurements have shown [74] that in urban areas mobile radio channels can be modelled appropriately for signal bandwidths up to 10 MHz and covered distances up to 30 m by so-called Gaussian wide-sense stationary uncorrelated scattering (GWSSUS) channels.

for all $n = 1, 2, \ldots, N_i$ $(i = 1, 2)$, where $\tau_{\max}$ is supposed to be given by $\tau_{\max} = T_i/4 = 1/(2\Delta f_i)$.

In the case of the limit $\Delta f_i \to 0$, one can show that from (5.18) the expression

$$c_{i,n} = \lim_{\Delta f_i \to 0} 2\sqrt{\Delta f_i S_{\mu_i\mu_i}(f_{i,n})} \tag{5.19}$$

follows, which is identical to the relation in (4.2a) given by Rice [16, 17]. Numerical investigations have shown that for $\Delta f_i > 0$, the formula

$$c_{i,n} = 2\sqrt{\Delta f_i S_{\mu_i\mu_i}(f_{i,n})}, \tag{5.20}$$

which can easily be evaluated, even then provides a quite useful approximation of the exact solution in (5.18) if the used number of sinusoids N_i is moderate.

We also want to show that $\tilde{r}_{\mu_i\mu_i}(\tau) \to r_{\mu_i\mu_i}(\tau)$ follows as $N_i \to \infty$ $(\Delta f_i \to 0)$. Substituting (5.1) and (5.18) in (4.11) and taking $\tau_{\max} = 1/(2\Delta f_i)$ into account, we may write

$$\begin{aligned}
\lim_{N_i\to\infty} \tilde{r}_{\mu_i\mu_i}(\tau) &= \lim_{N_i\to\infty} \sum_{n=1}^{N_i} \frac{c_{i,n}^2}{2} \cos(2\pi f_{i,n}\tau) \\
&= \lim_{N_i\to\infty} 4 \sum_{n=1}^{N_i} \int_0^{\frac{1}{2\Delta f_i}} r_{\mu_i\mu_i}(\tau') \cos(2\pi f_{i,n}\tau') \cos(2\pi f_{i,n}\tau)\, d\tau' \Delta f_i \\
&= 4 \int_0^{\infty}\int_0^{\infty} r_{\mu_i\mu_i}(\tau') \cos(2\pi f\tau') \cos(2\pi f\tau)\, d\tau'\, df \\
&= 2 \int_0^{\infty} S_{\mu_i\mu_i}(f) \cos(2\pi f\tau)\, df \\
&= r_{\mu_i\mu_i}(\tau).
\end{aligned} \tag{5.21}$$

Next, we will study the application of the mean-square-error method (MSEM) on the Jakes and the Gaussian power spectral densities.

Jakes power spectral density: When using the MSEM, the formula for the computation of the discrete Doppler frequencies $f_{i,n}$ is identical to the relation in (5.7), which has been obtained by applying the MED. For the corresponding path gains $c_{i,n}$, however, we obtain quite different expressions. After substituting (3.25) in (5.18), we find

$$c_{i,n} = 2\sigma_0 \sqrt{\frac{1}{\tau_{\max}} \int_0^{\tau_{\max}} J_0(2\pi f_{\max}\tau) \cos(2\pi f_{i,n}\tau)\, d\tau}, \tag{5.22}$$

where $\tau_{\max} = 1/(2\Delta f_i) = N_i/(2f_{\max})$. There is no closed-form solution for the definite integral appearing in (5.22), so that in this case a numerical integration technique has to be applied in order to calculate the path gains $c_{i,n}$.

As an example, we consider Figure 5.7, where the power spectral density $\tilde{S}_{\mu_i\mu_i}(f)$ and the corresponding autocorrelation function $\tilde{r}_{\mu_i\mu_i}(\tau)$ for $N_i = 25$ are depicted. For reasons of comparison, the autocorrelation function $r_{\mu_i\mu_i}(\tau)$ of the reference model [see (3.25)] is also shown in Figure 5.7(b). The undesired periodic behaviour of $\tilde{r}_{\mu_i\mu_i}(\tau)$, as a consequence of the equidistant discrete Doppler frequencies, is clearly visible.

The evaluation of the performance criteria $E_{r_{\mu_i\mu_i}}$ and $E_{p_{\mu_i}}$ [see (4.128) and (4.127), respectively] has been performed for the MSEM. The obtained results, pointing out the influence of the number of sinusoids N_i, are shown in Figures 5.8(a) and 5.8(b). For a better assessment of the performance of the MSEM, the results found before by using the MED as well as the results obtained by applying the approximate solution in (5.20) are likewise included in these figures.

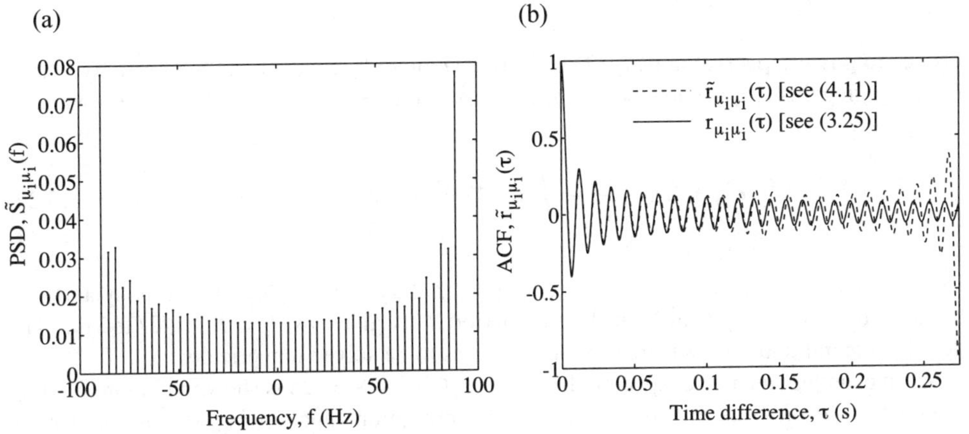

Figure 5.7 (a) Power spectral density $\tilde{S}_{\mu_i\mu_i}(f)$ and (b) autocorrelation function $\tilde{r}_{\mu_i\mu_i}(\tau)$ for $N_i = 25$ (MSEM, Jakes PSD, $f_{\text{max}} = 91$ Hz, $\sigma_0^2 = 1$).

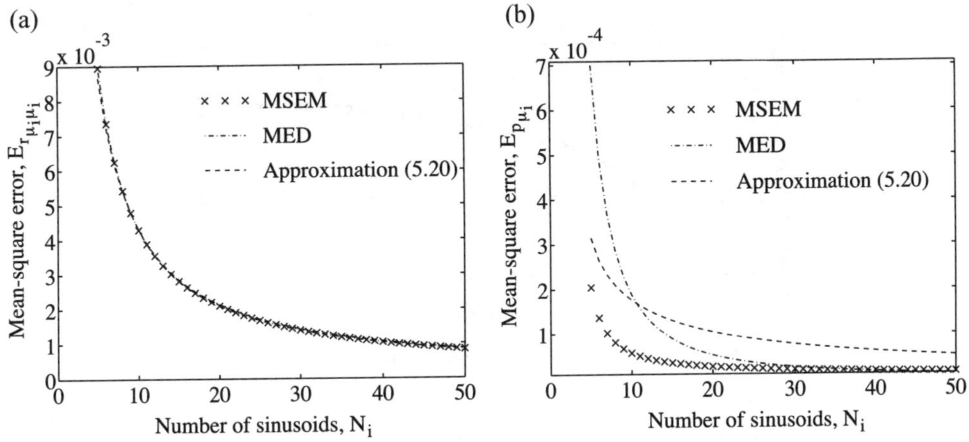

Figure 5.8 Mean-square errors: (a) $E_{r_{\mu_i\mu_i}}$ and (b) $E_{p_{\mu_i}}$ (MSEM, Jakes PSD, $f_{\text{max}} = 91$ Hz, $\sigma_0^2 = 1$, $\tau_{\text{max}} = N_i/(2f_{\text{max}})$).

In case of the MSEM, a simple solution for the model error $\Delta\beta_i$ does not exist. By means of (5.7), (5.22), (3.68), and (4.29), the following formula for $\tilde{\beta}_i$ is obtained after a short computation

$$\tilde{\beta}_i = \beta \frac{1}{N_i} \sum_{n=1}^{N_i} (2n-1)^2 \int_0^1 J_0(\pi N_i u) \cos\left[\frac{\pi}{2}(2n-1)u\right] du. \tag{5.23}$$

With this expression and by making use of $\beta = 2(\pi f_{\max}\sigma_0)^2$, the model error $\Delta\beta_i = \tilde{\beta}_i - \beta$ can be calculated. Figure 5.9 depicts the resulting relative model error $\Delta\beta_i/\beta$ in terms of N_i. This figure also shows the results which can be found when the approximate solution in (5.20) is used. For reasons of comparison, the graph of $\Delta\beta_i/\beta$ obtained by applying the MED is presented here once again.

Gaussian power spectral density: The discrete Doppler frequencies $f_{i,n}$ are given by (5.12). For the path gains $c_{i,n}$, we now obtain, after substituting (3.27) in (5.18), the expression

$$c_{i,n} = 2\sigma_0 \sqrt{\frac{1}{\tau_{\max}} \int_0^{\tau_{\max}} e^{-(\pi f_c \tau)^2/\ln 2} \cos(2\pi f_{i,n}\tau)\, d\tau} \tag{5.24}$$

for all $n = 1, 2, \ldots, N_i$ $(i = 1, 2)$, where $\tau_{\max} = 1/(2\Delta f_i) = N_i/(2\kappa_c f_c)$. Let the quantity κ_c be defined by $\kappa_c = 2\sqrt{2/\ln 2}$, so that the period T_i is given by $T_i = N_i/(\sqrt{2/\ln 2} f_c)$. The definite integral under the square root of (5.24) has to be solved numerically.

As an example, the power spectral density $\tilde{S}_{\mu_i\mu_i}(f)$ for $N_i = 25$ is shown in Figure 5.10(a). Figure 5.10(b) presents the corresponding autocorrelation function $\tilde{r}_{\mu_i\mu_i}(\tau)$ in comparison with the autocorrelation function $r_{\mu_i\mu_i}(\tau)$ of the reference model in the range $0 \leq \tau \leq T_i/2$.

The mean-square errors $E_{r_{\mu_i\mu_i}}$ and $E_{p_{\mu_i}}$ [see (4.128) and (4.127)], occurring when the MSEM is applied, are depicted in Figures 5.11(a) and 5.11(b), respectively. For comparison,

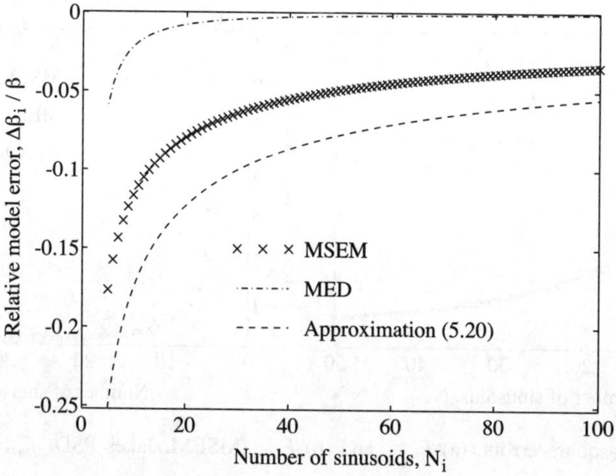

Figure 5.9 Relative model error $\Delta\beta_i/\beta$ (MSEM, Jakes PSD).

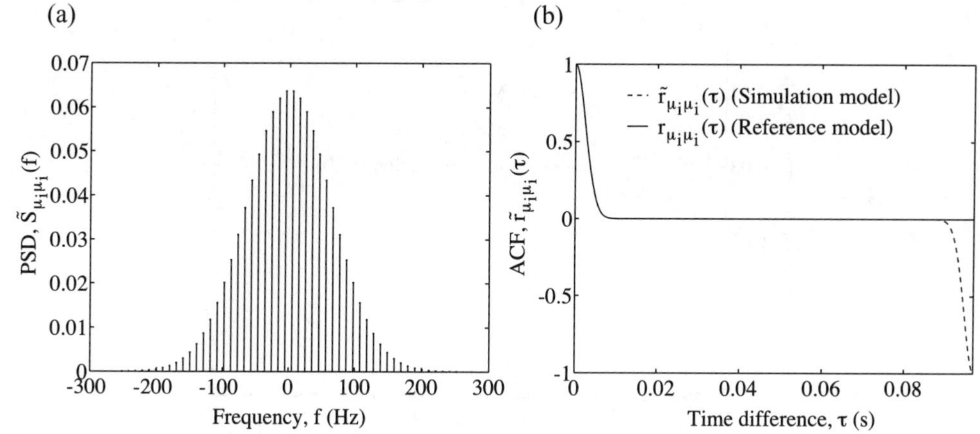

Figure 5.10 (a) Power spectral density $\tilde{S}_{\mu_i\mu_i}(f)$ and (b) autocorrelation function $\tilde{r}_{\mu_i\mu_i}(\tau)$ for $N_i = 25$ (MSEM, Gaussian PSD, $f_c = \sqrt{\ln 2} f_{\max}$, $f_{\max} = 91$ Hz, $\sigma_0^2 = 1$, $\kappa_c = 2\sqrt{2/\ln 2}$).

the results found previously for the MED and the results by using the approximation (5.20) are also shown in these figures.

We turn briefly to the model error $\Delta\beta_i$. Inserting (5.12) and (5.24) into the formula for $\tilde{\beta}_i$ [see (4.29)] and making use of (3.68), the expression

$$\tilde{\beta}_i = \beta \frac{\kappa_c^2 \ln 2}{N_i^2} \sum_{n=1}^{N_i} (2n-1)^2 \int_0^1 e^{-\left(\frac{\pi N_i}{2\kappa_c\sqrt{\ln 2}} u\right)^2} \cos\left[\frac{\pi}{2}(2n-1)u\right] du \tag{5.25}$$

follows, which enables the computation of the model error $\Delta\beta_i = \tilde{\beta}_i - \beta$. Figure 5.12 displays the resulting relative model error $\Delta\beta_i/\beta$ as a function of N_i. In addition, it includes the results

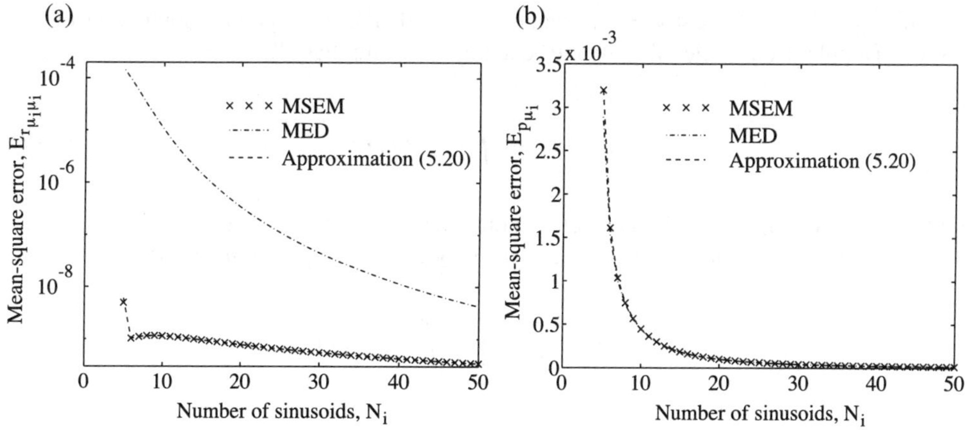

Figure 5.11 Mean-square errors: (a) $E_{r_{\mu_i\mu_i}}$ and (b) $E_{p_{\mu_i}}$ (MSEM, Gaussian PSD, $f_c = \sqrt{\ln 2} f_{\max}$, $f_{\max} = 91$ Hz, $\sigma_0^2 = 1$, $\tau_{\max} = N_i/(2\kappa_c f_c)$, $\kappa_c = 2\sqrt{2/\ln 2}$).

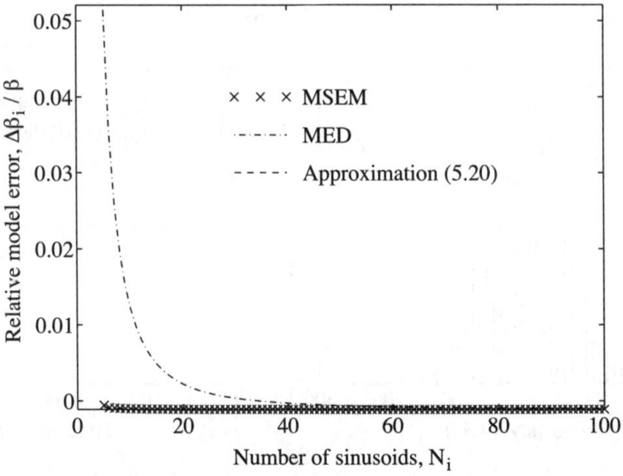

Figure 5.12 Relative model error $\Delta\beta_i/\beta$ (MSEM, Gaussian PSD, $\kappa_c = 2\sqrt{2/\ln 2}$).

which can be obtained by using the approximate solution in (5.20) derived for the path gains $c_{i,n}$. For comparison, this figure includes the graph obtained previously for the MED.

5.1.3 *Method of Equal Areas (MEA)*

The method of equal areas (MEA) [147] is characterized by the fact that the discrete Doppler frequencies $f_{i,n}$ are determined in such a way that the area under the Doppler power spectral density $S_{\mu_i\mu_i}(f)$ is equal to $\sigma_0^2/(2N_i)$ within the frequency range $f_{i,n-1} < f \leq f_{i,n}$, i.e.,

$$\int_{f_{i,n-1}}^{f_{i,n}} S_{\mu_i\mu_i}(f)\,df = \frac{\sigma_0^2}{2N_i} \tag{5.26}$$

for all $n = 1, 2, \ldots, N_i$ and $i = 1, 2$, where $f_{i,0} := 0$. For an explicit computation of the discrete Doppler frequencies $f_{i,n}$, the introduction of the auxiliary function

$$G_{\mu_i}(f_{i,n}) := \int_{-\infty}^{f_{i,n}} S_{\mu_i\mu_i}(f)\,df \tag{5.27}$$

turns out to be helpful. In the case of symmetrical Doppler power spectral densities, i.e., $S_{\mu_i\mu_i}(f) = S_{\mu_i\mu_i}(-f)$, and by using (5.26), we may express $G_{\mu_i}(f_{i,n})$ in the form

$$\begin{aligned} G_{\mu_i}(f_{i,n}) &= \frac{\sigma_0^2}{2} + \sum_{\nu=1}^{n} \int_{f_{i,\nu-1}}^{f_{i,\nu}} S_{\mu_i\mu_i}(f)\,df \\ &= \frac{\sigma_0^2}{2}\left(1 + \frac{n}{N_i}\right). \end{aligned} \tag{5.28}$$

If the inverse function of G_{μ_i}, denoted by $G_{\mu_i}^{-1}$, exists, then the discrete Doppler frequencies $f_{i,n}$ are given by

$$f_{i,n} = G_{\mu_i}^{-1}\left[\frac{\sigma_0^2}{2}\left(1+\frac{n}{N_i}\right)\right] \tag{5.29}$$

for all $n = 1, 2, \ldots, N_i$ and $i = 1, 2$.

The path gains $c_{i,n}$ are now determined by imposing on both the reference model and the simulation model that the mean power of the stochastic process $\mu_i(t)$ is identical to that of the deterministic process $\tilde{\mu}_i(t)$ within the frequency interval $I_{i,n} := (f_{i,n-1}, f_{i,n}]$, i.e.,

$$\int\limits_{f\in I_{i,n}} S_{\mu_i\mu_i}(f)\,df = \int\limits_{f\in I_{i,n}} \tilde{S}_{\mu_i\mu_i}(f)\,df. \tag{5.30}$$

From the equation above and by using the relations (4.18) and (5.26), the following simple formula for the path gains $c_{i,n}$ can be deduced

$$c_{i,n} = \sigma_0\sqrt{\frac{2}{N_i}}, \tag{5.31}$$

where $n = 1, 2, \ldots, N_i$ and $i = 1, 2$. Just as with the previous methods, we will apply this procedure to the Jakes and Gaussian power spectral density.

Jakes power spectral density: With the Jakes power spectral density in (3.23), we obtain the following expression for (5.27)

$$G_{\mu_i}(f_{i,n}) = \frac{\sigma_0^2}{2}\left[1+\frac{2}{\pi}\arcsin\left(\frac{f_{i,n}}{f_{\max}}\right)\right], \tag{5.32}$$

where $0 < f_{i,n} \leq f_{\max}$, $\forall n = 1, 2, \ldots, N_i$ and $i = 1, 2$. If we set up a relation between the right-hand side of (5.32) and (5.28), then the discrete Doppler frequencies $f_{i,n}$ can be computed explicitly. As a result, we find the equation

$$f_{i,n} = f_{\max}\sin\left(\frac{\pi n}{2N_i}\right), \tag{5.33}$$

which is valid for all $n = 1, 2, \ldots, N_i$ and $i = 1, 2$. The corresponding path gains $c_{i,n}$ are furthermore given by (5.31). Theoretically, for all relevant values of N_i, say $N_i \geq 5$, the greatest common divisor $F_i := \gcd\{f_{i,n}\}_{n=1}^{N_i}$ is equal to zero, and thus the period $T_i = 1/F_i$ is infinite. Hence, in this idealized case, the deterministic process $\tilde{\mu}_i(t)$ is nonperiodic. In practical applications, however, the discrete Doppler frequencies $f_{i,n}$ can only be calculated with a finite precision. Let us assume that the discrete Doppler frequencies $f_{i,n}$, according to (5.33), are representable up to the lth decimal place after the comma, then the greatest common divisor is equal to $F_i = \gcd\{f_{i,n}\}_{n=1}^{N_i} = 10^{-l}\ \text{s}^{-1}$. Consequently, the period T_i of the deterministic process $\tilde{\mu}_i(t)$ is $T_i = 1/F_i = 10^l$ s, so that $\tilde{\mu}_i(t)$ can be considered as *quasi-nonperiodic* if $l \geq 10$.

Deterministic processes $\tilde{\mu}_i(t)$ designed with (5.31) and (5.33) are characterized by the mean value $\tilde{m}_{\mu_i} = 0$ and the mean power

$$\tilde{\sigma}_{\mu_i}^2 = \tilde{r}_{\mu_i\mu_i}(0) = \sum_{n=1}^{N_i} \frac{c_{i,n}^2}{2} = \sigma_0^2. \tag{5.34}$$

When designing the complex deterministic processes $\tilde{\mu}(t) = \tilde{\mu}_1(t) + j\tilde{\mu}_2(t)$, the demand for uncorrelatedness of the real part and the imaginary part can be fulfilled sufficiently, if the number of sinusoids N_2 is defined by $N_2 := N_1 + 1$. However, the fact that $f_{1,N_1} = f_{2,N_2} = f_{\max}$ always holds for any chosen values of N_1 and N_2 has the consequence that $\tilde{\mu}_1(t)$ and $\tilde{\mu}_2(t)$ are not completely uncorrelated. But even for moderate values of N_i, the resulting correlation is very small, so that this effect will be ignored in order to simplify matters.

Let us choose $N_i = 25$, for example, then we obtain the results shown in Figures 5.13(a) for the power spectral density $\tilde{S}_{\mu_i\mu_i}(f)$. The corresponding autocorrelation function $\tilde{r}_{\mu_i\mu_i}(\tau)$ is presented in 5.13(b).

Without any difficulty, it can be proved that $\tilde{r}_{\mu_i\mu_i}(\tau) \rightarrow r_{\mu_i\mu_i}(\tau)$ holds as $N_i \rightarrow \infty$. To prove this property, we substitute (5.31) and (5.33) in (4.11), so that we may write

$$\begin{aligned}
\lim_{N_i\to\infty} \tilde{r}_{\mu_i\mu_i}(\tau) &= \lim_{N_i\to\infty} \sum_{n=1}^{N_i} \frac{c_{i,n}^2}{2} \cos(2\pi f_{i,n}\tau) \\
&= \lim_{N_i\to\infty} \sigma_0^2 \frac{1}{N_i} \sum_{n=1}^{N_i} \cos\left[2\pi f_{\max}\tau \sin\left(\frac{\pi n}{2N_i}\right)\right] \\
&= \sigma_0^2 \frac{2}{\pi} \int_0^{\pi/2} \cos(2\pi f_{\max}\tau \sin\alpha)\, d\alpha \\
&= \sigma_0^2 J_0(2\pi f_{\max}\tau) \\
&= r_{\mu_i\mu_i}(\tau).
\end{aligned} \tag{5.35}$$

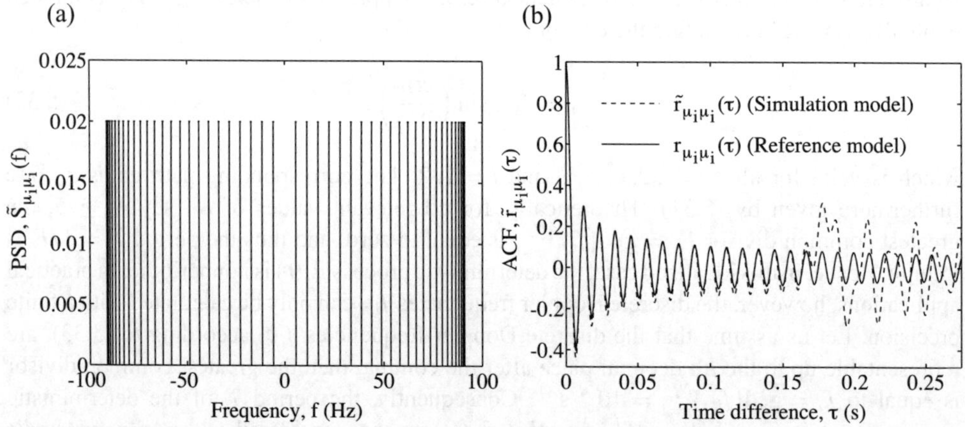

Figure 5.13 (a) Power spectral density $\tilde{S}_{\mu_i\mu_i}(f)$ and (b) autocorrelation function $\tilde{r}_{\mu_i\mu_i}(\tau)$ for $N_i = 25$ (MEA, Jakes PSD, $f_{\max} = 91$ Hz, $\sigma_0^2 = 1$).

In Subsection 4.3.1, we have proved that $\tilde{p}_{\mu_i}(x) \to p_{\mu_i}(x)$ holds if $N_i \to \infty$ and $c_{i,n} = \sigma_0\sqrt{2/N_i}$. Consequently, for an infinite number of sinusoids, the deterministic Gaussian process $\tilde{\mu}_i(t)$ represents a sample function of the stochastic Gaussian random process $\mu_i(t)$. Note that the same relation also exists between the deterministic Rice process $\tilde{\xi}(t)$ and the stochastic Rice process $\xi(t)$.

A deeper insight into the performance of the MEA can be gained by evaluating the performance criteria in (4.127) and (4.128). Both of the resulting mean-square errors $E_{r_{\mu_i\mu_i}}$ and $E_{p_{\mu_i}}$ are shown in Figures 5.14(a) and 5.14(b), respectively.

Next, let us analyze the model error $\Delta\beta_i$. With (5.31), (5.33), and (3.68), we first find the following expression for $\tilde{\beta}_i$ [see (4.29)]

$$\tilde{\beta}_i = \beta \frac{2}{N_i} \sum_{n=1}^{N_i} \sin^2\left(\frac{\pi n}{2N_i}\right) = \beta\left(1 + \frac{1}{N_i}\right). \tag{5.36}$$

Since $\tilde{\beta}_i$ has been introduced as $\tilde{\beta}_i = \beta + \Delta\beta_i$, we thus obtain a simple closed-form formula for the model error

$$\Delta\beta_i = \beta/N_i. \tag{5.37}$$

One may note that $\Delta\beta_i \to 0$ as $N_i \to \infty$. The convergence characteristic of the relative model error $\Delta\beta_i/\beta$ can be studied in Figure 5.15.

Furthermore, we look at the relative error of the level-crossing rate $\tilde{N}_\xi(r)$, to which we refer in the following as ϵ_{N_ξ}, i.e.,

$$\epsilon_{N_\xi} = \frac{N_\xi(r) - \tilde{N}_\xi(r)}{N_\xi(r)}. \tag{5.38}$$

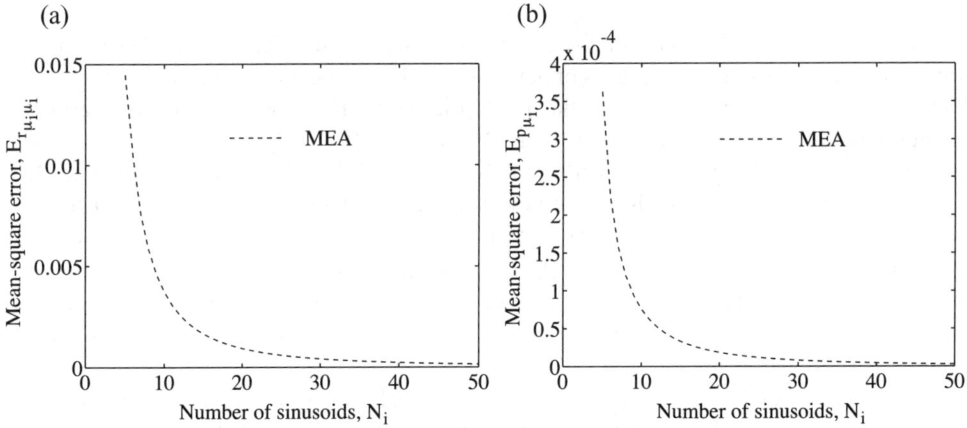

Figure 5.14 Mean-square errors: (a) $E_{r_{\mu_i\mu_i}}$ and (b) $E_{p_{\mu_i}}$ (MEA, Jakes PSD, $f_{\max} = 91$ Hz, $\sigma_0^2 = 1$, $\tau_{\max} = N_i/(2f_{\max})$).

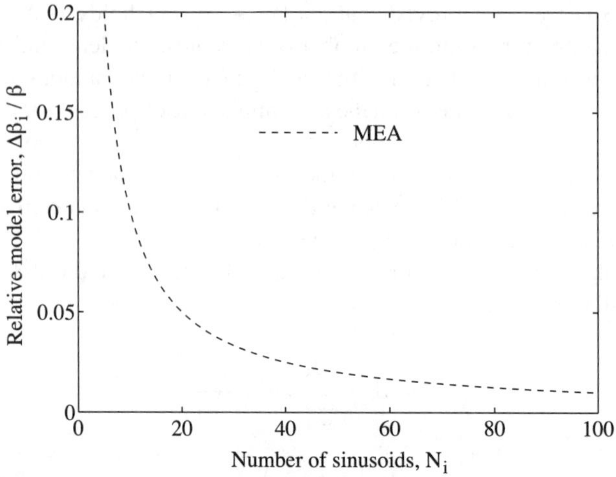

Figure 5.15 Relative model error $\Delta\beta_i/\beta$ (MEA, Jakes PSD).

By using (4.73) and (5.37), we can approximate the relative error ϵ_{N_ξ} in the present case as follows

$$\epsilon_{N_\xi} \approx -\frac{\Delta\beta}{2\beta} \approx -\frac{\Delta\beta_i}{2\beta} = -\frac{1}{2N_i}. \tag{5.39}$$

This result reveals that for a finite number of N_i, the level-crossing rate of the simulation model designed by using the MEA is always higher than the level-crossing rate of the reference model. Obviously, $\epsilon_{N_\xi} \to 0$ holds as $N_i \to \infty$.

In a similar way, one finds the following approximate solution for the relative error $\epsilon_{T_{\xi_-}}$ of the average duration of fades $\tilde{T}_{\xi_-}(r)$

$$\epsilon_{T_{\xi_-}} \approx \frac{\Delta\beta}{2\beta} \approx \frac{\Delta\beta_i}{2\beta} = \frac{1}{2N_i}. \tag{5.40}$$

The quasi-nonperiodic property of $\tilde{\mu}_i(t)$ now allows us to determine both the level-crossing rate and the average duration of fades of deterministic Rice processes by means of simulation. For this purpose, the parameters of the simulation model $\{c_{i,n}\}$ and $\{f_{i,n}\}$ were determined by applying the method of equal areas with $(N_1, N_2) = (10, 11)$. For the computation of the phases $\{\theta_{i,n}\}$, everything that was said at the beginning of this chapter also holds here. Just as in the previous examples, the Jakes power spectral density in (3.23) was characterized by $f_{\max} = 91$ Hz and $\sigma_0^2 = 1$. For the sampling interval T_s of the deterministic Rice process $\tilde{\xi}(kT_s)$, the value $T_s = 10^{-4}$s was chosen. The simulation time T_{sim} was determined for each individual signal level r such that always 10^6 fading intervals or downwards (upwards) level crossings could be evaluated. The results found under these conditions are presented in Figures 5.16(a) and 5.16(b).

These figures depict the analytical solutions previously found for the reference model and the simulation model. The quantities $\tilde{N}_\xi(r)$ and $\tilde{T}_{\xi_-}(r)$ were computed by using $\tilde{\beta} = \tilde{\beta}_1 = \beta(1 + 1/N_1)$ and by means of (4.71) and (4.76), respectively. These figures also demonstrate the excellent correspondence between the analytical expressions derived for the simulation

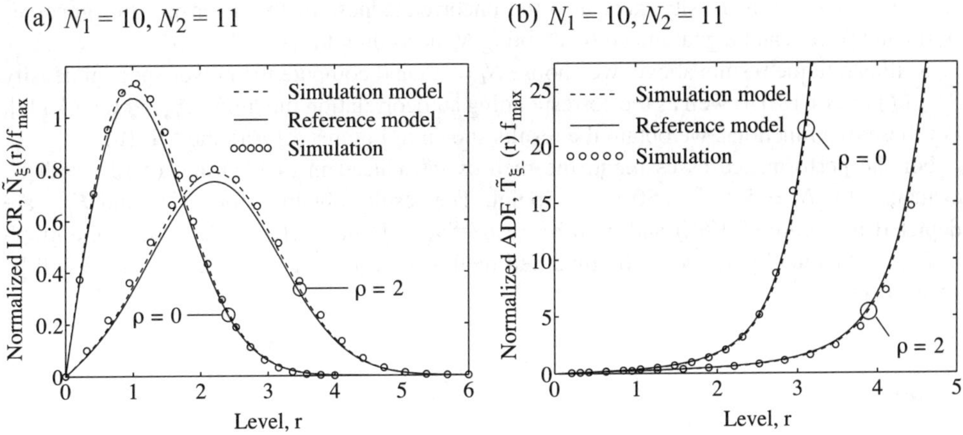

Figure 5.16 (a) Normalized level-crossing rate $\tilde{N}_\xi(r)/f_{\max}$ and (b) normalized average duration of fades $\tilde{T}_{\xi_-}(r) \cdot f_{\max}$ (MEA, Jakes PSD, $f_{\max} = 91$ Hz, $\sigma_0^2 = 1$).

model and the corresponding quantities determined from the measurement results of the simulated envelope behaviour. Unfortunately, the statistical deviations between the reference model and the simulation model are comparatively high, which prompts us to search for a better parameter computation technique. For example, under the conditions given here, where $N_1 = 10$ and $N_2 = 11$, the percentage of the relative error of the level-crossing rate $\tilde{N}_\xi(r)$ and the average duration of fades $\tilde{T}_{\xi_-}(r)$ is about $\epsilon_{N_\xi} \approx -5$ % and $\epsilon_{T_{\xi_-}} \approx +5$ %, respectively.

Gaussian power spectral density: With the Gaussian power spectral density in (3.26), we obtain the following expression for (5.27)

$$G_{\mu_i}(f_{i,n}) = \frac{\sigma_0^2}{2}\left[1 + \operatorname{erf}\left(\frac{f_{i,n}}{f_c}\sqrt{\ln 2}\right)\right] \tag{5.41}$$

for all $n = 1, 2, \ldots, N_i$ and $i = 1, 2$. Since the inverse function $\operatorname{erf}^{-1}(\cdot)$ of the Gaussian error function does not exist, we cannot express the discrete Doppler frequencies $f_{i,n}$ in closed form in this case. Nevertheless, from the difference of the expressions in (5.28) and (5.41), we obtain the equation

$$\frac{n}{N_i} - \operatorname{erf}\left(\frac{f_{i,n}}{f_c}\sqrt{\ln 2}\right) = 0, \quad \forall n = 1, 2, \ldots, N_i \quad (i = 1, 2), \tag{5.42}$$

from which the discrete Doppler frequencies $f_{i,n}$ can be determined by means of a proper numerical root-finding technique.

Since the difference between two neighbouring discrete Doppler frequencies $\Delta f_{i,n} = f_{i,n} - f_{i,n-1}$ depends on the index n via a strongly nonlinear relation, we can conclude that the greatest common divisor $F_i = \gcd\{f_{i,n}\}_{n=1}^{N_i}$ is very small, so that the period $T_i = 1/F_i$ of $\tilde{\mu}_i(t)$ becomes extremely large. We can therefore consider $\tilde{\mu}_i(t)$ as quasi-nonperiodic.

Moreover, for the corresponding path gains $c_{i,n}$, Equation (5.31) holds. Thus, the designed deterministic processes $\tilde{\mu}_i(t)$ have the mean power $\tilde{\sigma}_{\mu_i}^2 = \sigma_0^2$. In the same way as with the

Jakes power spectral density, here also the uncorrelatedness of the deterministic processes $\tilde{\mu}_1(t)$ and $\tilde{\mu}_2(t)$ can be guaranteed by defining N_2 according to $N_2 := N_1 + 1$.

To illustrate the results above, we choose $N_i = 25$ and compute the power spectral density $\tilde{S}_{\mu_i\mu_i}(f)$ [cf. (4.18)] as well as the corresponding autocorrelation function $\tilde{r}_{\mu_i\mu_i}(\tau)$ [cf. (4.11)]. For these two functions, we obtain the results shown in Figures 5.17(a) and 5.17(b).

For the performance assessment, the metrics introduced in (4.128) and (4.127) will be evaluated for $N_i = 5, 6, \ldots, 50$ at this point. The results obtained for $E_{r_{\mu_i\mu_i}}$ and $E_{p_{\mu_i}}$ are depicted in Figures 5.18(a) and 5.18(b), respectively. Figure 5.18(a) also shows the graph of $E_{r_{\mu_i\mu_i}}$ obtained by applying the modified method of equal areas (MMEA), which will be described below.

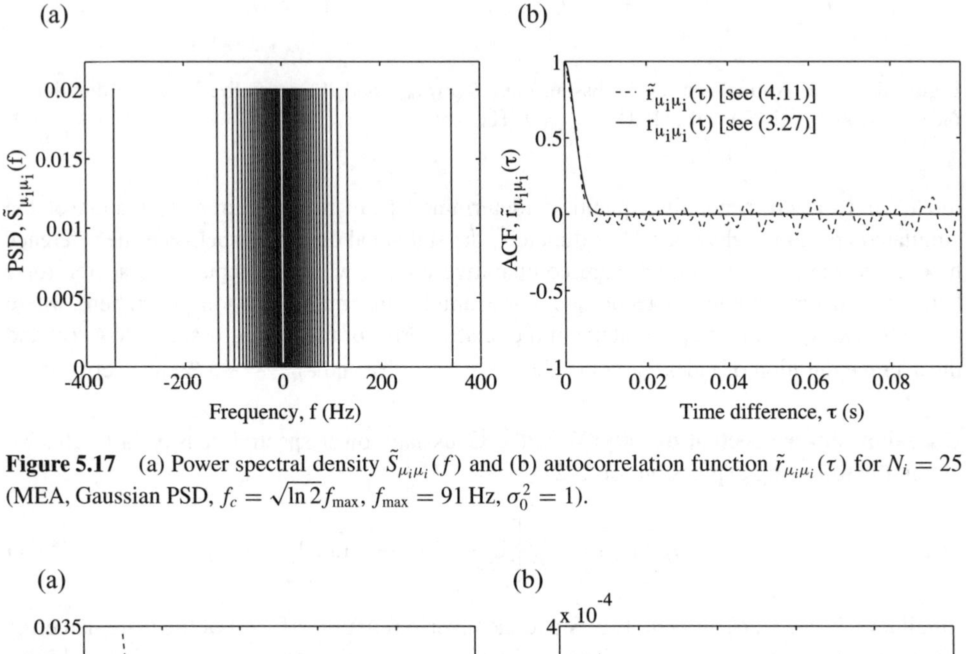

Figure 5.17 (a) Power spectral density $\tilde{S}_{\mu_i\mu_i}(f)$ and (b) autocorrelation function $\tilde{r}_{\mu_i\mu_i}(\tau)$ for $N_i = 25$ (MEA, Gaussian PSD, $f_c = \sqrt{\ln 2} f_{\max}$, $f_{\max} = 91$ Hz, $\sigma_0^2 = 1$).

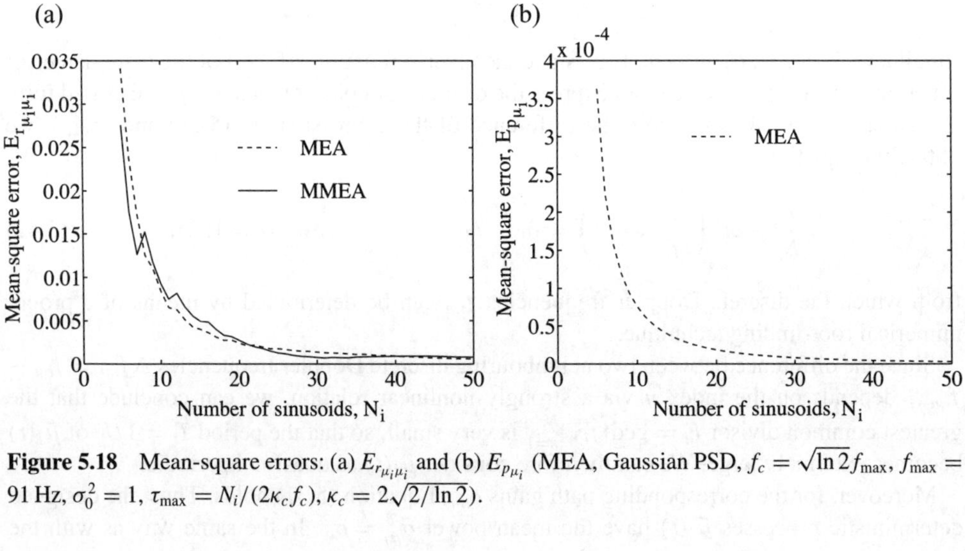

Figure 5.18 Mean-square errors: (a) $E_{r_{\mu_i\mu_i}}$ and (b) $E_{p_{\mu_i}}$ (MEA, Gaussian PSD, $f_c = \sqrt{\ln 2} f_{\max}$, $f_{\max} = 91$ Hz, $\sigma_0^2 = 1$, $\tau_{\max} = N_i/(2\kappa_c f_c)$, $\kappa_c = 2\sqrt{2/\ln 2}$).

Next, we focus on the analysis of the model error $\Delta\beta_i$. Since the discrete Doppler frequencies $f_{i,n}$ are not available in an explicit form, it is not possible to find a closed-form expression for the model error $\Delta\beta_i$. Therefore, we proceed as follows. First, the parameter sets $\{c_{i,n}\}$ and $\{f_{i,n}\}$ are computed by means of (5.31) and (5.42). Then, the quantity $\tilde{\beta}_i$ will be determined by using (4.29). In connection with $\beta = 2(\pi f_c \sigma_0)^2 / \ln 2$, we are now able to evaluate the model error $\Delta\beta_i = \tilde{\beta}_i - \beta$. The results obtained for the relative model error $\Delta\beta_i/\beta$ are presented in Figure 5.19 in dependence on the number of sinusoids N_i.

Taking Figure 5.19 into account, it turns out that the percentage of the relative model error $\Delta\beta_i/\beta$ is lower than 50 per cent only if $N_i \geq 49$. Hence, the MEA is totally unsuitable for the Gaussian power spectral density. Since the main reason for this is the bad positioning of the discrete Doppler frequency $f_{i,n}$ for the value $n = N_i$ [see also Figure 5.17(a)], this imperfect adaptation can be avoided by a simple modification of the procedure. Instead of computing the complete set $\{f_{i,n}\}_{n=1}^{N_i}$ of the discrete Doppler frequencies according to (5.42), as done before, we will now only use the root-finding algorithm for the computation of $\{f_{i,n}\}_{n=1}^{N_i-1}$ and determine the remaining discrete Doppler frequency f_{i,N_i} such that $\tilde{\beta}_i = \beta$ holds.

For this type of modified method of equal areas (MMEA), one obtains the following set of equations:

$$\frac{n}{N_i} - \operatorname{erf}\left(\frac{f_{i,n}}{f_c}\sqrt{\ln 2}\right) = 0, \quad \forall n = 1, 2, \ldots, N_i - 1, \tag{5.43a}$$

$$f_{i,N_i} = \sqrt{\frac{\beta N_i}{(2\pi\sigma_0)^2} - \sum_{n=1}^{N_i-1} f_{i,n}^2}. \tag{5.43b}$$

The corresponding path gains $c_{i,n}$ are of course still given by (5.31). The advantage of the modified method of equal areas is that the relative model error $\Delta\beta_i/\beta$ is always equal to zero for all given values of $N_i = 1, 2, \ldots$ $(i = 1, 2)$. This is demonstrated graphically in

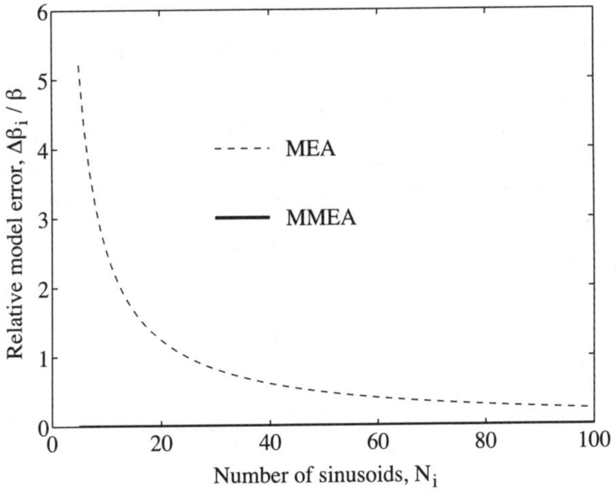

Figure 5.19 Relative model error $\Delta\beta_i/\beta$ (MEA, Gaussian PSD, $f_c = \sqrt{\ln 2} f_{\max}$, $f_{\max} = 91$ Hz, $\sigma_0^2 = 1$).

Figure 5.19. However, the effects on the mean-square error $E_{r_{\mu_i\mu_i}}$ are small, as can be seen from Figure 5.18(a).

For the determination of the level-crossing rate $\tilde{N}_\xi(r)$ and the average duration of fades $\tilde{T}_{\xi_-}(r)$, we proceed in the same way as described before for the Jakes power spectral density. The results obtained for $\tilde{N}_\xi(r)$ and $\tilde{T}_{\xi_-}(r)$ by choosing $(N_1, N_2) = (10, 11)$ are presented in Figures 5.20(a) and 5.20(b), respectively. Here, the modified method of equal areas was used for the computation of the model parameters.

Closed-form analytical expressions for the relative error of both the level-crossing rate $\tilde{N}_\xi(r)$ and the average duration of fades $\tilde{T}_{\xi_-}(r)$ cannot be derived for the MEA in the case of the Gaussian power spectral density. The reason for this is to be found in the implicit Equation (5.42) for the determination of the discrete Doppler frequencies $f_{i,n}$. For the MMEA, however, both relative errors ϵ_{N_ξ} and $\epsilon_{T_{\xi_-}}$ are equal to zero.

5.1.4 *Monte Carlo Method (MCM)*

The Monte Carlo method was first proposed in [100] for the stochastic modelling and the digital simulation of mobile radio channels. Based on this paper, a model for the equivalent discrete-time channel [155] in the complex baseband was introduced in [99, 156]. In the following, we will use this method for the design of deterministic processes whose resulting statistical properties will be analyzed afterwards.

The principle of the Monte Carlo method is based on the realization of the discrete Doppler frequencies $f_{i,n}$ according to a given probability density function $p_{\mu_i}(f)$, which is related to the power spectral density $S_{\mu_i\mu_i}(f)$ of the coloured Gaussian random process $\mu_i(t)$ by

$$p_{\mu_i}(f) = \frac{1}{\sigma_0^2} S_{\mu_i\mu_i}(f). \tag{5.44}$$

Again, σ_0^2 denotes the mean power (variance) of the Gaussian random process $\mu_i(t)$.

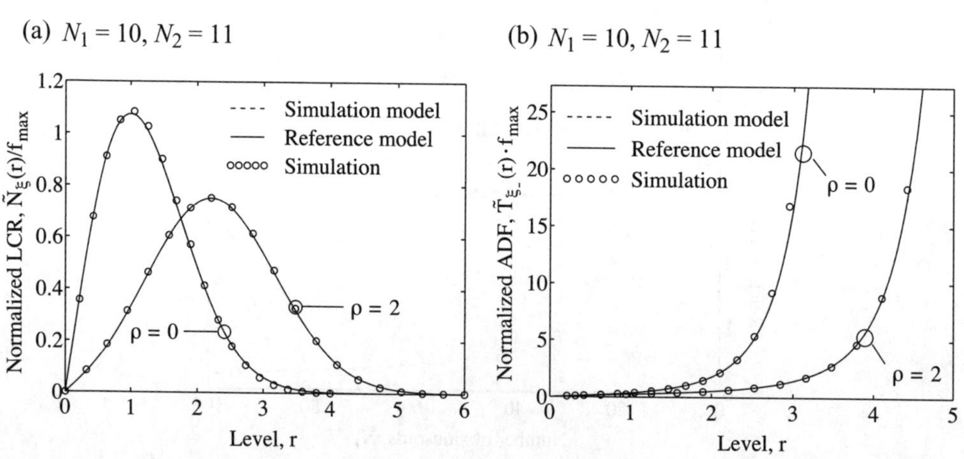

Figure 5.20 (a) Normalized level-crossing rate $\tilde{N}_\xi(r)/f_{\max}$ and (b) normalized average duration of fades $\tilde{T}_{\xi_-}(r) \cdot f_{\max}$ (MMEA, Gaussian PSD, $f_c = \sqrt{\ln 2} f_{\max}$, $f_{\max} = 91$ Hz, $\sigma_0^2 = 1$).

For the computation of the discrete Doppler frequencies $f_{i,n}$, we will be guided by the procedure presented in [99, 156]. Let $u_{i,n}$ be a random variable uniformly distributed over the interval (0,1]. Furthermore, let $g_{\mu_i}(u_{i,n})$ be a mapping that is chosen in such a way that the distribution of the discrete Doppler frequencies $f_{i,n} = g_{\mu_i}(u_{i,n})$ is equal to the desired cumulative distribution function

$$F_{\mu_i}(f_{i,n}) = \int_{-\infty}^{f_{i,n}} p_{\mu_i}(f)\, df. \tag{5.45}$$

According to [41], $g_{\mu_i}(u_{i,n})$ can then be identified with the inverse function of $F_{\mu_i}(f_{i,n}) = u_{i,n}$. Consequently, for the discrete Doppler frequencies $f_{i,n}$, the relation

$$f_{i,n} = g_{\mu_i}(u_{i,n}) = F_{\mu_i}^{-1}(u_{i,n}) \tag{5.46}$$

holds for all $n = 1, 2, \ldots, N_i$ $(i = 1, 2)$. Generally, we obtain positive as well as negative values for $f_{i,n}$. In cases where the probability density function $p_{\mu_i}(f)$ is an even function, i.e., $p_{\mu_i}(f) = p_{\mu_i}(-f)$, we can confine ourselves to positive values for $f_{i,n}$ without restriction of generality. This will be achieved by substituting the uniformly distributed random variable $u_{i,n} \in (0, 1]$ in (5.45) by $(1 + u_{i,n})/2 \in (0.5, 1]$.

As it follows from $u_{i,n} > 0$ that $f_{i,n} > 0$ holds, the time average of $\tilde{\mu}_i(t)$ is equal to zero, i.e., $\tilde{m}_{\mu_i} = m_{\mu_i} = 0$.

The path gains $c_{i,n}$ are chosen such that the mean power of $\tilde{\mu}_i(t)$ is identical to the variance of $\mu_i(t)$, i.e., $\tilde{\sigma}_0^2 = \tilde{r}_{\mu_i\mu_i}(0) = \sigma_0^2$, which is guaranteed by choosing $c_{i,n}$ according to (5.31). Hence, it follows

$$c_{i,n} = \sigma_0 \sqrt{\frac{2}{N_i}} \tag{5.47}$$

for all $n = 1, 2, \ldots, N_i$ $(i = 1, 2)$.

One may consider that with the Monte Carlo method, not only the phases $\theta_{i,n}$, but also the discrete Doppler frequencies $f_{i,n}$ are random variables. In principle, there is no difference whether statistical or deterministic methods are applied for the determination of the model parameters $(c_{i,n}, f_{i,n}, \theta_{i,n})$, because the resulting process $\tilde{\mu}_i(t)$ that matters here is by definition a deterministic process. (We refer to Section 4.1, where deterministic processes $\tilde{\mu}_i(t)$ have been introduced as sample functions or as realizations of stochastic processes $\hat{\mu}_i(t)$.) However, especially for a small number of sinusoids, the ergodic properties of the stochastic process $\hat{\mu}_i(t)$ are poor, if the Monte Carlo method is applied for the computation of the discrete Doppler frequencies $f_{i,n}$ [78]. The consequence is that many important characteristic quantities of the deterministic process $\tilde{\mu}_i(t)$, like the Doppler spread, the level-crossing rate, and the average duration of fades become random values, which in particular cases can deviate considerably from the desired characteristic quantities prescribed by the reference model. In the following, we want to put this into concrete terms with the example of the Jakes and Gaussian power spectral density.

Jakes power spectral density: The application of the Monte Carlo method in connection with the Jakes power spectral density in (3.23) results in the following expression for the discrete

Doppler frequencies $f_{i,n}$

$$f_{i,n} = f_{\max} \sin\left(\frac{\pi}{2} u_{i,n}\right), \tag{5.48}$$

where $u_{i,n} \in (0, 1]$ for all $n = 1, 2, \ldots, N_i$ $(i = 1, 2)$. For the path gains $c_{i,n}$, Equation (5.47) still holds true. It should be observed that the replacement of $u_{i,n}$ in (5.48) by the deterministic quantity n/N_i leads exactly to the relation (5.33) that we found for the method of equal areas.

Since the discrete Doppler frequencies $f_{i,n}$ are random variables, the greatest common divisor $F_i = \gcd\{f_{i,n}\}_{n=1}^{N_i}$ is a random variable as well. However, for a given realization of the set $\{f_{i,n}\}$ with N_i elements, the greatest common divisor F_i is a constant that can be determined by applying the Euclidian algorithm to $\{f_{i,n}\}$, where we have to take into account that the discrete Doppler frequencies $f_{i,n}$ are real numbers. Generally, one can assume that the greatest common divisor F_i is very small, and thus the period $T_i = 1/F_i$ is very large, so that $\tilde{\mu}_i(t)$ can be considered as a quasi-nonperiodical function. Needless to say that this holds even more the greater the number of sinusoids N_i is chosen.

The demand for uncorrelatedness of the real and the imaginary parts does not cause any difficulties for the Monte Carlo method when designing complex deterministic processes $\tilde{\mu}(t) = \tilde{\mu}_1(t) + j\tilde{\mu}_2(t)$. The reason for this is that even for $N_1 = N_2$, the realized sets $\{f_{1,n}\}$ and $\{f_{2,n}\}$ are in general mutually exclusive events, leading to the fact that $\tilde{\mu}_1(t)$ and $\tilde{\mu}_2(t)$ are uncorrelated in the time-average sense.

An example of the power spectral density $\tilde{S}_{\mu_i\mu_i}(f)$, obtained with $N_i = 25$ sinusoids, is shown in Figure 5.21(a). The autocorrelation function $\tilde{r}_{\mu_i\mu_i}(\tau)$, which was computed according to (4.11), is plotted in Figure 5.21(b) for two different realizations of the sets $\{f_{i,n}\}$.

Regarding Figure 5.21(b), one can see that even in the range $0 \leq \tau \leq \tau_{\max}$, the autocorrelation function $\tilde{r}_{\mu_i\mu_i}(\tau)$ of the deterministic process $\tilde{\mu}_i(t)$ can deviate considerably from the ideal autocorrelation function $r_{\mu_i\mu_i}(\tau)$ of the stochastic process $\mu_i(t)$.[2]

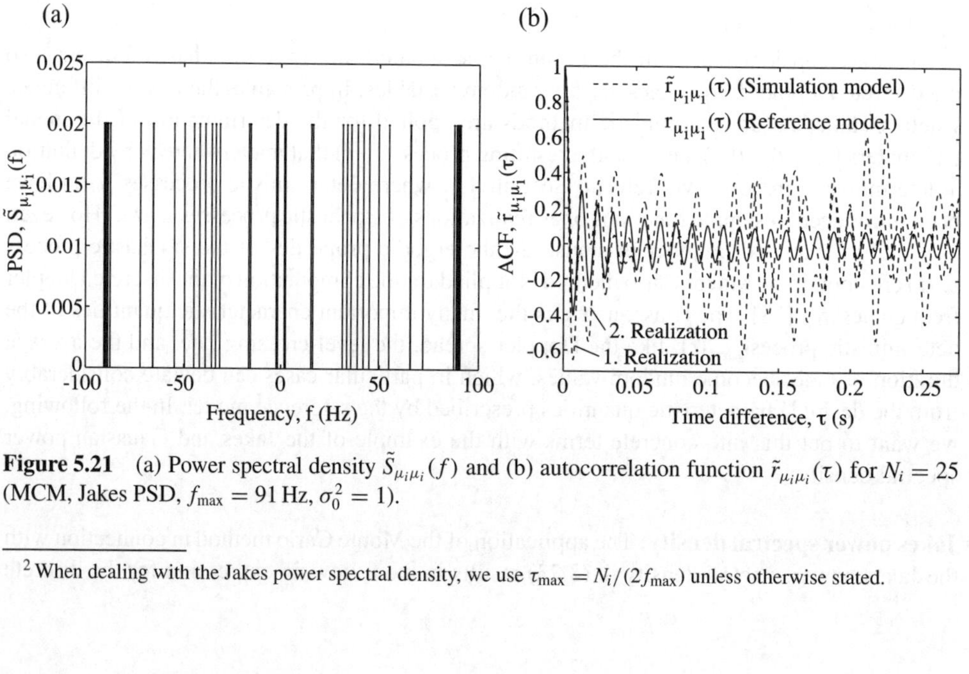

Figure 5.21 (a) Power spectral density $\tilde{S}_{\mu_i\mu_i}(f)$ and (b) autocorrelation function $\tilde{r}_{\mu_i\mu_i}(\tau)$ for $N_i = 25$ (MCM, Jakes PSD, $f_{\max} = 91$ Hz, $\sigma_0^2 = 1$).

[2] When dealing with the Jakes power spectral density, we use $\tau_{\max} = N_i/(2f_{\max})$ unless otherwise stated.

On the other hand, if we analyze the autocorrelation function $\hat{r}_{\mu_i\mu_i}(\tau)$ of the stochastic process $\hat{\mu}_i(t)$, then, by using (4.95), we obtain

$$\begin{aligned}\hat{r}_{\mu_i\mu_i}(\tau) &:= E\{\hat{\mu}_i(t)\hat{\mu}_i(t+\tau)\}\\ &= \sum_{n=1}^{N_i} \frac{c_{i,n}^2}{2} E\{\cos(2\pi f_{i,n}\tau)\}\\ &= \sum_{n=1}^{N_i} \frac{c_{i,n}^2}{2} J_0(2\pi f_{\max}\tau)\\ &= \sigma_0^2 J_0(2\pi f_{\max}\tau)\\ &= r_{\mu_i\mu_i}(\tau). \end{aligned} \tag{5.49}$$

Summarizing, we can say that the autocorrelation function $\hat{r}_{\mu_i\mu_i}(\tau)$ of the stochastic simulation model is equal to the ideal autocorrelation function $r_{\mu_i\mu_i}(\tau)$ of the reference model, whereas the autocorrelation function $\tilde{r}_{\mu_i\mu_i}(\tau)$ of the deterministic simulation model is different from the two preceding autocorrelation functions, i.e., $r_{\mu_i\mu_i}(\tau) = \hat{r}_{\mu_i\mu_i}(\tau) \neq \tilde{r}_{\mu_i\mu_i}(\tau)$ [78]. Due to $\hat{r}_{\mu_i\mu_i}(\tau) \neq \tilde{r}_{\mu_i\mu_i}(\tau)$, the stochastic process $\hat{\mu}_i(t)$ is therefore not ergodic with respect to the autocorrelation function (cf. Subsection 4.3.4).

The performance of the Monte Carlo method can be assessed more precisely with the help of the mean-square error $E_{r_{\mu_i\mu_i}}$ [see (4.128)]. Figure 5.22 illustrates the evaluation of $E_{r_{\mu_i\mu_i}}$ as a function of N_i for a single realization of the autocorrelation function $\tilde{r}_{\mu_i\mu_i}(\tau)$ as well as for the expected value obtained by averaging $E_{r_{\mu_i\mu_i}}$ over one thousand realizations of $\tilde{r}_{\mu_i\mu_i}(\tau)$.

Figure 5.22 also shows the results found for the method of equal areas, which are obviously better compared to the results obtained for the Monte Carlo method. The relation (5.47) for the

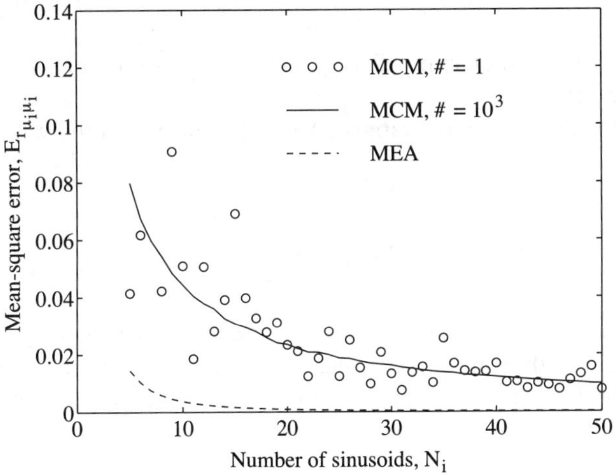

Figure 5.22 Mean-square error $E_{r_{\mu_i\mu_i}}$ (MCM, Jakes PSD, $f_{\max} = 91\,\text{Hz}$, $\sigma_0^2 = 1$, $\tau_{\max} = N_i/(2f_{\max})$).

computation of the path gains $c_{i,n}$ matches (5.31) exactly. Consequently, for the mean-square error $E_{p_{\mu_i}}$ [see (4.127)], we obtain the same results as presented in Figure 5.14(b).

The discussion on the model error $\Delta\beta_i$ follows. Let us start with (4.29). Then, by using (3.68) and (5.47), the quantity $\tilde{\beta}_i = \beta + \Delta\beta_i$ can be expressed as a function of the discrete Doppler frequencies $f_{i,n}$ as follows

$$\tilde{\beta}_i = \frac{2\beta}{f_{\max}^2 N_i} \sum_{n=1}^{N_i} f_{i,n}^2. \tag{5.50}$$

For the Monte Carlo method, the discrete Doppler frequencies $f_{i,n}$ are random variables, so that $\tilde{\beta}_i$ is also a random variable. In what follows, we will determine the probability density function of $\tilde{\beta}_i$.

Starting from the uniform distribution of $u_{i,n} \in (0, 1]$ and noting that the mapping from $u_{i,n}$ to $f_{i,n}$ is defined by (5.48), it follows that the probability density function of the discrete Doppler frequencies $f_{i,n}$ can be written as

$$p_{f_{i,n}}(f_{i,n}) = \begin{cases} \dfrac{2}{\pi f_{\max}\sqrt{1 - \left(\frac{f_{i,n}}{f_{\max}}\right)^2}}, & 0 < f \le f_{\max}, \\ 0, & \text{else}. \end{cases} \tag{5.51}$$

Now, with the probability density function of $f_{i,n}$, the density of $f_{i,n}^2$ can easily be computed, and in order to compute the density of the sum of these squares, we prefer to apply the concept of the characteristic function. After some straightforward computations, we obtain the result for the probability density function of $\tilde{\beta}_i$ in the following form [78]

$$p_{\tilde{\beta}_i}(\tilde{\beta}_i) = \begin{cases} 2\displaystyle\int_0^\infty \left[J_0\left(\frac{2\pi\beta\nu}{N_i}\right)\right]^{N_i} \cos[2\pi(\tilde{\beta}_i - \beta)\nu]\, d\nu, & \text{if } \tilde{\beta}_i \in (0, 2\beta], \\ 0, & \text{if } \tilde{\beta}_i \notin (0, 2\beta]. \end{cases} \tag{5.52}$$

By way of illustration, the probability density function $p_{\tilde{\beta}_i}(\tilde{\beta}_i)$ of $\tilde{\beta}_i$ is plotted in Figure 5.23 with N_i as a parameter.

The expected value $E\{\tilde{\beta}_i\}$ and the variance $\text{Var}\{\tilde{\beta}_i\}$ of $\tilde{\beta}_i$ are as follows:

$$E\{\tilde{\beta}_i\} = \beta, \tag{5.53a}$$

$$\text{Var}\{\tilde{\beta}_i\} = \frac{\beta^2}{2N_i}. \tag{5.53b}$$

It will also be shown that for large values of N_i, the random variable $\tilde{\beta}_i$ is approximately normally distributed with a mean value and a variance according to (5.53a) and (5.53b), respectively. Using the following approximation of the Bessel function of zeroth order [77, Equation (9.1.12)]

$$J_0(x) \approx 1 - \frac{x^2}{4} \tag{5.54}$$

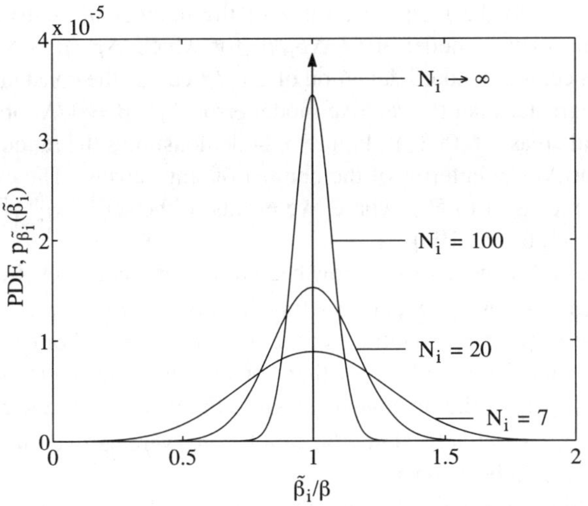

Figure 5.23 Probability density function $p_{\tilde{\beta}_i}(\tilde{\beta}_i)$ of $\tilde{\beta}_i$ by using the Monte Carlo method (Jakes PSD, $f_{\max} = 91\,\text{Hz}$, $\sigma_0^2 = 1$).

and taking into account that the relation [77, Equation (4.2.21)]

$$e^{-x} = \lim_{N_i \to \infty} \left(1 - \frac{x}{N_i}\right)^{N_i} \tag{5.55}$$

can be approximated by $e^{-x} \approx (1 - x/N_i)$ if N_i is sufficiently large, then we may express $p_{\tilde{\beta}_i}(\tilde{\beta}_i)$ [see (5.52)] approximately by

$$p_{\tilde{\beta}_i}(\tilde{\beta}_i) \approx \int_{-\infty}^{\infty} e^{-\frac{(\pi\beta\nu)^2}{N_i}} e^{-j2\pi(\tilde{\beta}_i-\beta)\nu}\, d\nu, \quad \tilde{\beta}_i \in (0, 2\beta]. \tag{5.56}$$

Finally, using the integral [23, Equation (3.323.2)]

$$\int_{-\infty}^{\infty} e^{-(ax)^2 \pm bx}\, dx = \frac{\sqrt{\pi}}{a} e^{(\frac{b}{2a})^2}, \quad a > 0, \tag{5.57}$$

the desired approximation follows directly in the form of

$$p_{\tilde{\beta}_i}(\tilde{\beta}_i) \approx \frac{1}{\sqrt{2\pi}\,\beta/\sqrt{2N_i}} e^{-\frac{(\tilde{\beta}_i-\beta)^2}{2\beta^2/(2N_i)}}, \quad \tilde{\beta}_i \in (0, 2\beta]. \tag{5.58}$$

Hence, for large values of N_i, the quantity $\tilde{\beta}_i$ is approximately normally distributed, and we may write $\tilde{\beta}_i \sim N(\beta, \beta^2/(2N_i))$ without causing a significant error. It should be observed that in the limit $N_i \to \infty$, it obviously follows $p_{\tilde{\beta}_i}(\tilde{\beta}_i) \to \delta(\tilde{\beta}_i - \beta)$. Evidently, the model error $\Delta\beta_i = \tilde{\beta}_i - \beta$ is likewise approximately normally distributed, i.e., $\Delta\beta_i \sim N(0, \beta^2/(2N_i))$, so that the random variable $\Delta\beta_i$ in fact has zero mean, but unfortunately its variance merely behaves proportionally to the reciprocal value of the number of sinusoids N_i. Finally, we also investigate the relative model error $\Delta\beta_i/\beta$, for which $\Delta\beta_i/\beta \sim N(0, 1/(2N_i))$ holds approximately. Hence, the standard deviation of $\Delta\beta_i/\beta$ equals the constant $1/\sqrt{2N_i}$, which is for $N_i > 2$ always greater than the relative model error $\Delta\beta_i/\beta = 1/N_i$ obtained by invoking the method of equal areas [cf. (5.37)]. Figure 5.24 demonstrates the random behaviour of the relative model error $\Delta\beta_i/\beta$ in terms of the number of sinusoids N_i. The evaluation of $\Delta\beta_i/\beta$ was performed by means of (5.50), where five events of the set $\{f_{i,n}\}_{n=1}^{N_i}$ were processed for every value of $N_i \in \{5, 6, \ldots, 100\}$.

Owing to (4.73), (4.77), and (4.85), it now becomes clear that the level-crossing rate $\tilde{N}_\xi(r)$, the average duration of fades $\tilde{T}_{\xi_-}(r)$, and the time intervals $\tilde{\tau}_q(r)$ likewise deviate in a random manner from the corresponding quantities of the reference model. For example, if we choose the pair (10, 11) for the 2-tuple (N_1, N_2), then for two different realizations of each of the sets $\{f_{1,n}\}_{n=1}^{N_1}$ and $\{f_{2,n}\}_{n=1}^{N_2}$, the mismatch shown in Figures 5.25(a) and 5.25(b) could occur for $\tilde{N}_\xi(r)$ and $\tilde{T}_{\xi_-}(r)$, respectively. Here, the simulations were carried out in the same way as described previously in Subsection 5.1.3.

With the Chebyshev inequality (2.24), one can show [see Appendix 5.C] that even if $N_i = 2500$ sinusoids are used, the probability that the absolute value of the relative model error $|\Delta\beta_i/\beta|$ exceeds a value of more than 2 per cent is merely less or equal to 50 per cent.

Gaussian power spectral density: If we apply the Monte Carlo method in connection with the Gaussian power spectral density in (3.26), then it is not possible to find a closed-form expression for the discrete Doppler frequencies $f_{i,n}$. In this case, however, the discrete Doppler frequencies $f_{i,n}$, are determined by the roots (zeros) of the following equation

$$u_{i,n} - \operatorname{erf}\left(\frac{f_{i,n}}{f_c}\sqrt{\ln 2}\right) = 0, \quad \forall n = 1, 2, \ldots, N_i \quad (i = 1, 2). \tag{5.59}$$

Here, the corresponding path gains $c_{i,n}$ are also available in the form of (5.47). A comparison between the equation above and (5.42) unveils the close relationship between the method

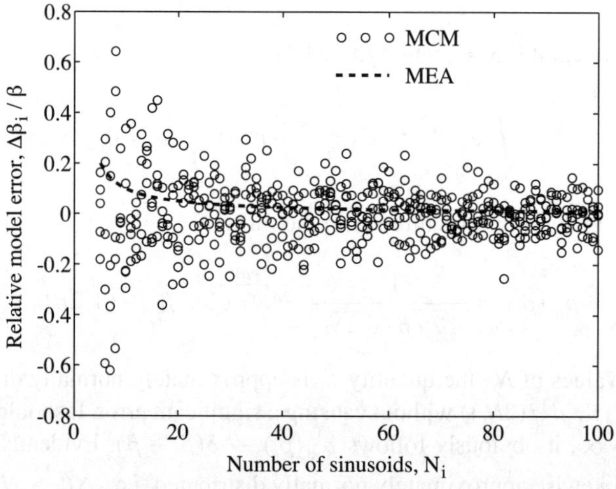

Figure 5.24 Relative model error $\Delta\tilde{\beta}_i/\beta$ (MCM, Jakes PSD).

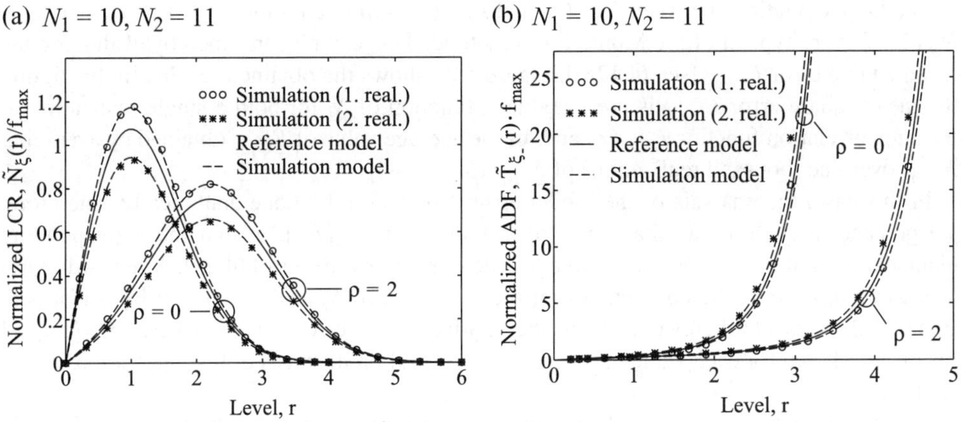

Figure 5.25 (a) Normalized level-crossing rate $\tilde{N}_\xi(r)/f_{\max}$ and (b) normalized average duration of fades $\tilde{T}_{\xi_-}(r) \cdot f_{\max}$ (MCM, Jakes PSD, $f_{\max} = 91$ Hz, $\sigma_0^2 = 1$).

of equal areas and the Monte Carlo method. If the uniformly distributed random variable $u_{i,n}$ is replaced by the deterministic quantity n/N_i, then the latter statistical procedure turns into the former deterministic one. For any (arbitrary) event $\{f_{i,n}\}$, it turns out that the mean value $\tilde{m}_{\mu_i}$, the mean power $\tilde{\sigma}^2_{\mu_i}$, and the period T_i of the deterministic process $\tilde{\mu}_i(t)$ have the same properties as described before when discussing the Jakes power spectral density. The same also holds for the cross-correlation properties of the deterministic processes $\tilde{\mu}_1(t)$ and $\tilde{\mu}_2(t)$.

For an event $\{f_{i,n}\}$ with $N_i = 25$ outcomes, the power spectral density $\tilde{S}_{\mu_i\mu_i}(f)$ is depicted in Figure 5.26(a). Likewise for $N_i = 25$, Figure 5.26(b) shows two possible realizations of the autocorrelation function $\tilde{r}_{\mu_i\mu_i}(\tau)$.

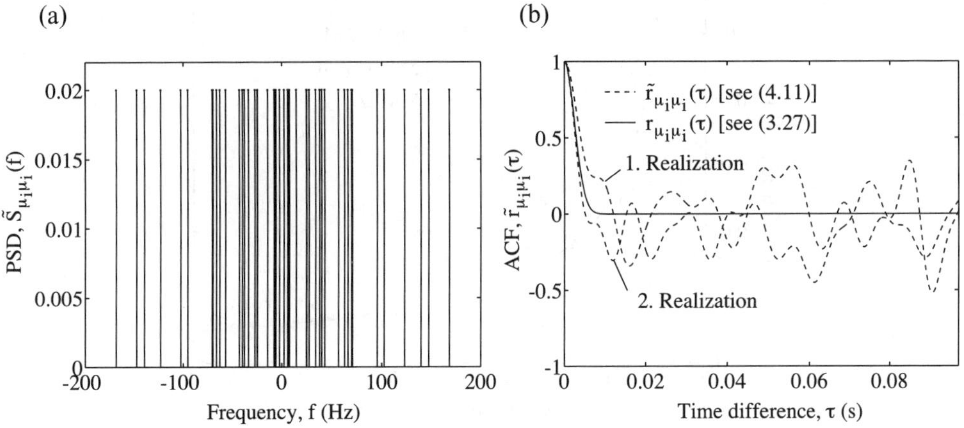

Figure 5.26 (a) Power spectral density $\tilde{S}_{\mu_i\mu_i}(f)$ and (b) autocorrelation function $\tilde{r}_{\mu_i\mu_i}(\tau)$ for $N_i = 25$ (MCM, Gaussian PSD, $f_c = \sqrt{\ln 2} f_{\max}$, $f_{\max} = 91$ Hz, $\sigma_0^2 = 1$).

The large deviations between $\tilde{r}_{\mu_i\mu_i}(\tau)$ and $r_{\mu_i\mu_i}(\tau)$ within the range $0 \leq \tau \leq \tau_{max}$ ($\tau_{max} = N_i/(2\kappa_c f_c)$) are typical of the Monte Carlo method. This can be confirmed by evaluating the mean-square error $E_{r_{\mu_i\mu_i}}$ [see (4.128)]. Figure 5.27 shows the obtained results. In this figure, the mean-square error $E_{r_{\mu_i\mu_i}}$ is presented as a function of N_i for both a single realization of the autocorrelation function $\tilde{r}_{\mu_i\mu_i}(\tau)$ and for the average value of $E_{r_{\mu_i\mu_i}}$ obtained by averaging $E_{r_{\mu_i\mu_i}}$ over one thousand realizations of $\tilde{r}_{\mu_i\mu_i}(\tau)$.

In this case, the analysis of the model error $\Delta\beta_i$ cannot be done analytically. Therefore, we proceed in such a way that for a given realization of $\{f_{i,n}\}$, at first the corresponding elementary event of the random variable $\tilde{\beta}_i$ is determined by means of (4.29). Afterwards, with $\beta = 2(\pi f_c \sigma_0)^2/\ln 2$, the computation of the model error $\Delta\beta_i = \tilde{\beta}_i - \beta$ will be performed. Figure 5.28 presents the evaluation of the relative model error $\Delta\beta_i/\beta$, where the obtained results are shown for each value of $N_i \in \{5, 6, \ldots, 100\}$ on the basis of five realizations of the set $\{f_{i,n}\}$.

The determination as well as the investigation of the properties of the level-crossing rate $\tilde{N}_\xi(r)$ and the average duration of fades $\tilde{T}_{\xi_-}(r)$ are also performed on the basis of several realizations of the set $\{f_{i,n}\}$. To illustrate the obtained results, we take a look at Figures 5.29(a) and 5.29(b), where two different realizations of $\tilde{N}_\xi(r)$ and $\tilde{T}_{\xi_-}(r)$ are shown, respectively. All the presented results have been obtained by using $(N_1, N_2) = (10, 11)$.

5.1.5 Jakes Method (JM)

The Jakes method (JM) [13] has been developed exclusively for the Jakes power spectral density. Not only for completeness, but also due to its widespread popularity, this classical method will be described here as well. Since the Jakes method and the resulting Jakes simulator are treated extensively in [13, p. 67ff.], a detailed description will not be given here. Instead, we will restrict ourselves mainly to the performance analysis of this popular method. Our

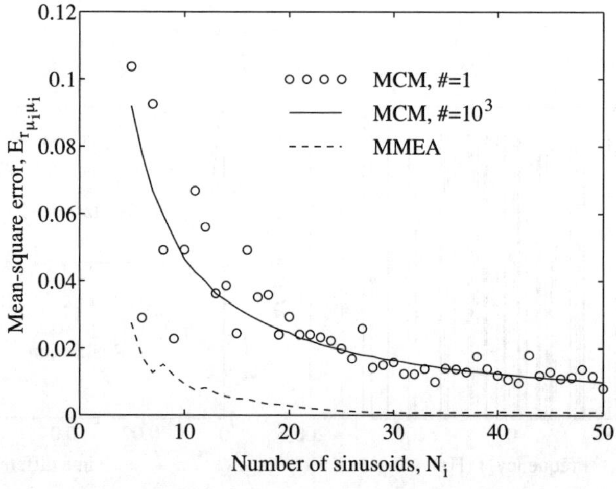

Figure 5.27 Mean-square error $E_{r_{\mu_i\mu_i}}$ (MCM, Gaussian PSD, $f_c = \sqrt{\ln 2} f_{max}$, $f_{max} = 91$ Hz, $\sigma_0^2 = 1$, $\tau_{max} = N_i/(2\kappa_c f_c)$, $\kappa_c = 2\sqrt{2/\ln 2}$).

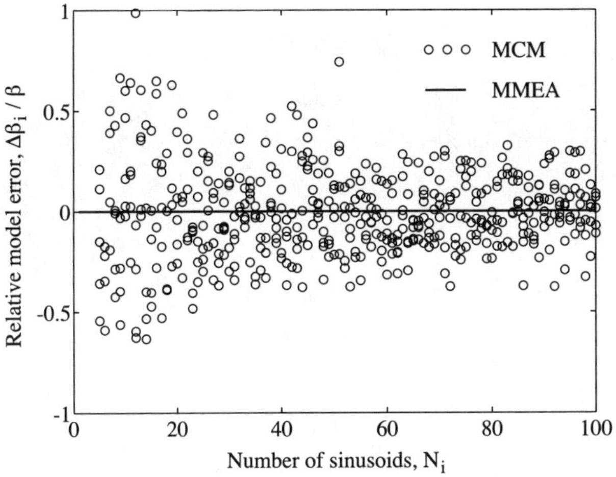

Figure 5.28 Relative model error $\tilde{\beta}_i/\beta$ (MCM, Gaussian PSD, $f_c = \sqrt{\ln 2} f_{\max}$, $f_{\max} = 91$ Hz, $\sigma_0^2 = 1$).

starting point is the structure of the original Jakes simulator [13, p. 70], which is shown in Figure 5.30(a) for the generation of Rayleigh fading in the complex baseband.

Jakes power spectral density: The structure of the Jakes simulator in Figure 5.30(a) can easily be brought into compliance with the standard structure of the deterministic sum-of-sinusoids simulation model shown in Figure 5.30(b). After some straightforward manipulations of the expressions for the parameters of the Jakes simulator and adopting our notation and terminology, the following relations hold for the path gains $c_{i,n}$, the discrete Doppler frequencies $f_{i,n}$,

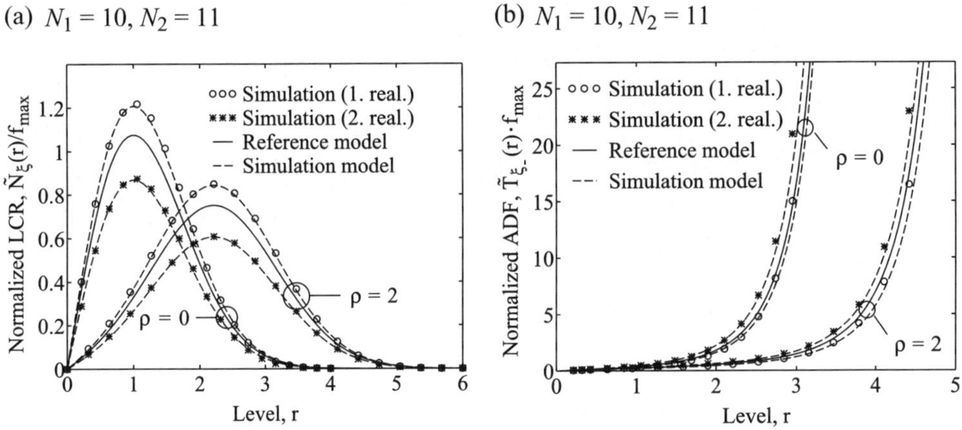

Figure 5.29 (a) Normalized level-crossing rate $\tilde{N}_\xi(r)/f_{\max}$ and (b) normalized average duration of fades $\tilde{T}_{\xi_-}(r) \cdot f_{\max}$ (MCM, Gaussian PSD, $f_c = \sqrt{\ln 2} f_{\max}$, $f_{\max} = 91$ Hz, $\sigma_0^2 = 1$).

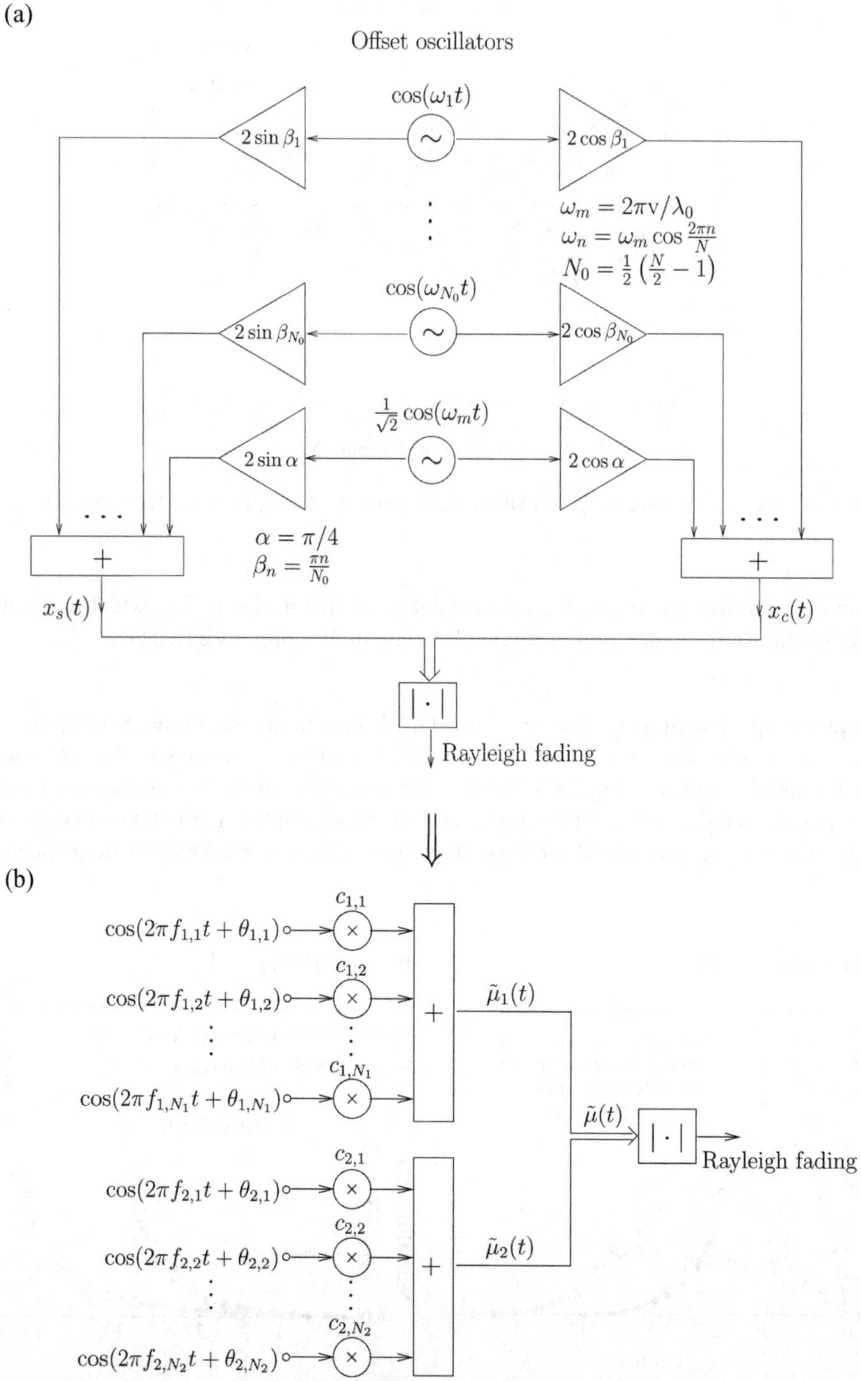

Figure 5.30 (a) Structure of the Jakes simulator and (b) the equivalent deterministic sum-of-sinusoids channel simulator for generating Rayleigh fading.

and the phases $\theta_{i,n}$ [157]:

$$c_{i,n} = \begin{cases} \dfrac{2\sigma_0}{\sqrt{N_i - \frac{1}{2}}} \sin\left(\dfrac{\pi n}{N_i - 1}\right), & n = 1, 2, \ldots, N_i - 1, \quad i = 1, \\ \dfrac{2\sigma_0}{\sqrt{N_i - \frac{1}{2}}} \cos\left(\dfrac{\pi n}{N_i - 1}\right), & n = 1, 2, \ldots, N_i - 1, \quad i = 2, \\ \dfrac{\sigma_0}{\sqrt{N_i - \frac{1}{2}}}, & n = N_i, \quad i = 1, 2, \end{cases} \tag{5.60}$$

$$f_{i,n} = \begin{cases} f_{\max} \cos\left(\dfrac{n\pi}{2N_i - 1}\right), & n = 1, 2, \ldots, N_i - 1, \quad i = 1, 2, \\ f_{\max}, & n = N_i, \quad i = 1, 2, \end{cases} \tag{5.61}$$

$$\theta_{i,n} = 0, \quad n = 1, 2, \ldots, N_i, \quad i = 1, 2. \tag{5.62}$$

where $N_1 = N_2$. Here, the path gains $c_{i,n}$ have been scaled such that the mean power $\tilde{\sigma}^2_{\mu_i}$ of $\tilde{\mu}_i(t)$ meets the relation $\tilde{\sigma}^2_{\mu_i} = \sigma_0^2$ for $i = 1, 2$. Due to $f_{i,n} \neq 0$, the following equation holds for the mean value: $\tilde{m}_{\mu_i} = m_{\mu_i} = 0$ ($i = 1, 2$).

The resulting power spectral densities $\tilde{S}_{\mu_1\mu_1}(f)$ and $\tilde{S}_{\mu_2\mu_2}(f)$ as well as the corresponding autocorrelation functions $\tilde{r}_{\mu_1\mu_1}(\tau)$ and $\tilde{r}_{\mu_2\mu_2}(\tau)$ are shown in Figures 5.31(a)–5.31(d) for $N_1 = N_2 = 9$. Even for small values of τ, it can be seen from Figures 5.31(b) and 5.31(d) that the autocorrelation functions of the deterministic processes $\tilde{\mu}_1(t)$ and $\tilde{\mu}_2(t)$ differ widely from the desired autocorrelation function $r_{\mu_i\mu_i}(\tau) = \sigma_0^2 J_0(2\pi f_{\max}\tau)$. On the other hand, as Figure 5.31(f) reveals, the autocorrelation function $\tilde{r}_{\mu\mu}(\tau)$ of the complex deterministic process $\tilde{\mu}(t)$ over the interval $\tau \in [0, \tau_{\max}]$ matches $r_{\mu\mu}(\tau) = 2\sigma_0^2 J_0(2\pi f_{\max}\tau)$ very well. Figure 5.31(e) shows the power spectral density $\tilde{S}_{\mu\mu}(f)$ that corresponds to the autocorrelation function $\tilde{r}_{\mu\mu}(\tau)$.

It is interesting to note that even for $N_i \to \infty$, the autocorrelation function $\tilde{r}_{\mu_i\mu_i}(\tau)$ does not tend to $r_{\mu_i\mu_i}(\tau)$. Instead, after substituting (5.60) and (5.61) into (4.11) and taking afterwards the limit $N_i \to \infty$, we instead obtain the expressions

$$\lim_{N_1 \to \infty} \tilde{r}_{\mu_1\mu_1}(\tau) = \frac{2\sigma_0^2}{\pi} \int_0^{\pi/2} \left[1 - \cos(4z)\right] \cos(2\pi f_{\max}\tau \cos z)\, dz \tag{5.63a}$$

and

$$\lim_{N_2 \to \infty} \tilde{r}_{\mu_2\mu_2}(\tau) = \frac{2\sigma_0^2}{\pi} \int_0^{\pi/2} \left[1 + \cos(4z)\right] \cos(2\pi f_{\max}\tau \cos z)\, dz, \tag{5.63b}$$

which, by making use of [23, Equation (3.715.19)]

$$\int_0^{\pi/2} \cos(z \cos x) \cos(2nx)\, dx = (-1)^n \cdot \frac{\pi}{2} J_{2n}(z), \tag{5.64}$$

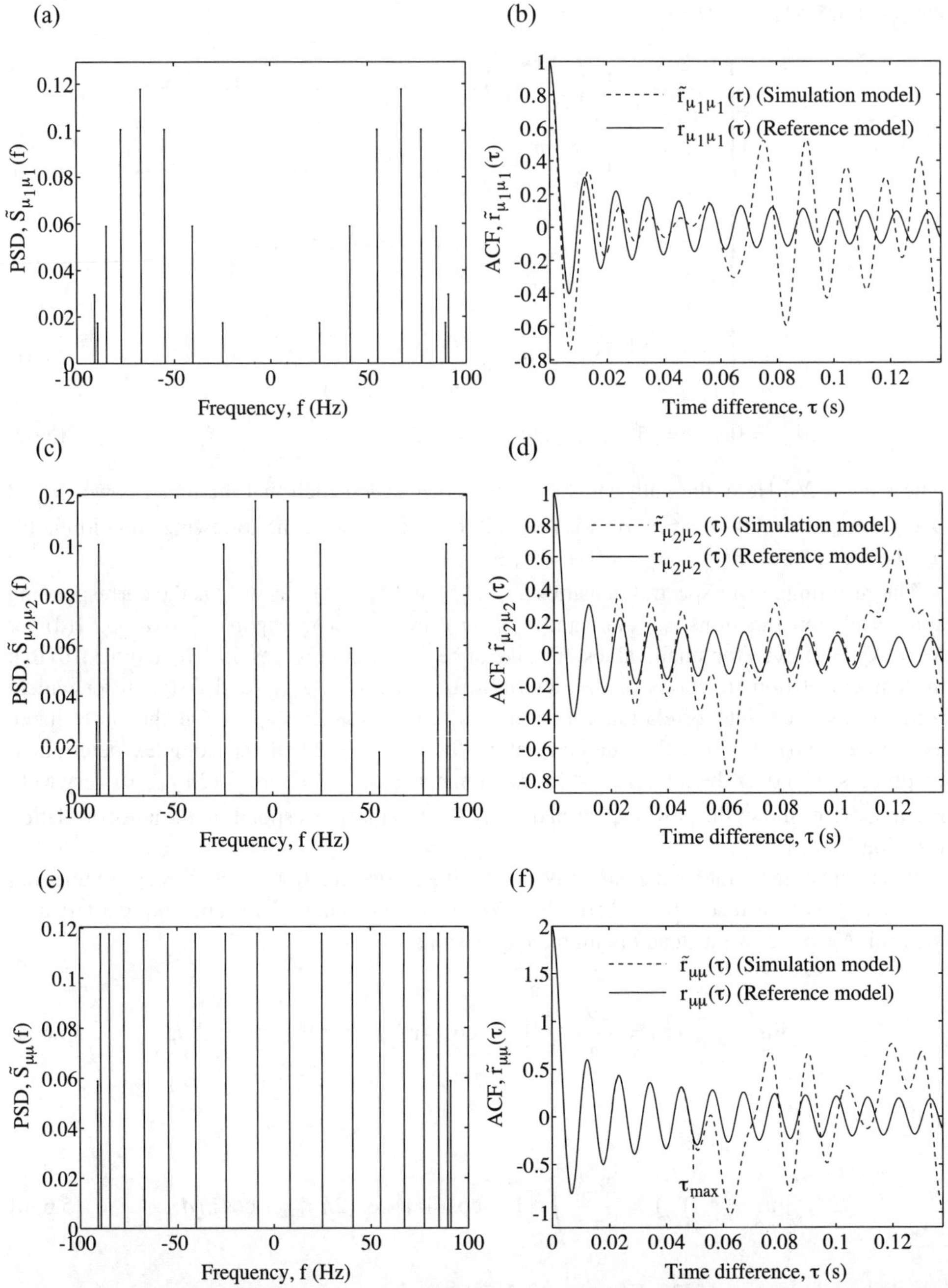

Figure 5.31 Power spectral densities and corresponding autocorrelation functions for $N_1 = N_2 = 9$: (a) $\tilde{S}_{\mu_1\mu_1}(f)$, (b) $\tilde{r}_{\mu_1\mu_1}(\tau)$, (c) $\tilde{S}_{\mu_2\mu_2}(f)$, (d) $\tilde{r}_{\mu_2\mu_2}(\tau)$, (e) $\tilde{S}_{\mu\mu}(f)$, and (f) $\tilde{r}_{\mu\mu}(\tau)$ (JM, Jakes PSD, $f_{\max} = 91$ Hz, $\sigma_0^2 = 1$).

can finally be presented in closed form as [157]

$$\lim_{N_1\to\infty} \tilde{r}_{\mu_1\mu_1}(\tau) = \sigma_0^2\big[J_0(2\pi f_{\max}\tau) - J_4(2\pi f_{\max}\tau)\big] \tag{5.65a}$$

and

$$\lim_{N_2\to\infty} \tilde{r}_{\mu_2\mu_2}(\tau) = \sigma_0^2\big[J_0(2\pi f_{\max}\tau) + J_4(2\pi f_{\max}\tau)\big]. \tag{5.65b}$$

Thus, the inequality $\tilde{r}_{\mu_i\mu_i}(\tau) \neq r_{\mu_i\mu_i}(\tau)$ $(i = 1, 2)$ holds even after taking the limit $N_i \to \infty$. On the contrary, the autocorrelation function $\tilde{r}_{\mu\mu}(\tau)$ of the complex deterministic process $\tilde{\mu}(t) = \tilde{\mu}_1(t) + j\tilde{\mu}_2(t)$ tends exactly to the autocorrelation function $r_{\mu\mu}(\tau)$ of the reference model as $N_i \to \infty$. This fact becomes evident immediately after substituting (5.65a) and (5.65b) into the general formula

$$\tilde{r}_{\mu\mu}(\tau) = \tilde{r}_{\mu_1\mu_1}(\tau) + \tilde{r}_{\mu_2\mu_2}(\tau) + j\big(\tilde{r}_{\mu_1\mu_2}(\tau) - \tilde{r}_{\mu_2\mu_1}(\tau)\big) \tag{5.66}$$

that can be derived from (2.143), and then making use of the relation $\tilde{r}_{\mu_1\mu_2}(\tau) = \tilde{r}_{\mu_2\mu_1}(\tau)$, which is valid here, as we will see below. Hence, one notices immediately that

$$\lim_{N_i\to\infty} \tilde{r}_{\mu\mu}(\tau) = r_{\mu\mu}(\tau) = 2\sigma_0^2 J_0(2\pi f_{\max}\tau) \tag{5.67}$$

holds.

Furthermore, we want to find the functions to which the power spectral densities $\tilde{S}_{\mu_1\mu_1}(f)$ and $\tilde{S}_{\mu_2\mu_2}(f)$ tend in the limits $N_1 \to \infty$ and $N_2 \to \infty$, respectively. Therefore, we transform (5.65a) and (5.65b) into the spectral domain by means of the Fourier transform and obtain

$$\lim_{N_1\to\infty} \tilde{S}_{\mu_1\mu_1}(f) = \begin{cases} \sigma_0^2 \cdot \dfrac{1 - \cos\big[4\arcsin(f/f_{\max})\big]}{\pi f_{\max}\sqrt{1-(f/f_{\max})^2}}, & |f| \le f_{\max}, \\ 0, & |f| > f_{\max}, \end{cases} \tag{5.68a}$$

$$\lim_{N_2\to\infty} \tilde{S}_{\mu_2\mu_2}(f) = \begin{cases} \sigma_0^2 \cdot \dfrac{1 + \cos\big[4\arcsin(f/f_{\max})\big]}{\pi f_{\max}\sqrt{1-(f/f_{\max})^2}}, & |f| \le f_{\max}, \\ 0, & |f| > f_{\max}. \end{cases} \tag{5.68b}$$

An alternative way to derive the two equations above would be to first substitute (5.60) and (5.61) in (4.18) and then to take the limit $N_i \to \infty$. If we now put the results given by (5.68a) and (5.68b) into the Fourier transform of the autocorrelation function $\tilde{r}_{\mu\mu}(\tau)$ in (5.66), then we obtain the Jakes power spectral density as expected, i.e.,

$$\lim_{N_i\to\infty} \tilde{S}_{\mu\mu}(f) = S_{\mu\mu}(f) = \begin{cases} \dfrac{2\sigma_0^2}{\pi f_{\max}\sqrt{1-(f/f_{\max})^2}}, & |f| \le f_{\max}, \\ 0, & |f| > f_{\max}. \end{cases} \tag{5.69}$$

Consequently, as $N_i \to \infty$, it follows $\tilde{S}_{\mu\mu}(f) \to S_{\mu\mu}(f)$ but not $\tilde{S}_{\mu_i\mu_i}(f) \to S_{\mu_i\mu_i}(f)$ $(i = 1, 2)$.

To illustrate the results given above, we study Figure 5.32, where the power spectral densities $\tilde{S}_{\mu_1\mu_1}(f)$, $\tilde{S}_{\mu_2\mu_2}(f)$, and $\tilde{S}_{\mu\mu}(f)$ are presented together with the corresponding autocorrelation functions for the limit $N_i \to \infty$.

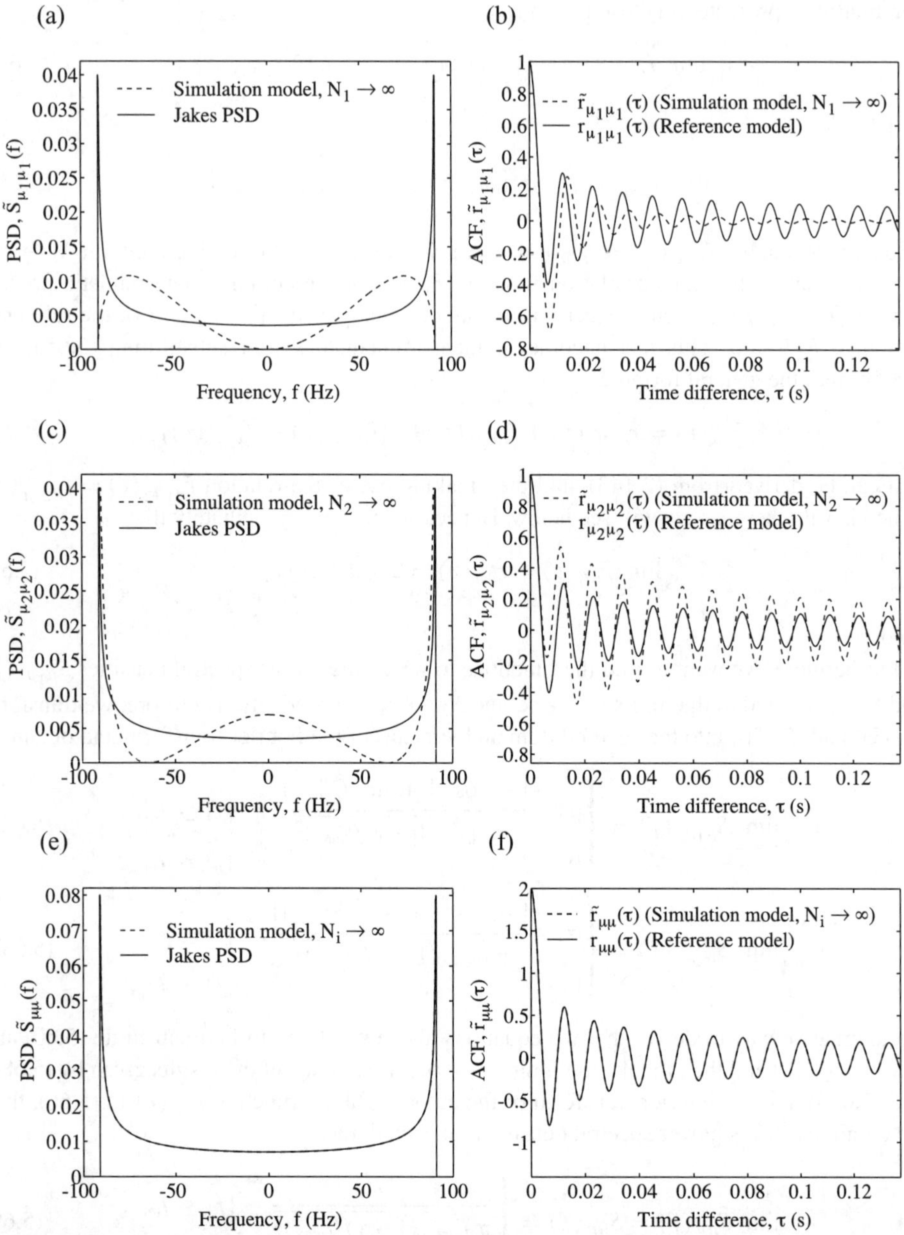

Figure 5.32 Power spectral densities and corresponding autocorrelation functions for $N_1 \to \infty$ and $N_2 \to \infty$: (a) $\tilde{S}_{\mu_1\mu_1}(f)$, (b) $\tilde{r}_{\mu_1\mu_1}(\tau)$, (c) $\tilde{S}_{\mu_2\mu_2}(f)$, (d) $\tilde{r}_{\mu_2\mu_2}(\tau)$, (e) $\tilde{S}_{\mu\mu}(f)$, and (f) $\tilde{r}_{\mu\mu}(\tau)$ (JM, Jakes PSD, $f_{\max} = 91$ Hz, $\sigma_0^2 = 1$).

When using the Jakes method, we have to factor in that the deterministic processes $\tilde{\mu}_1(t)$ and $\tilde{\mu}_2(t)$ are correlated, because $f_{1,n} = f_{2,n}$ holds according to (5.61) for all $n = 1, 2, \ldots, N_1$ ($N_1 = N_2$). After substituting (5.60)–(5.62) into (4.13), we find the following expression for the cross-correlation function $\tilde{r}_{\mu_1\mu_2}(\tau)$

$$\tilde{r}_{\mu_1\mu_2}(\tau) = \frac{\sigma_0^2}{N_i - \frac{1}{2}} \left\{ \sum_{n=1}^{N_i-1} \sin\left(\frac{2\pi n}{N_i - 1}\right) \cos\left[2\pi f_{\max} \cos\left(\frac{n\pi}{2N_i - 1}\right)\tau\right] + \frac{1}{2}\cos(2\pi f_{\max}\tau) \right\}. \tag{5.70}$$

Since $\tilde{r}_{\mu_1\mu_2}(\tau)$ is a real-valued and even function, it can be shown, by using (2.108), that $\tilde{r}_{\mu_2\mu_1}(\tau) = \tilde{r}^*_{\mu_1\mu_2}(-\tau) = \tilde{r}_{\mu_1\mu_2}(\tau)$ holds. An idea of the behaviour of the cross-correlation function $\tilde{r}_{\mu_1\mu_2}(\tau)$ computed according to (5.70) is provided by Figure 5.33(b). The adjacent Figure 5.33(a) shows the corresponding cross-power spectral density $\tilde{S}_{\mu_1\mu_2}(f)$ computed by using (4.20). The results illustrated in these figures have been obtained by choosing $N_1 = N_2 = 9$.

One can clearly see that there exists a strong correlation between $\tilde{\mu}_1(t)$ and $\tilde{\mu}_2(t)$. This problem was addressed in [158], where a modification of the Jakes method has been suggested, which is based essentially on an alteration of the gains $c_{i,n}$ in (5.60). However, this variant guarantees only that $\tilde{r}_{\mu_1\mu_2}(\tau)$ is zero at the origin $\tau = 0$. The only way to ensure that $\tilde{r}_{\mu_1\mu_2}(\tau) = 0$ holds for all values of τ is to design the deterministic processes $\tilde{\mu}_1(t)$ and $\tilde{\mu}_2(t)$ by using disjoint sets $\{f_{1,n}\}$ and $\{f_{2,n}\}$, i.e., $\{f_{1,n}\} \cap \{f_{2,n}\} = \emptyset$.

The question of whether the cross-correlation function $\tilde{r}_{\mu_1\mu_2}(\tau)$ vanishes for $N_i \to \infty$, will be addressed in the following. Therefore, we let N_i tend to infinity in (5.70), which provides the integral

$$\lim_{N_i\to\infty} \tilde{r}_{\mu_1\mu_2}(\tau) = \frac{2\sigma_0^2}{\pi} \int_0^{\pi/2} \sin(4z)\cos(2\pi f_{\max}\tau\cos z)\, dz, \tag{5.71}$$

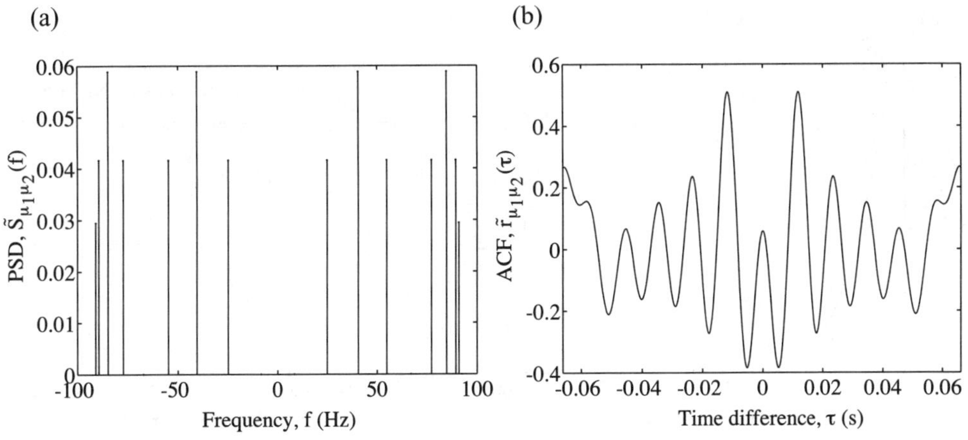

Figure 5.33 (a) Cross-power spectral density $\tilde{S}_{\mu_1\mu_2}(f)$ and (b) cross-correlation function $\tilde{r}_{\mu_1\mu_2}(\tau)$ for $N_1 = N_2 = 9$ (JM, Jakes PSD, $f_{\max} = 91$ Hz, $\sigma_0^2 = 1$).

that needs to be solved numerically. The result of the numerical integration is shown in Figure 5.34(b). Obviously, even for an infinite number of sinusoids N_i, the correlation between $\tilde{\mu}_1(t)$ and $\tilde{\mu}_2(t)$ does not vanish. Consequently, $\tilde{r}_{\mu_1\mu_2}(\tau) \to r_{\mu_1\mu_2}(\tau)$ does not hold for $N_i \to \infty$.

In the case of the limit $N_i \to \infty$, we obtain the following closed-form expression for the cross-power spectral density $\tilde{S}_{\mu_1\mu_2}(f)$ after performing the Fourier transform of (5.71)

$$\lim_{N_i\to\infty} \tilde{S}_{\mu_1\mu_2}(f) = \begin{cases} \sigma_0^2 \cdot \dfrac{\sin\left[4 \arccos(|f|/f_{\max})\right]}{\pi f_{\max}\sqrt{1-(f/f_{\max})^2}}, & |f| \leq f_{\max}, \\ 0, & |f| > f_{\max}. \end{cases} \tag{5.72}$$

The graph of this equation is presented in Figure 5.34(a). In contrast to the Jakes power spectral density [see (3.23)], which becomes singular at $f = \pm f_{\max}$, the cross-power spectral density in (5.72) takes on the finite value $4/(\pi f_{\max})$ at these frequencies, i.e., $\tilde{S}_{\mu_1\mu_2}(\pm f_{\max}) = 4/(\pi f_{\max})$ holds, which can readily be proved by applying l'Hôpital's rule [159].

Expediently, the mean-square error of the autocorrelation function defined in (4.128) will in the present case be evaluated with respect to $\tilde{r}_{\mu_1\mu_1}(\tau)$, $\tilde{r}_{\mu_2\mu_2}(\tau)$, and $\tilde{r}_{\mu\mu}(\tau)$. The obtained results are displayed in Figure 5.35(a) in terms of N_i. Using the Jakes method, the path gains $c_{i,n}$ differ quite considerably from the (quasi-)optimal quantities $c_{i,n} = \sigma_0\sqrt{2/N_i}$. This inevitably leads to an increase in the mean-square error $E_{p_{\mu_i}}$ [see (4.127)], which is clearly discernible in Figure 5.35(b).

For the Jakes method, we have: $N_1 = N_2$ and $f_{1,n} = f_{2,n}$ $\forall n = 1, 2, \ldots, N_1$ (N_2). On the other hand, $c_{1,n} \neq c_{2,n}$ still holds for almost all $n = 1, 2, \ldots, N_1$ (N_2). As a consequence, the model errors $\Delta\beta_1$ and $\Delta\beta_2$ are different for any given number of sinusoids N_i. To illustrate this fact, we consider Figure 5.36 in which the relative model errors $\Delta\beta_1/\beta$ and $\Delta\beta_2/\beta$ are presented as a function of N_i.

Owing to $\Delta\beta_1 \neq \Delta\beta_2$, we must use the expression (3.D.13) for the computation of the level-crossing rate $\tilde{N}_\xi(r)$, where β_1 and β_2 have to be replaced by $\tilde{\beta}_1 = \beta + \Delta\beta_1$ and $\tilde{\beta}_2 = \beta + \Delta\beta_2$, respectively. For the general case that $\tilde{\mu}_1(t)$ and $\tilde{\mu}_2(t)$ are correlated, we recall that the

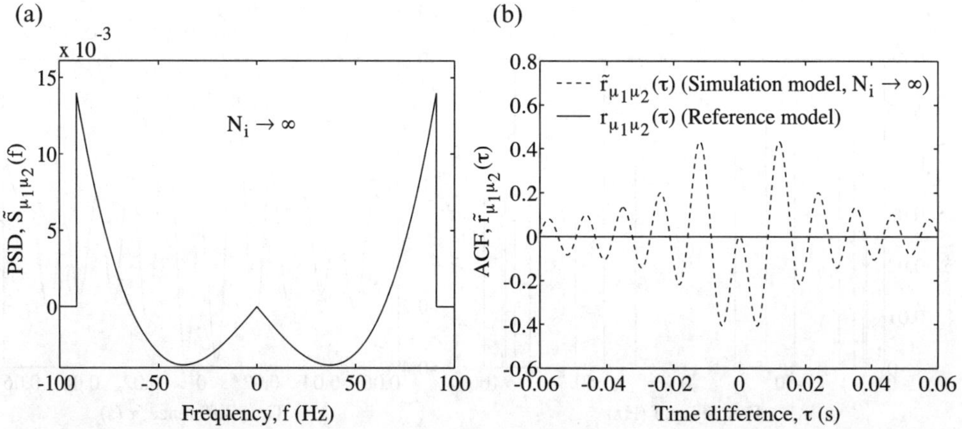

Figure 5.34 (a) Cross-power spectral density $\tilde{S}_{\mu_1\mu_2}(f)$ and (b) cross-correlation function $\tilde{r}_{\mu_1\mu_2}(\tau)$ for $N_i \to \infty$ (JM, Jakes PSD, $f_{\max} = 91$ Hz, $\sigma_0^2 = 1$).

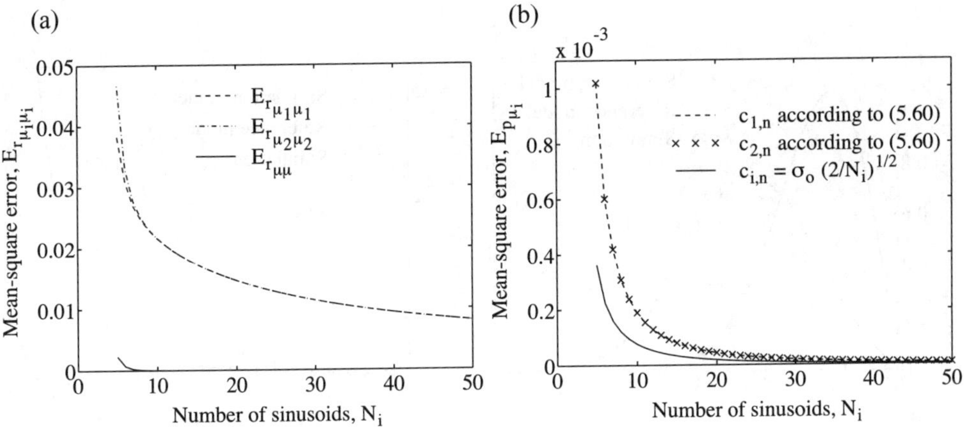

Figure 5.35 (a) Mean-square errors $E_{r_{\mu_1\mu_1}}$, $E_{r_{\mu_2\mu_2}}$, and $E_{r_{\mu\mu}}$ with $\tau_{\max} = N_i/(2f_{\max})$ and (b) mean-square error $E_{p_{\mu_i}}$ (JM, Jakes PSD, $f_{\max} = 91$ Hz, $\sigma_0^2 = 1$).

level-crossing rate $\tilde{N}_\xi(r)$ depends also on the quantities $\dot{\tilde{r}}_{\mu_1\mu_2}(0)$ and $\ddot{\tilde{r}}_{\mu_1\mu_2}(0)$. Since it follows immediately from (5.70) that $\dot{\tilde{r}}_{\mu_1\mu_2}(0)$ is equal to zero, and since, furthermore, the influence of $\ddot{\tilde{r}}_{\mu_1\mu_2}(0)$ on $\tilde{N}_\xi(r)$ is quite small, this dependency has a negligible effect, which will be ignored here. Figure 5.37(a) presents the analytical results for $N_\xi(r)/f_{\max}$ as well as for $\tilde{N}_\xi(r)/f_{\max}$ by choosing $N_1 = N_2 = 9$. This figure shows in addition the corresponding simulation results, which match the theoretical results for $N_\xi(r)/f_{\max}$ and $\tilde{N}_\xi(r)/f_{\max}$ very well.

When using the Jakes method, it seems that the relatively large deviations between the autocorrelation functions $\tilde{r}_{\mu_i\mu_i}(\tau)$ and $r_{\mu_i\mu_i}(\tau)$ as well as the non-zero cross-correlation function $\tilde{r}_{\mu_1\mu_2}(\tau)$ do not have a significant negative impact on $\tilde{N}_\xi(r)$. By the way, this also holds for the average duration of fades $\tilde{T}_{\xi_-}(r)$ [see Figure 5.37(b)] as well as for the probability density function $\tilde{p}_{0_-}(\tau_-; r)$ of the fading intervals τ_- at low signal levels r [see Figure 5.38].

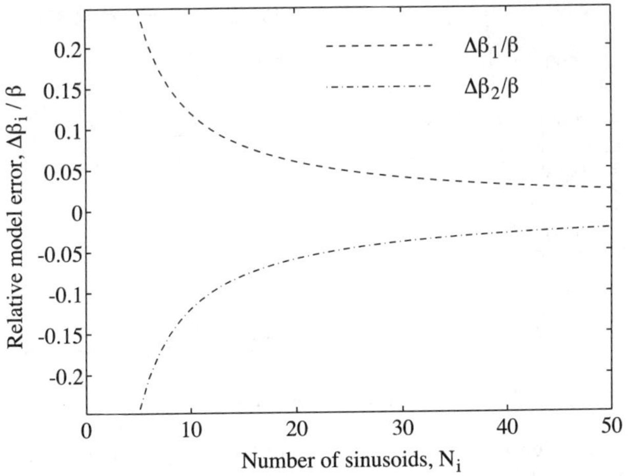

Figure 5.36 Relative model errors $\Delta\beta_1/\beta$ and $\Delta\beta_2/\beta$ (JM, Jakes PSD, $f_{\max} = 91$ Hz, $\sigma_0^2 = 1$).

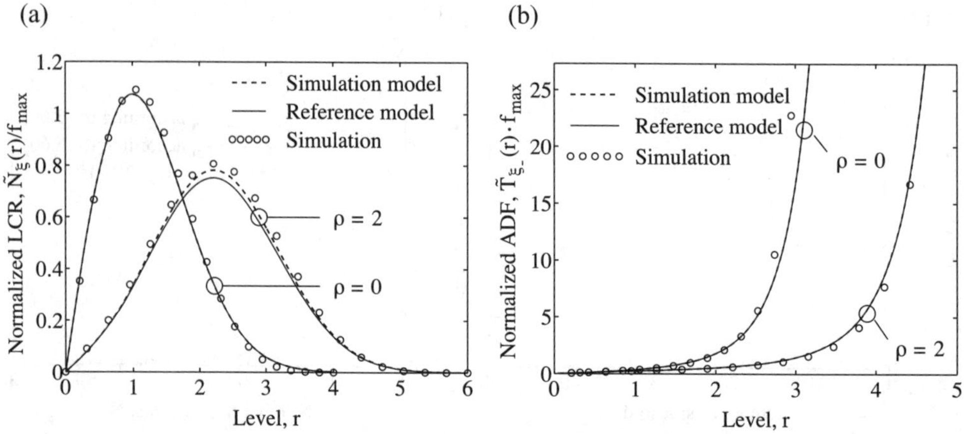

Figure 5.37 (a) Normalized level-crossing rate $\tilde{N}_\xi(r)/f_{\max}$ and (b) normalized average duration of fades $\tilde{T}_{\xi_-}(r) \cdot f_{\max}$ for $N_1 = N_2 = 9$ (JM, Jakes PSD, $f_{\max} = 91\,\text{Hz}$, $\sigma_0^2 = 1$).

This observation should not mislead us to the conclusion that the influence of the cross-correlation function $\tilde{r}_{\mu_1\mu_2}(\tau)$ on $\tilde{N}_\xi(r)$ and $\tilde{T}_{\xi_-}(r)$ can generally be ignored. It rather depends on the specific type of the cross-correlation function $\tilde{r}_{\mu_1\mu_2}(\tau)$ whether its influence can be neglected or not. In Chapter 6, we will see that there exist certain classes of cross-correlation functions, which not only affect the higher-order statistical properties, but also influence the probability density function of the signal envelope. By this means, it is possible to increase the flexibility of mobile fading channel models considerably.

Summarizing, we can say that the main disadvantage of the Jakes method is not to be seen in the non-zero cross-correlation function, but in the fact that the deterministic processes $\tilde{\mu}_1(t)$

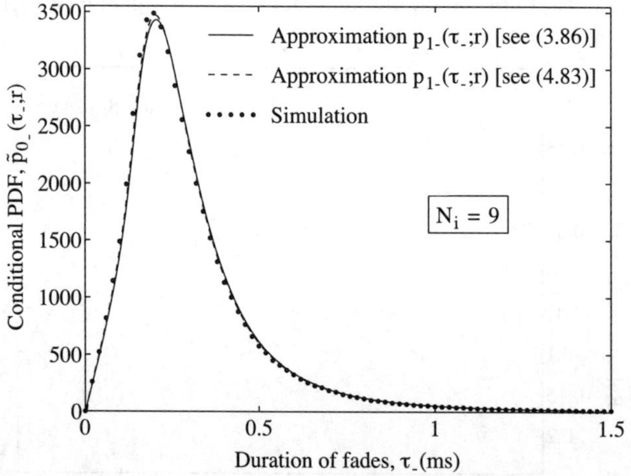

Figure 5.38 Probability density function $\tilde{p}_{0_-}(\tau_-; r)$ of the fading intervals τ_- at the signal level $r = 0.1$ (JM, Jakes PSD, $f_{\max} = 91\,\text{Hz}$, $\sigma_0^2 = 1$).

and $\tilde{\mu}_2(t)$ do not follow optimally the Gaussian distribution [cf. Figure 5.35(b)] for a given limited number of sinusoids N_i. Since the performance loss is not very significant and can easily be compensated by a slight increase in the number of sinusoids N_i, we can conclude that, all in all, the Jakes method is quite suitable for the modelling of Rayleigh and Rice processes described by the classical Doppler power spectral density in (3.23), provided that the chosen value for the number of sinusoids N_i is greater than or equal to nine [157]. Finally, it should also be noted that an implementation technique of the Jakes method on a signal processor has been described in [160, 161].

5.1.6 L_p-Norm Method (LPNM)

The L_p-norm method (LPNM) is based on the idea that the sets $\{c_{i,n}\}$ and $\{f_{i,n}\}$ are to be determined such that the following two postulations are fulfilled for a given number of sinusoids N_i [96, 149]:

(i) With respect to the L_p-norm

$$E_{p_{\mu_i}}^{(p)} := \left\{ \int_{-\infty}^{\infty} |p_{\mu_i}(x) - \tilde{p}_{\mu_i}(x)|^p \, dx \right\}^{1/p}, \quad p = 1, 2, \ldots, \tag{5.73}$$

the probability density function $\tilde{p}_{\mu_i}(x)$ of the deterministic process $\tilde{\mu}_i(t)$ will be an optimal approximation of the Gaussian distribution $p_{\mu_i}(x)$ of the stochastic process $\mu_i(t)$.

(ii) With respect to the L_p-norm

$$E_{r_{\mu_i\mu_i}}^{(p)} := \left\{ \frac{1}{\tau_{\max}} \int_{0}^{\tau_{\max}} |r_{\mu_i\mu_i}(\tau) - \tilde{r}_{\mu_i\mu_i}(\tau)|^p \, d\tau \right\}^{1/p}, \quad p = 1, 2, \ldots, \tag{5.74}$$

the autocorrelation function $\tilde{r}_{\mu_i\mu_i}(\tau)$ of the deterministic process $\tilde{\mu}_i(t)$ will be fitted as close as possible to a given (desired) autocorrelation function $r_{\mu_i\mu_i}(\tau)$ of the stochastic process $\mu_i(t)$, where $\tau_{\max}$ again defines an appropriate time-lag interval $[0, \tau_{\max}]$ over which the approximation of $r_{\mu_i\mu_i}(\tau)$ is of interest.

We first direct our attention to the Postulation (i). In view of (4.41) stating that $\tilde{p}_{\mu_i}(x)$ merely depends on the number of sinusoids N_i and the path gains $c_{i,n}$, we should raise the following question: Does an optimal solution exist for the set of path gains $\{c_{i,n}\}$ such that the L_p-norm $E_{p_{\mu_i}}^{(p)}$ becomes minimal? In seeking to answer this question, we first substitute (4.41) and (4.43) in (5.73), and afterwards perform a numerical optimization of the path gains $c_{i,n}$, so that $E_{p_{\mu_i}}^{(p)}$ becomes minimal. As a proper numerical optimization technique, the Fletcher-Powell algorithm [162], for example, is particularly well suited for solving this kind of problem. After the minimization of (5.73), the optimized path gains $c_{i,n} = c_{i,n}^{(opt)}$ are available for the realization of deterministic simulation models. Figure 5.39(a) shows the resulting probability density function $\tilde{p}_{\mu_i}(x)$ by using the optimized quantities $c_{i,n}^{(opt)}$. For the choice of suitable starting values for the path gains, we appropriately fall back to the quantities $c_{i,n} = \sigma_0\sqrt{2/N_i}$. For a better assessment of the obtained results, the probability density function $\tilde{p}_{\mu_i}(x)$, which is obtained by using the starting values $c_{i,n} = \sigma_0\sqrt{2/N_i}$ [cf. also Figure 4.4(a)], is again presented in Figure 5.39(b).

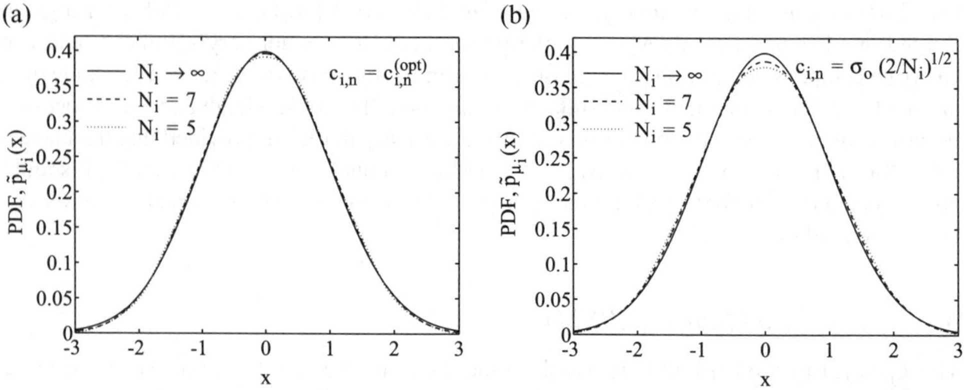

Figure 5.39 Probability density function $\tilde{p}_{\mu_i}(x)$ for $N_i \in \{5, 7, \infty\}$ by using: (a) $c_{i,n} = c_{i,n}^{(opt)}$ and (b) $c_{i,n} = \sigma_0\sqrt{2/N_i}$ ($\sigma_0^2 = 1$).

More expressive than the comparison of Figures 5.39(a) and 5.39(b) are the results of Figure 5.40, where the mean-square error $E_{p_{\mu_i}}$ [see (4.46)] is presented for $c_{i,n} = c_{i,n}^{(opt)}$ as well as for $c_{i,n} = \sigma_0\sqrt{2/N_i}$. One can clearly realize that the benefit of the optimization decreases strictly monotonously if the number of sinusoids N_i increases.

It is also worth mentioning that after the minimization of (5.73), all optimized path gains $c_{i,n}^{(opt)}$ are in fact identical (due to the central limit theorem). But for a finite number of sinusoids N_i, they are always smaller than the pre-set starting values, i.e., $c_{i,n}^{(opt)} < \sigma_0\sqrt{2/N_i}$, $\forall N_i = 1, 2, \ldots$ Since the optimized path gains $c_{i,n}^{(opt)}$ are identical, which is even the case when arbitrary starting values are chosen, it is probable that the L_p-norm in (5.73) has a global

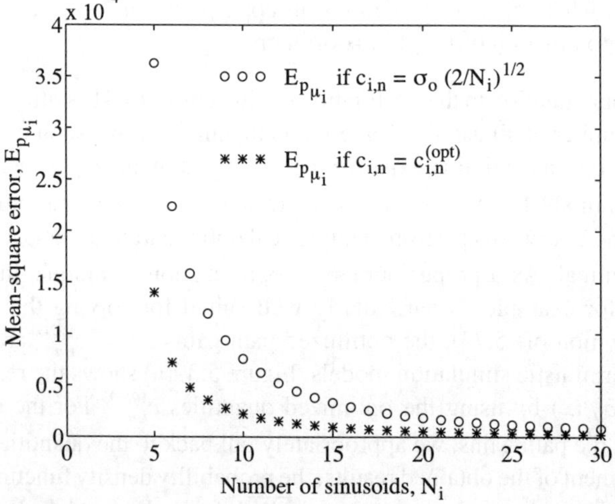

Figure 5.40 Mean-square error $E_{p_{\mu_i}}$, if $c_{i,n} = c_{i,n}^{(opt)}$ (∗∗∗) and $c_{i,n} = \sigma_0\sqrt{2/N_i}$ (∘∘∘) with $\sigma_0^2 = 1$.

minimum at $c_{i,n} = c_{i,n}^{(opt)}$, and thus the path gains $c_{i,n} = c_{i,n}^{(opt)}$ are optimal. For finite values of N_i, one may also take into account that due to $c_{i,n} = c_{i,n}^{(opt)} < \sigma_0\sqrt{2/N_i}$, the mean power of the deterministic process $\tilde{\mu}_i(t)$ is always smaller than the variance of the stochastic process $\mu_i(t)$, i.e., it holds $\tilde{\sigma}_0^2 < \sigma_0^2$. This can be realized by considering Figure 5.41.

Obviously, a compromise between the attainable approximation accuracies of $\tilde{p}_{\mu_i}(x) \approx p_{\mu_i}(x)$ and $\tilde{\sigma}_0^2 \approx \sigma_0^2$ has to be found here. Let us try to avoid this compromise by imposing the so-called power constraint defined by $\tilde{\sigma}_0^2 = \sigma_0^2$ on the simulation model. Then, we have to optimize, e.g., the first $N_i - 1$ path gains $c_{i,1}, c_{i,2}, \ldots, c_{i,N_i-1}$ and the remaining parameter c_{i,N_i} is determined such that the imposed power constraint $\tilde{\sigma}_0^2 = \sigma_0^2$ is always fulfilled. In this case, the optimization results in $c_{i,n}^{(opt)} = \sigma_0\sqrt{2/N_i}$ for all $n = 1, 2, \ldots, N_i$. Thus, by including the power constraint $\tilde{\sigma}_0^2 = \sigma_0^2$ in the parameter design, an optimal approximation of the Gaussian distribution $p_{\mu_i}(x)$ for any number of sinusoids N_i can only become possible if the path gains $c_{i,n}$ are given by $c_{i,n} = c_{i,n}^{(opt)} = \sigma_0\sqrt{2/N_i}$. Therefore, when modelling Gaussian random processes and other processes derivable from these, such as Rayleigh processes, Rice processes, and lognormal processes, we will usually make use of the relation $c_{i,n} = \sigma_0\sqrt{2/N_i}$ in the following.

The suggested method is still quite useful and advantageous for the approximation of probability density functions which are in general not derivable from Gaussian distributions, such as the Nakagami-*m* distribution [27] introduced in (2.58). The Nakagami-*m* distribution is more flexible than the frequently used Rayleigh or Rice distribution and often enables a closer match to probability density functions that follow from experimental measurement results [26].

In order to be able to determine the set of path gains $\{c_{i,n}\}$ such that the probability density function of the deterministic simulation model approximates the Nakagami-*m* distribution, we perform the optimization of the path gains in a similar manner as that described previously in connection with the normal distribution. The only difference is that in (5.73) we have to substitute the Gaussian distribution $p_{\mu_i}(x)$ by the Nakagami-*m* distribution $p_\omega(z)$ [see (2.58)]

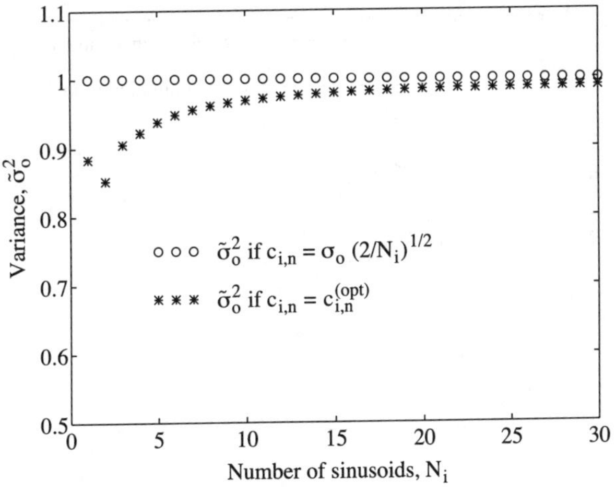

Figure 5.41 Mean power $\tilde{\sigma}_0^2$ of the deterministic process $\tilde{\mu}_i(t)$, if $c_{i,n} = c_{i,n}^{(opt)}$ (∗∗∗) and $c_{i,n} = \sigma_0\sqrt{2/N_i}$ (∘∘∘) with $\sigma_0^2 = 1$.

and $\tilde{p}_{\mu_i}(x)$ has to be replaced by $\tilde{p}_{\xi}(z)$ given by (4.57). Some optimization results obtained for various values of the parameter m are shown in Figure 5.42, where $N_1 = N_2 = 10$ sinusoids have been used in all cases.

Further details on the derivation and simulation of Nakagami fading channels can be found in [163, 164]. Results of the analysis of both the level-crossing rate and the average duration of fades of Nakagami processes were first published in [165].

In [38], the Weibull distribution was suggested for the approximation of the probability density function of real-world mobile radio channels in the 900 MHz frequency range. As is well known, the Weibull distribution can be derived by means of a nonlinear transformation of a uniformly distributed random variable [166]. Since the uniform distribution can be determined from a further nonlinear transformation of two Gaussian distributed random variables [166], the problem of modelling the Weibull distribution can thus be reduced to the problem of modelling Gaussian random processes, which we have already discussed. Therefore, we do not expect any essential new discoveries from further analysis of this matter.

Let us now consider the Postulation (ii) [see (5.74)]. According to (4.11), the autocorrelation function $\tilde{r}_{\mu_i\mu_i}(\tau)$ depends on both the path gains $c_{i,n}$ and the discrete Doppler frequencies $f_{i,n}$. Since the path gains $c_{i,n}$ were already determined so that the probability density function $\tilde{p}_{\mu_i}(x)$ of the deterministic process $\tilde{\mu}_i(t)$ approximates the Gaussian distribution $p_{\mu_i}(x)$ of the stochastic process $\tilde{\mu}_i(t)$ as good as possible, only the discrete Doppler frequencies $f_{i,n}$ can be used for the minimization of the L_p-norm $E^{(p)}_{r_{\mu_i\mu_i}}$ defined by (5.74). The discrete Doppler frequencies $f_{i,n}$ can now be optimized, e.g., by applying the Fletcher-Powell algorithm, so that $E^{(p)}_{r_{\mu_i\mu_i}}$ becomes as small as possible, and, hence, the autocorrelation function $\tilde{r}_{\mu_i\mu_i}(\tau)$ of the deterministic process $\tilde{\mu}_i(t)$ approximates the given autocorrelation function $r_{\mu_i\mu_i}(\tau)$ of the stochastic process $\mu_i(t)$ within the interval $[0, \tau_{\max}]$. In general, we cannot guarantee that the Fletcher-Powell algorithm — like any other optimization algorithm suitable for this problem — finds the global minimum of $E^{(p)}_{r_{\mu_i\mu_i}}$, so that in most cases, we have to be satisfied with a local minimum. This property, which at first glance seems to be a drawback, can easily be

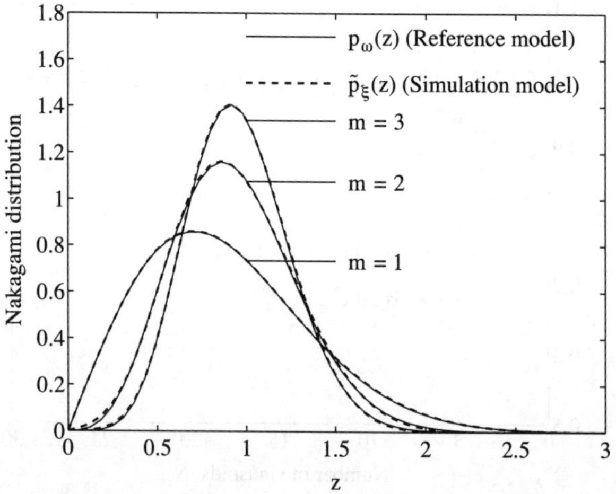

Figure 5.42 Approximation of the Nakagami-m distribution by using deterministic processes with $N_1 = N_2 = 10$ ($\Omega = 1$).

turned into an advantage if we take into account that various local minima lead to various sets of discrete Doppler frequencies $\{f_{i,n}\}$. For the generation of uncorrelated deterministic processes $\tilde{\mu}_1(t)$ and $\tilde{\mu}_2(t)$, we are therefore no longer restricted to the previous convention $N_2 := N_1 + 1$, as we can now guarantee that the processes $\tilde{\mu}_1(t)$ and $\tilde{\mu}_2(t)$ are also uncorrelated for $N_1 = N_2$. However, the latter property can likewise be obtained by carrying out the optimizations with different values for the parameter p or by using different starting values for the discrete Doppler frequencies $f_{i,n}$. This concept unfolds its full potential when it is applied to the problem of designing multiple uncorrelated Rayleigh fading waveforms.

In the following, we will apply the L_p-norm method to the Jakes and the Gaussian power spectral densities. When taking the power constraint $\tilde{\sigma}_0^2 = \sigma_0^2$ into account, the Postulation (i) is already fulfilled by $c_{i,n} = c_{i,n}^{(opt)} = \sigma_0\sqrt{2/N_i}$, and therefore only the Postulation (ii) needs to be investigated in more detail.

Jakes power spectral density: By substituting (3.25) and (4.11) into (5.74), we obtain an optimized set $\{f_{i,n}^{(opt)}\}$ for the discrete Doppler frequencies after the numerical minimization of the L_p-norm $E_{r_{\mu_i\mu_i}}^{(p)}$. Suitable starting values for the discrete Doppler frequencies $f_{i,n}$ are, e.g., the quantities $f_{i,n} = f_{\max}\sin[n\pi/(2N_i)], \forall\, n = 1, 2, \ldots, N_i\ (i = 1, 2)$, derived by using the method of equal areas. For the Jakes power spectral density, the upper limit of the integral in (5.74) is given by the relation $\tau_{\max} = N_i/(2f_{\max})$, which we already know from Subsection 5.1.1. All of the following optimization results are based on the L_p-norm $E_{r_{\mu_i\mu_i}}^{(p)}$ with $p = 2$.

Generally valid statements on the greatest common divisor $F_i = \gcd\{f_{i,n}^{(opt)}\}_{n=1}^{N_i}$ cannot be made here. Numerical investigations, however, have shown that F_i is usually zero or at least extremely small. Therefore, the deterministic processes $\tilde{\mu}_i(t)$ designed with the L_p-norm method are nonperiodical or quasi-nonperiodical. For the time average $\tilde{m}_{\mu_i}$ and the mean power $\tilde{\sigma}_{\mu_i}^2$, again the relations $\tilde{m}_{\mu_i} = m_{\mu_i} = 0$ and $\tilde{\sigma}_{\mu_i}^2 = \sigma_0^2$ follow, respectively.

As for the preceding methods, the power spectral density $\tilde{S}_{\mu_i\mu_i}(f)$ and the autocorrelation function $\tilde{r}_{\mu_i\mu_i}(\tau)$ have been evaluated using the example of $N_i = 25$, which leads to the results shown in Figures 5.43(a) and 5.43(b), respectively.

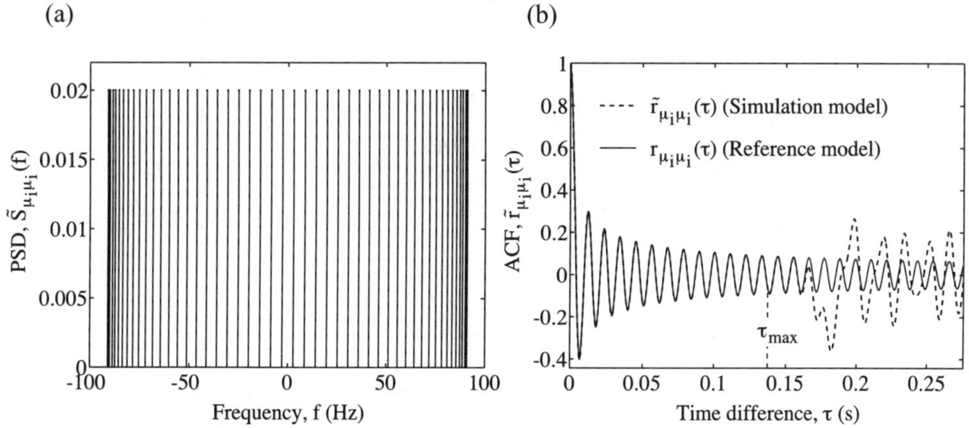

Figure 5.43 (a) Power spectral density $\tilde{S}_{\mu_i\mu_i}(f)$ and (b) autocorrelation function $\tilde{r}_{\mu_i\mu_i}(\tau)$ for $N_i = 25$ (LPNM, Jakes PSD, $f_{\max} = 91$ Hz, $\sigma_0^2 = 1$).

Since the path gains $c_{i,n}$ have been chosen as $c_{i,n} = \sigma_0\sqrt{2/N_i}$, we obtain the graph presented in Figure 5.14(b) for the mean-square error $E_{p_{\mu_i}}$ [see (4.127)]. The results of the evaluation of $E_{r_{\mu_i\mu_i}}$ [see (4.128)] are shown in Figure 5.44, which also includes the corresponding graph obtained by applying the method of equal areas.

In order to compute the model error $\Delta\beta_i = \tilde{\beta}_i - \beta$, the expression (4.29) has to be evaluated for $c_{i,n} = \sigma_0\sqrt{2/N_i}$ and $f_{i,n} = f_{i,n}^{(opt)}$. In comparison with the method of equal areas, we then obtain the graphs illustrated in Figure 5.45 for the relative model error $\Delta\beta_i/\beta$.

The simulation of the level-crossing rate and the average duration of fades is carried out in the same way as described in Subsection 5.1.3. For reasons of uniformity, we again choose here the pair (10, 11) for the 2-tuple (N_1, N_2). The simulation results for the normalized

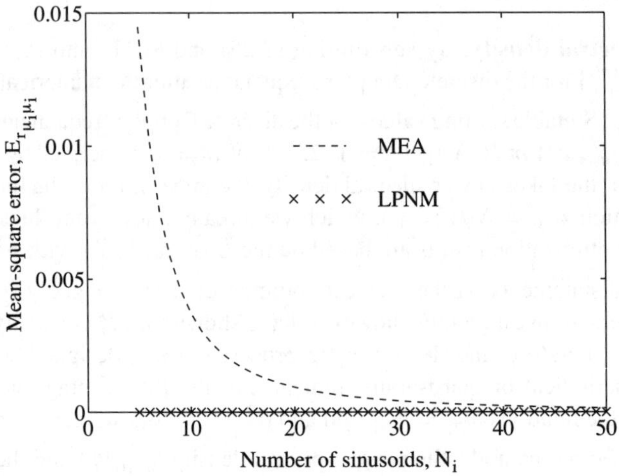

Figure 5.44 Mean-square error $E_{r_{\mu_i\mu_i}}$ (LPNM, Jakes PSD, $f_{max} = 91$ Hz, $\sigma_0^2 = 1$, $\tau_{max} = N_i/(2f_{max})$).

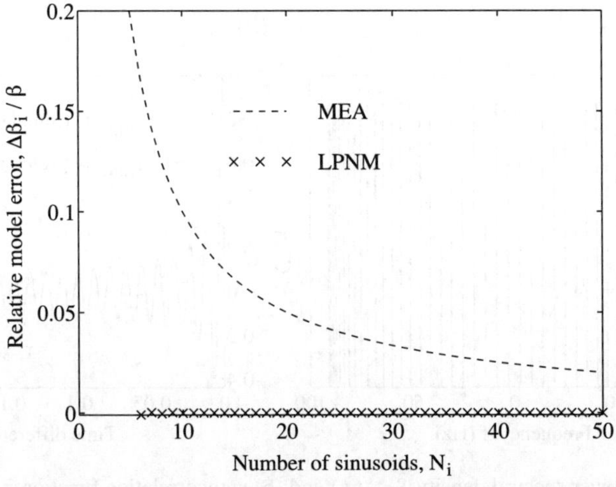

Figure 5.45 Relative model error $\Delta\beta_i/\beta$ (LPNM, Jakes PSD).

level-crossing rate $\tilde{N}_\xi(r)/f_{\max}$ and the normalized average duration of fades $\tilde{T}_{\xi_-}(r) \cdot f_{\max}$ are visualized in Figures 5.46(a) and 5.46(b), respectively. The analytical results one finds for the reference model and the simulation model are also illustrated in these figures. Since the relative model errors $\Delta\beta_1$ and $\Delta\beta_2$ are extremely small for both cases $N_1 = 10$ and $N_2 = 11$, the individual curves can no longer be distinguished from each other in the presented graphs.

Gaussian power spectral density: The previously analyzed methods for the determination of the model parameters of the deterministic processes have made it quite clear that the Gaussian power spectral density causes more severe problems than the Jakes power spectral density. In this subsection, we show how these problems can be overcome by using the L_p-norm method. Therefore, we fully exploit the degrees of freedom which this method has to offer. All in all, this leads us to three fundamental variants of the L_p-norm method. In the following, these variants will be briefly described and afterwards analyzed with respect to their performances.

First variant of the L_p-norm method (LPNM I): In the first variant, the path gains $c_{i,n}$ are again computed according to the equation $c_{i,n} = \sigma_0\sqrt{2/N_i}$ for all $n = 1, 2, \ldots, N_i$, whereas the discrete Doppler frequencies $f_{i,n}$ are optimized for $n = 1, 2, \ldots, N_i - 1$ in such a way that the L_p-norm $E^{(p)}_{r_{\mu_i\mu_i}}$ [see (5.74)] results in a (local) minimum, i.e.,

$$E^{(p)}_{r_{\mu_i\mu_i}}(\boldsymbol{f}_i) = \text{Min!}, \tag{5.75}$$

where $\boldsymbol{f}_i$ stands for the parameter vector $\boldsymbol{f}_i = (f_{i,1}, f_{i,2}, \ldots, f_{i,N_i-1})^T \in \mathbb{R}^{N_i-1}$. Boundary conditions, like the restriction that the components of the parameter vector $\boldsymbol{f}_i$ shall be positive, do not need to be imposed on the procedure, since the Gaussian power spectral density is symmetrical. The remaining discrete Doppler frequency f_{i,N_i} is defined by

$$f_{i,N_i} := \sqrt{\frac{\beta N_i}{(2\pi\sigma_0)^2} - \sum_{n=1}^{N_i-1} f_{i,n}^2}, \tag{5.76}$$

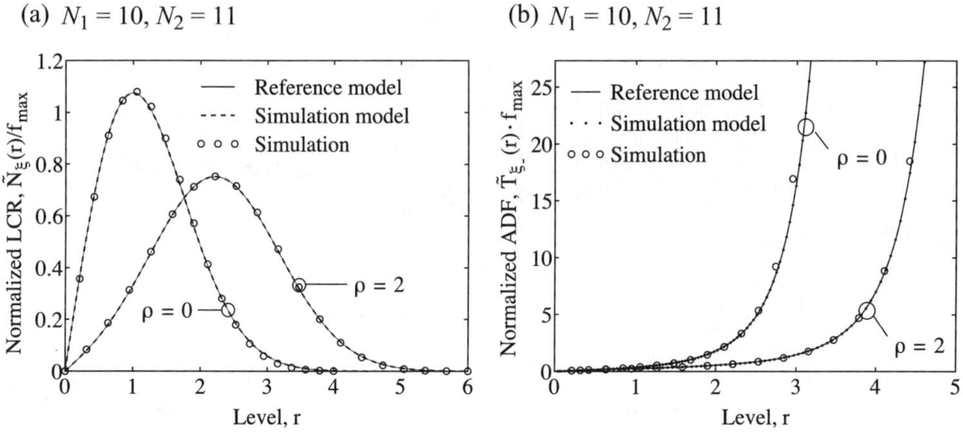

Figure 5.46 (a) Normalized level-crossing rate $\tilde{N}_\xi(r)/f_{\max}$ and (b) normalized average duration of fades $\tilde{T}_\xi(r) \cdot f_{\max}$ (LPNM, Jakes PSD, $f_{\max} = 91\,\text{Hz}$, $\sigma_0^2 = 1$).

so that we have guaranteed in a simple manner that the model error $\Delta\beta_i$ is always zero for all chosen values of $N_i = 1, 2, \ldots$ $(i = 1, 2)$. With the corresponding quantity β, we can of course make use of this possibility when dealing with the Jakes power spectral density (or any other given power spectral density) as well. Suitable starting values for the optimization of the involved discrete Doppler frequencies are the quantities found by employing the method of equal areas [cf. Subsection 5.1.3]. For the evaluation of the L_p-norm $E^{(p)}_{r_{\mu_i\mu_i}}$, it is sufficient for our purpose to restrict ourselves to the case of $p = 2$. Concerning the parameter $\tau_{\max}$ in the L_p-norm $E^{(p)}_{r_{\mu_i\mu_i}}$, we revert to the relation $\tau_{\max} = N_i/(2\kappa_c f_c)$ with $\kappa_c = 2\sqrt{2/\ln 2}$ and $f_c = \sqrt{\ln 2} f_{\max}$, which has already been applied several times before. For the quantities F_i, $\tilde{m}_{\mu_i}$, and $\tilde{\sigma}^2_{\mu_i}$, the statements made for the Jakes power spectral density are still valid in the present case. Uncorrelated deterministic processes $\tilde{\mu}_1(t)$ and $\tilde{\mu}_2(t)$ can also be obtained for $N_1 = N_2$ by optimizing the parameter vectors $\boldsymbol{f}_1$ and $\boldsymbol{f}_2$ under different conditions. Therefore, it is sufficient to change, for example, $\tau_{\max}$ or p slightly and then to repeat the optimization once again.

We choose $N_i = 25$ and with the first variant of the L_p-norm method, we obtain the power spectral density $\tilde{S}_{\mu_i\mu_i}(f)$ presented in Figure 5.47(a), while Figure 5.47(b) shows the corresponding autocorrelation function $\tilde{r}_{\mu_i\mu_i}(\tau)$.

Second variant of the L_p-norm method (LPNM II): With the second variant of the L_p-norm method, our purpose is to fit the autocorrelation function $\tilde{r}_{\mu_i\mu_i}(\tau)$ within the interval $[0, \tau_{\max}]$ far closer to $r_{\mu_i\mu_i}(\tau)$ than it is possible by using the LPNM I. Therefore, we combine all parameters determining the behaviour of $\tilde{r}_{\mu_i\mu_i}(\tau)$ into the parameter vectors $\boldsymbol{c}_i = (c_{i,1}, c_{i,2}, \ldots, c_{i,N_i})^T \in \mathbb{R}^{N_i}$ and $\boldsymbol{f}_i = (f_{i,1}, f_{i,2}, \ldots, f_{i,N_i})^T \in \mathbb{R}^{N_i}$. Now, the task is actually to optimize the parameter vectors $\boldsymbol{c}_i$ and $\boldsymbol{f}_i$ in such a way that the L_p-norm $E^{(p)}_{r_{\mu_i\mu_i}}$ becomes minimal, i.e.,

$$E^{(p)}_{r_{\mu_i\mu_i}}(\boldsymbol{c}_i, \boldsymbol{f}_i) = \text{Min!}. \tag{5.77}$$

Also in this case, we do not need to impose any boundary conditions on the components of the parameter vectors $\boldsymbol{c}_i$ and $\boldsymbol{f}_i$.

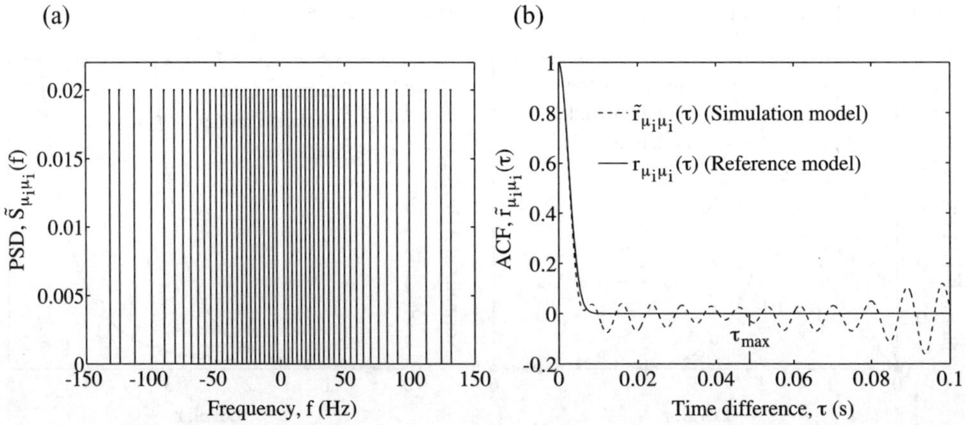

Figure 5.47 (a) Power spectral density $\tilde{S}_{\mu_i\mu_i}(f)$ and (b) autocorrelation function $\tilde{r}_{\mu_i\mu_i}(\tau)$ for $N_i = 25$ (LPNM I, Gaussian PSD, $f_c = \sqrt{\ln 2} f_{\max}$, $f_{\max} = 91$ Hz, $\sigma_0^2 = 1$).

An example of the resulting power spectral density $\tilde{S}_{\mu_i\mu_i}(f)$ is depicted in Figure 5.48(a), where again $N_i = 25$ is chosen. Additionally, Figure 5.48(b) shows the graph of the corresponding autocorrelation function $\tilde{r}_{\mu_i\mu_i}(\tau)$.

It does not go without notice that the approximation $r_{\mu_i\mu_i}(\tau) \approx \tilde{r}_{\mu_i\mu_i}(\tau)$ for $\tau \in [0, \tau_{\max}]$ is exceptionally good. However, in order to obtain this advantage, we have to accept some disadvantages. Thus, for example, the power constraint $\tilde{\sigma}^2_{\mu_i} = \sigma_0^2$ is only fulfilled approximately; besides, the model error $\Delta\beta_i$ is unequal to zero. In general, the obtained approximations $\tilde{\sigma}^2_{\mu_i} \approx \sigma_0^2$ and $\tilde{\beta}_i \approx \beta$ or $\Delta\beta_i \approx 0$ are still very good and absolutely sufficient for most practical applications. A problem which should be considered as serious, however, occurs for the LPNM II when optimizing the path gains $c_{i,n}$. The degradation of the probability density function $\tilde{p}_{\mu_i}(x)$, to which this problem leads, will be discussed further below. At this point, it is sufficient to mention that all these disadvantages can be avoided with the third variant.

Third variant of the L_p-norm method (LPNM III): The third variant is aiming at optimizing both the autocorrelation function $\tilde{r}_{\mu_i\mu_i}(\tau)$ and the probability density function $\tilde{p}_{\mu_i}(x)$. An error function suitable for this purpose has the form

$$E^{(p)}(\boldsymbol{c}_i, \boldsymbol{f}_i) = W_1 \cdot E^{(p)}_{r_{\mu_i\mu_i}}(\boldsymbol{c}_i, \boldsymbol{f}_i) + W_2 \cdot E^{(p)}_{p_{\mu_i}}(\boldsymbol{c}_i), \tag{5.78}$$

where $E^{(p)}_{r_{\mu_i\mu_i}}(\cdot)$ and $E^{(p)}_{p_{\mu_i}}(\cdot)$ denote the L_p-norms introduced by (5.74) and (5.73), respectively. The quantities W_1 and W_2 are appropriate weighting factors, which will be defined by $W_1 = 1/4$ and $W_2 = 3/4$ in the sequel. To guarantee that both boundary conditions $\tilde{\sigma}^2_{\mu_i} = \sigma_0^2$ and $\tilde{\beta}_i = \beta$ can now be fulfilled exactly, we define the parameter vectors $\boldsymbol{c}_i$ and $\boldsymbol{f}_i$ by

$$\boldsymbol{c}_i = (c_{i,1}, c_{i,2}, \ldots, c_{i,N_i-1})^T \in \mathbb{R}^{N_i-1} \tag{5.79a}$$

and

$$\boldsymbol{f}_i = (f_{i,1}, f_{i,2}, \ldots, f_{i,N_i-1})^T \in \mathbb{R}^{N_i-1}, \tag{5.79b}$$

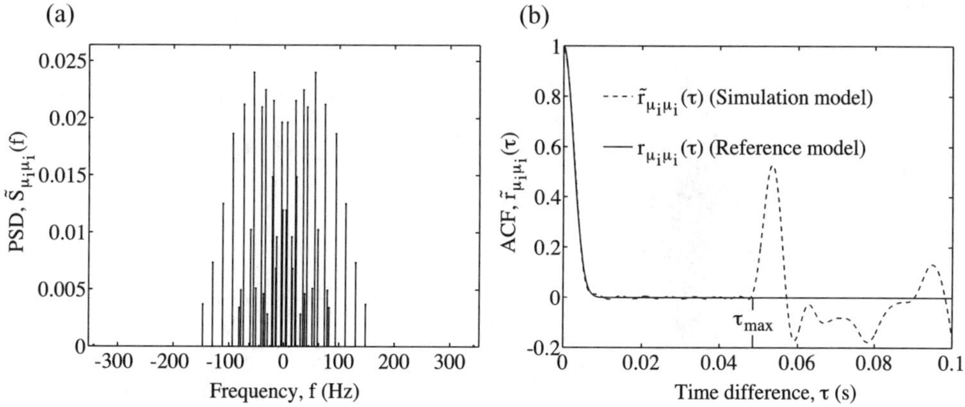

Figure 5.48 (a) Power spectral density $\tilde{S}_{\mu_i\mu_i}(f)$ and (b) autocorrelation function $\tilde{r}_{\mu_i\mu_i}(\tau)$ for $N_i = 25$ (LPNM II, Gaussian PSD, $f_c = \sqrt{\ln 2} f_{\max}$, $f_{\max} = 91$ Hz, $\sigma_0^2 = 1$).

respectively, and calculate the remaining model parameters c_{i,N_i} and f_{i,N_i} as follows:

$$c_{i,N_i} = \sqrt{2\sigma_0^2 - \sum_{n=1}^{N_i-1} c_{i,n}^2}, \tag{5.80a}$$

$$f_{i,N_i} = \frac{1}{c_{i,N_i}} \sqrt{\frac{\beta}{2\pi^2} - \sum_{n=1}^{N_i-1} (c_{i,n} f_{i,n})^2}, \tag{5.80b}$$

where $\beta = -\ddot{r}_{\mu_i\mu_i}(0) = 2(\pi f_c \sigma_0)^2 / \ln 2$ $(i = 1, 2)$. Correlations between the deterministic processes $\tilde{\mu}_1(t)$ and $\tilde{\mu}_2(t)$ can now be avoided for $N_1 = N_2$ by performing the minimization of the error function in (5.78) for $i = 1$ and $i = 2$ with different weighting factors of the respective L_p-norms $E_{r_{\mu_i\mu_i}}^{(p)}$ and $E_{p_{\mu_i}}^{(p)}$.

As in the preceding examples, we choose $N_i = 25$ and observe the resulting power spectral density $\tilde{S}_{\mu_i\mu_i}(f)$ in Figure 5.49(a). The corresponding autocorrelation function $\tilde{r}_{\mu_i\mu_i}(\tau)$ is plotted in Figure 5.49(b).

Finally, we will also analyze the performance of these three variants of the L_p-norm method. Here, we are especially interested in the mean-square errors $E_{r_{\mu_i\mu_i}}$ and $E_{p_{\mu_i}}$ [see (4.128) and (4.127)], both of which are shown as a function of N_i in Figures 5.50(a) and 5.50(b), respectively. For the starting values, the parameters to be optimized were in all cases computed by employing the method of equal areas. When studying Figure 5.50(a), it becomes striking that the quality of the approximation $r_{\mu_i\mu_i}(\tau) \approx \tilde{r}_{\mu_i\mu_i}(\tau)$ can be improved significantly if the discrete Doppler frequencies $f_{i,n}$ and the path gains $c_{i,n}$ are included in the optimization procedure, as it is the objective of the LPNM II and III. It should be noted that among the three variants of the L_p-norm method introduced here, the LPNM I has in fact the largest mean-square error $E_{r_{\mu_i\mu_i}}$ [see Figure 5.50(a)], while on the other hand the mean-square error $E_{p_{\mu_i}}$ [see Figure 5.50(b)] is the smallest. Exactly the opposite statement is true for the LPNM II. Only the LPNM III provides a proven recipe for finding a good compromise between the

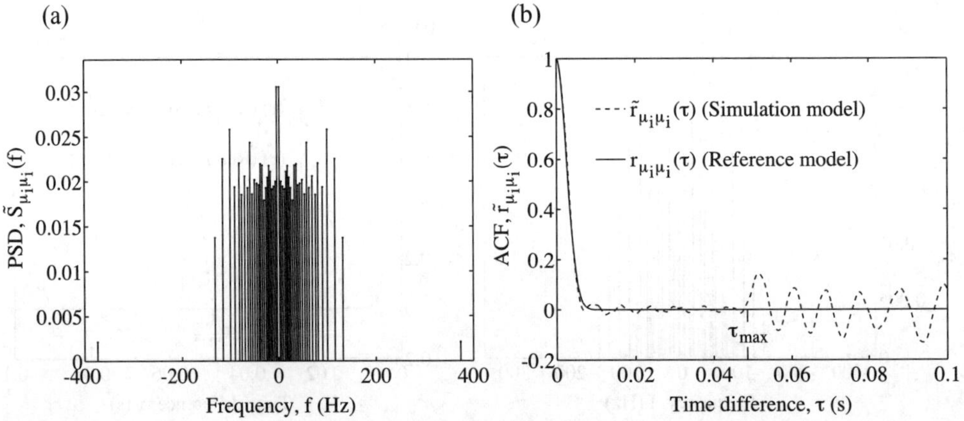

Figure 5.49 (a) Power spectral density $\tilde{S}_{\mu_i\mu_i}(f)$ and (b) autocorrelation function $\tilde{r}_{\mu_i\mu_i}(\tau)$ for $N_i = 25$ (LPNM III, Gaussian PSD, $f_c = \sqrt{\ln 2} f_{max}$, $f_{max} = 91$ Hz, $\sigma_0^2 = 1$).

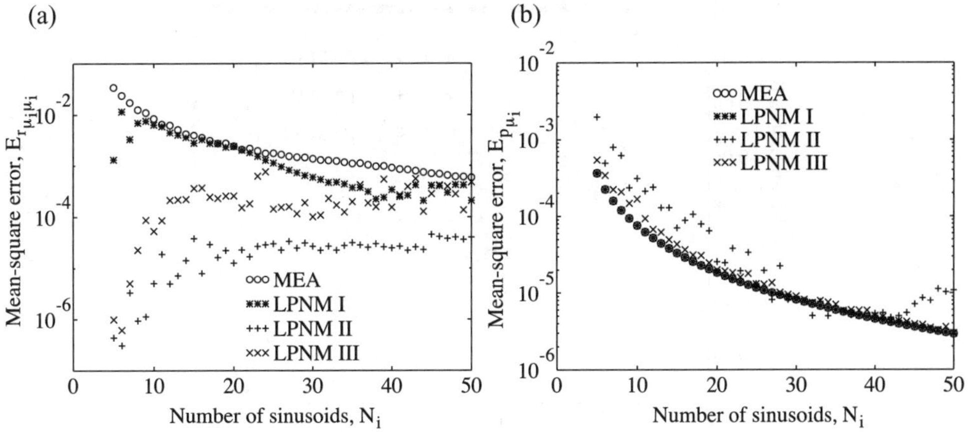

Figure 5.50 Mean-square errors: (a) $E_{r_{\mu_i\mu_i}}$ and (b) $E_{p_{\mu_i}}$ (LPNM I–III, Gaussian PSD, $f_c = \sqrt{\ln 2} f_{\max}$, $f_{\max} = 91$ Hz, $\sigma_0^2 = 1$, $\tau_{\max} = N_i/(2\kappa_c f_c)$, $\kappa_c = 2\sqrt{2/\ln 2}$).

minimization of the error norms $E_{r_{\mu_i\mu_i}}$ and $E_{p_{\mu_i}}$. By a skilful choice of the weighting factors in (5.78), this method always allows the minimization of $E^{(p)}_{r_{\mu_i\mu_i}}$ in a manner such that we do not have to accept noticeable degradations concerning the error norm $E^{(p)}_{p_{\mu_i}}$. Owing not only to this property, but also because the boundary conditions $\tilde{\sigma}^2_{\mu_i} = \sigma_0^2$ and $\tilde{\beta}_i = \beta$ can be fulfilled exactly by the LPNM III, this variant of the L_p-norm method is without doubt the most efficient.

Concerning the evaluation of the model error $\Delta\beta_i = \tilde{\beta}_i - \beta$ for the three variants of the L_p-norm method, we recall that during the introduction of the LPNM I and III, we have set great store by the fact that the model error $\Delta\beta_i$ is always equal to zero, which is fully guaranteed by (5.76) and (5.80b), respectively. In order to find the model error for the LPNM II, we substitute the optimized path gains $c_{i,n} = c^{(opt)}_{i,n}$ and the optimized discrete Doppler frequencies $f_{i,n} = f^{(opt)}_{i,n}$ into (4.29). This leads to the results of the relative model error $\Delta\beta_i/\beta$ shown in Figure 5.51. It can be seen that the model error $\Delta\beta_i$ corresponding to the LPNM II is different from zero. In the present case, the autocorrelation function $\tilde{r}_{\mu_i\mu_i}(\tau)$ was optimized over the interval $[0, \tau_{\max}]$ with a constant weighting factor. If the approximation error of $\tilde{r}_{\mu_i\mu_i}(\tau)$ is weighted higher in an infinitesimal ϵ-interval around $\tau = 0$, then the model error $\Delta\beta_i$ can be reduced further. However, as can be seen clearly in Figure 5.51, the relative model error $\Delta\beta_i/\beta$ is sufficiently small, so that we will be content with the results reported for this procedure and continue with the analysis of the level-crossing rate and the average duration of fades.

For the analysis of the level-crossing rate $\tilde{N}_\xi(r)$ and the average duration of fades $\tilde{T}_{\xi_-}(r)$, we will confine ourselves to the LPNM III. Again, the simulation of the quantities $\tilde{N}_\xi(r)$ and $\tilde{T}_{\xi_-}(r)$ will be performed under the conditions described in Subsection 5.1.3. For the normalized level-crossing rate $\tilde{N}_\xi(r)/f_{\max}$, the simulation results as well as the analytical results are depicted in Figure 5.52(a), where, exactly as in the preceding examples, the pair (10, 10) was chosen for the 2-tuple (N_1, N_2). The adjacent Figure 5.52(b) shows the corresponding normalized average duration of fades $\tilde{T}_{\xi_-}(r) f_{\max}$.

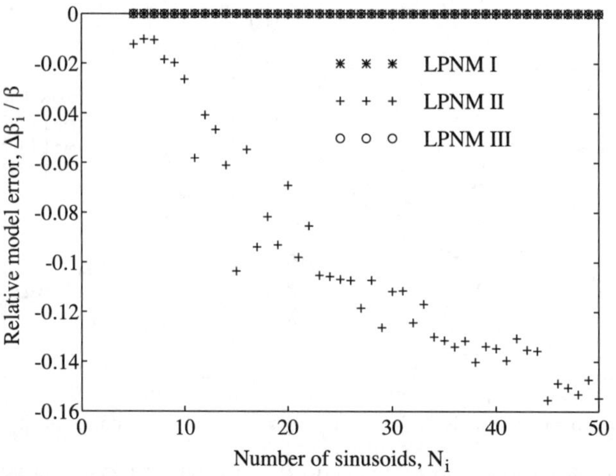

Figure 5.51 Relative model error $\Delta\beta_i/\beta$ (LPNM I–III, Gaussian PSD, $f_c = \sqrt{\ln 2} f_{\max}$, $f_{\max} = 91$ Hz, $\sigma_0^2 = 1$).

We want to close this subsection with some general remarks on the L_p-norm method. The obvious advantage of this method lies in its potential to design deterministic processes $\tilde{\mu}_i(t)$ or $\tilde{\xi}(t)$ such that they have the ability to reproduce the characteristic properties of snapshot measurements taken from real-world mobile radio channels. This feature is important for the design of measurement-based channel simulators, which can be accomplished directly by replacing the probability density function $p_{\mu_i}(x)$ in (5.73) and the autocorrelation function $r_{\mu_i\mu_i}(\tau)$ in (5.74) by the corresponding measured quantities. The optimization can then be performed using the same techniques as described above. Compared to other methods, the only drawback of the L_p-norm method is its relatively high numerical complexity, but this is

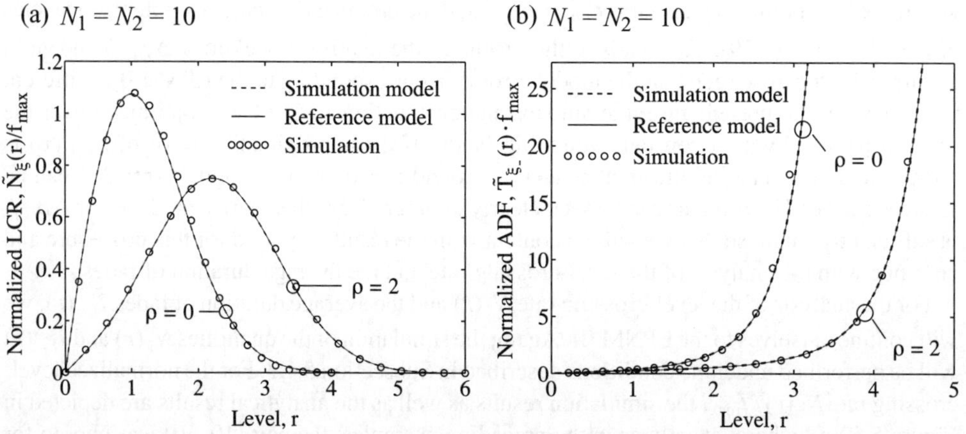

Figure 5.52 (a) Normalized level-crossing rate $\tilde{N}_\xi(r)/f_{\max}$ and (b) normalized average duration of fades $\tilde{T}_{\xi_-}(r) \cdot f_{\max}$ (LPNM III, Gaussian PSD, $f_c = \sqrt{\ln 2} f_{\max}$, $f_{\max} = 91$ Hz, $\sigma_0^2 = 1$).

no longer a serious problem with modern computers and databases. In some specific cases, however, the expenditure is not worth it, at least not when attempting to model the Jakes power spectral density, because there exists a quite simple, elegant, and quasi-optimal solution, which will be presented in the next subsection.

5.1.7 *Method of Exact Doppler Spread (MEDS)*

The method of exact Doppler spread (MEDS) was first introduced in [149] and has been developed especially for the often used Jakes power spectral density. Despite its simplicity, the method is distinguished by its high performance and enables a quasi-optimal approximation of the autocorrelation function corresponding to the Jakes power spectral density. In the following, we will first derive the method of exact Doppler spread in connection with the Jakes power spectral density and afterwards, we will investigate to what extent the method also offers advantages for the application to the Gaussian power spectral density.

Jakes power spectral density: Let us start with the integral presentation of the Bessel function of zeroth order [77, Equation (9.1.18)]

$$J_0(z) = \frac{2}{\pi} \int_0^{\pi/2} \cos(z \sin \alpha)\, d\alpha, \tag{5.81}$$

which can be expressed in form of an infinite series as

$$J_0(z) = \lim_{N_i \to \infty} \frac{2}{\pi} \sum_{n=1}^{N_i} \cos(z \sin \alpha_n) \Delta\alpha, \tag{5.82}$$

where $\alpha_n = \pi(2n-1)/(4N_i)$ and $\Delta\alpha = \pi/(2N_i)$. Hence, for $r_{\mu_i\mu_i}(\tau)$ in (3.25), we can write alternatively

$$r_{\mu_i\mu_i}(\tau) = \lim_{N_i \to \infty} \frac{\sigma_0^2}{N_i} \sum_{n=1}^{N_i} \cos\left\{2\pi f_{\max} \sin\left[\frac{\pi}{2N_i}\left(n - \frac{1}{2}\right)\right] \cdot \tau\right\}. \tag{5.83}$$

This relation describes the autocorrelation function of the stochastic reference model for a Gaussian random process $\mu_i(t)$, whose power spectral density is given by the Jakes power spectral density. Now, if we do not take the limit $N_i \to \infty$, then the stochastic reference model turns into the stochastic simulation model, as described in Section 4.1. Hence, the autocorrelation function of the stochastic simulation model for the process $\hat{\mu}_i(t)$ is

$$\hat{r}_{\mu_i\mu_i}(\tau) = \frac{\sigma_0^2}{N_i} \sum_{n=1}^{N_i} \cos\left\{2\pi f_{\max} \sin\left[\frac{\pi}{2N_i}\left(n - \frac{1}{2}\right)\right] \cdot \tau\right\}. \tag{5.84}$$

The stochastic process $\hat{\mu}_i(t)$ will be ergodic with respect to the autocorrelation function. Then, regarding Subsection 4.3.4, it follows that $\hat{r}_{\mu_i\mu_i}(\tau) = \tilde{r}_{\mu_i\mu_i}(\tau)$ holds. Consequently, for the autocorrelation function of the deterministic process $\tilde{\mu}_i(t)$, we obtain the equation

$$\tilde{r}_{\mu_i\mu_i}(\tau) = \frac{\sigma_0^2}{N_i} \sum_{n=1}^{N_i} \cos\left\{2\pi f_{\max} \sin\left[\frac{\pi}{2N_i}\left(n - \frac{1}{2}\right)\right] \cdot \tau\right\}. \tag{5.85}$$

If we now compare the relation above with the general expression (4.11), then the path gains $c_{i,n}$ and the discrete Doppler frequencies $f_{i,n}$ can be identified with the equations

$$c_{i,n} = \sigma_0 \sqrt{\frac{2}{N_i}} \tag{5.86}$$

and

$$f_{i,n} = f_{\max} \sin\left[\frac{\pi}{2N_i}\left(n - \frac{1}{2}\right)\right], \tag{5.87}$$

respectively, for all $n = 1, 2, \ldots, N_i$ $(i = 1, 2)$. A deterministic process $\tilde{\mu}_i(t)$ designed with these parameters has the time average $\tilde{m}_{\mu_i} = m_{\mu_i} = 0$ and the mean power $\tilde{\sigma}^2_{\mu_i} = \sigma_0^2$. For all relevant values of N_i, the greatest common divisor $F_i = \gcd\{f_{i,n}\}_{n=1}^{N_i}$ is equal to zero (or very small), so that the period $T_i = 1/F_i$ becomes infinite (or very large). The uncorrelatedness of two deterministic processes $\tilde{\mu}_1(t)$ and $\tilde{\mu}_2(t)$ is again guaranteed by the convention $N_2 := N_1 + 1$.

The autocorrelation function $\tilde{r}_{\mu_i\mu_i}(\tau)$ computed according to (5.85) is presented in Figure 5.53(a) for $N_i = 7$ and in Figure 5.53(b) for $N_i = 21$.

In connection with the Jakes power spectral density, the following rule of thumb applies: Let $\tilde{\mu}_1(t)$ be a deterministic process designed by using the method of exact Doppler spread with N_i sinusoids, then the approximation $r_{\mu_i\mu_i}(\tau) \approx \tilde{r}_{\mu_i\mu_i}(\tau)$ is excellent at least up to the N_ith zero-crossing of $r_{\mu_i\mu_i}(\tau)$.

As the equation $c_{i,n} = \sigma_0\sqrt{2/N_i}$, which has been used numerous times before, holds here again for the path gains, we may conclude that the mean-square error $E_{p_{\mu_i}}$ [see (4.127)] follows the trend shown in Figure 5.14(b). The evaluation of the mean-square error $E_{r_{\mu_i\mu_i}}$ [see (4.128)] in terms of N_i results in the graph depicted in Figure 5.54. As shown in this figure, the comparison with the L_p-norm method clearly demonstrates that even by applying numerical optimization techniques, only minor improvements can be achieved.

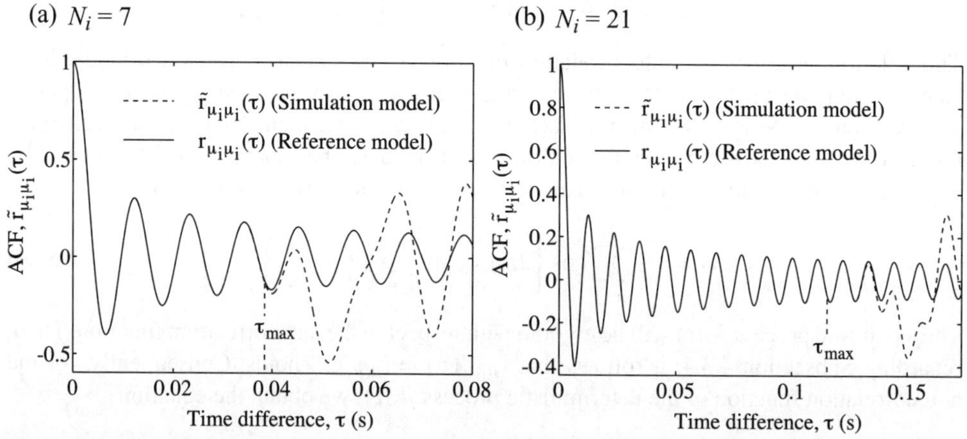

Figure 5.53 Autocorrelation function $\tilde{r}_{\mu_i\mu_i}(\tau)$ for (a) $N_i = 7$ and (b) $N_i = 21$ (MEDS, Jakes PSD, $f_{\max} = 91$ Hz, $\sigma_0^2 = 1$).

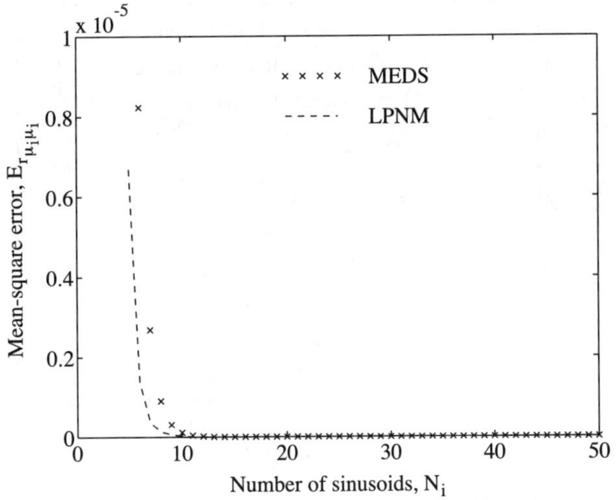

Figure 5.54 Mean-square error $E_{r_{\mu_i\mu_i}}$ (MEDS, Jakes PSD, $f_{\max} = 91$ Hz, $\sigma_0^2 = 1$, $\tau_{\max} = N_i/(2f_{\max})$).

Substituting Equations (5.86) and (5.87) in (4.29) and making use of relation (3.68), we can easily show that $\tilde{\beta}_i = \beta$ holds. This means that the model error $\Delta\beta_i$ is equal to zero for all $N_i \in \mathbb{N}\backslash\{0\}$. Since we have $\tilde{\sigma}^2_{\mu_i} = \sigma_0^2$ and $\tilde{\beta}_i = \tilde{\beta} = \beta$ in the present case, it follows from (3.30b) and (4.32) that

$$\tilde{B}^{(2)}_{\mu\mu} = \tilde{B}^{(2)}_{\mu_i\mu_i} = B^{(2)}_{\mu_i\mu_i} = B^{(2)}_{\mu\mu} \tag{5.88}$$

holds. Hence, the Doppler spread of the simulation model is identical to that of the reference model. This is exactly the reason why this procedure is called the 'method of exact Doppler spread'.

To study the level-crossing rate of the channel simulator by means of time-domain simulations of the envelope, we choose the pair (5, 6) for the 2-tuple (N_1, N_2) and proceed otherwise exactly in the same manner as described in Subsection 5.1.3. Even when such small values are chosen for the numbers of sinusoids, the simulation results match the analytical results very well, as can be seen from an inspection of the results shown in Figure 5.55.

Gaussian power spectral density: By comparing (5.33) and (5.87), we can realize that the latter equation follows immediately from the former if we replace n by $n - 1/2$. This indicates that there exists obviously a close relationship between the method of equal areas and the method of exact Doppler spread. We will return briefly to this relation at the end of this subsection. At first glance, it seems obvious to make an attempt to apply the mapping $n \to n - 1/2$ likewise to (5.43a) and (5.43b), so that the discrete Doppler frequencies $f_{i,n}$ can now be computed by means of the relations

$$\frac{2n-1}{2N_i} - \text{erf}\left(\frac{f_{i,n}}{f_c}\sqrt{\ln 2}\right) = 0, \quad \forall n = 1, 2, \ldots, N_i - 1, \tag{5.89a}$$

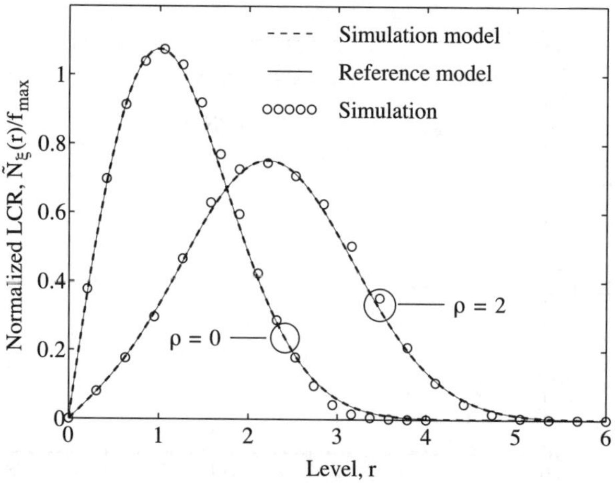

Figure 5.55 Normalized level-crossing rate $\tilde{N}_\xi(r)/f_{\max}$ for $N_1 = 5$ and $N_2 = 6$ (MEDS, Jakes PSD, $f_{\max} = 91$ Hz, $\sigma_0^2 = 1$).

and

$$f_{i,N_i} = \sqrt{\frac{\beta N_i}{(2\pi\sigma_0)^2} - \sum_{n=1}^{N_i-1} f_{i,n}^2}, \tag{5.89b}$$

where the latter equation again guarantees that the model error $\Delta\beta_i$ is equal to zero for all $N_i = 1, 2, \ldots$ $(i = 1, 2)$. For the path gains $c_{i,n}$, the expression (5.86) still remains valid.

The autocorrelation function $\tilde{r}_{\mu_i\mu_i}(\tau)$ can be computed according to (4.11) with the model parameters obtained by this scheme. Figures 5.56(a) and 5.56(b) give us an impression of the behaviour of $\tilde{r}_{\mu_i\mu_i}(\tau)$ for $N_i = 7$ and $N_i = 21$, respectively.

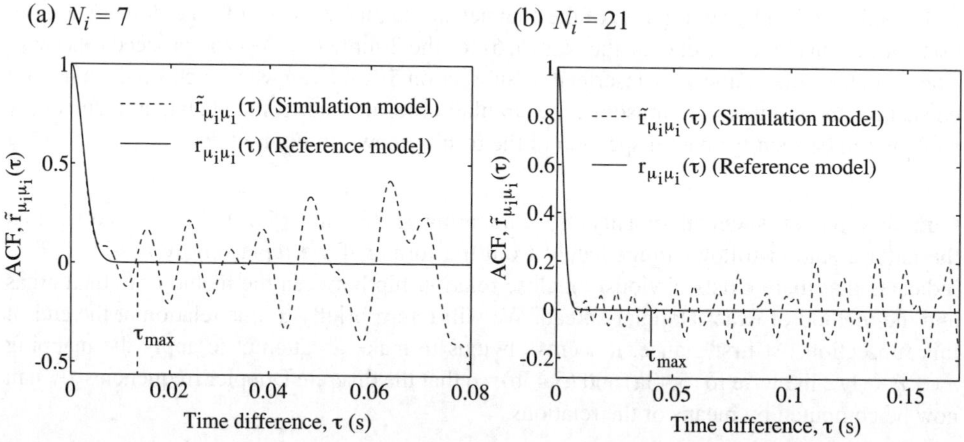

Figure 5.56 Autocorrelation function $\tilde{r}_{\mu_i\mu_i}(\tau)$ for (a) $N_i = 7$ and (b) $N_i = 21$ (MEDS, Gaussian PSD, $f_c = \sqrt{\ln 2} f_{\max}$, $f_{\max} = 91$ Hz, $\sigma_0^2 = 1$).

The mean-square error $E_{r_{\mu_i \mu_i}}$ [see (4.128)], which results from the application of the present method and the standard L_p-norm method, is illustrated in Figure 5.57 as a function of N_i. Unlike the case of the Jakes power spectral density, for a small number of sinusoids N_i, the method of exact Doppler spread leads to significantly higher values of $E_{r_{\mu_i \mu_i}}$ than the L_p-norm method does. However, if $N_i \geq 25$, then intrinsic improvements are no longer achievable by means of numerical optimization techniques.

Owing to $\tilde{\sigma}^2_{\mu_i} = \sigma_0^2$ and $\Delta\beta_i = 0$, i.e., $\tilde{\beta}_i = \beta$, Equation (5.88) holds again here.

It is certainly worth mentioning that the analytical results of the level-crossing rate can be confirmed very precisely by simulation, even when very small values are chosen for N_1 and N_2, e.g., $(N_1, N_2) = (5, 6)$. The following Figure 5.58 serves to illustrate the good match between analytical and simulation results.

As mentioned above, the method of equal areas is closely related to the method of exact Doppler spread. In fact, the former method can be transformed into the latter and vice versa. For example, if we replace the right-hand side of (5.26) by $\sigma_0^2/(4N_i)$ and $f_{i,n}$ by $f_{i,2n-1}$ in (5.27), then we obtain (5.29) in which we remove n and put $n - 1/2$ in its place. Consequently, for (5.33) and (5.43a), we obtain exactly Equations (5.87) and (5.89a), respectively. A similar relationship exists between the Monte Carlo method and the method of exact Doppler spread. For example, if we substitute the random variable $u_{i,n} \in (0, 1]$ in (5.48) by the deterministic quantity $(n - 1/2)/N_i$ for all $n = 1, 2, \ldots, N_i$ $(i = 1, 2)$, then we again obtain (5.87).

5.1.8 Randomized Method of Exact Doppler Spread (RMEDS)

A combination of the Monte Carlo method and the method of exact Doppler spread has been proposed in [111], which was originally published in [153]. This method can be interpreted as a randomized version of the method of exact Doppler spread, which will therefore be called

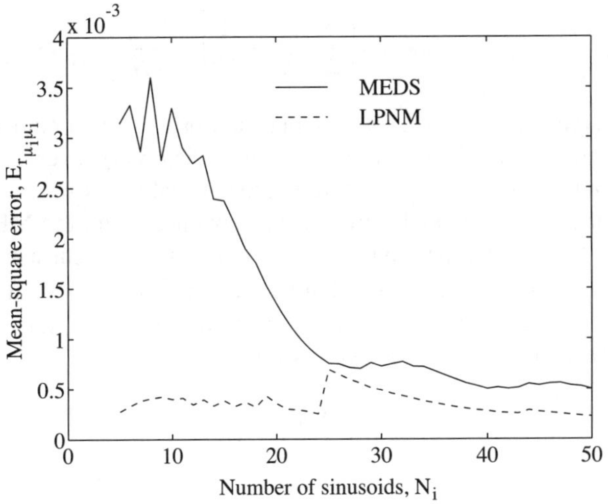

Figure 5.57 Mean-square error $E_{r_{\mu_i \mu_i}}$ (MEDS, Gaussian PSD, $f_c = \sqrt{\ln 2} f_{\max}$, $f_{\max} = 91$ Hz, $\sigma_0^2 = 1$, $\tau_{\max} = N_i/(2\kappa_c f_c)$, $\kappa_c = 2\sqrt{2/\ln 2}$).

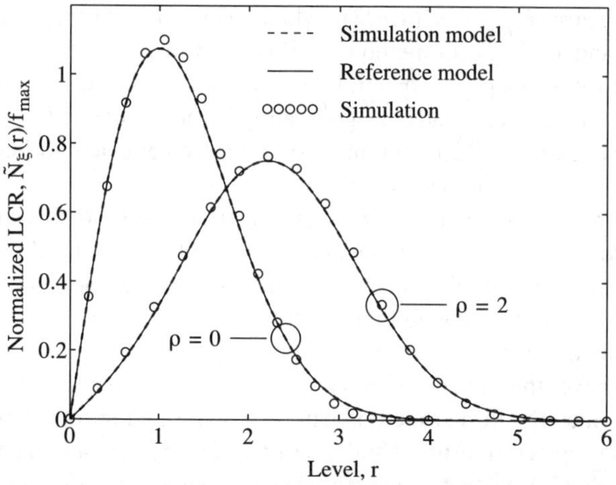

Figure 5.58 Normalized level-crossing rate $\tilde{N}_\xi(r)/f_{\max}$ for $N_1 = 5$ and $N_2 = 6$ (MEDS, Gaussian PSD, $f_c = \sqrt{\ln 2} f_{\max}$, $f_{\max} = 91$ Hz, $\sigma_0^2 = 1$).

the *randomized method of exact Doppler spread (RMEDS)*. In the following, we will present this method briefly by focussing on the modelling of the Jakes power spectral density.

Jakes power spectral density. Although not mentioned explicitly, the basic idea behind the scheme in [111] was obviously to use the method of exact Doppler spread as a basis and then to randomize the angles-of-arrival by adding a small random variable. According to the method in [111], the path gains $c_{i,n}$ are the same as the deterministic quantities in (5.47) and (5.86). The discrete Doppler frequencies $f_{i,n}$ are defined by

$$f_{i,n} = f_{\max} \cos\left[\frac{\pi}{2N_i}\left(n - \frac{1}{2}\right) + \frac{u_{i,n}}{4N_i}\right], \tag{5.90}$$

where $u_{i,n}$ are i.i.d. random variables, each having a uniform distribution over $[-\pi, \pi)$. From the fact that the random variables $u_{i,n}$ are uniformly distributed over $[-\pi, \pi)$, it follows that the discrete Doppler frequencies $f_{i,n}$ are random variables as well. Their mean values can be related to the discrete Doppler frequencies obtained using the MEDS by $E\{f_{i,n}\} = \mathrm{sinc}(\pi/(4N_i)) \cdot f_{i,n}^{(MEDS)}$, where the symbol $f_{i,n}^{(MEDS)}$ represents the right-hand side of Equation (5.87) in which we replace the sinus function by a cosine function.[3] Analyzing the relative model error $\Delta\beta_i/\beta$ by following a similar reasoning as in Appendix 5.A, reveals that the mean value and the variance of $\Delta\beta_i/\beta$ can be expressed as

$$E\left\{\frac{\Delta\beta_i}{\beta}\right\} = 0 \tag{5.91}$$

[3] Here, the sinc function $\mathrm{sinc}(x)$ is defined as $\mathrm{sinc}(x) := \sin(x)/x$.

and

$$\mathrm{Var}\left\{\frac{\Delta\beta_i}{\beta}\right\} = \begin{cases} \dfrac{1}{2}, & \text{if} \quad N_i = 1, \\ \dfrac{1}{2N_i}\left[1 - \mathrm{sinc}^2\left(\dfrac{\pi}{2N_i}\right)\right], & \text{if} \quad N_i \geq 2, \end{cases} \tag{5.92}$$

respectively. In comparison to the Monte Carlo method [see (5.A.9)], we observe that the variance of the relative model error decreases faster to zero with increasing values of N_i.

Computing the autocorrelation function $\tilde{r}_{\mu_i\mu_i}(\tau)$ by using (5.85) results in the graphs shown in Figures 5.59(a) and 5.59(b) for $N_i = 7$ and $N_i = 21$, respectively. As it was expected, the RMEDS performs better than the Monte Carlo method, but a certain degree of uncontrolled randomness inherent in all Monte Carlo procedures still remains. The next subsection presents an ultimate method that outperforms the RMEDS with respect to both single trials and averaging over multiple trials.

5.1.9 *Method of Exact Doppler Spread with Set Partitioning (MEDS-SP)*

The method of exact Doppler spread with set partitioning (MEDS-SP) was first described in [150]. As the name indicates, this procedure incorporates the idea of set partitioning into the MEDS. The concept of set partitioning was originally introduced in Ungerboeck's seminal paper [167] to design trellis-coded modulation schemes. Set partitioning has been described as the "key that cracked the problem of constructing efficient coded modulation techniques for bandlimited channels" [168]. In [150], it has been demonstrated that set partitioning is also a quite useful approach for mobile fading channel modelling. The proposed procedure enables the generation of multiple uncorrelated deterministic waveforms, each with a specific autocorrelation function. We will see that the sample mean of the obtained autocorrelation

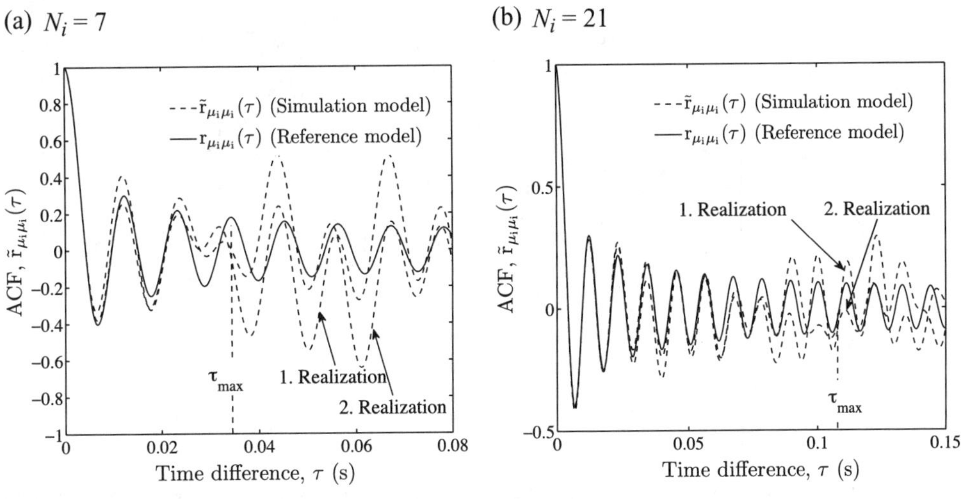

Figure 5.59 Autocorrelation function $\tilde{r}_{\mu_i\mu_i}(\tau)$ for (a) $N_i = 7$ and (b) $N_i = 21$ (RMEDS, Jakes PSD, $f_{\max} = 91$ Hz, $\sigma_0^2 = 1$).

functions tends to the desired autocorrelation function described by the reference model, as the number of waveforms increases.

The frequency-nonselective reference model described by (3.15) can be derived from the well-known geometrical one-ring model, which assumes that an infinite number of local scatterers are uniformly distributed around a ring centred on the mobile station [13]. The simulation model, however, is related to a geometrical model by assuming that there are only a few scatterers around the mobile station. The problem is then to place a minimum number of scatterers on a ring in such a way that the simulation model has nearly the same statistical properties as the reference model.

To visualize the location of the scatterers and to understand the proposed concept of set partitioning, we consider the scattering diagram shown in Figure 5.60. Note that all scatterers used in the simulation model are located in the first quadrant; the other scatterers are redundant, which follows from the symmetry properties of the autocorrelation function $\tilde{r}_{\mu_i\mu_i}(\tau)$ in (4.11). The simple example in Figure 5.60 illustrates that a constellation consisting of $N_i' = 8$ relevant scatterers can be partitioned into two subconstellations, each of which consists of

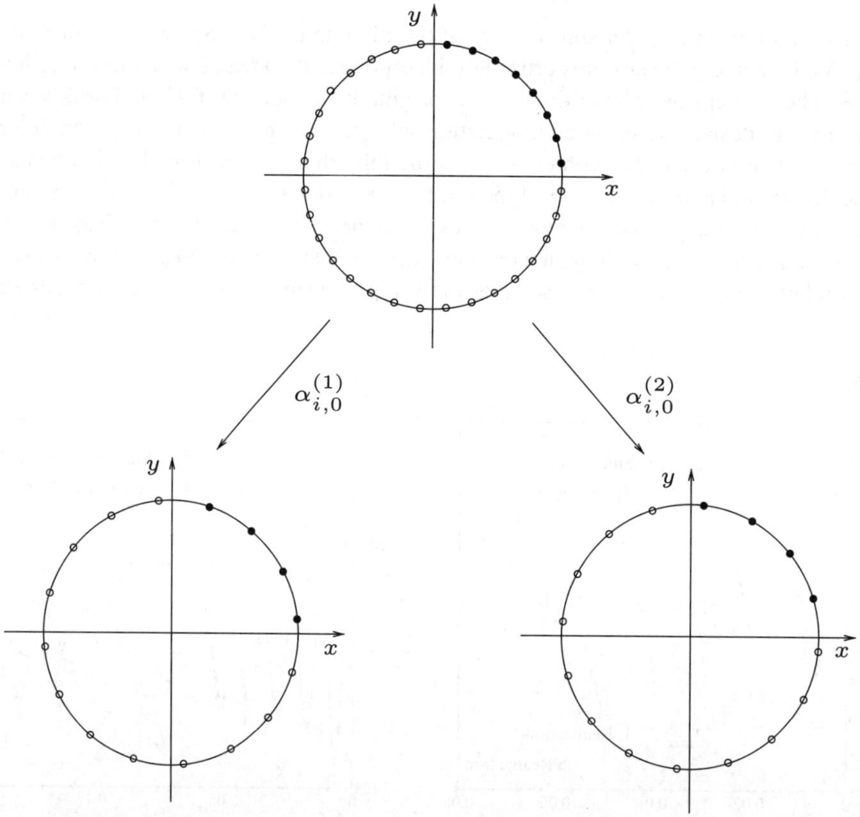

Figure 5.60 Scattering diagram illustrating the set partitioning of an 8-scatterer constellation into two subconstellations, each consisting of four relevant scatterers ("•" relevant scatterers, "∘" redundant scatterers).

$N_i = N_i'/2 = 4$ scatterers. The corresponding sets of angles-of-arrival $\{\alpha_{i,n}^{(1)}\}$ and $\{\alpha_{i,n}^{(2)}\}$ are given by

$$\alpha_{i,n}^{(k)} = \frac{\pi}{2N_i}\left(n - \frac{1}{2}\right) + \alpha_{i,0}^{(k)}, \quad \forall n = 1, 2, \ldots, N_i \quad (i = 1, 2,) \tag{5.93}$$

with $\alpha_{i,0}^{(1)} = -\pi/(8N_i)$ and $\alpha_{i,0}^{(2)} = \pi/(8N_i)$, respectively, where $\alpha_{i,0}^{(k)}$ is called the angle-of-rotation.

The following generalization of the principle above establishes the MEDS-SP. Let K be an integer number. Given is a scattering diagram consisting of $N_i' = K \cdot N_i$ relevant scatterers, where the corresponding angles-of-arrival $\{\alpha_{i,n}\}_{in=1}^{N'}$ are determined by the original MEDS, i.e.,

$$\alpha_{i,n} = \frac{\pi}{2N_i'}\left(n - \frac{1}{2}\right), \quad n = 1, 2, \ldots, N_i'. \tag{5.94}$$

Next, we partition the constellation with $N_i' = K \cdot N_i$ scatterers into K subconstellations, each of which consists of N_i scatterers. The angles-of-arrival $\{\alpha_{i,n}^{(k)}\}_{n=1}^{N_i}$ of the kth subconstellation are determined by

$$\alpha_{i,n}^{(k)} = \frac{\pi}{2N_i}\left(n - \frac{1}{2}\right) + \frac{\pi}{2KN_i}\left(k - \frac{K+1}{2}\right), \tag{5.95}$$

where $k = 1, 2, \ldots, K$ and $n = 1, 2, \ldots, N_i$ $(i = 1, 2)$. A comparison with (5.93) shows that the angle-of-rotation $\alpha_{i,0}^{(k)}$ can be identified as $\alpha_{i,0}^{(k)} = \pi[k - (K+1)/2]/(2KN_i)$. The discrete Doppler frequencies $f_{i,n}^{(k)}$ can now be obtained using $f_{i,n}^{(k)} = f_{\max}\cos(\alpha_{i,n}^{(k)})$.

The path gains $c_{i,n}$ are the same as in (5.86). The phases $\theta_{i,n}$ are i.i.d. random variables, each of which is uniformly distributed over $[0, 2\pi)$. A specific realization of the random variable $\theta_{i,n}$ will be denoted by $\theta_{i,n}^{(k)}$. Since the phases $\theta_{i,n}$ are random variables, it follows that the sum-of-sinusoids process $\hat{\mu}_i(t)$ in (4.3) is a stochastic process. A specific sample function of $\hat{\mu}_i(t)$ is defined by the sets $\{\alpha_{i,n}^{(k)}\}_{n=1}^{N_i}$ and $\{\theta_{i,n}^{(k)}\}_{n=1}^{N_i}$ according to

$$\tilde{\mu}_i^{(k)}(t) = \sigma_0\sqrt{\frac{2}{N_i}}\sum_{n=1}^{N_i}\cos\left(2\pi f_{\max}\cos(\alpha_{i,n}^{(k)})t + \theta_{i,n}^{(k)}\right), \quad k = 1, 2, \ldots, K, \tag{5.96}$$

where we have used (5.86). To emphasize the fact that each sample function $\tilde{\mu}_i^{(k)}(t)$ $(k = 1, 2, \ldots, K)$ is a deterministic function of time t, we call $\tilde{\mu}_i^{(k)}(t)$ henceforth the *kth deterministic process* (or *kth waveform*). In accordance with [41, p. 373], a stochastic process $\hat{\mu}_i(t)$ can be interpreted as a family (or an ensemble) of deterministic processes $\tilde{\mu}_i^{(k)}(t)$, i.e., $\hat{\mu}_i(t) = \{\tilde{\mu}_i^{(1)}(t), \tilde{\mu}_i^{(2)}(t), \ldots\}$. Theoretically, the stochastic process $\hat{\mu}_i(t)$ is defined by an infinite number of deterministic processes. In practice, however, one usually simulates only a few waveforms $\tilde{\mu}_i^{(1)}(t), \tilde{\mu}_i^{(2)}(t), \ldots, \tilde{\mu}_i^{(K)}(t)$ and evaluates the statistical properties of the sum-of-sinusoids channel simulator by computing sample averages.

Owing to the deterministic nature of the waveforms $\tilde{\mu}_i^{(k)}(t)$, the correlation properties of $\tilde{\mu}_i^{(k)}(t)$ have to be determined by means of time averages. The resulting autocorrelation function of $\tilde{\mu}_i^{(k)}(t)$ $(i = 1, 2)$ and the cross-correlation function of $\tilde{\mu}_1^{(k)}(t)$ and $\tilde{\mu}_2^{(k)}(t)$ can be expressed as

$$\tilde{r}^{(k)}_{\mu_i\mu_i}(\tau) = \frac{\sigma_0^2}{N_i} \sum_{n=1}^{N_i} \cos\left(2\pi f_{\max} \cos(\alpha_{i,n}^{(k)})\tau\right), \tag{5.97}$$

$$\tilde{r}^{(k)}_{\mu_1\mu_2}(\tau) = 0 \quad \text{if} \quad \alpha_{1,m}^{(k)} \pm \alpha_{2,n}^{(k)} \neq l\pi, \tag{5.98}$$

respectively, where $m = 1, 2, \ldots, N_1$, $n = 1, 2, \ldots, N_2$, and $l = 0, \pm 1, \pm 2, \ldots$. We mention that the condition in (5.98) is always fulfilled by applying (5.95) with N_2 defined as $N_2 := N_1 + 1$. This means that the MEDS-SP enables the generation of multiple uncorrelated deterministic waveforms. In the following, the sample mean $\bar{r}_{\mu_i\mu_i}(\tau)$ of the K autocorrelation functions $\tilde{r}^{(1)}_{\mu_i\mu_i}(\tau), \tilde{r}^{(2)}_{\mu_i\mu_i}(\tau), \ldots, \tilde{r}^{(K)}_{\mu_i\mu_i}(\tau)$, which is defined as

$$\bar{r}_{\mu_i\mu_i}(\tau) := \frac{1}{K} \sum_{k=1}^{K} \tilde{r}^{(k)}_{\mu_i\mu_i}(\tau) \tag{5.99}$$

will be of central importance.

A direct consequence of the proposed set partitioning concept is that instead of simulating a single deterministic process $\tilde{\mu}_i(t)$ with $N_i' = K \cdot N_i$ terms, we can alternatively simulate K mutually uncorrelated deterministic processes $\tilde{\mu}_i^{(1)}(t), \tilde{\mu}_i^{(2)}(t), \ldots, \tilde{\mu}_i^{(K)}(t)$ with reduced complexity determined by $N_i = N_i'/K$. In practice, it is recommended to simulate the deterministic processes $\tilde{\mu}_i^{(k)}(t)$ $(k = 1, 2, \ldots, K)$ successively by computing from time to time, e.g., after every 100 000 generated samples, a new set of phases $\{\theta_{i,n}^{(k)}\}_{n=1}^{N_i}$ and a new set of angles-of-arrival $\{\alpha_{i,n}^{(k)}\}_{n=1}^{N_i}$ by using (5.95). It should be mentioned that the sample mean autocorrelation function $\bar{r}_{\mu_i\mu_i}(\tau)$ [see (5.99)] of $\tilde{r}^{(k)}_{\mu_i\mu_i}(\tau)$ equals the autocorrelation function $\tilde{r}_{\mu_i\mu_i}(\tau)$ of a single deterministic process $\tilde{\mu}_i(t)$ determined by applying the original MEDS with $N_i' = K \cdot N_i$. Numerical studies have shown that, by replacing $\tilde{r}_{\mu_i\mu_i}(\tau)$ by $\bar{r}_{\mu_i\mu_i}(\tau)$ in (4.128), we obtain always the minimum value for the mean-square error $E_{r_{\mu_i\mu_i}}$ of the autocorrelation function $\bar{r}_{\mu_i\mu_i}(\tau)$. In Appendix 5.B, it is shown that $\bar{r}_{\mu_i\mu_i}(\tau)$ approaches $r_{\mu_i\mu_i}(\tau)$ as $K \to \infty$. This statement holds for all values of $N_i \geq 1$. Similarly one can show that $\bar{r}_{\mu_i\mu_i}(\tau) \to r_{\mu_i\mu_i}(\tau)$ for all values of $K \geq 1$ as $N_i \to \infty$.

Performance analysis: The MEDS-SP results in an excellent approximation $\bar{r}_{\mu_i\mu_i}(\tau) \approx r_{\mu_i\mu_i}(\tau)$ for all values of τ ranging from 0 to $KN_i/(2f_{\max})$. Figure 5.61(a) illustrates this assertion by two examples in which small values have been chosen for K ($K \in \{2, 4\}$). Obviously, by keeping N_i constant and doubling K, the range over which a quasi-perfect fitting can be achieved increases by a factor of two. This property does not hold for the RMEDS, as can be seen in Figure 5.61(b), where degradations even within the interval $[0, N_i/(2f_{\max})]$ can be observed. Only in the case of the MEDS-SP, the resulting sample mean autocorrelation function of the simulation model is extremely close to the autocorrelation function of the reference model over a domain, which increases linearly with the product of the number of generated waveforms and the number of sinusoids per waveform. Another performance comparison is illustrated in Figures 5.61(c) and 5.61(d) for a larger number of waveforms ($K = 20$). It can clearly be seen that the MEDS-SP performs better than the RMEDS. To further compare the

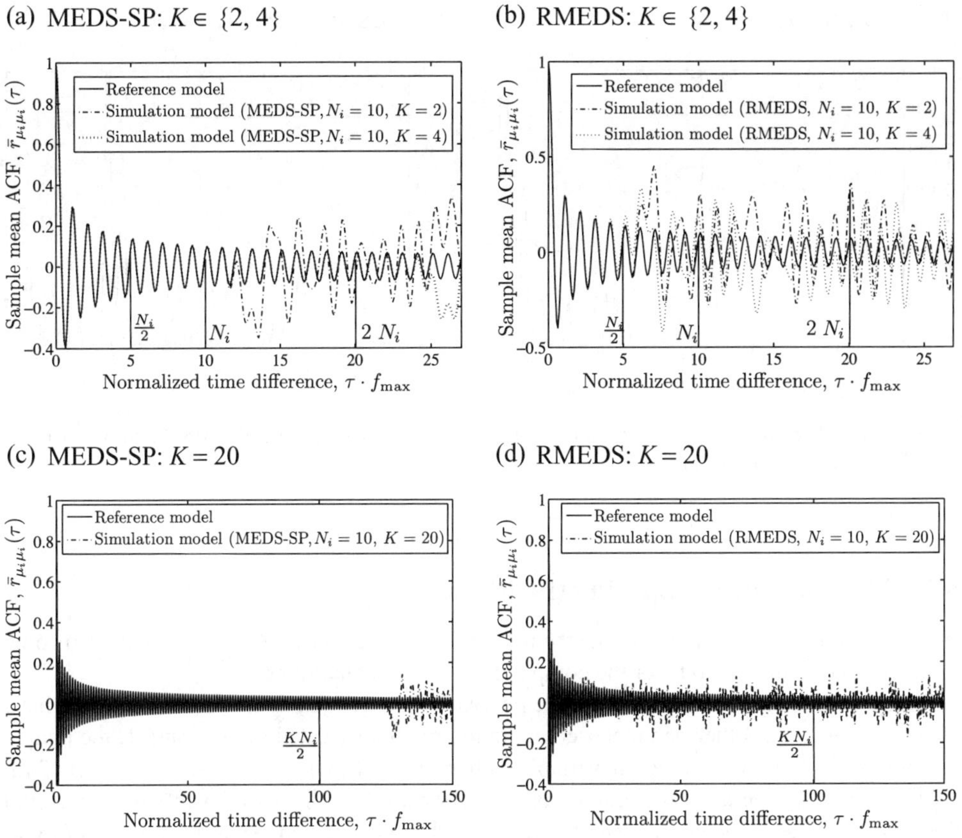

Figure 5.61 Sample mean autocorrelation function $\bar{r}_{\mu_i\mu_i}(\tau)$ by using (a) the MEDS-SP and (b) the RMEDS both for $K \in \{2, 4\}$, as well as (c) the MEDS-SP and (d) the RMEDS by choosing $K = 20$ (Jakes PSD, $\sigma_0^2 = 1$).

performance of the MEDS-SP and the RMEDS, we consider the square error $e_2(\tau)$ of the sample mean autocorrelation function $\bar{r}_{\mu_i\mu_i}(\tau)$, defined as

$$e_2(\tau) = |r_{\mu_i\mu_i}(\tau) - \bar{r}_{\mu_i\mu_i}(\tau)|^2. \tag{5.100}$$

The evaluation of this error is shown in Figure 5.62 for $N_i = 10$ and various values of K. From the results presented in Figures 5.61 and 5.62, we conclude that the deterministic MEDS-SP outperforms the stochastic RMEDS by far with respect to the sample mean autocorrelation function $\bar{r}_{\mu_i\mu_i}(\tau)$. Since the MEDS-SP includes the original MEDS as a special case ($K = 1$), we can also conclude that the MEDS-SP outperforms the RMEDS with respect to both single and multiple trials.

From the study in [112], it is known that the RMEDS is the best among the Monte-Carlo-based parameter computation methods. From this fact and the fact that the MEDS-SP outperforms the RMEDS as well as the original MEDS, we may conclude that the MEDS-SP provides the best solution ever for the design of Rayleigh fading channel simulators.

(a) MEDS-SP: $K \in \{2, 4\}$ (b) RMEDS: $K \in \{2, 4\}$

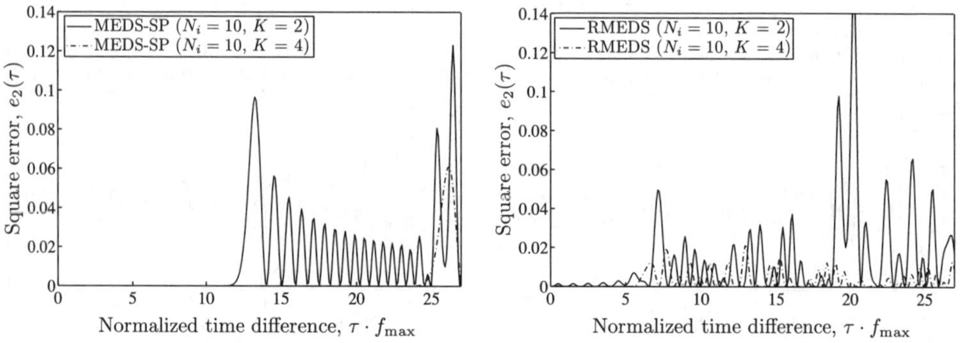

Figure 5.62 The square error $e_2(\tau) = |r_{\mu_i\mu_i}(\tau) - \tilde{r}_{\mu_i\mu_i}(\tau)|^2$ by using (a) the MEDS-SP and (b) the RMEDS (Jakes PSD, $\sigma_0^2 = 1$).

5.2 Methods for Computing the Phases

In this section, we will briefly outline the importance of the phases $\theta_{i,n}$, and we will also come up with some basic methods for the computation of these quantities.

Except for the Jakes method, where the phases $\theta_{i,n}$ are per definition equal to zero, we have assumed that for all other parameter computation methods treated in Section 5.1, the phases $\theta_{i,n}$ are realizations of a random variable uniformly distributed over the interval $(0, 2\pi]$. Without loss of generality, we assume in the following that the set of path gains $\{c_{i,n}\}$ and the set of discrete Doppler frequencies $\{f_{i,n}\}$ have been computed using the method of exact Doppler spread. For two specific realizations of the sets $\{\theta_{1,n}\}_{n=1}^{N_1}$ and $\{\theta_{2,n}\}_{n=1}^{N_2}$ of sizes $N_1 = 7$ and $N_2 = 8$, the pseudo-random behaviour of the resulting deterministic Rayleigh process $\tilde{\zeta}(t)$ is demonstrated in Figure 5.63(a). Here, it has to be taken into account that different events $\{\theta_{i,n}\}_{n=1}^{N_i}$ always result in different realizations of $\tilde{\zeta}(t)$. However, all of these different realizations have the same statistical properties, because the underlying stochastic processes $\hat{\mu}_1(t)$ and $\hat{\mu}_2(t)$ are ergodic with respect to the autocorrelation function. Moreover, the method of exact Doppler spread guarantees that due to the definition $N_2 := N_1 + 1$, the relation $f_{1,n} \neq \pm f_{2,m}$ holds for all $n = 1, 2, \ldots, N_1$ and $m = 1, 2, \ldots, N_2$, so that the cross-correlation function $\tilde{r}_{\mu_1\mu_2}(\tau)$ [cf. (4.13)], which in general depends on $\theta_{i,n}$, is equal to zero. Since the phases $\theta_{i,n}$ have no influence on the statistical properties of $\tilde{\zeta}(t)$ if the underlying deterministic Gaussian processes $\tilde{\mu}_1(t)$ and $\tilde{\mu}_2(t)$ are uncorrelated, we are inclined to set the phases $\theta_{i,n}$ equal to zero. In this case, however, we obtain $\tilde{\mu}_i(0) = \sigma_0\sqrt{2N_i}$ $(i = 1, 2)$, which means that the deterministic Rayleigh process $\tilde{\zeta}(t)$ takes its maximum value $2\sigma_0\sqrt{N_1 + 1/2}$ always at the origin $t = 0$, i.e., $\tilde{\zeta}(0) = 2\sigma_0\sqrt{N_1 + 1/2}$. This leads to the typical transient behaviour depicted in Figure 5.63(b). As we can see in Figure 5.63(c), a similar effect is also obtained, if the phases $\theta_{i,n}$ are computed deterministically, according to $\theta_{i,n} = 2\pi n/N_i$ $(n = 1, 2, \ldots, N_i$ and $i = 1, 2)$. A simple solution to avoid the transient behaviour around the origin is to replace the time variable t by $t + T_0$, where T_0 is a positive real-valued quantity, which has to be chosen sufficiently large. It should therefore be noted that the substitution $t \to t + T_0$ is equivalent to

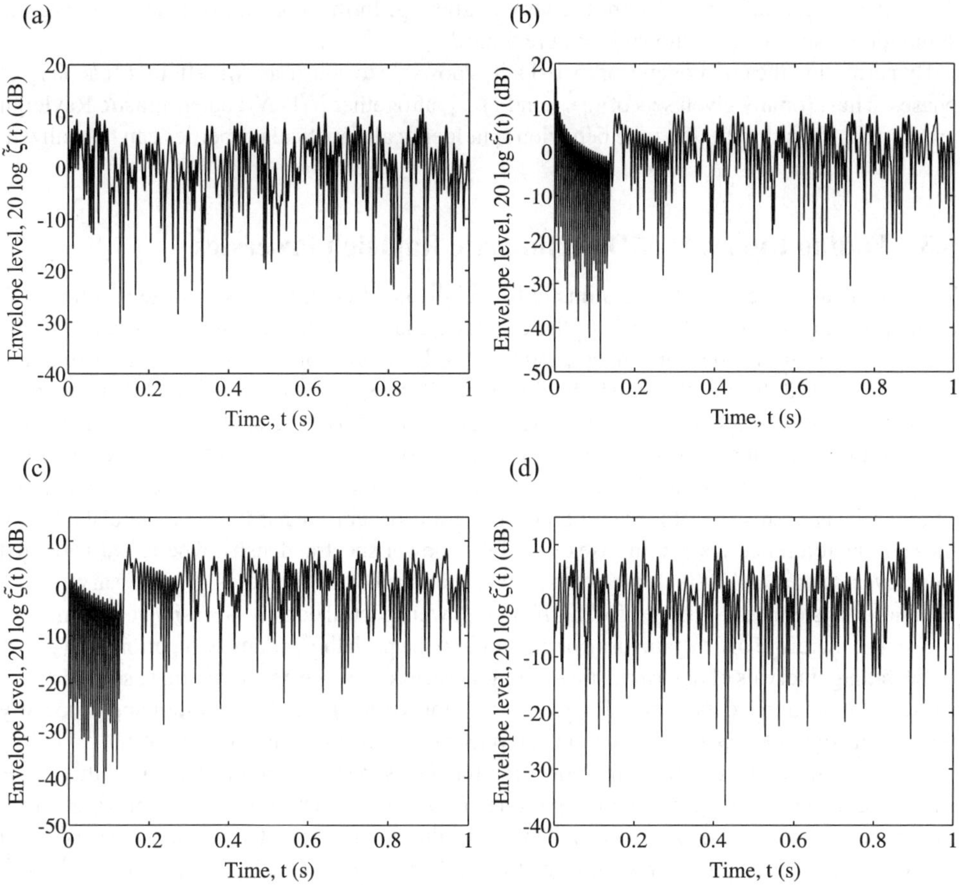

Figure 5.63 Influence of the phases $\theta_{i,n}$ on the transient behaviour of $\tilde{\zeta}(t)$ around the origin: (a) random phases $\theta_{i,n} \in (0, 2\pi]$, (b) $\theta_{i,n} = 0$, (c) $\theta_{i,n} = 2\pi n/N_i$ $(n = 1, 2, \ldots, N_i)$, and (d) permuted phases (MEDS, Jakes PSD, $f_{\max} = 91$ Hz, $\sigma_0^2 = 1$, $N_1 = 21$, $N_2 = 22$).

the substitution $\theta_{i,n} \to \theta_{i,n} + 2\pi f_{i,n} T_0$, which leads to the desired property that the transformed phases are not in a rational ratio for different values of n.

A further possibility would be to introduce a standard phase vector $\vec{\Theta}_i$ with N_i deterministic components according to [96]

$$\vec{\Theta}_i = \left(2\pi \frac{1}{N_i + 1}, 2\pi \frac{2}{N_i + 1}, \ldots, 2\pi \frac{N_i}{N_i + 1}\right) \tag{5.101}$$

and to regard the phases $\theta_{i,n}$ as components of the so-called phase vector

$$\vec{\theta}_i = (\theta_{i,1}, \theta_{i,2}, \ldots, \theta_{i,N_i}). \tag{5.102}$$

By identifying the components of the phase vector $\vec{\theta}_i$ with the permuted components of the standard phase vector $\vec{\Theta}_i$, the transient behaviour observed in the vicinity of the origin of the time axis can be avoided from the very beginning. In this context, one may observe the simulation results of $\tilde{\zeta}(t)$ shown in Figure 5.63(d).

By permuting the components of (5.101), it is possible to construct $N_i!$ different sets $\{\theta_{i,n}\}$ of phases. Thus, for any given sets of $\{c_{i,n}\}$ and $\{f_{i,n}\}$, altogether $N_1! \cdot N_2!$ deterministic Rayleigh processes $\tilde{\zeta}(t)$ with different time behaviour but identical statistical properties can be realized.

5.3 Fading Intervals of Deterministic Rayleigh Processes

The statistical properties of deterministic Rayleigh and Rice processes analyzed so far, such as the probability density function of the envelope and phase, the level-crossing rate, and the average duration of fades, are independent of the behaviour of the autocorrelation function $\tilde{r}_{\mu_i\mu_i}(\tau)$ $(i = 1, 2)$ for $\tau > 0$. In the following, we will investigate the question: Which statistical properties depend at all on $\tilde{r}_{\mu_i\mu_i}(\tau)$ $(i = 1, 2)$ for $\tau > 0$? Closely related to this question is the open problem of determining the size of the interval $[0, \tau_{\max}]$ over which the approximation of $r_{\mu_i\mu_i}(\tau)$ by $\tilde{r}_{\mu_i\mu_i}(\tau)$ is of interest. The task at hand is to find a proper value for $\tau_{\max}$ such that further relevant statistical properties of the simulation system can hardly be distinguished from those of the reference system. In case of the Jakes power spectral density, where $\tau_{\max}$ is related to N_i over $\tau_{\max} = N_i/(2f_{\max})$, we will see that the number of sinusoids N_i necessary for the simulation system can — at least for this kind of power spectral density — easily be determined.

We will therefore once again turn our attention to the probability density function $\tilde{p}_{0_-}(\tau_-; r)$ of the fading intervals of deterministic Rayleigh processes. Since an approximate solution with sufficient precision exists neither for $\tilde{p}_{0_-}(\tau_-; r)$ nor for $p_{0_-}(\tau_-; r)$ at medium and especially not at high signal levels r, this problem can only be solved by means of simulation.

We will first carry out the simulation for the Jakes power spectral density with $f_{\max} =$ 91 Hz and $\sigma_0^2 = 1$ and determine the parameters of the simulation model by making use of the method of exact Doppler spread. Owing to the advantages of this method (very good approximation of the autocorrelation function $r_{\mu_i\mu_i}(\tau) = \sigma_0^2 J_0(2\pi f_{\max}\tau)$ from $\tau = 0$ to $\tau = \tau_{\max} = N_i/(2f_{\max})$, no model error, no correlation between $\tilde{\mu}_1(t)$ and $\tilde{\mu}_2(t)$, and, last but not least, the very good periodicity properties), the resulting deterministic simulation model meets all essential requirements. In this specific case, we may consider the simulation model designed by using the 2-tuple $(N_1, N_2) = (100, 101)$ as the reference model. The simulation of the discrete deterministic process $\tilde{\zeta}(kT)$ has been carried out by setting the sampling interval T_s to $0.5 \cdot 10^{-4}$ s. The generated samples of $\tilde{\zeta}(kT_s)$ have then been used to measure the probability density function $\tilde{p}_{0_-}(\tau_-; r)$ at a low signal level $(r = 0.1)$, a medium signal level $(r = 1)$, and a high signal level $(r = 2.5)$. All obtained results are shown in Figures 5.64(a)–(c) for various 2-tuples (N_1, N_2), where 10^7 fading intervals τ_- have been used for the determination of each probability density function $\tilde{p}_{0_-}(\tau_-; r)$. As can be seen from Figure 5.64(a), there is an exceptionally good agreement between the simulation results obtained for $\tilde{p}_{0_-}(\tau_-; r)$ and the theoretical approximation $p_{1_-}(\tau_-; r)$ [cf. (3.86)] at the low signal level $r = 0.1$. That was to be expected, because the probability that a long fading interval occurs is very low at deep fading levels. Consequently, the probability that further level-crossings can be observed in the interval between t_1 and $t_2 = t_1 + \tau_-$ is negligible. Exactly for this case, the approximation $p_{0_-}(\tau_-; r) \approx p_{1_-}(\tau_-; r)$ turns out to be very useful. Figures 5.64(a) and (b) clearly demonstrate that by choosing $N_1 = 7$ and $N_2 = 8$, the numbers of sinusoids are sufficiently large, so that at least at low and medium signal levels, the obtained probability

(a) Low level: $r = 0.1$

(b) Medium level: $r = 1$

(c) High level: $r = 2.5$

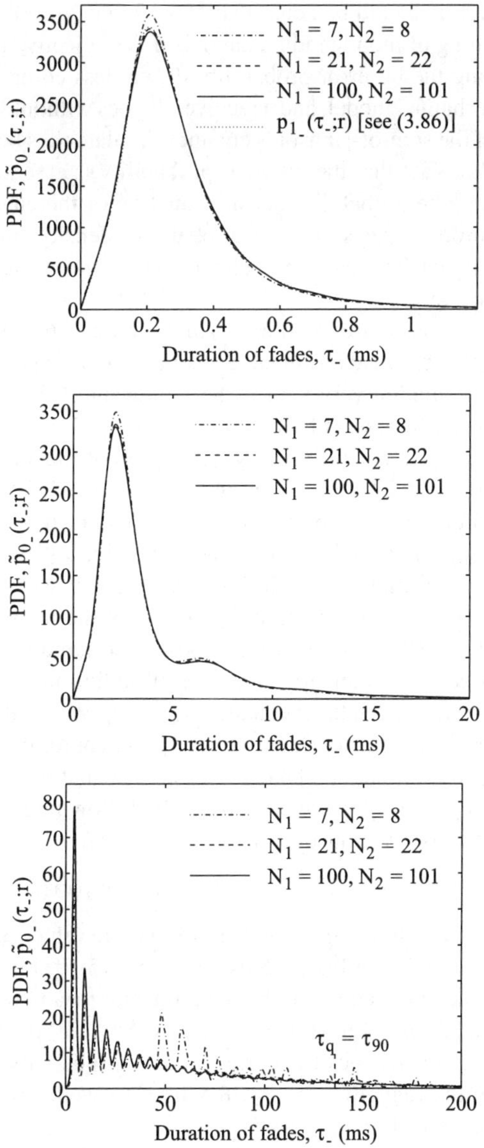

Figure 5.64 Probability density function $\tilde{p}_{0_-}(\tau_-; r)$ of the fading intervals of deterministic Rayleigh processes $\tilde{\zeta}(t)$: (a) $r = 0.1$, (b) $r = 1$, and (c) $r = 2.5$ (MEDS, Jakes PSD, $f_{\max} = 91$ Hz, $\sigma_0^2 = 1$).

density functions $\tilde{p}_{0_-}(\tau_-; r)$ are hardly to be distinguished from those of the reference model ($N_1 = 100, N_2 = 101$). As shown in Figure 5.64(c), significant differences in comparison with the reference model first occur if the signal level r is high ($r = 2.5$), and if the simulation model is designed with $N_1 = 7$ and $N_2 = 8$ sinusoids. However, if the deviations from the reference model shall be negligible at this level too, then at least $N_1 = 21$ and $N_2 = 22$ sinusoids are required. A further increase in N_i is not meaningful!

At this point, it should be noted that $N_1 = 7$ and $N_2 = 8$ sinusoids are in general sufficient for the modelling of mobile radio channels if the objective is to employ the channel simulator for determining the bit error probability of wireless communication systems consisting of a transmitter, a channel model, and a receiver. It goes without saying that $N_1 = 7$ and $N_2 = 8$ will suffice only if the sum-of-sinusoids channel simulator has been designed correctly. This can be attributed to the fact that the bit error probability is essentially determined by the statistical properties (i.e., the probability density function of the envelope, the level-crossing rate, the average duration of fades, and the probability density function of the fading intervals) of $\tilde{\zeta}(t)$ at low signal levels r. The behaviour of $\tilde{\zeta}(t)$ at high signal levels is in this case of minor importance.

A comparison of Figures 3.10(a)–(c) and Figures 5.64(a)–(c) shows that the theoretical approximation $p_{1\_}(\tau_-; r)$ at all signal levels r only fits the probability density function $\tilde{p}_0(\tau_-; r)$ obtained by simulation very well if the fading intervals τ_- are short. One should also note that for $\tau_- \to \infty$, it always follows $\tilde{p}_{0\_}(\tau_-; r) \to 0 \; \forall \, r \in \{0.1, 1, 2.5\}$. However, this convergence property is not fulfilled by $p_{1\_}(\tau_-; r)$ at the signal levels $r = 1$ and $r = 2.5$ [see Figures 3.10(b) and 3.10(c)]. Given the convergence behaviour of $\tilde{p}_{0\_}(\tau_-; r)$, we have now the chance to determine approximately the interval $[0, \tau_{max}]$ over which the approximation $r_{\mu_i\mu_i}(\tau) \approx \tilde{r}_{\mu_i\mu_i}(\tau)$ has to be as accurate as possible. We will therefore take advantage of the quantity $\tau_q = \tau_q(r)$, introduced in Subsection 3.4.4, where we replace $p_{0\_}(\tau_-; r)$ in (3.88) by the probability density function $\tilde{p}_{0\_}(\tau_-; r)$ of the fading intervals of the reference model ($N_1 = 100$, $N_2 = 101$), and then we choose q so large that the probability density function $\tilde{p}_{0\_}(\tau_-; r)$ becomes sufficiently small for all fading intervals $\tau_- \geq \tau_q$. Furthermore, we demand that τ_{max} must fulfil the inequality $\tau_{max} \geq \tau_q$. We recall that by using the method of exact Doppler spread, $\tilde{r}_{\mu_i\mu_i}(\tau)$ represents a very good approximation for $r_{\mu_i\mu_i}(\tau)$ within the range $0 \leq \tau \leq \tau_{max}$, where τ_{max} is related to N_i via the equation $\tau_{max} = N_i/(2f_{max})$. Hence, by using $\tau_{max} = N_i/(2f_{max}) \geq \tau_q(r)$, we can obtain the following simple formula for the estimation of the required number of sinusoids N_i

$$N_i \geq \lceil 2 f_{max} \tau_q(r) \rceil . \tag{5.103}$$

For example, if we choose $q = 90$, then we find the value 135.7 ms for $\tau_{90} = \tau_{90}(r)$ at the high signal level $r = 2.5$ [see Figure 5.64(c)]. From (5.103) it then follows $N_i \geq 25$. This result matches the one obtained before by experimental means very well. Now, in reverse order, let us assume that N_i is given (for example, by $N_i \geq 7$), then the resulting probability density function $\tilde{p}_0(\tau_-; r)$ matches the corresponding probability density function of the reference model within the range $0 \leq \tau_- \leq 38.5$ ms very well. This is also confirmed by the results shown Figure 5.64(c).

For low and medium signal levels, where usually $\tau_{90} < 1/f_{max}$ holds, (5.103) does not provide any admissible values for N_i, because in such cases the obtained values fall below the lower limit $N_i = 7$, which is considered as the minimum number of sinusoids guaranteeing a sufficient approximation of the Gaussian probability density function $p_{\mu_i}(x)$. As a corollary, we obtain finally the following useful estimate for the required number of sinusoids

$$N_i \geq \max\{7, \lceil 2 f_{max} \tau_q(r) \rceil\}, \tag{5.104}$$

which is valid for all signal levels $r \geq 0$.

Next, we will study the statistics of the fading intervals of deterministic Rayleigh processes $\tilde{\zeta}(t)$ for the case that the underlying sum-of-sinusoids processes $\tilde{\mu}_1(t)$ and $\tilde{\mu}_2(t)$ are characterized by Gaussian power spectral densities. Let the shape of the Gaussian power spectral density in (3.26) be determined by the parameters $f_c = \sqrt{\ln 2} f_{\max}$, $f_{\max} = 91$ Hz, and $\sigma_0^2 = 1$. The parameters of the simulation model are again computed by means of the method of exact Doppler spread. Exactly as in the preceding case, we consider the simulation model with the 2-tuple $(N_1, N_2) = (100, 101)$ as the reference model. The repetition of the experimental determination of the probability density function $\tilde{p}_{0_-}(\tau_-; r)$ leads, after the evaluation of the generated samples of $\tilde{\zeta}(kT_s)$ at low, medium, and high signal levels r, to the results shown in Figures 5.65(a)–(c). For all signal levels, the sampling interval T_s has been empirically set to $0.5 \cdot 10^{-4}$ s. Here again 10^7 fading intervals τ_- have been evaluated for determining each of the presented probability density functions $\tilde{p}_{0_-}(\tau_-; r)$.

From the comparison of Figures 5.65(a) and 5.64(a), it follows that the corresponding probability density functions $\tilde{p}_{0_-}(\tau_-; r)$ are identical. That was to be expected, as the exact shape of the power spectral density of the processes $\tilde{\mu}_1(t)$ and $\tilde{\mu}_2(t)$ has no influence on the density $\tilde{p}_0(\tau_-; r)$ at low signal levels r. Only the values of the quantities $\tilde{\sigma}_0^2 = \tilde{r}_{\mu_i\mu_i}(0)$ and $\tilde{\beta}_i = -\ddot{\tilde{r}}_{\mu_i\mu_i}(0)$ are of importance here. In the present case, they are identical for the Jakes and the Gaussian power spectral density. Only with an increasing signal level r, the behaviour of $\tilde{r}_{\mu_i\mu_i}(\tau)$ for $\tau > 0$ exerts more and more influence on the density $\tilde{p}_{0_-}(\tau_-; r)$. For clarification of this issue, it is recommended to compare Figures 5.65(b) and 5.65(c) with Figures 5.64(b) and 5.64(c), respectively. Obviously, the following fundamental relation exists between $\tilde{p}_{0_-}(\tau_-; r)$ and $\tilde{r}_{\mu\mu}(\tau)$: The probability density function $\tilde{p}_{0_-}(\tau_-; r)$ has several maxima if and only if this also holds for the autocorrelation function $\tilde{r}_{\mu\mu}(\tau)$ of the underlying complex sum-of-sinusoids process $\tilde{\mu}(t) = \tilde{\mu}_1(t) + j\tilde{\mu}_2(t)$.

In what follows, we will study the two-dimensional joint probability density function of the fading and connecting intervals, which will be denoted here by $\tilde{p}_{0_{-+}}(\tau_-, \tau_+; r)$. The function $\tilde{p}_{0_{-+}}(\tau_-, \tau_+; r)$ describes the density of the joint probability that the fading interval τ_- and the connecting interval τ_+ occur in pairs at the signal level r. More specifically, this is the probability density of the joint event that the deterministic Rayleigh process $\tilde{\zeta}(t)$ crosses a constant signal level r upwards for the first time after the time duration τ_- within the interval $(t + \tau_-, t + \tau_- + d\tau_-)$, and afterwards the process $\tilde{\zeta}(t)$ falls again below the same signal level r for the first time after the duration τ_+ within the interval $(t + \tau_- + \tau_+, t + \tau_- + \tau_+ + d\tau_+)$, provided that a level-crossing through r from up to down has been occurred at the time instant t.

Some simulation results for the two-dimensional joint probability density function $\tilde{p}_{0_{-+}}(\tau_-, \tau_+; r)$ are shown in Figures 5.66(a)–(c) and 5.67(a)–(c) for the case that the underlying deterministic processes $\tilde{\mu}_i(t)$ are characterized by the Jakes and the Gaussian power spectral density, respectively. The numerical integration of the joint probability density function $\tilde{p}_{0_{-+}}(\tau_-, \tau_+; r)$ over the connecting interval τ_+ leads to the marginal density $\tilde{p}_{0_-}(\tau_-; r) = \int_0^\infty \tilde{p}_{0_{-+}}(\tau_-, \tau_+; r)\, d\tau_+$, which has already been illustrated in Figures 5.64(a)–(c) and 5.65(a)–(c).

At the end of this chapter, we will once more return to the Monte Carlo method and the Jakes method. We repeat the previously described simulations necessary to determine the probability density function $\tilde{p}_{0_-}(\tau_-; r)$, where we now compute the parameters of the simulation model first by employing the Monte Carlo method and afterwards by means of the Jakes method. For brevity, we will here only apply both methods on the Jakes power spectral density in (3.23) with the parameters $f_{\max} = 91$ Hz and $\sigma_0^2 = 1$. The probability density functions $\tilde{p}_{0_-}(\tau_-; r)$, which were found by using the Monte Carlo method, are shown in Figures 5.68(a)–(c) for various signal levels r and for two different realizations of the sets of discrete Doppler frequencies

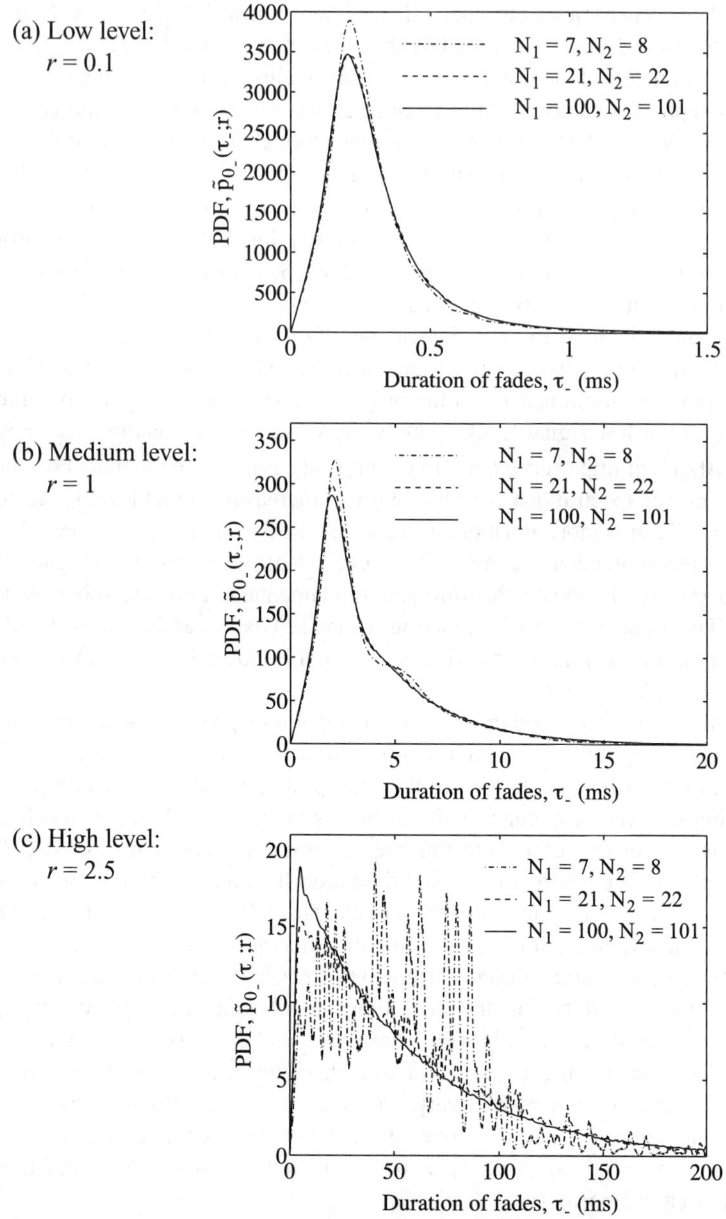

Figure 5.65 Probability density function $\tilde{p}_{0-}(\tau_-; r)$ of the fading intervals of deterministic Rayleigh processes $\tilde{\zeta}(t)$: (a) $r = 0.1$, (b) $r = 1$, and (c) $r = 2.5$ (MEDS, Gaussian PSD, $f_c = \sqrt{\ln 2} f_{\max}$, $f_{\max} = 91\,\text{Hz}$, $\sigma_0^2 = 1$).

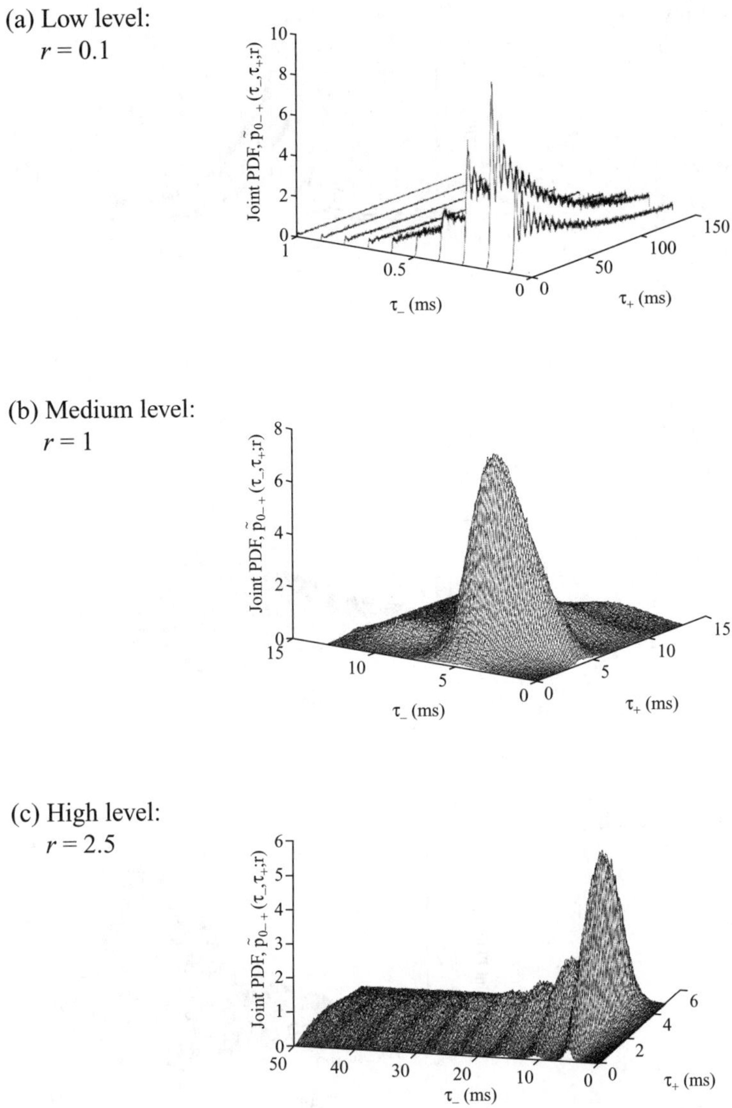

Figure 5.66 Joint probability density function $\tilde{p}_{0_{-+}}(\tau_-, \tau_+; r)$ of the fading and connecting intervals of deterministic Rayleigh processes $\tilde{\zeta}(t)$: (a) $r = 0.1$, (b) $r = 1$, and (c) $r = 2.5$ (MEDS, Jakes PSD, $f_{\max} = 91\,\text{Hz}$, $\sigma_0^2 = 1$).

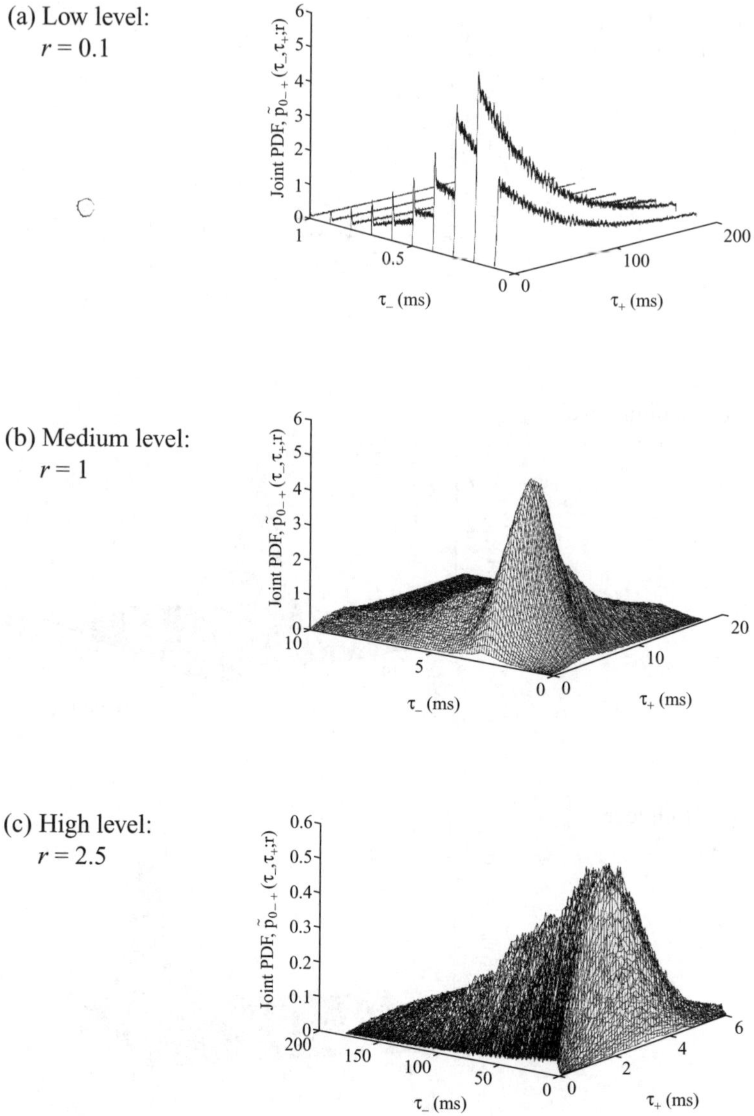

Figure 5.67 Joint probability density function $\tilde{p}_{0_-+}(\tau_-, \tau_+; r)$ of the fading and connecting intervals of deterministic Rayleigh processes $\tilde{\zeta}(t)$: (a) $r = 0.1$, (b) $r = 1$, and (c) $r = 2.5$ (MEDS, Gaussian PSD, $f_c = \sqrt{\ln 2} f_{\max}$, $f_{\max} = 91\,\text{Hz}$, $\sigma_0^2 = 1$).

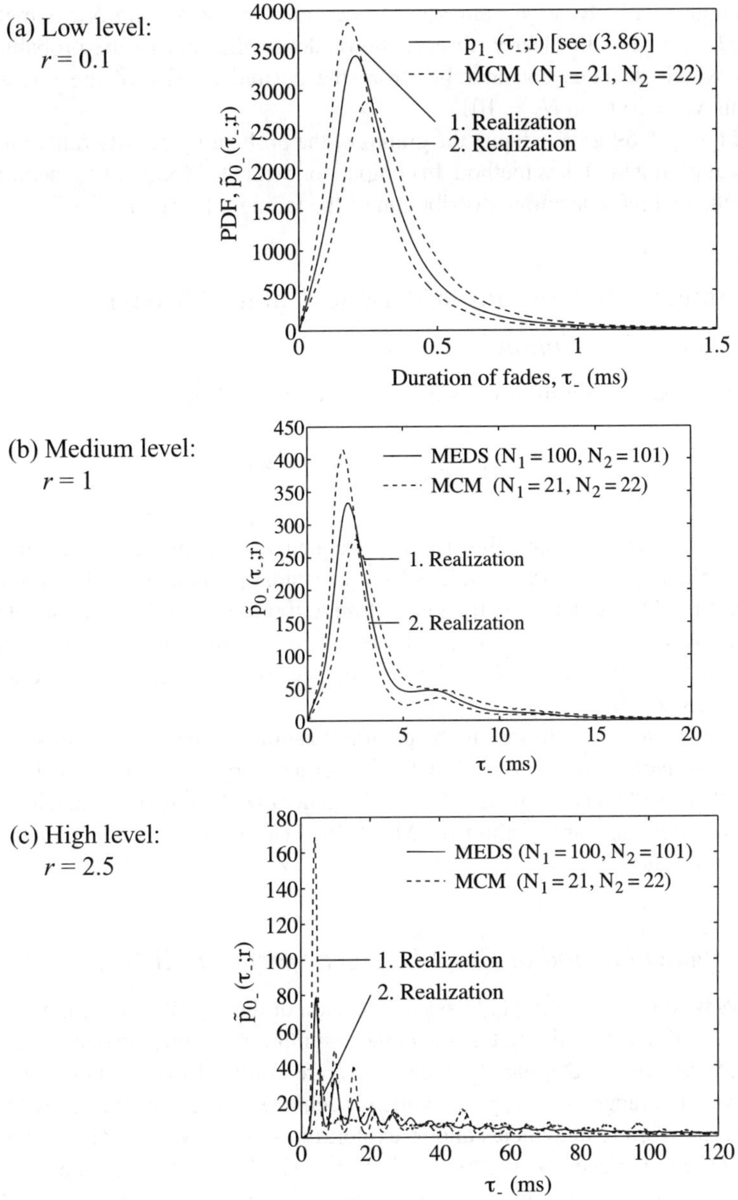

Figure 5.68 Probability density function $\tilde{p}_{0_-}(\tau_-; r)$ of the fading intervals of deterministic Rayleigh processes $\tilde{\zeta}(t)$: (a) $r = 0.1$, (b) $r = 1$, and (c) $r = 2.5$ (MCM, Jakes PSD, $f_{\max} = 91$ Hz, $\sigma_0^2 = 1$).

$\{f_{i,n}\}$. Although relatively large values have been chosen for N_1 and N_2, namely $N_1 = 21$ and $N_2 = 22$, one can still clearly observe the random behaviour of the probability density $\tilde{p}_{0\_}(\tau_-; r)$, which deviates considerably from the desired density of the reference model (MEDS with $N_1 = 100$ and $N_2 = 101$).

Finally, Figures 5.69(a)–(c) depict the graphs of the probability density function $\tilde{p}_{0\_}(\tau_-; r)$ obtained by applying the Jakes method. In comparison with the Monte Carlo method, the Jakes method results in a more accurate distribution of the fading intervals τ_-.

5.4 Parametrization of Sum-of-Cisoids Channel Models

5.4.1 Problem Description

Let us consider a stochastic sum-of-cisoids (SOC) process of the form

$$\hat{\mu}(t) = \sum_{n=1}^{N} c_n \, e^{j(2\pi f_n t + \theta_n)}, \tag{5.105}$$

where c_n, f_n, and θ_n are the path gains, Doppler frequencies, and phases of the nth propagation path. The problem is to find proper values for the cisoids' parameters such that the statistical properties of the SOC model are sufficiently close to those of the reference model for a given number of cisoids N. Supposing that the phases θ_n are i.i.d. random variables, each of which is uniformly distributed over $(0, 2\pi]$, the problem lies basically in finding the gains c_n and Doppler frequencies f_n.

In the literature, several solutions to the parametrization problem of SOC models have been proposed. These include the extended method of exact Doppler spread (EMEDS) [169], the L_p-norm method (LPNM) (see Section 5.1.6), the generalized method of equal areas (GMEA) [151], and the Riemann sum method (RSM) [170]. The first three of these methods will be described in the following.

5.4.2 Extended Method of Exact Doppler Spread (EMEDS)

The EMEDS was introduced in [171] as an extension of the MEDS for computing the model parameters of SOC channel simulators under the assumption of isotropic scattering. In contrast to the MEDS, where the Doppler frequencies f_n are confined to the interval $(0, f_{\max})$, the EMEDS covers the range $(-f_{\max}, f_{\max})$, which is achieved by replacing $\pi/2$ by 2π in (5.87). Furthermore, to avoid redundancies and to exploit the symmetry properties of the cisoids in (5.105), we have to replace $n - 1/2$ by $n - 1/4$ in (5.87). Hence, for the Jakes power spectral density, the EMEDS specifies the path gains c_n and Doppler frequencies f_n as [171]:

$$c_n = \sigma_0 \sqrt{\frac{2}{N}}, \tag{5.106}$$

$$f_n = f_{\max} \cos\left[\frac{2\pi}{N}\left(n - \frac{1}{4}\right)\right], \tag{5.107}$$

where $n = 1, 2, \ldots, N$. One can verify, by substituting (5.106) and (5.107) into (4.109), that the EMEDS exactly reproduces the Doppler spread of isotropic scattering channels for any

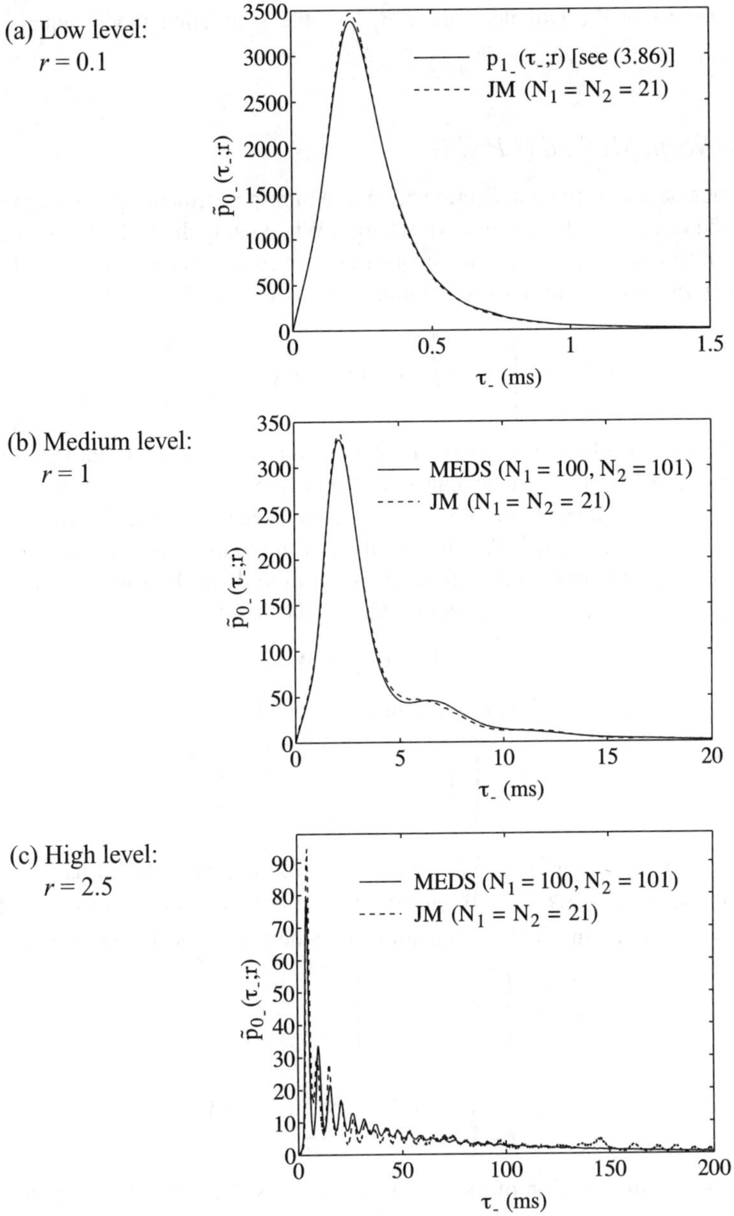

Figure 5.69 Probability density function $\tilde{p}_{0-}(\tau_-; r)$ of the fading intervals of deterministic Rayleigh processes $\tilde{\zeta}(t)$: (a) $r = 0.1$, (b) $r = 1$, and (c) $r = 2.5$ (JM, Jakes PSD, $f_{\max} = 91$ Hz, $\sigma_0^2 = 1$).

$N \geq 1$, meaning that the Doppler spread $\hat{B}^{(2)}_{\mu\mu}$ of the simulation model equals that of the reference model $B^{(2)}_{\mu\mu}$, i.e., $\hat{B}^{(2)}_{\mu\mu} = B^{(2)}_{\mu\mu} = f_{\max}/\sqrt{2}, \forall N \geq 1$.

5.4.3 L_p-Norm Method (LPNM)

The LPNM has been described in Section 5.1.6 as a high-performance parameter computation method for the design of SOS channel simulators. When using the LPNM in connection with stochastic SOC processes $\hat{\mu}(t)$, then the path gains c_n of $\hat{\mu}(t)$ are defined as in (5.106), whereas the Doppler frequencies f_n have to be computed in such a way that the L_p-norm

$$E^{(p)}_{r_{\mu\mu}} = \left\{ \frac{1}{\tau_{\max}} \int_0^{\tau_{\max}} \left| r_{\mu\mu}(\tau) - \hat{r}_{\mu\mu}(\tau) \right|^p d\tau \right\}^{1/p} \tag{5.108}$$

results in a minimum, where p is a positive integer, and $\tau_{\max} > 0$ determines the length of the interval $[0, \tau_{\max}]$ inside of which the approximation $r_{\mu\mu}(\tau) \approx \hat{r}_{\mu\mu}(\tau)$ is of interest.

There exist other versions of the LPNM, which have been surveyed in Section 5.1.6 for SOS models. In the following, we will describe briefly how they can be used for the parametrization of SOC models. For the first variant under consideration, both the gains c_n and the Doppler frequencies f_n of $\hat{\mu}(t)$ are to be computed such that the cost function

$$E^{(p)} = W_1 \cdot E^{(p)}_{r_{\mu\mu}} + W_2 \cdot E^{(p)}_{p_\zeta} \tag{5.109}$$

is minimized, where W_1 and W_2 are weighting factors and

$$E^{(p)}_{p_\zeta} = \left\{ \int_0^{\infty} \left| p_\zeta(z) - \hat{p}_\zeta(z) \right|^p dz \right\}^{1/p} . \tag{5.110}$$

The second variant aims to minimize $E^{(p)}$ too, but it considers only $N - 1$ pairs of parameters c_n, f_n, where $n = 1, 2, \ldots, N - 1$. To ensure that the boundary conditions $\hat{r}_{\mu\mu}(0) = r_{\mu\mu}(0)$ and $\ddot{\hat{r}}_{\mu\mu}(0) = \ddot{r}_{\mu\mu}(0)$ are fulfilled, the remaining parameters c_N and f_N are obtained as follows:

$$c_N = \left\{ 2\sigma_0^2 - \sum_{n=1}^{N-1} c_n^2 \right\}^{1/2} , \tag{5.111}$$

$$f_N = \frac{1}{c_N} \left\{ \frac{-\ddot{r}_{\mu\mu}(0)}{4\pi^2} - \sum_{n=1}^{N-1} (c_n f_n)^2 \right\}^{1/2} . \tag{5.112}$$

We refer to the original version of the LPNM as the LPNM I, while the first and the second variants described above will be called the LPNM II and LPNM III, respectively. In accordance with the analysis presented in Section 5.1.6, the LPNM III is the most efficient variant of the LPNM.

The EMEDS and the LPNM have widely been in use to simulate SISO and MIMO fading channels, see, e.g., [105, 169, 172]. The performance of both methods is remarkable indeed, as one may conclude from the analysis presented in [169] and [105]. Unfortunately, the applicability of the EMEDS is restricted to the design of channel simulators characterized by the symmetrical Jakes power spectral density (isotropic scattering), while the optimization technique inherent to the LPNM makes the determination of the model parameters a time-consuming task. To avoid these constraints, we propose the GMEA as an efficient and general solution to the parametrization problem of SOC processes.

5.4.4 *Generalized Method of Equal Areas (GMEA)*

In this subsection, we present the generalized method of equal areas (GMEA) [151], which is well suited for the design of sum-of-cisoids (SOC) simulators for Rayleigh fading channels characterized by any (symmetrical or asymmetrical) specified type of Doppler power spectral density. Unlike the original MEA, its generalized version is not restricted to the modelling of symmetrical Doppler power spectral densities. In fact, we will see that the GMEA is quite useful for the design of simulation models characterized by asymmetrical Doppler power spectral densities. This is an important feature, since the simulation of such scenarios is of great importance for the laboratory analysis of mobile communication systems under non-isotropic scattering conditions.

The complex Gaussian process $\mu(t)$ acts in the following as reference model, which is characterized by its autocorrelation function $r_{\mu\mu}(\tau)$. Let $p_\alpha(\alpha)$ be any given distribution function of the angle-of-arrival α, then $r_{\mu\mu}(\tau)$ can be expressed by means of $r_{\mu\mu}(\tau) = 2\sigma_0^2 E\{\exp\{j2\pi f_{\max}\cos(\alpha)\tau\}\}$ as

$$r_{\mu\mu}(\tau) = 4\sigma_0^2 \int_0^\pi g_\alpha(\alpha)\, e^{j2\pi f_{\max}\cos(\alpha)\tau}\, d\alpha, \tag{5.113}$$

where $g_\alpha(\alpha) := [p_\alpha(\alpha) + p_\alpha(-\alpha)]/2$ is the even part of the probability density function $p_\alpha(\alpha)$ of α. In addition, one can show that the autocorrelation function $r_{\mu\mu}(\tau)$ satisfies the relations

$$r_{\mu\mu}(\tau) = 2\left[r_{\mu_1\mu_1}(\tau) + j\, r_{\mu_1\mu_2}(\tau)\right] = 2\left[r_{\mu_2\mu_2}(\tau) - j\, r_{\mu_2\mu_1}(\tau)\right], \tag{5.114}$$

where

$$r_{\mu_1\mu_1}(\tau) = r_{\mu_2\mu_2}(\tau) = 2\sigma_0^2 \int_0^\pi g_\alpha(\alpha) \cos\left(2\pi f_{\max}\cos(\alpha)\tau\right) d\alpha \tag{5.115}$$

and

$$r_{\mu_1\mu_2}(\tau) = -r_{\mu_2\mu_1}(\tau) = 2\sigma_0^2 \int_0^\pi g_\alpha(\alpha) \sin\left(2\pi f_{\max}\cos(\alpha)\tau\right) d\alpha \tag{5.116}$$

are the autocorrelation functions and cross-correlation functions of the inphase and quadrature components of $\mu(t) = \mu_1(t) + j\mu_2(t)$. It follows from (5.114)–(5.116) that if $\mu_1(t)$ and $\mu_2(t)$ are uncorrelated, i.e., $r_{\mu_1\mu_2}(\tau) = r_{\mu_2\mu_1}(\tau) = 0$, then $r_{\mu\mu}(\tau)$ will be a real-valued even function; otherwise, $r_{\mu\mu}(\tau)$ will be complex-valued and Hermitian symmetric.

The channel model described by $\mu(t)$ can alternatively be characterized by its Doppler power spectral density $S_{\mu\mu}(f)$, which is defined by the Fourier transform of the autocorrelation function $r_{\mu\mu}(\tau)$ [see (2.111)], i.e., $S_{\mu\mu}(f) = \int_{-\infty}^{\infty} r_{\mu\mu}(\tau)\exp\{-j2\pi f\tau\}d\tau$. Substituting (5.113) in (2.111) results in

$$S_{\mu\mu}(f) = 4\sigma_0^2 \cdot \frac{g_\alpha(\arccos(f/f_{\max}))}{f_{\max}\sqrt{1-(f/f_{\max})^2}} \tag{5.117}$$

for $|f| < f_{\max}$. From the properties of the Fourier transformation [173, Section 3.6], we know that if the inphase and quadrature components of $\mu(t)$ are uncorrelated, meaning that $r_{\mu\mu}(\tau)$ is even and real-valued, then $S_{\mu\mu}(f)$ is symmetrical with respect to the origin. In the opposite case, if $\mu_1(t)$ and $\mu_2(t)$ are correlated, then $S_{\mu\mu}(f)$ will be asymmetrical.

Concerning the computation of the discrete Doppler frequencies f_n, we recall that the autocorrelation function of $\mu(t)$ is given in terms of the distribution $p_\alpha(\alpha)$ of the random angles-of-arrival α_n, or equivalently, in terms of the even part $g_\alpha(\alpha)$ of $p_\alpha(\alpha)$ [see (5.113)]. On the other hand, the autocorrelation function $\hat{r}_{\mu\mu}(\tau)$ of $\hat{\mu}(t)$ is completely determined by the parameters c_n, f_n , and N [see (4.104)].

In this regard, it is reasonable to think that in order for $\hat{r}_{\mu\mu}(\tau)$ to resemble the autocorrelation function $r_{\mu\mu}(\tau)$ of the reference model for a given value of N, the gains and Doppler frequencies of $\hat{\mu}(t)$ should provide information about the function $g_\alpha(\alpha)$. Following this line of reasoning and considering that the gains in (5.106) are blind to the angle-of-arrival statistics, we will compute the Doppler frequencies of $\hat{\mu}(t)$ such that the deterministic angles-of-arrival α_n satisfy the equation

$$\int_{\alpha_{n-1}}^{\alpha_n} g_\alpha(\alpha)d\alpha = \frac{1}{2N}, \qquad n = 2, 3, \ldots, N, \tag{5.118}$$

where $\alpha_n \in (0, \pi)$. We recall that the uncorrelatedness between the inphase and quadrature components of $\hat{\mu}(t)$ is a fundamental feature for the design of simulation models for fading channels characterized by symmetrical Doppler power spectral densities. It is shown in the Appendix 5.C that if the Doppler power spectral density of $\mu(t)$ is symmetrical and the angles-of-arrival α_n satisfy (5.118), then the inphase and quadrature components of $\hat{\mu}(t)$ are uncorrelated if and only if

$$\int_{0}^{\alpha_1} g_\alpha(\alpha)d\alpha = \frac{1}{4N}. \tag{5.119}$$

We will consider (5.119) as an initial condition for computing the angles-of-arrival α_n to ensure that the parameter computation method presented in this section can be applied to the simulation of fading channels characterized by symmetrical Doppler power spectral densities.

Experiments have shown that if $S_{\mu\mu}(f) = S_{\mu\mu}(-f)$, then the initial condition given in (5.119) minimizes the error $E_{r_{\mu\mu}}^{(2)}$ between $r_{\mu\mu}(\tau)$ and $r_{\hat{\mu}\hat{\mu}}(\tau)$ for $\tau \in [0, N/(4f_{\max})]$. However, this is in general not the case if the channel Doppler power spectral density is asymmetrical, i.e., if $S_{\mu\mu}(f) \neq S_{\mu\mu}(-f)$.

It has been pointed out in [151] that the simulation model described by $\hat{\mu}(t)$ is mean-ergodic and autocorrelation-ergodic if the following inequalities are fulfilled: (i) $f_n \neq f_m$ for $n \neq m$ and (ii) $f_n \neq 0$ for all $n = 1, 2, \ldots, N$. An important implication of the requirement stated in (5.118) is that the angles-of-arrival α_n satisfy $\alpha_n \neq \alpha_m$ for $n \neq m$, and $\alpha_n > 0\ \forall n$. Thus, by demanding the angles-of-arrival α_n to meet (5.118), we implicitly guarantee the fulfillment of the inequality $f_n \neq f_m$, since $f_n = f_m$ holds for some $n \neq m$ if and only if $\alpha_n = \pm\alpha_m$. Furthermore, $f_n = 0$ for some n if and only if $\alpha_n = \pi/2$, which is rather unlikely when the Doppler power spectral density of $\mu(t)$ is asymmetrical, and is never the case when $S_{\mu\mu}(f)$ is symmetrical and the number of cisoids N is even.

On the basis of (5.118) and (5.119), we can compute the angles-of-arrival α_n by employing numerical root-finding techniques to solve

$$\int_0^{\alpha_n} g_\alpha(\alpha)\,d\alpha = \frac{1}{2N}\left(n - \frac{1}{2}\right) \tag{5.120}$$

for $n = 1, 2, \ldots, N$. The function $g_\alpha(\alpha)$ may itself be regarded as being a probability density function, the corresponding cumulative distribution function is given by $F_\alpha(\alpha) := \int_{-\infty}^{\alpha} g_\alpha(x)\,dx$. The evaluation of $F_\alpha(\alpha)$ for $\alpha \leq \alpha_n$ results in $F_\alpha(\alpha_n) = (n + N - 1/2)/(2N)$. Hence, if a closed-form solution exists for the inverse function $F_\alpha^{-1}(\cdot)$ of $F_\alpha(\cdot)$, then the angles-of-arrival α_n can be computed by means of

$$\alpha_n = F_\alpha^{-1}\left(\frac{1}{2N}\left[n + N - \frac{1}{2}\right]\right) \tag{5.121}$$

for $n = 1, 2, \ldots, N$.

Once the angles-of-arrival α_n are known, the discrete Doppler frequencies f_n can easily be obtained by using the relation $f_n = f_{\max}\cos(\alpha_n)$. Often, however, the channel autocorrelation function and/or Doppler power spectral density are introduced without giving any explicit information about the distribution of the random angles-of-arrival α_n, such as in [174]. For those cases, we observe that

$$\int_0^{\alpha_n} g_\alpha(\alpha)\,d\alpha = \frac{1}{4\sigma_0^2}\int_{f_n}^{f_{\max}} S_{\mu\mu}(f)\,df. \tag{5.122}$$

Taking into account the result presented in (5.120), we can write

$$\int_{-f_{\max}}^{f_n} S_{\mu\mu}(f)\,df = \frac{2\sigma_0^2}{N}\left(N - n + \frac{1}{2}\right) \tag{5.123}$$

for $n = 1, 2, \ldots, N$, where $f_n \in (-f_{\max}, f_{\max})$. The integral in (5.123) describes a sort of cumulative power function, which is defined as $P_S(f_n) := \int_{-\infty}^{f_n} S_{\mu\mu}(f)\,df$. Thus, for the special case where the inverse $P_S^{-1}(\cdot)$ of $P_S(\cdot)$ exists, the Doppler frequencies f_n of the SOC process $\hat{\mu}(t)$ in (5.105) can be computed by evaluating

$$f_n = P_S^{-1}\left(\frac{2\sigma_0^2}{N}\left[N - n + \frac{1}{2}\right]\right) \tag{5.124}$$

for $n = 1, 2, \ldots, N$. In case the inverse of $P_S(\cdot)$ does not exist, then the Doppler frequencies f_n have to be computed by solving (5.123) using numerical root-finding techniques.

The GMEA defines the path gains c_n as in (5.106) to allow for a proper emulation of the reference model's envelope distribution $p_\zeta(z)$ regardless of the underlying Doppler power spectral density and angle-of-arrival statistics of $\mu(t)$. In this respect, it is important to mention that the numerical results presented in [135] indicate that the envelope distribution $\hat{p}_\zeta(z)$ of the SOC simulation model is closely in line with the Rayleigh distribution when the path gains

c_n are given as in (5.106) and $N \geq 10$. In fact, it is shown in [135] that the root mean square error $E^{(2)}_{p_\zeta}$ between $p_\zeta(z)$ and $\hat{p}_\zeta(z)$ is around 0.02 if $N = 10$, and smaller than 0.01 if $N > 20$.

5.4.5 *Performance Analysis*

To demonstrate the performance of the LPNM and the GMEA, it will be necessary to specify a concrete distribution for the random angles-of-arrival α. For this purpose serves in the following the von Mises distribution [175]. The usefulness of the von Mises distribution regarding the modelling of the angle-of-arrival statistics of mobile fading channels was first proposed and confirmed by measured data in [176].

The von Mises distribution of the angle-of-arrival α and its even part are given by

$$p_\alpha(\alpha) = \frac{e^{\kappa\cos(\alpha - m_\alpha)}}{2\pi I_0(\kappa)}, \tag{5.125}$$

$$g_\alpha(\alpha) = \frac{e^{\kappa\cos(\alpha)\cos(m_\alpha)}}{2\pi I_0(\kappa)} \cdot \cosh\left(\kappa \sin(\alpha)\sin(m_\alpha)\right), \tag{5.126}$$

respectively, where $\alpha \in [-\pi, \pi)$, $m_\alpha \in [-\pi, \pi)$ denotes the mean angle-of-arrival, $\kappa \geq 0$ is a concentration parameter controlling the angular spread, and $I_0(\cdot)$ is the modified Bessel function of the first kind of order zero. By setting $\kappa = 0$ in (5.125), the von Mises distribution reduces to the uniform distribution $p_\alpha(\alpha) = 1/(2\pi)$ characterizing isotropic scattering. For the von Mises distribution, one can readily verify by using (5.117) that the Doppler power spectral density $S_{\mu\mu}(f)$ of $\mu(t)$ is equal to [176]

$$S_{\mu\mu}(f) = \frac{2\sigma_0^2\, e^{\kappa\cos(m_\alpha) f/f_{\max}}}{\pi f_{\max} I_0(\kappa)\sqrt{1-(f/f_{\max})^2}} \cdot \cosh\left(\kappa \sin(m_\alpha)\sqrt{1-\left(\frac{f}{f_{\max}}\right)^2}\right) \tag{5.127}$$

for $|f| < f_{\max}$. Similarly, by using (5.113), one may demonstrate that [176]

$$r_{\mu\mu}(\tau) = \frac{2\sigma_0^2}{I_0(\kappa)} I_0\left(\left\{\kappa^2 - (2\pi f_{\max}\tau)^2 + j4\pi\kappa f_{\max}\cos(m_\alpha)\tau\right\}^{1/2}\right). \tag{5.128}$$

It is worth noticing that, for $\kappa = 0$, the Doppler power spectral density $S_{\mu\mu}(f)$ shown in (5.127) reduces to the Jakes power spectral density $S_{\mu\mu}(f) = 2\sigma_0^2/(\pi f_{\max}\sqrt{1-(f/f_{\max})^2})$ characterizing isotropic scattering channels, while the autocorrelation function in (5.128) simplifies to $r_{\mu\mu}(\tau) = 2\sigma_0^2 J_0(2\pi f_{\max}\tau)$.

Furthermore, by following (3.28a) and (3.28b), we find that the average Doppler shift $B^{(1)}_{\mu\mu}$ and the Doppler spread $B^{(2)}_{\mu\mu}$ of $\mu(t)$ are given by

$$B^{(1)}_{\mu\mu} = \frac{f_{\max}\cos(m_\alpha) I_1(\kappa)}{I_0(\kappa)} \tag{5.129}$$

$$B^{(2)}_{\mu\mu} = \left\{\frac{f_{\max}^2}{\kappa I_0(\kappa)}\left[\frac{\kappa[I_0(\kappa) + I_2(\kappa)]\cos^2(m_\alpha)}{2} + I_1(\kappa)\sin^2(m_\alpha)\right] - \left(B^{(1)}_{\mu\mu}\right)^2\right\}^{1/2}. \tag{5.130}$$

Notice that if $\kappa = 0$, then $B^{(1)}_{\mu\mu} = 0$ and $B^{(2)}_{\mu\mu} = f_{\max}/\sqrt{2}$.

Next, we analyze the performance of the GMEA and the LPNM for four different propagation scenarios, where the angle-of-arrival statistics are characterized by the von Mises distribution with the following pairs of parameters: ($m_\alpha = 0°$, $\kappa = 0$), ($m_\alpha = 0°$, $\kappa = 5$), ($m_\alpha = 30°$, $\kappa = 10$), and ($m_\alpha = 90°$, $\kappa = 10$). The first and the last of such pairs of parameters are related to channels characterized by symmetrical Doppler power spectral densities, whereas the other two pairs are associated with channels with asymmetrical Doppler power spectral densities. For the performance comparison of the GMEA with the three variants of the LPNM, we choose $\sigma_0^2 = 1/2$, $f_{\max} = 91$ Hz, $p = 2$, and $\tau_{\max} = N/4f_{\max}$. Following the experiments performed in Subsection 5.1.6, we set the underlying weighting factors W_1 and W_2 of $E^{(p)}$ to $W_1 = 1/4$ and $W_2 = 3/4$. We use the Doppler frequencies computed by the GMEA as initial values to minimize the L_p-norms $E^{(p)}_{r_{\mu\mu}}$ and $E^{(p)}$. Notice that neither the cumulative distribution function $F_\alpha(\alpha)$ of the von Mises distribution, nor the cumulative power function $P_S(f)$ of the resulting Doppler power spectral density $S_{\mu\mu}(f)$ [see (5.127)] can be computed analytically. For this reason, the application of the GMEA to the von Mises distribution requires solving (5.120) (or (5.123)) numerically, which can be done easily with the `fzero` function of MATLAB®.

Emulation of the envelope distribution: Figure 5.70 presents a comparison between the envelope distribution $p_\zeta(z)$ of the reference model and the envelope distribution $\hat{p}_\zeta(z)$ of $\hat{\mu}(t)$ by applying the GMEA with $N = 10$. Both analytical and empirical graphs of $\hat{p}_\zeta(z)$ are depicted in this figure. The analytical graphs were generated by numerically evaluating (4.121) using the MATLAB® `trapz` function. The empirical graphs were obtained by evaluating $50 \cdot 10^6$ samples of $\hat{\zeta}(t)$. Such samples were collected at the same time instant $t = t_i$, where the value of t_i was chosen at random over a time interval of two hours. Notice that the GMEA results in the same envelope distribution regardless of the angle-of-arrival statistics of the

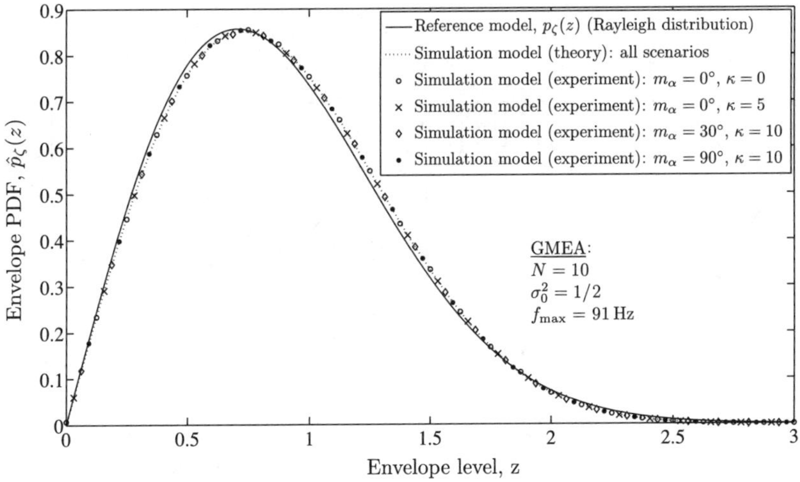

Figure 5.70 Comparison between the envelope distribution $p_\zeta(z)$ of the reference model and the envelope distribution $p_{\hat{\zeta}}(z)$ of the simulation model by applying the GMEA to the von Mises distribution of the angle-of-arrival α.

simulated channel. This is because the envelope distribution of the simulation model is solely influenced by the set of path gains $\{c_n\}$ and the number of cisoids N [see (4.121)], and the GMEA defines a unique set of path gains for each value of N irrespective of the angle-of-arrival distribution [see (5.106)]. We can see in the figure that the analytical graphs of $\hat{p}_\zeta(z)$ match perfectly the empirical ones, so that one can presume the correctness of the analytical solution of $\hat{p}_\zeta(z)$ given by (4.121). Moreover, we can observe from Figure 5.70 that the graphs of $\hat{p}_\zeta(z)$ are in a very good agreement with the curve described by the envelope distribution $p_\zeta(z)$ of the reference model, which demonstrates the accuracy of the GMEA in approximating the Rayleigh distribution.

Next, we evaluate the root mean square (RMS) error $E_{p_\zeta}^{(2)}$ between $p_\zeta(z)$ and $\hat{p}_\zeta(z)$ [see (5.110) with $p = 2$] by considering the three variants of the LPNM and the GMEA for $N \in \{10, 20, 30, 40, 50\}$. The results are presented in Figure 5.71. Notice that the error caused by the GMEA and the LPNM I is exactly the same. This is because the distribution of $\hat{\zeta}(t)$ is the same for both methods, as they specify the path gains c_n in the same way. The obtained results leave no doubt that the LPNM II is the method that generates the smallest error under isotropic scattering conditions, while the GMEA and the LPNM I, III have basically the same performance. On the other hand, the GMEA and the LPNM I outperform the LPNM II and the LPNM III regarding the emulation of the reference model's envelope distribution under non-isotropic scattering conditions.

Emulation of the autocorrelation function: Figure 5.72 shows a comparison between the absolute value of the autocorrelation function $r_{\mu\mu}(\tau)$ of the reference model [see (5.128)] and the absolute value of the autocorrelation function $\hat{r}_{\mu\mu}(\tau)$ of $\hat{\mu}(t)$ [see (4.104)] by applying the methods under consideration with $N = 20$. In addition to the theoretical curves of $|\hat{r}_{\mu\mu}(\tau)|$, Figure 5.72 presents empirical graphs of the time-averaged autocorrelation function $\tilde{r}_{\mu\mu}(\tau)$ of the sample functions $\tilde{\mu}(t)$ of $\hat{\mu}(t)$. All of these graphs were generated from a single realization of $\hat{\mu}(t)$. It can be seen that the graphs of $|\tilde{r}_{\mu\mu}(\tau)|$ superimpose the ones of $|\hat{r}_{\mu\mu}(\tau)|$ in all cases,

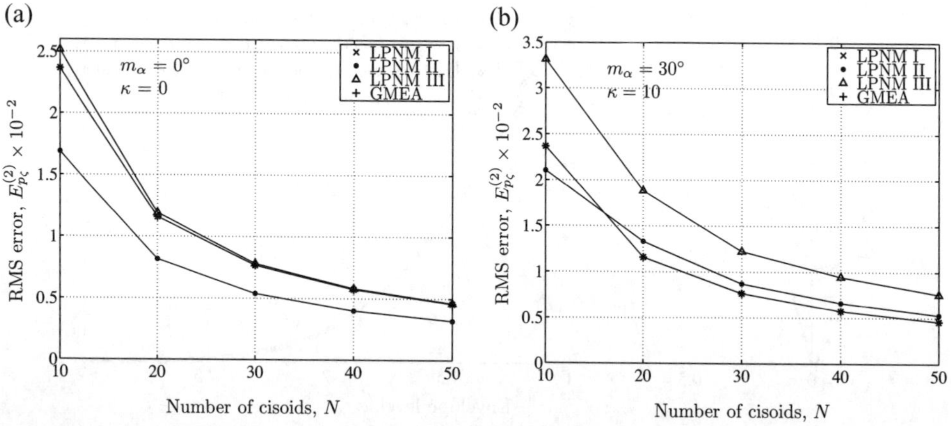

Figure 5.71 Root mean square error $E_{p_\zeta}^{(2)}$ of the envelope distribution of the simulation model designed by applying the LPNM and the GMEA to the von Mises distribution for (a) isotropic scattering and (b) non-isotropic scattering.

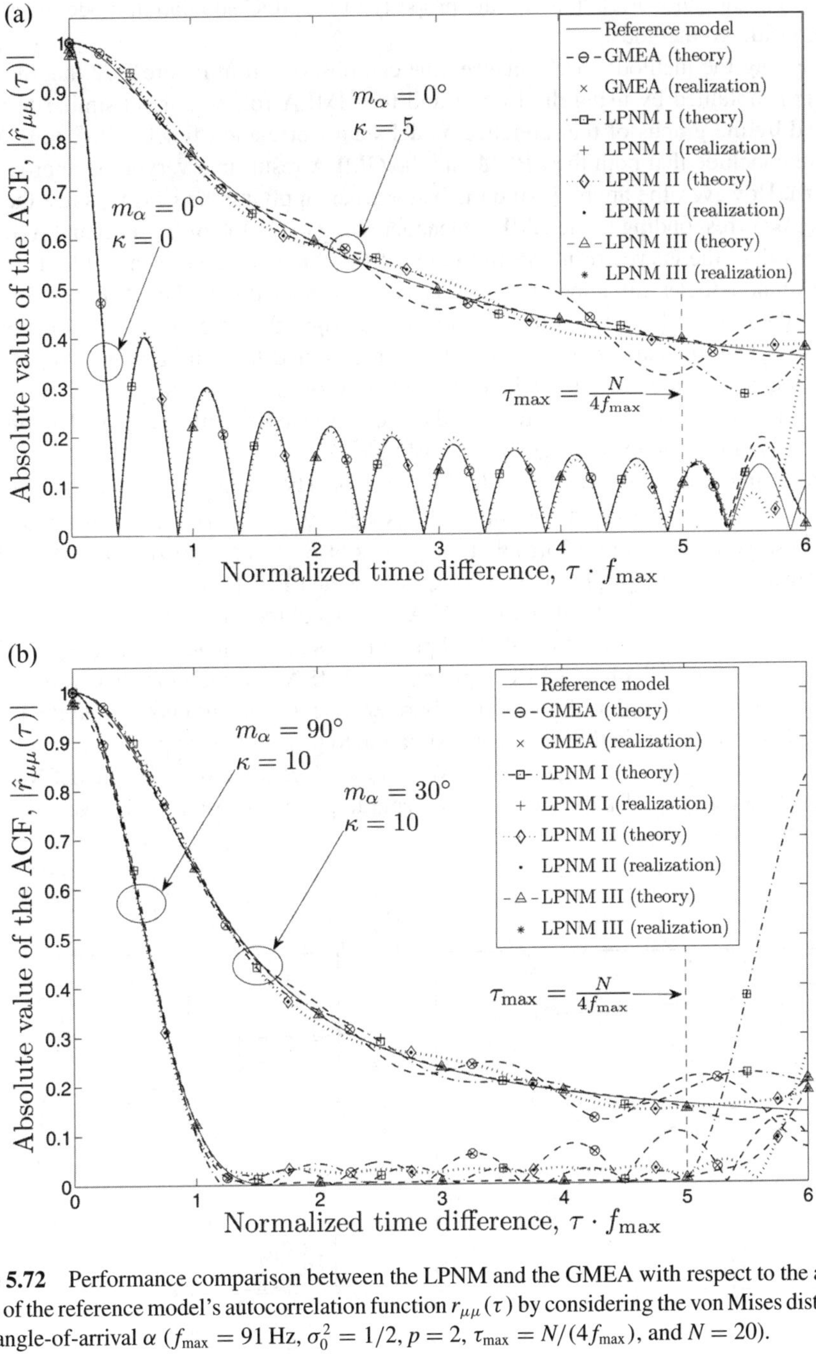

Figure 5.72 Performance comparison between the LPNM and the GMEA with respect to the approximation of the reference model's autocorrelation function $r_{\mu\mu}(\tau)$ by considering the von Mises distribution of the angle-of-arrival α ($f_{\max} = 91$ Hz, $\sigma_0^2 = 1/2$, $p = 2$, $\tau_{\max} = N/(4f_{\max})$, and $N = 20$).

meaning that the autocorrelation-ergodic property of the SOC simulation model is preserved by the considered methods.

Concerning the methods' performance, one can observe from Figure 5.72 that the graphs of $|\hat{r}_{\mu\mu}(\tau)|$ obtained by using the LPNM and the GMEA follow a trend similar to the one described by the graphs of the reference model's autocorrelation function $|r_{\mu\mu}(\tau)|$. We can therefore conclude that both the LPNM and the GMEA result in a very good approximation to $r_{\mu\mu}(\tau)$. However, the approximation is better when applying the LPNM, since the curves of $|\hat{r}_{\mu\mu}(\tau)|$ corresponding to the GMEA meander along the graph of $|r_{\mu\mu}(\tau)|$ and show larger deviations than the curves resulting from any of the three variants of the LPNM. It should be noticed, nonetheless, that the graphs of $|\hat{r}_{\mu\mu}(\tau)|$ associated with the GMEA are in general closer to the ones of $|r_{\mu\mu}(\tau)|$ at the vicinity of the origin than the graphs obtained by using the LPNM. This indicates that the GMEA is better suited than the LPNM to approximate the statistical quantities of $\mu(t)$, which depend only on the value, slope, and curvature of $r_{\mu\mu}(\tau)$ at $\tau = 0$. Such quantities are, e.g., the average Doppler shift, the Doppler spread, the level-crossing rate, and the average duration of fades.

The resulting root mean square error $E^{(2)}_{r_{\mu\mu}}$ [see (5.108)] between $\hat{r}_{\mu\mu}(\tau)$ and $r_{\mu\mu}(\tau)$ is shown in Figure 5.73 for $N \in \{10, 20, 30, 40, 50\}$. Especially in case of non-isotropic scattering, we can observe from Figure 5.73(b) that the graphs of $E^{(2)}_{r_{\mu\mu}}$ corresponding to the GMEA are almost always above the graphs obtained by applying the LPNM. This confirms the hypothesis that the LPNM performs better than the GMEA in terms of the emulation of $r_{\mu\mu}(\tau)$. However, Figure 5.73(b) shows that only the LPNM II provides a significant advantage over the GMEA, since the values of $E^{(2)}_{r_{\mu\mu}}$ obtained by applying the GMEA and the other two variants of the LPNM are in the same order of magnitude. Furthermore, one should notice that the LPNM II is the method that results in the worst approximation to $|r_{\mu\mu}(\tau)|$ in the proximity of the origin, as one can readily see from Figure 5.72. We can therefore anticipate that the LPNM II will be the method that provides the worst results regarding the emulation of the average Doppler shift and the Doppler spread.

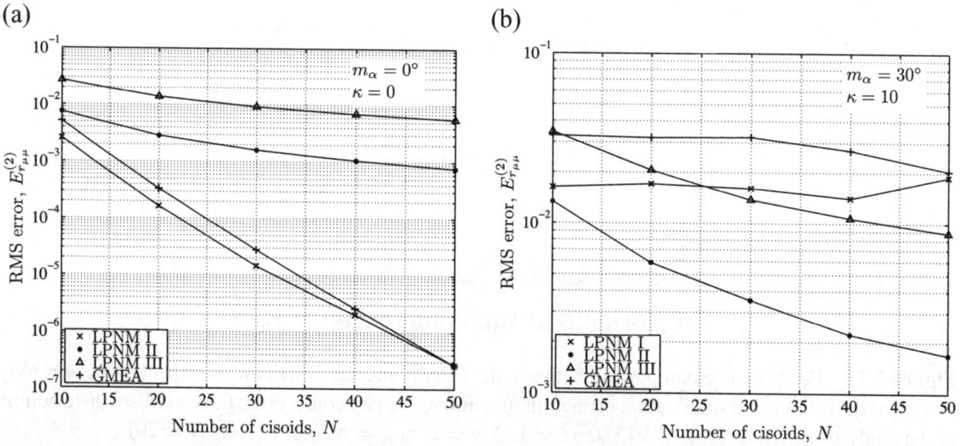

Figure 5.73 Root mean square error $E^{(2)}_{r_{\mu\mu}}$ of the autocorrelation function $\hat{r}_{\mu\mu}(\tau)$ of the stochastic SOC simulation model designed by applying the LPNM and the GMEA to the von Mises distribution for (a) isotropic scattering and (b) non-isotropic scattering ($\tau_{\max} = N/(4f_{\max})$).

Emulation of the average Doppler shift and Doppler spread: With respect to the emulation of the average Doppler shift $B_{\mu\mu}^{(1)}$ of $\mu(t)$, we plot in Figure 5.74 the absolute error

$$e_{B_{\mu\mu}^{(1)}} = |B_{\mu\mu}^{(1)} - \hat{B}_{\mu\mu}^{(1)}| \tag{5.131}$$

of the average Doppler shift $\hat{B}_{\mu\mu}^{(1)}$ of the simulation model as a function of N. In addition, the results obtained for the relative error

$$e_{B_{\mu\mu}^{(2)}} = \frac{|B_{\mu\mu}^{(2)} - \hat{B}_{\mu\mu}^{(2)}|}{B_{\mu\mu}^{(2)}} \tag{5.132}$$

of the Doppler spread $\hat{B}_{\mu\mu}^{(2)}$ are plotted in Figure 5.75. Figures 5.74 and 5.75 show clearly that in case of isotropic scattering, the smallest values of $e_{B_{\mu\mu}^{(1)}}$ and $e_{B_{\mu\mu}^{(2)}}$ are obtained by applying the GMEA, while the LPNM III performs similar to the GMEA if non-isotropic scattering is considered. Recall that we mentioned in Section 3.3 that the average Doppler shift of $\mu(t)$ is equal to zero if the Doppler power spectral density $S_{\mu\mu}(f)$ is a symmetric function. From the graphs of $e_{B_{\mu\mu}^{(1)}}$ corresponding to channels with symmetrical Doppler power spectral densities [Figures 5.74(a)], we can observe that the GMEA is the only method for which the simulation model's average Doppler shift $\hat{B}_{\mu\mu}^{(1)}$ equals approximately zero. Thus, in view of the results presented in Figures 5.74 and 5.75, we can conclude that the GMEA is better suited than the LPNM to emulate the average Doppler shift and the Doppler spread. Notice that the LPNM II is the method that produced the largest values of $e_{B_{\mu\mu}^{(1)}}$ and $e_{B_{\mu\mu}^{(2)}}$. This illustrates the observations we made above when we discussed the emulation of the autocorrelation function.

Summary of the performance analysis: In this subsection, we evaluated the performance of the three different variants of the LPNM in comparison to the GMEA with respect to their capability to approximate the Rayleigh distribution, the autocorrelation function, the average Doppler shift, and the Doppler spread of a given reference channel model. This

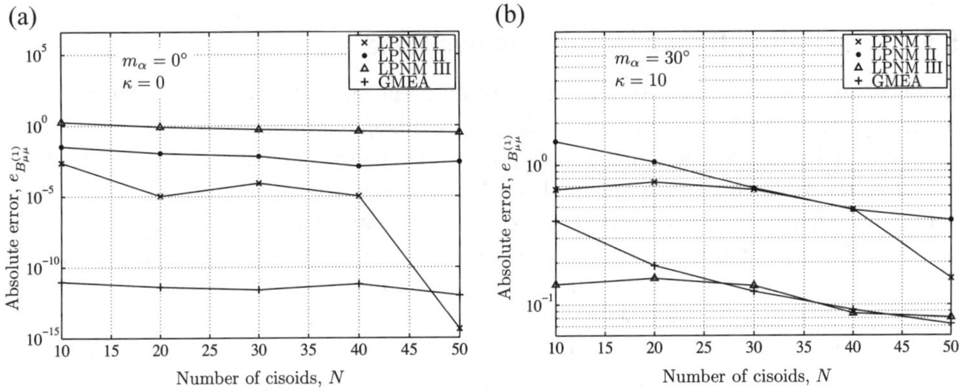

Figure 5.74 Absolute error $e_{B_{\mu\mu}^{(1)}}$ of the average Doppler shift of the simulation model designed by applying the LPNM and the GMEA to the von Mises distribution for (a) isotropic scattering and (b) non-isotropic scattering.

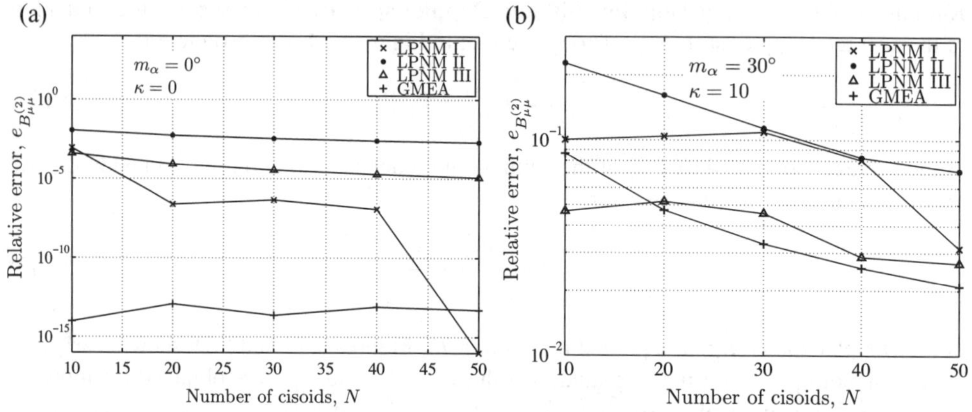

Figure 5.75 Relative error $e_{B^{(2)}_{\mu\mu}}$ of the Doppler spread of the simulation model designed by applying the LPNM and the GMEA to the von Mises distribution for (a) isotropic scattering and (b) non-isotropic scattering.

study showed that the GMEA and the LPNM I are equivalent to each other regarding the approximation of the Rayleigh distribution. The LPNM II performed slightly better than the other methods when the scattering is isotropic, but the GMEA (LPNM I) outperformed the LPNM II and the LPNM III regarding the approximation of the envelope distribution under non-isotropic scattering conditions. The three variants of the LPNM perform better than the GMEA with respect to the emulation of the autocorrelation function, but the GMEA proves to be more accurate concerning the emulation of the channel's average Doppler shift and Doppler spread. This characteristic of the GMEA, together with its good performance, make this method a suitable tool for the design of SOC channel simulators to enable the performance analysis of modern mobile communication systems under isotropic and non-isotropic scattering conditions.

5.5 Concluding Remarks and Further Reading

In this chapter, we have discussed a variety of different methods for the computation of the primary parameters of the simulation model (discrete Doppler frequencies $f_{i,n}$ and path gains $c_{i,n}$). Exactly like the original Rice method [16, 17], the method of equal distances [147, 148], as well as the mean-square-error method [148] are characterized by the fact that the distances between two neighbouring discrete Doppler frequencies are equidistant. These three methods merely differ in their way of how the path gains are adapted to the desired Doppler power spectral density. Owing to the equidistant property of the discrete Doppler frequencies, all three procedures have one decisive disadvantage in common, namely the comparatively small period of the designed deterministic Gaussian processes, and thus of the resulting simulation model. This disadvantage can be avoided, e.g., by using the method of equal areas [147, 148], which has an acceptable performance when applied to the Jakes power spectral density. However, this method fails or leads to a comparatively high realization complexity, if the procedure is used in connection with Gaussian shaped power spectral densities. First in Germany and later worldwide, the Monte Carlo method [99, 100] has become quite popular. In comparison with other methods, however, the performance of this method is poor [78, 148] if the approximation accuracy of the autocorrelation function of the resulting deterministic Gaussian processes is invoked as performance criterion. The principle of the Monte Carlo method is that the discrete Doppler frequencies of the stochastic simulation system are obtained from the mapping of a uniformly distributed random variable onto a random variable with a distribution proportional to the desired Doppler power spectral density. Consequently, the discrete Doppler frequencies themselves are random variables. The realization of a set $\{f_{i,n}\}$ of discrete Doppler frequencies can thus result in a deterministic Gaussian process $\tilde{\mu}_i(t)$, whose statistical properties may largely deviate from the desired properties of the (ideal) stochastic Gaussian random process $\mu_i(t)$. This even holds if the chosen number of sinusoids N_i is very large, let us say $N_i = 100$ [78]. A quasi-optimal procedure is the method of exact Doppler spread [96, 149]. This method is almost perfectly suitable for Jakes shaped power spectral densities. The performance of the method of exact Doppler spread can only be outperformed by the L_p-norm method [96, 149]. Unfortunately, the numerical complexity of this method is comparatively high, so that an application, especially in connection with the Jakes and Gaussian power spectral density, is often not worth the effort. The L_p-norm method only attains its full performance when the statistical properties of the deterministic simulation model have to be adapted to snapshot measurements of real-world mobile radio channels. A further design method is the Jakes method [13], which, however, does not fulfil the often imposed requirement that the real and imaginary part of the complex Gaussian random processes describing the Rayleigh (Rice) process should be uncorrelated.

The classification scheme introduced in Section 4.4 achieves its full advantage when it is applied to the presented parameter computation methods. Starting with the original Rice method [16, 17], we realize by considering (4.2a) and (4.2b) that the gains $c_{i,n}$ and frequencies $f_{i,n}$ are constant quantities. Supposing that the phases $\theta_{i,n}$ are random variables, it follows from Table 4.1 that the resulting channel simulator belongs to Class II. Such a channel simulator enables the generation of stochastic processes which are not only first order stationary but also mean and autocorrelation ergodic. On the other hand, if the Monte Carlo method [99, 100] or the randomized method of exact Doppler spread [111, 153] is applied, then the gains $c_{i,n}$ are constant quantities and the frequencies $f_{i,n}$ and phases $\theta_{i,n}$ are random variables. Consequently, the resulting channel simulator can be identified as a Class IV channel simulator, which is first order stationary and mean ergodic, but unfortunately non-autocorrelation ergodic. For a given parameter computation method, the stationary and ergodic properties of the resulting channel simulator can be concluded directly from the results presented in Table 4.1. It goes without saying that the above concept is easy to handle and can be applied to any given parameter computation method. For some selected methods [13, 16, 17, 96, 97, 99, 100, 102, 111, 148, 150], the obtained results are presented in Table 5.1.

The Jakes method [13] has achieved great popularity. When this method is used, then the gains $c_{i,n}$, frequencies $f_{i,n}$, and phases $\theta_{i,n}$ are constant quantities. Consequently, the Jakes channel simulator is per definition completely deterministic. To obtain the underlying stochastic channel simulator, it is advisable to replace the constant phases[4] $\theta_{i,n}$ by random phases. According to Table 4.1, the replacement of constant phases $\theta_{i,n}$ by random variables transforms a Class I type channel simulator into a Class II type one, which is first order stationary, mean ergodic, and autocorrelation ergodic. It should be pointed out here that this statement is in contrast to the analysis in [177], where it has been claimed that Jakes' simulator

[4] Recall that the phases $\theta_{i,n}$ in Jakes' channel simulator are equal to 0 for all $i = 1, 2$ and $n = 1, 2, \ldots, N_i$ [see (5.62)].

is nonstationary. However, this is not surprising if one takes into account that the proof in [177] is based on the assumption that the gains $c_{i,n}$ are random variables. But this assumption cannot be justified in the sense of the original Jakes method, where all model parameters are constant quantities.

Nevertheless, the Jakes simulator has some disadvantages as discussed above, which can be avoided, e.g., by using the method of exact Doppler spread [96] or the even more powerful L_p-norm method [96]. Both methods enable the design of deterministic channel simulators, where all parameters are fixed including the phases $\theta_{i,n}$. The investigation of the stationary and ergodic properties of deterministic processes is meaningless, since the concept of stationarity and ergodicity is only applicable to stochastic processes. A deterministic channel simulator can be interpreted as an emulator for sample functions of the underlying stochastic channel simulator. According to the concept of deterministic channel modelling, the corresponding stochastic simulation model is obtained by replacing the constant phases $\theta_{i,n}$ by random phases (see Figure 4.3). It is now important to realize that all stochastic channel simulators derived in this way from deterministic channels simulators are Class II channel simulators having the statistical properties listed in Table 4.1. This must be taken into account when considering the results shown in Table 5.1.

Appendix 5.A Analysis of the Relative Model Error by Using the Monte Carlo Method

We consider the relative model error

$$\frac{\Delta\beta_i}{\beta} = \frac{\tilde{\beta}_i - \beta}{\beta}, \quad i = 1, 2, \tag{5.A.1}$$

Table 5.1 Overview of parameter computation methods for sum-of-sinusoids processes and their stationary and ergodic properties [134].

Parameter computation method	Class	First-order stationary	Wide-sense stationary	Mean ergodic	Auto-correlation ergodic
Rice method [16, 17]	II	Yes	Yes	Yes	Yes
Monte Carlo method [99, 100]	IV	Yes	Yes	Yes	No
Jakes method [13] (with random phases)	II	Yes	Yes	Yes	Yes
Harmonic decomposition technique [102]	IV	Yes	Yes	Yes	No
Method of equal distances [148]	II	Yes	Yes	Yes	Yes
Method of equal areas [148]	II	Yes	Yes	Yes	Yes
Mean-square-error method [148]	II	Yes	Yes	Yes	Yes
Method of exact Doppler spread [96]	II	Yes	Yes	Yes	Yes
L_p-Norm method [96]	II	Yes	Yes	Yes	Yes
Randomized method of exact Doppler spread [111, 153]	IV	Yes	Yes	Yes	No
Improved method by Zheng and Xiao [97]	VIII	Yes	Yes	Yes	No
Method of exact Doppler spread with set partitioning [150]	II	Yes	Yes	Yes	Yes

where the quantities β and $\tilde{\beta}_i$ are in case of the Jakes power spectral density given by

$$\beta = 2(\pi\sigma_0 f_{\max})^2 \tag{5.A.2}$$

and

$$\tilde{\beta}_i = \frac{2\beta}{f_{\max}^2 N_i} \sum_{n=1}^{N_i} f_{i,n}^2, \tag{5.A.3}$$

respectively. If we use the Monte Carlo method for the computation of the model parameters, then the discrete Doppler frequencies $f_{i,n}$ are independent, identically distributed random variables characterized by the probability density function

$$p_{f_{i,n}}(f_{i,n}) = \begin{cases} \dfrac{2}{\pi f_{\max}\sqrt{1-(f_{i,n}/f_{\max})^2}}, & 0 < f \leq f_{\max}, \\ 0, & \text{else.} \end{cases} \tag{5.A.4}$$

With the Chebyshev inequality (2.24), the relation

$$P\left\{\left|\frac{\Delta\beta_i}{\beta} - E\left\{\frac{\Delta\beta_i}{\beta}\right\}\right| \geq \varepsilon\right\} \leq \frac{\text{Var}\{\Delta\beta_i/\beta\}}{\varepsilon^2} \tag{5.A.5}$$

holds for all $\varepsilon > 0$. Using (5.A.4), we find

$$E\left\{f_{i,n}^2\right\} = \frac{f_{\max}^2}{2} \tag{5.A.6}$$

and

$$\begin{aligned} \text{Var}\left\{f_{i,n}^2\right\} &= E\left\{f_{i,n}^4\right\} - \left(E\left\{f_{i,n}^2\right\}\right)^2 \\ &= \frac{3}{8} f_{\max}^4 - \frac{f_{\max}^4}{4} \\ &= \frac{f_{\max}^4}{8}. \end{aligned} \tag{5.A.7}$$

Hence, by using (5.A.1) and (5.A.3), we obtain the following expression for the mean value and the variance of the relative model error $\Delta\beta_i/\beta$

$$E\left\{\frac{\Delta\beta_i}{\beta}\right\} = 0 \tag{5.A.8}$$

and

$$\begin{aligned} \text{Var}\left\{\frac{\Delta\beta_i}{\beta}\right\} &= \text{Var}\left\{\frac{\beta_i}{\beta}\right\} \\ &= \left(\frac{2}{f_{\max}^2 N_i}\right)^2 \text{Var}\left\{\sum_{n=1}^{N_i} f_{i,n}^2\right\} \\ &= \left(\frac{2}{f_{\max}^2 N_i}\right)^2 \sum_{n=1}^{N_i} \text{Var}\left\{f_{i,n}^2\right\} \\ &= \frac{1}{2N_i}, \end{aligned} \tag{5.A.9}$$

respectively. Thus, with the Chebyshev inequality (5.A.5) the relation

$$P\left\{\left|\frac{\Delta\beta_i}{\beta}\right| \geq \varepsilon\right\} \leq \frac{1}{2N_i\varepsilon^2} \tag{5.A.10}$$

follows. For example, let $\varepsilon = 0.02$ and $N_i = 2500$ (!), then the inequality above can be interpreted as follows: The probability that the absolute value of the relative model error $|\Delta\beta_i/\beta|$ is higher than or equal to 2 per cent is smaller than or equal to 50 per cent.

Appendix 5.B Proof of the Convergence of the Sample Mean Autocorrelation Function by Using the MEDS-SP

In this appendix, we prove that the sample mean autocorrelation function $\bar{r}_{\mu_i\mu_i}(\tau)$ approaches the autocorrelation function $r_{\mu_i\mu_i}(\tau)$ of the reference model for any value of $N_i \geq 1$ as $K \to \infty$. To show this, we substitute (5.97) in (5.99) and make use of (5.95). Hence, we obtain

$$\begin{aligned}
\bar{r}_{\mu_i\mu_i}(\tau) &= \frac{1}{K}\sum_{k=1}^{K} \tilde{r}^{(k)}_{\mu_i\mu_i}(\tau) \\
&= \frac{\sigma_0^2}{KN_i}\sum_{k=1}^{K}\sum_{n=1}^{N_i} \cos(2\pi f_{\max}\cos(\alpha_{i,n}^{(k)})\,\tau) \\
&= \frac{\sigma_0^2}{KN_i}\sum_{k=1}^{K}\sum_{n=1}^{N_i} \cos\left\{2\pi f_{\max}\cos\left[\frac{\pi}{2N_i}\left(n-\frac{1}{2}\right)+\frac{\pi(2k-1-K)}{4KN_i}\right]\tau\right\}.
\end{aligned} \tag{5.B.1}$$

In the limit $K \to \infty$, we obtain

$$\begin{aligned}
\lim_{K\to\infty} \bar{r}_{\mu_i\mu_i}(\tau) &= \sigma_0^2\,\frac{2}{\pi}\sum_{n=1}^{N_i}\int_{-\frac{\pi}{4N_i}}^{\frac{\pi}{4N_i}} \cos\left\{2\pi f_{\max}\tau\cos\left[\frac{\sigma_0^2}{2N_i}\left(n-\frac{1}{2}\right)+\alpha\right]\right\}d\alpha \\
&= \sigma_0^2\,\frac{2}{\pi}\sum_{n=1}^{N_i}\int_{\frac{\pi(n-1)}{2N_i}}^{\frac{\pi n}{2N_i}} \cos(2\pi f_{\max}\tau\cos\alpha)\,d\alpha \\
&= \sigma_0^2\,\frac{2}{\pi}\int_{0}^{\pi/2} \cos(2\pi f_{\max}\tau\cos\alpha)\,d\alpha.
\end{aligned} \tag{5.B.2}$$

The integral above can be identified as the integral representation of the zeroth order Bessel function of the first kind [77, Equation (9.1.18)]. Thus,

$$\begin{aligned}
\lim_{K\to\infty} \bar{r}_{\mu_i\mu_i}(\tau) &= \sigma_0^2 J_0(2\pi f_{\max}\tau) \\
&= r_{\mu_i\mu_i}(\tau)
\end{aligned} \tag{5.B.3}$$

for $i = 1, 2$. Note that this result holds for any value of $N_i \geq 1$.

Appendix 5.C Proof of the Condition for Uncorrelated Inphase and Quadrature Components of SOC Processes

Theorem: *Let the Doppler power spectral density $S_{\mu\mu}(f)$ of $\mu(t)$ be a symmetric continuous function on the frequency domain $(-f_{\max}, f_{\max}]$. Suppose that the path gains c_n and the angles-of-arrival α_n of a stochastic SOC process $\hat{\mu}(t)$ are given by (5.106) and (5.118), respectively. Then, the inphase and quadrature components of $\hat{\mu}(t) = \hat{\mu}_1(t) + j\hat{\mu}_2(t)$ are uncorrelated if and only if the Equation (5.119) holds.*

Proof: To prove this theorem, we start by noticing from (5.118) that the constant angles-of-arrival α_n fulfill the relationships $0 < \alpha_n < \pi \ \forall n$ and $\alpha_n \neq \alpha_m \ \forall n \neq m$. Consequently, the Doppler frequencies $f_n = f_{\max}\cos(\alpha_n)$ of $\hat{\mu}(t)$ satisfy $f_n \neq f_m$ for all $n \neq m$. Now, if $f_n \neq f_m$ and $c_n = c_m$ for all $n \neq m$, then $\hat{\mu}_1(t)$ and $\hat{\mu}_2(t)$ are uncorrelated if and only if

$$f_n = -f_{N-n+1}, \qquad n = 1, 2, \ldots, N. \tag{5.C.1}$$

For notational convenience and without loss of generality, we have assumed that the Doppler frequencies f_n are indexed such that $f_n < f_m$ for $n < m$.

The equality $f_n = -f_{N-n+1}$ implies that the underlying angles-of-arrival satisfy $\alpha_n = \pi - \alpha_{N-n+1}$, where $\alpha_n \in (0, \pi)$. In turn, the symmetry of $S_{\mu\mu}(f)$ implies that $g_\alpha(\alpha) = g_\alpha(\pi - \alpha)$ for $\alpha \in (0, \pi]$. Thus, if $S_{\mu\mu}(f) = S_{\mu\mu}(-f)$ and the Doppler frequencies f_n satisfy (5.C.1), then

$$\int_0^{\alpha_n} g_\alpha(\alpha)\, d\alpha = \int_{\alpha_{N-n+1}}^{\pi} g_\alpha(\alpha)\, d\alpha \tag{5.C.2}$$

for all $n = 1, 2, \ldots, N$. Hence, to prove the theorem, it is sufficient to demonstrate that (5.C.2) holds if and only if (5.119) is met. This can be proved easily by noticing that [cf. (5.118)]

$$\int_0^{\alpha_n} g_\alpha(\alpha)\, d\alpha = \frac{n-1}{2N} + \int_0^{\alpha_1} g_\alpha(\alpha)\, d\alpha \tag{5.C.3}$$

$$\int_{\alpha_{N-n+1}}^{\pi} g_\alpha(\alpha)\, d\alpha = \frac{n}{2N} - \int_0^{\alpha_1} g_\alpha(\alpha)\, d\alpha. \tag{5.C.4}$$

It is evident that (5.C.3) equals (5.C.4) if and only if $\int_0^{\hat{\alpha}_1} g_\alpha(\alpha)\, d\alpha = 1/(4N)$.

6

Frequency-Nonselective Channel Models

Frequency-nonselective or flat fading channels are characterized by time-invariant transfer functions whose absolute values are independent of frequency. This property holds true if the propagation delay differences of the reflected and scattered signal components at the receiver antenna are negligible in comparison to the symbol duration. Under this condition, the random fluctuations of the received signal can be modelled by multiplying the transmitted signal with a suitable stochastic process. The problem of finding and describing suitable stochastic processes and their fitting to real-world channels has been a subject of research [26, 178–181] for a considerable time.

The simplest stochastic processes that can be used to model frequency-nonselective channels for terrestrial, cellular mobile radio systems and land mobile satellite systems are Rayleigh and Rice processes which have been described in detail in Chapter 3. The flexibility of these models is, however, relatively limited and often not sufficient to enable a good fitting to the statistics of real-world channels. For frequency-nonselective land mobile radio channels, it has turned out that the *Suzuki process* [26, 182] is a more suitable stochastic model in many cases. The Suzuki process is a product process of a Rayleigh process and a lognormal process. Thereby, the Rayleigh process models as usual the fast fading behaviour of the received signal. The slow fading variations of the local mean value of the received signal are modelled by the lognormal process. In Suzuki channels, it is assumed that no line-of-sight component exists due to shadowing. Usually, it is also assumed that the two narrow-band real-valued Gaussian random processes from which the Rayleigh process is derived are uncorrelated. Dropping the last assumption leads to the so-called *modified Suzuki process* analyzed in [70, 183].

Although the Suzuki process and its modified version were originally suggested as a model for the terrestrial, cellular mobile radio channel, these stochastic processes are also quite suitable for the modelling of land mobile satellite channels in urban regions, where the assumption that the line-of-sight signal component is blocked is justified most of the time. Suburban and rural regions or even open areas with partial or no shadowing of the line-of-sight component, however, make further model extensions necessary. A contribution to this

Mobile Radio Channels, Second Edition. Matthias Pätzold.

topic was made in [181]. The stochastic model introduced there is based on the product of a Rice process and a lognormal process. Such a product process is suitable for modelling a large class of environments (urban, suburban, rural, and open areas). Here, the two real-valued Gaussian random processes producing the Rice process are again assumed to be uncorrelated. If this assumption is dropped, then the flexibility of this model can be improved considerably with respect to the statistics of higher order. Depending on the type of cross-correlation, we distinguish between *extended Suzuki processes of Type I* [75] and those of *Type II* [184].

Moreover, a so-called *generalized Suzuki process* was suggested in [185], which contains the classical Suzuki process [26, 182], the modified Suzuki process [70, 183], as well as the two extended Suzuki processes [75, 184] of Type I and Type II as special cases. The first and second order statistical properties of generalized Suzuki processes are very flexible and can therefore be fitted very well to measurements of real-world channels.

A further stochastic model was introduced by Loo [178, 179, 186, 187]. Loo's model is designated for a satellite mobile radio channel in rural environments, where a line-of-sight component between the satellite and the vehicle exists for most of the transmission time. The model is based on a Rayleigh process with constant mean power for the absolute value of the sum of all scattered multipath components. For the line-of-sight component, it is assumed that its amplitude follows the statistics of a lognormal process. In this way, the slow amplitude variations of the line-of-sight component caused by shadowing are taken into account.

All the stochastic channel models described above have in common that they are stationary in the sense that they are based on stationary stochastic processes with constant parameters. A nonstationary model, which is valid for very large areas, was introduced by Lutz et al. [180] for frequency-nonselective land mobile satellite channels. This model distinguishes between good channel states and bad channel states. Lutz's model can be represented by using a two-state Markov model. If the Markov model is in the good channel state, then the fading envelope follows the Rice process. Otherwise, if the Markov model is in the bad channel state, then the classical Suzuki process serves as a proper model for the fading envelope. This procedure can be generalized, leading to an M-state Markov model [188, 189]. Experimental measurements have shown that a four-state model is sufficient for most channels [190]. A concept for embedding a highly flexible stationary model in a dynamic two-state channel model was proposed in [191]. In this paper, a measurement-based channel model was developed by assigning, for each channel state, a specific set of parameters to one and the same stationary channel model. A change of a channel state corresponds thus to a new configuration of a universal stationary channel model.

This chapter deals with the description of the extended Suzuki process of Type I (Section 6.1) and of Type II (Section 6.2) as well as with the generalized Suzuki process (Section 6.3). We will also get to know a modified version of the Loo model (Section 6.4), which includes the classical Loo model as a special case. Moreover, in Section 6.5, several methods for the modelling of nonstationary land mobile satellite channels will be introduced. Starting from the Lutz model, we will see how the underlying concept can be extended and applied to real-world satellite channels. Each section starts with a description of the respective reference model. Afterwards, the corresponding simulation model will be presented. To demonstrate the usefulness of the suggested reference models, the main statistical properties such as the probability density function of the envelope, the level-crossing rate, and the average duration of fades are fitted generally to measurement results available in the literature.

The achieved correspondence between the reference model, the simulation model, and the underlying measurements is in most cases extremely good, as will be demonstrated clearly by various examples.

6.1 The Extended Suzuki Process of Type I

As mentioned at the beginning of this chapter, the product process of a Rayleigh process and a lognormal process is said to be a Suzuki process. For this kind of process, an extension is suggested in the text that follows. The Rayleigh process is in this case substituted by a Rice process taking the influence of a line-of-sight component into account. In the proposed model, the line-of-sight component can be Doppler shifted. Moreover, a cross-correlation between the two real-valued Gaussian random processes determining the Rice process is admitted. In this way, the number of degrees of freedom increases, which in fact increases the mathematical complexity, but in the end clearly improves the flexibility of the stochastic model. The resulting product process of a Rice process with cross-correlated underlying Gaussian random processes and a lognormal process was introduced as the *extended Suzuki process (of Type I)* in [75, 192]. This process is suitable as a stochastic model for a large class of satellite and land mobile radio channels in environments, where a direct line-of-sight connection between the transmitter and the receiver cannot be ignored.

The description of the reference model and the derivation of the statistical properties are carried out here by using the (complex) baseband notation as usual. First, we will deal with the Rice process, which is used for the modelling of short-term fading.

6.1.1 Modelling and Analysis of Short-Term Fading

For the modelling of short-term fading, also called fast fading, we will consider the Rice process (3.21), i.e.,

$$\xi(t) = |\mu_\rho(t)| = |\mu(t) + m(t)|, \tag{6.1}$$

where the line-of-sight component $m(t)$ will again be described by (3.17), and $\mu(t)$ is the narrow-band complex-valued Gaussian random process introduced by (3.16), whose real and imaginary parts have zero-mean and identical variances $\sigma_{\mu_1}^2 = \sigma_{\mu_2}^2 = \sigma_0^2$.

Until now, we have assumed that the angles-of-arrival of the electromagnetic waves arriving at the receiving antenna are uniformly distributed over the interval $[0, 2\pi)$ and that the antenna has a circular-symmetrical radiation pattern. Hence, the Doppler power spectral density $S_{\mu\mu}(f)$ of the complex-valued process $\mu(t)$ has a symmetrical form (see (3.23)), and consequently, the two real-valued Gaussian random processes $\mu_1(t)$ and $\mu_2(t)$ are uncorrelated. In the following, we will drop this assumption. Instead, it is supposed that, e.g., by spatially limited obstacles, no electromagnetic waves with angles-of-arrival within the interval from α_0 to $2\pi - \alpha_0$ can arrive at the receiver, where α_0 shall be restricted to the interval $[\pi/2, 3\pi/2]$. This assumption is also justified if we use directional antennas or sector antennas, i.e., antennas with asymmetrical radiation patterns. The resulting asymmetrical Doppler power spectral density $S_{\mu\mu}(f)$ is then described as follows

$$S_{\mu\mu}(f) = \begin{cases} \dfrac{2\sigma_0^2}{\pi f_{\max}\sqrt{1-(f/f_{\max})^2}}, & -f_{\min} \leq f \leq f_{\max}, \\ 0, & \text{else,} \end{cases} \tag{6.2}$$

where $f_{\max}$ again denotes the maximum Doppler frequency, and $f_{\min} = -f_{\max}\cos\alpha_0$ lies within the range $0 \leq f_{\min} \leq f_{\max}$. Only for the special case $\alpha_0 = \pi$, i.e., $f_{\min} = f_{\max}$, we obtain the symmetrical Jakes power spectral density. In general, however, the shape of $S_{\mu\mu}(f)$ in (6.2) is asymmetrical, which results in a cross-correlation of the real-valued Gaussian random processes $\mu_1(t)$ and $\mu_2(t)$. In the following, we denote the Doppler power spectral density according to (6.2) as *left-sided restricted Jakes power spectral density*. With a given value for $f_{\max}$ and a suitable choice of $f_{\min}$, one can often achieve a better fitting to the Doppler spread of measured fading signals than with the conventional Jakes power spectral density whose Doppler spread is often too large in comparison with reality (see Subsection 6.1.5).

Figure 6.1 depicts the reference model for the Rice process $\xi(t)$, whose underlying complex-valued Gaussian random process is characterized by the left-sided restricted Jakes power spectral density introduced in (6.2).

From this figure, we conclude the relations

$$\mu_1(t) = \nu_1(t) + \nu_2(t) \tag{6.3}$$

and

$$\mu_2(t) = \check{\nu}_1(t) - \check{\nu}_2(t), \tag{6.4}$$

where $\nu_i(t)$ represents a coloured Gaussian random process, and its Hilbert transform is denoted by $\check{\nu}_i(t)$ $(i = 1, 2)$. Here, the spectral shaping of $\nu_i(t)$ is based on filtering of white Gaussian noise $n_i(t) \sim N(0, 1)$ by using an ideal filter whose transfer function is given by $H_i(f) = \sqrt{S_{\nu_i\nu_i}(f)}$. In the following, we assume that the white Gaussian random processes $n_1(t)$ and $n_2(t)$ are uncorrelated.

The autocorrelation function of $\mu(t) = \mu_1(t) + j\mu_2(t)$, which is generally defined by (2.107), can be expressed in terms of the autocorrelation and cross-correlation functions of $\mu_1(t)$ and

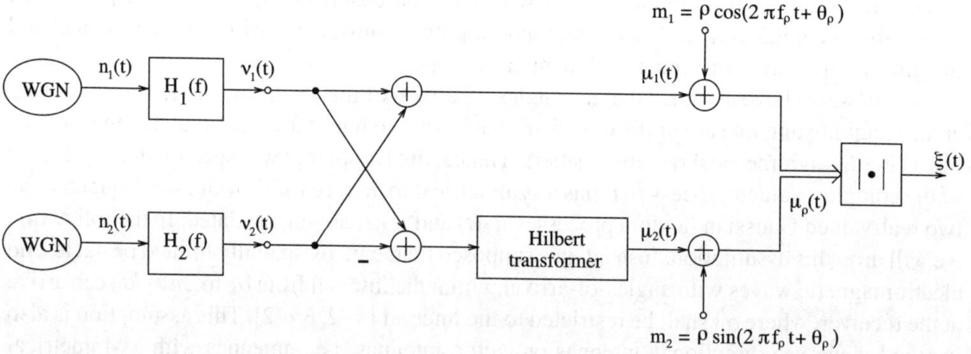

Figure 6.1 Reference model for Rice processes $\xi(t)$ with cross-correlated underlying Gaussian random processes $\mu_1(t)$ and $\mu_2(t)$.

$\mu_2(t)$ as follows

$$r_{\mu\mu}(\tau) = r_{\mu_1\mu_1}(\tau) + r_{\mu_2\mu_2}(\tau) + j(r_{\mu_1\mu_2}(\tau) - r_{\mu_2\mu_1}(\tau)). \tag{6.5}$$

Using the relations $r_{\nu_i\nu_i}(\tau) = r_{\check{\nu}_i\check{\nu}_i}(\tau)$ and $r_{\nu_i\check{\nu}_i}(\tau) = r_{\check{\nu}_i\nu_i}(-\tau) = -r_{\check{\nu}_i\nu_i}(\tau)$ (cf. also (2.135e) and (2.135c), respectively), we may write:

$$r_{\mu_1\mu_1}(\tau) = r_{\nu_1\nu_1}(\tau) + r_{\nu_2\nu_2}(\tau) = r_{\mu_2\mu_2}(\tau), \tag{6.6a}$$

$$r_{\mu_1\mu_2}(\tau) = r_{\nu_1\check{\nu}_1}(\tau) - r_{\nu_2\check{\nu}_2}(\tau) = -r_{\mu_2\mu_1}(\tau), \tag{6.6b}$$

so that (6.5) can be expressed as

$$r_{\mu\mu}(\tau) = 2[r_{\nu_1\nu_1}(\tau) + r_{\nu_2\nu_2}(\tau) + j(r_{\nu_1\check{\nu}_1}(\tau) - r_{\nu_2\check{\nu}_2}(\tau))]. \tag{6.7}$$

After the Fourier transform of (6.5) and (6.7), we obtain the following expressions for the Doppler power spectral density

$$S_{\mu\mu}(f) = S_{\mu_1\mu_1}(f) + S_{\mu_2\mu_2}(f) + j(S_{\mu_1\mu_2}(f) - S_{\mu_2\mu_1}(f)), \tag{6.8a}$$

$$S_{\mu\mu}(f) = 2[S_{\nu_1\nu_1}(f) + S_{\nu_2\nu_2}(f) + j(S_{\nu_1\check{\nu}_1}(f) - S_{\nu_2\check{\nu}_2}(f))]. \tag{6.8b}$$

For the Doppler power spectral densities $S_{\nu_i\nu_i}(f)$ and $S_{\nu_i\check{\nu}_i}(f)$ as well as for the corresponding autocorrelation functions $r_{\nu_i\nu_i}(\tau)$ and $r_{\nu_i\check{\nu}_i}(\tau)$, the following relations hold:

$$S_{\nu_1\nu_1}(f) = \frac{\sigma_0^2}{2\pi f_{\max}\sqrt{1-(f/f_{\max})^2}}, \tag{6.9a}$$

$$\bullet\!-\!\circ$$

$$r_{\nu_1\nu_1}(\tau) = \frac{\sigma_0^2}{2} J_0(2\pi f_{\max}\tau), \tag{6.9b}$$

$$S_{\nu_2\nu_2}(f) = \mathrm{rect}\,(f/f_{\min}) \cdot S_{\nu_1\nu_1}(f), \tag{6.9c}$$

$$\bullet\!-\!\circ$$

$$r_{\nu_2\nu_2}(\tau) = f_{\min}\sigma_0^2 J_0(2\pi f_{\max}\tau) * \mathrm{sinc}\,(2\pi f_{\min}\tau), \tag{6.9d}$$

$$S_{\nu_1\check{\nu}_1}(f) = -j\,\mathrm{sgn}\,(f) \cdot S_{\nu_1\nu_1}(f), \tag{6.9e}$$

$$\bullet\!-\!\circ$$

$$r_{\nu_1\check{\nu}_1}(\tau) = \frac{\sigma_0^2}{2} H_0(2\pi f_{\max}\tau), \tag{6.9f}$$

$$S_{\nu_2\check{\nu}_2}(f) = -j\,\mathrm{sgn}\,(f) \cdot S_{\nu_2\nu_2}(f), \tag{6.9g}$$

$$\bullet\!-\!\circ$$

$$r_{\nu_2\check{\nu}_2}(\tau) = f_{\min}\sigma_0^2 H_0(2\pi f_{\max}\tau) * \mathrm{sinc}\,(2\pi f_{\min}\tau), \tag{6.9h}$$

where $J_0(\cdot)$ and $H_0(\cdot)$ denote the zeroth order Bessel function of the first kind and the Struve function of zeroth order, respectively.[1] If we now substitute (6.9e) and (6.9g) into (6.8b), we can express $S_{\mu\mu}(f)$ in terms of $S_{\nu_i\nu_i}(f)$ as follows

$$S_{\mu\mu}(f) = 2[(1 + \text{sgn}\,(f)) \cdot S_{\nu_1\nu_1}(f) + (1 - \text{sgn}\,(f)) \cdot S_{\nu_2\nu_2}(f)]. \tag{6.10}$$

Figure 6.2 illustrates the shapes of $S_{\nu_1\nu_1}(f)$ and $S_{\nu_2\nu_2}(f)$ as well as the corresponding left-sided restricted Jakes power spectral density $S_{\mu\mu}(f)$.

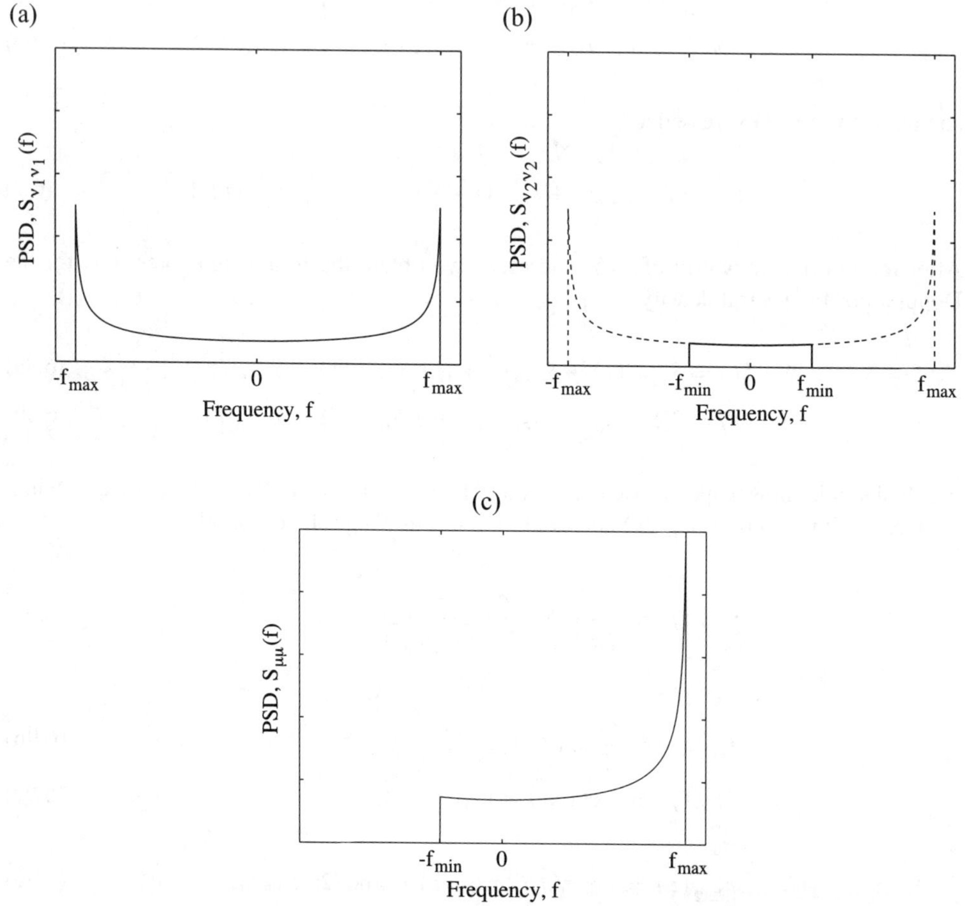

Figure 6.2 Doppler power spectral densities: (a) $S_{\nu_1\nu_1}(f)$, (b) $S_{\nu_2\nu_2}(f)$, and (c) the resulting left-sided restricted Jakes power spectral density.

[1] The rectangular function used in (6.9c) is defined by

$$\text{rect}(x) = \begin{cases} 1 & \text{for} \quad |x| < 1 \\ 1/2 & \text{for} \quad x = \pm 1 \\ 0 & \text{for} \quad |x| > 1 \end{cases}$$

and $\text{sinc}\,(x) = \sin(x)/x$ in (6.9d) denotes the sinc function.

In the following derivation of the statistical properties of $\xi(t) = |\mu_\rho(t)|$ and $\vartheta(t) = \arg\{\mu_\rho(t)\}$, we often make use of the abbreviations

$$\psi_0^{(n)} := \left. \frac{d^n}{d\tau^n} r_{\mu_1\mu_1}(\tau) \right|_{\tau=0} = \left. \frac{d^n}{d\tau^n} r_{\mu_2\mu_2}(\tau) \right|_{\tau=0} \tag{6.11a}$$

and

$$\phi_0^{(n)} := \left. \frac{d^n}{d\tau^n} r_{\mu_1\mu_2}(\tau) \right|_{\tau=0} \tag{6.11b}$$

for $n = 0, 1, 2$. Using (6.6) and (6.9), these characteristic quantities can be expressed as follows:

$$\psi_0^{(0)} = \psi_0 = \frac{\sigma_0^2}{2}\left[1 + \frac{2}{\pi}\arcsin(\kappa_0)\right], \tag{6.12a}$$

$$\psi_0^{(1)} = \dot{\psi}_0 = 0, \tag{6.12b}$$

$$\psi_0^{(2)} = \ddot{\psi}_0 = -(\pi\sigma_0 f_{\max})^2\left\{1 + \frac{2}{\pi}\left[\arcsin(\kappa_0) - \frac{1}{2}\sin(2\arcsin(\kappa_0))\right]\right\}, \tag{6.12c}$$

$$\phi_0^{(0)} = \phi_0 = 0, \tag{6.12d}$$

$$\phi_0^{(1)} = \dot{\phi}_0 = 2\sigma_0^2 f_{\max}\sqrt{1-\kappa_0^2}, \tag{6.12e}$$

$$\phi_0^{(2)} = \ddot{\phi}_0 = 0, \tag{6.12f}$$

where the overdot indicates the time derivative, and the parameter κ_0 denotes the frequency ratio

$$\kappa_0 = f_{\min}/f_{\max}, \quad 0 \leq \kappa_0 \leq 1. \tag{6.13}$$

One should note that the shape of $S_{\mu\mu}(f)$ is only symmetrical for the special case $\kappa_0 = 1$. In this case, the processes $\mu_1(t)$ and $\mu_2(t)$ are uncorrelated, and from (6.12a)–(6.12f) we obtain the relations $\psi_0 = \sigma_0^2$, $\ddot{\psi}_0 = -2(\pi\sigma_0 f_{\max})^2$, and $\dot{\phi}_0 = 0$, which we already know from Subsection 3.4.3.

A starting point for the derivation of the statistical properties of Rice processes $\xi(t)$ with asymmetrical Doppler power spectral densities is given by the joint probability density function of the processes $\mu_{\rho_1}(t)$, $\mu_{\rho_2}(t)$, $\dot{\mu}_{\rho_1}(t)$, and $\dot{\mu}_{\rho_2}(t)$ [see (3.19)] at the same point within the time t. This joint probability density function will here be denoted by $p_{\mu_{\rho_1}\mu_{\rho_2}\dot{\mu}_{\rho_1}\dot{\mu}_{\rho_2}}(x_1, x_2, \dot{x}_1, \dot{x}_2)$. It should be noted that $\mu_{\rho_i}(t)$ is a real-valued Gaussian random process with the time variant mean value $E\{\mu_{\rho_i}(t)\} = m_i(t)$ and the variance $\mathrm{Var}\{\mu_{\rho_i}(t)\} = \mathrm{Var}\{\mu_i(t)\} = r_{\mu_i\mu_i}(0) = \psi_0$. Consequently, its time derivative $\dot{\mu}_{\rho_i}(t)$ is a real-valued Gaussian random process too. However, this process is characterized by the mean value $E\{\dot{\mu}_{\rho_i}(t)\} = \dot{m}_i(t)$ and the variance $\mathrm{Var}\{\dot{\mu}_{\rho_i}(t)\} = \mathrm{Var}\{\dot{\mu}_i(t)\} = r_{\dot{\mu}_i\dot{\mu}_i}(0) = -\ddot{r}_{\mu_i\mu_i}(0) = -\ddot{\psi}_0$. It is also worth mentioning that the processes $\mu_{\rho_i}(t)$ and $\dot{\mu}_{\rho_i}(t)$ are correlated in pairs at the same time instant t. The joint probability density function $p_{\mu_{\rho_1}\mu_{\rho_2}\dot{\mu}_{\rho_1}\dot{\mu}_{\rho_2}}(x_1, x_2, \dot{x}_1, \dot{x}_2)$ can therefore be expressed by the multivariate Gaussian distribution in (2.36), i.e.,

$$p_{\mu_{\rho_1}\mu_{\rho_2}\dot{\mu}_{\rho_1}\dot{\mu}_{\rho_2}}(x_1, x_2, \dot{x}_1, \dot{x}_2) = \frac{e^{-\frac{1}{2}(\boldsymbol{x}-\boldsymbol{m})^T \boldsymbol{C}_{\mu_\rho}^{-1}(\boldsymbol{x}-\boldsymbol{m})}}{(2\pi)^2\sqrt{\det \boldsymbol{C}_{\mu_\rho}}}, \tag{6.14}$$

where $\boldsymbol{x}$ and $\boldsymbol{m}$ are the column vectors defined by

$$\boldsymbol{x} = \begin{pmatrix} x_1 \\ x_2 \\ \dot{x}_1 \\ \dot{x}_2 \end{pmatrix} \tag{6.15}$$

and

$$\boldsymbol{m} = \begin{pmatrix} E\{\mu_{\rho_1}(t)\} \\ E\{\mu_{\rho_2}(t)\} \\ E\{\dot{\mu}_{\rho_1}(t)\} \\ E\{\dot{\mu}_{\rho_2}(t)\} \end{pmatrix} = \begin{pmatrix} m_1(t) \\ m_2(t) \\ \dot{m}_1(t) \\ \dot{m}_2(t) \end{pmatrix} = \begin{pmatrix} \rho\cos(2\pi f_\rho t + \theta_\rho) \\ \rho\sin(2\pi f_\rho t + \theta_\rho) \\ -2\pi f_\rho \rho \sin(2\pi f_\rho t + \theta_\rho) \\ 2\pi f_\rho \rho \cos(2\pi f_\rho t + \theta_\rho) \end{pmatrix}, \tag{6.16}$$

respectively, and det $\mathbf{C}_{\mu_\rho}$ ($\boldsymbol{C}_{\mu_\rho}^{-1}$) denotes the determinant (inverse) of the covariance matrix

$$\boldsymbol{C}_{\mu_\rho} = \begin{pmatrix} C_{\mu_{\rho_1}\mu_{\rho_1}} & C_{\mu_{\rho_1}\mu_{\rho_2}} & C_{\mu_{\rho_1}\dot{\mu}_{\rho_1}} & C_{\mu_{\rho_1}\dot{\mu}_{\rho_2}} \\ C_{\mu_{\rho_2}\mu_{\rho_1}} & C_{\mu_{\rho_2}\mu_{\rho_2}} & C_{\mu_{\rho_2}\dot{\mu}_{\rho_1}} & C_{\mu_{\rho_2}\dot{\mu}_{\rho_2}} \\ C_{\dot{\mu}_{\rho_1}\mu_{\rho_1}} & C_{\dot{\mu}_{\rho_1}\mu_{\rho_2}} & C_{\dot{\mu}_{\rho_1}\dot{\mu}_{\rho_1}} & C_{\dot{\mu}_{\rho_1}\dot{\mu}_{\rho_2}} \\ C_{\dot{\mu}_{\rho_2}\mu_{\rho_1}} & C_{\dot{\mu}_{\rho_2}\mu_{\rho_2}} & C_{\dot{\mu}_{\rho_2}\dot{\mu}_{\rho_1}} & C_{\dot{\mu}_{\rho_2}\dot{\mu}_{\rho_2}} \end{pmatrix}. \tag{6.17}$$

The entries of the covariance matrix $\boldsymbol{C}_{\mu_\rho}$ can be calculated as follows

$$\boldsymbol{C}_{\mu_{\rho_i}^{(k)}\mu_{\rho_j}^{(\ell)}} = \boldsymbol{C}_{\mu_{\rho_i}^{(k)}\mu_{\rho_j}^{(\ell)}}(t_i, t_j) \tag{6.18a}$$

$$= E\{\left(\mu_{\rho_i}^{(k)}(t_i) - m_i^{(k)}(t_i)\right)\left(\mu_{\rho_j}^{(\ell)}(t_j) - m_j^{(\ell)}(t_j)\right)\} \tag{6.18b}$$

$$= E\{\mu_i^{(k)}(t_i)\mu_j^{(\ell)}(t_j)\} \tag{6.18c}$$

$$= r_{\mu_i^{(k)}\mu_j^{(\ell)}}(t_i, t_j) \tag{6.18d}$$

$$= r_{\mu_i^{(k)}\mu_j^{(\ell)}}(\tau), \tag{6.18e}$$

for all $i, j = 1, 2$ and $k, \ell = 0, 1$. The result in (6.18e) follows from (6.18d) due to the fact that the Gaussian random processes $\mu_i(t)$ and $\dot{\mu}_i(t)$ are wide-sense stationary. As a consequence, the autocorrelation and cross-correlation functions only depend on the time difference $\tau = t_j - t_i$, i.e., $r_{\mu_i^{(k)}\mu_j^{(\ell)}}(t_i, t_j) = r_{\mu_i^{(k)}\mu_j^{(\ell)}}(t_i, t_i + \tau) = r_{\mu_i^{(k)}\mu_j^{(\ell)}}(\tau)$. Studying Equations (6.17) and (6.18e), it now becomes clear that the covariance matrix $\boldsymbol{C}_{\mu_\rho}$ of the processes $\mu_{\rho_1}(t)$, $\mu_{\rho_2}(t)$, $\dot{\mu}_{\rho_1}(t)$, and $\dot{\mu}_{\rho_2}(t)$ is identical to the correlation matrix $\boldsymbol{R}_\mu$ of the processes $\mu_1(t)$, $\mu_2(t)$, $\dot{\mu}_1(t)$, and $\dot{\mu}_2(t)$, i.e., we may write

$$C_{\mu_\rho}(\tau) = R_\mu(\tau) = \begin{pmatrix} r_{\mu_1\mu_1}(\tau) & r_{\mu_1\mu_2}(\tau) & r_{\mu_1\dot{\mu}_1}(\tau) & r_{\mu_1\dot{\mu}_2}(\tau) \\ r_{\mu_2\mu_1}(\tau) & r_{\mu_2\mu_2}(\tau) & r_{\mu_2\dot{\mu}_1}(\tau) & r_{\mu_2\dot{\mu}_2}(\tau) \\ r_{\dot{\mu}_1\mu_1}(\tau) & r_{\dot{\mu}_1\mu_2}(\tau) & r_{\dot{\mu}_1\dot{\mu}_1}(\tau) & r_{\dot{\mu}_1\dot{\mu}_2}(\tau) \\ r_{\dot{\mu}_2\mu_1}(\tau) & r_{\dot{\mu}_2\mu_2}(\tau) & r_{\dot{\mu}_2\dot{\mu}_1}(\tau) & r_{\dot{\mu}_2\dot{\mu}_2}(\tau) \end{pmatrix}. \tag{6.19}$$

Concerning the entries of the correlation matrix $R_\mu(\tau)$, the following relations hold [41]:

$$r_{\mu_j\mu_i}(\tau) = r_{\mu_i\mu_j}(-\tau), \; r_{\mu_i\dot{\mu}_j}(\tau) = \dot{r}_{\mu_i\mu_j}(\tau), \tag{6.20a,b}$$

$$r_{\dot{\mu}_i\mu_j}(\tau) = -\dot{r}_{\mu_i\mu_j}(\tau), \; r_{\dot{\mu}_i\dot{\mu}_j}(\tau) = -\ddot{r}_{\mu_i\mu_j}(\tau), \tag{6.20c,d}$$

for all $i, j = 1, 2$.

For the derivation of the level-crossing rate and the average duration of fades, we have to consider the correlation properties of the processes $\mu_{\rho_i}^{(k)}(t_i)$ and $\mu_{\rho_j}^{(\ell)}(t_j)$ at the same time instant, i.e., $t_i = t_j$, and thus the time-difference variable $\tau = t_j - t_i$ is equal to zero. Therefore, in connection with (6.12a)–(6.12f), we can profit from the notation (6.11) enabling us to present the covariance matrix and the correlation matrix (6.19) as follows

$$C_{\mu_\rho}(0) = R_\mu(0) = \begin{pmatrix} \psi_0 & 0 & 0 & \dot{\phi}_0 \\ 0 & \psi_0 & -\dot{\phi}_0 & 0 \\ 0 & -\dot{\phi}_0 & -\ddot{\psi}_0 & 0 \\ \dot{\phi}_0 & 0 & 0 & -\ddot{\psi}_0 \end{pmatrix}. \tag{6.21}$$

After substituting (6.21) into the relation (6.14), we can now express the joint probability density function $p_{\mu_{\rho_1}\mu_{\rho_2}\dot{\mu}_{\rho_1}\dot{\mu}_{\rho_2}}(x_1, x_2, \dot{x}_1, \dot{x}_2)$ in terms of the quantities (6.12a)–(6.12f). For our purposes, however, it is advisable to first transform the Cartesian coordinates (x_1, x_2) into polar coordinates (z, θ). For that purpose, we consider the following system of equations:

$$z = \sqrt{x_1^2 + x_2^2}\,, \quad \dot{z} = \frac{x_1\dot{x}_1 + x_2\dot{x}_2}{\sqrt{x_1^2 + x_2^2}}, \tag{6.22a}$$

$$\theta = \arctan\left(\frac{x_2}{x_1}\right), \quad \dot{\theta} = \frac{x_1\dot{x}_2 - x_2\dot{x}_1}{x_1^2 + x_2^2}. \tag{6.22b}$$

For $z > 0$, $|\dot{z}| < \infty$, $|\theta| \le \pi$, and $|\dot{\theta}| < \infty$, this system of equations has the real-valued solutions

$$x_1 = z\cos\theta\,, \quad \dot{x}_1 = \dot{z}\cos\theta - \dot{\theta}z\sin\theta, \tag{6.23a}$$

$$x_2 = z\sin\theta\,, \quad \dot{x}_2 = \dot{z}\sin\theta + \dot{\theta}z\cos\theta. \tag{6.23b}$$

Applying the transformation rule (2.87) leads to the joint probability density function

$$p_{\xi\dot{\xi}\vartheta\dot{\vartheta}}(z, \dot{z}, \theta, \dot{\theta}) = |J|^{-1} p_{\mu_{\rho_1}\mu_{\rho_2}\dot{\mu}_{\rho_1}\dot{\mu}_{\rho_2}}(z\cos\theta,\ z\sin\theta,\ \dot{z}\cos\theta - \dot{\theta}z\sin\theta,\ \dot{z}\sin\theta + \dot{\theta}z\cos\theta), \tag{6.24}$$

where J denotes the Jacobian determinant

$$J = \begin{vmatrix} \frac{\partial z}{\partial x_1} & \frac{\partial z}{\partial x_2} & \frac{\partial z}{\partial \dot{x}_1} & \frac{\partial z}{\partial \dot{x}_2} \\ \frac{\partial \dot{z}}{\partial x_1} & \frac{\partial \dot{z}}{\partial x_2} & \frac{\partial \dot{z}}{\partial \dot{x}_1} & \frac{\partial \dot{z}}{\partial \dot{x}_2} \\ \frac{\partial \theta}{\partial x_1} & \frac{\partial \theta}{\partial x_2} & \frac{\partial \theta}{\partial \dot{x}_1} & \frac{\partial \theta}{\partial \dot{x}_2} \\ \frac{\partial \dot{\theta}}{\partial x_1} & \frac{\partial \dot{\theta}}{\partial x_2} & \frac{\partial \dot{\theta}}{\partial \dot{x}_1} & \frac{\partial \dot{\theta}}{\partial \dot{x}_2} \end{vmatrix} = \begin{vmatrix} \frac{\partial x_1}{\partial z} & \frac{\partial x_1}{\partial \dot{z}} & \frac{\partial x_1}{\partial \theta} & \frac{\partial x_1}{\partial \dot{\theta}} \\ \frac{\partial x_2}{\partial z} & \frac{\partial x_2}{\partial \dot{z}} & \frac{\partial x_2}{\partial \theta} & \frac{\partial x_2}{\partial \dot{\theta}} \\ \frac{\partial \dot{x}_1}{\partial z} & \frac{\partial \dot{x}_1}{\partial \dot{z}} & \frac{\partial \dot{x}_1}{\partial \theta} & \frac{\partial \dot{x}_1}{\partial \dot{\theta}} \\ \frac{\partial \dot{x}_2}{\partial z} & \frac{\partial \dot{x}_2}{\partial \dot{z}} & \frac{\partial \dot{x}_2}{\partial \theta} & \frac{\partial \dot{x}_2}{\partial \dot{\theta}} \end{vmatrix}^{-1} = -\frac{1}{z^2}. \tag{6.25}$$

After some further algebraic manipulations, we are now in the position to bring the desired joint probability density function $p_{\xi\dot{\xi}\vartheta\dot{\vartheta}}(z, \dot{z}, \theta, \dot{\theta})$ into the following form [75]

$$\begin{aligned} p_{\xi\dot{\xi}\vartheta\dot{\vartheta}}(z, \dot{z}, \theta, \dot{\theta}) = {} & \frac{z^2}{(2\pi)^2 \psi_0 \beta} e^{-\frac{z^2+\rho^2}{2\psi_0}} \cdot e^{\frac{z\rho}{\psi_0}\cos(\theta - 2\pi f_\rho t - \theta_\rho)} \\ & \cdot e^{-\frac{1}{2\beta}\left[\dot{z} - \sqrt{2\beta}\alpha\rho\sin(\theta - 2\pi f_\rho t - \theta_\rho)\right]^2} \\ & \cdot e^{-\frac{z^2}{2\beta}\left\{\dot{\theta} - \frac{\dot{\phi}_0}{\psi_0} - \sqrt{2\beta}\frac{\alpha\rho}{z}\cos(\theta - 2\pi f_\rho t - \theta_\rho)\right\}^2} \end{aligned} \tag{6.26}$$

for $z \geq 0$, $|\dot{z}| < \infty$, $|\theta| \leq \pi$, and $|\dot{\theta}| < \infty$, where

$$\alpha = \left(2\pi f_\rho - \frac{\dot{\phi}_0}{\psi_0}\right) \Big/ \sqrt{2\beta}, \tag{6.27}$$

$$\beta = -\ddot{\psi}_0 - \dot{\phi}_0^2/\psi_0. \tag{6.28}$$

The joint probability density function (6.26) represents a fundamental equation. With it, we will first determine the probability density function of the envelope and the phase of the process $\mu_\rho(t)$ in the following subsection, and then we proceed with the derivation of the level-crossing rate and the average duration of fades of the process $\xi(t) = |\mu_\rho(t)|$ by again making use of (6.26).

6.1.1.1 Probability Density Function of the Envelope and the Phase

Employing the rule (2.89) now allows us to calculate the probability density $p_\xi(z)$ of the process $\xi(t)$ from the joint probability density function $p_{\xi\dot{\xi}\vartheta\dot{\vartheta}}(z, \dot{z}, \theta, \dot{\theta})$. We therefore consider the threefold integral

$$p_\xi(z) = \int_{-\infty}^{\infty}\int_{-\pi}^{\pi}\int_{-\infty}^{\infty} p_{\xi\dot{\xi}\vartheta\dot{\vartheta}}(z, \dot{z}, \theta, \dot{\theta})\, d\dot{\theta}\, d\theta\, d\dot{z}, \quad z \geq 0. \tag{6.29}$$

Putting (6.26) in the above expression results in the well-known Rice distribution

$$p_\xi(z) = \begin{cases} \frac{z}{\psi_0} e^{-\frac{z^2+\rho^2}{2\psi_0}} I_0\left(\frac{z\rho}{\psi_0}\right), & z \geq 0, \\ 0, & z < 0. \end{cases} \tag{6.30}$$

Owing to the correlation of the processes $\mu_1(t)$ and $\mu_2(t)$, this result cannot be regarded as a matter of course, as we will see later in Subsection 6.2. Since the probability density in (6.30) is independent of the quantity $\dot{\phi}_0$ in the present case, it follows that the correlation between the processes $\mu_1(t)$ and $\mu_2(t)$ has no influence on the probability density function of the envelope $\xi(t)$. However, one should note that the parameter κ_0 determining the Doppler bandwidth exerts an influence on the variance ψ_0 of the processes $\mu_1(t)$ and $\mu_2(t)$ [cf. (6.12a)] and consequently determines the behaviour of (6.30) decisively.

The probability density function of the phase $\vartheta(t)$, denoted by $p_\vartheta(\theta)$, can be calculated in a similar way. Substituting (6.26) into

$$p_\vartheta(\theta) = \int_0^\infty \int_{-\infty}^\infty \int_{-\infty}^\infty p_{\xi\dot{\xi}\vartheta\dot{\vartheta}}(z,\dot{z},\theta,\dot{\theta})\,d\dot{\theta}\,d\dot{z}\,dz, \quad -\pi \le \theta \le \pi, \tag{6.31}$$

results in

$$p_\vartheta(\theta) = p_\vartheta(\theta;t) = \frac{e^{-\frac{\rho^2}{2\psi_0}}}{2\pi}\left\{1 + \sqrt{\frac{\pi}{2\psi_0}}\rho\cos(\theta - 2\pi f_\rho t - \theta_\rho)\cdot e^{\frac{\rho^2\cos^2(\theta-2\pi f_\rho t-\theta_\rho)}{2\psi_0}}\right.$$
$$\left.\left[1 + \operatorname{erf}\left(\frac{\rho\cos(\theta - 2\pi f_\rho t - \theta_\rho)}{\sqrt{2\psi_0}}\right)\right]\right\}, \quad -\pi \le \theta \le \pi. \tag{6.32}$$

Even in this instance, the cross-correlation function $r_{\mu_1\mu_2}(\tau)$ has no influence on the probability density function $p_\vartheta(\theta)$, since $p_\vartheta(\theta)$ is independent of $\dot{\phi}_0$. For the special case $\kappa_0 = 1$, we have $\psi_0 = \sigma_0^2$, and thus from (6.32) it follows (3.56). The investigation of further special cases, for example: (i) $f_\rho = 0$, (ii) $\rho \to 0$, and (iii) $\rho \to \infty$ leads to the statements made below Equation (3.56), which will not be repeated here for the sake of brevity.

6.1.1.2 Level-Crossing Rate and Average Duration of Fades

The derivation of the level-crossing rate using

$$N_\xi(r) = \int_0^\infty \dot{z}\,p_{\xi\dot{\xi}}(r,\dot{z})\,d\dot{z} \tag{6.33}$$

requires the knowledge of the joint probability density function $p_{\xi\dot{\xi}}(z,\dot{z})$ of the stationary processes $\xi(t)$ and $\dot{\xi}(t)$ at the same time instant t at the signal level $z = r$. For the joint probability density function $p_{\xi\dot{\xi}}(z,\dot{z})$, one finds, after substituting (6.26) into

$$p_{\xi\dot{\xi}}(z,\dot{z}) = \int_{-\pi}^{\pi}\int_{-\infty}^{\infty} p_{\xi\dot{\xi}\vartheta\dot{\vartheta}}(z,\dot{z},\theta,\dot{\theta})\,d\dot{\theta}\,d\theta, \quad z \ge 0, \quad |\dot{z}| < \infty, \tag{6.34}$$

the result

$$p_{\xi\dot{\xi}}(z,\dot{z}) = \frac{z}{\psi_0\sqrt{\beta}(2\pi)^{3/2}}\cdot e^{-\frac{z^2+\rho^2}{2\psi_0}}\int_{-\pi}^{\pi} e^{\frac{z\rho}{\psi_0}\cos\theta}$$
$$\cdot e^{-\frac{1}{2\beta}[\dot{z}-\sqrt{2\beta}\alpha\rho\sin\theta]^2}\,d\theta, \quad z \ge 0, \quad |\dot{z}| < \infty, \tag{6.35}$$

where α, β, and ψ_0 are the quantities introduced by (6.27), (6.28), and (6.12a), respectively. Obviously, the processes $\xi(t)$ and $\dot{\xi}(t)$ are in general statistically dependent, because $p_{\xi\dot{\xi}}(z, \dot{z}) \neq p_\xi(z) \cdot p_{\dot{\xi}}(\dot{z})$ holds. Only for the special case $\alpha = 0$, i.e., (i) if the two real-valued Gaussian random processes $\mu_1(t)$ and $\mu_2(t)$ are uncorrelated and f_ρ is equal to zero, or (ii) if f_ρ and $\dot{\phi}_0$ are related by $f_\rho = \dot{\phi}_0/(2\pi\psi_0)$, we obtain statistically independent processes $\xi(t)$ and $\dot{\xi}(t)$, since from (6.35) it follows

$$p_{\xi\dot{\xi}}(z, \dot{z}) = p_\xi(z) \cdot p_{\dot{\xi}}(\dot{z})$$
$$= \frac{z}{\psi_0} e^{-\frac{z^2+\rho^2}{2\psi_0}} I_0\left(\frac{z\rho}{\psi_0}\right) \cdot \frac{e^{-\frac{\dot{z}^2}{2\beta}}}{\sqrt{2\pi\beta}}, \tag{6.36}$$

where β in this case again represents $\beta = -\ddot{\psi}_0 - \dot{\phi}_0^2/\psi_0 \geq 0$. Hence, for $\alpha = 0$, the joint probability density function $p_{\xi\dot{\xi}}(z, \dot{z})$ is equal to the product of the probability density functions of the stochastic processes $\xi(t)$ and $\dot{\xi}(t)$, which are Rice and Gaussian distributed, respectively.

With the joint probability density function (6.35), we are now able to calculate the level-crossing rate of Rice processes whose underlying complex-valued Gaussian process has cross-correlated inphase and quadrature components. In this way, after substituting (6.35) in the definition (6.33) and performing some tedious algebraic manipulations, we finally obtain the result [75]

$$N_\xi(r) = \frac{r\sqrt{2\beta}}{\pi^{3/2}\psi_0} e^{-\frac{r^2+\rho^2}{2\psi_0}} \int_0^{\pi/2} \cosh\left(\frac{r\rho}{\psi_0}\cos\theta\right)$$
$$\cdot \left\{e^{-(\alpha\rho\sin\theta)^2} + \sqrt{\pi}\alpha\rho\sin(\theta)\,\mathrm{erf}\,(\alpha\rho\sin\theta)\right\} d\theta, \quad r \geq 0, \tag{6.37}$$

where the characteristic quantities α, β, and ψ_0 are given in the form (6.27), (6.28), and (6.12a), respectively. Further simplifications are not possible; the remaining integral has to be solved numerically. Let us again consider the special case $\kappa_0 = 1$. Then, we obtain: $\alpha = 2\pi f_\rho/\sqrt{2\beta}$, $\beta = -\ddot{\psi}_0 = -\ddot{r}_{\mu_i\mu_i}(0)$, and $\psi_0 = \sigma_0^2$, so that the level-crossing rate $N_\xi(r)$ given above results in the expression introduced in (3.63), as was to be expected.

Let us assume that the line-of-sight component tends to zero, i.e., $\rho \to 0$, which leads to $\xi(t) \to \zeta(t)$. Then, it follows that (6.37) tends to

$$N_\xi(r) = \sqrt{\frac{\beta}{2\pi}} \cdot \frac{r}{\psi_0} e^{-\frac{r^2}{2\psi_0}}, \quad r \geq 0, \tag{6.38}$$

where the quantity β is given by (6.28). The above result demonstrates that the level-crossing rate is proportional to the Rayleigh distribution. This property has also been mentioned in [70]. Owing to (6.28), the proportionality factor $\sqrt{\beta/(2\pi)}$ is not only determined by the curvature of the autocorrelation function at the origin $\tau = 0$ ($\ddot{\psi}_0 = \ddot{r}_{\mu_i\mu_i}(0)$), but also decisively by the gradient of the cross-correlation function at $\tau = 0$ ($\dot{\phi}_0 = \dot{r}_{\mu_1\mu_2}(0)$).

Now, let $\rho \neq 0$ and $f_\rho = \dot{\phi}_0/(2\pi\psi_0)$. Then, it follows $\alpha = 0$ [see (6.27)], and from (6.37) we obtain the level-crossing rate $N_\xi(r)$ according to (3.66), if σ_0^2 is replaced by ψ_0 in that equation, i.e.,

$$N_\xi(r) = \sqrt{\frac{\beta}{2\pi}} \cdot \frac{r}{\psi_0} e^{-\frac{r^2+\rho^2}{2\phi_0}} I_0\left(\frac{r\rho}{\psi_0}\right), \quad r \geq 0, \tag{6.39}$$

where β is again given by (6.28).

In connection with the Jakes power spectral density, the level-crossing rate $N_\xi(r)$ described by (6.37) is always proportional to the maximum Doppler frequency $f_{\max}$. The normalization of $N_\xi(r)$ to $f_{\max}$ therefore eliminates the influence of both the velocity of the vehicle and the carrier frequency. The influence of the parameters κ_0 and σ_0^2 on the normalized level-crossing rate $N_\xi(r)/f_{\max}$ is illustrated in Figure 6.3(a) and in Figure 6.3(b), respectively.

For the calculation of the average duration of fades $T_{\xi_-}(r)$, we will be guided by the basic relation (2.120), i.e.,

$$T_{\xi_-}(r) = \frac{F_{\xi_-}(r)}{N_\xi(r)}, \tag{6.40}$$

where $F_{\xi_-}(r)$ denotes the cumulative distribution function of the Rice process $\xi(t)$, which states the probability that $\xi(t)$ takes on a value which is lower than or equal to the signal level r. Using (6.30), the following integral expression can be derived for $F_{\xi_-}(r)$

$$\begin{aligned} F_{\xi_-}(r) &= P\{\xi(t) \leq r\} \\ &= \int_0^r p_\xi(z)\, dz \\ &= \frac{e^{-\frac{\rho^2}{2\psi_0}}}{\psi_0} \int_0^r z e^{-\frac{z^2}{2\psi_0}} I_0\left(\frac{z\rho}{\psi_0}\right) dz. \end{aligned} \tag{6.41}$$

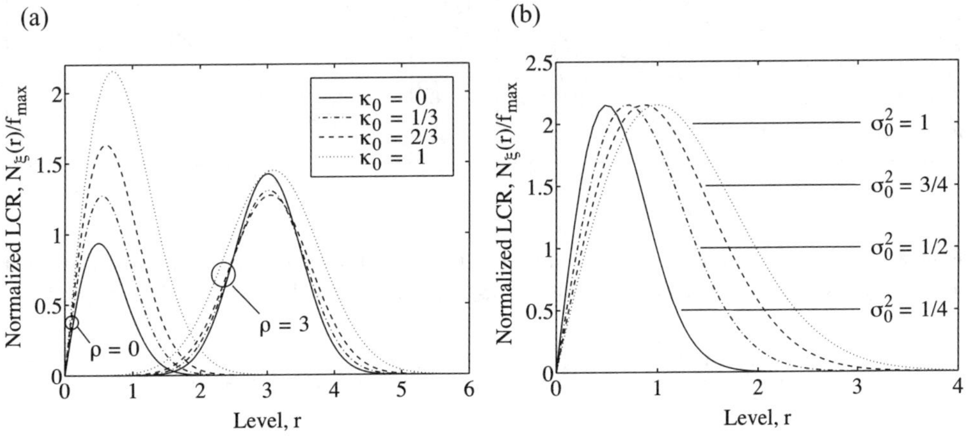

Figure 6.3 Normalized level-crossing rate $N_\xi(r)/f_{\max}$ of Rice processes (with cross-correlated underlying Gaussian random processes): (a) $\kappa_0 = f_{\min}/f_{\max}$ ($\sigma_0^2 = 1$) and (b) σ_0^2 ($\rho = 0, \kappa_0 = 1$).

The average duration of fades of Rice processes $\xi(t)$ with cross-correlated inphase and quadrature components $\mu_1(t)$ and $\mu_2(t)$ is, thus, the quotient (6.40) of the integral expressions (6.41) and (6.37), which have to be solved numerically.

The influence of the parameters κ_0 and σ_0^2 on the normalized average duration of fades $T_{\xi_-}(r) \cdot f_{\max}$ is depicted in Figures 6.4(a) and 6.4(b), respectively.

6.1.2 Modelling and Analysis of Long-Term Fading

Measurements have shown that the statistical properties of slow fading are quite similar to those of a lognormal process [63, 193, 194]. With such a process, the slow fluctuations of the local mean value of the received signal, which are determined by shadowing effects, can be reproduced. In the following, we will denote lognormal processes by $\lambda(t)$. Lognormal processes can be derived by means of the nonlinear transform

$$\lambda(t) = e^{\sigma_3 \nu_3(t) + m_3} \tag{6.42}$$

from a third real-valued Gaussian random process $\nu_3(t)$ with the expected value $E\{\nu_3(t)\} = 0$ and the variance $\text{Var}\{\nu_3(t)\} = 1$. Fitting the model behaviour to the statistics of real-world channels, the model parameters m_3 and σ_3 can be used in connection with the parameters of the Rice process $(\sigma_0^2, f_{\max}, f_{\min}, \rho, f_\rho)$. We assume henceforth that the stochastic process $\nu_3(t)$ is statistically independent of the processes $\nu_1(t)$ and $\nu_2(t)$. Figure 6.5 illustrates the reference model for the lognormal process $\lambda(t)$ introduced by the Equation (6.42).

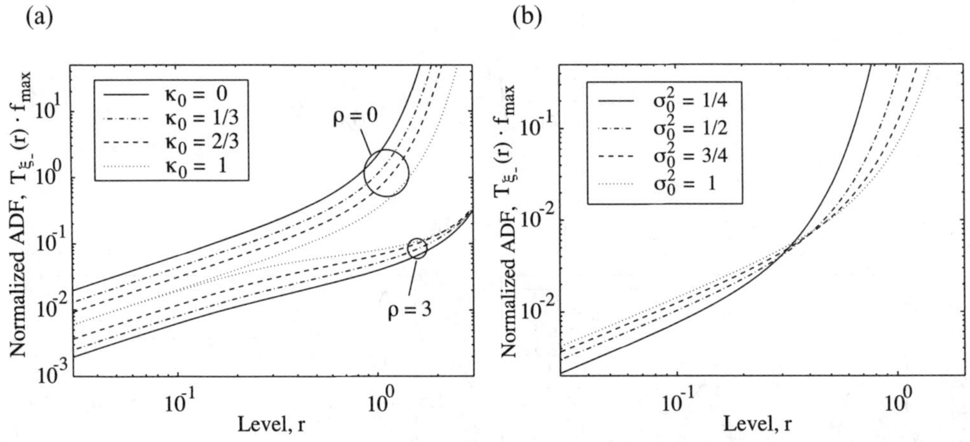

Figure 6.4 Normalized average duration of fades $T_\xi(r) \cdot f_{\max}$ of Rice processes (with cross-correlated underlying Gaussian random processes): (a) $\kappa_0 = f_{\min}/f_{\max}$ $(\sigma_0^2 = 1)$ and (b) σ_0^2 $(\rho = 0, \kappa_0 = 1)$.

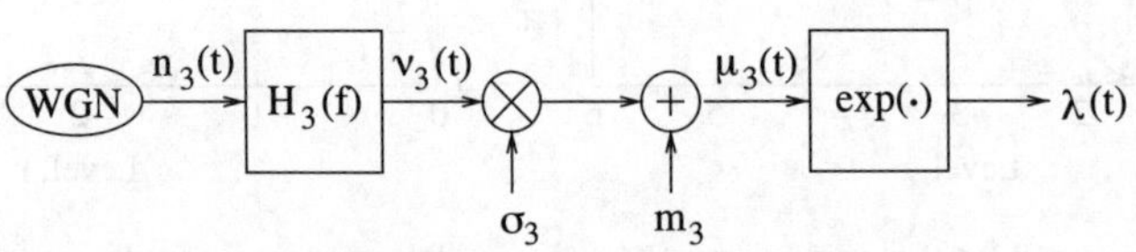

Figure 6.5 Reference model for lognormal processes $\lambda(t)$.

In this figure, the process $\nu_3(t)$ is obtained by filtering white Gaussian noise $n_3(t) \sim N(0, 1)$ with a real-valued lowpass filter, whose transfer function $H_3(f)$ is related to the power spectral density $S_{\nu_3\nu_3}(f)$ of the process $\nu_3(t)$ according to (2.131f), i.e., $H_3(f) = \sqrt{S_{\nu_3\nu_3}(f)}$. For $S_{\nu_3\nu_3}(f)$, the Gaussian power spectral density is assumed in the form [cf. also (3.26)]

$$S_{\nu_3\nu_3}(f) = \frac{1}{\sqrt{2\pi}\sigma_c} e^{-\frac{f^2}{2\sigma_c^2}}, \tag{6.43}$$

where the 3-dB-cut-off frequency $f_c = \sigma_c\sqrt{2\ln 2}$ is in general much smaller than the maximum Doppler frequency $f_{\max}$. In order to simplify the notation, we introduce the symbol κ_c for the frequency ratio $f_{\max}/f_c$, i.e., $\kappa_c = f_{\max}/f_c$. A study on modified Suzuki processes has shown [70] that both the parameter κ_c as well as the exact shape of the power spectral density of $\nu_3(t)$ have no considerable influence on the relevant statistical properties of modified Suzuki processes if $\kappa_c > 10$. Other types of power spectral densities $S_{\nu_3\nu_3}(f)$ than that presented in (6.43) have been introduced, for example, in [70, 183] and [179], where RC-lowpass filters and Butterworth filters of third order, respectively, have been proposed for shaping the spectral characteristics of $\nu_3(t)$.

The autocorrelation function $r_{\nu_3\nu_3}(\tau)$ of the process $\nu_3(t)$ can be described after calculating the inverse Fourier transform of (6.43) by

$$r_{\nu_3\nu_3}(\tau) = e^{-2(\pi\sigma_c\tau)^2}. \tag{6.44}$$

Next, let us consider the lognormal process $\lambda(t)$ [see (6.42)]. The autocorrelation function $r_{\lambda\lambda}(\tau)$ of this process can be expressed in terms of $r_{\nu_3\nu_3}(\tau)$ as follows

$$\begin{aligned} r_{\lambda\lambda}(\tau) &= E\{\lambda(t)\cdot\lambda(t+\tau)\} \\ &= E\{e^{2m_3+\sigma_3[\nu_3(t)+\nu_3(t+\tau)]}\} \\ &= \int_{-\infty}^{\infty}\int_{-\infty}^{\infty} e^{2m_3+\sigma_3(x_1+x_2)} \cdot p_{\nu_3\nu_3'}(x_1, x_2)\, dx_1\, dx_2, \end{aligned} \tag{6.45}$$

where

$$p_{\nu_3\nu_3'}(x_1, x_2) = \frac{1}{2\pi\sqrt{1 - r^2_{\nu_3\nu_3}(\tau)}} e^{-\frac{x_1^2 - 2r_{\nu_3\nu_3}(\tau)x_1x_2 + x_2^2}{2[1-r^2_{\nu_3\nu_3}(\tau)]}} \tag{6.46}$$

describes the joint probability density function of the Gaussian random process $\nu_3(t)$ at two different time instants $t_1 = t$ and $t_2 = t + \tau$. After substituting (6.46) in (6.45) and solving the double integral, the autocorrelation function $r_{\lambda\lambda}(\tau)$ can be expressed in closed form by

$$r_{\lambda\lambda}(\tau) = e^{2m_3+\sigma_3^2[1+r_{\nu_3\nu_3}(\tau)]}. \tag{6.47}$$

With this relation, the mean power of the lognormal process $\lambda(t)$ can easily be determined. We obtain $r_{\lambda\lambda}(0) = e^{2(m_3+\sigma_3^2)}$.

The power spectral density $S_{\lambda\lambda}(f)$ of the lognormal process $\lambda(t)$ can now be expressed in terms of the power spectral density $S_{\nu_3\nu_3}(f)$ of $\nu_3(t)$ as follows [195]

$$
\begin{aligned}
S_{\lambda\lambda}(f) &= \int_{-\infty}^{\infty} r_{\lambda\lambda}(\tau)\, e^{-j2\pi f\tau} d\tau \\
&= e^{2m_3+\sigma_3^2} \cdot \left\{ \delta(f) + \int_{-\infty}^{\infty} \left(e^{\sigma_3^2 r_{\nu_3\nu_3}(\tau)} - 1 \right) e^{-j2\pi f\tau}\, d\tau \right\} \\
&= e^{2m_3+\sigma_3^2} \cdot \left[\delta(f) + \sum_{n=1}^{\infty} \frac{\sigma_3^{2n}}{n!} \cdot \frac{S_{\nu_3\nu_3}\left(\frac{f}{\sqrt{n}}\right)}{\sqrt{n}} \right].
\end{aligned}
\tag{6.48}
$$

This result shows that the power spectral density $S_{\lambda\lambda}(f)$ of the lognormal process $\lambda(t)$ consists of a weighted Dirac delta function at the origin $f = 0$ and of an infinite sum of strictly monotonously decreasing power spectral densities $S_{\nu_3\nu_3}(f/\sqrt{n})/\sqrt{n}$. It should be noted that $S_{\nu_3\nu_3}(f/\sqrt{n})/\sqrt{n}$ follows directly from (6.43), if the quantity σ_c is replaced by $\sqrt{n}\sigma_3$ there.

The probability density function $p_\lambda(y)$ of the lognormal process $\lambda(t)$ is described by the lognormal distribution (2.51), i.e.,

$$
p_\lambda(y) = \begin{cases} \dfrac{1}{\sqrt{2\pi}\sigma_3 y} e^{-\frac{(\ln y - m_3)^2}{2\sigma_3^2}}, & y \geq 0, \\ 0, & y < 0, \end{cases} \tag{6.49}
$$

with the expected value and the variance according to (2.53) and (2.54), respectively.

For the calculation of the level-crossing rate and the average duration of fades of (extended) Suzuki processes, we require the knowledge of the joint probability density function of the lognormal process $\lambda(t)$ and its corresponding time derivative $\dot{\lambda}(t)$ at the same time instant t. This joint probability density function, denoted by $p_{\lambda\dot{\lambda}}(y, \dot{y})$, will be derived briefly in the following. We start from the underlying Gaussian random process $\nu_3(t)$ and its time derivative $\dot{\nu}_3(t)$. For the cross-correlation function of these two processes, it follows that $r_{\nu_3\dot{\nu}_3}(0) = 0$ holds, i.e., $\nu_3(t_1)$ and $\dot{\nu}_3(t_2)$ are uncorrelated at the same time instant $t = t_1 = t_2$. Since $\nu_3(t)$ and, hence, also $\dot{\nu}_3(t)$ are Gaussian random processes, it follows from the uncorrelatedness that these processes are statistically independent. For the joint probability density function $p_{\nu_3\dot{\nu}_3}(x, \dot{x})$ of the processes $\nu_3(t)$ and $\dot{\nu}_3(t)$, we can therefore write

$$
p_{\nu_3\dot{\nu}_3}(x, \dot{x}) = p_{\nu_3}(x) \cdot p_{\dot{\nu}_3}(\dot{x}) = \frac{e^{-\frac{x^2}{2}}}{\sqrt{2\pi}} \cdot \frac{e^{-\frac{\dot{x}^2}{2\gamma}}}{\sqrt{2\pi\gamma}}, \tag{6.50}
$$

where

$$
\gamma = r_{\dot{\nu}_3\dot{\nu}_3}(0) = -\ddot{r}_{\nu_3\nu_3}(0) = (2\pi\sigma_c)^2 \tag{6.51}
$$

denotes the variance of the process $\dot{\nu}_3(t)$.

Similar to the scheme described in detail in Subsection 6.1.1, we can take $p_{\nu_3\dot{\nu}_3}(x, \dot{x})$ as our starting point to determine the desired joint probability density function $p_{\lambda\dot{\lambda}}(y, \dot{y})$. The nonlinear mapping (6.42) in connection with the following substitution of variables

$$x = \frac{\ln y - m_3}{\sigma_3}, \quad \dot{x} = \frac{\dot{y}}{\sigma_3 y} \tag{6.52a,b}$$

yields the expression $J = (\sigma_3 y)^2$ for the Jacobian determinant (6.25). With the transformation rule (2.87), we then obtain the following result for the joint probability density function $p_{\lambda\dot{\lambda}}(y, \dot{y})$

$$p_{\lambda\dot{\lambda}}(y, \dot{y}) = \frac{e^{-\frac{(\ln y - m_3)^2}{2\sigma_3^2}}}{\sqrt{2\pi}\,\sigma_3 y} \cdot \frac{e^{-\frac{\dot{y}^2}{2\gamma(\sigma_3 y)^2}}}{\sqrt{2\pi\gamma}\,\sigma_3 y}. \tag{6.53}$$

This result shows that the processes $\lambda(t)$ and $\dot{\lambda}(t)$ are statistically dependent, although the underlying Gaussian processes $\nu_3(t)$ and $\dot{\nu}_3(t)$ are statistically independent.

6.1.3 The Stochastic Extended Suzuki Process of Type I

The extended Suzuki process (Type I), denoted by $\eta(t)$, was introduced in [75] as a product process of a Rice process $\xi(t)$ [see (6.1)] with cross-correlated underlying Gaussian random processes $\mu_1(t)$ and $\mu_2(t)$ and a lognormal process $\lambda(t)$ [see (6.42)], i.e.,

$$\eta(t) = \xi(t) \cdot \lambda(t). \tag{6.54}$$

Figure 6.6 shows the structure of the reference model described by the extended Suzuki process (Type I) $\eta(t)$ for a frequency-nonselective mobile radio channel.

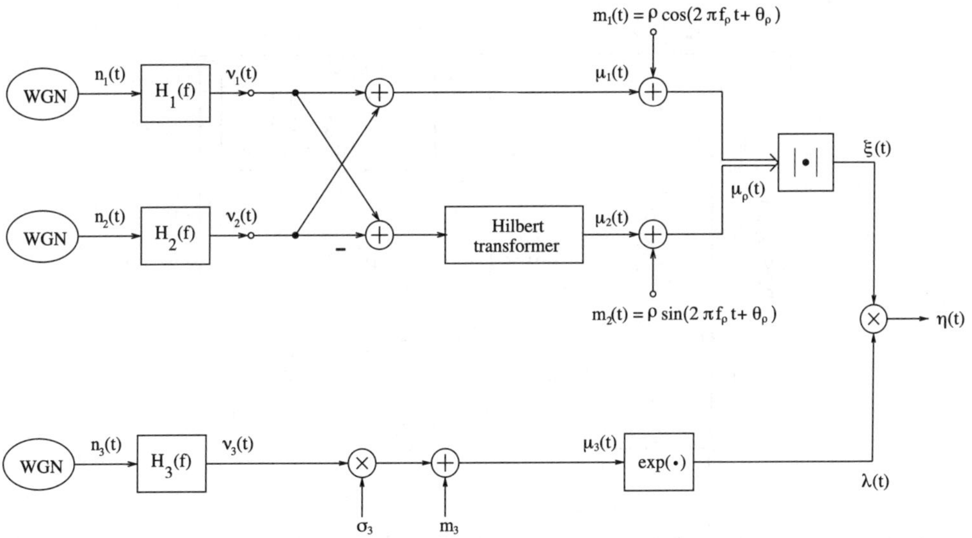

Figure 6.6 Reference model for extended Suzuki processes (Type I).

The probability density function $p_\eta(z)$ of the extended Suzuki process $\eta(t)$ can be calculated by means of the relation [41]

$$p_\eta(z) = \int_{-\infty}^{\infty} \frac{1}{|y|} \, p_{\xi\lambda}\left(\frac{z}{y}, y\right) dy, \tag{6.55}$$

where $p_{\xi\lambda}(x, y)$ is the joint probability density function of the processes $\xi(t)$ and $\lambda(t)$ at the same time instant t. According to our assumption, the coloured Gaussian random processes $\nu_1(t)$, $\nu_2(t)$, and $\nu_3(t)$ are mutually statistically independent. Consequently, the Rice process $\xi(t)$ and the lognormal process $\lambda(t)$ are also statistically independent, so that for the joint probability density function $p_{\xi\lambda}(x, y)$ it follows: $p_{\xi\lambda}(x, y) = p_\xi(x) \cdot p_\lambda(y)$. Hence, the multiplicative relation between the processes $\xi(t)$ and $\lambda(t)$ leads to the following integral equation for the probability density function of extended Suzuki processes

$$p_\eta(z) = \frac{z}{\sqrt{2\pi}\,\psi_0\sigma_3} \int_0^{\infty} \frac{1}{y^3} \, e^{-\frac{(z/y)^2+\rho^2}{2\psi_0}} \, I_0\left(\frac{z\rho}{y\psi_0}\right) e^{-\frac{(\ln y - m_3)^2}{2\sigma_3^2}} \, dy, \quad z \geq 0. \tag{6.56}$$

For $\rho = 0$, it should be noted that the probability density function (6.56) can be reduced to the (classical) Suzuki distribution (2.55) introduced in [26]. The influence of the parameters ρ and σ_3 on the behaviour of $p_\xi(z)$ can be concluded from Figure 6.7.

Studying (6.56), one can clearly see that $p_\eta(z)$ solely depends on the quantities ψ_0, ρ, σ_3, and m_3. Accordingly, the exact shape of the power spectral density of the complex-valued Gaussian random process $\mu(t)$ and especially the cross-correlation of the processes $\mu_1(t)$ and $\mu_2(t)$ have no influence on the probability density function of the extended Suzuki process. For the fitting of $p_\eta(z)$ according to (6.56) to a given measured probability density function by merely optimizing these model parameters, there is a risk that the statistics of real-world channels are reproduced insufficiently by the channel model.

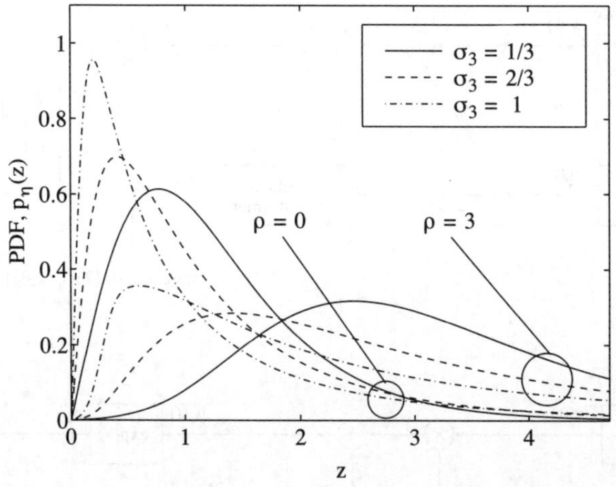

Figure 6.7 Probability density function $p_\eta(z)$ for various values of the parameters ρ and σ_3 ($\psi_0 = 1$, $m_3 = -\sigma_3^2/2$).

In the following, we therefore study the level-crossing rate $N_\eta(r)$ of the process $\eta(t)$, i.e.,

$$N_\eta(r) = \int_0^\infty \dot{z}\, p_{\eta\dot{\eta}}(r, \dot{z})\, d\dot{z}, \tag{6.57}$$

which requires the knowledge of the joint probability density function $p_{\eta\dot{\eta}}(z, \dot{z})$ of the process $\eta(t)$ and its time derivative $\dot{\eta}(t)$ at the same time instant t. This joint probability density can be derived by substituting the Equations (6.35) and (6.53) obtained for $p_{\xi\dot{\xi}}(x, \dot{x})$ and $p_{\lambda\dot{\lambda}}(y, \dot{y})$, respectively, into the relation [183]

$$p_{\eta\dot{\eta}}(z, \dot{z}) = \int_0^\infty \int_{-\infty}^\infty \frac{1}{y^2} p_{\xi\dot{\xi}}\left(\frac{z}{y}, \frac{\dot{z}}{y} - \frac{z}{y^2}\dot{y}\right) p_{\lambda\dot{\lambda}}(y, \dot{y})\, d\dot{y}\, dy, \quad z \geq 0,\ |\dot{z}| < \infty. \tag{6.58}$$

Hence, after some tedious algebraic manipulations, we find the expression

$$p_{\eta\dot{\eta}}(z, \dot{z}) = \frac{z}{(2\pi)^{\frac{3}{2}} \psi_0 \sqrt{\beta}} \int_0^\infty \frac{e^{-\frac{(z/y)^2+\rho^2}{2\psi_0}}}{y^3 K(z, y)} \cdot \frac{e^{-\frac{(\ln y - m_3)^2}{2\sigma_3^2}}}{\sqrt{2\pi}\, \sigma_3 y} \cdot$$
$$\int_0^{2\pi} e^{\frac{z\rho}{y\psi_0}\cos\theta} \cdot e^{-\frac{\left(\dot{z} - \sqrt{2\beta}\alpha y \rho \sin\theta\right)^2}{2\beta y^2 K^2(z,y)}}\, d\theta\, dy, \quad z \geq 0, \quad |\dot{z}| < \infty, \tag{6.59}$$

where

$$K(z, y) = \sqrt{1 + \frac{\gamma}{\beta}\left(\frac{z\sigma_3}{y}\right)^2}. \tag{6.60}$$

After substituting (6.59) into (6.57), we then obtain the following final result for the level-crossing rate $N_\eta(r)$ of the extended Suzuki process of Type I

$$N_\eta(r) = \frac{r\sqrt{2\beta}}{\pi^{3/2}\psi_0} \cdot \int_0^\infty \frac{K(r, y)}{y} \cdot \frac{e^{-\frac{(\ln y - m_3)^2}{2\sigma_3^2}}}{\sqrt{2\pi}\, \sigma_3 y} \cdot e^{-\frac{(r/y)^2+\rho^2}{2\psi_0}} \cdot \int_0^{\pi/2} \cosh\left(\frac{r\rho}{y\psi_0}\cos\theta\right)$$
$$\left\{ e^{-\left(\alpha\rho\frac{\sin\theta}{K(r,y)}\right)^2} + \sqrt{\pi}\,\alpha\rho \frac{\sin\theta}{K(r, y)} \operatorname{erf}\left[\alpha\rho\frac{\sin\theta}{K(r, y)}\right]\right\} d\theta\, dy, \tag{6.61}$$

where α, β, and γ again are the quantities introduced by (6.27), (6.28), and (6.51), respectively, and ψ_0 is determined by (6.12a). Exactly due to α and β, the influence of the shape of the Doppler power spectral density is now taken into consideration, because α depends on $\dot{\phi}_0$ and β is a function of $\dot{\phi}_0$ and $\ddot{\psi}_0$. A detailed analysis of (6.61) here also shows that $N_\eta(r)$ is again proportional to the maximum Doppler frequency and thus to the speed of the vehicle as well.

Moreover, we are interested in some special cases. Assuming $\sigma_3 \to 0$, then the lognormal distribution (6.49) converges to the probability density function $p_\lambda(y) = \delta(y - e^{m_3})$. Consequently, especially in case $m_3 = 0$, the level-crossing rate $N_\eta(r)$ tends to $N_\xi(r)$ according to (6.37).

In the case of a missing line-of-sight component, i.e., $\rho = 0$, the level-crossing rate of modified Suzuki processes follows from (6.61)

$$N_\eta(r)\big|_{\rho=0} = \sqrt{\frac{\beta}{2\pi}}\frac{r}{\psi_0}\int_0^\infty \frac{K(r,y)}{y}\, p_\lambda(y)\, e^{-\frac{r^2}{2\psi_0 y^2}}\, dy$$

$$= \sqrt{\frac{\beta}{2\pi}}\int_0^\infty K(r,y)\, p_\zeta(r/y)\, p_\lambda(y)\, dy \qquad (6.62)$$

as stated in [70, 183]. It should also be mentioned that for $\rho \neq 0$, the two cases

$$\text{(i)}\quad f_\rho = \dot{\phi}_0/(2\pi\psi_0), \qquad (6.63a)$$

$$\text{(ii)}\quad f_\rho = 0 \quad\text{and}\quad \dot{\phi}_0 = 0 \qquad (6.63b)$$

are equivalent with respect to the level-crossing rate $N_\eta(r)$, because we then have $\alpha = 0$ due to (6.27), which allows us to deduce from (6.61) in both cases the same expression

$$N_\eta(r)\big|_{\alpha=0} = \sqrt{\frac{\beta}{2\pi}}\frac{r}{\psi_0}\int_0^\infty \frac{K(r,y)}{y}\, p_\lambda(y)\, e^{-\frac{(r/y)^2+\rho^2}{2\psi_0}}\, I_0\left(\frac{r\rho}{y\psi_0}\right) dy$$

$$= \sqrt{\frac{\beta}{2\pi}}\int_0^\infty K(r,y)\, p_\xi(r/y)\, p_\lambda(y)\, dy. \qquad (6.64)$$

One should note, however, that the cases (i) and (ii) result in different values for β. Under the condition (i), the general relation (6.28) is valid for β, whereas this equation can be simplified in case of (ii) to $\beta = -\ddot{\psi}_0$.

At the end of this subsection, we also derive the cumulative distribution function $F_{\eta_-}(r) = P\{\eta(t) \leq r\}$, which is required for the calculation of the average duration of fades of extended Suzuki processes of Type I

$$T_{\eta_-}(r) = \frac{F_{\eta_-}(r)}{N_\eta(r)}. \qquad (6.65)$$

Thus, using (6.56), we obtain

$$F_{\eta_-}(r) = \int_0^r p_\eta(z)\, dz$$

$$= \frac{1}{\sqrt{2\pi}\,\psi_0\sigma_3}\int_0^\infty\int_0^r \frac{z}{y^3} e^{-\frac{(z/y)^2+\rho^2}{2\psi_0}}\, I_0\left(\frac{z\rho}{y\psi_0}\right) e^{-\frac{(\ln y - m_3)^2}{2\sigma_3^2}}\, dz\, dy$$

$$= 1 - \int_0^\infty Q_1\left(\frac{\rho}{\sqrt{\psi_0}}, \frac{r}{y\sqrt{\psi_0}}\right) p_\lambda(y)\, dy, \qquad (6.66)$$

where $Q_1(.,.)$ (see [11, p. 44]) is the *generalized Marcum Q-function* defined by

$$Q_m(a,b) = \int_b^{\infty} z \left(\frac{z}{a}\right)^{m-1} e^{-\frac{z^2+a^2}{2}} I_{m-1}(az)\,dz, \quad m = 1, 2, \ldots \tag{6.67}$$

In order to illustrate the results found in this section, let us consider the parameter study shown in Figures 6.8(a)–6.8(d). Figures 6.8(a) and 6.8(b) depict the normalized level-crossing rate $N_\eta(r)/f_{\max}$ calculated according to (6.61) for several values of the parameters m_3 and σ_3. The graphs of the corresponding normalized average duration of fades $T_{\eta_-}(r) \cdot f_{\max}$ are presented in Figures 6.8(c) and 6.8(d).

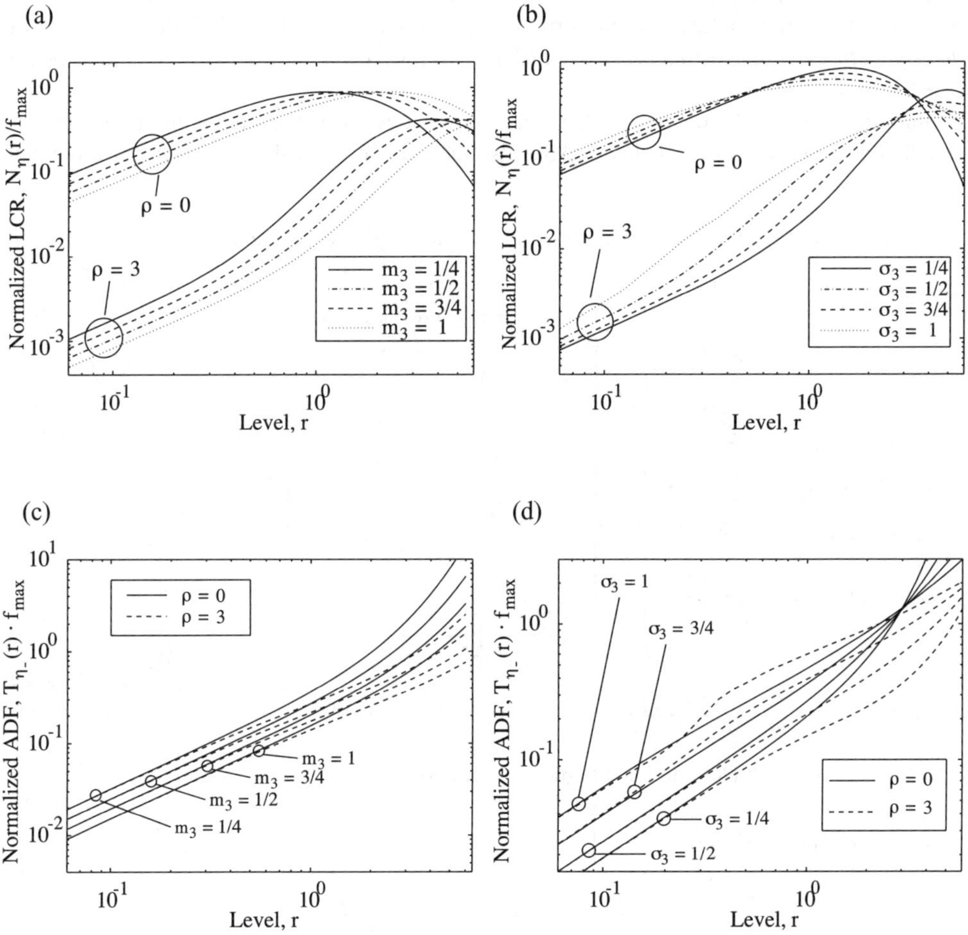

Figure 6.8 Normalized level-crossing rate $N_\eta(r)/f_{\max}$ of extended Suzuki processes (Type I) for several values of (a) m_3 ($\sigma_3 = 1/2$) and (b) σ_3 ($m_3 = 1/2$) as well as (c) and (d) the corresponding normalized average duration of fades $T_{\eta_-}(r) \cdot f_{\max}$ ($\kappa_0 = 1$, $\psi_0 = 1$).

6.1.4 The Deterministic Extended Suzuki Process of Type I

In the preceding subsection, we have seen that the reference model for the extended Suzuki process of Type I is based on the use of three real-valued coloured Gaussian random processes $\nu_i(t)$ or $\mu_i(t)$ $(i = 1, 2, 3)$ (see Figure 6.6). We now make use of the principle of deterministic channel modelling explained in Section 4.1, and approximate the ideal Gaussian random processes $\nu_i(t)$ by

$$\tilde{\nu}_i(t) = \sum_{n=1}^{N_i} c_{i,n} \cos(2\pi f_{i,n} t + \theta_{i,n}), \quad i = 1, 2, 3. \tag{6.68}$$

In the following, we therefore assume that the processes $\tilde{\nu}_1(t)$, $\tilde{\nu}_2(t)$, and $\tilde{\nu}_3(t)$ are uncorrelated in pairs. The uncorrelatedness condition can easily be fulfilled by nearly all of the parameter computation methods discussed in Section 5.1. After a few elementary network transformations, the continuous-time structure shown in Figure 6.9 enabling the simulation of *deterministic extended Suzuki processes of Type I* follows from the stochastic reference model (see Figure 6.6).

Studying Figure 6.9, we notice that not only the design of the digital filters, which are usually employed for spectral shaping, but also the realization of the Hilbert transformer can

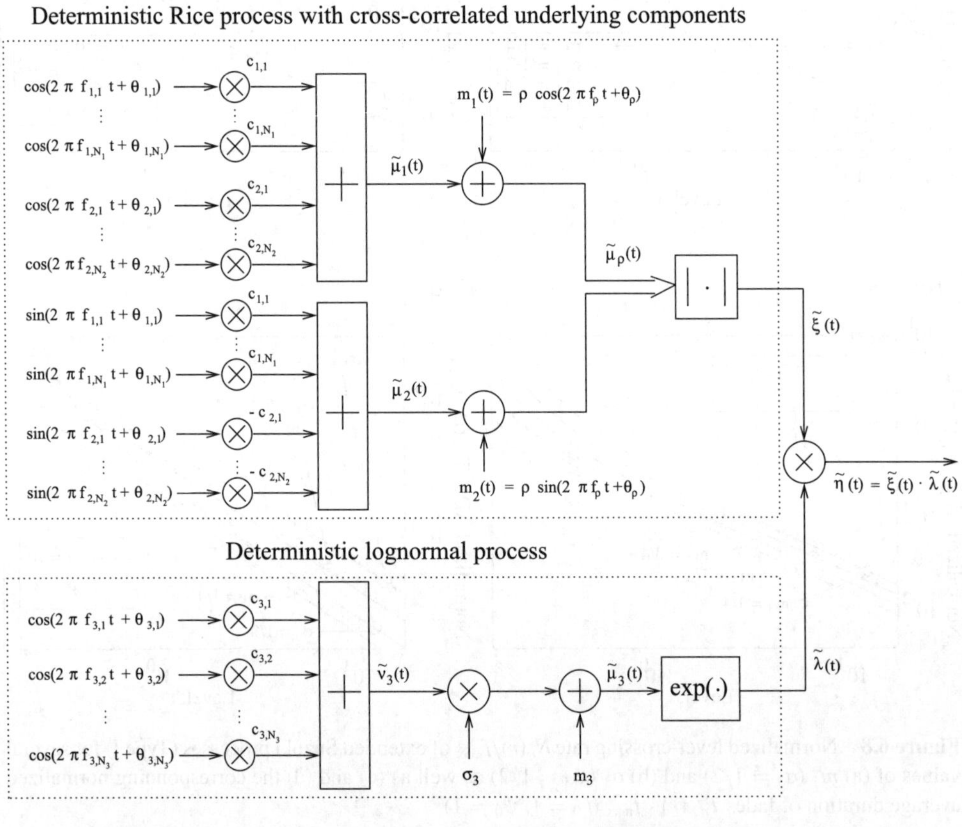

Figure 6.9 Deterministic simulation model for extended Suzuki processes (Type I).

be avoided. Moreover, deterministic simulation models offer the advantage that all relations derived for the reference model before such as, for instance, the expressions for the probability density function $p_\eta(z)$, the level-crossing rate $N_\eta(r)$, and the average duration of fades $T_{\eta_-}(r)$ can be used to describe approximately the behaviour of the deterministic extended Suzuki process $\tilde{\eta}(t)$. In all those expressions, which are of interest for us, we therefore only have to replace the characteristic quantities of the reference model ψ_0, $\ddot{\psi}_0$, and $\dot{\phi}_0$ by the corresponding quantities of the simulation model, i.e.,

$$\tilde{\psi}_0 = \tilde{r}_{\mu_1\mu_1}(0) = \tilde{r}_{\nu_1\nu_1}(0) + \tilde{r}_{\nu_2\nu_2}(0) = \tilde{r}_{\mu_2\mu_2}(0), \tag{6.69a}$$

$$\ddot{\tilde{\psi}}_0 = \ddot{\tilde{r}}_{\mu_1\mu_1}(0) = \ddot{\tilde{r}}_{\nu_1\nu_1}(0) + \ddot{\tilde{r}}_{\nu_2\nu_2}(0) = \ddot{\tilde{r}}_{\mu_2\mu_2}(0), \tag{6.69b}$$

$$\dot{\tilde{\phi}}_0 = \dot{\tilde{r}}_{\mu_1\mu_2}(0) = \dot{\tilde{r}}_{\nu_1\check{\nu}_1}(0) - \dot{\tilde{r}}_{\nu_2\check{\nu}_2}(0) = -\dot{\tilde{r}}_{\mu_2\mu_1}(0), \tag{6.69c}$$

where the tilde ($\sim$) refers to the fact that the underlying processes are deterministic processes. These quantities determine the statistical behaviour of $\tilde{\eta}(t)$ decisively and can be calculated explicitly in a simple way. With the autocorrelation function

$$\tilde{r}_{\nu_i\nu_i}(\tau) = \sum_{n=1}^{N_i} \frac{c_{i,n}^2}{2} \cos(2\pi f_{i,n}\tau), \quad i = 1, 2, 3, \tag{6.70}$$

and the property (2.135a), it then follows from (6.69a)–(6.69c) [192]:

$$\tilde{\psi}_0 = \sum_{n=1}^{N_1} \frac{c_{1,n}^2}{2} + \sum_{n=1}^{N_2} \frac{c_{2,n}^2}{2}, \tag{6.71a}$$

$$\ddot{\tilde{\psi}}_0 = -2\pi^2 \left[\sum_{n=1}^{N_1} (c_{1,n} f_{1,n})^2 + \sum_{n=1}^{N_2} (c_{2,n} f_{2,n})^2 \right], \tag{6.71b}$$

$$\dot{\tilde{\phi}}_0 = \pi \left[\sum_{n=1}^{N_1} c_{1,n}^2 f_{1,n} - \sum_{n=1}^{N_2} c_{2,n}^2 f_{2,n} \right]. \tag{6.71c}$$

Throughout Chapter 6, we will employ exclusively the method of exact Doppler spread described in detail in Subsection 5.1.7 for the computation of the model parameters $c_{i,n}$ and $f_{i,n}$. The phases $\theta_{i,n} \in (0, 2\pi]$ are assumed to be realizations (outcomes) of a uniformly distributed random generator. For the method of exact Doppler spread, however, we have to take into account that this procedure was derived originally for the classical Jakes power spectral density ($\kappa_0 = 1$). Its application to the restricted Jakes power spectral density ($\kappa_0 \leq 1$) requires a slight modification. For the discrete Doppler frequencies $f_{i,n}$, we now have [75]

$$f_{i,n} = \begin{cases} f_{\max} \sin\left[\dfrac{\pi}{2N_1}\left(n - \dfrac{1}{2}\right)\right], & i = 1, \quad n = 1, 2, \ldots, N_1, \\ f_{\max} \sin\left[\dfrac{\pi}{2N_2'}\left(n - \dfrac{1}{2}\right)\right], & i = 2, \quad n = 1, 2, \ldots, N_2, \end{cases} \tag{6.72}$$

where

$$N_2' = \left\lceil \frac{N_2}{\frac{2}{\pi}\arcsin(\kappa_0)} \right\rceil \tag{6.73}$$

is an auxiliary variable that depends on the frequency ratio $\kappa_0 = f_{\min}/f_{\max}$. In connection with (6.72), the quantity N'_2 restricts the discrete Doppler frequencies $f_{2,n}$ to the relevant interval $(0, f_{\min}]$. It should be pointed out that the actual required number of sinusoids $N_2\ (\leq N'_2)$, which is necessary for the realization of $\tilde{\nu}_2(t)$, is still to be defined by the user. We therefore call the auxiliary variable N'_2 the *virtual number of sinusoids* of $\tilde{\nu}_2(t)$. Moreover, the path gains $c_{i,n}$ are also affected by this modification, particularly since a power adaptation is necessary. Now, the path gains read as follows

$$c_{i,n} = \begin{cases} \sigma_0\sqrt{1/N_1}, & i = 1, \quad n = 1, 2, \ldots, N_1, \\ \sigma_0\sqrt{1/N'_2}, & i = 2, \quad n = 1, 2, \ldots, N_2. \end{cases} \tag{6.74}$$

The computation of the discrete Doppler frequencies $f_{3,n}$ of the third deterministic Gaussian process $\tilde{\nu}_3(t)$, whose power spectral density is Gaussian shaped, can be accomplished by means of (5.89a) and (5.89b). After the adaptation of these equations to the notation used here, we obtain the following set of equations

$$\frac{2n-1}{2N_3} - \operatorname{erf}\left(\frac{f_{3,n}}{\sqrt{2}\sigma_c}\right) = 0, \quad \forall n = 1, 2, \ldots, N_3 - 1, \tag{6.75a}$$

and

$$f_{3,N_3} = \sqrt{\frac{\gamma N_3}{(2\pi)^2} - \sum_{n=1}^{N_3-1} f_{3,n}^2}, \tag{6.75b}$$

where the meaning of $\sigma_c = f_{\max}/(\kappa_c\sqrt{2\ln 2})$ follows from (6.43), and the parameter γ is defined by (6.51). Due to $\nu_3(t) \sim N(0, 1)$, we compute $c_{3,n}$ according to the formula $c_{3,n} = \sqrt{2/N_3}$ for all $n = 1, 2, \ldots, N_3$.

When using the method of exact Doppler spread, we obtain the results shown in Figure 6.10 as a function of $N_1 = N_2 = N_i$ for the convergence behaviour and for the approximation quality of the normalized characteristic quantities $\ddot{\tilde{\psi}}_0/f_{\max}^2$ and $\dot{\tilde{\phi}}_0/f_{\max}$. Figures 6.10(a)

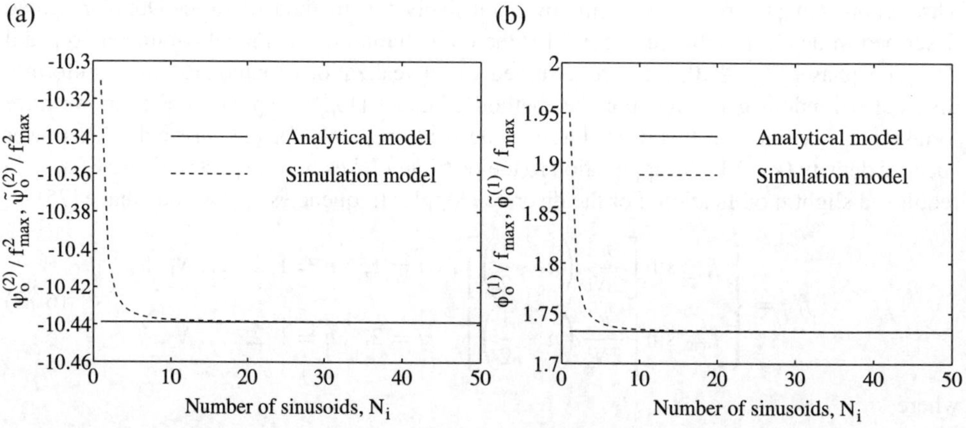

Figure 6.10 Illustration of (a) $\ddot{\psi}_0/f_{\max}^2$ and $\ddot{\tilde{\psi}}_0/f_{\max}^2$ as well as (b) $\dot{\phi}_0/f_{\max}$ and $\dot{\tilde{\phi}}_0/f_{\max}$ (MEDS, $\sigma_0^2 = 1$, $\kappa_0 = 1/2$).

and 6.10(b) also show us that in all cases where $N_i \geq 10$ holds, the deviations between the presented characteristic quantities of the simulation model and the ones of the reference model can be ignored.

Let $N_i \geq 7$, then (6.61) can be considered as an excellent approximation for the level-crossing rate of the simulation model $\tilde{N}_\eta(r)$, if the characteristic quantities of the reference model $(\psi_0, \ddot{\psi}_0, \dot{\phi}_0)$ and (α, β, γ) are replaced by the corresponding quantities of the simulation model $(\tilde{\psi}_0, \ddot{\tilde{\psi}}_0, \dot{\tilde{\phi}}_0)$ and $(\tilde{\alpha}, \tilde{\beta}, \tilde{\gamma})$, respectively. The same of course also holds for the average duration of fades $\tilde{T}_{\eta_-}(r)$ of the simulation model. Hence, $\tilde{N}_\eta(r)$ and $\tilde{T}_{\eta_-}(r)$ must not necessarily be determined by lengthy and time-consuming simulation runs, but they can be determined directly by solving the integral equation (6.61) and making use of (6.65) in conjunction with (6.71a)–(6.71c). Nevertheless, if $\tilde{N}_\eta(r)$ $(\tilde{T}_{\eta_-}(r))$ is determined subsequently by means of simulation of the fading envelope $\tilde{\eta}(t)$, then this only serves to confirm the correctness of the obtained theoretical results. In the following subsection, we will see that the deviations between $\tilde{N}_\eta(r)$ and $N_\eta(r)$ are in fact extremely small, so that a deeper study of the derivations does not seem to be appropriate at this point.

6.1.5 *Applications and Simulation Results*

In this subsection, we show how the statistics of the channel model can be adapted to the statistics of real-world channels by optimizing the relevant parameters of the reference model. As we are not satisfied with an adaptation of the first order statistics only, we include the statistics of the second order in the design procedure as well. Starting from the fitted reference model, the parameters of the corresponding deterministic simulation model will be determined afterwards. At the end of this subsection, the proposed procedure will be verified by means of simulations.

The measurement results of the complementary cumulative distribution function[2] $F^\star_{\eta_+}(r)$ [Figure 6.11(a)] and the level-crossing rate $N^\star_\eta(r)$ [Figure 6.11(b)] considered here were taken from the literature [196]. For the measurement experiments carried out therein, a helicopter equipped with an 870 MHz transmitter and a vehicle with a receiver were used to simulate a real-world satellite channel. Concerning the relative location of the helicopter and the mobile receiver, the elevation angle was held constant at 15°. One test route led through regions, in which the line-of-sight component was heavily shadowed, another through regions with light shadowing. The measurement results of this so-called *equivalent satellite channel* have also been used in [179]. Therefore, they offer a suitable basis for a fair comparison of the procedures. Further reports on measurement results of real-world satellite channels can be found, e.g., in [197–200].

Now let us combine all relevant model parameters, which determine decisively the statistical properties of the extended Suzuki process (Type I), into a parameter vector denoted and defined by $\mathbf{\Omega} := (\sigma_0, \kappa_0, \rho, f_\rho, \sigma_3, m_3)$. In practice, the frequency ratio $\kappa_c = f_{\max}/f_c$ is in general greater than 10, so that according to the statements made in Subsection 6.1.2, this parameter exerts no influence on the first and second order statistics of $\eta(t)$. This is the reason, why κ_c has not been included in the parameter vector $\mathbf{\Omega}$. Without restriction of generality, we will therefore arbitrarily set the value of the frequency ratio κ_c to 20 and choose *a priori* $\theta_\rho = 0$.

---

[2] Recall that the complementary cumulative distribution function $F_{\eta_+}(r) = P\{\eta(t) > r\}$ and the cumulative distribution function $F_{\eta_-}(r) = P\{\eta(t) \leq r\}$ are related by $F_{\eta_+}(r) = 1 - F_{\eta_-}(r)$.

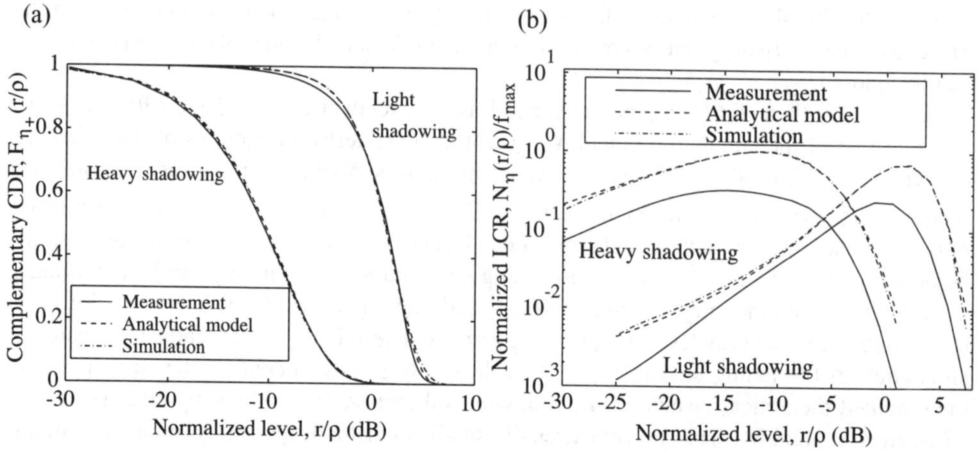

Figure 6.11 (a) Complementary cumulative distribution function $F_{\eta_+}(r/\rho)$ and (b) normalized level-crossing rate $N_\eta(r/\rho)/f_{\max}$ for regions with heavy and light shadowing (without optimization of κ_0 and f_ρ). The measurement results were adopted from [196].

As a suitable measure of the deviations between the complementary distribution functions $F_{\eta_+}(r/\rho)$ and $F^\star_{\eta_+}(r/\rho)$ as well as between the normalized level-crossing rates $N_\eta(r/\rho)/f_{\max}$ and $N^\star_\eta(r/\rho)/f_{\max}$, we introduce the following error function

$$E_2(\boldsymbol{\Omega}) := \left\{ \sum_{m=1}^{M_r} \left[W_1\left(\frac{r_m}{\rho}\right) \left(F^\star_{\eta_+}\left(\frac{r_m}{\rho}\right) - F_{\eta_+}\left(\frac{r_m}{\rho}\right) \right) \right]^2 \right\}^{1/2} + \frac{1}{f_{\max}} \left\{ \sum_{m=1}^{M_r} \left[W_2\left(\frac{r_m}{\rho}\right) \left(N^\star_\eta\left(\frac{r_m}{\rho}\right) - N_\eta\left(\frac{r_m}{\rho}\right) \right) \right]^2 \right\}^{1/2}, \qquad (6.76)$$

where M_r is the number of different signal levels r_m at which the measurements were taken. In addition, $W_1(\cdot)$ and $W_2(\cdot)$ denote two weighting functions, which are chosen to be proportional to the reciprocals of $F^\star_{\eta_+}(\cdot)$ and $N^\star_\eta(\cdot)$, respectively. The optimization of the components of the parameter vector $\boldsymbol{\Omega}$ is carried out numerically by applying the quasi-Newton procedure according to the Fletcher-Powell algorithm [162].

We first perform the optimization by using the classical Jakes power spectral density. Therefore, we keep the parameter $\kappa_0 = f_{\min}/f_{\max}$ constant at the value $\kappa_0 = 1$ during the minimization. Furthermore, we also fix f_ρ to the value $f_\rho = 0$, so that the extended Suzuki model simplifies to the conventional Rice-lognormal model. Now, we are confronted with the problem that there are no free parameters available for the optimization of the normalized level-crossing rate $N_\eta(r/\rho)/f_{\max}$, because all the remaining model parameters (σ_0, ρ, σ_3, m_3) are used completely for the optimization of the complementary cumulative distribution function $F_{\eta_+}(r/\rho)$. In other words, a better approximation of the second order statistics is only possible at the expense of a worse approximation of the first order statistics. We will not make this compromise at this point. For the moment, we are content with the approximation of $F_{\eta_+}(r/\rho)$, and temporarily set $W_2(r/\rho)$ equal to zero. The results of the parameter optimization for regions with light and heavy shadowing are listed in Table 6.1.

Table 6.1 The optimized parameters of the reference model for areas with heavy and light shadowing (without optimization of κ_0 and f_ρ).

Shadowing	σ_0	ρ	σ_3	m_3
Heavy	0.1847	0.0554	0.1860	0.3515
Light	0.3273	0.9383	0.0205	0.1882

Studying Figure 6.11(a), where the resulting complementary cumulative distribution function $F_{\eta_+}(r/\rho)$ is depicted, one can see that this function can be fitted very closely to the given measurement results. However, it is apparent from Figure 6.11(b) that these satisfying results cannot be obtained for the normalized level-crossing rate $N_\eta(r/\rho)/f_{\max}$. The deviations from the measurement results are partly more than 300 per cent in this case. The deeper reason of this mismatching is due to the far too high Doppler spread of the Jakes power spectral density.

For comparison and in order to confirm the results found, Figures 6.11(a) and 6.11(b) also show the results obtained from a discrete-time simulation of the extended Suzuki process. The deterministic processes $\tilde{\nu}_1(t)$, $\tilde{\nu}_2(t)$, and $\tilde{\nu}_3(t)$ were in this case designed by applying the techniques described in the preceding Subsection 6.1.4 (MEDS with $N_1 = 15$, $N_2 = 16$, and $N_3 = 15$).

The next step is to enable a reduction of the Doppler bandwidth and thus of the Doppler spread as well, by including the parameter κ_0 in the optimization procedure. In order to exploit the full flexibility of the channel model, the optimization of the parameter f_ρ will now also be permitted within the range $-f_{\min} \leq f_\rho \leq f_{\max}$. The numerical minimization of the error function (6.76) then results in the components of the parameter vector $\mathbf{\Omega}$, as presented in Table 6.2. With these parameters, which have been optimized with respect to both $F_{\eta_+}(r/\rho)$ and $N_\eta(r/\rho)/f_{\max}$, the behaviour of $F_{\eta_+}(r/\rho)$ remains almost unchanged (cf. Figures 6.11(a) and 6.12(a).) However, the actual advantages of the extended Suzuki model of Type I first become apparent by studying the statistics of second order. Observe that due to the model extension, the normalized level-crossing rate $N_\eta(r/\rho)/f_{\max}$ of the reference model can now obviously be fitted to the measurement results better than in the case $\kappa_0 = 1$ and $f_\rho = 0$ (cf. Figures 6.11(b) and 6.12(b)).

It should also be mentioned that the Rice factor (3.53) of the extended Suzuki model (Type I) now reads as

$$c_R = \frac{\rho^2}{2\psi_0} = \frac{\rho^2}{\sigma_0^2[1 + \frac{2}{\pi}\arcsin(\kappa_0)]}. \tag{6.77}$$

Using the parameters listed in Table 6.2, we obtain the values $c_R = -5.15$ dB (heavy shadowing) and $c_R = 6.82$ dB (light shadowing) for the Rice factor.

Table 6.2 The optimized parameters of the reference channel model for areas with heavy and light shadowing (with optimization of κ_0 and f_ρ).

Shadowing	σ_0	κ_0	ρ	σ_3	m_3	$f_\rho/f_{\max}$
Heavy	0.2022	4.4E-11	0.1118	0.1175	0.4906	0.6366
Light	0.4497	5.9E-08	0.9856	0.0101	0.0875	0.7326

The verification of the analytical results is again established by means of simulations. Owing to the fact that $\kappa_0 = f_{\min}/f_{\max}$ is very small in both cases determined by light and heavy shadowing (see Table 6.2), the influence of $\nu_2(t)$ or $\tilde{\nu}_2(t)$ can be neglected, and, consequently, N_2 can be set to zero, which is synonymous with an additional drastic reduction concerning the realization complexity. The other processes $\tilde{\nu}_1(t)$ and $\tilde{\nu}_3(t)$ are again realized by employing the method of exact Doppler spread with $N_1 = 15$ and $N_3 = 15$ cosine functions, respectively. The simulation results are also depicted in Figures 6.12(a) and 6.12(b). From these figures, it can be realized that there is almost an exact correspondence between the reference model and the simulation model.

In order to illustrate the results, Figures 6.13(a) and 6.13(b) both show us a part of the simulated sequence of the deterministic extended Suzuki process $\tilde{\eta}(t)$ for regions with heavy and light shadowing, respectively. One recognizes that for a heavily shadowed line-of-sight component (see Figure 6.13(a)), the average signal level is, all in all, obviously smaller than for an only lightly shadowed line-of-sight component (see Figure 6.13(b)). Also, the deep fades for heavy shadowing are much deeper than for light shadowing.

6.2 The Extended Suzuki Process of Type II

In Section 6.1, it has been shown how a higher model class can be created by introducing a correlation between the two Gaussian random processes determining the Rice process. In this way, the flexibility of the statistical properties of the second order could be increased. On the other hand, the statistical properties of the first order were not influenced. The model described in Section 6.1, however, is not the only possible one for which cross-correlated Gaussian random processes can be used. A further possibility, which was first introduced in [184], will be discussed in this section. We will see that a special type can be found for the cross-correlation function of the real part and the imaginary part of a

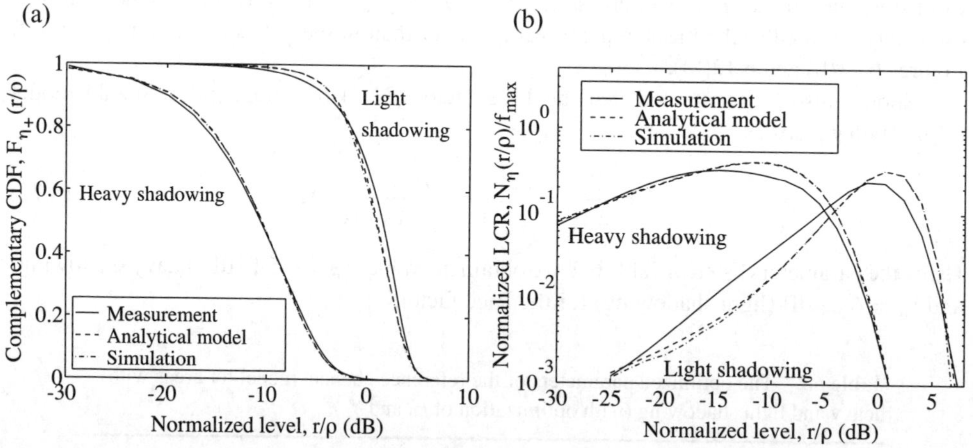

Figure 6.12 (a) Complementary cumulative distribution function $F_{\eta_+}(r/\rho)$ and (b) normalized level-crossing rate $N_\eta(r/\rho)/f_{\max}$ for regions with heavy and light shadowing (with optimization of κ_0 and f_ρ). The measurement results were adopted from [196].

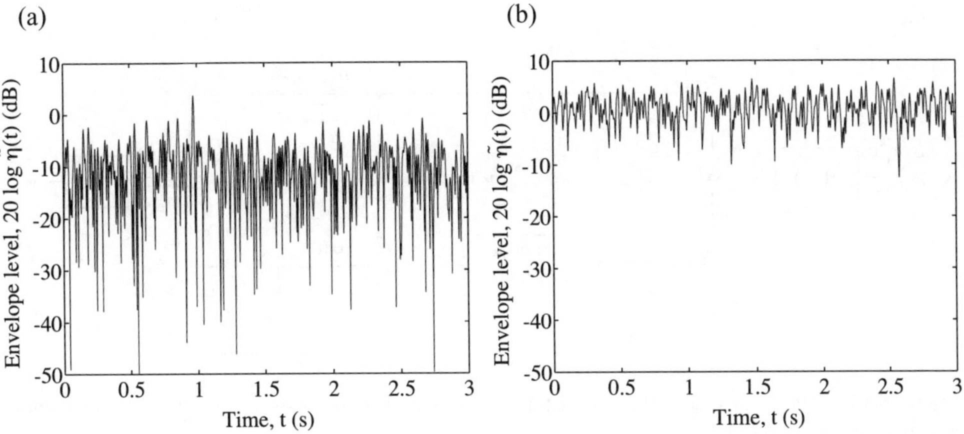

Figure 6.13 Simulation of deterministic extended Suzuki processes $\tilde{\eta}(t)$ of Type I for regions with (a) heavy shadowing and (b) light shadowing (MEDS, $N_1 = 15$, $N_2 = 0$, $N_3 = 15$, $f_{\max} = 91$ Hz, $\kappa_c = 20$).

complex-valued Gaussian random process. This type of cross-correlation function increases not only the flexibility of the statistical properties of the second order of the stochastic model for modelling of short-term fading, but also those of the first order. This model includes the Rice, Rayleigh, and one-sided Gaussian random processes as special cases. The long-term fading is again modelled by means of a lognormal process. The product of both processes, which is useful for modelling short-term and long-term fading, is called the *extended Suzuki process of Type II*.

The aim of this section is to describe the extended Suzuki process (of Type II) and to analyze its first- and second-order statistical properties. To pursue this aim, we will first deal with the modelling and analysis of short-term fading.

6.2.1 Modelling and Analysis of Short-Term Fading

The modelling of short-term fading will be performed by considering the reference model depicted in Figure 6.14. In the following, this model will be described.

Regarding this figure, one should notice that the complex-valued Gaussian random process

$$\mu(t) = \mu_1(t) + j\mu_2(t) \tag{6.78}$$

with the cross-correlated components $\mu_1(t)$ and $\mu_2(t)$ is derived from a single real-valued zero-mean Gaussian random process $\nu_0(t)$. In order to simplify the model, we assume in the following that the Doppler frequency of the line-of-sight component is equal to zero, and thus the line-of-sight component is described by the time-invariant expression (3.18), i.e.,

$$m = m_1 + jm_2 = \rho e^{j\theta_\rho}. \tag{6.79}$$

As for the preceding models, we will also derive a further stochastic process for this one by taking the absolute value of the complex-valued Gaussian random process $\mu_\rho(t) = \mu(t) + m$,

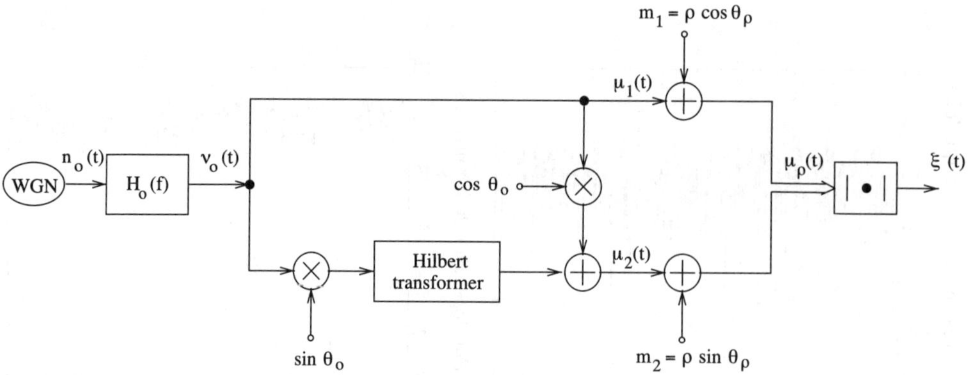

Figure 6.14 Reference model for stochastic processes $\xi(t)$ with cross-correlated Gaussian random processes $\mu_1(t)$ and $\mu_2(t)$.

i.e.,

$$\xi(t) = |\mu_\rho(t)| = \sqrt{(\mu_1(t) + m_1)^2 + (\mu_2(t) + m_2)^2}. \tag{6.80}$$

We will see in Subsection 6.2.1.1 that Rice, Rayleigh, and one-sided Gaussian random processes are merely special cases of this process. To do justice to this property, the output process of the model shown in Figure 6.14 will be called *extended Rice process* in the following.

The Doppler power spectral density $S_{\nu_0\nu_0}(f)$ of the process $\nu_0(t)$ is described by the function (see Figure 6.15(a))

$$S_{\nu_0\nu_0}(f) = \begin{cases} \dfrac{\sigma_0^2}{\pi f_{\max}\sqrt{1-(f/f_{\max})^2}}, & |f| \leq \kappa_0 \cdot f_{\max}, \\ 0, & |f| > \kappa_0 \cdot f_{\max}, \end{cases} \tag{6.81}$$

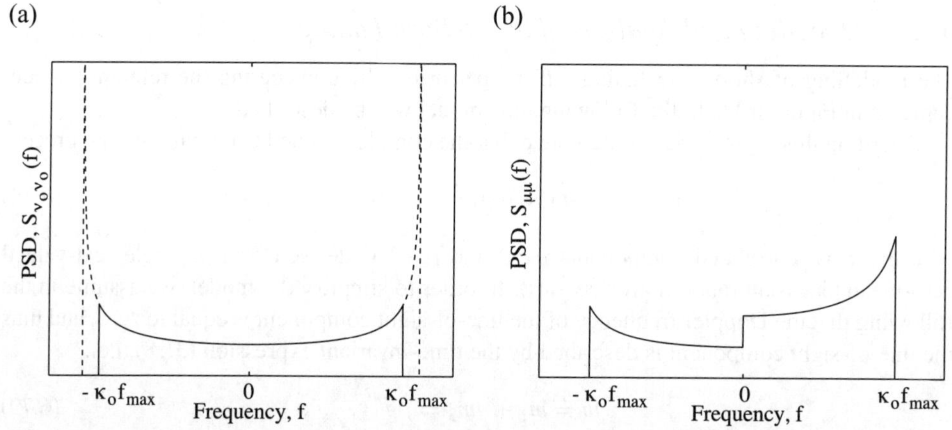

Figure 6.15 Doppler power spectral densities: (a) restricted Jakes power spectral density $S_{\nu_0\nu_0}(f)$ and (b) $S_{\mu\mu}(f)$ ($\theta_0 = 19.5°$).

where $0 < \kappa_0 \leq 1$. The symmetrical Doppler power spectral density $S_{\nu_0\nu_0}(f)$ as defined above is called the *restricted Jakes power spectral density*. We note that for the special case $\kappa_0 = 1$, the (classical) Jakes power spectral density in (3.23) follows from the restricted Jakes power spectral introduced in (6.81). The underlying physical model of the restricted Jakes power spectral density is based on the simplified assumption that due to the presence of spatially limited obstacles or if sector antennas are used, the contribution of the electromagnetic waves to the received signal can be neglected if their angles-of-arrival lie within the interval $(-\alpha_0, \alpha_0)$ and $(\pi - \alpha_0, \pi + \alpha_0)$. Here, α_0 will be restricted to the range $(0, \pi/2]$. Furthermore, α_0 can be related to the parameter κ_0 via the equation $\kappa_0 = f_{\min}/f_{\max} = \cos\alpha_0$. All angles-of-arrival, which do not lie in any of the intervals just mentioned, are again assumed to be uniformly distributed. The actual reason for introducing the restricted Jakes power spectral density in our model is not to be found in the fitting of the theoretical Doppler power spectral density to power spectral densities rarely encountered in practice. Instead of this, the variable κ_0 provides us with a very simple but effective way to reduce the Doppler spread of the Jakes power spectral density, which is often too large compared with practice.

From Figure 6.14, we can deduce the relations

$$\mu_1(t) = \nu_0(t) \tag{6.82}$$

and

$$\mu_2(t) = \cos\theta_0 \cdot \nu_0(t) + \sin\theta_0 \cdot \check{\nu}_0(t), \tag{6.83}$$

where the parameter θ_0 will be kept restricted to the interval $[-\pi, \pi)$, and $\check{\nu}_0(t)$ denotes the Hilbert transform of the coloured Gaussian random process $\mu_0(t)$. The spectral shaping of $\nu_0(t)$ in the reference model is obtained by filtering of white Gaussian noise $n_0(t) \sim N(0, 1)$, where we again assume that the filter is real-valued and completely described by the transfer function $H_0(f) = \sqrt{S_{\nu_0\nu_0}(f)}$.

The autocorrelation functions $r_{\mu_1\mu_1}(\tau)$ and $r_{\mu_2\mu_2}(\tau)$, as well as the cross-correlation functions $r_{\mu_1\mu_2}(\tau)$ and $r_{\mu_2\mu_1}(\tau)$ can be expressed in terms of the autocorrelation function $r_{\nu_0\nu_0}(\tau)$ of the process $\nu_0(t)$ and the cross-correlation function $r_{\check{\nu}_0\nu_0}(\tau)$ of the processes $\check{\nu}_0(t)$ and $\nu_0(t)$ as follows:

$$r_{\mu_1\mu_1}(\tau) = r_{\mu_2\mu_2}(\tau) = r_{\nu_0\nu_0}(\tau), \tag{6.84a}$$

$$r_{\mu_1\mu_2}(\tau) = \cos\theta_0 \cdot r_{\nu_0\nu_0}(\tau) - \sin\theta_0 \cdot r_{\check{\nu}_0\nu_0}(\tau), \tag{6.84b}$$

$$r_{\mu_2\mu_1}(\tau) = \cos\theta_0 \cdot r_{\nu_0\nu_0}(\tau) + \sin\theta_0 \cdot r_{\check{\nu}_0\nu_0}(\tau). \tag{6.84c}$$

One should be aware of the influence of the parameter θ_0. Note that this parameter does not have any influence on the autocorrelation functions $r_{\mu_1\mu_1}(\tau)$ and $r_{\mu_2\mu_2}(\tau)$, but on the cross-correlation functions $r_{\mu_1\mu_2}(\tau)$ and $r_{\mu_2\mu_1}(\tau)$.

Substituting the relations (6.84a)–(6.84c) into (6.5), we obtain the following expression for the autocorrelation function $r_{\mu\mu}(\tau)$ of the complex-valued process $\mu(t) = \mu_1(t) + j\mu_2(t)$

$$r_{\mu\mu}(\tau) = 2r_{\nu_0\nu_0}(\tau) - j2\sin\theta_0 \cdot r_{\check{\nu}_0\nu_0}(\tau). \tag{6.85}$$

The Fourier transform of the above result gives us the power spectral density in the form

$$S_{\mu\mu}(f) = 2S_{\nu_0\nu_0}(f) - j2\sin\theta_0 \cdot S_{\check{\nu}_0\nu_0}(f). \tag{6.86}$$

From (2.135b) and (2.135d), we obtain the relation $S_{\check{\nu}_0\nu_0}(f) = j\,\text{sgn}\,(f) \cdot S_{\nu_0\nu_0}(f)$, so that $S_{\mu\mu}(f)$ can now be expressed in terms of the restricted Jakes power spectral density $S_{\nu_0\nu_0}(f)$ as follows

$$S_{\mu\mu}(f) = 2[1 + \text{sgn}\,(f)\sin\theta_0] \cdot S_{\nu_0\nu_0}(f)\,. \tag{6.87}$$

Note that $S_{\mu\mu}(f)$ is an asymmetric function for all values of $\theta_0 \in (-\pi, \pi)\backslash\{0\}$. An example of the power spectral density $S_{\mu\mu}(f)$ is depicted in Figure 6.15(b), where the value 19.5° has been chosen for the parameter θ_0.

When deriving the statistical properties of $\xi(t) = |\mu_\rho(t)|$ and $\vartheta(t) = \arg\{\mu_\rho(t)\}$, we again make use of the abbreviations (6.11a) and (6.11b). Therefore, we substitute (6.84a) into (6.11a) and (6.84b) into (6.11b), so that after some lengthy but simple algebraic computations, the characteristic quantities of the extended Rice model can be written as follows:

$$\psi_0^{(0)} = \psi_0 = \frac{2}{\pi}\sigma_0^2 \arcsin(\kappa_0), \tag{6.88a}$$

$$\psi_0^{(1)} = \dot{\psi}_0 = 0, \tag{6.88b}$$

$$\psi_0^{(2)} = \ddot{\psi}_0 = -\psi_0 \cdot 2(\pi f_{\max})^2 \left\{1 - \frac{\sin[2\arcsin(\kappa_0)]}{2\arcsin(\kappa_0)}\right\}, \tag{6.88c}$$

$$\phi_0^{(0)} = \phi_0 = \psi_0 \cdot \cos\theta_0, \tag{6.88d}$$

$$\phi_0^{(1)} = \dot{\phi}_0 = 4\sigma_0^2 f_{\max}(1 - \sqrt{1-\kappa_0^2}\,) \cdot \sin\theta_0, \tag{6.88e}$$

$$\phi_0^{(2)} = \ddot{\phi}_0 = \ddot{\psi}_0 \cdot \cos\theta_0, \tag{6.88f}$$

where $0 < \kappa_0 \leq 1$ and $-\pi \leq \theta_0 < \pi$. A comparison between Equations (6.88a)–(6.88f) and (6.12a)–(6.12f) shows us that for the present model even the quantities Φ_0 and $\ddot{\phi}_0$ are in general different from zero. Only for the special case $\theta_0 = \pm\pi/2$, we obtain $\phi_0 = \ddot{\phi}_0 = 0$. Hence, there are reasons for supposing that the statistical properties of the extended Rice process are different from those of the classical Rice process.

The starting point, which enables the analysis of the statistical properties of extended Rice processes, is again the multivariate Gaussian distribution of the processes $\mu_{\rho_1}(t)$, $\mu_{\rho_2}(t)$, $\dot{\mu}_{\rho_1}(t)$, and $\dot{\mu}_{\rho_2}(t)$ at the same time instant t [see (6.14)]. For the present model, where it was assumed for simplification that $f_\rho = 0$, the multivariate Gaussian distribution (6.14) is completely described by the column vectors

$$\boldsymbol{x} = \begin{pmatrix} x_1 \\ x_2 \\ \dot{x}_1 \\ \dot{x}_2 \end{pmatrix} \quad \text{and} \quad \boldsymbol{m} = \begin{pmatrix} m_1 \\ m_2 \\ \dot{m}_1 \\ \dot{m}_2 \end{pmatrix} = \begin{pmatrix} \rho\cos\theta_\rho \\ \rho\sin\theta_\rho \\ 0 \\ 0 \end{pmatrix} \tag{6.89a,b}$$

as well as by the covariance or correlation matrix

$$C_{\mu_\rho}(0) = R_\mu(0) = \begin{pmatrix} \psi_0 & \phi_0 & 0 & \dot{\phi}_0 \\ \phi_0 & \psi_0 & -\dot{\phi}_0 & 0 \\ 0 & -\dot{\phi}_0 & -\ddot{\psi}_0 & -\ddot{\phi}_0 \\ \dot{\phi}_0 & 0 & -\ddot{\phi}_0 & -\ddot{\psi}_0 \end{pmatrix}. \tag{6.90}$$

Employing the relations (6.88d) and (6.88f) results in

$$C_{\mu_\rho}(0) = R_\mu(0) = \begin{pmatrix} \psi_0 & \psi_0 \cos\theta_0 & 0 & \dot{\phi}_0 \\ \psi_0 \cos\theta_0 & \psi_0 & -\dot{\phi}_0 & 0 \\ 0 & -\dot{\phi}_0 & -\ddot{\psi}_0 & -\ddot{\psi}_0 \cos\theta_0 \\ \dot{\phi}_0 & 0 & -\ddot{\psi}_0 \cos\theta_0 & -\ddot{\psi}_0 \end{pmatrix}. \tag{6.91}$$

Now, after substituting (6.89a,b) and (6.91) into (6.14), the desired joint probability density function $p_{\mu_{\rho_1}\mu_{\rho_2}\dot{\mu}_{\rho_1}\dot{\mu}_{\rho_2}}(x_1, x_2, \dot{x}_1, \dot{x}_2)$ of our model can be calculated. We then transform the Cartesian coordinates (x_1, x_2) of this density to polar coordinates (z, θ) by means of (6.22a,b). After some further algebraic manipulations, we succeed in converting the joint probability density function (6.24) to the following form [184]

$$\begin{aligned} p_{\xi\dot{\xi}\vartheta\dot{\vartheta}}(z, \dot{z}, \theta, \dot{\theta}) = {} & \frac{z^2}{(2\pi)^2 \beta \psi_0 \sin^2\theta_0} \cdot e^{-\frac{1}{2\psi_0 \sin^2\theta_0}[z^2+\rho^2-2z\rho\cos(\theta-\theta_\rho)]} \\ & \cdot e^{\frac{\cos\theta_0}{2\psi_0 \sin^2\theta_0}[z^2 \sin 2\theta + \rho^2 \sin 2\theta_\rho - 2z\rho \sin(\theta+\theta_\rho)]} \\ & \cdot e^{-\frac{1}{2\beta(1+\cos\theta_0 \cdot \sin 2\theta)} \cdot \left\{\dot{z} + \frac{\dot{\phi}_0[\rho\sin(\theta-\theta_\rho) - \cos\theta_0 (z\cos 2\theta - \rho\cos(\theta+\theta_\rho))]}{\psi_0 \sin^2\theta_0}\right\}^2} \\ & \cdot e^{-\frac{z^2(1+\cos\theta_0 \cdot \sin 2\theta)}{2\beta \sin^2\theta_0} \cdot \left\{\dot{\theta} + \frac{\dot{\phi}_0[\rho\cos(\theta-\theta_\rho) - z] - \psi_0 \dot{z} \cos\theta_0 \cdot \cos 2\theta}{\psi_0 z (1+\cos\theta_0 \cdot \sin 2\theta)}\right\}^2} \end{aligned} \tag{6.92}$$

for $z \geq 0$, $|\dot{z}| < \infty$, $|\theta| \leq \pi$, and $|\dot{\theta}| < \infty$. Here, one should note that the quantity β in (6.92) is no longer given by (6.28), but is defined by the extended expression

$$\beta = -\ddot{\psi}_0 - \frac{\dot{\phi}_0^2}{\psi_0 \sin^2\theta_0}. \tag{6.93}$$

In the next subsection, we will derive the probability density function of the envelope $\xi(t)$ and the phase $\vartheta(t)$ from the joint probability density function presented in (6.92). After that follows the derivation and analysis of the level-crossing rate and the average duration of fades of $\xi(t)$.

6.2.1.1 Probability Density Function of the Envelope and Phase

For the probability density function of the extended Rice process $\xi(t)$, denoted by $p_\xi(z)$, we obtain the following result after substituting (6.92) into (6.29)

$$p_\xi(z) = \frac{z}{2\pi\,\psi_0|\sin\theta_0|}\, e^{-\frac{z^2+\rho^2}{2\psi_0\sin^2\theta_0}} \cdot \int_{-\pi}^{\pi} e^{\frac{z\rho\cos(\theta-\theta_\rho)}{\psi_0\sin^2\theta_0}} \cdot e^{\frac{\cos\theta_0}{2\psi_0\sin^2\theta_0}[z^2\sin 2\theta+\rho^2\sin 2\theta_\rho-2z\rho\sin(\theta+\theta_\rho)]}\, d\theta, \quad z \geq 0. \tag{6.94}$$

Just as for conventional Rice processes, the probability density function $p_\xi(z)$ in this case also depends on the mean power of the processes $\mu_1(t)$ and $\mu_2(t)$, i.e., ψ_0, as well as on the amplitude ρ of the line-of-sight component. Moreover, the density of the extended Rice process is also determined by the parameter θ_0 and — surprising at first sight — by the phase θ_ρ of the line-of-sight component. We will understand this property as soon as we have derived the corresponding simulation model (see Subsection 6.2.3). In order to illustrate the results, we study Figures 6.16(a) and 6.16(b), where the probability density function (6.94) is shown for various values of the parameters θ_0 and θ_ρ, respectively. It should be pointed out that even for this model, the density $p_\xi(z)$ neither depends on the first and second time derivative of the autocorrelation function (6.11a), i.e., $\dot{\psi}_0$ and $\ddot{\psi}_0$, nor on the first and second time derivative of the cross-correlation function (6.11b), i.e., $\dot{\phi}_0$ and $\ddot{\phi}_0$.

In the following, we study some special cases. For example, if $\theta_0 = \pm\pi/2$, then the integral in (6.94) can be solved analytically. As a result, we obtain the Rice distribution

$$p_\xi(z) = \frac{z}{\psi_0}\, e^{-\frac{z^2+\rho^2}{2\psi_0}}\, I_0\!\left(\frac{z\rho}{\psi_0}\right), \quad z \geq 0, \tag{6.95}$$

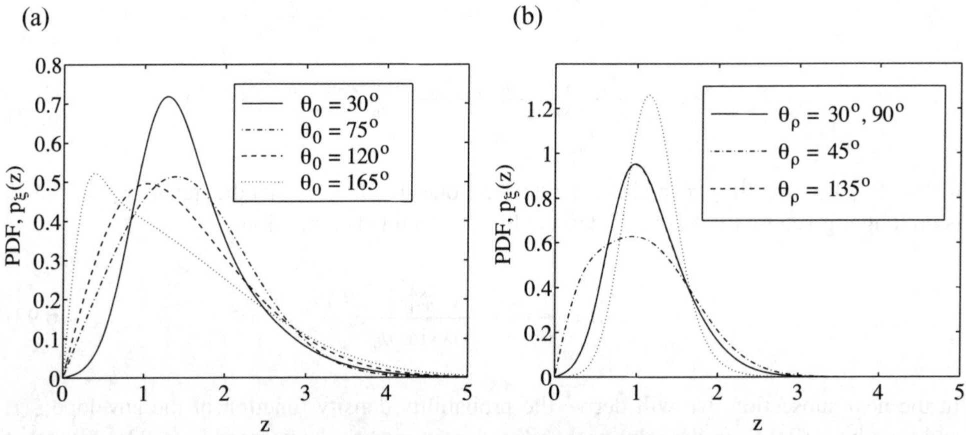

Figure 6.16 Probability density function $p_\xi(z)$ of extended Rice processes $\xi(t)$ for various values of the parameters (a) θ_0 ($\psi_0 = 1$, $\rho = 1$, $\theta_\rho = 127°$) and (b) θ_ρ ($\psi_0 = 1$, $\rho = 1$, $\theta_0 = 45°$).

where ψ_0 is given by (6.88a). For a shadowed line-of-sight component, i.e., $\rho = 0$, but especially for arbitrary values of $\theta_0 \in [-\pi, \pi)$, Equation (6.94) reduces to the expression

$$p_\xi(z) = \frac{z}{\psi_0 |\sin\theta_0|} e^{-\frac{z^2}{2\psi_0 \sin^2\theta_0}} I_0\left(\frac{z^2\cos\theta_0}{2\psi_0 \sin^2\theta_0}\right), \quad z \geq 0, \tag{6.96}$$

from which the Rayleigh distribution

$$p_\xi(z) = \frac{z}{\psi_0} e^{-\frac{z^2}{2\psi_0}}, \quad z \geq 0, \tag{6.97}$$

follows if $\theta_0 = \pm\pi/2$, and in the limit $\theta_0 \to 0$, we obtain the one-sided Gaussian distribution

$$p_\xi(z) = \frac{1}{\sqrt{\pi\psi_0}} e^{-\frac{z^2}{4\psi_0}}, \quad z \geq 0. \tag{6.98}$$

Consequently, the Rice distribution, the Rayleigh distribution, and the one-sided Gaussian distribution are special cases of the extended Rice distribution in (6.94).

For the probability density function of the phase $\vartheta(t)$, denoted by $p_\vartheta(\theta)$, we obtain, after substituting (6.92) into (6.31), the expression

$$p_\vartheta(\theta) = \frac{|\sin\theta_0|}{2\pi(1 - \cos\theta_0 \cdot \sin 2\theta)} \cdot e^{-\frac{\rho^2(1-\cos\theta_0 \cdot \sin 2\theta_\rho)}{2\psi_0 \sin^2\theta_0}}$$
$$\cdot \left\{1 + \sqrt{\pi} f(\theta) e^{f^2(\theta)} [1 + \text{erf}(f(\theta))]\right\}, \quad -\pi \leq \theta \leq \pi, \tag{6.99}$$

where

$$f(\theta) = \frac{\rho[\cos(\theta - \theta_\rho) - \cos\theta_0 \cdot \sin(\theta + \theta_\rho)]}{|\sin\theta_0| \sqrt{2\psi_0(1 - \cos\theta_0 \cdot \sin 2\theta)}}. \tag{6.100}$$

Just like the probability density function of the envelope [see (6.94)], the probability density function of the phase merely depends on the parameters ψ_0, ρ, θ_0, and θ_ρ, and not on the quantities $\dot{\psi}_0$, $\ddot{\psi}_0$, ϕ_0, and $\dot{\phi}_0$.

The same probability density function of the phase, which we became acquainted with during the analysis of Rice processes with uncorrelated Gaussian random processes $\mu_1(t)$ and $\mu_2(t)$ in Subsection 3.4.1 [see (3.57)], also follows from (6.99) for the special case $\theta_0 = \pm\pi/2$. If the parameters ρ and θ_0 are determined by $\rho = 0$ and $\theta_0 = \pm\pi/2$, then the phase $\vartheta(t)$ is uniformly distributed over the interval $[-\pi, \pi]$.

Finally, the influence of the parameters θ_0 and θ_ρ on the behaviour of the density $p_\vartheta(\theta)$ shall be further clarified by Figures 6.17(a) and 6.17(b), respectively.

6.2.1.2 Level-Crossing Rate and Average Duration of Fades

For the calculation of the level-crossing rate $N_\xi(r)$, the joint probability density function $p_{\xi\dot{\xi}}(z, \dot{z})$ of the stochastic processes $\xi(t)$ and $\dot{\xi}(t)$ has to be known at the same time

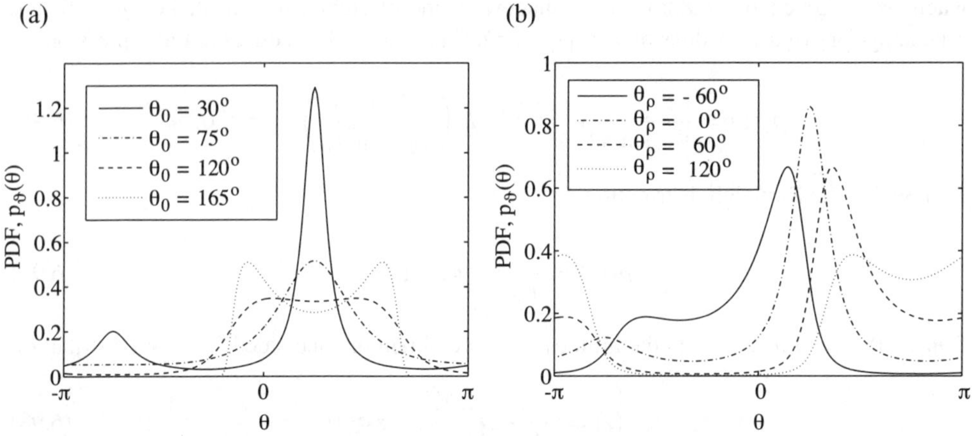

Figure 6.17 Probability density function $p_\vartheta(\theta)$ of the phase $\vartheta(t)$ for various values of the parameters (a) θ_0 ($\psi_0 = 1$, $\rho = 1$, $\theta_\rho = 45°$) and (b) θ_ρ ($\psi_0 = 1$, $\rho = 1$, $\theta_0 = 45°$).

instant t. For this density, we obtain the following integral expression after substituting (6.92) into (6.34)

$$p_{\xi\dot{\xi}}(z,\dot{z}) = \frac{z}{(2\pi)^{3/2}\psi_0\sqrt{\beta}|\sin\theta_0|} e^{-\frac{z^2+\rho^2}{2\psi_0 \sin^2\theta_0}} \int_{-\pi}^{\pi} \frac{1}{\sqrt{1+\cos\theta_0\cdot\sin 2\theta}} \cdot e^{\frac{z\rho\cos(\theta-\theta_\rho)}{\psi_0\sin^2\theta_0}} \cdot e^{\frac{\cos\theta_0}{2\psi_0\sin^2\theta_0}[z^2\sin 2\theta+\rho^2\sin 2\theta_\rho - 2z\rho\sin(\theta+\theta_\rho)]} \cdot e^{-\frac{1}{2\beta(1+\cos\theta_0\cdot\sin 2\theta)}\left\{\dot{z}+\frac{\dot{\phi}_0[\rho\sin(\theta-\theta_\rho)-\cos\theta_0(z\cos 2\theta-\rho\cos(\theta+\theta_\rho))]}{\psi_0\sin^2\theta_0}\right\}^2} d\theta, \qquad (6.101)$$

for $z \geq 0$ and $|\dot{z}| < \infty$. Here, ψ_0, $\dot{\phi}_0$, and β again are the quantities defined by (6.88a), (6.88e), and (6.93), respectively. Within the interval $(-\pi, \pi)\backslash\{0\}$, no value can be found for the parameter θ_0 in such a way that the stochastic processes $\xi(t)$ and $\dot{\xi}(t)$ become statistically independent, because $p_{\xi\dot{\xi}}(z,\dot{z}) \neq p_\xi(z)\cdot p_{\dot{\xi}}(\dot{z})$ always holds. Even for the special case $\theta_0 = \pm\pi/2$, Equation (6.35) may follow from (6.101), but here it has to be taken into consideration that now relations (6.88a)–(6.88f) hold, so that $\dot{\phi}_0 \neq 0$ ($\alpha \neq 0$) follows, and thus (6.101) can never be brought into the form (6.36).

With the joint probability density function (6.101), all assumptions for the derivation of the level-crossing rate $N_\xi(r)$ of extended Rice processes $\xi(t)$ are made. We substitute (6.101) into the definition (6.33), and, after some algebraic manipulations, obtain the result

$$N_\xi(r) = \frac{r\sqrt{\beta}}{(2\pi)^{3/2}\psi_0|\sin\theta_0|} \cdot e^{-\frac{r^2+\rho^2}{2\psi_0\sin^2\theta_0}} \int_{-\pi}^{\pi} \sqrt{1+\cos\theta_0\cdot\sin 2\theta} \cdot e^{\frac{r\rho\cos(\theta-\theta_\rho)}{\psi_0\sin^2\theta_0}} e^{\frac{\cos\theta_0}{2\psi_0\sin^2\theta_0}[r^2\sin 2\theta+\rho^2\sin 2\theta_\rho-2r\rho\sin(\theta+\theta_\rho)]} \cdot \left\{e^{-g^2(r,\theta)} + \sqrt{\pi}g(r,\theta)[1+\mathrm{erf}\,(g(r,\theta))]\right\} d\theta, \quad r \geq 0, \qquad (6.102)$$

where the function $g(r, \theta)$ stands for

$$g(r, \theta) = -\frac{\dot{\phi}_0\{\rho \sin(\theta - \theta_\rho) - \cos\theta_0[r\cos 2\theta - \rho\cos(\theta + \theta_\rho)]\}}{\psi_0 \sin^2\theta_0 \sqrt{2\beta(1 + \cos\theta_0 \cdot \sin 2\theta)}}. \tag{6.103}$$

The quantities ψ_0, $\dot{\phi}_0$, and β are again defined by (6.88a), (6.88e), and (6.93), respectively. It should be noted that we have made use of the integral [23, Equation (3.462.5)]

$$\int_0^\infty x e^{-ax^2-2bx} dx = \frac{1}{2a}\left\{1 - b\sqrt{\frac{\pi}{a}}\, e^{\frac{b^2}{a}}\left[1 - \operatorname{erf}\left(\frac{b}{\sqrt{a}}\right)\right]\right\}, \quad a > 0, \tag{6.104}$$

for the derivation of (6.102). Using (6.88a)–(6.88f), we easily find out that (6.102) is proportional to the maximum Doppler frequency f_{max}, i.e., the normalized level-crossing rate $N_\xi(r)/f_{max}$ is independent of the speed of the vehicle and the carrier frequency, just as in the previous case. A brief parameter study, which illustrates the influence of the parameters κ_0, ρ, θ_0, and θ_ρ on the normalized level-crossing rate $N_\xi(r)/f_{max}$, is depicted in Figures 6.18(a)–6.18(d). The variation of κ_0 (Figure 6.18(a)) and ρ (Figure 6.18(b)) leads to graphs, which are in principle similar to those shown in Figure 6.3(a) and Figure 3.8(b), respectively. A further, more powerful parameter exists, namely θ_0, which has a decisive influence on the behaviour of $N_\xi(r)/f_{max}$, as can be seen in Figure 6.18(c). For the results shown in Figure 6.18(d), the value of the quantity θ_ρ is only of secondary importance if the other parameters (ψ_0, κ_0, ρ, θ_0) are chosen as described in the figure caption.

Now, attention is given to some special cases. On the assumption that $\theta_0 = \pm\pi/2$ holds, the level-crossing rate described by (6.37) follows from (6.102). If, in addition, $\rho = 0$ holds, then $N_\xi(r)$ becomes directly proportional to the Rayleigh distribution and can be brought into the form of (6.38). Moreover, for the special case $\rho = 0$ and $\theta_0 \to 0°$, one can show that the level-crossing rate of one-sided Gaussian random processes follows from (6.102), i.e.,

$$N_\xi(r) = \frac{\sqrt{\beta}}{\pi\sqrt{\psi_0}}\, e^{-\frac{r^2}{4\psi_0}}, \quad r \geq 0, \tag{6.105}$$

where β is given by $\beta = -\ddot{\psi}_0 > 0$ in the present case. Further special cases such as, e.g., $\rho = 0$ in connection with arbitrary values for $\theta_0 \in [-\pi, \pi)$ can also be easily analyzed by means of (6.102).

When calculating the average duration of fades $T_{\xi_-}(r)$ [see (6.40)], we need to know the level-crossing rate $N_\xi(r)$ and the cumulative distribution function $F_{\xi_-}(r)$ of the extended Rice process $\xi(t)$. For the cumulative distribution function $F_{\xi_-}(r)$, we obtain the following double integral by using the probability density function in (6.94)

$$\begin{aligned} F_{\xi_-}(r) &= \int_0^r p_\xi(z)\, dz \\ &= \int_0^r \frac{z}{2\pi\psi_0|\sin\theta_0|}\, e^{-\frac{z^2+\rho^2}{2\psi_0\sin^2\theta_0}} \cdot \int_{-\pi}^{\pi} e^{\frac{z\rho\cos(\theta-\theta_\rho)}{\psi_0\sin^2\theta_0}} \\ &\quad \cdot e^{\frac{\cos\theta_0}{2\psi_0\sin^2\theta_0}\cdot[z^2\sin 2\theta + \rho^2\sin 2\theta_\rho - 2z\rho\sin(\theta+\theta_\rho)]}\, d\theta\, dz, \quad r \geq 0. \end{aligned} \tag{6.106}$$

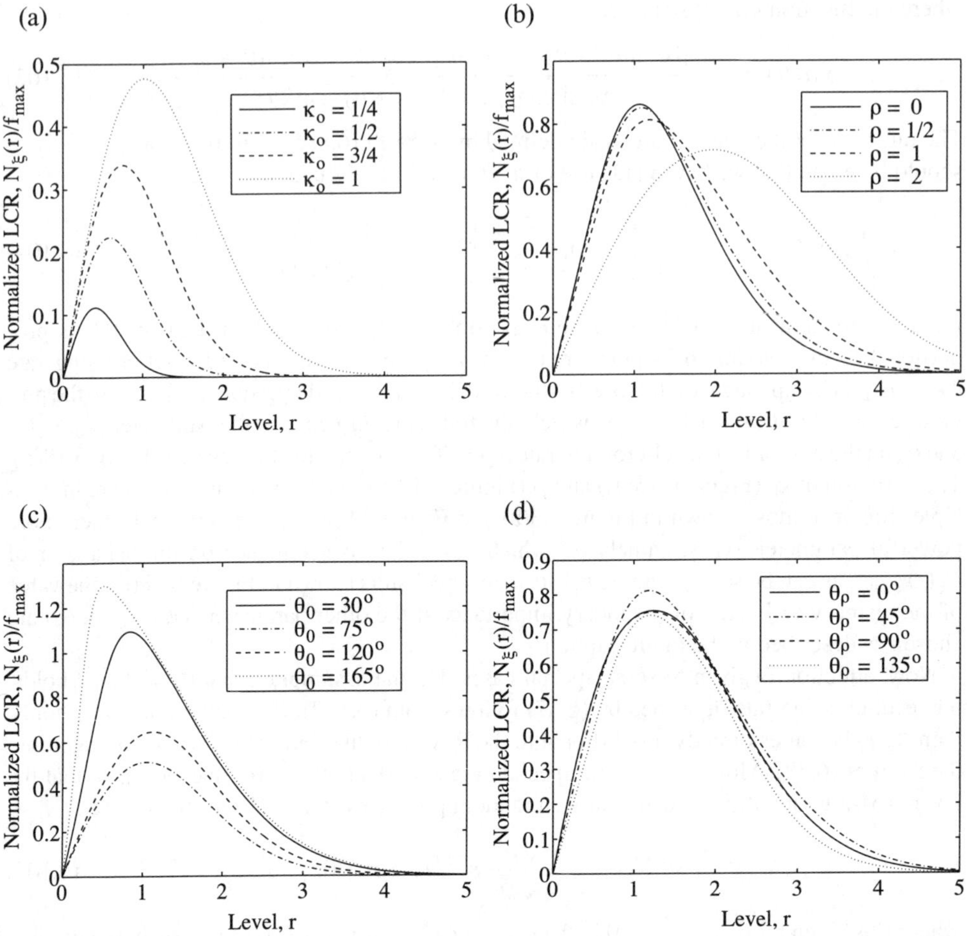

Figure 6.18 Normalized level-crossing rate $N_\xi(r)/f_{\max}$ of extended Rice processes (Type II) depending on: (a) κ_0 ($\sigma_0^2 = 1, \rho = 0, \theta_0 = 45°$), (b) ρ ($\psi_0 = 1, \kappa_0 = 1, \theta_\rho = 45°, \theta_0 = 45°$), (c) θ_0 ($\psi_0 = 1, \kappa_0 = 1, \rho = 0$), and (d) θ_ρ ($\psi_0 = 1, \kappa_0 = 1, \rho = 1, \theta_0 = 45°$).

According to (6.40), the average duration of fades $T_{\xi_-}(r)$ of extended Rice processes $\xi(t)$ can be obtained from the quotient of (6.106) and (6.102).

Figures 6.19(a) to 6.19(d) show clearly the influence of the parameters κ_0, ρ, θ_0, and θ_ρ on the normalized average duration of fades $T_{\xi_-}(r) \cdot f_{\max}$. The model parameters that lead to the results shown in Figures 6.18(a) to 6.18(d) were also used here for the calculation of $T_{\xi_-}(r) \cdot f_{\max}$. By varying the parameter κ_0, we recognize similar effects on $T_{\xi_-}(r) \cdot f_{\max}$ in Figure 6.19(a) as in Figure 6.4(a). Figure 6.19(b) shows that even at low signal levels r, an increase in ρ results in a reduction in the normalized average duration of fades $T_{\xi_-}(r) \cdot f_{\max}$. This is obviously in contrast to the results depicted in Figure 3.9(b), where no considerable effect on $T_{\xi_-}(r) \cdot f_{\max}$ can be observed at low signal levels r by a variation of ρ. From Figure 6.19(c), it can be realized that the parameter θ_0 affects the behaviour of $T_{\xi_-}(r) \cdot f_{\max}$

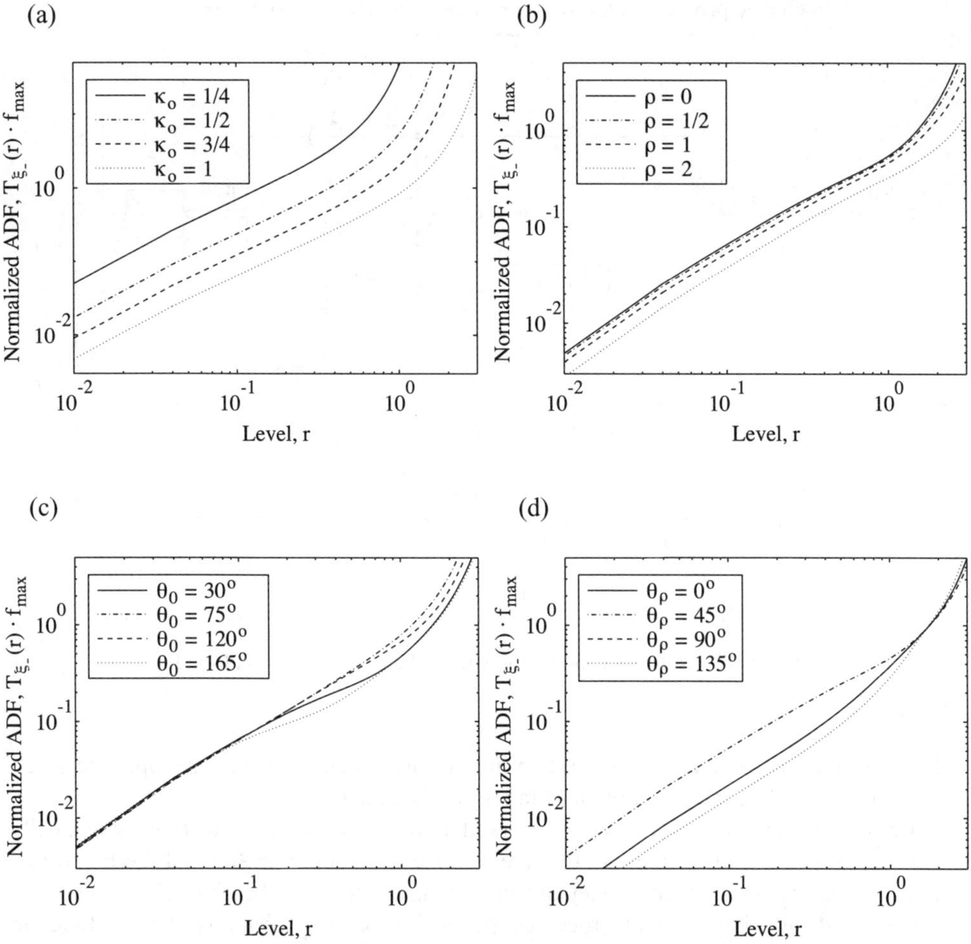

Figure 6.19 Normalized average duration of fades $T_{\xi_-}(r) \cdot f_{\max}$ of extended Rice processes (Type II) depending on: (a) κ_0 ($\sigma_0^2 = 1$, $\rho = 0$, $\theta_0 = 45^\circ$), (b) ρ ($\psi_0 = 1$, $\kappa_0 = 1$, $\theta_\rho = 45^\circ$, $\theta_0 = 45^\circ$), (c) θ_0 ($\psi_0 = 1$, $\kappa_0 = 1$, $\rho = 0$), and (d) θ_ρ ($\psi_0 = 1$, $\kappa_0 = 1$, $\rho = 1$, $\theta_0 = 45^\circ$).

at medium and high signal levels r, whereas its influence at low signal levels r can be ignored (at least if the parameters are chosen as in the present example: $\psi_0 = 1$, $\kappa_0 = 1$, and $\rho = 0$). Analogously, the opposite statement holds for the results shown in Figure 6.19(d) illustrating the influence of the parameter θ_ρ.

6.2.2 *The Stochastic Extended Suzuki Process of Type II*

In [184], the extended Suzuki process of Type II, denoted by $\eta(t)$, was introduced as a product process of the extended Rice process $\xi(t)$ studied before and the lognormal process $\lambda(t)$ described in Subsection 6.1.2, i.e., $\eta(t) = \xi(t) \cdot \lambda(t)$. The structure of the reference model corresponding to this process is depicted in Figure 6.20.

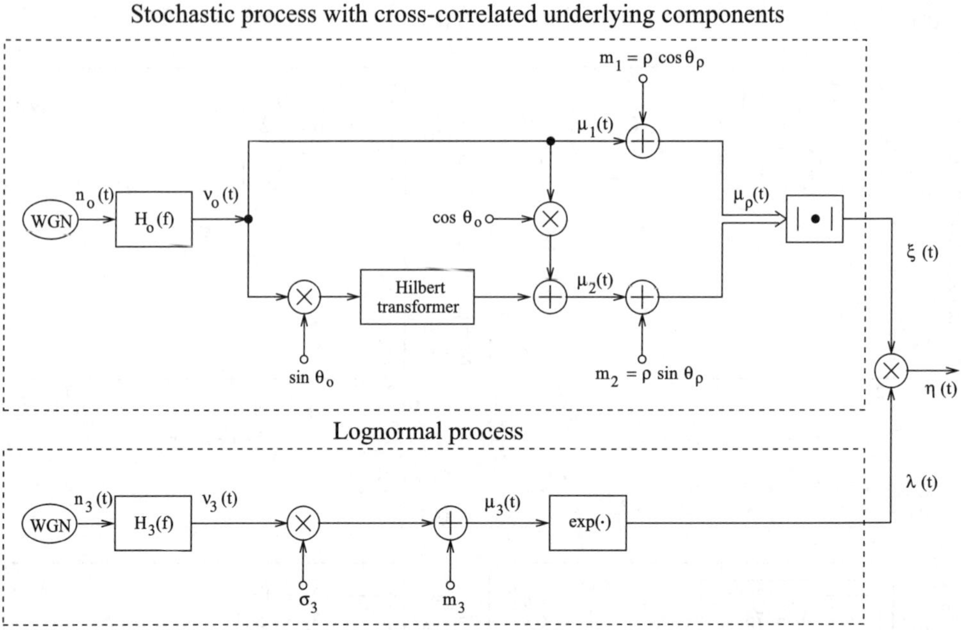

Figure 6.20 Reference model for extended Suzuki processes (Type II).

In the following, we analyze the probability density function of the envelope, the level-crossing rate, and the average duration of fades of this model.

Let us assume that the coloured Gaussian random processes $\nu_0(t)$ and $\nu_3(t)$ are statistically independent, which leads to the fact that the extended Rice process $\xi(t)$ and the lognormal process $\lambda(t)$ are also statistically independent. Owing to the multiplicative relation between the two statistically independent processes $\xi(t)$ and $\lambda(t)$, the probability density function $p_\eta(z)$ of the extended Suzuki process of Type II can be derived by using (6.94) and (6.49) as follows:

$$p_\eta(z) = \int_{-\infty}^{\infty} \frac{1}{|y|} p_\xi\left(\frac{z}{y}\right) \cdot p_\lambda(y)\, dy, \tag{6.107a}$$

$$= \int_{-\infty}^{\infty} \frac{1}{|y|} p_\xi(y) \cdot p_\lambda\left(\frac{z}{y}\right) dy, \tag{6.107b}$$

$$= \frac{1}{2\pi\psi_0 |\sin\theta_0|} \int_0^{\infty} \frac{e^{-\frac{[\ln(z/y)-m_3]^2}{2\sigma_3^2}}}{\sqrt{2\pi}\sigma_3 (z/y)} \cdot e^{-\frac{y^2+\rho^2}{2\psi_0 \sin^2\theta_0}} \cdot \int_{-\pi}^{\pi} e^{\frac{y\rho\cos(\theta-\theta_\rho)}{\psi_0 \sin^2\theta_0}} \cdot e^{\frac{\cos\theta_0}{2\psi_0 \sin^2\theta_0}[y^2\sin 2\theta + \rho^2 \sin 2\theta_\rho - 2y\rho\sin(\theta+\theta_\rho)]}\, d\theta\, dy, \quad z \geq 0. \tag{6.107c}$$

Here, we deliberately preferred the relation (6.107b) to (6.107a), because the solution of (6.107c) can then be performed more advantageously by means of numerical integration techniques. For $\sigma_3 \to 0$ and $m_3 \to 0$, it follows $p_\lambda(z/y) \to |y|\,\delta(z-y)$ and thus $p_\eta(z) \to p_\xi(z)$, where $p_\xi(z)$ is described by (6.94). In general, the probability density function (6.107c) depends on the mean power ψ_0, the parameters σ_3, m_3, ρ, θ_ρ, and, last but not least, on θ_0. Figures 6.21(a) and 6.21(b) give an idea of the influence of the parameters σ_3 and m_3, respectively, on the behaviour of the probability density function $p_\eta(z)$.

Next, we will calculate the level-crossing rate $N_\eta(r)$ of extended Suzuki processes (Type II). Since the joint probability density function $p_{\eta\dot{\eta}}(z,\dot{z})$ of the processes $\eta(t)$ and $\dot{\eta}(t)$ at the same time t is required for our purpose, we first substitute the relations (6.101) and (6.53) found for $p_{\xi\dot{\xi}}(z,\dot{z})$ and $p_{\lambda\dot{\lambda}}(y,\dot{y})$, respectively, into (6.58). Thus,

$$p_{\eta\dot{\eta}}(z,\dot{z}) = \frac{1}{(2\pi)^{3/2}\psi_0\sqrt{\beta}|\sin\theta_0|} \cdot \int_0^\infty \frac{e^{-\frac{[\ln(z/y)-m_3]^2}{2\sigma_3^2}}}{\sqrt{2\pi}\sigma_3(z/y)^2} \cdot e^{-\frac{y^2+\rho^2}{2\psi_0\sin^2\theta_0}}$$

$$\cdot \int_{-\pi}^{\pi} \frac{e^{\frac{y\rho\cos(\theta-\theta_\rho)}{\psi_0\sin^2\theta_0}}\, e^{\frac{\cos\theta_0}{2\psi_0\sin^2\theta_0}[y^2\sin 2\theta+\rho^2\sin 2\theta_\rho-2y\rho\sin(\theta+\theta_\rho)]}}{h(y,\theta)\sqrt{1+\cos\theta_0\cdot\sin 2\theta}}$$

$$\cdot e^{-\frac{\left\{\dot{z}+\frac{\dot{\phi}_0(z/y)[\rho\sin(\theta-\theta_\rho)-\cos\theta_0(y\cos 2\theta-\rho\cos(\theta+\theta_\rho))]}{\psi_0\sin^2\theta_0}\right\}}{2\beta(z/y)^2h^2(y,\theta)(1+\cos\theta_0\cdot\sin 2\theta)}}\, d\theta\, dy, \quad z \geq 0, \quad |\dot{z}| < \infty, \tag{6.108}$$

where

$$h(y,\theta) = \sqrt{1 + \frac{\gamma(\sigma_3 y)^2}{\beta(1+\cos\theta_0\cdot\sin 2\theta)}}. \tag{6.109}$$

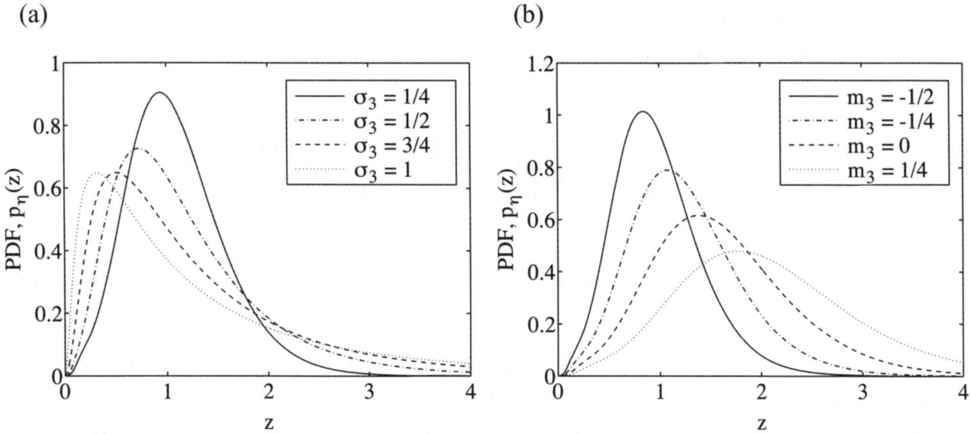

Figure 6.21 Probability density function $p_\eta(z)$ of extended Suzuki processes (Type II) for various values of the parameters (a) σ_3 ($m_3 = 1$, $\psi_0 = 0.0412$, $\rho = 0.918$, $\theta_\rho = 86°$, $\theta_0 = 97°$) and (b) m_3 ($\sigma_3 = 0.5$, $\psi_0 = 0.0412$, $\rho = 0.918$, $\theta_\rho = 86°$, $\theta_0 = 97°$).

Similarly as before, ψ_0, $\dot{\phi}_0$, β, and γ are again the quantities introduced in (6.88a), (6.88e), (6.93), and (6.51), respectively. If we furthermore substitute (6.108) into (6.57), we obtain the result for the level-crossing rate $N_\eta(r)$ of extended Suzuki processes of Type II as follows

$$N_\eta(r) = \frac{\sqrt{\beta}}{(2\pi)^2\sigma_3\psi_0|\sin\theta_0|} \cdot \int_0^\infty e^{-\frac{[\ln(r/y)-m_3]^2}{2\sigma_3^2}} \cdot e^{-\frac{y^2+\rho^2}{2\psi_0\sin^2\theta_0}}$$

$$\cdot \int_{-\pi}^{\pi} h(y,\theta)\sqrt{1+\cos\theta_0\cdot\sin 2\theta}$$

$$\cdot e^{\frac{y\rho\cos(\theta-\theta_\rho)}{\psi_0\sin^2\theta_0}} \cdot e^{\frac{\cos\theta_0}{2\psi_0\sin^2\theta_0}[y^2\sin 2\theta+\rho^2\sin 2\theta_\rho-2y\rho\sin(\theta+\theta_\rho)]}$$

$$\cdot \left\{ e^{-\left[\frac{g(y,\theta)}{h(y,\theta)}\right]^2} + \sqrt{\pi}\frac{g(y,\theta)}{h(y,\theta)}\left[1+\operatorname{erf}\left(\frac{g(y,\theta)}{h(y,\theta)}\right)\right]\right\} d\theta\, dy, \quad r \geq 0, \qquad (6.110)$$

where the functions $g(y,\theta)$ and $h(y,\theta)$ are given by (6.103) and (6.109), respectively.

In the case $\sigma_3 \to 0$ and $m_3 \to 0$, it follows $p_\lambda(r/y) \to |y|\,\delta(r-y)$ and $h(y,\theta) \to 1$, so that $N_\eta(r)$, according to (6.110), converges towards the expression (6.102), which describes the level-crossing rate of extended Rice processes. This result was to be expected. Furthermore, it should be taken into account that although (6.110) can be brought into the form (6.61) for the special case $\theta_0 = \pm\pi/2$, however, the definitions (6.88a)–(6.88f) still hold and not (6.12a)–(6.12f), so that, generally speaking, the level-crossing rate of extended Suzuki processes of Type II cannot be mapped exactly onto that of Type I. The maximum Doppler frequency $f_{\max}$ is again proportional to the level-crossing rate $N_\eta(r)$, as can easily be shown by substituting (6.88a)–(6.88f) into (6.110).

Concerning the computation of the average duration of fades $T_{\eta_-}(r)$, we make use of the definition (6.65). For the necessary cumulative distribution function $F_{\eta_-}(r) = P\{\eta(t) \leq r\}$, we obtain the following double integral by means of (6.107c)

$$F_{\eta_-}(r) = \int_0^r p_\eta(z)\,dz$$

$$= \frac{1}{2\pi\psi_0|\sin\theta_0|}\int_0^\infty \frac{y}{2}\left\{1+\operatorname{erf}\left[\frac{\ln(r/y)-m_3}{\sigma_3}\right]\right\} \cdot e^{-\frac{y^2+\rho^2}{2\psi_0\sin^2\theta_0}}$$

$$\cdot \int_{-\pi}^{\pi} e^{\frac{y\rho\cos(\theta-\theta_\rho)}{\psi_0\sin^2\theta_0}} \cdot e^{\frac{\cos\theta_0}{2\psi_0\sin^2\theta_0}[y^2\sin 2\theta+\rho^2\sin 2\theta_\rho-2y\rho\sin(\theta+\theta_\rho)]} d\theta\, dy. \qquad (6.111)$$

According to (6.65), the quotient of (6.111) and (6.110) results in the average duration of fades $T_{\eta_-}(r)$ of extended Suzuki processes of Type II.

Some examples illustrating the results found for $N_\eta(r)$ and $T_{\eta_-}(r)$ are depicted in Figures 6.22(a)–6.22(d). Figures 6.22(a) and 6.22(b) show the normalized level-crossing rate $N_\eta(r)/f_{\max}$, calculated according to (6.110) for various values of the parameter m_3 and $\kappa_c = f_{\max}/f_c$, respectively. In the logarithmic representation of Figure 6.22(a), we notice that a change of the parameter m_3 essentially causes a horizontal shift of the normalized

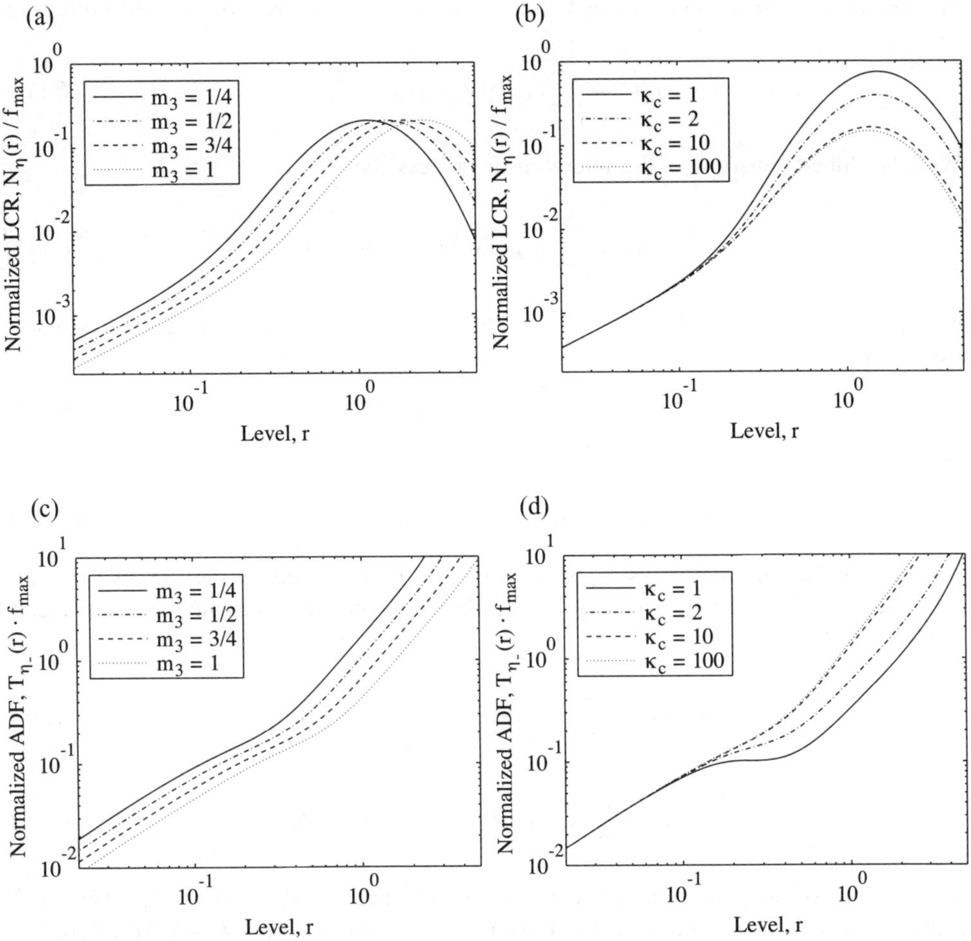

Figure 6.22 Normalized level-crossing rate $N_\eta(r)/f_{\max}$ of extended Suzuki processes (Type II) for various values of the parameters: (a) m_3 ($\kappa_c = 5$) and (b) κ_c ($m_3 = 0.5$), as well as (c) and (d) the corresponding normalized average duration of fades $T_{\eta_-}(r) \cdot f_{\max}$ ($\psi_0 = 0.0412$, $\kappa_0 = 0.4553$, $\rho = 0.918$, $\theta_\rho = 86°$, $\theta_0 = 97°$, $\sigma_3 = 0.5$).

level-crossing rate. Figure 6.22(b) reveals that the influence of the parameter κ_c is absolutely negligible, if κ_c takes on realistic values, i.e., $\kappa_c > 10$. The normalized average duration of fades $T_{\eta_-}(r) \cdot f_{\max}$, which was calculated according to (6.65), is depicted in Figures 6.22(c) and 6.22(d) for different values of m_3 and κ_c, respectively.

6.2.3 The Deterministic Extended Suzuki Process of Type II

Referring to the stochastic model of the extended Suzuki process of Type II described in the subsection before, we now derive the corresponding deterministic model. Therefore, we again make use of the principle of deterministic channel modelling (see Section 4.1), and approximate the coloured zero-mean Gaussian random process $\nu_0(t)$ by a finite sum of sinusoidal functions

$$\tilde{\nu}_0(t) = \sum_{n=1}^{N_1} c_{1,n} \cos(2\pi f_{1,n} t + \theta_{1,n}). \tag{6.112}$$

With the Hilbert transform of the deterministic process above

$$\check{\tilde{\nu}}_0(t) = \sum_{n=1}^{N_1} c_{1,n} \sin(2\pi f_{1,n} t + \theta_{1,n}), \tag{6.113}$$

we can transform the two relations (6.82) and (6.83) into the deterministic model. Accordingly, we obtain

$$\tilde{\mu}_1(t) = \tilde{\nu}_0(t) \tag{6.114}$$

and

$$\tilde{\mu}_2(t) = \cos\theta_0 \cdot \tilde{\nu}_0(t) + \sin\theta_0 \cdot \check{\tilde{\nu}}_0(t). \tag{6.115}$$

If we now replace the deterministic process $\tilde{\nu}_0(t)$ and its Hilbert transform $\check{\tilde{\nu}}_0(t)$ by the right-hand side of (6.112) and (6.113), respectively, then the generating deterministic components can be written as follows:

$$\tilde{\mu}_1(t) = \sum_{n=1}^{N_1} c_{1,n} \cos(2\pi f_{1,n} t + \theta_{1,n}), \tag{6.116}$$

$$\tilde{\mu}_2(t) = \sum_{n=1}^{N_1} c_{1,n} \cos(2\pi f_{1,n} t + \theta_{1,n} - \theta_0). \tag{6.117}$$

Now, the role of the parameter θ_0 becomes clear: The quantity θ_0 describes the phase shift between the elementary sinusoidal functions $\tilde{\mu}_{1,n}(t)$ and $\tilde{\mu}_{2,n}(t)$ [see (4.34)]. Therefore, the phases $\theta_{2,n}$ of the second deterministic process $\tilde{\mu}_2(t)$ depend on the phases $\theta_{1,n}$ of the first deterministic process $\tilde{\mu}_1(t)$, because $\theta_{2,n} = \theta_{1,n} - \theta_0$ holds.

One may also take into account that for the path gains $c_{i,n}$ and the Doppler frequencies $f_{i,n}$ the relations $c_{1,n} = c_{2,n}$ and $f_{1,n} = f_{2,n}$ hold. In particular, for the special case $\theta_0 = \pm 90°$, the complex-valued deterministic process $\tilde{\mu}(t) = \tilde{\mu}_1(t) + j\tilde{\mu}_2(t)$ can be represented as a sum-of-cisoids

$$\tilde{\mu}(t) = \sum_{n=1}^{N_1} c_{1,n}\, e^{\pm j(2\pi f_{1,n} t + \theta_{1,n})}. \tag{6.118}$$

The deterministic lognormal process $\tilde{\lambda}(t)$, which models the slow fading, can be realized using the structure shown in the bottom part of Figure 6.9. Accordingly, a further deterministic process $\tilde{\nu}_3(t)$ is necessary, which has to be designed in such a way that it does not correlate with the process $\tilde{\nu}_0(t)$. Since these two processes are (approximately) Gaussian distributed, the statistical independence of $\tilde{\nu}_0(t)$ and $\tilde{\nu}_3(t)$ follows from the uncorrelatedness. As a result, the deterministic processes $\tilde{\xi}(t)$ and $\tilde{\lambda}(t)$ derived from these are also statistically independent.

By using (6.116) and (6.117), the stochastic reference model for the extended Suzuki process of Type II (see Figure 6.20) can now easily be transformed into the deterministic simulation model shown in Figure 6.23.

The statistical properties of deterministic extended Suzuki processes $\tilde{\eta}(t)$ of Type II can be described approximately by the relations $p_\eta(z)$, $N_\eta(r)$, and $T_{\eta_-}(r)$ derived for the reference model before, if the characteristic quantities in (6.88a)–(6.88f) are there replaced by those corresponding to the simulation model. In the following, we will derive the characteristic quantities of the simulation model. We therefore need the autocorrelation functions of the processes $\tilde{\mu}_1(t)$ and $\tilde{\mu}_2(t)$

$$\tilde{r}_{\mu_1\mu_1}(\tau) = \tilde{r}_{\mu_2\mu_2}(\tau) = \sum_{n=1}^{N_1} \frac{c_{1,n}^2}{2} \cos(2\pi f_{1,n}\tau) \tag{6.119}$$

as well as the cross-correlation function calculated according to (4.13)

$$\tilde{r}_{\mu_1\mu_2}(\tau) = \tilde{r}_{\mu_2\mu_1}(-\tau) = \sum_{n=1}^{N_1} \frac{c_{1,n}^2}{2} \cos(2\pi f_{1,n}\tau - \theta_0). \tag{6.120}$$

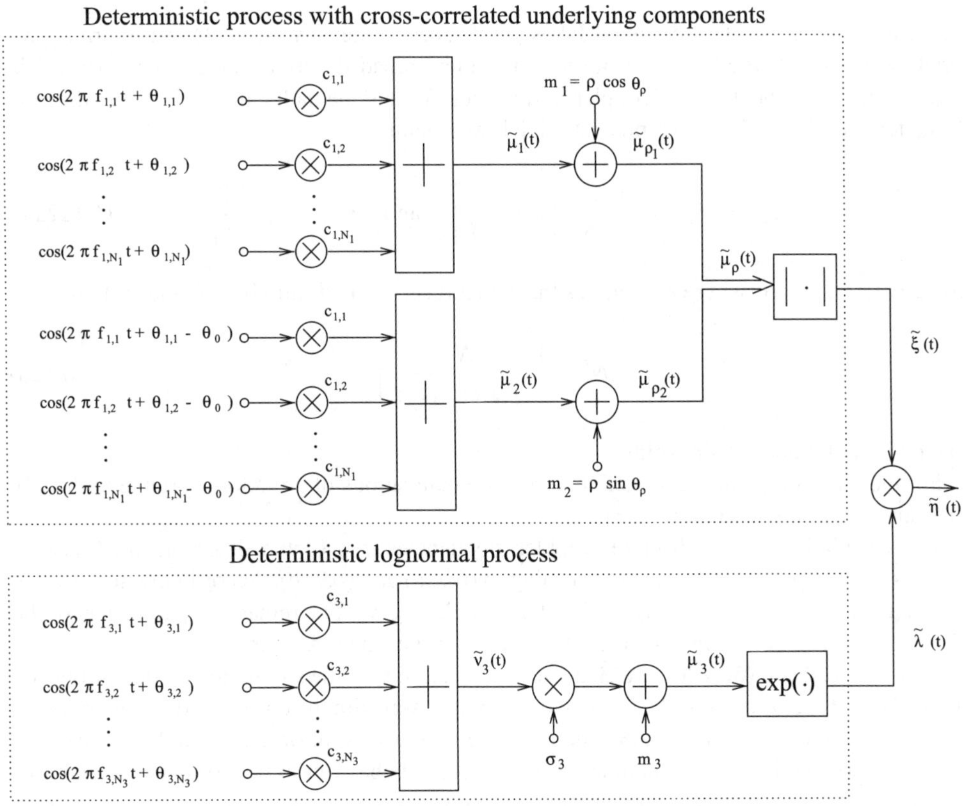

Figure 6.23 Deterministic simulation model for extended Suzuki processes (Type II).

With these two functions, the characteristic quantities of the simulation model $\tilde{\psi}_0^{(n)} = \tilde{r}_{\mu_1\mu_1}^{(n)}(0)$ and $\tilde{\phi}_0^{(n)} = \tilde{r}_{\mu_1\mu_2}^{(n)}(0)$ can easily be determined for $n = 0, 1, 2$. We immediately obtain the following closed-form expressions:

$$\tilde{\psi}_0^{(0)} = \tilde{\psi}_0 = \sum_{n=1}^{N_1} \frac{c_{1,n}^2}{2}, \tag{6.121a}$$

$$\tilde{\psi}_0^{(1)} = \dot{\tilde{\psi}}_0 = 0, \tag{6.121b}$$

$$\tilde{\psi}_0^{(2)} = \ddot{\tilde{\psi}}_0 = -2\pi^2 \sum_{n=1}^{N_1} (c_{1,n} f_{1,n})^2, \tag{6.121c}$$

$$\tilde{\phi}_0^{(0)} = \tilde{\phi}_0 = \tilde{\psi}_0 \cdot \cos\theta_0, \tag{6.121d}$$

$$\tilde{\phi}_0^{(1)} = \dot{\tilde{\phi}}_0 = \pi \sum_{n=1}^{N_1} (c_{1,n}^2 f_{1,n}) \cdot \sin\theta_0, \tag{6.121e}$$

$$\tilde{\phi}_0^{(2)} = \ddot{\tilde{\phi}}_0 = \ddot{\tilde{\psi}}_0 \cdot \cos\theta_0. \tag{6.121f}$$

Since this model uses the restricted Jakes power spectral density ($\kappa_0 \leq 1$), we return appropriately to the modified method of exact Doppler spread described in Subsection 6.1.4 in order to calculate the discrete Doppler frequencies $f_{1,n}$ and the path gains $c_{1,n}$. After adjusting Equations (6.72)–(6.74) to the present model, we obtain

$$f_{1,n} = f_{\max} \sin\left[\frac{\pi}{2N_1'}\left(n - \frac{1}{2}\right)\right] \quad \text{and} \quad c_{1,n} = \sigma_0 \sqrt{\frac{2}{N_1'}} \tag{6.122a,b}$$

for $n = 1, 2, \ldots, N_1$, where N_1 denotes the actual (user defined) number of sinusoids and

$$N_1' = \left\lceil \frac{N_1}{\frac{2}{\pi}\arcsin(\kappa_0)} \right\rceil \tag{6.123}$$

is the virtual number of sinusoids.

For the phases $\theta_{1,n}$, it is assumed that they are realizations of a random variable uniformly distributed over the interval $(0, 2\pi]$.

The calculation of the discrete Doppler frequencies $f_{3,n}$ of the deterministic Gaussian process $\tilde{\nu}_3(t)$ is performed exactly according to (6.75a) and (6.75b). Accordingly, for $c_{3,n}$, the formula $c_{3,n} = \sqrt{2/N_3}$ is used again for all $n = 1, 2, \ldots, N_3$. The remaining parameters of the simulation model (ρ, θ_ρ, m_3, σ_3) are identical to those of the reference model.

With (6.122a) and (6.122b), the characteristic quantities of the simulation model introduced in (6.121a)–(6.121f) can now be evaluated. A comparison with the corresponding quantities of the reference model (6.88a)–(6.88f) then provides the desired information on the precision of the simulation model. As an example, the convergence behaviour of the normalized quantities $\ddot{\tilde{\psi}}_0/f_{\max}^2$ and $\dot{\tilde{\phi}}_0/f_{\max}$ is depicted in Figures 6.24(a) and 6.24(b), respectively. Just as in Figures 6.10(a) and 6.10(b), one can here as well see that the deviations of the depicted quantities

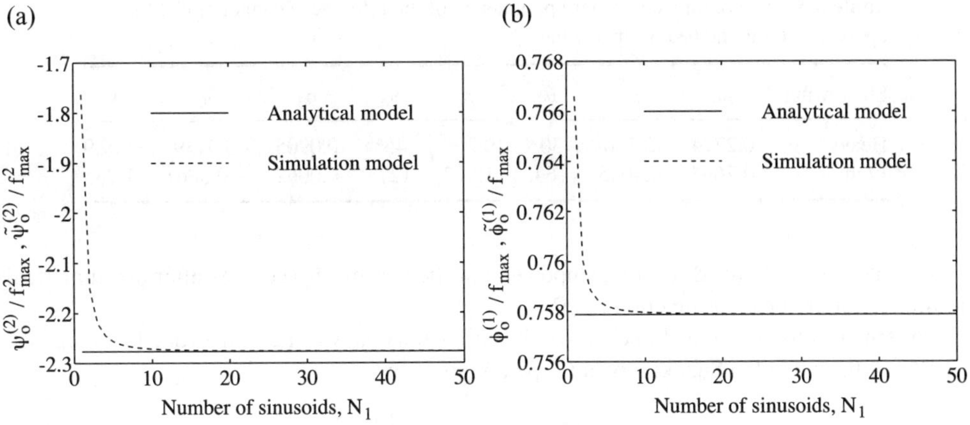

Figure 6.24 Illustration of (a) $\ddot{\psi}_0/f_{\max}^2$ and $\ddot{\tilde{\psi}}_0/f_{\max}^2$ as well as (b) $\dot{\phi}_0/f_{\max}$ and $\dot{\tilde{\phi}}_0/f_{\max}$ as a function of N_1 (MEDS, $\sigma_0^2 = 2$, $\kappa_0 = 1/2$, $\theta_0 = 45°$).

of the simulation model are negligible compared to the reference model for all cases relevant in practice (i.e., $N_1 \geq 7$).

For the level-crossing rate $\tilde{N}_\eta(r)$ and the average duration of fades $\tilde{T}_{\eta_-}(r)$ of the simulation model the statements made in Subsection 6.1.4 are also valid in the present case.

6.2.4 Applications and Simulation Results

This subsection describes how the statistical properties of stochastic and deterministic extended Suzuki processes of Type II can be brought in accordance with those of measured channels. This is again performed by optimizing the primary model parameters. The basis for the approach is provided by the measurement results found in the literature [196] ($F_{\eta_+}^\star(r)$, $N_\eta^\star(r)$, $T_{\eta_-}^\star(r)$), which we have already introduced in Subsection 6.1.5. Only in this way, can a fair performance comparison be made between the extended Suzuki processes of Type I and those of Type II.

In the present case, the parameter vector $\mathbf{\Omega}$ is defined by

$$\mathbf{\Omega} := (\sigma_0, \kappa_0, \theta_0, \rho, \theta_\rho, \sigma_3, m_3, \kappa_c). \tag{6.124}$$

This time, the vector $\mathbf{\Omega}$ contains all primary model parameters of the extended Suzuki process (Type II), including κ_c, although exactly this parameter has no significant influence on the first and second order statistics of the process $\eta(t)$, if κ_c exceeds the value 10. It will be left to the optimization procedure to find a suitable value for this quantity.

Since the error function $E_2(\mathbf{\Omega})$ [see (6.76)] has proved to be particularly useful for our previous applications, we will also make use of this function in the present minimization problem, which can again be solved by applying the Fletcher-Powell algorithm [162]. Of course, concerning the evaluation of (6.76), it has to be taken into account that the complementary cumulative distribution function $F_{\eta_+}(r/\rho) = 1 - F_{\eta_-}(r/\rho)$ has now to be calculated by means of (6.111) and that the level-crossing rate $N_\eta(r/\rho)$ is defined by (6.110). Table 6.3

Table 6.3 The optimized primary parameters of the reference channel model for areas with light and heavy shadowing.

Shadowing	σ_0	κ_0	θ_0	ρ	θ_ρ	σ_3	m_3	κ_c
Heavy	0.2774	0.506	30°	0.269	45°	0.0905	0.0439	119.9
Light	0.7697	0.4045	164°	1.567	127°	0.0062	−0.3861	1.735

shows the results obtained for the components of the parameter vector $\mathbf{\Omega}$ after the numerical minimization of the error function $E_2(\mathbf{\Omega})$.

With the results shown in Table 6.3 for the parameters σ_0, κ_0, and ρ, the Rice factor c_R [see (3.53)] of the extended Suzuki model (Type II), given by

$$c_R = \frac{\rho^2}{2\psi_0} = \frac{\pi}{4} \cdot \frac{\rho^2}{\sigma_0^2 \arcsin(\kappa_0)}, \tag{6.125}$$

takes on the values $c_R = 1.43$ dB (heavy shadowing) and $c_R = 8.93$ dB (light shadowing).

Figure 6.25(a) shows the complementary cumulative distribution function $F_{\eta_+}(r/\rho)$ of the reference model in comparison with that of the real-world channel $F^{\star}_{\eta_+}(r/\rho)$. At heavy shadowing, we obtain minor deviations at low (normalized to ρ) signal levels r/ρ. The deviations almost disappear as soon as r/ρ takes on medium or even large values. At light shadowing, on the other hand, the deviations are largest at medium signal levels, whereas they can be ignored at low levels.

Figure 6.25(b) shows the normalized level-crossing rate $N_\eta(r/\rho)/f_{\max}$ of the reference model and that of the measured channel $N^{\star}_\eta(r/\rho)/f_{\max}$. One can see that the two level-crossing rates match each other astonishingly well over the whole depicted domain of the signal level.

A comparison between the corresponding normalized average duration of fades is shown in Figure 6.25(c). The results presented there are quite good already, but it stands to reason that there is still room for further improvement, which we can indeed achieve by a further model extension, as we will see in the next section.

At this point, a comparison of the performance between the two extended Suzuki processes (Type I and Type II) is indicated. With regard to the complementary cumulative distribution function, both model types provide the same good results to a certain extent (compare Figure 6.25(a) with Figure 6.12(a)). However, the flexibility of the level-crossing rate of the extended Suzuki process of Type II seems to be higher than that of Type I, which would explain the clearly better results of Figure 6.25(b) compared to those of Figure 6.12(b). To be fair though, we have to add that the higher flexibility comes along with a greater complexity of the reference model. Since the achievable improvements can only be reached with a higher numerical computational complexity, the user himself has to decide from case to case, i.e., in our terminology "from channel to channel", whether the achievable improvements justify a higher analytical and computational complexity or not.

However, if the parameters of the reference model have been determined, then the determination of the parameters of the corresponding deterministic simulation model can be regarded as trivial due to the closed-form formulas derived here.

If we once again study the simulation models depicted in Figures 6.9 and 6.23, it becomes apparent that the structure corresponding to the model Type II is, generally speaking, the more

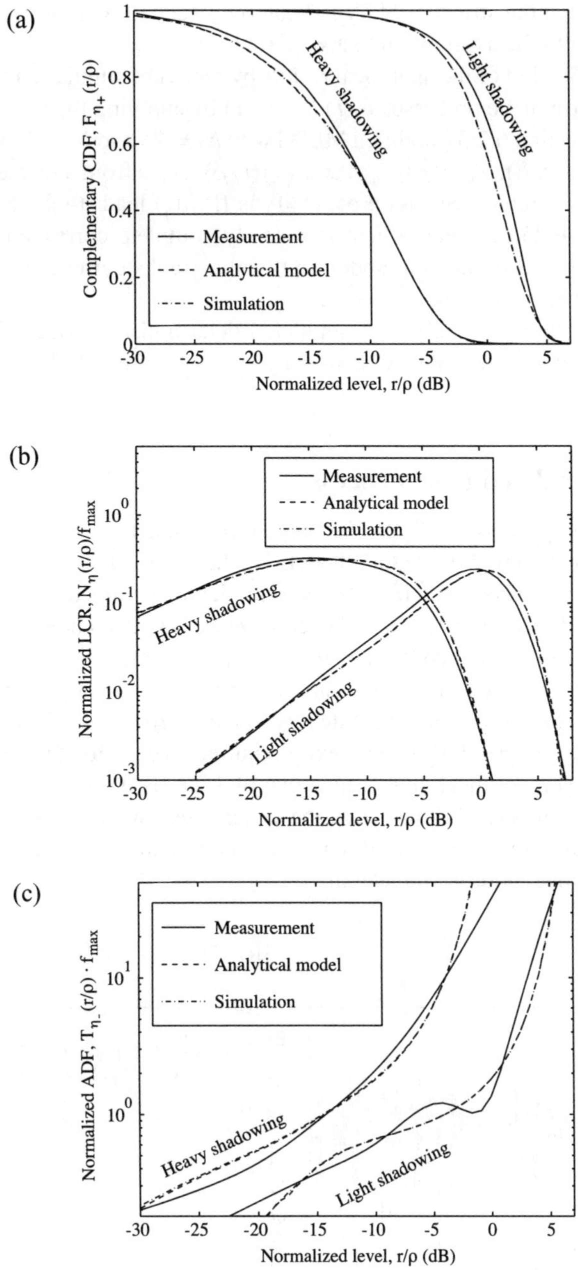

Figure 6.25 (a) Complementary cumulative distribution function $F_{\eta_+}(r/\rho)$, (b) normalized level-crossing rate $N_\eta(r/\rho)/f_{\max}$, and (c) normalized average duration of fades $T_{\eta_-}(r/\rho) \cdot f_{\max}$ for areas with heavy and light shadowing. The measurement results were adopted from [196].

efficient one, and that the structure of Type I can only keep up with it, if N_2 is equal to zero, which is equivalent to the assumption that $\kappa_0 = 0$ holds.

Finally, the verification of the analytical results by means of simulation remains. Therefore, we design the deterministic processes $\tilde{\nu}_0(t)$ and $\tilde{\nu}_3(t)$ by applying the techniques described in the preceding Subsection 6.2.3 (modified MEDS with $N_1 = 25$ and $N_3 = 15$). The measurement of the functions $\tilde{F}_{\eta_+}(r/\rho)$, $\tilde{N}_\eta(r/\rho)/f_{\max}$, and $\tilde{T}_{\eta_-}(r/\rho) \cdot f_{\max}$ from a discrete-time simulation of the deterministic extended Suzuki process (Type II) $\tilde{\eta}(t)$ leads to the curves also depicted in Figures 6.25(a)–6.25(c). Again, there is a nearly complete correspondence between the reference model and the simulation model, so that the graphs corresponding to these models can hardly be distinguished from each other.

A small part of the sequence of the simulated deterministic process $\tilde{\eta}(t)$ is depicted in Figure 6.26(a) for an area with heavy shadowing and in Figure 6.26(b) for an area with light shadowing.

6.3 The Generalized Rice Process

The extended Suzuki processes of Type I and Type II represent two classes of stochastic models with different statistical properties. Both models are, however, identical if in the former model the parameter κ_0 is set to zero, and if in the latter model the parameters κ_0 and θ_0 are given by $\kappa_0 = 1$ and $\theta_0 = \pi/2$, respectively. But in general, we can say that neither the extended Suzuki process of Type I is completely covered by that of Type II nor that the reverse is true. In [185], it has been pointed out that both models can be combined to a single model. This so-called *generalized Suzuki model* contains the extended Suzuki processes of Type I and of Type II as special cases. The mathematical complexity required to describe the generalized model is considerable, however, not much higher than that of Type II. Without the lognormal process, the *generalized Rice process* follows from the generalized Suzuki process. The generalized Rice process is considerably easier to describe and is in many cases sufficient for modelling frequency-nonselective mobile radio channels.

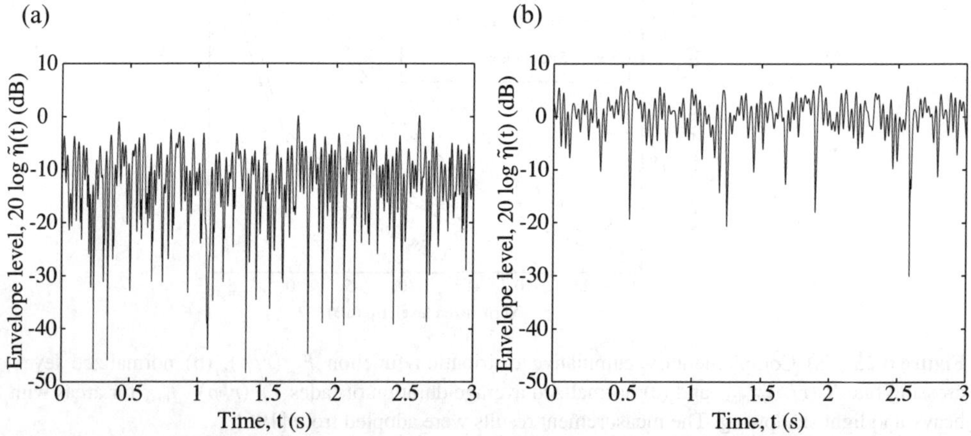

Figure 6.26 Simulation of deterministic extended Suzuki processes $\tilde{\eta}(t)$ of Type II for areas with (a) heavy shadowing and (b) light shadowing (MEDS, $N_1 = 25$, $N_3 = 15$, $f_{\max} = 91$ Hz).

This section deals with the description and the analysis of stochastic generalized Rice processes. Here, just as in previous sections, we will generally be concerned with the probability density function of the envelope, the level-crossing rate, and the average duration of fades. Since the derivation of these quantities is performed analogously to the procedure described in Subsection 6.1.1, we will be considerably briefer here. However, the comprehensibility of the derived results will still be maintained for the reader. Starting from the stochastic generalized Rice model, it then follows the derivation of the corresponding deterministic simulation model. Finally, the section closes with the fitting of the stochastic reference model and the deterministic simulation model to a real-world channel.

6.3.1 The Stochastic Generalized Rice Process

Let us study the reference model for a generalized Rice process $\xi(t)$ as depicted in Figure 6.27. The directly visible parameters of this model are θ_0, ρ, and θ_ρ, which are already known to us. We demand from the coloured real-valued Gaussian random processes $\nu_1(t)$ and $\nu_2(t)$ that they are zero-mean and statistically independent. For the Doppler power spectral density $S_{\nu_i\nu_i}(f)$ of the Gaussian random processes $\nu_i(t)$ $(i = 1, 2)$ it holds

$$S_{\nu_i\nu_i}(f) = \begin{cases} \dfrac{\sigma_i^2}{2\pi f_{\max}\sqrt{1-(f/f_{\max})^2}}, & |f| \leq \kappa_i f_{\max}, \\ 0, & |f| > \kappa_i f_{\max}, \end{cases} \tag{6.126}$$

where $f_{\max}$ again denotes the maximum Doppler frequency, and κ_i is a positive constant determining the Doppler bandwidth. Note that κ_i, together with the quantity σ_i^2, determines the variance of $\nu_i(t)$. In order to ensure that the chosen notation remains homogeneous, we make the following agreements: $\kappa_1 = 1$ and $\kappa_2 = \kappa_0$ with $\kappa_0 \in [0, 1]$, so that $S_{\nu_1\nu_1}(f)$ corresponds to the classical Jakes power spectral density in (6.9a), and $S_{\nu_2\nu_2}(f)$ is identical to the restricted Jakes power spectral introduced in (6.9c).

The reference model shown in Figure 6.27 includes two special cases:

$$\text{(i)}\ \sigma_1^2 = \sigma_2^2 = \sigma_0^2 \quad \text{and} \quad \theta_0 = \pi/2, \tag{6.127a}$$

$$\text{(ii)}\quad \sigma_1^2 = 0 \quad \text{and} \quad \sigma_2^2 = 2\sigma_0^2. \tag{6.127b}$$

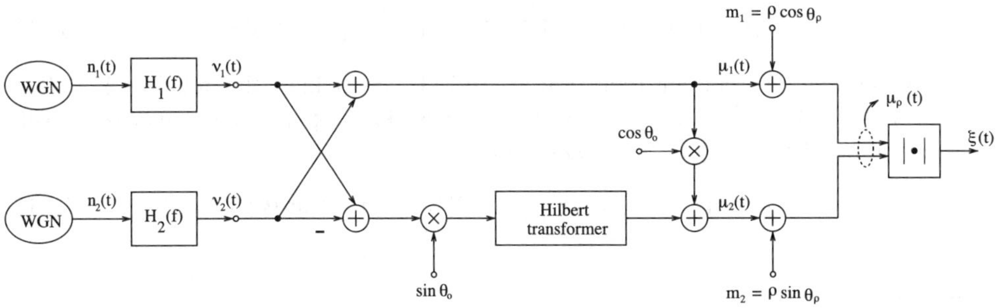

Figure 6.27 Reference model for generalized Rice processes $\xi(t)$.

In case (i), the very Rice process, depicted in Figure 6.1, whose underlying complex-valued Gaussian process is described by the left-sided restricted Jakes power spectral density in (6.2), follows from the generalized Rice process. If we disregard the missing minus sign in the lower branch of the structure shown in Figure 6.14,[3] then in case (ii) the extended Rice process (Figure 6.14) follows from the generalized Rice process (Figure 6.27).

Next, we are interested in the autocorrelation function $r_{\mu\mu}(\tau)$ and the Doppler power spectral density $S_{\mu\mu}(f)$ of the complex-valued process $\mu(t) = \mu_1(t) + j\mu_2(t)$. From Figure 6.27, we first read the relations

$$\mu_1(t) = \nu_1(t) + \nu_2(t) \tag{6.128}$$

and

$$\mu_2(t) = [\nu_1(t) + \nu_2(t)] \cos\theta_0 + \left[\check{\nu}_1(t) - \check{\nu}_2(t)\right] \sin\theta_0 . \tag{6.129}$$

From these equations, we obtain the following relations for the autocorrelation functions $r_{\mu_1\mu_1}(\tau)$ and $r_{\mu_2\mu_2}(\tau)$, as well as for the cross-correlation functions $r_{\mu_1\mu_2}(\tau)$ and $r_{\mu_2\mu_1}(\tau)$

$$r_{\mu_1\mu_1}(\tau) = r_{\mu_2\mu_2}(\tau) = r_{\nu_1\nu_1}(\tau) + r_{\nu_2\nu_2}(\tau), \tag{6.130a}$$

$$r_{\mu_1\mu_2}(\tau) = [r_{\nu_1\nu_1}(\tau) + r_{\nu_2\nu_2}(\tau)] \cos\theta_0 + [r_{\nu_1\check{\nu}_1}(\tau) - r_{\nu_2\check{\nu}_2}(\tau)] \sin\theta_0, \tag{6.130b}$$

$$r_{\mu_2\mu_1}(\tau) = [r_{\nu_1\nu_1}(\tau) + r_{\nu_2\nu_2}(\tau)] \cos\theta_0 - [r_{\nu_1\check{\nu}_1}(\tau) - r_{\nu_2\check{\nu}_2}(\tau)] \sin\theta_0, \tag{6.130c}$$

where $r_{\nu_i\nu_i}(\tau)$ $(i = 1, 2)$ denotes the inverse Fourier transform of (6.126), i.e,

$$r_{\nu_i\nu_i}(\tau) = \sigma_i^2 \frac{2}{\pi} \int_0^{\arcsin(\kappa_i)} \cos(2\pi f_{\max}\tau \sin\varphi)\, d\varphi, \tag{6.131}$$

and $r_{\nu_i\check{\nu}_i}(\tau)$, due to (2.135a), denotes the Hilbert transform of $r_{\nu_i\nu_i}(\tau)$, so that

$$r_{\nu_i\check{\nu}_i}(\tau) = \sigma_i^2 \frac{2}{\pi} \int_0^{\arcsin(\kappa_i)} \sin(2\pi f_{\max}\tau \sin\varphi)\, d\varphi \tag{6.132}$$

holds. By using the relation (6.5), the desired autocorrelation function $r_{\mu\mu}(\tau)$ can now be written as

$$r_{\mu\mu}(\tau) = 2[r_{\nu_1\nu_1}(\tau) + r_{\nu_2\nu_2}(\tau)] + j2[r_{\nu_1\check{\nu}_1}(\tau) - r_{\nu_2\check{\nu}_2}(\tau)] \sin\theta_0 . \tag{6.133}$$

After performing the Fourier transform of (6.133) and taking the relation $S_{\nu_i\check{\nu}_i}(f) = -j\,\text{sgn}\,(f) S_{\nu_i\nu_i}(f)$ into account, we can then express the Doppler power spectral density $S_{\mu\mu}(f)$ in terms of $S_{\nu_i\nu_i}(f)$ [cf. (6.126)] as follows

$$S_{\mu\mu}(f) = 2[1 + \text{sgn}\,(f) \sin\theta_0]\, S_{\nu_1\nu_1}(f) + 2[1 - \text{sgn}\,(f) \sin\theta_0]\, S_{\nu_2\nu_2}(f). \tag{6.134}$$

---

[3] It should be noted that the minus sign has no influence on the statistics of $\xi(t)$. On the other hand, the minus sign can easily be incorporated by substituting $-\theta_0$ for θ_0.

An example of this in general asymmetrical Doppler power spectral density is depicted in Figure 6.28. For the two special cases (6.127a) and (6.127b), it is obvious that Figure 6.28 converts to Figures 6.2(c) and 6.15(b), respectively. Also, the Doppler power spectral density in (6.134) includes the classical Jakes power spectral density according to (3.22) as a further special case, because we obtain the latter with the parameter constellation $\sigma_1^2 = \sigma_2^2 = \sigma_0^2$, $\kappa_1 = \kappa_2 = 1$, and $\theta_0 = \pi/2$.

Next follows the derivation of the characteristic quantities $\psi_0^{(n)}$ and $\phi_0^{(n)}$ $(n = 0, 1, 2)$. Therefore, we substitute (6.130a) into (6.11a) and (6.130b) into (6.11b), which leads to the following expressions:

$$\psi_0^{(0)} = \psi_0 = \frac{\sigma_2^2}{2}\left[\left(\frac{\sigma_1}{\sigma_2}\right)^2 + \frac{2}{\pi}\arcsin(\kappa_0)\right], \tag{6.135a}$$

$$\psi_0^{(1)} = \dot{\psi}_0 = 0, \tag{6.135b}$$

$$\psi_0^{(2)} = \ddot{\psi}_0 = -(\pi\sigma_2 f_{\max})^2\left\{\left(\frac{\sigma_1}{\sigma_2}\right)^2 + \frac{2}{\pi}\left[\arcsin(\kappa_0) - \frac{1}{2}\sin(2\arcsin(\kappa_0))\right]\right\}, \tag{6.135c}$$

$$\phi_0^{(0)} = \phi_0 = \psi_0 \cdot \cos\theta_0, \tag{6.135d}$$

$$\phi_0^{(1)} = \dot{\phi}_0 = 2\sigma_2^2 f_{\max}\left[\left(\frac{\sigma_1}{\sigma_2}\right)^2 - \left(1 - \sqrt{1-\kappa_0^2}\right)\right] \cdot \sin\theta_0, \tag{6.135e}$$

$$\phi_0^{(2)} = \ddot{\phi}_0 = \ddot{\psi}_0 \cdot \cos\theta_0, \tag{6.135f}$$

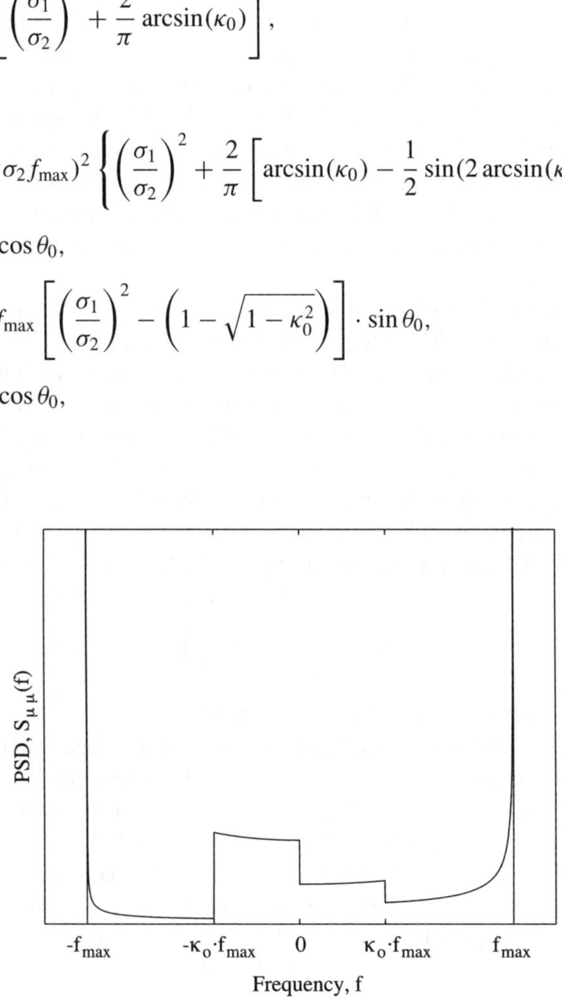

Figure 6.28 Asymmetrical Doppler power spectral density $S_{\mu\mu}(f)$ ($\sigma_1^2 = 0.25$, $\sigma_2^2 = 1$, $\theta_0 = 15°$, $\kappa_0 = 0.4$).

where $0 \leq \kappa_0 \leq 1$ and $-\pi \leq \theta_0 < \pi$. Notice that in the special case (i) described by (6.127a), the quantities presented above exactly result in Equations (6.12a)–(6.12f). On the other hand, the special case (ii) [cf. (6.127b)] leads to the formulas[4] (6.88a)–(6.88f).

With the characteristic quantities described by (6.135a)–(6.135f), the covariance matrix $\boldsymbol{C}_{\mu_\rho}(\tau)$ of the vector process $\boldsymbol{\mu}_\rho(t) = (\mu_{\rho_1}(t), \mu_{\rho_2}(t), \dot{\mu}_{\rho_1}(t), \dot{\mu}_{\rho_2}(t))$ at the same time t, i.e., $\tau = 0$, is completely determined. It holds

$$\boldsymbol{C}_{\mu_\rho}(0) = \boldsymbol{R}_\mu(0) = \begin{pmatrix} \psi_0 & \psi_0 \cos\theta_0 & 0 & \dot{\phi}_0 \\ \psi_0 \cos\theta_0 & \psi_0 & -\dot{\phi}_0 & 0 \\ 0 & -\dot{\phi}_0 & -\ddot{\psi}_0 & -\ddot{\psi}_0 \cos\theta_0 \\ \dot{\phi}_0 & 0 & -\ddot{\psi}_0 \cos\theta_0 & -\ddot{\psi}_0 \end{pmatrix}. \tag{6.136}$$

It is important to recognize here that the covariance matrix in (6.136) has the same form as that in (6.91). As a consequence of this, we again obtain the joint probability density function $p_{\xi\dot{\xi}\vartheta\dot{\vartheta}}(z, \dot{z}, \theta, \dot{\theta})$ described by (6.92), where we have to replace the quantities ψ_0, $\ddot{\psi}_0$, and $\dot{\phi}_0$ by the equations derived above, i.e., (6.135a), (6.135c), and (6.135e), respectively. Consequently, all relations derivable from these, as for example those for $p_\xi(z)$, $p_\vartheta(\theta)$, $N_\xi(r)$, and $T_{\xi_-}(r)$, lead to the results already found in Subsection 6.2.1. In the formulas given there, we merely have to replace ψ_0, $\ddot{\psi}_0$, and $\dot{\phi}_0$ with (6.135a), (6.135c), and (6.135e), respectively. Therefore, all further calculations for the description of generalized Rice processes can at this point be omitted.

This fact, however, should not lead us to conclude that extended Rice processes and generalized Rice processes are two different ways of describing one and the same stochastic process. The flexibility of generalized Rice processes is definitely higher than that of extended Rice processes. The reason for this lies in the additional primary model parameter σ_1^2, which is zero per definition for the extended Rice process and which contributes to a further de-coupling of the secondary model parameters $(\psi_0, \dot{\psi}_0, \ddot{\psi}_0, \phi_0, \dot{\phi}_0, \ddot{\phi}_0)$ of the generalized Rice process. To illustrate this point, we consider (6.88e). Obviously, there exists no real number for the parameter κ_0 in the interval $(0, 1]$, so that $\dot{\phi}_0 = 0$ holds. On the other hand, the quantity $\dot{\phi}_0$ according to (6.135e) behaves differently. Let $\sigma_1^2 \in [\sigma_2^2, 2\sigma_2^2)$, then a real-valued number

$$\kappa_0 = \frac{\sigma_1}{\sigma_2}\sqrt{2 - \left(\frac{\sigma_1}{\sigma_2}\right)^2} \tag{6.137}$$

always exists in the interval $(0, 1]$, so that $\dot{\phi}_0 = 0$ holds.

The multiplication of the generalized Rice process with a lognormal process results in the so-called *generalized Suzuki process* suggested in [185]. The generalized Suzuki process contains the classical Suzuki process [26], the modified Suzuki process [70], as well as the two extended Suzuki processes of Type I [75] and of Type II [184] as special cases. This product process is described by the probability density function (6.107c), where we have to use Equation (6.135a) for ψ_0. Similarly, for the level-crossing rate one finds the expression (6.110). Now, however, it has to be emphasized that the entries $(\psi_0, \dot{\psi}_0, \ddot{\psi}_0, \phi_0, \dot{\phi}_0, \ddot{\phi}_0)$ of the covariance matrix $\boldsymbol{C}_{\mu_\rho}$ are defined by (6.135a)–(6.135f).

[4] Due to the minus sign in the lower part of the signal flow diagram shown in Figure 6.27, it has to be taken into account that Equations (6.88e) and (6.135e) have different signs.

A detailed discussion of generalized Rice processes or Suzuki processes is not necessary for our purposes. Instead, we will continue with the design of deterministic generalized Rice processes.

6.3.2 *The Deterministic Generalized Rice Process*

We proceed again in such a way that first each of the coloured zero-mean Gaussian random processes $\nu_1(t)$ and $\nu_2(t)$ is replaced by a finite sum of N_i sinusoids of the form

$$\tilde{\nu}_i(t) = \sum_{n=1}^{N_i} c_{i,n} \cos(2\pi f_{i,n} t + \theta_{i,n}), \quad i = 1, 2. \tag{6.138}$$

When designing the deterministic processes (6.138), it has to be taken into account that $\tilde{\nu}_1(t)$ and $\tilde{\nu}_2(t)$ have to be uncorrelated, i.e., $f_{1,n} \neq f_{2,m}$ must hold for all $n = 1, 2, \ldots, N_1$ and $m = 1, 2, \ldots, N_2$. With the deterministic processes designed in this way and the corresponding Hilbert transforms, i.e.,

$$\check{\tilde{\nu}}_i(t) = \sum_{n=1}^{N_i} c_{i,n} \sin(2\pi f_{i,n} t + \theta_{i,n}), \quad i = 1, 2, \tag{6.139}$$

we can directly replace the stochastic processes $\mu_1(t)$ and $\mu_2(t)$ [cf. (6.128) and (6.129), respectively] with the corresponding deterministic processes $\tilde{\mu}_1(t)$ and $\tilde{\mu}_2(t)$. Thus, the latter processes can be expressed as follows:

$$\tilde{\mu}_1(t) = \sum_{n=1}^{N_1} c_{1,n} \cos(2\pi f_{1,n} t + \theta_{1,n}) + \sum_{n=1}^{N_2} c_{2,n} \cos(2\pi f_{2,n} t + \theta_{2,n}), \tag{6.140}$$

$$\tilde{\mu}_2(t) = \sum_{n=1}^{N_1} c_{1,n} \cos(2\pi f_{1,n} t + \theta_{1,n} - \theta_0) + \sum_{n=1}^{N_2} c_{2,n} \cos(2\pi f_{2,n} t + \theta_{2,n} + \theta_0). \tag{6.141}$$

As a result, the *deterministic generalized Rice process* is completely determined, and we obtain the simulation system in the continuous-time representation form depicted in Figure 6.29.

Now, let $\theta_0 = \pi/2$, then the structure of the deterministic Rice process with cross-correlated underlying components (cf. Figure 6.9) follows from Figure 6.29. Moreover, in the special case $\sigma_2^2 = 0$, i.e., $N_2 = 0$, we obtain the deterministic extended Rice process depicted in the upper part of Figure 6.23.

In the following, we will derive the characteristic quantities of the simulation model, i.e., $\tilde{\psi}_0^{(n)} = \tilde{r}^{(n)}_{\mu_1\mu_1}(0) = \tilde{r}^{(n)}_{\mu_2\mu_2}(0)$ and $\tilde{\phi}_0^{(n)} = \tilde{r}^{(n)}_{\mu_1\mu_2}(0)$ for $n = 0, 1, 2$. The necessary autocorrelation functions $\tilde{r}_{\mu_1\mu_1}(\tau)$ and $\tilde{r}_{\mu_2\mu_2}(\tau)$ can be expressed as

$$\begin{aligned}\tilde{r}_{\mu_1\mu_1}(\tau) &= \tilde{r}_{\mu_2\mu_2}(\tau)\\ &= \sum_{n=1}^{N_1} \frac{c_{1,n}^2}{2} \cos(2\pi f_{1,n}\tau) + \sum_{n=1}^{N_2} \frac{c_{2,n}^2}{2} \cos(2\pi f_{2,n}\tau),\end{aligned} \tag{6.142}$$

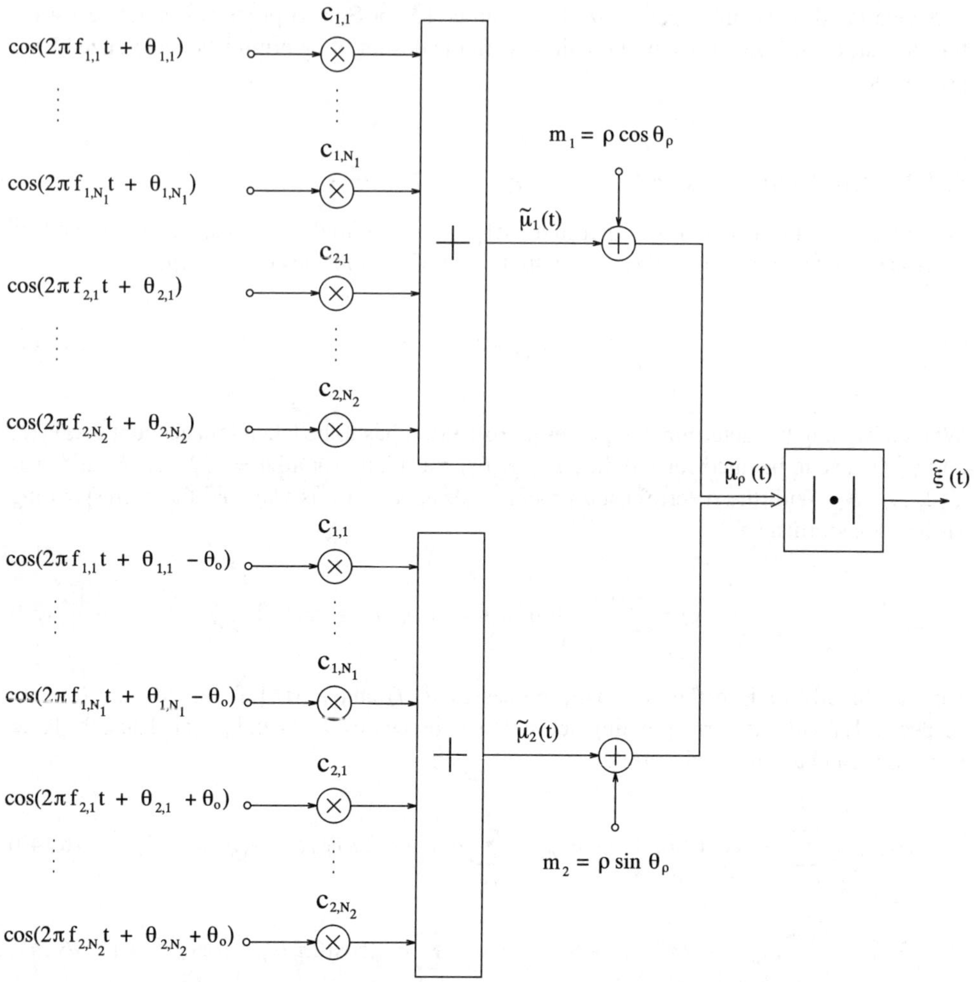

Figure 6.29 Deterministic simulation model for generalized Rice processes.

and for the cross-correlation function $\tilde{r}_{\mu_1\mu_2}(\tau)$, calculated according to (4.13), it holds

$$\begin{aligned}\tilde{r}_{\mu_1\mu_2}(\tau) &= \tilde{r}_{\mu_2\mu_1}(-\tau)\\ &= \sum_{n=1}^{N_1}\frac{c_{1,n}^2}{2}\cos(2\pi f_{1,n}\tau-\theta_0)+\sum_{n=1}^{N_2}\frac{c_{2,n}^2}{2}\cos(2\pi f_{2,n}\tau+\theta_0). \qquad (6.143)\end{aligned}$$

Thus, we obtain the following expressions for the characteristic quantities of the deterministic simulation model:

$$\tilde{\psi}_0^{(0)} = \tilde{\psi}_0 = \sum_{n=1}^{N_1}\frac{c_{1,n}^2}{2}+\sum_{n=1}^{N_2}\frac{c_{2,n}^2}{2}, \qquad (6.144a)$$

$$\tilde{\psi}_0^{(1)} = \dot{\tilde{\psi}}_0 = 0, \qquad (6.144b)$$

$$\tilde{\psi}_0^{(2)} = \ddot{\tilde{\psi}}_0 = -2\pi^2 \left[\sum_{n=1}^{N_1} (c_{1,n} f_{1,n})^2 + \sum_{n=1}^{N_2} (c_{2,n} f_{2,n})^2 \right], \quad (6.144c)$$

$$\tilde{\phi}_0^{(0)} = \tilde{\phi}_0 = \tilde{\psi}_0 \cdot \cos\theta_0, \quad (6.144d)$$

$$\tilde{\phi}_0^{(1)} = \dot{\tilde{\phi}}_0 = \pi \left[\sum_{n=1}^{N_1} (c_{1,n}^2 f_{1,n}) - \sum_{n=1}^{N_2} (c_{2,n}^2 f_{2,n}) \right] \cdot \sin\theta_0, \quad (6.144e)$$

$$\tilde{\phi}_0^{(2)} = \ddot{\tilde{\phi}}_0 = \ddot{\tilde{\psi}}_0 \cdot \cos\theta_0. \quad (6.144f)$$

With these quantities, $\tilde{\beta}$ is also determined, because

$$\tilde{\beta} = -\ddot{\tilde{\psi}}_0 - \frac{\dot{\tilde{\phi}}_0^2}{\tilde{\psi}_0 \sin^2\theta_0} \quad (6.145)$$

holds.

The calculation of the model parameters $f_{i,n}$ and $c_{i,n}$ is performed according to the method of exact Doppler spread. As described in Subsection 6.1.4, however, this procedure must be modified slightly due to $\kappa_2 = \kappa_0 \in (0, 1]$. Therefore, the formula (6.72) is also valid for computing the discrete Doppler frequencies $f_{i,n}$ in the present case, where it should be noted that (6.72) must be seen in connection with (6.73). Similarly, the calculation of the path gains $c_{i,n}$ is performed by using (6.74), where σ_0 has to be replaced by σ_i. Finally, for the phases $\theta_{i,n}$, it is assumed that these quantities are realizations (outcomes) of a random variable uniformly distributed over $(0, 2\pi]$.

Analyzing the characteristic quantities of the simulation model, we restrict ourselves to $\tilde{\psi}_0$ and $\tilde{\beta}/f_{\max}^2$. If these quantities are calculated according to (6.144a) and (6.145) by means of (6.72) and (6.74), then the convergence behaviour in terms of N_i ($N_1 = N_2$) becomes apparent as shown in Figures 6.30(a) and 6.30(b). The presented results are based on the primary model parameters σ_1^2, σ_2^2, and κ_0, as they are listed in Table 6.4 in the following subsection.

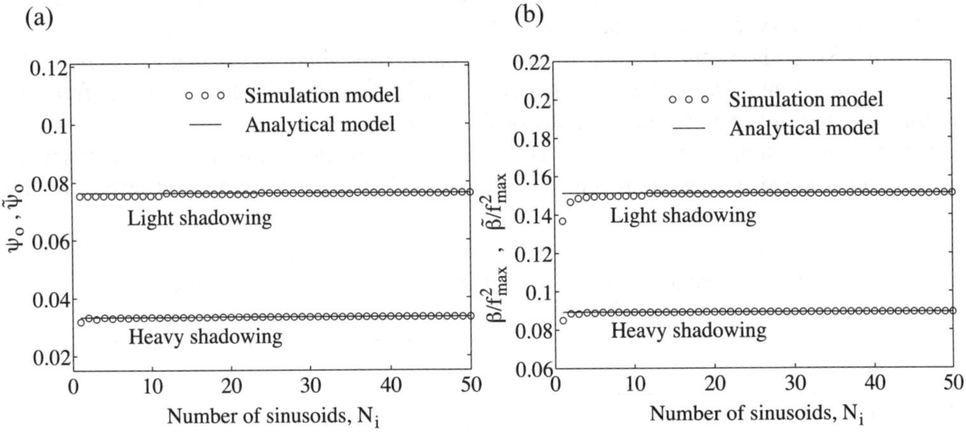

Figure 6.30 Illustration of (a) ψ_0 and $\tilde{\psi}_0$ as well as (b) $\beta/f_{\max}^2$ and $\tilde{\beta}/f_{\max}^2$ (MEDS, σ_i^2 and κ_2 according to Table 6.4).

Since the deviations between $\tilde{\psi}_0$ and ψ_0 as well as between $\tilde{\beta}/f_{max}^2$ and β/f_{max}^2 are negligible for all cases relevant in practice ($N_i \geq 7$), it follows that the probability density function $\tilde{p}_\xi(z)$, the level-crossing rate $\tilde{N}_\xi(r)$, and the average duration of fades $\tilde{T}_{\xi_-}(r)$ of the simulation model are extremely close to the corresponding quantities of the reference model.

6.3.3 Applications and Simulation Results

In this subsection, it will be shown that the statistical properties of stochastic and deterministic generalized Rice processes can be brought into surprisingly good agreement with real-world measurement results, even without multiplying the Rice process with a lognormal process. Since a fair comparison of the performance between different channel models is intended, we use again the measurement results for $F_{\xi_+}^\star(r)$, $N_\xi^\star(r)$, and $T_{\xi_-}^\star(r)$ from [196], which were also the basis for the experiments described in Subsections 6.1.5 and 6.2.4.

In the present case, the parameter vector $\mathbf{\Omega}$ contains all six primary model parameters. Thus, $\mathbf{\Omega}$ is defined by

$$\mathbf{\Omega} := (\sigma_1, \sigma_2, \kappa_0, \theta_0, \ \rho, \theta_\rho). \tag{6.146}$$

The optimization of the components of $\mathbf{\Omega}$ is performed as described in Subsection 6.1.5 by minimizing the error function $E_2(\mathbf{\Omega})$ [cf. (6.76)] by means of the Fletcher-Powell algorithm [162]. The optimization results found are presented in Table 6.4.

The Rice factor c_R [see (3.53)], i.e.,

$$c_R = \frac{\rho^2}{2\psi_0} = \frac{\rho^2}{\sigma_2^2\left[\left(\frac{\sigma_1}{\sigma_2}\right)^2 + \frac{2}{\pi}\arcsin(\kappa_2)\right]}, \tag{6.147}$$

of the present model takes on the values $c_R = 0.134$ dB (heavy shadowing) and $c_R = 8.65$ dB (light shadowing), which are about as large as the Rice factors determined for the extended Suzuki model of Type II (cf. Subsection 6.2.4).

Figure 6.31(a) shows the complementary cumulative distribution function $F_{\xi_+}(r/\rho)$ of the reference model and that of the measured channel $F_{\xi_+}^\star(r/\rho)$. Significant deviations from the results depicted in Figure 6.25(a) can obviously not be observed.

On the other hand, especially when considering the channel with heavy shadowing, we are able to achieve further improvement in respect of fitting the normalized level-crossing

Table 6.4 The optimized primary model parameters of the reference model for areas with heavy and light shadowing.

Shadowing	σ_1	σ_2	κ_0	θ_0	ρ	θ_ρ
Heavy	0.0894	0.7468	0.1651	0.3988	0.2626	30.3°
Light	0.1030	0.9159	0.2624	0.3492	1.057	53.1°

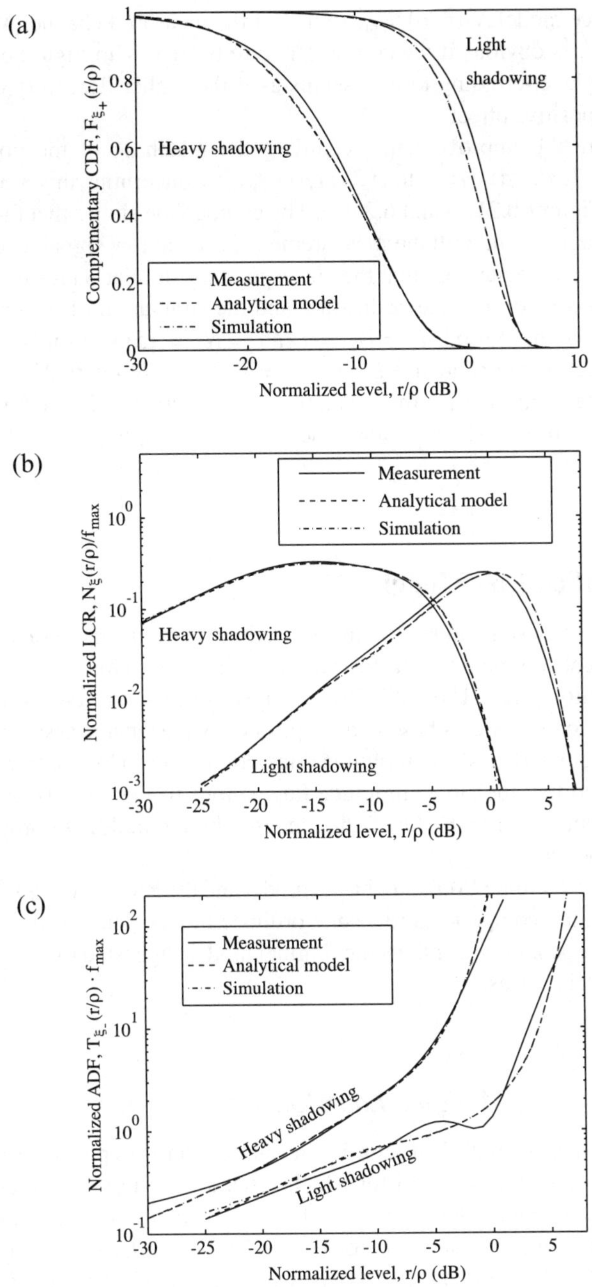

Figure 6.31 (a) Complementary cumulative distribution function $F_{\xi_+}(r/\rho)$, (b) normalized level-crossing rate $N_\xi(r/\rho)/f_{\max}$, and (c) normalized average duration of fades $T_{\xi_-}(r/\rho) \cdot f_{\max}$ for areas with heavy and light shadowing. The measurement results were adopted from [196].

rate of the reference model $N_\xi(r/\rho)/f_{\max}$ to that of the measured channel $N_\xi^\star(r/\rho)/f_{\max}$. This immediately becomes obvious if we compare Figure 6.31(b) with Figure 6.25(b). Especially with regard to the level-crossing rate, it seems as if the higher flexibility of the generalized Rice model has a positive effect.

Also evident are the improvements regarding the matching of the normalized average duration of fades $T_{\xi_-}(r/\rho) \cdot f_{\max}$ to $T_{\xi_-}^\star(r/\rho) \cdot f_{\max}$. Concerning this statement, one may compare the two Figures 6.31(c) and 6.25(c). The comparison shows that the present model is now in very good agreement with the measurements, even at low signal levels.

Finally, it should be pointed out that the corresponding simulation results are also depicted in Figures 6.31(a)–6.31(c). For the realization and the simulation of the channel with heavy (light) shadowing, we have used $N_1 = N_2 = 7$ ($N_1 = N_2 = 15$) sinusoids. In each of the two situations, a channel output sequence $\tilde{\xi}(kT_s)$ ($k = 1, 2, \ldots, N_s$) with $N_s = 3 \cdot 10^6$ sampling values was generated and used for the evaluation of the statistics. For the maximum Doppler frequency $f_{\max}$, the value 91 Hz was chosen here, and the sampling interval T_s was prescribed by $T_s = 0.3$ ms.

6.4 The Modified Loo Model

Loo developed a stochastic model for the modelling of frequency-nonselective terrestrial mobile radio channels on the basis of measurements in [178]. This model was also the topic of further investigations [179, 186, 187, 201], the results of which were summarized in a later publication [202]. Loo's model is based on the physically reasonable assumption that the line-of-sight component underlies slow amplitude fluctuations caused by shadowing effects. In this model, it is assumed that the slow amplitude fluctuations of the line-of-sight component are lognormally distributed, while the fast fading, caused by the multipath propagation, behaves like a Rayleigh process.

In this section, we combine Loo's stochastic model and the Rice model with cross-correlated inphase and quadrature components to a superordinate model. The resulting model, which is called the *modified Loo model*, contains the original model suggested by Loo and the extended Rice process as special cases.

6.4.1 *The Stochastic Modified Loo Model*

The model with which we will deal with in this subsection is depicted in Figure 6.32. This figure shows the modified Loo model for which $\nu_1(t)$, $\nu_2(t)$, and $\nu_3(t)$ are uncorrelated zero-mean real-valued Gaussian random processes. Let the Doppler power spectral density $S_{\nu_i\nu_i}(f)$ of the Gaussian random processes $\nu_i(t)$ for $i = 1, 2$ be given by the restricted Jakes power spectral density in (6.126) with $\kappa_i \in [0, 1]$, whereas we again use the Gaussian power spectral density according to (6.43) for $S_{\nu_3\nu_3}(f)$.

In this model, the fast signal fluctuations caused by the multipath propagation are modelled in the equivalent complex baseband by a complex-valued Gaussian random process

$$\mu(t) = \mu_1(t) + j\mu_2(t), \tag{6.148}$$

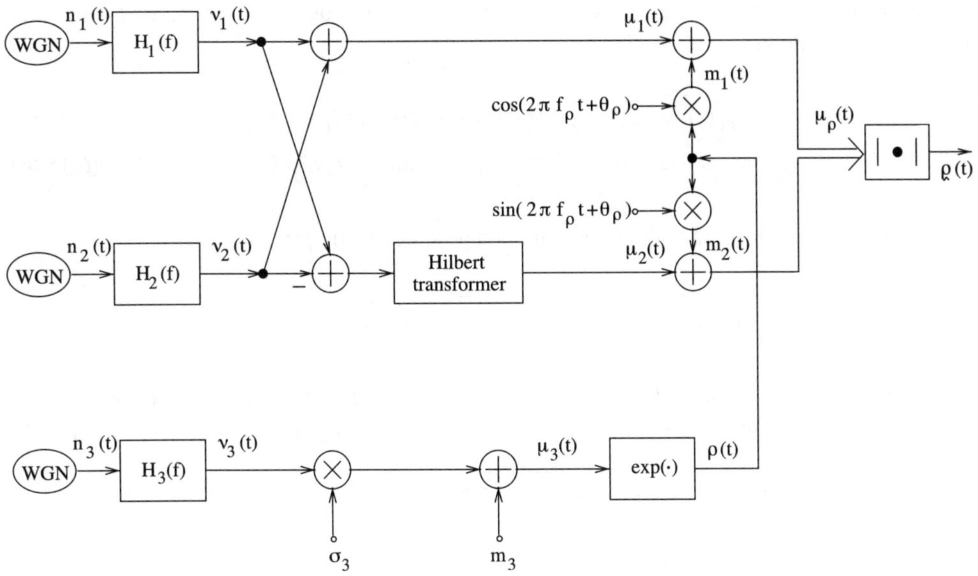

Figure 6.32 The modified Loo model (reference model).

where its real and imaginary part

$$\mu_1(t) = \nu_1(t) + \nu_2(t), \tag{6.149a}$$

$$\mu_2(t) = \check{\nu}_1(t) - \check{\nu}_2(t), \tag{6.149b}$$

are statistically uncorrelated. Here, the symbol $\check{\nu}_i(t)$ $(i = 1, 2)$ designates the Hilbert transform of $\nu_i(t)$.

For the line-of-sight component $m(t) = m_1(t) + jm_2(t)$, we read from Figure 6.32 that

$$m(t) = \rho(t) \cdot e^{j(2\pi f_\rho t + \theta_\rho)} \tag{6.150}$$

holds, where f_ρ and θ_ρ again denote the Doppler frequency and the phase of the line-of-sight component, respectively, and

$$\rho(t) = e^{\sigma_3 \nu_3(t) + m_3} \tag{6.151}$$

designates a lognormal process with which the slow amplitude fluctuations of the line-of-sight component are modelled. For the spectral and statistical properties of the lognormal process (6.151), the statements made in Subsection 6.1.2 hold. Let us assume that the bandwidth of the Gaussian random process $\nu_3(t)$ is very small in comparison with the bandwidth of $\mu(t)$, so that, consequently, the amplitude $\rho(t)$ of the line-of-sight component (6.150) only varies relatively slowly compared to the fast signal fading.

The sum of the scattered component and the line-of-sight component results in the complex-valued Gaussian random process

$$\mu_\rho(t) = \mu(t) + m(t), \tag{6.152}$$

whose real and imaginary part can be expressed — by using (6.149a), (6.149b), and (6.150) — as follows:

$$\mu_{\rho_1}(t) = \nu_1(t) + \nu_2(t) + \rho(t) \cdot \cos(2\pi f_\rho t + \theta_\rho), \tag{6.153a}$$

$$\mu_{\rho_2}(t) = \check{\nu}_1(t) - \check{\nu}_2(t) + \rho(t) \cdot \sin(2\pi f_\rho t + \theta_\rho). \tag{6.153b}$$

The absolute value of (6.152) finally results in a new stochastic process

$$\varrho(t) = \sqrt{\left[\mu_1(t) + \rho(t)\cos(2\pi f_\rho t + \theta_\rho)\right]^2 + \left[\mu_2(t) + \rho(t)\sin(2\pi f_\rho t + \theta_\rho)\right]^2}, \tag{6.154}$$

which is called the *modified Loo process*. This process will be used in the following as a stochastic model to describe the fading behaviour of frequency-nonselective satellite mobile radio channels.

The modified Loo model introduced here contains the following three special cases:

$$\text{(i)}\;\; \sigma_1^2 = \sigma_2^2 = \sigma_0^2, \quad \kappa_1 = \kappa_2 = 1, \quad \text{and} \quad f_\rho = 0, \tag{6.155a}$$

$$\text{(ii)}\;\; \sigma_2^2 = 0 \quad \text{or} \quad \kappa_2 = 0, \tag{6.155b}$$

$$\text{(iii)}\;\; \sigma_1^2 = \sigma_2^2 = \sigma_0^2, \quad \kappa_1 = 1, \kappa_2 = \kappa_0, \quad \text{and} \quad \sigma_3^2 = 0. \tag{6.155c}$$

Later on, we will see that in the special case (i), the power spectral density $S_{\mu\mu}(f)$ of the complex-valued Gaussian random process $\mu(t)$ [see (6.148)] is equal to the Jakes power spectral density. Since the Gaussian random processes $\mu_1(t)$ and $\mu_2(t)$ are uncorrelated due to the symmetry of the Jakes power spectral density, the modified Loo model (Figure 6.32) can be reduced to the classical Loo model [178, 179] depicted in Figure 6.33. One should take into consideration that also $f_\rho = 0$ holds, so that the power spectral density of the line-of-sight component does not experience a frequency shift (Doppler shift) in this model. For the second special case (ii), where σ_2^2 or κ_2 is equal to zero, the coloured Gaussian random process $\nu_2(t)$ can just as well be removed, and thus, one obtains the channel model proposed in [195], which stands out against the general variant due to its considerably lower realization complexity. Finally, the third special case (iii) leads to the Rice process depicted in Figure 6.1, for which the underlying Gaussian random processes $\mu_1(t)$ and $\mu_2(t)$ are, admittedly, also correlated, but for which the absolute value of the line-of-sight component $m(t)$ is time-independent, i.e., it then holds $|m(t)| = \rho(t) = \rho = e^{m_3}$.

6.4.1.1 Autocorrelation Function and Doppler Power Spectral Density

We are now interested in the autocorrelation function $r_{\mu_\rho\mu_\rho}(\tau)$ and in the corresponding Doppler power spectral density $S_{\mu_\rho\mu_\rho}(f)$ of the complex-valued random process $\mu_\rho(t)$ introduced in (6.152). Therefore, we first calculate the autocorrelation function $r_{\mu_{\rho_i}\mu_{\rho_i}}(\tau)$ $(i = 1, 2)$ of the processes $\mu_{\rho_i}(t)$ as well as the cross-correlation function $r_{\mu_{\rho_1}\mu_{\rho_2}}(\tau)$ of the processes $\mu_{\rho_1}(t)$ and $\mu_{\rho_2}(t)$. By using (6.153a) and (6.153b), we obtain the following relations for these

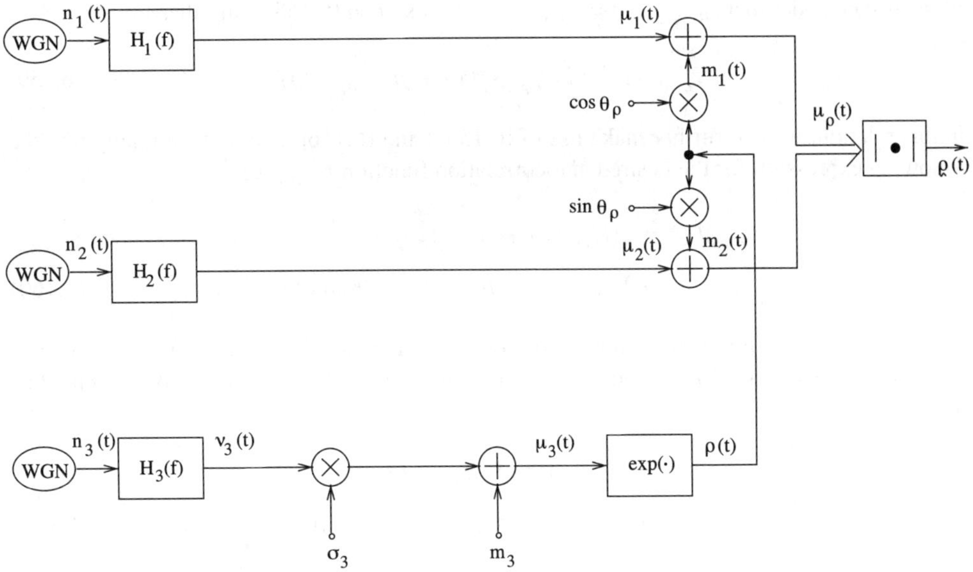

Figure 6.33 The classical Loo model (reference model).

correlation functions:

$$r_{\mu_{\rho_1}\mu_{\rho_1}}(\tau) = r_{\mu_{\rho_2}\mu_{\rho_2}}(\tau) = r_{\nu_1\nu_1}(\tau) + r_{\nu_2\nu_2}(\tau) + \frac{1}{2} r_{\rho\rho}(\tau) \cdot \cos(2\pi f_\rho \tau), \tag{6.156a}$$

$$r_{\mu_{\rho_1}\mu_{\rho_2}}(\tau) = r^*_{\mu_{\rho_2}\mu_{\rho_1}}(-\tau) = r_{\nu_1\check{\nu}_1}(\tau) - r_{\nu_2\check{\nu}_2}(\tau) + \frac{1}{2} r_{\rho\rho}(\tau) \cdot \sin(2\pi f_\rho \tau), \tag{6.156b}$$

where $r_{\nu_i\nu_i}(\tau)$ ($i = 1, 2$) describes the autocorrelation function of the Gaussian random process $\nu_i(t)$, and $r_{\nu_i\check{\nu}_i}(\tau)$ ($i = 1, 2$) denotes the cross-correlation function of $\nu_i(t)$ and $\check{\nu}_i(t)$. Recall that $r_{\nu_i\nu_i}(\tau)$ and $r_{\nu_i\check{\nu}_i}(\tau)$ are already known to us due to (6.131) and (6.132), respectively. Furthermore, $r_{\rho\rho}(\tau)$ describes the autocorrelation function of $\rho(t)$ [cf. (6.151)] in (6.156a) and in (6.156b). Notice that $\rho(t)$ has been introduced as a lognormal process in this section. That is why the autocorrelation function $r_{\rho\rho}(\tau)$ of $\rho(t)$ can be identified directly with the right-hand side of (6.47). Hence, we can therefore immediately write

$$r_{\rho\rho}(\tau) = e^{2m_3 + \sigma_3^2(1 + r_{\nu_3\nu_3}(\tau))}. \tag{6.157}$$

Following the relation in (6.5), the autocorrelation function $r_{\mu_\rho\mu_\rho}(\tau)$ of the complex-valued process $\mu_\rho(t) = \mu_{\rho_1}(t) + j\mu_{\rho_2}(t)$ can be expressed in terms of the autocorrelation functions and the cross-correlation functions of $\mu_{\rho_1}(t)$ and $\mu_{\rho_2}(t)$ as

$$r_{\mu_\rho\mu_\rho}(\tau) = r_{\mu_{\rho_1}\mu_{\rho_1}}(\tau) + r_{\mu_{\rho_2}\mu_{\rho_2}}(\tau) + j\big(r_{\mu_{\rho_1}\mu_{\rho_2}}(\tau) - r_{\mu_{\rho_2}\mu_{\rho_1}}(\tau)\big). \tag{6.158}$$

When studying (6.157) and taking (6.44) into account, we notice that $r_{\rho\rho}(\tau)$ is a real and even function in τ. From the relation (6.132), we can on the other hand conclude that $r_{\nu_i\check{\nu}_i}(\tau)$ is real and odd, so that from (6.156b) the relation $r_{\mu_{\rho_1}\mu_{\rho_2}}(\tau) = r^*_{\mu_{\rho_2}\mu_{\rho_1}}(-\tau) = -r_{\mu_{\rho_2}\mu_{\rho_1}}(\tau)$ follows.

If we also consider that $r_{\mu_{\rho_1}\mu_{\rho_1}}(\tau) = r_{\mu_{\rho_2}\mu_{\rho_2}}(\tau)$ holds, then (6.158) simplifies to

$$r_{\mu_\rho\mu_\rho}(\tau) = 2\big(r_{\mu_{\rho_1}\mu_{\rho_1}}(\tau) + jr_{\mu_{\rho_1}\mu_{\rho_2}}(\tau)\big). \tag{6.159}$$

In this relation, we can further make use of (6.156a) and (6.156b), so that we finally find the following expression for the desired autocorrelation function $r_{\mu_\rho\mu_\rho}(\tau)$

$$\begin{aligned} r_{\mu_\rho\mu_\rho}(\tau) &= 2\big(r_{\nu_1\nu_1}(\tau) + jr_{\nu_1\check{\nu}_1}(\tau)\big) \\ &\quad +2\big(r_{\nu_2\nu_2}(\tau) - jr_{\nu_2\check{\nu}_2}(\tau)\big) + r_{\rho\rho}(\tau)e^{j2\pi f_\rho \tau}. \end{aligned} \tag{6.160}$$

After performing the Fourier transform of (6.160) and using the relation $S_{\nu_i\check{\nu}_i}(f) = -j\,\mathrm{sgn}\,(f) \cdot S_{\nu_i\nu_i}(f)$, we obtain the Doppler power spectral density $S_{\mu_\rho\mu_\rho}(f)$, which can be presented as follows

$$\begin{aligned} S_{\mu_\rho\mu_\rho}(f) &= 2\big(1 + \mathrm{sgn}\,(f)\big)S_{\nu_1\nu_1}(f) \\ &\quad +2\big(1 - \mathrm{sgn}\,(f)\big)S_{\nu_2\nu_2}(f) + S_{\rho\rho}(f - f_\rho), \end{aligned} \tag{6.161}$$

where $S_{\nu_i\nu_i}(f)$ $(i = 1, 2)$ is again given by (6.126), and $S_{\rho\rho}(f - f_\rho)$ can be identified with the right-hand side of (6.48) if the frequency variable f is replaced by $f - f_\rho$ there, i.e.,

$$S_{\rho\rho}(f - f_\rho) = e^{2m_3+\sigma_3^2} \cdot \left[\delta(f - f_\rho) + \sum_{n=1}^{\infty} \frac{\sigma_3^{2n}}{n!} \cdot \frac{S_{\nu_3\nu_3}\left(\frac{f-f_\rho}{\sqrt{n}}\right)}{\sqrt{n}}\right], \tag{6.162}$$

where $S_{\nu_3\nu_3}(f)$ denotes the Gaussian power spectral density according to (6.43).

Figures 6.34(a)–6.34(f) show symbolically how the generally asymmetrical Doppler power spectral density $S_{\mu_\rho\mu_\rho}(f)$ is composed of the individual power spectral densities $S_{\nu_1\nu_1}(f)$, $S_{\nu_2\nu_2}(f)$, and $S_{\nu_3\nu_3}(f)$. The spectra shown are valid for the following parameters: $\sigma_1^2 = \sigma_2^2 = 1$, $\kappa_1 = 0.8$, $\kappa_2 = 0.6$, $\sigma_3^2 = 0.01$, $m_3 = 0$, $f_\rho = 0.4 f_{\max}$, $f_c = 0.13 f_{\max}$, and $\sigma_c^2 = 100$.

From the general representation (6.161), we can easily derive the power spectral densities determined by the special cases (i)–(iii) according to (6.155a)–(6.155c), respectively. For example, on condition that (6.155a) holds, the Doppler power spectral density $S_{\mu_\rho\mu_\rho}(f)$ of the classical Loo model (see Figure 6.33) can be derived from (6.161) in the form of

$$S_{\mu_\rho\mu_\rho}(f) = S_{\mu\mu}(f) + S_{\rho\rho}(f), \tag{6.163}$$

where $S_{\mu\mu}(f)$ denotes the Jakes power spectral density according to (3.22), and $S_{\rho\rho}(f)$ represents the power spectral density of the lognormal process $\rho(t)$. An example of the shape of $S_{\rho\rho}(f)$ is depicted in Figure 6.34(d). For the special case (ii) determined by (6.155b), the Doppler power spectral density $S_{\mu_\rho\mu_\rho}(f)$ disappears for negative frequencies in Figure 6.34(f). Finally, in the special case (iii), $S_{\rho\rho}(f - f_\rho)$ only delivers a contribution to $S_{\mu_\rho\mu_\rho}(f)$ according to (6.161), which is characterized by a weighted Dirac delta function at the point $f = f_\rho$.

Next, we calculate the characteristic quantities $\psi_0^{(n)}$ and $\phi_0^{(n)}$ $(n = 0, 1, 2)$ valid for the modified Loo model. Therefore, we substitute $r_{\mu_1\mu_1}(\tau) = r_{\nu_1\nu_1}(\tau) + r_{\nu_2\nu_2}(\tau)$ and $r_{\mu_1\mu_2}(\tau) = r_{\nu_1\check{\nu}_1}(\tau) - r_{\nu_2\check{\nu}_2}(\tau)$ into (6.11a) and (6.11b), respectively, and obtain the following expressions

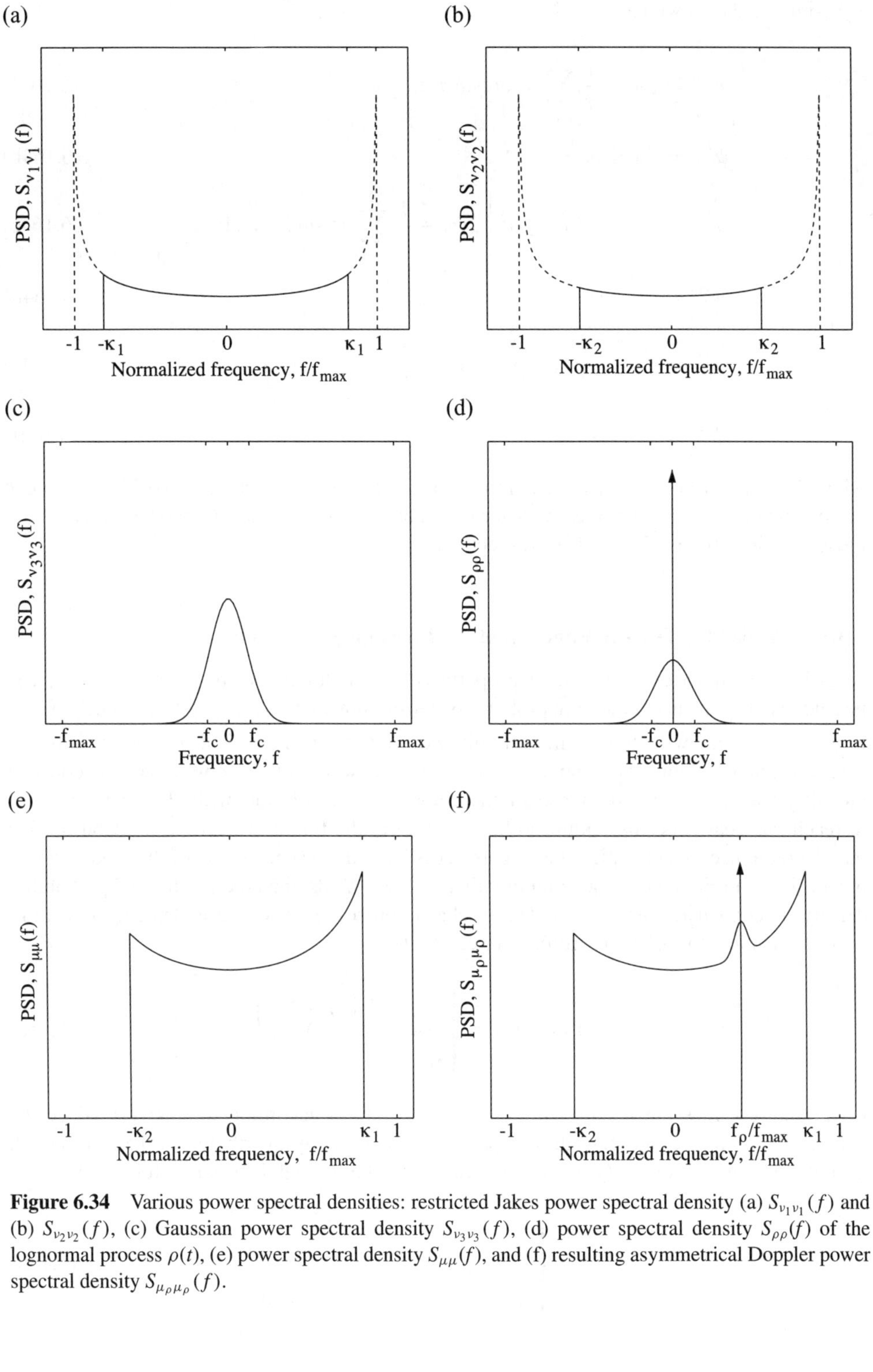

Figure 6.34 Various power spectral densities: restricted Jakes power spectral density (a) $S_{\nu_1\nu_1}(f)$ and (b) $S_{\nu_2\nu_2}(f)$, (c) Gaussian power spectral density $S_{\nu_3\nu_3}(f)$, (d) power spectral density $S_{\rho\rho}(f)$ of the lognormal process $\rho(t)$, (e) power spectral density $S_{\mu\mu}(f)$, and (f) resulting asymmetrical Doppler power spectral density $S_{\mu_\rho\mu_\rho}(f)$.

by using (6.131) as well as (6.132):

$$\psi_0^{(0)} = \psi_0 = \frac{1}{\pi}\sum_{i=1}^{2}\sigma_i^2 \arcsin(\kappa_i), \tag{6.164a}$$

$$\psi_0^{(1)} = \dot{\psi}_0 = 0, \tag{6.164b}$$

$$\psi_0^{(2)} = \ddot{\psi}_0 = -(\pi f_{\max})^2\left[2\psi_0 - \frac{1}{\pi}\sum_{i=1}^{2}\sigma_i^2 \sin\left(2\arcsin(\kappa_i)\right)\right], \tag{6.164c}$$

$$\phi_0^{(0)} = \phi_0 = 0, \tag{6.164d}$$

$$\phi_0^{(1)} = \dot{\phi}_0 = -2f_{\max}\sum_{i=1}^{2}(-1)^i\sigma_i^2\left(1-\sqrt{1-\kappa_i^2}\right), \tag{6.164e}$$

$$\phi_0^{(2)} = \ddot{\phi}_0 = 0, \tag{6.164f}$$

where $0 \leq \kappa_i \leq 1$ holds for $i = 1, 2$. In the special case (iii) described by (6.155c), one can easily convince oneself that the characteristic quantities in (6.164a)–(6.164f) are identical to those described by (6.12a)–(6.12f), respectively.

6.4.1.2 Probability Density Function of the Envelope and the Phase

As in the previous cases, the statistical properties of the modified Loo process $\varrho(t) = |\mu_\rho(t)|$ can be calculated by means of the joint probability density function $p_{\mu_{\rho_1}\mu_{\rho_2}\dot{\mu}_{\rho_1}\dot{\mu}_{\rho_2}}(x_1, x_2, \dot{x}_1, \dot{x}_2)$ or $p_{\varrho\dot{\varrho}\vartheta\dot{\vartheta}}(z, \dot{z}, \theta, \dot{\theta})$. Owing to the time variability of $\rho(t)$, the mathematical complexity is in this case much higher than for the models analyzed before, where $\rho(t) = \rho$ was always a constant quantity. Therefore, we will choose another, more elegant, solution method, which allows us to reach our goal faster by taking profit from the results found in Section 6.1. Considering that the reference model depicted in Figure 6.1 is basically a special case of the modified Loo model shown in Figure 6.32 on condition that $\rho(t) = \rho$ holds, then the conditional probability density function $p_\varrho(z\,|\,\rho(t) = \rho)$ of the stochastic process $\varrho(t)$, which is defined by (6.154), must be identical to (6.30). Therefore, we can write

$$p_\varrho(z\,|\,\rho(t) = \rho) = p_\xi(z) = \begin{cases} \dfrac{z}{\psi_0}\, e^{-\frac{z^2+\rho^2}{2\psi_0}} I_0\left(\dfrac{z\rho}{\psi_0}\right), & z \geq 0, \\ 0, & z < 0, \end{cases} \tag{6.165}$$

where ψ_0 describes the mean power of $\mu_i(t)$ $(i = 1, 2)$ according to (6.164a). Since the amplitude $\rho(t)$ of the line-of-sight component is lognormally distributed in the Loo model, meaning that the density $p_\rho(y)$ of $\rho(t)$ is given by the lognormal distribution [cf. (2.51)]

$$p_\rho(y) = \begin{cases} \dfrac{1}{\sqrt{2\pi}\sigma_3 y}\, e^{-\frac{(\ln y - m_3)^2}{2\sigma_3^2}}, & y \geq 0, \\ 0, & y < 0, \end{cases} \tag{6.166}$$

the probability density function $p_\varrho(z)$ of the modified Loo process $\varrho(t)$ can be derived from the joint probability density function $p_{\varrho\rho}(z, y)$ of the stochastic processes $\varrho(t)$ and $\rho(t)$ as follows:

$$\begin{aligned} p_\varrho(z) &= \int_0^\infty p_{\varrho\rho}(z, y)\, dy \\ &= \int_0^\infty p_\varrho(z \,|\, \rho(t) = y) \cdot p_\rho(y)\, dy \\ &= \int_0^\infty p_\xi(z; \rho = y) \cdot p_\rho(y)\, dy, \quad z \geq 0. \end{aligned} \tag{6.167}$$

If we now substitute (6.30) (or (6.165)) and (6.166) into (6.167), then we obtain the following expression for the probability density function $p_\varrho(z)$ of the modified Loo process $\varrho(t)$

$$p_\varrho(z) = \frac{z}{\sqrt{2\pi}\,\psi_0\sigma_3} \int_0^\infty \frac{1}{y}\, e^{-\frac{z^2+y^2}{2\psi_0}}\, I_0\!\left(\frac{zy}{\psi_0}\right) e^{-\frac{(\ln y - m_3)^2}{2\sigma_3^2}}\, dy, \quad z \geq 0, \tag{6.168}$$

where ψ_0 is given by (6.164a). We notice that the probability density function $p_\varrho(z)$ depends on three parameters, namely ψ_0, σ_3, and m_3. In connection with (6.155a), it now becomes apparent that (6.168) also holds for the classical Loo model, if we disregard the influences of the parameters σ_i^2 and κ_i on ψ_0. The same statement also holds for the special case (ii) introduced by (6.155b). Therefore, it is not surprising if one also finds the probability density function $p_\varrho(z)$ in the form (6.168), e.g., in [178, 179, 202] and [195]. Differences, however, do occur for the level-crossing rate and the average duration of fades, as we will see in Subsection 6.4.1.3. For completeness, we will also briefly study the effects of the special case (iii) [see (6.155c)]. In the limit $\sigma_3^2 \to 0$, the lognormal distribution (6.166) converges to $p_\rho(y) = \delta(y - \rho)$, where $\rho = e^{m_3}$. In this case, the Rice distribution (6.165) follows directly from (6.167), where it has to be taken into account that ρ is equal to e^{m_3}.

In order to illustrate the probability density function $p_\varrho(z)$ of the modified Loo process $\varrho(t)$, we study Figures 6.35(a) and 6.35(b), which allow the influence of the parameters σ_3 and m_3, respectively, to stand out.

Next, we analyze the probability density function $p_\vartheta(\theta)$ of the phase $\vartheta(t) = \arg\{\mu_\rho(t)\}$ of the modified Loo model. Here, we proceed in a similar way as for the computation of $p_\varrho(z)$. In the present situation this means that we exploit the fact that the probability density function $p_\vartheta(\theta)$ is identical to the right-hand side of (6.32) if $\rho(t) = \rho = const$. Hence, we have

$$p_\vartheta(\theta; t \,|\, \rho(t) = \rho) = \frac{e^{-\frac{\rho^2}{2\psi_0}}}{2\pi} \left\{ 1 + \sqrt{\frac{\pi}{2\psi_0}}\, \rho \cos(\theta - 2\pi f_\rho t - \theta_\rho)\, e^{\frac{\rho^2 \cos^2(\theta - 2\pi f_\rho t - \theta_\rho)}{2\psi_0}} \left[1 + \operatorname{erf}\left(\frac{\rho \cos(\theta - 2\pi f_\rho t - \theta_\rho)}{\sqrt{2\psi_0}} \right) \right] \right\}, \quad -\pi \leq \theta \leq \pi. \tag{6.169}$$

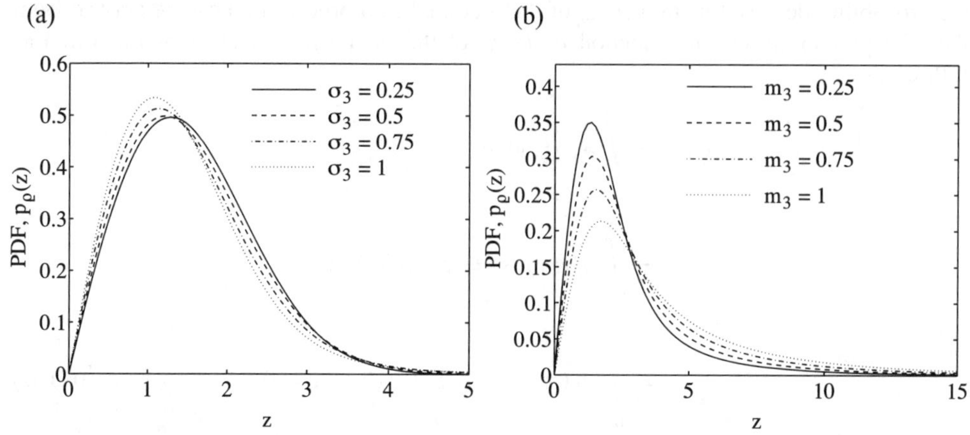

Figure 6.35 Probability density function $p_\varrho(z)$ of the envelope $\varrho(t)$ of modified and classical Loo processes depending on: (a) σ_3 ($\psi_0 = 1, m_3 = -\sigma_3^2$) and (b) m_3 ($\psi_0 = 1, \sigma_3^2 = 1$).

Since the conditional probability density function of the phase $\vartheta(t)$ for $f_\rho \neq 0$ is always a function of the time t according to this equation, we first perform an averaging of the expression above with respect to the time t. This leads to the uniform distribution

$$\begin{aligned} p_\vartheta(\theta \mid \rho(t) = \rho) &= \lim_{T\to\infty} \frac{1}{2T} \int_{-T}^{T} p_\vartheta(\theta; t \mid \rho(t) = \rho)\, dt \\ &= \frac{1}{2\pi}, \qquad -\pi \leq \theta \leq \pi. \end{aligned} \tag{6.170}$$

The desired probability density function $p_\vartheta(\theta)$ of the phase $\vartheta(t) = \arg\{\mu_\rho(t)\}$ can now be determined by means of the joint probability density function $p_{\vartheta\rho}(\theta, y)$ of $\vartheta(t)$ and $\rho(t)$ as follows:

$$\begin{aligned} p_\vartheta(\theta) &= \int_0^\infty p_{\vartheta\rho}(\theta, y)\, dy \\ &= \int_0^\infty p_\vartheta(\theta \mid \rho(t) = y) \cdot p_\rho(y)\, dy \\ &= \frac{1}{2\pi} \int_0^\infty p_\rho(y)\, dy \\ &= \frac{1}{2\pi}, \qquad -\pi \leq \theta \leq \pi. \end{aligned} \tag{6.171}$$

Thus, it is proven that the phase $\vartheta(t)$ of $\mu_\rho(t)$ is uniformly distributed in the interval $[-\pi, \pi]$, if the Doppler frequency f_ρ of the line-of-sight component $m(t)$ is not equal to zero. Similarly, an expression for $p_\vartheta(\theta)$ can be derived for the case $f_\rho = 0$. We will, however, refrain from a presentation of the resulting formula at this point.

6.4.1.3 Level-Crossing Rate and Average Duration of Fades

The derivation of the level-crossing rate $N_\varrho(r)$ of the modified Loo process $\varrho(t)$ is performed by means of the fundamental relation (6.33). Since the knowledge of the joint probability density function $p_{\varrho\dot{\varrho}}(z, \dot{z})$ of the processes $\varrho(t)$ and $\dot{\varrho}(t)$ at the same time t is necessary again, we will first derive this quantity. Therefore, we write

$$\begin{aligned} p_{\varrho\dot{\varrho}}(z, \dot{z}) &= \int_0^\infty \int_{-\infty}^\infty p_{\varrho\dot{\varrho}\rho\dot{\rho}}(z, \dot{z}, y, \dot{y})\, d\dot{y}\, dy \\ &= \int_0^\infty \int_{-\infty}^\infty p_{\varrho\dot{\varrho}}(z, \dot{z}\,|\,\rho(t) = y, \dot{\rho}(t) = \dot{y}) \cdot p_{\rho\dot{\rho}}(y, \dot{y})\, d\dot{y}\, dy. \end{aligned} \tag{6.172}$$

In the latter expression, $p_{\rho\dot{\rho}}(y, \dot{y})$ denotes the joint probability density function of $\rho(t)$ and $\dot{\rho}(t)$ at the same time t. Since the process $\rho(t)$ is lognormally distributed in the modified Loo model, we can readily identify $p_{\rho\dot{\rho}}(y, \dot{y})$ with the expression in (6.53), which allows us to write

$$p_{\rho\dot{\rho}}(y, \dot{y}) = \frac{e^{-\frac{(\ln y - m_3)^2}{2\sigma_3^2}}}{\sqrt{2\pi}\,\sigma_3 y} \cdot \frac{e^{-\frac{\dot{y}^2}{2\gamma(\sigma_3 y)^2}}}{\sqrt{2\pi\gamma}\,\sigma_3 y}, \tag{6.173}$$

where $\gamma = -\ddot{r}_{\nu_3\nu_3}(0) = (2\pi\sigma_c)^2$. At the beginning of Subsection 6.4.1, we assumed that the amplitude $\rho(t)$ of the line-of-sight component changes only very slowly. Therefore, $\dot{\rho}(t) \approx 0$ must hold approximately, so that the probability density function $p_{\dot{\rho}}(\dot{y})$ of $\dot{\rho}(t)$ can be approximated by $p_{\dot{\rho}}(\dot{y}) \approx \delta(\dot{y})$. Since this always holds if γ is sufficiently small or if the frequency ratio $\kappa_c = f_{\max}/f_c$ is sufficiently large, we can in this case replace (6.173) by the approximation

$$p_{\rho\dot{\rho}}(y, \dot{y}) \approx p_\rho(y) \cdot \delta(\dot{y}), \tag{6.174}$$

where $p_\rho(y)$ again denotes the lognormal distribution according to (6.166). Regarding the sifting property of the Dirac delta function, we now substitute (6.174) into (6.172) and obtain the approximation

$$\begin{aligned} p_{\varrho\dot{\varrho}}(z, \dot{z}) &\approx \int_0^\infty \int_{-\infty}^\infty p_{\varrho\dot{\varrho}}(z, \dot{z}\,|\,\rho(t) = y, \dot{\rho}(t) = \dot{y}) \cdot p_\rho(y)\, \delta(\dot{y})\, d\dot{y}\, dy \\ &= \int_0^\infty p_{\varrho\dot{\varrho}}(z, \dot{z}\,|\,\rho(t) = y, \dot{\rho}(t) = 0) \cdot p_\rho(y)\, dy. \end{aligned} \tag{6.175}$$

With this relation, we can now approximate the level-crossing rate $N_\varrho(r)$ [cf. (6.33)] as follows:

$$\begin{aligned} N_\varrho(r) &= \int_0^\infty \dot{z}\, p_{\varrho\dot{\varrho}}(r, \dot{z})\, d\dot{z} \\ &\approx \int_0^\infty \int_0^\infty \dot{z}\, p_{\varrho\dot{\varrho}}(z, \dot{z}\,|\,\rho(t) = y, \dot{\rho}(t) = 0) \cdot p_\rho(y)\, dy\, d\dot{z} \\ &= \int_0^\infty N_\varrho(r\,|\,\rho(t) = y, \dot{\rho}(t) = 0) \cdot p_\rho(y)\, dy. \end{aligned} \tag{6.176}$$

Here, we have to take into account that the level-crossing rate $N_\varrho(r \,|\, \rho(t) = \rho, \dot{\rho}(t) = 0)$ appearing under the integral of (6.176) corresponds exactly to the relation (6.37) derived in Subsection 6.1.1.2. If we now substitute this equation together with (6.166) into (6.176), then we obtain the following approximation for the level-crossing rate of the modified Loo process

$$N_\varrho(r) \approx \int_0^\infty \frac{e^{-\frac{(\ln y - m_3)^2}{2\sigma_3^2}}}{\sqrt{2\pi}\sigma_3 y} \cdot \frac{r\sqrt{2\beta}}{\pi^{3/2}\psi_0} e^{-\frac{r^2+y^2}{2\psi_0}} \cdot \int_0^{\pi/2} \cosh\left(\frac{ry}{\psi_0}\cos\theta\right)$$
$$\cdot \left[e^{-(\alpha y \sin\theta)^2} + \sqrt{\pi}\alpha y \sin(\theta)\,\mathrm{erf}\,(\alpha y \sin\theta)\right] d\theta\, dy, \qquad (6.177)$$

where the relations (6.27) and (6.28) hold for α and β, respectively, if there the formulas (6.164a), (6.164c), and (6.164e) are used for the characteristic quantities ψ_0, $\ddot{\psi}_0$, and $\dot{\phi}_0$, respectively.

The investigation of the special case (i) [see (6.155a)] first of all provides $\alpha = 0$. This leads to the fact that the approximation (6.177) can be simplified considerably. Thus, provided that $\alpha = 0$ holds, the level-crossing rate $N_\varrho(r)$ of the modified Loo model simplifies to that of the classical Loo model, which can be approximated as follows:

$$N_\varrho(r)|_{\alpha=0} \approx \sqrt{\frac{\beta}{2\pi}} \cdot \frac{r}{\psi_0} \int_0^\infty \frac{e^{-\frac{(\ln y - m_3)^2}{2\sigma_3^2}}}{\sqrt{2\pi}\sigma_3 y} \cdot e^{-\frac{r^2+y^2}{2\psi_0}} I_0\left(\frac{ry}{\psi_0}\right) dy$$
$$= \sqrt{\frac{\beta}{2\pi}} \int_0^\infty p_\xi(r; \rho = y) \cdot p_\rho(y)\, dy, \qquad (6.178)$$

where the quantities β and ψ_0 are in this case given by $\beta = -2(\pi\sigma_0 f_{\max})^2$ and $\psi_0 = \sigma_0^2$, respectively, and $p_\xi(r; \rho = y)$ denotes the Rice distribution in (2.44) if ρ is replaced there by y. Studying (6.178) and (6.167) it becomes clear that on condition that $\alpha = 0$ holds, the level-crossing rate $N_\varrho(r)$ is again proportional to the probability density function $p_\varrho(r)$. This is always the case if the Doppler power spectral density is symmetrical, which often does not correspond to reality. The special case (ii) [see (6.155b)] does not lead to a simplification of (6.177). Here, however, the characteristic quantities ψ_0, $\ddot{\psi}_0$, and $\dot{\phi}_0$ are more tightly coupled to each other, which restricts the flexibility of $N_\varrho(r)$. Finally, we will investigate the consequences which the special case (iii) [see (6.155c)] has for the level-crossing rate $N_\varrho(r)$. In the limit $\sigma_3^2 \to 0$, we obtain $p_\rho(y) = \delta(y - \rho)$ with $\rho = e^{m_3}$, so that (6.37) again follows from the right-hand side of (6.176). By the way, (6.174) is then exactly fulfilled, so that we can replace the approximation sign by an equals sign in (6.176) without hesitation.

In order to be able to calculate the average duration of fades

$$T_{\varrho_-}(r) = \frac{F_{\varrho_-}(r)}{N_\varrho(r)} \qquad (6.179)$$

of the modified Loo process, we still need an expression for the cumulative distribution function $F_{\varrho_-}(r) = P\{\varrho(t) \leq r\}$ of the stochastic process $\varrho(t)$. For the derivation of $F_{\varrho_-}(r)$, we use (6.168) and obtain

$$F_{\varrho_-}(r) = \int_0^r p_\varrho(z)\,dz$$

$$= \frac{1}{\sqrt{2\pi}\,\psi_0\sigma_3}\int_0^r\int_0^\infty \frac{z}{y}\,e^{-\frac{z^2+y^2}{2\psi_0}}\,I_0\left(\frac{zy}{\psi_0}\right)e^{-\frac{(\ln y-m_3)^2}{2\sigma_3^2}}\,dy\,dz$$

$$= 1 - \int_0^\infty Q_1\left(\frac{y}{\sqrt{\psi_0}}, \frac{r}{\sqrt{\psi_0}}\right) p_\rho(y)\,dy, \tag{6.180}$$

where $Q_1(\cdot,\cdot)$ is the generalized Marcum Q-function defined by (6.67).

In order to illustrate the results found for the level-crossing rate $N_\varrho(r)$ and the average duration of fades $T_{\varrho_-}(r)$, we study the graphs depicted in Figures 6.36(a)–6.36(d). In Figures 6.36(a) and 6.36(b), the normalized level-crossing rate $N_\varrho(r)/f_{\max}$, calculated according to (6.177), is presented for various values of the parameters m_3 and σ_3, respectively. The figures below, Figures 6.36(c) and 6.36(d), show the behaviour of the corresponding normalized average duration of fades $T_{\varrho_-}(r)\cdot f_{\max}$.

6.4.2 *The Deterministic Modified Loo Model*

For the derivation of a proper simulation model for modified Loo processes, we proceed as in Subsection 6.1.4. That means, we replace the three stochastic Gaussian random processes $\nu_i(t)$ $(i = 1, 2, 3)$ by deterministic Gaussian processes $\tilde{\nu}_i(t)$ in the form of (6.68). When constructing the sets $\{f_{1,n}\}$, $\{f_{2,n}\}$, and $\{f_{3,n}\}$, one has to take care that they are mutually disjoint (mutually exclusive), which leads to the fact that the resulting deterministic Gaussian processes $\tilde{\nu}_1(t)$, $\tilde{\nu}_2(t)$, and $\tilde{\nu}_3(t)$ are in pairs uncorrelated. The substitution $\nu_i(t) \to \tilde{\nu}_i(t)$ leads to $\mu_i(t) \to \tilde{\mu}_i(t)$, where the deterministic Gaussian processes $\tilde{\mu}_i(t)$ $(i = 1, 2, 3)$ can be expressed after a simple auxiliary calculation as follows:

$$\tilde{\mu}_1(t) = \sum_{n=1}^{N_1} c_{1,n}\cos(2\pi f_{1,n}t + \theta_{1,n}) + \sum_{n=1}^{N_2} c_{2,n}\cos(2\pi f_{2,n}t + \theta_{2,n}), \tag{6.181a}$$

$$\tilde{\mu}_2(t) = \sum_{n=1}^{N_1} c_{1,n}\sin(2\pi f_{1,n}t + \theta_{1,n}) - \sum_{n=1}^{N_2} c_{2,n}\sin(2\pi f_{2,n}t + \theta_{2,n}), \tag{6.181b}$$

$$\tilde{\mu}_3(t) = \sigma_3\sum_{n=1}^{N_3} c_{3,n}\cos(2\pi f_{3,n}t + \theta_{3,n}) + m_3. \tag{6.181c}$$

With these relations, the stochastic reference model (see Figure 6.32) can be directly transformed into the *deterministic Loo model*, shown in Figure 6.37. The output process $\tilde{\varrho}(t)$ of this model is mnemonically named *deterministic modified Loo process*.

For the special case (i), introduced in (6.155a), the structure of the so-called deterministic classical Loo model follows from Figure 6.37. Here, each of the two deterministic processes

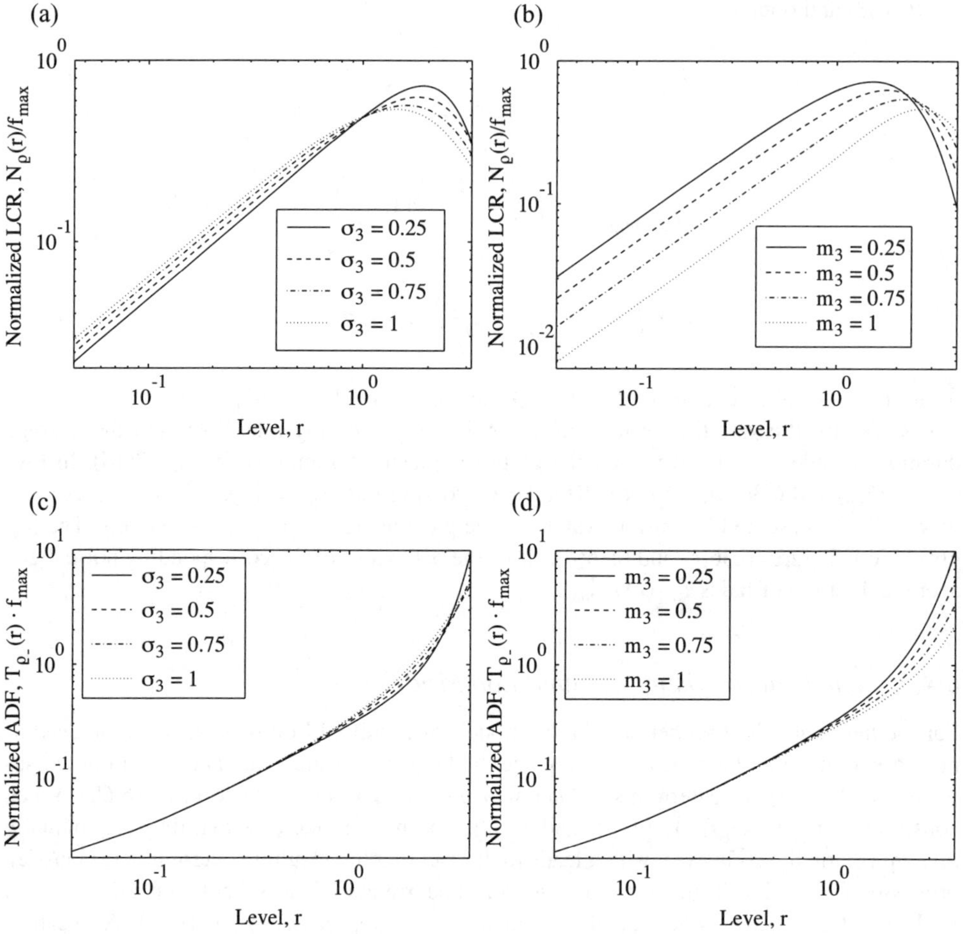

Figure 6.36 Normalized level-crossing rate $N_\varrho(r)/f_{\max}$ of the modified Loo model for various values of (a) m_3 ($\sigma_3 = 1/2$) and (b) σ_3 ($m_3 = 1/2$) as well as (c) and (d) the corresponding normalized average duration of fades $T_{\varrho_-}(r) \cdot f_{\max}$ ($\kappa_1 = \kappa_2 = 1$, $\psi_0 = 1$, $f_\rho = 0$).

$\tilde{\mu}_1(t)$ and $\tilde{\mu}_2(t)$, given by (6.181a) and (6.181b), respectively, can be replaced by the sum-of-sinusoids model introduced in (4.4), as a result of which the realization complexity of the model reduces considerably. The special case (ii) [see (6.155b)] also leads to a simplification of the structure of the simulation model, because $\sigma_2^2 = 0$ is equivalent to $N_2 = 0$. Under this condition, one obtains the simulation model introduced in [195]. Finally, we want to point out that in the special case (iii) described by (6.155c), it follows from Figure 6.37 the structure of the deterministic Rice process with cross-correlated components, which we know already from the upper part of Figure 6.9.

On condition that $N_i \geq 7$ holds, Equations (6.168), (6.177), and (6.179) derived for the reference model in Subsection 6.4.1 also hold approximately for deterministic modified Loo processes $\tilde{\varrho}(t)$, if the substitutions $\psi_0 \to \tilde{\psi}_0$, $\ddot{\psi}_0 \to \ddot{\tilde{\psi}}_0$, and $\dot{\phi}_0 \to \dot{\tilde{\phi}}_0$ are performed in the

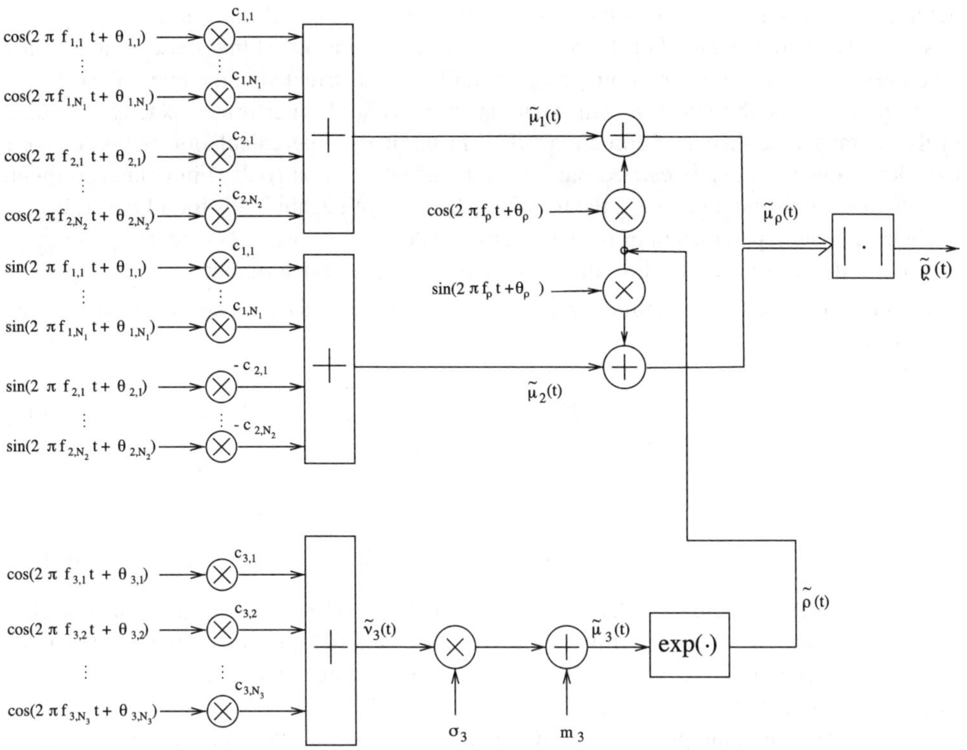

Figure 6.37 The deterministic modified Loo model (simulation model).

respective formulas. Here, the characteristic quantities $\tilde{\psi}_0$, $\ddot{\tilde{\psi}}_0$, and $\dot{\tilde{\phi}}_0$ of the simulation model are given by the relations (6.71a), (6.71b), and (6.71c) derived in Subsection 6.1.4, respectively. This is not particularly surprising because here as well as in Subsection 6.1.4, the deterministic Gaussian processes $\tilde{\mu}_1(t)$ and $\tilde{\mu}_2(t)$ are based on the same expressions. Differences, however, only occur in the calculation of the model parameters $f_{i,n}$ and $c_{i,n}$ for $i = 1, 2$. In the present case, we have to take into account that the Jakes power spectral density is in general left-sided restricted as well as right-sided restricted, due to $\kappa_1 \in [0, 1]$ and $\kappa_2 \in [0, 1]$. If we take this fact into account, when calculating the model parameters $f_{i,n}$ and $c_{i,n}$, by means of the method of exact Doppler spread (MEDS), then the following expressions hold for the deterministic modified Loo model:

$$f_{i,n} = f_{\max} \sin\left[\frac{\pi}{2N_i'}\left(n - \frac{1}{2}\right)\right], \quad n = 1, 2, \ldots, N_i \quad (i = 1, 2), \tag{6.182a}$$

$$c_{i,n} = \frac{\sigma_i}{\sqrt{N_i'}}, \quad n = 1, 2, \ldots, N_i \quad (i = 1, 2), \tag{6.182b}$$

where

$$N_i' = \left\lceil \frac{N_i}{\frac{2}{\pi}\arcsin(\kappa_i)} \right\rceil, \quad i = 1, 2, \tag{6.183}$$

describes the virtual number of sinusoids, and N_i denotes the actual number, i.e., the number of sinusoids set by the user. For the phases $\theta_{i,n}$, we assume as usual that these quantities are outcomes (realizations) of a random generator uniformly distributed in the interval $(0, 2\pi]$.

The design of the third deterministic Gaussian process $\tilde{\nu}_3(t)$ is performed exactly according to the method described in Subsection 6.1.4. In particular, the calculation of the discrete Doppler frequencies $f_{3,n}$ is carried out by means of the relation (6.75a) in connection with (6.75b), and for the path gains $c_{i,n}$ the formula $c_{3,n} = \sqrt{2/N_3}$ again holds for all $n = 1, 2, \ldots, N_3$. The remaining parameters of the simulation model $(f_\rho, \theta_\rho, m_3, \sigma_3)$ of course correspond to those of the reference model, so that all paramcters are now determined.

With the characteristic quantities $\tilde{\psi}_0$, $\ddot{\tilde{\psi}}_0$, and $\dot{\tilde{\phi}}_0$, the secondary model parameters of the simulation model

$$\tilde{\alpha} = \left(2\pi f_\rho - \frac{\dot{\tilde{\phi}}_0}{\tilde{\psi}_0}\right) \Big/ \sqrt{2\tilde{\beta}} \tag{6.184}$$

and

$$\tilde{\beta} = -\ddot{\tilde{\psi}}_0 - \dot{\tilde{\phi}}_0^2/\tilde{\psi}_0 \tag{6.185}$$

can be calculated explicitly similarly to (6.27) and (6.28). The convergence behaviour of $\tilde{\alpha}$ and $\tilde{\beta}/f_{\max}^2$ is depicted in terms of N_i in Figure 6.38(a) and 6.38(b), respectively. The curves were drawn based upon the values of the primary model parameters σ_1, σ_2, κ_1, κ_2, and f_ρ as they are presented in Table 6.5.

Figure 6.39 is an example of the time behaviour of the deterministic Loo process $\tilde{\varrho}(t)$ (continuous line), where the values $N_1 = N_2 = N_3 = 13$ were chosen for the number of sinusoids N_i $(i = 1, 2, 3)$, and the maximum Doppler frequency $f_{\max}$ was again determined by $f_{\max} = 91$ Hz. This figure also illustrates the behaviour of the deterministic lognormal process $\tilde{\rho}(t)$ (dotted line).

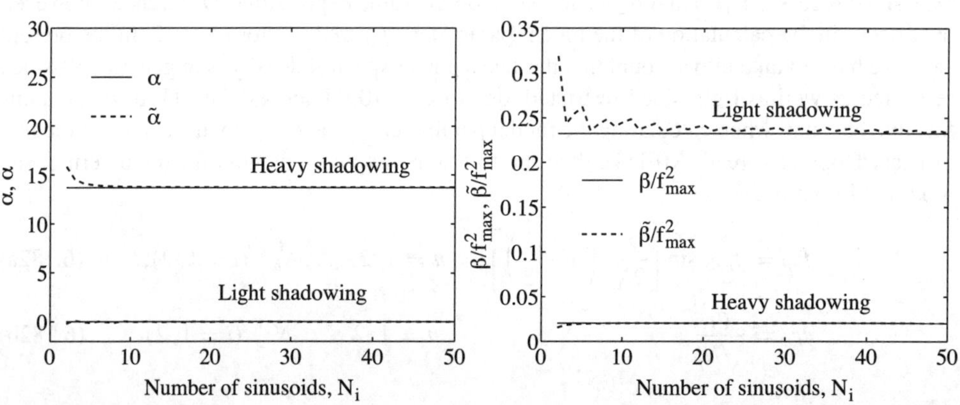

Figure 6.38 Illustration of (a) α and $\tilde{\alpha}$ as well as (b) $\beta/f_{\max}^2$ and $\tilde{\beta}/f_{\max}^2$ when using the MEDS with $N_1 = N_2$ but $N_1' \neq N_2'$ (σ_i, κ_i, and f_ρ according to Table 6.5).

Table 6.5 The optimized primary model parameters of the modified Loo model for areas with light and heavy shadowing.

Shadowing	σ_1	σ_2	κ_1	κ_2	σ_3	m_3	$f_\rho/f_{\max}$
Heavy	0	0.3856	0	0.499	0.5349	−1.593	0.1857
Light	0.404	0.4785	0.6223	0.4007	0.2628	−0.0584	0.0795

A comparison between the statistical properties of the reference model and those of the simulation model is shown in Figures 6.40(a)–6.40(c). Except for the parameter $\kappa_c = f_{\max}/f_c$, whose influence will be investigated here, all parameters of the simulation model and of the reference model were chosen exactly as in the previous example. The sampling interval T_s of the discrete deterministic Loo process $\tilde{\varrho}(kT_s)$ $(k = 1, 2, \ldots, K)$ was given by $T_s = 1/(36.63 f_{\max})$. Altogether $K = 3 \cdot 10^7$ sampling values of the process $\tilde{\varrho}(kT_s)$ $(k = 1, 2, \ldots, K)$ have been simulated and used for the determination of the probability density function $\tilde{p}_\varrho(z)$ [see Figure 6.40(a)], the normalized level-crossing rate $\tilde{N}_\varrho(r)/f_{\max}$ [see Figure 6.40(b)], and the normalized average duration of fades $\tilde{T}_{\varrho_-}(r) \cdot f_{\max}$ [see Figure 6.40(c)] of the simulation model.

Figure 6.40(a) clearly indicates that the behaviour of the probability density function $\tilde{p}_\varrho(z)$ is not influenced by the quantity κ_c. This result was to be expected, because $\tilde{p}_\varrho(z)$ is, according to (6.168), independent of the bandwidth of the process $\nu_3(t)$, which explains completely the missing influence of the frequency ratio $\kappa_c = f_{\max}/f_c$. The minor differences that can be observed between $p_\varrho(z)$ and $\tilde{p}_\varrho(z)$, are due to the limited number of sinusoids, which are here equal to $N_i = 13$ for $i = 1, 2, 3$. It goes without saying that these deviations decrease if N_i increases, and that they converge against zero as $N_i \to \infty$.

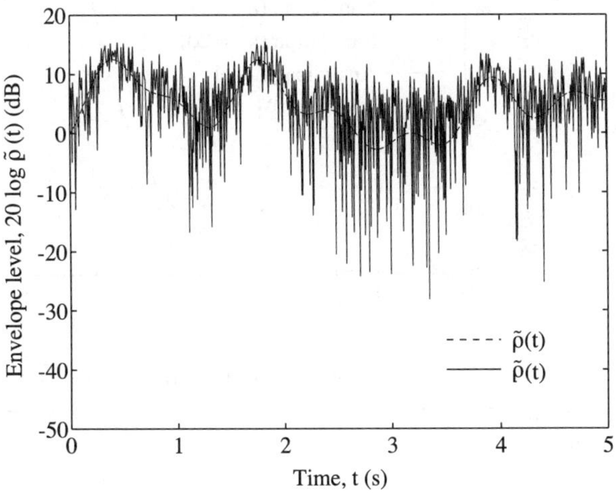

Figure 6.39 The deterministic processes $\tilde{\varrho}(t)$ and $\tilde{\rho}(t)$ $(\sigma_1^2 = \sigma_2^2 = 1$, $\kappa_1 = 0.8$, $\kappa_2 = 0.5$, $\sigma_3 = 0.5$, $m_3 = 0.25$, $f_\rho = 0.2 f_{\max}$, $\theta_\rho = 0$, $\kappa_c = 50$, and $f_{\max} = 91$ Hz).

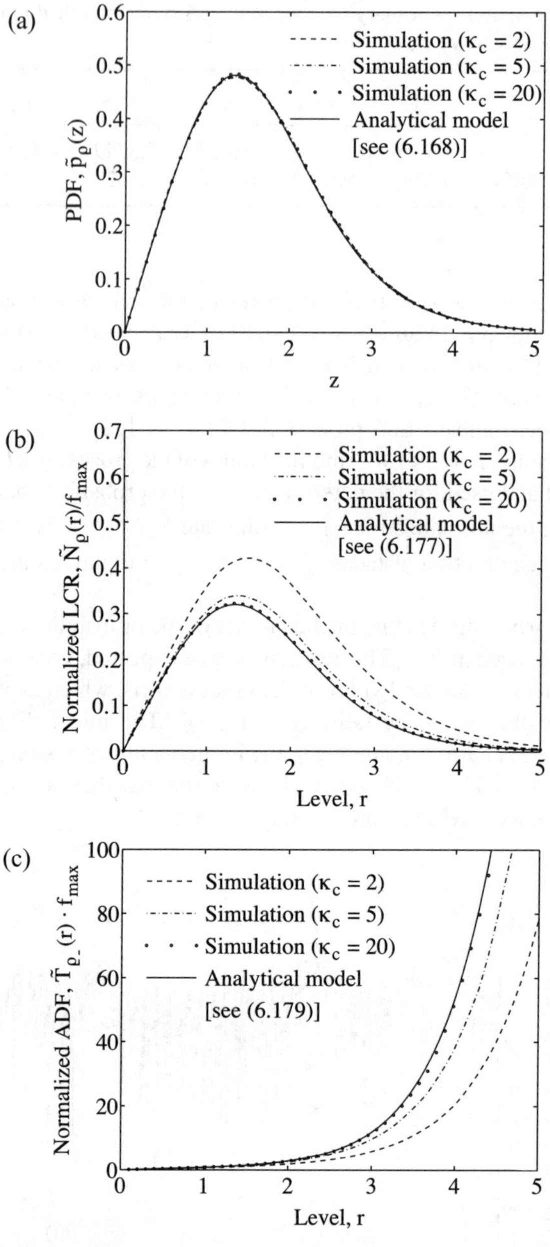

Figure 6.40 Comparisons between: (a) $p_\varrho(z)$ and $\tilde{p}_\varrho(z)$, (b) $N_\varrho(r)/f_{\max}$ and $\tilde{N}_\varrho(r)/f_{\max}$, as well as (c) $T_{\varrho_-}(r) \cdot f_{\max}$ and $\tilde{T}_{\varrho_-}(r) \cdot f_{\max}$ ($\sigma_1^2 = \sigma_2^2 = 1$, $\kappa_1 = 0.8$, $\kappa_2 = 0.5$, $\sigma_3 = 0.5$, $m_3 = 0.25$, $f_\rho = 0.2 f_{\max}$, $\theta_\rho = 0$, and $f_{\max} = 91$ Hz).

The results of Figure 6.40(b) show us that the deviations between the level-crossing rate of the reference model and that of the simulation model are only relatively high for unrealistically small values of κ_c, i.e., $\kappa_c \leq 5$. On the contrary, for $\kappa_c \geq 20$ the differences between the analytical approximate solution (6.177) and the corresponding simulation results can be ignored.

Studying Figure 6.40(c), we notice that the same statements also hold for the average duration of fades. Consequently, the approximate solutions derived for this model for the level-crossing rate and the average duration of fades are very exact, provided that the frequency ratio $\kappa_c = f_{\max}/f_c$ is greater than or equal to 20, i.e., if the amplitude of the line-of-sight component changes relatively slowly compared to the envelope variations of the scattered component. This means also that for all practically relevant cases, there is no need to be concerned about the restriction imposed by the boundary condition $\kappa_c \geq 20$, as $\kappa_c \gg 1$ holds anyway in real-world channels.

6.4.3 Applications and Simulation Results

In this subsection, we want to fit the statistical properties of the modified Loo model to the statistics of real-world channels. Just as for the extended Suzuki process of Type I and Type II as well as for the generalized Rice process, we here also use the measurement results presented in [196] as a basis for the complementary cumulative distribution function, the level-crossing rate, and the average duration of fades.

In the following, we will choose the realistic value $\kappa_c = 20$ for the frequency ratio $\kappa_c = f_{\max}/f_c$, so that the level-crossing rate $N_\varrho(r)$ of the Loo model is approximated by (6.177) very well. Without restriction of generality, we will also set the phase θ_ρ of the line-of-sight component to the arbitrary value $\theta_\rho = 0$.

The remaining free model parameters of the modified Loo model are the quantities σ_1, σ_2, κ_1, κ_2, σ_3, m_3, and f_ρ, which are set for the model fitting procedure. With these primary model parameters, we defined the parameter vector

$$\mathbf{\Omega} := \left(\sigma_1,\ \sigma_2,\ \kappa_1,\ \kappa_2,\ \sigma_3,\ m_3,\ f_\rho\right), \tag{6.186}$$

whose components are to be optimized according to the scheme described in Subsection 6.1.5. In order to minimize the error function $E_2(\mathbf{\Omega})$ [cf. (6.76)], we again make use of the Fletcher-Powell algorithm [162]. The optimized components of the parameter vector $\mathbf{\Omega}$ obtained in this way are presented in Table 6.5 for areas with light and heavy shadowing.

For the modified Loo model, the Rice factor c_R is calculated as follows:

$$\begin{aligned} c_R &= \frac{E\left\{|m(t)|^2\right\}}{E\left\{|\mu(t)|^2\right\}} = \frac{E\left\{\varrho^2(t)\right\}}{2E\left\{\mu_i^2(t)\right\}} \qquad (i = 1, 2) \\ &= \frac{r_{\varrho\varrho}(0)}{2\psi_0} = \frac{\pi}{2} \cdot \frac{e^{2(m_3+\sigma_3^2)}}{\sum\limits_{i=1}^{2} \sigma_i^2 \arcsin(\kappa_i)}. \end{aligned} \tag{6.187}$$

Thus, with the parameters taken from Table 6.5, the Rice factor c_R is $c_R = 1.7$ dB for heavy shadowing and $c_R = 8.96$ dB for light shadowing.

In Figure 6.41(a), the complementary cumulative distribution function $F_{\varrho+}(r) = 1 - F_{\varrho-}(r)$ of the modified Loo model is depicted together with that of the measured channel for the areas

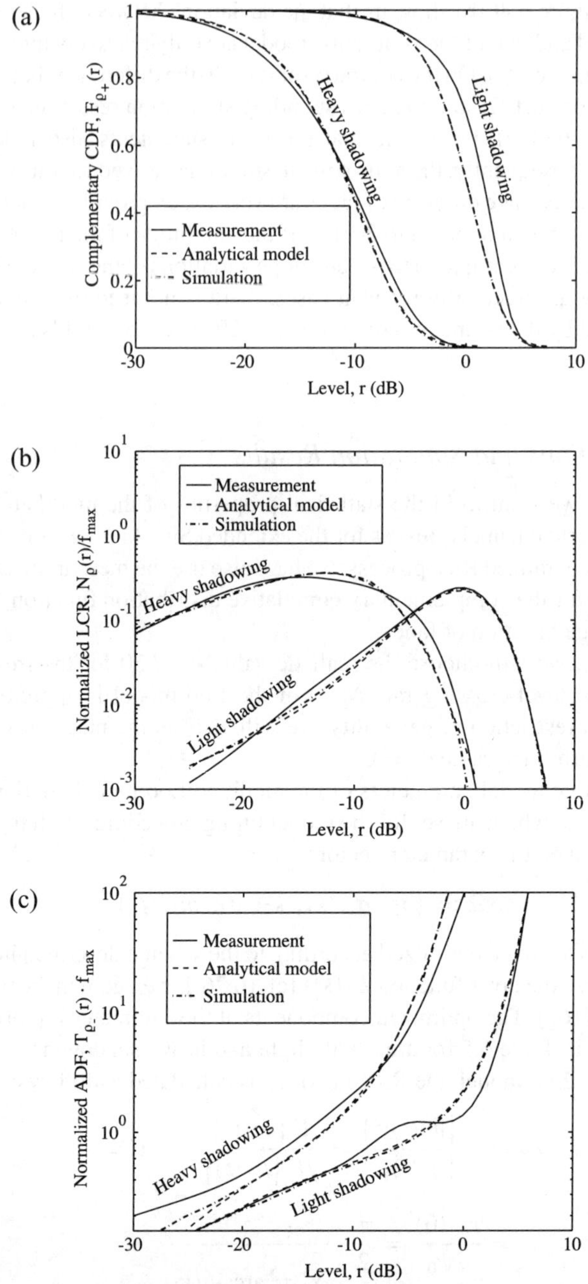

Figure 6.41 (a) Complementary cumulative distribution function $F_{\varrho+}(r)$, (b) normalized level-crossing rate $N_\varrho(r)/f_{\max}$, and (c) normalized average duration of fades $T_{\varrho-}(r/\rho) \cdot f_{\max}$ for areas with light and heavy shadowing. The measurement results were adopted from [196].

with light shadowing and with heavy shadowing. Figure 6.41(b) makes it clear that the differences between the normalized level-crossing rate $N_\varrho(r)/f_{\max}$ of the modified Loo model and the measured normalized level-crossing rate used here are acceptable. Finally, Figure 6.41(c) shows the corresponding normalized average duration of fades. Here, an excellent agreement is again observable between the reference model and the measured channel.

For the verification of the analytical results, the corresponding simulation results are also depicted in Figures 6.41(a)–6.41(c). The simulation model was based on the structure shown in Figure 6.37, where the underlying deterministic Gaussian processes $\tilde{\nu}_1(t)$, $\tilde{\nu}_2(t)$, and $\tilde{\nu}_3(t)$ have been designed by applying the method described in Subsection 6.4.2 using $N_1 = N_2 = N_3 = 15$ cosine functions.

Finally, the deterministic modified Loo process $\tilde{\varrho}(t)$ is depicted in Figures 6.42(a) and 6.42(b) for areas with light shadowing and for areas with heavy shadowing, respectively.

6.5 Modelling of Nonstationary Land Mobile Satellite Channels

A measurement campaign initiated by the European Space Agency (ESA) was carried out by the German Aerospace Center (DLR)[5] in 1995 aiming to investigate the propagation properties of land mobile satellite (LMS) channels in typical environments and under different elevation angles [203]. The evaluation of a vast amount of measurement data has revealed the following two characteristic properties of real-world LMS channels. First, the level-crossing rate is in general not proportional to the distribution of the fading envelope; and second, the level-crossing rate and the probability density function have in most cases more than one maximum value.

The missing proportionality between the level-crossing rate and the probability density function might be an indication of the fact that the inphase and quadrature components of

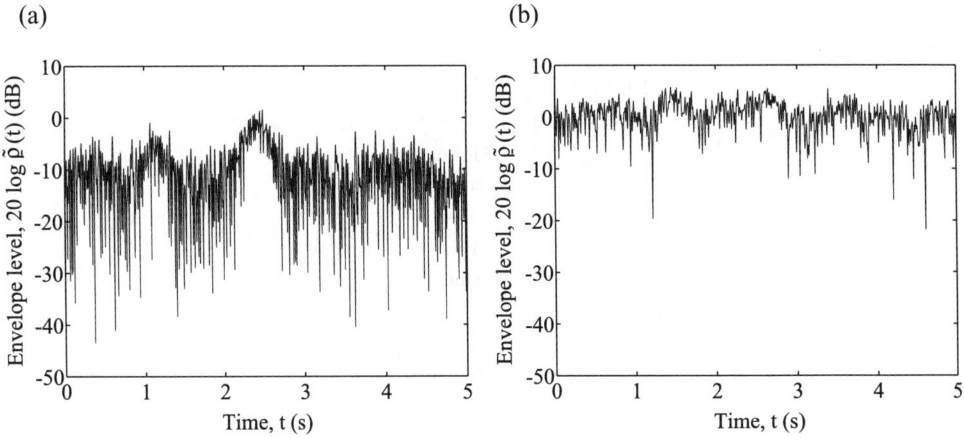

Figure 6.42 Simulation of deterministic modified Loo processes $\tilde{\varrho}(t)$ for areas with (a) heavy shadowing and (b) light shadowing (MEDS, $N_1 = N_2 = N_3 = 15$, $f_{\max} = 91$ Hz, $\kappa_c = 20$).

[5] German: **D**eutsches Zentrum für **L**uft- und **R**aumfahrt e.V.

the scattered component are correlated, which is equivalent to an asymmetric Doppler power spectral density. This property, which is obviously typical for real-world LMS channels, is often not considered by popular channel models such as the Loo model [202] and the model proposed by Lutz et al. [180]. The special merits of the extended Suzuki processes of Type I and II are their high flexibility and the desirable feature that the level-crossing rate is in general not proportional to the probability density function of the fading envelope. However, the extended Suzuki processes of Type I and II are both stationary, as they are derived from coloured Gaussian noise processes which are by their nature stationary.

Real-world LMS channels always become nonstationary if the observation time interval increases. Consequently, the modelling of real-world channels by using stationary processes is limited to those cases where the environment is extremely homogeneous, or, where the duration of the snapshot measurements is not sufficiently long to reveal the typical phenomena of nonstationary channels. The following subsections will guide us along the path from modelling to simulation of nonstationary LMS channels.

6.5.1 *Lutz's Two-State Channel Model*

A nonstationary model was introduced by Lutz et al. [180]. This model, which is shown in Figure 6.43, is valid for very large areas and has been developed especially for frequency-nonselective LMS channels. One distinguishes between regions, in which the line-of-sight component is present (good channel state), and regions, in which the line-of-sight component is shadowed (bad channel state). The switching between good and bad channel states is controlled by a two-state Markov chain. In the Lutz model, the envelope of the fading signal is described by a Rice process if the Markov chain is in the good channel state. Otherwise, if the Markov chain is in the bad channel state, then the fading envelope is modelled by the classical Suzuki process (Rayleigh-lognormal process). The underlying complex Gaussian

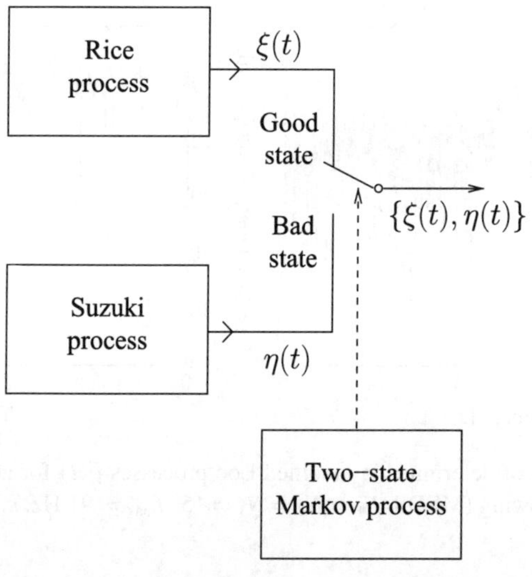

Figure 6.43 The Lutz model with two states for modelling of frequency-nonselective LMS channels.

process from which the Rice (Rayleigh) process is derived has the classical Jakes Doppler spectrum.

A Markov chain is a special kind of a Markov process, where the model can occupy a finite or at most an infinite countable number of states. With the Markov concept, the next state is dependent only upon the present state and not on how the process arrived in that state. Markov chains can be represented by state transition diagrams. The state transition diagram for a two-state Markov model is illustrated in Figure 6.44.

In the Lutz model, it is assumed that the bad channel state and the good channel state are represented by S_1 and S_2, respectively. The quantities p_{ij} $(i,j = 1,2)$ presented in Figure 6.44 are called the *transition probabilities*. For $i \neq j$, the transition probability p_{ij} signifies the probability that the Markov chain moves from state S_i to state S_j. Analogously, for $i = j$, the transition probability p_{ii} denotes the probability that the Markov chain remains in state S_i. According to the Markov model, the sum of the transition probabilities leaving a state must be equal to 1. Thus,

$$p_{11} = 1 - p_{12} \quad \text{and} \quad p_{22} = 1 - p_{21}. \qquad (6.188\text{a,b})$$

In digital transmission systems, each state transition has a duration that equals the symbol duration. The probability of staying in state S_i for more than n data symbols equals p_{ii}^n $(i = 1,2)$. Let D_i denote the mean number of symbol durations the system spends in state S_i $(i = 1,2)$. Then, the mean number of symbol durations D_i can be related to the transition probabilities p_{ij} by [180]

$$D_1 = \frac{1}{p_{12}} \quad \text{and} \quad D_2 = \frac{1}{p_{21}}. \qquad (6.189\text{a,b})$$

Furthermore, let A_1 be the *time-share factor*, which is a measure for the proportion of time for which the channel is in state S_1 (bad state), then A_1 can be expressed in terms of D_1 and D_2 as [180]

$$A_1 = \frac{D_1}{D_1 + D_2} = \frac{p_{21}}{p_{12} + p_{21}}. \qquad (6.190)$$

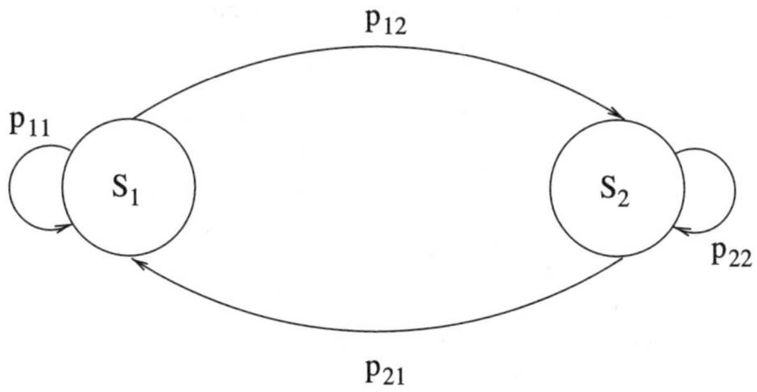

Figure 6.44 State transition diagram of the two-state Markov model.

With the time-share factor A_1, the overall probability density function $p_{Lutz}(z)$ of the envelope of the Lutz model can be written as

$$p_{Lutz}(z) = A_1\, p_\eta(z) + (1 - A_1)\, p_\xi(z), \tag{6.191}$$

where $p_\eta(z)$ denotes the Suzuki distribution [see (2.55)] and $p_\xi(z)$ is the Rice distribution [see (2.44)].

For city and highway environments, the parameters A_1, D_1, D_2, as well as the parameters characterizing the Rice and Suzuki processes have been determined from measurements at 1.54 GHz (L-band) under different satellite elevation angles using a least-square curve-fitting procedure in [180] (see there Table II). An improved method based on a physical-statistical approach for determining the time-share factor A_1 from measurements has been proposed in [204].

6.5.2 *M*-State Channel Models

The concept of Lutz's model can easily be generalized, leading to an M-state Markov process, where each state is represented by a specific stationary stochastic process. In this sense, the fading behaviour of nonstationary channels can be approximated by M stationary channel models [188, 189]. Experimental measurements have shown that a four-state Markov model is sufficient for most channels [190].

Instead of using a mixture of M channel models, it is also possible to use one and the same stationary channel model with different parameter settings related to each state [191]. The proposed dynamic model with M states for modelling of nonstationary LMS channels is presented in Figure 6.45. The usefulness of this dynamic model depends mainly on the flexibility of the embedded stochastic process. A proper choice for a highly flexible channel model is, e.g., the extended Suzuki process of Type I (see Section 6.1).

For each of the M channel states, a specific set of model parameters — combined to parameter vectors $\mathbf{\Omega}^{(m)}$ ($m = 1, 2, \ldots, M$) — has to be determined and properly assigned to the channel model. A change of the channel state from one state to another leads thus to a new configuration of the embedded stationary channel model. As shown in Figure 6.45, the switching of the parameter vectors is controlled by a discrete M-state Markov process.

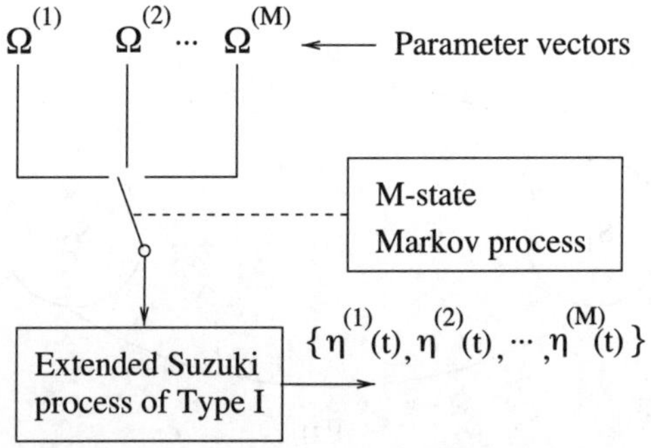

Figure 6.45 Dynamic model with *M* states for modelling of nonstationary LMS channels.

In this way, a large class of real-world (nonstationary) LMS channels can be approximated by embedding a stationary channel model in an M-state Markov process that controls the dynamic exchange of the model parameters. The dynamic M-state channel model can also be interpreted as a generalization of Lutz's LMS channel model. It allows the generation of fading envelopes characterized by probability density functions and level-crossing rates having up to M local maximum values.

6.5.3 *Modelling of Nonstationary Real-World LMS Channels*

This section deals with the modelling of nonstationary LMS channels using the dynamic M-state model presented in Figure 6.45. We show how the statistical properties of this channel model can be adapted to the statistics of real-world LMS channels in different environments. The extended Suzuki process of Type I will be embedded in an M-state Markov process such that a transition from one state to another corresponds to an exchange of all those model parameters which characterize the statistics of the Suzuki process. This means that in each channel state, a specific set of model parameters has to be assigned to the extended Suzuki process of Type I. Such a dynamic M-state model has been proposed in [191] as a universal model that allows the modelling of a large class of nonstationary LMS channels characterized by probability density functions and level-crossing rates having more than one maximum.

For the important case $M = 2$, a fitting procedure will be described, and it is demonstrated that the primary statistical properties (probability density function, level-crossing rate, average duration of fades) of the proposed dynamic channel model are nearly identical with given measured values of equivalent satellite channels in different environments. Moreover, starting from the dynamic M-state model, which is considered here as reference model, an efficient simulation model is derived by replacing all coloured Gaussian noise processes by finite sums of sinusoids. With the objective to produce gentle transitions between states, we introduce so-called handover processes as a sophisticated technique to interpolate the simulated fading envelope between consecutive states. Finally, it is demonstrated by several theoretical and simulation results that the performance of the reference model and the corresponding simulation model are excellent with respect to the probability density function of the fading envelope, level-crossing rate, and average duration of fades.

6.5.3.1 Review of the Extended Suzuki Process of Type I

For ease of understanding, we will briefly review the extended Suzuki process of Type I, which has been described in detail in Section 6.1. The extended Suzuki process of Type I, denoted by $\eta(t)$, is a product process of a Rice process $\xi(t)$ with underlying cross-correlated inphase and quadrature components and a lognormal process $\lambda(t)$, i.e.,

$$\eta(t) = \xi(t) \cdot \lambda(t). \tag{6.192}$$

The Rice process $\xi(t)$ is obtained from a complex Gaussian noise process $\mu(t)$ and a line-of-sight component $m(t)$ according to

$$\xi(t) = |\mu(t) + m(t)|. \tag{6.193}$$

The inphase and quadrature components of the complex Gaussian noise process

$$\mu(t) = \mu_1(t) + j\mu_2(t) \tag{6.194}$$

are derived from two zero-mean real Gaussian noise processes $\nu_1(t)$ and $\nu_2(t)$ as follows

$$\mu_1(t) = \nu_1(t) + \nu_2(t), \tag{6.195}$$

$$\mu_2(t) = \check{\nu}_1(t) - \check{\nu}_2(t), \tag{6.196}$$

where $\check{\nu}_i(t)$ denotes the Hilbert transform of $\nu_i(t)$ $(i = 1, 2)$. The Doppler power spectral density $S_{\mu\mu}(f)$ of the complex Gaussian noise process $\mu(t)$ can be expressed as

$$S_{\mu\mu}(f) = 2[(1 + \text{sgn}\,(f)) \cdot S_{\nu_1\nu_1}(f) + (1 - \text{sgn}\,(f)) \cdot S_{\nu_2\nu_2}(f)], \tag{6.197}$$

where

$$S_{\nu_1\nu_1}(f) = \frac{\sigma_0^2}{2\pi f_{\max}\sqrt{1 - (f/f_{\max})^2}}, \qquad |f| \leq f_{\max}, \tag{6.198a}$$

and

$$S_{\nu_2\nu_2}(f) = \text{rect}\left(\frac{f}{\kappa_0 f_{\max}}\right) \cdot S_{\nu_1\nu_1}(f), \quad 0 \leq \kappa_0 \leq 1, \tag{6.198b}$$

are the Doppler power spectral densities of $\nu_1(t)$ and $\nu_2(t)$, respectively. In the above expressions, $f_{\max}$ denotes the maximum Doppler frequency and $\sigma_0^2/2$ is the variance of $\nu_1(t)$. It should be observed that the Doppler power spectral density $S_{\mu\mu}(f)$ has an asymmetrical shape if $\kappa_0 < 1$. In this case, the inphase and quadrature components [see (6.195) and (6.196)] are correlated. On the other hand, the shape of $S_{\mu\mu}(f)$ is symmetrical if $\kappa_0 = 1$, and, consequently, $\mu_1(t)$ and $\mu_2(t)$ are uncorrelated.

The line-of-sight component $m(t)$ in (6.193) is a time-variant function having the form of

$$m(t) = \rho\, e^{j(2\pi f_\rho t + \theta_\rho)}, \tag{6.199}$$

where the parameters ρ, f_ρ, and θ_ρ are called the amplitude, Doppler frequency, and phase of the line-of-sight component, respectively. The lognormal process $\lambda(t)$ introduced in (6.192) is derived from a third zero-mean real Gaussian noise process $\nu_3(t)$ with Gaussian power spectral density characteristics and unit variance according to

$$\lambda(t) = e^{\sigma_3 \nu_3(t) + m_3}, \tag{6.200}$$

where σ_3 and m_3 are parameters describing the extent of the variation of the local mean.

Next, we combine the primary model parameters introduced above to a parameter vector

$$\mathbf{\Omega} := \left(\sigma_0, \kappa_0, \sigma_3, m_3, \rho, f_\rho, f_{\max}\right). \tag{6.201}$$

The expressions (6.192)–(6.201) describe the extended Suzuki process $\eta(t)$ of Type I as a single-state model. If such a process is embedded in an M-state Markov model (see Figure 6.45), then we will use notations such as $\eta^{(m)}(t)$ and $\mathbf{\Omega}^{(m)}$ for all $m = 1, 2, \ldots, M$. In particular, we write $\mathbf{\Omega}^{(m)} \mapsto \eta^{(m)}(t)$ in order to emphasize that the parameter vector $\mathbf{\Omega}^{(m)}$ is assigned to the extended Suzuki process $\eta^{(m)}(t)$ as long as the Markov process is in state S_m. A transition from state S_m to state S_n at time $t = t_0$ results in $\mathbf{\Omega}^{(m)} \rightarrow \mathbf{\Omega}^{(n)}$, and thus $\eta^{(m)}(t) \rightarrow \eta^{(n)}(t)$.

Thereby, we require that the generated output process fulfills the steadiness condition $\eta^{(m)}(t) = \eta^{(n)}(t)$ at $t = t_0$.

6.5.3.2 The Simulation Model

An efficient simulation model for the extended Suzuki process $\eta(t)$ of Type I can be obtained by replacing the real-valued Gaussian noise processes $\nu_i(t)$ by the following sum of N_i sinusoids

$$\tilde{\nu}_i(t) = \sum_{n=1}^{N_i} c_{i,n} \cos(2\pi f_{i,n} t + \theta_{i,n}), \quad i = 1, 2, 3, \tag{6.202}$$

where the quantities $c_{i,n}$, $f_{i,n}$, and $\theta_{i,n}$ are called gains, discrete Doppler frequencies, and phases, respectively. These parameters can easily be computed, e.g., by using the modified method of exact Doppler spread (MEDS). According to this method, the gains $c_{i,n}$ and discrete Doppler frequencies $f_{i,n}$ of the sum-of-sinusoids $\tilde{\nu}_i(t)$ are given by

$$c_{i,n} = \begin{cases} \dfrac{\sigma_0}{\sqrt{N_1}}, & i = 1, \\ \dfrac{\sigma_0}{\sqrt{N_2'}}, & i = 2, \end{cases} \tag{6.203a}$$

$$f_{i,n} = \begin{cases} f_{\max} \sin\left[\dfrac{\pi\left(n-\frac{1}{2}\right)}{2N_1}\right], & i = 1, \\ f_{\max} \sin\left[\dfrac{\pi\left(n-\frac{1}{2}\right)}{2N_2'}\right], & i = 2, \end{cases} \tag{6.203b}$$

for all $n = 1, 2, \ldots, N_i$, where N_2' denotes the virtual number of sinusoids, which is defined by $N_2' = \left\lceil \frac{N_2}{\frac{2}{\pi}\arcsin(\kappa_0)} \right\rceil$.

The discrete Doppler frequencies $f_{3,n}$ of the third sum-of-sinusoids $\tilde{\nu}_3(t)$ are obtained numerically by computing the zeros of

$$\frac{2n-1}{2N_3} - \operatorname{erf}\left(\frac{f_{3,n}}{\sqrt{2}\sigma_c}\right) = 0 \tag{6.204}$$

for all $n = 1, 2, \ldots, N_3 - 1$, where $\operatorname{erf}(\cdot)$ denotes the error function and $\sigma_c = f_{\max}/(20\sqrt{2\ln 2})$. The corresponding gains $c_{3,n}$ are given by $c_{3,n} = \sqrt{2/N_3}$.

The phases $\theta_{i,n}$ are outcomes of a random generator which is uniformly distributed over $(0, 2\pi]$ for all $n = 1, 2, \ldots, N_i$ $(i = 1, 2, 3)$. The remaining parameters of the simulation model, i.e., σ_3, m_3, ρ, f_ρ, and θ_ρ, are identical to those of the reference model introduced in Subsection 6.5.3.1.

By analogy to the reference model, we combine the primary parameters of the simulation model to a parameter vector defined by

$$\tilde{\mathbf{\Omega}} := \left(f_{1,1}, \ldots, c_{1,1}, \ldots, \theta_{1,1}, \ldots, m_3, \sigma_3, \rho, f_\rho\right). \tag{6.205}$$

Using the notation introduced in the previous subsection, we write for the single-state simulation model $\tilde{\mathbf{\Omega}} \mapsto \tilde{\eta}(t)$ to express the fact that the envelope $\tilde{\eta}(t)$ of the simulation model is completely defined for all time t by the parameter vector $\tilde{\mathbf{\Omega}}$. In the case where that the simulation model is embedded in an M-state Markov model, we may write $\tilde{\mathbf{\Omega}}^{(m)}$ and $\tilde{\eta}^{(m)}(t)$ ($m = 1, 2, \ldots, M$), and hence $\tilde{\mathbf{\Omega}}^{(m)} \mapsto \tilde{\eta}^{(m)}(t)$. It is important to note that the resulting simulation model for the dynamic M-state reference model violates in its present form the steadiness condition, because it follows from $\tilde{\mathbf{\Omega}}^{(m)} \neq \tilde{\mathbf{\Omega}}^{(n)}$ ($m \neq n$) that $\tilde{\eta}^{(m)}(t) \neq \tilde{\eta}^{(n)}(t)$ holds at $t = t_0$. Fortunately, the problem of unsteadiness can be solved by using handover processes, as we will see in the following subsection.

6.5.3.3 Handover Processes

In the dynamic M-state channel model considered above, a transition from state S_m to another state S_n is obtained instantaneously at random instants $t = t_0$. In practice, however, this is not true, as measurements have shown that the transitions from shadowed to unshadowed situations and vice versa are smooth [205]. To avoid discontinuous fading envelopes caused by instantaneous transitions, we introduce so-called *handover states* S_0 such that S_m passes over to S_n via S_0, i.e., $S_m \to S_0 \to S_n$. An example of the resulting modified M-state channel model is shown in Figure 6.46 for the important case of $M = 2$.

If the dynamic model moves from state S_m to the handover state S_0, it remains in that state for the period Δ_0. A reasonable assumption is that the quantity Δ_0, called the *transition time interval*, is a random variable which is uniformly distributed over the interval $[\Delta_{\min}, \Delta_{\max}]$. We refer to Subsection 6.5.3.4, where it is explained in detail how the quantities $\Delta_{\min}$ and $\Delta_{\max}$ can be determined from measurements.

If the modified Markov model is in the handover state S_0, then the envelope $\tilde{\eta}^{(0)}(t)$ of the simulation model is determined by the parameter vector $\tilde{\mathbf{\Omega}}^{(0)}$ having time-variant components, i.e., $\tilde{\mathbf{\Omega}}^{(0)} = \tilde{\mathbf{\Omega}}^{(0)}(t)$. By analogy with (6.205), we can write formally

$$\tilde{\mathbf{\Omega}}^{(0)}(t) := \left(f_{1,1}^{(0)}(t), \ldots, c_{1,1}^{(0)}(t), \ldots, \theta_{1,1}^{(0)}(t), \ldots, m_3^{(0)}(t), \sigma_3^{(0)}(t), \rho^{(0)}(t), f_\rho^{(0)}(t)\right). \tag{6.206}$$

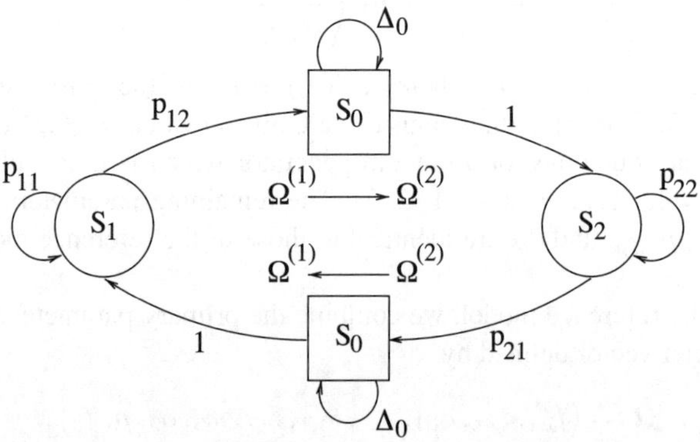

Figure 6.46 State transition diagram of the modified (2+2)-state Markov model by using handover states S_0.

The time-variant components of $\tilde{\mathbf{\Omega}}^{(0)}(t)$ are directly related to the fixed components of the parameter vectors $\tilde{\mathbf{\Omega}}^{(m)}$ and $\tilde{\mathbf{\Omega}}^{(n)}$. By way of example, we consider the amplitude of the line-of-sight component $\rho^{(0)}(t)$ that interpolates the values between the constant quantities $\rho^{(m)}$ and $\rho^{(n)}$ in state S_0 by

$$\rho^{(0)}(t) = \frac{\rho^{(m)} + \rho^{(n)}}{2} + \frac{\rho^{(m)} - \rho^{(n)}}{2} \cos\left(\pi \frac{t - t_0}{\Delta_0}\right) \tag{6.207}$$

for $t \in [t_0, t_0 + \Delta_0]$. From the expression above, it follows $\rho^{(0)}(t_0) = \rho^{(m)}$ and $\rho^{(0)}(t_0 + \Delta_0) = \rho^{(n)}$. Figure 6.47 illustrates the behaviour of the amplitude of the line-of-sight component $\rho^{(m)}(t)$ if the channel moves from state S_1 to state S_2 via the handover state S_0.

In an analogous manner, we proceed with all other components of $\tilde{\mathbf{\Omega}}^{(0)}(t)$ in (6.206). As a result, at the time instants $t = t_0$ and $t = t_0 + \Delta_0$, we have $\tilde{\mathbf{\Omega}}^{(0)}(t_0) = \tilde{\mathbf{\Omega}}^{(m)}$ and $\tilde{\mathbf{\Omega}}^{(0)}(t_0 + \Delta_0) = \tilde{\mathbf{\Omega}}^{(n)}$, respectively. From these equations and from the mapping $\tilde{\mathbf{\Omega}}^{(0)}(t) \mapsto \tilde{\eta}^{(0)}(t)$, it follows now $\tilde{\eta}^{(0)}(t_0) = \tilde{\eta}^{(m)}(t_0)$ and $\tilde{\eta}^{(0)}(t_0 + \Delta_0) = \tilde{\eta}^{(n)}(t_0 + \Delta_0)$, i.e., the envelope of the modified M-state simulation model fulfills the steadiness condition at $t = t_0$ and $t = t_0 + \Delta_0$. Moreover, due to (6.207), it can be shown that the time derivative of the simulated envelope fulfills the steadiness condition as well, i.e, $\dot{\tilde{\eta}}^{(0)}(t_0) = \dot{\tilde{\eta}}^{(m)}(t_0)$ and $\dot{\tilde{\eta}}^{(0)}(t_0 + \Delta_0) = \dot{\tilde{\eta}}^{(n)}(t_0 + \Delta_0)$, where the overdot indicates the time derivative. The task of the time variant parameter vector $\tilde{\mathbf{\Omega}}^{(0)}(t)$ is to hand over the components of $\tilde{\mathbf{\Omega}}^{(m)}$ to the components of $\tilde{\mathbf{\Omega}}^{(n)}$ within the time interval $\Delta_0 \in [\Delta_{\min}, \Delta_{\max}]$. For that reason, the process $\tilde{\eta}^{(0)}(t)$ defined by the mapping $\tilde{\mathbf{\Omega}}^{(0)}(t) \mapsto \tilde{\eta}^{(0)}(t)$ is called *handover process*. An example that illustrates the solution of the unsteadiness problem by using a handover process is shown in Figure 6.48. This figure shows the gentle transitions of

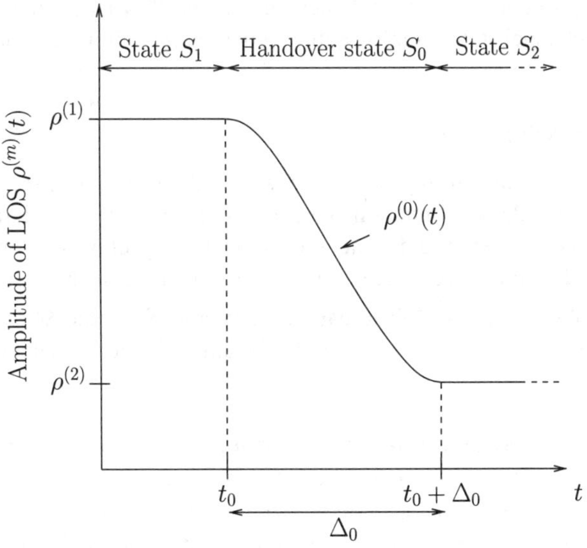

Figure 6.47 Behaviour of the amplitude of the line-of-sight component $\rho^{(m)}(t)$ if the channel is in state S_m $(m = 0, 1, 2)$.

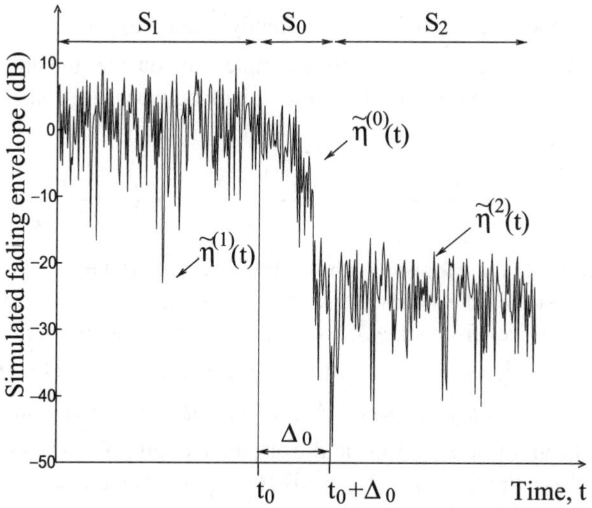

Figure 6.48 Solving the unsteadiness problem by using handover processes $\tilde{\eta}^{(0)}(t)$.

the simulated envelope when the modified (2+2)-state Markov model is moving from the good channel state S_1 to the bad channel state S_2 via the handover state S_0. It should be mentioned that in practice the mean value of the time interval Δ_0 is relatively short in comparison with the average duration of being in state S_1 or S_2, so that the influence of $\tilde{\eta}^{(0)}(t)$ on the overall statistics of the dynamic channel model can be neglected.

An alternative way to model continuous transitions from one state to another is to introduce extra states, as proposed in [205], where it has been shown that at least three states are needed to model accurately smooth transitions. Other techniques allowing to create gentle transitions between different channel states employ cubic spline interpolation techniques [206].

6.5.3.4 Fitting Procedure

In this section, we present a fitting procedure which allows us to fit the relevant statistical properties (probability density function, level-crossing rate, average duration of fades) of the dynamic M-state reference and simulation models very closely to those of real-world satellite channels. The proposed procedure can be organized in three steps. In Step I, we determine the number of states M. The parameter vectors $\mathbf{\Omega}^{(m)}$ and $\tilde{\mathbf{\Omega}}^{(m)}$ are determined in Step II. Finally, the transition probabilities of the dynamic M-state model are then computed in Step III.

Step I — Determination of the number of states: From the measurement data of the envelope, denoted by $\eta^{\star}(t)$, we compute the level-crossing rate $N_{\eta}^{\star}(r)$, which is the expected number of crossings at which the envelope $\eta^{\star}(t)$ crosses a given signal level r with positive (or negative) slope within one second. The number of states M is chosen to be equal to the number of local maxima of $N_{\eta}^{\star}(r)$. The evaluation of a large amount of measurement data [203] collected in different areas under various elevation angles has revealed that the great majority of all measured level-crossing rates $N_\eta^\star(r)$ shows two local maxima. Only in some cases, the measured quantity $N_\eta^\star(r)$ showed one maximum or three local maxima. Hence, if $N_\eta^\star(r)$ reveals only one maximum, then we set $M = 1$. This means that a single flexible channel model with fixed parameters is sufficient, and we can proceed with the fitting procedure described in Subsection 6.1.5. Otherwise, if the measured level-crossing rate $N_\eta^\star(r)$ has two local maxima, then a dynamic channel model with $M = 2$ states is adequate.

Step II — Determination of the parameter vectors: The components of the parameter vector $\mathbf{\Omega}^{(m)}$ determine the shape of the complementary cumulative distribution function $F_{\eta+}^{(m)}(r)$ and the level-crossing rate $N_\eta^{(m)}(r)$ of the M-state model when it is in state S_m for $m \in \{1, 2, \ldots, M\}$. Exact formulas for $F_{\eta-}^{(m)}(r) = 1 - F_{\eta+}^{(m)}(r)$ and $N_\eta^{(m)}(r)$ have been derived for the single-state model in Section 6.1.3 [see (6.66) and (6.61), respectively]. Using these results for the important case of $M = 2$ and neglecting the influence of the handover processes, we can express the overall complementary cumulative distribution function $F_{\eta+}(r)$, the overall level-crossing rate $N_\eta(r)$, and the overall average duration of fades $T_{\eta-}(r)$ of the 2-state model in terms of $F_{\eta+}^{(m)}(r)$ and $N_\eta^{(m)}(r)$ as follows:

$$F_{\eta+}(r) = A_1 F_{\eta+}^{(1)}(r) + (1 - A_1) F_{\eta+}^{(2)}(r), \tag{6.208a}$$

$$N_\eta(r) = A_1 N_\eta^{(1)}(r) + (1 - A_1) N_\eta^{(2)}(r), \tag{6.208b}$$

$$T_{\eta-}(r) = [1 - F_{\eta+}(r)]/N_\eta(r). \tag{6.208c}$$

The parameter vectors $\mathbf{\Omega}^{(1)}$ and $\mathbf{\Omega}^{(2)}$, as well as the time-share factor A_1, have to be computed numerically by minimizing the error function

$$E_2\left(\mathbf{\Omega}^{(1)}, \mathbf{\Omega}^{(2)}, A_1\right) := \left\{\sum_{m=1}^{M_r} \left[W_1(r_m)\left(F_{\eta+}^\star(r_m) - F_{\eta+}(r_m)\right)\right]^2\right\}^{1/2} + \left\{\sum_{m=1}^{M_r} \left[W_2(r_m)\left(N_\eta^\star(r_m) - N_\eta(r_m)\right)\right]^2\right\}^{1/2}, \tag{6.209}$$

where M_r is the number of signal levels r_m at which the measurements were taken, $W_1(\cdot)$ and $W_2(\cdot)$ are appropriate weighting functions, $F_{\eta+}^\star(\cdot)$ denotes the complementary cumulative distribution function of the measured envelope $\eta^\star(t)$, and $N_\eta^\star(\cdot)$ designates the level-crossing rate of $\eta^\star(t)$. The minimization of the error norm $E_2(\cdot)$ in (6.209) can be performed by using any elaborated numerical optimization technique, such as the well-known Fletcher-Powell algorithm [162, 207]. Once the parameter vectors $\mathbf{\Omega}^{(1)}$ and $\mathbf{\Omega}^{(2)}$ of the reference model are determined, one has to choose proper values for the number of sinusoids N_i ($i = 1, 2, 3$), and then one can compute directly the corresponding quantities of the simulation model $\tilde{\mathbf{\Omega}}^{(1)}$, $\tilde{\mathbf{\Omega}}^{(2)}$, and $\tilde{\mathbf{\Omega}}^{(0)}$ by applying the procedures described in Sections 6.5.3.2 and 6.5.3.3.

Step III — Determination of the transition probabilities: First of all, we determine the transition time interval Δ_0. Let us therefore consider Figure 6.49(a), where a typical example

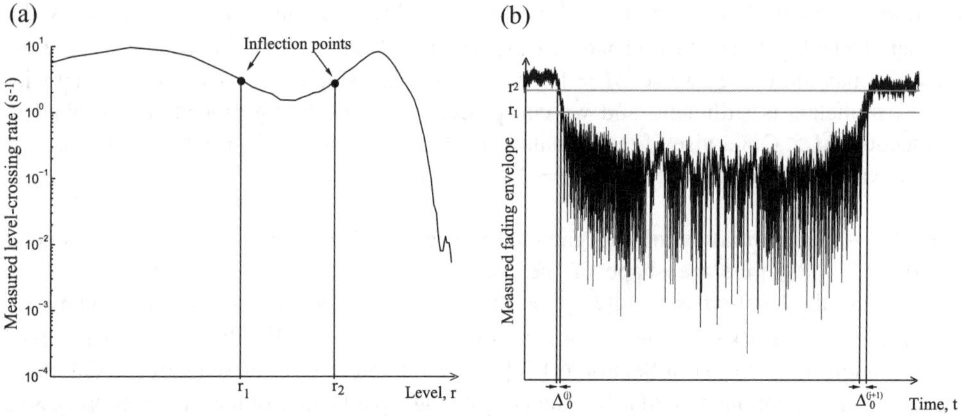

Figure 6.49 (a) Inflection points defining the location of the two thresholds r_1 and r_2, and (b) determination of the transition time intervals $\Delta_0^{(j)}$ for $j = 1, 2, \ldots, \mathcal{J}$.

of a measured level-crossing rate $N_\eta^\star(r)$ of an LMS channel is shown. It can be observed that $N_\eta^\star(r)$ has two local maxima, so that we choose $M = 2$. Next, we search for the inflection points of $N_\eta^\star(r)$, which are located at the signal levels r_1 and r_2 in Figure 6.49(a). These levels are defining the two thresholds shown in Figure 6.49(b) in which a short section of a measured envelope $\eta^\star(t)$ is presented. The transition time interval $\Delta_0^{(j)}$ for downward crossings is the duration between the first downward crossing of $\eta^\star(t)$ through the signal level r_1 at time $t_0 + \Delta_0^{(j)}$ given the last downward crossing of $\eta^\star(t)$ through the signal level r_2 at t_0. Analogously, the transition time interval for upward crossings can be defined, as illustrated in Figure 6.49(b). The evaluation of the measured data results finally in the set $\{\Delta_0^{(1)}, \Delta_0^{(2)}, \ldots, \Delta_0^{(\mathcal{J})}\}$, where $\mathcal{J}$ is the number of observed events defined by the number of transitions from state S_1 to S_2 and vice versa. From this set, we determine $\Delta_{\min} = \min\{\Delta_0^{(j)}\}_{j=1}^{\mathcal{J}}$ and $\Delta_{\max} = \max\{\Delta_0^{(j)}\}_{j=1}^{\mathcal{J}}$. In our resulting dynamic (2+2)-state simulation model, we assume that the transition time interval Δ_0 is a random variable which is — for simplicity — uniformly distributed over the interval $[\Delta_{\min}, \Delta_{\max}]$.

With the number of observed events $\mathcal{J}$ and taking into account that the sum of the transition probabilities leaving a state must be equal to one, the transition probabilities p_{11}, p_{12}, p_{21}, and p_{22} can now be computed as follows [191]:

$$p_{11} = 1 - p_{12}, \tag{6.210a}$$

$$p_{12} = \frac{\mathcal{J}}{2A_1 K} \frac{T_s}{T_s^\star}, \tag{6.210b}$$

$$p_{21} = p_{12} \frac{A_1}{1 - A_1}, \tag{6.210c}$$

$$p_{22} = 1 - p_{21}, \tag{6.210d}$$

where T_s ($T_s^\star$) is the sampling interval of the simulation model (measured signal), K designates the number of measured samples $\eta^\star(kT_s^\star)$ ($k = 1, 2, \ldots, K$), and A_1 is the time-share factor that accounts for the proportion of time for which the channel is in the channel state S_1.

6.5.3.5 Applications to Measured LMS Channels

In this section, we apply the proposed fitting procedure described in Subsection 6.5.3.4 to the measurement results of an equivalent (aircraft) LMS channel [203]. The measurement campaign was performed by the DLR at 1.82 GHz. The measurement data of the fading envelope $\eta^\star(t)$ considered here has been collected in a suburban and in an urban area under the elevation angles of 15° and 25°, respectively. The results of the evaluation of the measured envelope $\eta^\star(t)$ with respect to the complementary cumulative distribution function $F_{\eta+}^\star(r)$, level-crossing rate $N_\eta^\star(r)$, and average duration of fades $T_{\eta-}^\star(r)$ are shown in Figures 6.50(a)–(c), respectively. From Figure 6.50(b), it can be observed that the level-crossing rate $N_\eta^\star(r)$ has in both scenarios (suburban and urban) two local maxima, and thus, it follows from Step I of the fitting procedure that the number of states M is equal to 2. In Step II, we use $F_{\eta+}^\star(r)$ and $N_\eta^\star(r)$ as object functions in (6.209). The minimization of the error norm E_2 provides the parameter vectors $\mathbf{\Omega}^{(m)}$ ($m = 1, 2$) and the time-share factor A_1. The found parameters of the reference model are listed in Table 6.6, while the time-share factor A_1 is included in Table 6.7. Finally, in Step III, we determine the transition probabilities p_{ij} ($i, j = 1, 2$) of the dynamic 2-state model and compute the transition time intervals $\Delta_{\min}$ and $\Delta_{\max}$ from the measured fading envelope $\eta^\star(t)$. The resulting quantities are provided in Table 6.7

The resulting complementary cumulative distribution function $F_{\eta+}(r)$, level-crossing rate $N_\eta(r)$, and average duration of fades $T_{\eta-}(r)$ of the reference model are shown in Figures 6.50(a)–(c), respectively. These figures also present the corresponding simulation results

Table 6.6 Optimized components of the parameter vectors $\mathbf{\Omega}^{(m)}$ ($m = 1, 2$) of the reference model for an equivalent satellite channel in a suburban and an urban area [191].

Area	m	$\sigma_0^{(m)}$	$\kappa_0^{(m)}$	$\sigma_3^{(m)}$	$m_3^{(m)}$	$\rho^{(m)}$	$f_\rho^{(m)}$ (Hz)	$f_{\max}^{(m)}$ (Hz)
Suburban	1	1.1133	0.5976	0.37967	−1.3428	0.75684	7.3062	14.175
	2	0.2717	0.7305	0.00833	0.3826	0.73937	1.9173	30.1185
Urban	1	0.1392	0.9997	0.52197	−0.3438	0.0631	6.3	20.1323
	2	0.43121	0.9934	0.09586	−0.2033	1.4767	3.9359	23.6614

Table 6.7 Parameters of the dynamic (2+2)-state model for an equivalent satellite channel in suburban and urban areas [191].

Area	$\Delta_{\min}$ (s)	$\Delta_{\max}$ (s)	A_1	p_{11}	p_{12}	p_{21}	p_{22}
Suburban	0.016	1.066	0.40	0.99873	0.00127	0.000847	0.99915
City	0.02	1.082	0.55	0.99931	0.00069	0.000843	0.99916

(a)

(b)

(c)

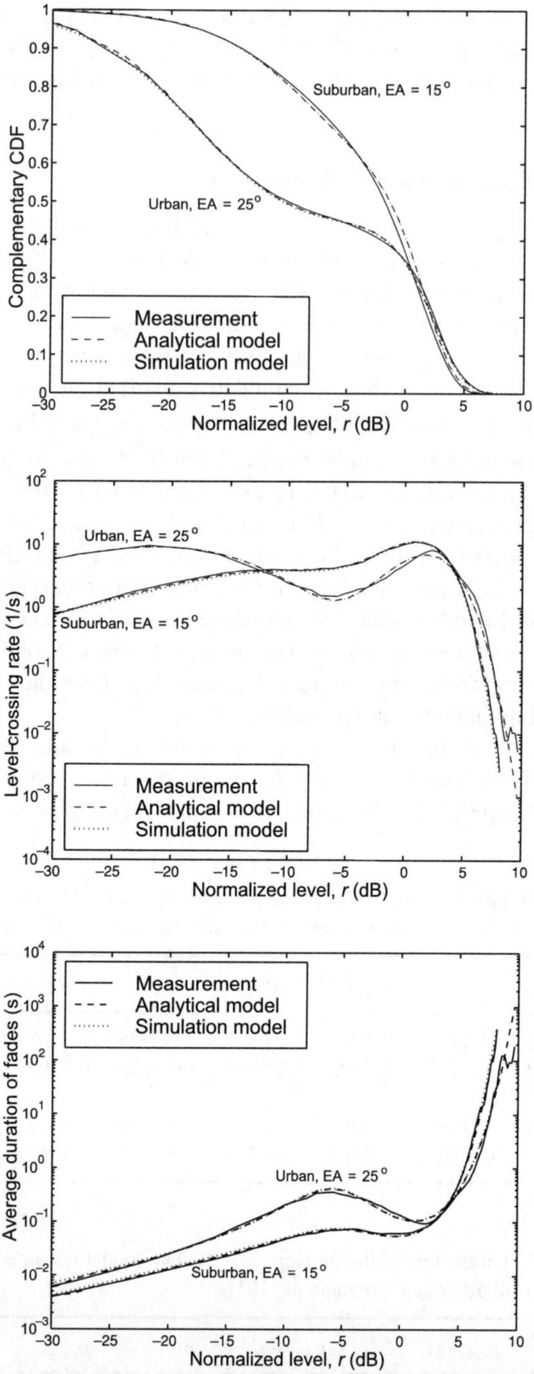

Figure 6.50 (a) Complementary cumulative distribution function, (b) level-crossing rate, and (c) average duration of fades of the fading envelope in a suburban area (EA $= 15^\circ$) and in an urban area (EA $= 25^\circ$) obtained by embedding the extended Suzuki processes of Type I in the dynamic (2+2)-state model (EA: elevation angle). The measurement results were adopted from [203].

which have been obtained by designing the simulation model with the modified MEDS (see Section 6.5.3.2) using $N_1 = N_2 = N_3 = 15$ sinusoids. The sampling frequency f_s of the simulation model was defined for the suburban and urban areas by $f_s = 40 \cdot \max\{f_{\max}^{(m)}\}$, where $f_{\max}^{(m)}$ is given in Table 6.6. Figures 6.50(a)–(c) show clearly that both the theoretical and the simulation results are in line with the corresponding measurements. Finally, Figure 6.51 illustrates the measured and the simulated fading envelope for the suburban environment.

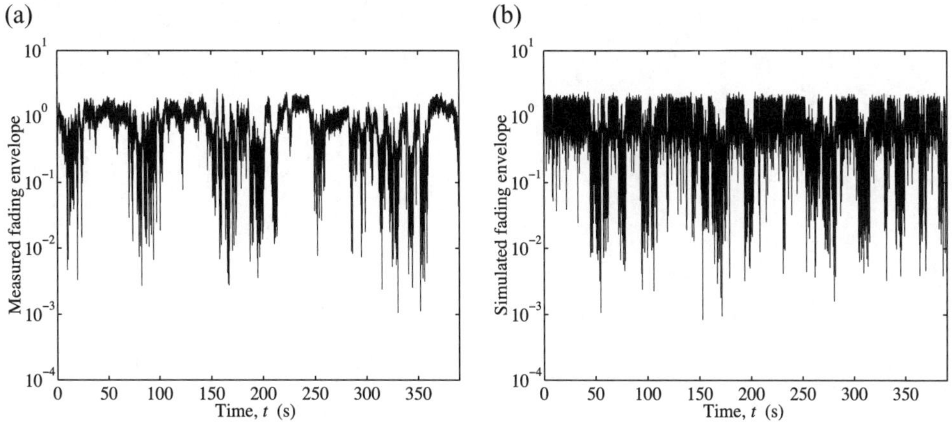

Figure 6.51 (a) Measured fading envelope of an equivalent LMS channel in a suburban area (15° elevation angle) and (b) the corresponding simulated fading envelope obtained by using the dynamic (2+2)-state model with embedded extended Suzuki processes of Type I.

7

Frequency-Selective Channel Models

The mobile radio channel is a combination of propagation paths, each with its own attenuation, phase distortion, and temporal dispersion which manifests itself as frequency-selective fading. The impact of the frequency selectivity of the channel on the transmitted signal increases with the transmitted data rate. In order to characterize temporal dispersive outdoor and indoor wireless channels, the so-called delay spread is commonly used. The delay spread is a measure of the spread of the propagation delays around the mean propagation delay. In time division multiple access (TDMA) systems, this quantity, for instance, plays a key role in determining the degree of intersymbol interference, and thus determines whether an adaptive equalizer is required at the receiver. In code division multiple access (CDMA) systems, the delay spread also determines the number of resolvable paths and the effectiveness of the RAKE receiver. In other words, understanding the behaviour of frequency-selective channels is of primary importance to radio system designers as the channel's temporal dispersion has a great impact on the performance of the wireless communication systems.

So far, we have been concerned exclusively with the modelling of frequency-nonselective mobile radio channels, which are characterized by the fact that the delay spread of the received multipath components is small in comparison to the symbol duration. This condition is less and less justified the shorter the symbol duration or the higher the data rate becomes. Channels whose delay spread is in the order of the symbol duration or even larger represent a further important class of channels, namely the class of *frequency-selective channels* which are also commonly called *wideband channels*. The stochastic and deterministic modelling of frequency-selective channels is the topic of this chapter.

Bello's [18] wide-sense stationary uncorrelated scattering (WSSUS) model has been accepted widely as an appropriate stochastic model for time-variant frequency-selective wireless channels. This model, which is valid for most radio channels, assumes that the channel is stationary in the wide sense and that the scattered components with different propagation delays are statistically uncorrelated. The description of WSSUS channels is the primary aim of this chapter.

Mobile Radio Channels, Second Edition. Matthias Pätzold.

Another aim is the design of efficient simulation models for frequency-selective mobile radio channels. We will see that frequency-selective channel simulators can be realized by using the tapped-delay-line structure [65]. The tapped-delay-line structure can be interpreted as a transversal filter of order $\mathcal{L}$ with time-varying tap gains. The main problem here is to determine the parameters of the simulation model such that the statistical properties of the channel simulator are, for a given order $\mathcal{L}$, as close as possible to the statistics of a specified or measured channel. Several solutions to this problem will be discussed in this chapter. Moreover, we present the concept of perfect channel modelling. This concept enables the design of channel simulators, which are perfectly fitted to measured wideband mobile radio channels. It thus facilitates the simulation of measurement-based mobile radio channels on a computer or on a reconfigurable hardware platform to enable the performance analysis of mobile communication systems under real-world propagation conditions in specific environments.

Chapter 7 is structured as follows. In order to illustrate the path geometry of multipath fading channels with different propagation delays, we will at first present in Section 7.1 the ellipse model introduced originally by Parsons and Bajwa [208]. In Section 7.2, we are concerned with the system theoretical description of frequency-selective channels. In this context, we will discuss four system functions introduced originally by Bello [18]. It will be pointed out how these system functions will lead to specific insights into the input-output behaviour of linear time-variant systems. Section 7.3 contains a description of the theory of frequency-selective stochastic channel models, which also goes back to Bello [18]. Here, the well-known WSSUS channel model is of central importance. In particular, stochastic system functions as well as the characteristic quantities derivable from these functions will be introduced to characterize the statistical properties of WSSUS channel models. These models are especially well suited for the modelling of the mobile radio channels specified by the European working group COST 207 [19]. The description of the COST 207 channel models is the topic of Subsection 7.3.3. The HIPERLAN/2 channel models [20] have attracted interest in recent years, and will therefore be described in Subsection 7.3.4. Section 7.4 is dedicated to the design of frequency-selective sum-of-sinusoids channel models. The mathematical tools used there can be considered as an extension of the theory of sum-of-sinusoids introduced in Chapter 4. It is shown how these techniques can be applied to the design of tapped-delay-line-based deterministic simulation models for frequency-selective channels. Section 7.5 discusses five stochastic and deterministic parameter computation methods for the modelling of given (specified or measured) power delay profiles. Section 7.6 introduces the concept of perfect channel modelling. Finally, Chapter 7 ends with a further reading section.

7.1 The Ellipse Model of Parsons and Bajwa

During the transmission of data, the emitted electromagnetic waves are influenced by a multitude of various obstacles. Depending on the geometric dimensions and the electromagnetic properties of these obstacles, one can distinguish between reflected waves, scattered waves, diffracted waves, and absorbed waves. For our purposes, a strict distinction between reflection, scattering, and diffraction is not as useful as the exact knowledge of the location and the consistency of each individual obstacle. Here, it is sufficient to merely speak of scattering, and – for the sake of simplicity – to introduce elliptical scattering zones, which lead us to the ellipse model of Parsons and Bajwa [208] (see also [209] and [65]) shown in Figure 7.1. All ellipses are confocal, i.e., they have common focal points (foci) Tx and Rx, which in our

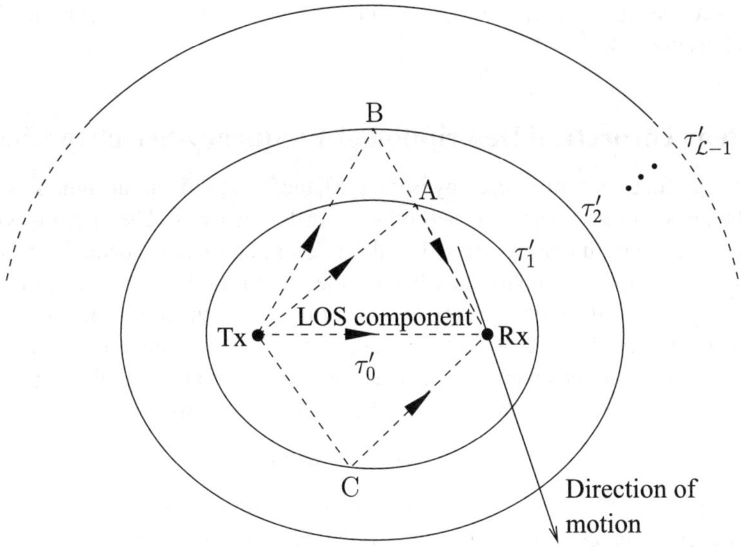

Figure 7.1 The ellipse model describing the path geometry according to Parsons and Bajwa [208].

case coincide with the position of the transmitter (Tx) and the receiver (Rx). In geometry, an ellipse is the set of all points in a plan such that the sum of the distances from any point on the ellipse to the two focal points Tx and Rx is a positive constant. Referring to Figure 7.1, this means the propagation paths Tx–A–Rx and Tx–C–Rx have the same path length. However, the respective angles-of-arrival are different, and, consequently, the corresponding Doppler frequencies, caused by the movement of the transmitter (receiver), are also different. The exact opposite statement holds for the (multipath) propagation paths Tx–A–Rx and Tx–B–Rx, where the path lengths are different, but the angles-of-arrival are equal, and thus the Doppler frequencies are equal, too.

The path length of each wave determines the propagation delay and essentially also the average power of the wave at the receiving antenna. Every wave in the scattering zone characterized by the ℓth ellipses undergoes the same *discrete propagation delay*

$$\tau'_\ell = \tau'_0 + \ell\,\Delta\tau', \quad \ell = 0, 1, \ldots, \mathcal{L} - 1, \tag{7.1}$$

where τ'_0 is the propagation delay of the line-of-sight (LOS) component, $\Delta\tau'$ is an infinitesimal propagation delay, and $\mathcal{L}$ denotes the number of paths with different propagation delays. It is evident that the ellipse model increases in precision if $\mathcal{L}$ increases and $\Delta\tau'$ becomes smaller. In the limit $\mathcal{L} \to \infty$ and $\Delta\tau' \to 0$, the discrete propagation delay τ'_ℓ results in the *continuous propagation delay* τ' restricted to the interval $[\tau'_0, \tau'_{max}]$. Here, τ'_{max} characterizes the maximum propagation delay, which depends on the environment. The maximum propagation delay τ'_{max} is chosen such that the energy contributions of the wave components with propagation delays τ' greater than τ'_{max} are negligibly small.

In the following discussion, we will see that the ellipse model forms to a certain extent the physical basis for the modelling of frequency-selective channels. In particular, the number of paths $\mathcal{L}$ with different propagation delays exactly corresponds to the number of delay elements required for the tapped-delay-line structure of the time-variant filter used for modelling frequency-selective channels. In view of obtaining a low-complexity realization, $\mathcal{L}$ should be kept as small as possible.

7.2 System Theoretical Description of Frequency-Selective Channels

Using the system functions introduced by Bello [18], the input and output signals of frequency-selective channels can be related to each other in different ways. The starting point for the derivation of the system functions is based on the assumption that the channel can be modelled in the equivalent complex baseband as a linear time-variant system. In time-variant systems, the impulse response — denoted by $h_0(t_0, t)$ — is a function of the time t_0 at which the channel has been excited by the impulse $\delta(t - t_0)$, and the time t at which the effect of the impulse is observed at the output of the channel. The relationship between the impulse $\delta(t - t_0)$ and the corresponding impulse response $h_0(t_0, t)$ can therefore be expressed by

$$\delta(t - t_0) \rightarrow h_0(t_0, t). \tag{7.2}$$

Since every physical channel is causal, the impulse cannot produce an effect before the impulse has excited the channel. This is the so-called *law of causality*, which can be stated mathematically as

$$h_0(t_0, t) = 0 \quad \text{for} \quad t < t_0 . \tag{7.3}$$

Using the impulse response $h_0(t_0, t)$, we now want to compute the output signal $y(t)$ of the channel for an arbitrary input signal $x(t)$. For this purpose, we first represent $x(t)$ as an infinite densely superposition of weighted Dirac delta functions. By applying the sifting property of the Dirac delta function, this allows us to write

$$x(t) = \int_{-\infty}^{\infty} x(t_0)\, \delta(t - t_0)\, dt_0 . \tag{7.4}$$

Alternatively, we can also use the expression

$$x(t) = \lim_{\Delta t_0 \to 0} \sum_{t_0} x(t_0)\, \delta(t - t_0) \Delta t_0 . \tag{7.5}$$

Since the channel was assumed to be linear, we may employ the principle of superposition [210]. Consequently, by using the relation in (7.2), the response to the sum in (7.5) can be written as

$$\sum_{t_0} x(t_0)\, \delta(t - t_0) \Delta t_0 \rightarrow \sum_{t_0} x(t_0)\, h_0(t_0, t) \Delta t_0 . \tag{7.6}$$

Hence, for the desired relationship

$$x(t) \rightarrow y(t), \tag{7.7}$$

we obtain the following result from (7.6) in the limit $\Delta t_0 \rightarrow 0$

$$\int_{-\infty}^{\infty} x(t_0)\, \delta(t - t_0)\, dt_0 \rightarrow \int_{-\infty}^{\infty} x(t_0)\, h_0(t_0, t)\, dt_0 . \tag{7.8}$$

If we now make use of the causality property (7.3), then the output signal is given by

$$y(t) = \int_{-\infty}^{t} x(t_0)\, h_0(t_0, t)\, dt_0 \,. \tag{7.9}$$

Next, we substitute the variable t_0 by the propagation delay

$$\tau' = t - t_0, \tag{7.10}$$

which defines the time elapsed from the moment at which the channel was excited by the impulse to the moment at which the response was observed at the output of the channel. Substituting t_0 by $t - \tau'$ in (7.9) results in

$$y(t) = \int_{0}^{\infty} x(t - \tau')\, h(\tau', t)\, d\tau' \,. \tag{7.11}$$

In order to simplify the notation, the time-variant impulse response $h_0(t - \tau', t)$ has been replaced by $h(\tau', t) := h_0(t - \tau', t)$. Physically, the time-variant impulse response $h(\tau', t)$ can be interpreted as the response of the channel at the time t to a Dirac delta impulse that stimulated the channel at the time $t - \tau'$. Considering (7.10), the causality property (7.3) can be expressed by

$$h(\tau', t) = 0 \quad \text{for} \quad \tau' < 0, \tag{7.12}$$

which means that the impulse response equals zero for negative propagation delays.

From (7.11), we now directly obtain the tapped-delay-line model shown in Figure 7.2 of a frequency-selective channel with the time-variant impulse response $h(\tau', t)$. Note that the tapped-delay-line model can be interpreted as a transversal filter with time-variant coefficients.

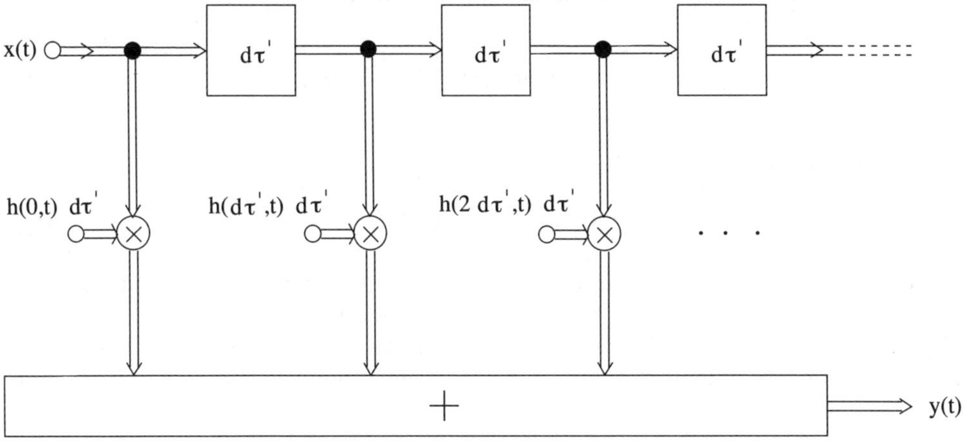

Figure 7.2 Tapped-delay-line representation of a frequency-selective and time-variant channel in the equivalent complex baseband.

Modelling mobile radio channels by using a tapped-delay-line structure with time-variant coefficients gives a deep insight into the channel distortions caused by scattering components with different propagation delays. For example, it is recognizable that the received signal is composed of an infinite number of delayed and weighted replicas of the transmitted signal. In digital data transmission this causes interference between adjacent received symbols. This effect, known as intersymbol interference (ISI), has to be eliminated as far as possible at the receiver, e.g., by using an equalizer. Moreover, the close relationship between the tapped-delay-line structure of the channel model and the geometrical ellipse model, described in the previous section, becomes obvious.

The *time-variant transfer function* $H(f', t)$ of the channel is defined by the Fourier transform of the time-variant impulse response $h(\tau', t)$ with respect to the propagation delay variable τ', i.e.,

$$H(f', t) := \int_0^\infty h(\tau', t)\, e^{-j2\pi f'\tau'} d\tau' . \tag{7.13}$$

This relation can be expressed symbolically by $h(\tau', t) \overset{\tau' \quad f'}{\circ\!\!-\!\!\!-\!\!\bullet} H(f', t)$. Notice that $H(f', t)$ only fulfils the condition $H^*(f', t) = H(-f', t)$ if $h(\tau', t)$ is a real-valued function. Starting from (7.11) and using $H(f', t)$, we can now establish the input-output relationship as follows

$$y(t) = \int_{-\infty}^\infty X(f')\, H(f', t)\, e^{j2\pi f' t} df', \tag{7.14}$$

where $X(f')$ is the Fourier transform of the input signal $x(t)$ at $f = f'$.

For the special case that $x(t)$ is a complex sinusoidal function of the form

$$x(t) = A\, e^{j2\pi f' t}, \tag{7.15}$$

where A denotes a complex-valued constant, it follows from (7.11)

$$y(t) = A \int_0^\infty h(\tau', t)\, e^{j2\pi f'(t-\tau')} d\tau' . \tag{7.16}$$

Using (7.13), we may also write

$$y(t) = A\, H(f', t)\, e^{j2\pi f' t} . \tag{7.17}$$

Therefore, the response of the channel can in this case be represented by the input signal $x(t)$, as given by (7.15), weighted by the time-variant transfer function $H(f', t)$, i.e., $y(t) = H(f', t)\, x(t)$. The relation in (7.17) makes clear that the time-variant transfer function $H(f', t)$ can be measured directly by sweeping a sinusoidal signal continuously through the relevant frequency range.

Neither the time-variant impulse response $h(\tau', t)$ nor the corresponding transfer function $H(f', t)$ allows an insight into the phenomena caused by the Doppler effect. In order to eliminate this disadvantage, we apply the Fourier transform to $h(\tau', t)$ with respect to the time variable t. In this way, we obtain a further system function

$$s(\tau', f) := \int_{-\infty}^{\infty} h(\tau', t)\, e^{-j2\pi f t} dt, \tag{7.18}$$

which is called the *Doppler-variant impulse response*.

Instead of (7.18), we may also write $h(\tau', t) \circ\!\!\overset{\tau \quad f}{\text{———}}\!\!\bullet\, s(\tau', f)$. Expressing the time-variant impulse response $h(\tau', t)$ by the inverse Fourier transform of $s(\tau', f)$, allows the representation of (7.11) in the form

$$y(t) = \int_{0}^{\infty} \int_{-\infty}^{\infty} x(t - \tau')\, s(\tau', f)\, e^{j2\pi f t} df\, d\tau' . \tag{7.19}$$

This relation shows that the output signal $y(t)$ can be represented by an infinite sum of delayed, weighted, and Doppler shifted replicas of the input signal $x(t)$. Signals delayed during transmission in the range of $[\tau', \tau' + d\tau')$ and affected by a Doppler shift within $[f, f + df)$ are weighted by the differential part $s(\tau', f)\, df\, d\tau'$. The Doppler-variant impulse response $s(\tau', f)$ thus describes explicitly the dispersive behaviour of the channel as a function of both the propagation delays τ' and the Doppler frequencies f. Consequently, the physical interpretation of $s(\tau', f)$ leads directly to the ellipse model shown in Figure 7.1.

A further system function, the so-called *Doppler-variant transfer function* $T(f', f)$, is defined by the two-dimensional Fourier transform of the time-variant impulse response $h(\tau', t)$ according to

$$T(f', f) := \int_{-\infty}^{\infty} \int_{0}^{\infty} h(\tau', t)\, e^{-j2\pi(ft + f'\tau')} d\tau' dt . \tag{7.20}$$

Due to (7.13) and (7.18), we may also write $T(f', f) \circ\!\!\overset{f \quad t}{\text{———}}\!\!\bullet\, H(f', t)$ or $T(f', f) \circ\!\!\overset{f' \quad \tau'}{\text{———}}\!\!\bullet\, s(\tau', f)$ for (7.20).

The computation of the Fourier transform of (7.11) with respect to the time variable t allows the representation of the spectrum $Y(f)$ of the output signal $y(t)$ in the form

$$Y(f) = \int_{-\infty}^{\infty} X(f - f')\, T(f - f', f')\, df' . \tag{7.21}$$

Finally, we exchange the frequency variables f and f' and obtain

$$Y(f') = \int_{-\infty}^{\infty} X(f' - f)\, T(f' - f, f)\, df . \tag{7.22}$$

This equation shows how a relation between the spectrum of the output signal and the input signal can be established by making use of the Doppler-variant transfer function $T(f', f)$. Regarding (7.22), it becomes apparent that the spectrum of the output signal can be interpreted as a superposition of an infinite number of Doppler shifted and filtered replicas of the spectrum of the input signal.

At the end of this section, we keep in mind that the four system functions $h(\tau', t)$, $H(f', t)$, $s(\tau', f)$, and $T(f', f)$ are related in pairs by the Fourier transform. The Fourier transform relationships established above are illustrated in Figure 7.3.

7.3 Frequency-Selective Stochastic Channel Models

7.3.1 Correlation Functions

In the following, we consider the channel as a stochastic system. In this case, the four functions $h(\tau', t)$, $H(f', t)$, $s(\tau', f)$, and $T(f', f)$ are stochastic system functions. Generally, these stochastic system functions can be described by the following autocorrelation functions:

$$r_{hh}(\tau_1', \tau_2'; t_1, t_2) := E\{h^*(\tau_1', t_1)\, h(\tau_2', t_2)\}, \tag{7.23a}$$

$$r_{HH}(f_1', f_2'; t_1, t_2) := E\{H^*(f_1', t_1)\, H(f_2', t_2)\}, \tag{7.23b}$$

$$r_{ss}(\tau_1', \tau_2'; f_1, f_2) := E\{s^*(\tau_1', f_1)\, s(\tau_2', f_2)\}, \tag{7.23c}$$

$$r_{TT}(f_1', f_2'; f_1, f_2) := E\{T^*(f_1', f_1)\, T(f_2', f_2)\}. \tag{7.23d}$$

Since the system functions are related by the Fourier transform, it is not surprising that analog relationships can also be derived for the autocorrelation functions. For example, (7.23a) and

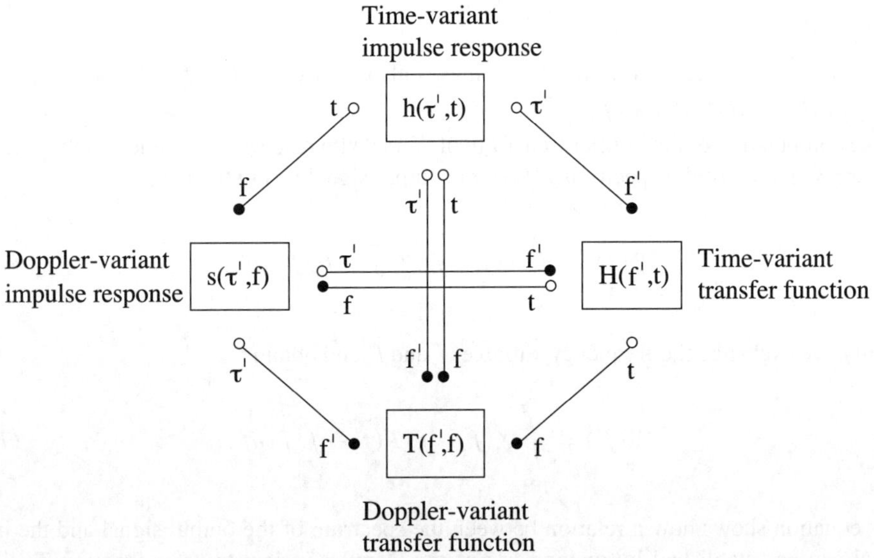

Figure 7.3 Fourier transform relationships between the system functions according to Bello [18].

(7.23b) are related by

$$
\begin{aligned}
r_{HH}(f_1', f_2'; t_1, t_2) &:= E\{H^*(f_1', t_1)\, H(f_2', t_2)\} \\
&= E\left\{ \int_{-\infty}^{\infty} h^*(\tau_1', t_1)\, e^{j2\pi f_1' \tau_1'}\, d\tau_1' \int_{-\infty}^{\infty} h(\tau_2', t_2)\, e^{-j2\pi f_2' \tau_2'}\, d\tau_2' \right\} \\
&= \int_{-\infty}^{\infty} \int_{-\infty}^{\infty} E\{h^*(\tau_1', t_1)\, h(\tau_2', t_2)\}\, e^{j2\pi (f_1'\tau_1' - f_2'\tau_2')}\, d\tau_1'\, d\tau_2' \\
&= \int_{-\infty}^{\infty} \int_{-\infty}^{\infty} r_{hh}(\tau_1', \tau_2'; t_1, t_2)\, e^{j2\pi (f_1'\tau_1' - f_2'\tau_2')}\, d\tau_1'\, d\tau_2' \,. \qquad (7.24)
\end{aligned}
$$

Finally, we replace the variable f_1' by $-f_1'$ on both sides of the last equation, to make clear that $r_{HH}(-f_1', f_2'; t_1, t_2)$ is the two-dimensional Fourier transform of $r_{hh}(\tau_1', \tau_2'; t_1, t_2)$ with respect to the two propagation delay variables τ_1' and τ_2'. This can be expressed symbolically by the notation $r_{hh}(\tau_1', \tau_2'; t_1, t_2) \circ\!\!\xrightarrow{\tau_1', \tau_2', f_1', f_2'}\!\!\bullet\, r_{HH}(-f_1', f_2'; t_1, t_2)$. The Fourier transform relationships between all the other pairs of (7.23a)–(7.23d) can be derived in a similar way. As a result, one finds the relationships between the autocorrelation functions of the stochastic system functions as shown in Figure 7.4.

In order to describe the input-output relationship of the stochastic channel, we assume that the input signal $x(t)$ is a stochastic process with the known autocorrelation function $r_{xx}(t_1, t_2) := E\{x^*(t_1)\, x(t_2)\}$. Since (7.11) is valid for deterministic systems as well as for stochastic systems, we can express the autocorrelation function $r_{yy}(t_1, t_2)$ of the output signal $y(t)$ in terms of

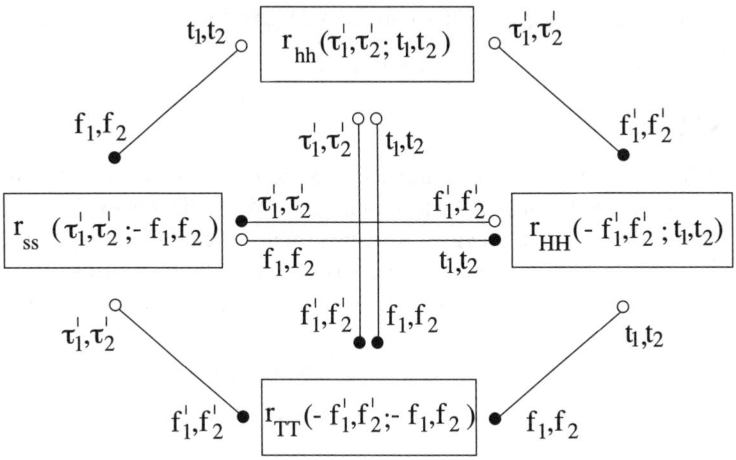

Figure 7.4 Fourier transform relationships between the autocorrelation functions of the stochastic system functions.

$r_{xx}(t_1, t_2)$ and $r_{hh}(\tau_1', \tau_2'; t_1, t_2)$ as follows:

$$
\begin{aligned}
r_{yy}(t_1, t_2) &:= E\{y^*(t_1)\, y(t_2)\} \\
&= E\left\{\int_0^\infty \int_0^\infty x^*(t_1 - \tau_1')\, x(t_2 - \tau_2')\, h^*(\tau_1', t_1)\, h(\tau_2', t_2)\, d\tau_1'\, d\tau_2'\right\} \\
&= \int_0^\infty \int_0^\infty E\{x^*(t_1 - \tau_1')\, x(t_2 - \tau_2')\}\, E\{h^*(\tau_1', t_1)\, h(\tau_2', t_2)\}\, d\tau_1'\, d\tau_2' \\
&= \int_0^\infty \int_0^\infty r_{xx}(t_1 - \tau_1'; t_2 - \tau_2')\, r_{hh}(\tau_1', \tau_2'; t_1, t_2)\, d\tau_1'\, d\tau_2' .
\end{aligned}
\tag{7.25}
$$

In the derivation above, we have assumed implicitly that the time-variant impulse response $h(\tau', t)$ of the channel and the input signal $x(t)$ are statistically independent.

Significant simplifications can be made by assuming that the time-variant impulse response $h(\tau', t)$ is stationary in the wide sense with respect to t, and that the scattering components with different propagation delays are statistically uncorrelated. Basing on these assumptions, Bello introduced the so-called WSSUS model in his fundamental work [18] on stochastic time-variant linear systems. The description of the WSSUS model is the topic of the following subsection.

7.3.2 The WSSUS Model According to Bello

The WSSUS model enables the statistical description of the input-output relationship of mobile radio channels for the transmission of bandpass signals in the equivalent complex baseband for observation periods in which the stationarity of the channel is ensured in the wide sense. According to empirical studies [208], the channel can be considered as wide-sense stationary as long as the mobile unit covers a distance in the order of a few tens of the wavelength of the carrier signal.

7.3.2.1 WSS Models

A channel model with a wide-sense stationary impulse response is called the *WSS channel model* (WSS: wide-sense stationary). Instead of the term WSS channel model, we also use the short form *WSS model*, since it is evident that this model is exclusively used for modelling channels. The assumption of wide-sense stationarity leads to the fact that the two autocorrelation functions in (7.23a) and (7.23b) are invariant with respect to a translation in time, i.e., the autocorrelation functions $r_{hh}(\tau_1', \tau_2'; t_1, t_2)$ and $r_{HH}(f_1', f_2'; t_1, t_2)$ merely depend on the time difference $\tau := t_2 - t_1$. With $t_1 = t$ and $t_2 = t + \tau$, we can therefore write in case of WSS models:

$$r_{hh}(\tau_1', \tau_2'; t, t + \tau) = r_{hh}(\tau_1', \tau_2'; \tau), \tag{7.26a}$$

$$r_{HH}(f_1', f_2'; t, t + \tau) = r_{HH}(f_1', f_2'; \tau). \tag{7.26b}$$

The restricted behaviour of these two autocorrelation functions certainly has consequences for the remaining autocorrelation functions in (7.23c) and (7.23d). To clarify this, we look at the Fourier transform relationship between $r_{hh}(\tau_1', \tau_2'; t_1, t_2)$ and $r_{ss}(\tau_1', \tau_2'; f_1, f_2)$ which can be formulated by considering Figure 7.4 as follows

$$r_{ss}(\tau_1', \tau_2'; f_1, f_2) = \int_{-\infty}^{\infty}\int_{-\infty}^{\infty} r_{hh}(\tau_1', \tau_2'; t_1, t_2)\, e^{j2\pi(t_1 f_1 - t_2 f_2)}\, dt_1\, dt_2\,. \tag{7.27}$$

The substitutions of the variables $t_1 \to t$ and $t_2 \to t + \tau$, in connection with (7.26a), result in

$$r_{ss}(\tau_1', \tau_2'; f_1, f_2) = \int_{-\infty}^{\infty} e^{-j2\pi(f_2-f_1)t}\, dt \int_{-\infty}^{\infty} r_{hh}(\tau_1', \tau_2'; \tau)\, e^{-j2\pi f_2 \tau}\, d\tau\,. \tag{7.28}$$

The first integral on the right-hand side of (7.28) can be identified with the Dirac delta function $\delta(f_2 - f_1)$. Consequently, $r_{ss}(\tau_1', \tau_2'; f_1, f_2)$ can be expressed by

$$r_{ss}(\tau_1', \tau_2'; f_1, f_2) = \delta(f_2 - f_1)\, S_{ss}(\tau_1', \tau_2'; f_1), \tag{7.29}$$

where $S_{ss}(\tau_1', \tau_2'; f_1)$ is the Fourier transform of the autocorrelation function $r_{hh}(\tau_1', \tau_2'; \tau)$ with respect to the time separation variable τ. The assumption that the time-variant impulse response $h(\tau', t)$ is wide-sense stationary therefore leads to the fact that the system functions $s(\tau_1', f_1)$ and $s(\tau_2', f_2)$ are statistically uncorrelated if the Doppler frequencies f_1 and f_2 are different.

It can be shown in a similar way that (7.23d) can be represented in the form

$$r_{TT}(f_1', f_2'; f_1, f_2) = \delta(f_2 - f_1)\, S_{TT}(f_1', f_2'; f_1), \tag{7.30}$$

where the symbol $S_{TT}(f_1', f_2'; f_1)$ denotes the Fourier transform of the autocorrelation function $r_{HH}(f_1', f_2'; \tau)$ with respect to τ. From (7.30), we can conclude that the system functions $T(f_1', f_1)$ and $T(f_2', f_2)$ are statistically uncorrelated for different Doppler frequencies f_1 and f_2.

Since the time-variant impulse response $h(\tau', t)$ results from a superposition of a multitude of scattered components, it can generally be stated that the WSS assumption leads to the fact that scattering components with different Doppler frequencies or different angles-of-arrival are statistically uncorrelated.

7.3.2.2 US Models

A second important class of channel models is obtained by assuming that scattering components with different propagation delays are statistically uncorrelated. These channel models are called *US channel models* or *US models* (US: **u**ncorrelated **s**cattering). The autocorrelation functions $r_{hh}(\tau_1', \tau_2'; t_1, t_2)$ and $r_{ss}(\tau_1', \tau_2'; f_1, f_2)$ of US models can first of all be described formally by

$$r_{hh}(\tau_1', \tau_2'; t_1, t_2) = \delta(\tau_2' - \tau_1')\, S_{hh}(\tau_1'; t_1, t_2), \tag{7.31a}$$

$$r_{ss}(\tau_1', \tau_2'; f_1, f_2) = \delta(\tau_2' - \tau_1')\, S_{ss}(\tau_1'; f_1, f_2). \tag{7.31b}$$

The singular behaviour of the autocorrelation function (7.31a) has significant consequences for the tapped-delay-line model shown in Figure 7.2, because the time-variant coefficients of this model are now uncorrelated as a result of the US assumption. In practice, the coefficients of the tapped-delay-line model are realized almost exclusively by using coloured Gaussian random processes. We recall that uncorrelated Gaussian random processes are also statistically independent.

Formal expressions for the autocorrelation functions of the remaining stochastic system functions $H(f', t)$ and $T(f', f)$ can easily be determined by using the relations illustrated in Figure 7.4. With the substitutions $f_1' \to f'$ and $f_2' \to f' + \upsilon'$, the following equations can be found:

$$r_{HH}(f', f' + \upsilon'; t_1, t_2) = r_{HH}(\upsilon'; t_1, t_2), \tag{7.32a}$$

$$r_{TT}(f', f' + \upsilon'; f_1, f_2) = r_{TT}(\upsilon'; f_1, f_2). \tag{7.32b}$$

Obviously, the autocorrelation functions of the system functions $H(f', t)$ and $T(f', f)$ only depend on the frequency difference $\upsilon' := f_2' - f_1'$. As a consequence of the US assumption, we can now conclude that US models are wide-sense stationary with respect to the frequency f'.

By comparing the above mentioned autocorrelation functions of US models with those derived for WSS models, it turns out that they are dual to each other. Therefore, we can say that the class of US models stands in a duality relationship to the class of WSS models.

7.3.2.3 WSSUS Models

The most important class of stochastic time-variant linear channel models is represented by models belonging to both the class of WSS models as well as to the class of US models. These channel models with wide-sense stationary impulse responses and uncorrelated scattering components are called WSSUS channel models or simply WSSUS models. Owing to their simplicity, they are of great practical importance and are nowadays employed almost exclusively for the modelling of frequency-selective mobile radio channels.

Under the WSSUS assumption, the autocorrelation function of the time-variant impulse response $h(\tau', t)$ has to satisfy the condition in (7.26a) as well as in (7.31a). Hence, we may write formally

$$r_{hh}(\tau_1', \tau_2'; t, t + \tau) = \delta(\tau_2' - \tau_1')\, S_{hh}(\tau_1', \tau), \tag{7.33}$$

where $S_{hh}(\tau_1', \tau)$ is called the *delay cross-power spectral density*. With this representation it becomes evident that the time-variant impulse response $h(\tau', t)$ of WSSUS models displays on the one hand the characteristic properties of uncorrelated white noise with respect to the propagation delay τ', and on the other $h(\tau', t)$ is also wide-sense stationary with respect to the time t.

By analogy, we can directly obtain the autocorrelation function of $T(f', f)$ by combining the properties in (7.30) and (7.32b). Thus, for WSSUS models, it holds

$$r_{TT}(f', f' + \upsilon'; f_1, f_2) = \delta(f_2 - f_1)\, S_{TT}(\upsilon', f_1), \tag{7.34}$$

where $S_{TT}(\upsilon', f_1)$ is called the *Doppler cross-power spectral density*. This result shows that the system function $T(f', f)$ of WSSUS models behaves like uncorrelated white noise with respect to the Doppler frequency f and like a wide-sense stationary stochastic process with respect to the frequency f'.

Furthermore, we can combine the relations (7.29) and (7.31b) and obtain the autocorrelation function of $s(\tau', f)$ in the form

$$r_{ss}(\tau_1', \tau_2'; f_1, f_2) = \delta(f_2 - f_1)\,\delta(\tau_2' - \tau_1')\,S(\tau_1', f_1)\,. \tag{7.35}$$

From this result, we conclude that the system function $s(\tau', f)$ of WSSUS models has the characteristic properties of uncorrelated white noise with respect to both τ' and f. In [18], Bello called the function $S(\tau_1', f_1)$ appearing in (7.35) the *scattering function*.

Finally, by combining (7.26b) and (7.32a), it follows the relation for the autocorrelation function of $H(f', t)$

$$r_{HH}(f', f' + \upsilon'; t, t + \tau) = r_{HH}(\upsilon', \tau)\,. \tag{7.36}$$

The autocorrelation function $r_{HH}(\upsilon', \tau)$ is called the *time-frequency correlation function*. Regarding (7.36), it becomes obvious that the system function $H(f', t)$ of WSSUS models has the properties of wide-sense stationary stochastic processes with respect to f' and t.

Figure 7.4 shows the universally valid relationships between the autocorrelation functions of the four system functions. With the expressions (7.33)–(7.36), it is now possible to derive the specific relations valid for WSSUS models. One may therefore study Figure 7.5, where the relationships between the delay cross-power spectral density $S_{hh}(\tau', \tau)$, the time-frequency correlation function $r_{HH}(\upsilon', \tau)$, the Doppler cross-power spectral density $S_{TT}(\upsilon', f)$, and the scattering function $S(\tau', f)$ are shown. We note that the substitutions $f_1 \to f$ and $\tau_1' \to \tau'$ have been carried out to simplify the notation.

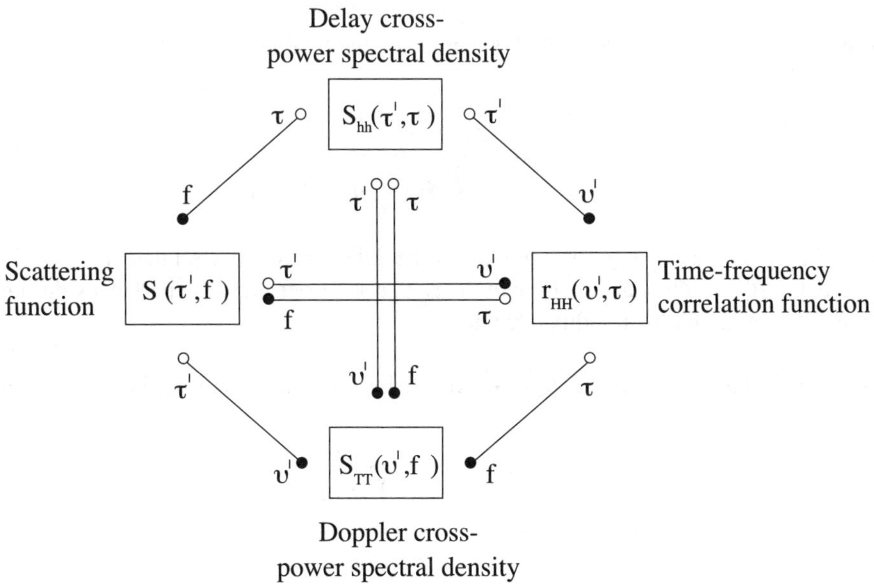

Figure 7.5 Fourier transform relationships between the delay cross-power spectral density $S_{hh}(\tau', \tau)$, the time-frequency correlation function $r_{HH}(\upsilon', \tau)$, the Doppler cross-power spectral density $S_{TT}(\upsilon', f)$, and the scattering function $S(\tau', f)$ of WSSUS models.

Figure 7.5 makes clear that the knowledge of one of the four depicted functions is sufficient to calculate the remaining three. For example, from the scattering function $S(\tau', f)$, we can obtain directly the delay cross-power spectral density $S_{hh}(\tau', \tau)$ by computing the inverse Fourier transform with respect to the Doppler frequency f, i.e.,

$$S_{hh}(\tau', \tau) = \int_{-\infty}^{\infty} S(\tau', f)\, e^{j2\pi f \tau}\, df, \tag{7.37}$$

where $\tau = t_2 - t_1$.

Delay power spectral density: The delay cross-power spectral density $S_{hh}(\tau', \tau)$ at $\tau = 0$ defines the so-called *delay power spectral density* $S_{\tau'}(\tau')$, which is more popularly known as the *power delay profile*. Due to (7.37), the delay power spectral density $S_{\tau'}(\tau')$ is related to the scattering function $S(\tau', f)$ according to

$$S_{\tau'}(\tau') := S_{hh}(\tau', 0) = \int_{-\infty}^{\infty} S(\tau', f)\, df\,. \tag{7.38}$$

The delay power spectral density $S_{\tau'}(\tau')$ is a measure of the average power of the multipath components associated with the propagation delays τ' in the interval $[\tau', \tau' + d\tau']$. It can easily be shown that $S_{\tau'}(\tau')$ is proportional to the probability density function $p_{\tau'}(\tau')$ of the propagation delays τ', i.e., $p_{\tau'}(\tau') \sim S_{\tau'}(\tau')$. From the delay power spectral density $S_{\tau'}(\tau')$, two important characteristic quantities for the characterization of WSSUS models can be derived: the *average delay* and the *delay spread*.

Average delay: The average delay $B_{\tau'}^{(1)}$ is defined by the first moment of $S_{\tau'}(\tau')$, i.e.,

$$B_{\tau'}^{(1)} := \frac{\int_{-\infty}^{\infty} \tau' S_{\tau'}(\tau')\, d\tau'}{\int_{-\infty}^{\infty} S_{\tau'}(\tau')\, d\tau'}\,. \tag{7.39}$$

It can be interpreted as the centre of gravity of the delay power spectral density $S_{\tau'}(\tau')$. The average delay $B_{\tau'}^{(1)}$ is the statistical mean delay that a carrier signal experiences during the transmission over a multipath fading channel.

Delay spread: The delay spread $B_{\tau'}^{(2)}$ is defined by the square root of the second central moment of $S_{\tau'}(\tau')$, i.e.,

$$B_{\tau'}^{(2)} := \sqrt{\frac{\int_{-\infty}^{\infty} \left(\tau' - B_{\tau'}^{(1)}\right)^2 S_{\tau'}(\tau')\, d\tau'}{\int_{-\infty}^{\infty} S_{\tau'}(\tau')\, d\tau'}}\,. \tag{7.40}$$

The delay spread $B_{\tau'}^{(2)}$ provides us with a measure of the time spread of an impulse passed through a multipath fading channel.

From Figure 7.5, we realize that the Doppler cross-power spectral density $S_{TT}(\upsilon', f)$ is the Fourier transform of the scattering function $S(\tau', f)$ with respect to the propagation delay τ', i.e., the relation

$$S_{TT}(\upsilon', f) = \int_{-\infty}^{\infty} S(\tau', f)\, e^{-j2\pi\upsilon'\tau'}\, d\tau' \tag{7.41}$$

holds, where $\upsilon' = f_2' - f_1{}'$.

Doppler power spectral density: For $\upsilon' = 0$, the already known Doppler power spectral density $S_{\mu\mu}(f)$ follows from the Doppler cross-power spectral density $S_{TT}(\upsilon', f)$, because

$$S_{\mu\mu}(f) := S_{TT}(0, f) = \int_{-\infty}^{\infty} S(\tau', f)\, d\tau' \tag{7.42}$$

holds. The Doppler power spectral density $S_{\mu\mu}(f)$ is a measure for the average power of the scattering components associated with the Doppler frequency f. In Appendix 3.A, it is shown that $S_{\mu\mu}(f)$ is proportional to the probability density function of the Doppler frequencies f. Remember that two important characteristic quantities can be derived from the Doppler power spectral density $S_{\mu\mu}(f)$, namely, the average Doppler shift $B_{\mu\mu}^{(1)}$ [cf. (3.28a)] and the Doppler spread $B_{\mu\mu}^{(2)}$ [cf. (3.28b)].

According to Figure 7.5, the time-frequency correlation function $r_{HH}(\upsilon', \tau)$ can be calculated from the scattering function $S(\tau', f)$ as follows

$$r_{HH}(\upsilon', \tau) = \int_{-\infty}^{\infty}\int_{-\infty}^{\infty} S(\tau', f)\, e^{-j2\pi(\upsilon'\tau' - f\tau)}\, d\tau'\, df, \tag{7.43}$$

where $\upsilon' = f_2' - f_1'$ and $\tau = t_2 - t_1$. Alternatively, we could have calculated $r_{HH}(\upsilon', \tau)$ by applying the Fourier transform to $S_{hh}(\tau', \tau)$ with respect to the propagation delay τ' or via the inverse Fourier transform of $S_{TT}(\upsilon', f)$ with respect to the Doppler frequency f.

From the time-frequency correlation function $r_{HH}(\upsilon', \tau)$, two further correlation functions can be derived. They are called *frequency correlation function* and *time correlation function.* From each of these functions, a further important characteristic quantity can be derived: the *coherence bandwidth* and the *coherence time.*

Frequency correlation function: The *frequency correlation function* $r_{\tau'}(\upsilon')$ is defined by the time-frequency correlation function $r_{HH}(\upsilon', \tau)$ at $\tau = t_2 - t_1 = 0$, i.e.,

$$\begin{aligned} r_{\tau'}(\upsilon') &:= r_{HH}(\upsilon', 0) \\ &= \int_{-\infty}^{\infty}\int_{-\infty}^{\infty} S(\tau', f)\, e^{-j2\pi\upsilon'\tau'}\, d\tau'\, df \\ &= \int_{-\infty}^{\infty} S_{\tau'}(\tau')\, e^{-j2\pi\upsilon'\tau'}\, d\tau'. \end{aligned} \tag{7.44}$$

Obviously, the frequency correlation function $r_{\tau'}(\upsilon')$ is the Fourier transform of the delay power spectral density $S_{\tau'}(\tau')$. Since the delay power spectral density is a real-valued function, the frequency correlation function exhibits the Hermitian symmetry property, i.e., $r_{\tau'}(\upsilon') = r^*_{\tau'}(-\upsilon')$. Furthermore, since $S_{\tau'}(\tau') = 0$ for $\tau' < 0$, the real and imaginary parts of $r_{\tau'}(\upsilon')$ are related to each other by the Hilbert transform, i.e., $\text{Re}\,\{r_{\tau'}(\upsilon')\} = \mathcal{H}\{\,\text{Im}\,\{r_{\tau'}(\upsilon')\}\}$, where $\mathcal{H}\{\cdot\}$ denotes the Hilbert transform operator. With reference to (7.23b) and (7.36), it turns out that $r_{\tau'}(\upsilon')$ can be computed alternatively by using the definition $r_{\tau'}(\upsilon') := E\{H^*(f',t)H(f'+\upsilon',t)\}$. This definition allows the following physical interpretation: The frequency correlation function measures the statistical correlation of the time-variant transfer functions $H(f_1', t)$ and $H(f_2', t)$ of the channel as a function of the frequency separation variable $\upsilon' = f_2' - f_1'$.

Coherence bandwidth: The frequency separation variable $\upsilon' = B_C$ that fulfils the condition

$$|r_{\tau'}(B_C)| = \frac{1}{2}|r_{\tau'}(0)| \tag{7.45}$$

is called the *coherence bandwidth*. The coherence bandwidth is a measure of the frequency interval over which the frequency response of the channel does not change significantly.

Since, referring to (7.44), the frequency correlation function $r_{\tau'}(\upsilon')$ and the delay power spectral density $S_{\tau'}(\tau')$ form a Fourier transform pair, the coherence bandwidth B_C is, according to the uncertainty principle of communications engineering [210], approximately reciprocally proportional to the delay spread $B_{\tau'}^{(2)}$, i.e., $B_C \approx 1/B_{\tau'}^{(2)}$. If the ratio of the signal bandwidth to the coherence bandwidth increases, then the complexity of the channel equalizer at the receiver grows, as the effect of ISI becomes more severe. An important special case occurs if the coherence bandwidth B_C is much greater than the symbol rate f_{sym}, i.e., if

$$B_C \gg f_{\text{sym}} \quad \text{or} \quad B_{\tau'}^{(2)} \ll T_{\text{sym}} \tag{7.46a,b}$$

holds, where $T_{\text{sym}} = 1/f_{\text{sym}}$ denotes the symbol interval. In this case, the effect of the impulse dispersion can be neglected and the time-variant impulse response $h(\tau', t)$ of the channel can approximately be represented by

$$h(\tau', t) = \delta(\tau') \cdot \mu(t), \tag{7.47}$$

where $\mu(t)$ is a proper complex stochastic process. Using (7.11), the output signal $y(t)$ may be expressed as

$$y(t) = \mu(t) \cdot x(t)\,. \tag{7.48}$$

Owing to the multiplicative relation between $\mu(t)$ and $x(t)$, we speak of *multiplicative fading* and we call the channel a *multiplicative channel* in this context. After substituting (7.47) in (7.13), we obtain the following expression for the time-variant transfer function $H(f', t)$ of the channel

$$H(f', t) = \mu(t). \tag{7.49}$$

In this case, the time-variant transfer function is obviously independent of the frequency f'. Thus, the channel is said to be *frequency-nonselective*, because all frequency components of the transmitted signal are subjected to the same temporal variations. A frequency-nonselective modelling of the mobile radio channel is always adequate if the delay spread $B_{\tau'}^{(2)}$ does not exceed 10 per cent to 20 per cent of the symbol interval T_{sym}. Otherwise, the channel is considered as *frequency-selective* if the conditions in (7.46a,b) are not fulfilled.

Time correlation function: The *time correlation function* $r_{\mu\mu}(\tau)$ is defined by the time-frequency correlation function $r_{HH}(\upsilon', \tau)$ at $\upsilon' = f_2' - f_1' = 0$, i.e.,

$$\begin{aligned} r_{\mu\mu}(\tau) &:= r_{HH}(0, \tau) \\ &= \int_{-\infty}^{\infty}\int_{-\infty}^{\infty} S(\tau', f)\, e^{j2\pi f\tau}\, d\tau'\, df \\ &= \int_{-\infty}^{\infty} S_{\mu\mu}(f)\, e^{j2\pi f\tau}\, df\,. \end{aligned} \tag{7.50}$$

This correlation function describes the correlation properties of the received scattered components as a function of the time difference $\tau = t_2 - t_1$.

Coherence time: The time interval $\tau = T_C$ that fulfils the condition

$$|r_{\mu\mu}(T_C)| = \frac{1}{2}|r_{\mu\mu}(0)| \tag{7.51}$$

is called the *coherence time.*

According to (7.50), the time correlation function $r_{HH}(0, \tau)$ and the Doppler power spectral density $S_{\mu\mu}(f)$ form a Fourier transform pair. Consequently, the coherence time T_C behaves approximately reciprocally proportional to the Doppler spread $B_{\mu\mu}^{(2)}$, i.e., $T_C \approx 1/B_{\mu\mu}^{(2)}$. The smaller the ratio of the coherence time T_C and the symbol interval T_{sym} is, the higher are the demands on the tracking capability of the channel estimator in the receiver. If the coherence time T_C is much larger than the symbol interval T_{sym}, i.e.,

$$T_C \gg T_{\text{sym}} \quad \text{or} \quad B_{\mu\mu}^{(2)} \ll f_{\text{sym}}\,, \tag{7.52a,b}$$

then the envelope and phase change of the channel may be regarded as approximately constant over the duration of one data symbol. In this case, we speak of *slow fading*. Contrarily, *fast fading* occurs if the coherence time T_C is relatively small compared to the symbol interval, i.e., $T_C < T_{\text{sym}}$. In this regime, the channel's envelope and phase may vary considerably during the transmission of one data symbol.

Figure 7.6 once more presents the relationships between the correlation functions and the power spectral densities introduced in this subsection in conjunction with the characteristic quantities of WSSUS models derived therefrom. This figure shows us vividly that the knowledge of the scattering function $S(\tau', f)$ is sufficient for WSSUS models to determine all correlation functions and power spectral densities as well as the characteristic quantities such as the delay spread and Doppler spread.

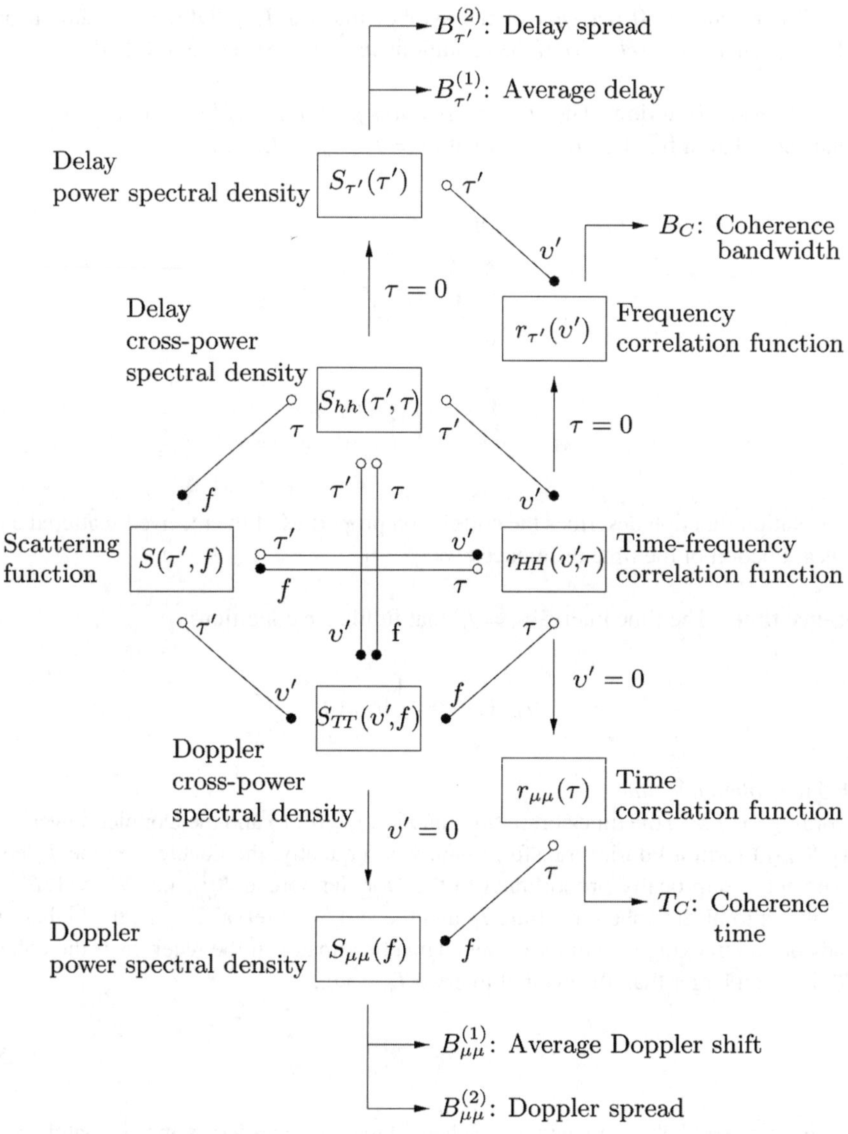

Figure 7.6 Relationships between the correlation functions, power spectral densities, and characteristic quantities of WSSUS models.

7.3.3 The COST 207 Channel Models

In 1984, the European working group COST[1] 207 was established by CEPT[2]. At that time, this working group developed suitable channel models for typical propagation environments

[1] COST: European Cooperation in the Field of Scientific and Technical Research.
[2] CEPT: Conference of European Postal and Telecommunications Administrations.

in view of the planned pan-European mobile communication system GSM[3]. The typical propagation environments were classified into areas with rural character (RA: Rural Area), areas typical for cities and suburbs (TU: Typical Urban), densely built urban areas with bad propagation conditions (BU: Bad Urban), and hilly terrains (HT: Hilly Terrain). Basing on the WSSUS assumption, the working group COST 207 developed specifications for the delay power spectral density and the Doppler power spectral density for these four classes of propagation environments [19]. The main results will be presented subsequently.

The specification of typical delay power spectral densities $S_{\tau'}(\tau')$ is based on the assumption that the corresponding probability density function $p_{\tau'}(\tau')$, which is proportional to $S_{\tau'}(\tau')$, can be represented by one or several negative exponential functions. The delay power spectral density functions $S_{\tau'}(\tau')$ of the channel models according to COST 207 are given in Table 7.1 and their graphs are plotted in Figure 7.7. The real-valued constant quantities c_{RA}, c_{TU}, c_{BU}, and c_{HT} introduced in in Table 7.1 can in principle be chosen arbitrarily. Hence, they have been determined such that the average delay power is equal to one, i.e., $\int_0^\infty S_{\tau'}(\tau')\,d\tau' = 1$. In this case, it holds:

$$c_{RA} = \frac{9.2}{1 - e^{-6.44}}, \quad c_{TU} = \frac{1}{1 - e^{-7}}, \tag{7.53a,b}$$

$$c_{BU} = \frac{2}{3(1 - e^{-5})}, \quad c_{HT} = \frac{1}{(1 - e^{-7})/3.5 + (1 - e^{-5})/10}. \tag{7.53c,d}$$

In the GSM system, the symbol interval T_{sym} is defined by $T_{\text{sym}} = 3.7\mu s$. If we bring T_{sym} in relation to the delay spread $B_{\tau'}^{(2)}$, which is listed in the last column of Table 7.1, then we realize that the condition in (7.46b) is only fulfilled for the RA channel. Consequently, the RA channel belongs to the class of frequency-nonselective channels, whereas the other channels (TU, BU, HT) are frequency selective.

Table 7.1 Specification of typical delay power spectral densities $S_{\tau'}(\tau')$ according to COST 207 [19].

Propagation area	Delay power spectral density $S_{\tau'}(\tau')$		Delay spread $B_{\tau'}^{(2)}$
Rural Area (RA)	$c_{RA}e^{-9.2\tau'/\mu s}$,	$0 \le \tau' < 0.7\mu s$	$0.1\ \mu s$
	0,	else	
Typical Urban (TU)	$c_{TU}e^{-\tau'/\mu s}$,	$0 \le \tau' < 7\mu s$	$0.98\ \mu s$
	0,	else	
Bad Urban (BU)	$c_{BU}e^{-\tau'/\mu s}$,	$0 \le \tau' < 5\mu s$	
	$c_{BU}\frac{1}{2}e^{(5-\tau'/\mu s)}$,	$5\mu s \le \tau' < 10\mu s$	$2.53\ \mu s$
	0,	else	
Hilly Terrain (HT)	$c_{HT}e^{-3.5\tau'/\mu s}$,	$0 \le \tau' < 2\mu s$	
	$c_{HT}0.1e^{(15-\tau'/\mu s)}$,	$15\mu s \le \tau' < 20\mu s$	$6.88\ \mu s$
	0,	else	

[3] GSM: Global System for Mobile Communications.

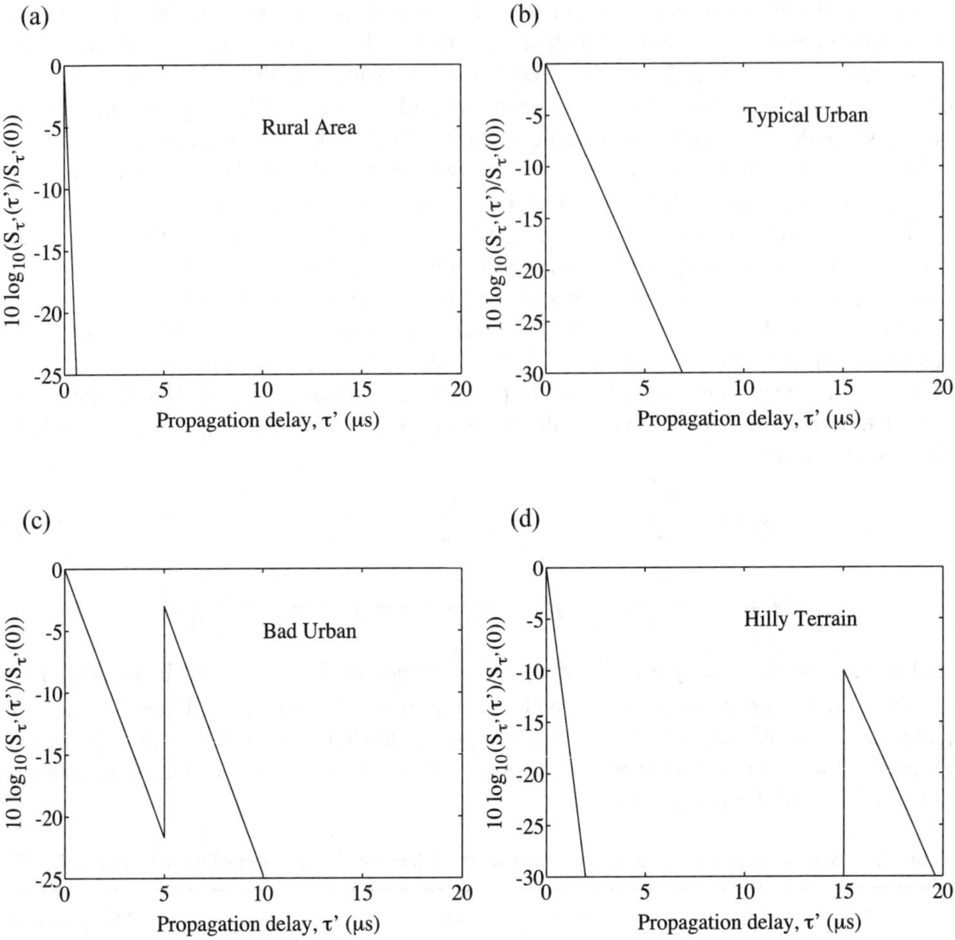

Figure 7.7 Delay power spectral densities $S_{\tau'}(\tau')$ of the channel models according to COST 207 [19].

Table 7.2 shows the four types of Doppler power spectral densities $S_{\mu\mu}(f)$ specified by COST 207. They are also presented in Figure 7.8 for better visualization. For the real-valued constants A_1 and A_2, preferably the values $A_1 = 50/(\sqrt{2\pi}3f_{\max})$ and $A_2 = 10^{1.5}/[\sqrt{2\pi}(\sqrt{10}+0.15)f_{\max}]$ are chosen, since it is then ensured that $\int_{-\infty}^{\infty} S_{\mu\mu}(f)\,df$ is normalized to unity. The classical Jakes power spectral density only occurs in the case of very short propagation delays ($\tau' \leq 0.5\,\mu$s) [see Figures 7.8(a) and 7.8(d)]. Only in this case, the assumptions that the amplitudes of the scattering components are homogeneous and the angles-of-arrival are uniformly distributed between 0 and 2π are justified. For scattering components with medium and long propagation delays τ', however, it is assumed that the corresponding Doppler frequencies are normally distributed, which results in a Doppler power spectral density with a Gaussian shape [see Figures 7.8(b) and 7.8(c)]. This had already been pointed out by Cox [74] at a very early stage after evaluating channel measurements conducted in urban regions.

Table 7.2 Specification of typical Doppler power spectral densities $S_{\mu\mu}(f)$ according to COST 207 [19], where $G(A_i, f_i, s_i)$ is defined by $G(A_i, f_i, s_i) := A_i \exp\{-(f - f_i)^2/(2s_i^2)\}$.

Type	Doppler power spectral density $S_{\mu\mu}(f)$	Propagation delay τ'	Doppler spread $B_{\mu\mu}^{(2)}$
"Jakes"	$\frac{1}{\pi f_{\max}\sqrt{1-(f/f_{\max})^2}}$	$0 \leq \tau' \leq 0.5\mu s$	$f_{\max}/\sqrt{2}$
"Gauss I"	$G(A_1, -0.8f_{\max}, 0.05f_{\max}) + G(A_1/10, 0.4f_{\max}, 0.1f_{\max})$	$0.5\mu s \leq \tau' \leq 2\mu s$	$0.45f_{\max}$
"Gauss II"	$G(A_2, 0.7f_{\max}, 0.1f_{\max}) + G(A_2/10^{1.5}, -0.4f_{\max}, 0.15f_{\max})$	$\tau' \geq 2\mu s$	$0.25f_{\max}$
"Rice"	$\frac{0.41^2}{\pi f_{\max}\sqrt{1-(f/f_{\max})^2}} + 0.91^2\delta(f - 0.7f_{\max})$	$\tau' = 0\mu s$	$0.39f_{\max}$

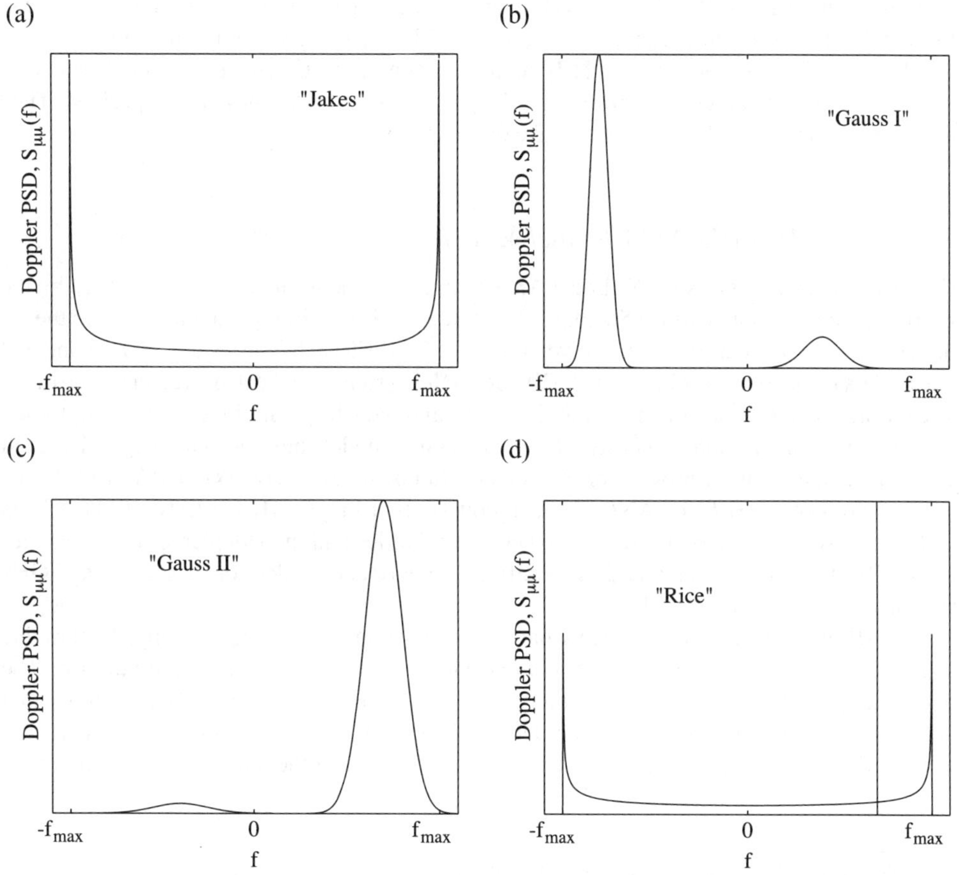

Figure 7.8 Doppler power spectral densities $S_{\mu\mu}(f)$ of the channel models according to COST 207 [19].

From Tables 7.1 and 7.2, it can be seen that the delay power spectral density $S_{\tau'}(\tau')$ is independent of the Doppler frequencies f, but the propagation delays τ' have a decisive influence on the shape of the Doppler power spectral density $S_{\mu\mu}(f)$. However, this is not valid for rural areas, where only the classical Jakes power spectral density is used. In this special case, the scattering function $S(\tau', f)$ can be represented as the product of the delay power spectral density and the Doppler power spectral density, i.e.,

$$S(\tau', f) = S_{\tau'}(\tau') \cdot S_{\mu\mu}(f). \tag{7.54}$$

Channels with a scattering function of the form (7.54) are called *independent time dispersive and frequency dispersive channels.* For this class of channels, the physical mechanism causing the propagation delays is independent of that which is responsible for the Doppler effect [211].

Regarding the design of hardware or software simulation models for frequency-selective channels, a discretization of the delay power spectral density $S_{\tau'}(\tau')$ has to be performed. In particular, the propagation delays τ' must be discretized and adapted to the sampling interval. This might be the reason why discrete $\mathcal{L}$-path channel models have been specified in [19] for the four propagation areas (RA, TU, BU, HT). Some of these specified $\mathcal{L}$-path channel models are listed in Table 7.3 for the cases $\mathcal{L} = 4$ and $\mathcal{L} = 6$. The resulting scattering functions $S(\tau', f)$ are shown in Figures 7.9(a)–(d). In [19], moreover, alternative 6-path channel models as well as more complex, but therefore more exact, 12-path channel models have been specified. They are presented in Appendix 7.A for the sake of completeness.

7.3.4 The HIPERLAN/2 Channel Models

HIPERLAN[4] is a wireless LAN (Local Area Network) standard that has been defined by the European Telecommunications Standards Institute (ETSI) as a European alternative to the set of IEEE 802.11 standards. Its second version, called HIPERLAN/2, uses the 5 GHz band and allows data rates up to 54 Mbit/s [212]. HIPERLAN/2 systems can be deployed in a wide range of environments, such as offices, exhibition halls, and industrial buildings. To account for the various environments, altogether five different channel models have been developed by ETSI within a standardization project for Broadband Radio Access Networks (BRAN) [20]. The specifications of the HIPERLAN/2 channel models according to ETSI BRAN [20] have been included in Appendix 7.B. There, the power delay profiles and the Doppler spectral densities of the HIPERLAN/2 channel models for different indoor and outdoor environments at 5 GHz are listed in Tables 7.B.1 to 7.B.3.

The HIPERLAN/2 channel models have been developed by using the tapped-delay-line approach. To reduce the number of required taps, the delay spacing is non-uniform. The power delay profile declines exponentially with increasing propagation delays resulting in a denser spacing for shorter propagation delays. Except for the first tap, which is in one case characterized by a Rice process with a Rice factor c_R of 10, all other taps have Rayleigh fading

[4] HIPERLAN: High Performance Radio Local Area Network.

Table 7.3 Specification of the $\mathcal{L}$-path channel models according to COST 207 [19], where $\mathcal{L} = 4$ (RA) and $\mathcal{L} = 6$ (TU, BU, HT).

Path no. ℓ	Propagation delay τ'_ℓ	Path power (linear)	Path power (dB)	Category of the Doppler power spectral density	Delay spread $B^{(2)}_{\tau'}$
		(a) Rural Area (RA)			
0	0.0 μs	1	0	"Rice"	0.1 μs
1	0.2 μs	0.63	−2	"Jakes"	
2	0.4 μs	0.1	−10	"Jakes"	
3	0.6 μs	0.01	−20	"Jakes"	
		(b) Typical Urban (TU)			
0	0.0 μs	0.5	−3	"Jakes"	1.1 μs
1	0.2 μs	1	0	"Jakes"	
2	0.6 μs	0.63	−2	"Gauss I"	
3	1.6 μs	0.25	−6	"Gauss I"	
4	2.4 μs	0.16	−8	"Gauss II"	
5	5.0 μs	0.1	−10	"Gauss II"	
		(c) Bad Urban (BU)			
0	0.0 μs	0.5	−3	"Jakes"	2.4 μs
1	0.4 μs	1	0	"Jakes"	
2	1.0 μs	0.5	−3	"Gauss I"	
3	1.6 μs	0.32	−5	"Gauss I"	
4	5.0 μs	0.63	−2	"Gauss II"	
5	6.6 μs	0.4	−4	"Gauss II"	
		(d) Hilly Terrain (HT)			
0	0.0 μs	1	0	"Jakes"	5.0 μs
1	0.2 μs	0.63	−2	"Jakes"	
2	0.4 μs	0.4	−4	"Jakes"	
3	0.6 μs	0.2	−7	"Jakes"	
4	15.0 μs	0.25	−6	"Gauss II"	
5	17.2 μs	0.06	−12	"Gauss II"	

statistics ($c_R = 0$). The classical (Jakes) Doppler spectrum is assumed for all taps and the terminal speed was set to 3 m/s.

7.4 Frequency-Selective Sum-of-Sinusoids Channel Models

In this section, we will deal with the derivation and the analysis of frequency-selective deterministic sum-of-sinusoids (SOS) channel models. For this purpose, we extend the principle of deterministic channel modelling introduced in Section 4.1 with respect to frequency selectivity.

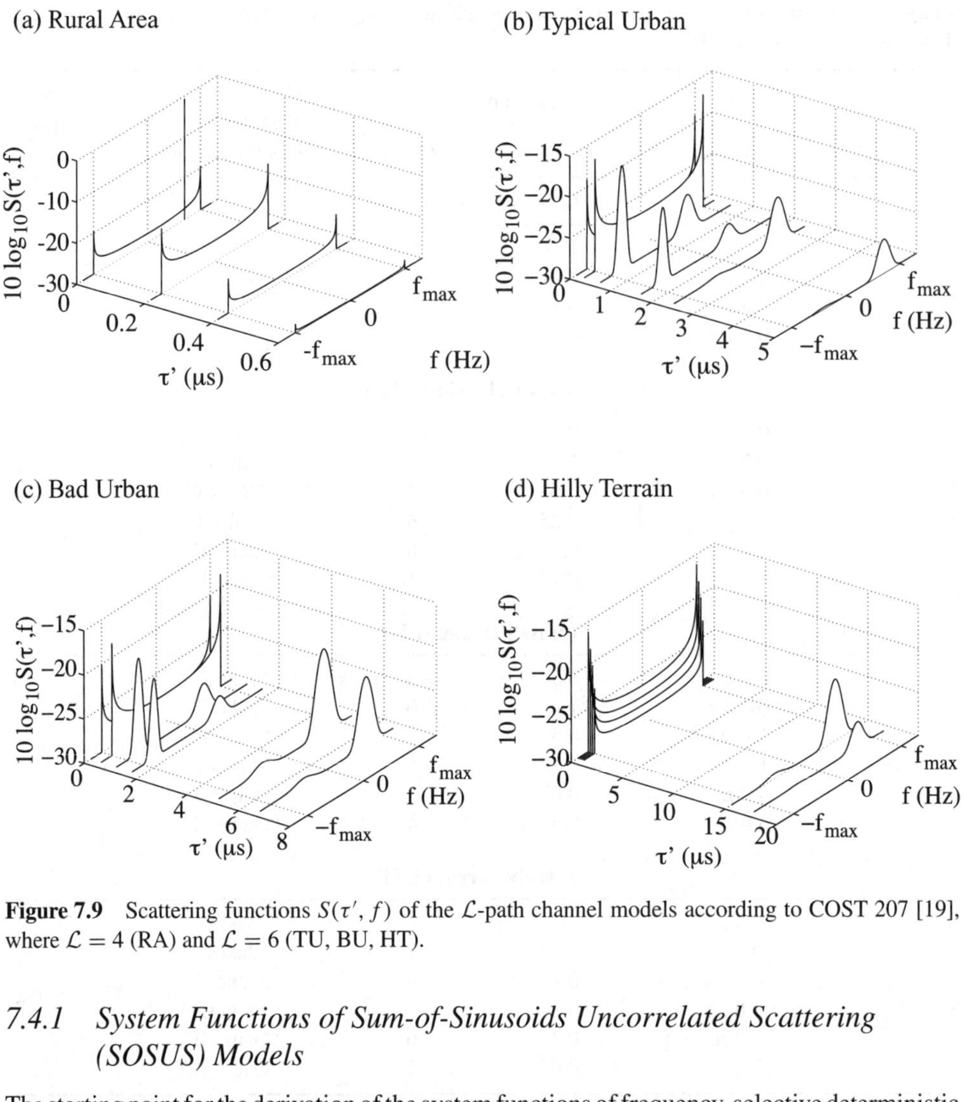

Figure 7.9 Scattering functions $S(\tau', f)$ of the $\mathcal{L}$-path channel models according to COST 207 [19], where $\mathcal{L} = 4$ (RA) and $\mathcal{L} = 6$ (TU, BU, HT).

7.4.1 System Functions of Sum-of-Sinusoids Uncorrelated Scattering (SOSUS) Models

The starting point for the derivation of the system functions of frequency-selective deterministic SOS channel models is the time-variant impulse response consisting of a sum of $\mathcal{L}$ discrete propagation paths

$$\tilde{h}(\tau', t) = \sum_{\ell=0}^{\mathcal{L}-1} \tilde{a}_\ell \, \tilde{\mu}_\ell(t) \, \delta(\tau' - \tilde{\tau}'_\ell) \, . \tag{7.55}$$

The symbols $\tilde{a}_\ell$ and $\tilde{\tau}'_\ell$ in (7.55) are real-valued constants which are called the *path gains* and the *discrete propagation delays*, respectively. As we will see later on, both the path gains $\tilde{a}_\ell$ and the discrete propagation delays $\tilde{\tau}'_\ell$ determine the delay power spectral density of frequency-selective deterministic channel models. Strictly speaking, the path gain $\tilde{a}_\ell$ is a measure of the square root of the average path power that is associated with the ℓth discrete propagation path. In general, one can say that the path gains $\tilde{a}_\ell$ and the discrete propagation delays $\tilde{\tau}'_\ell$ determine the frequency-selective behaviour of the channel, which can be attributed to the multipath effect. In the present case, it is assumed that elliptical scattering zones with different axes are the reason for multipath propagation. The fading of the channel, which is a result of the Doppler effect caused by the motion of the receiver (transmitter), is modelled in (7.55) according to the principle of deterministic channel modelling by complex deterministic SOS processes

$$\tilde{\mu}_\ell(t) = \tilde{\mu}_{1,\ell}(t) + j\tilde{\mu}_{2,\ell}(t), \quad \ell = 0, 1, \ldots, \mathcal{L} - 1, \tag{7.56a}$$

where

$$\tilde{\mu}_{i,\ell}(t) = \sum_{n=1}^{N_{i,\ell}} c_{i,n,\ell} \cos(2\pi f_{i,n,\ell} t + \theta_{i,n,\ell}), \quad i = 1, 2. \tag{7.56b}$$

Here, $N_{i,\ell}$ denotes the number of sinusoids assigned to the real part ($i = 1$) or the imaginary part ($i = 2$) of the ℓth propagation path. In (7.56b), $c_{i,n,\ell}$ is the gain of the nth component of the ℓth propagation path, and the remaining model parameters $f_{i,n,\ell}$ and $\theta_{i,n,\ell}$ are called the Doppler frequency and the phase, respectively.

Figure 7.10 shows the structure of the complex deterministic SOS process $\tilde{\mu}_\ell(t)$ in the continuous-time representation. To ensure that the frequency-selective channel simulator derived below has the same striking properties as the US model, the complex deterministic SOS processes $\tilde{\mu}_\ell(t)$ must be uncorrelated for different propagation paths. Therefore, it is inevitable that the deterministic SOS processes $\tilde{\mu}_\ell(t)$ and $\tilde{\mu}_\lambda(t)$ are designed such that they are uncorrelated for $\ell \neq \lambda$, where $\ell, \lambda = 0, 1, \ldots, \mathcal{L} - 1$. This demand can easily be fulfilled. One merely has to ensure that the discrete Doppler frequencies $f_{i,n,\ell}$ are designed such that the resulting sets $\{\pm f_{i,n,\ell}\}$ are disjoint (mutually exclusive) for different propagation paths. Regarding the simulation model, the demand for US propagation can therefore be expressed

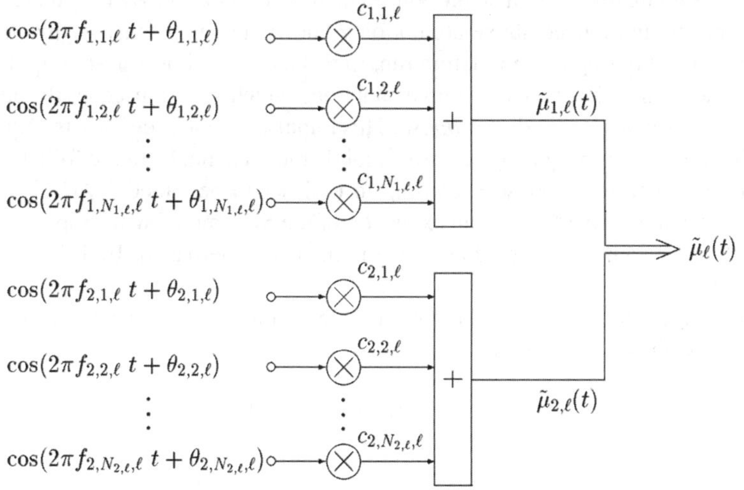

Figure 7.10 Simulation model for complex deterministic SOS processes $\tilde{\mu}_\ell(t)$.

by the following three equivalent conditions:

$$\text{(i) US} \quad \Longleftrightarrow \quad \tilde{\mu}_\ell(t) \;\text{ and }\; \tilde{\mu}_\lambda(t) \;\text{ are uncorrelated for } \ell \neq \lambda, \tag{7.57a}$$

$$\text{(ii) US} \quad \Longleftrightarrow \quad \{f_{i,n,\ell}\} \cap \{\pm f_{k,m,\lambda}\} = \emptyset \quad \text{for } \ell \neq \lambda, \tag{7.57b}$$

$$\text{(iii) US} \quad \Longleftrightarrow \quad f_{i,n,\ell} \neq \pm f_{k,m,\lambda} \quad \text{for } \ell \neq \lambda, \tag{7.57c}$$

where $i, k = 1, 2$, $n = 1, 2, \ldots, N_{i,\ell}$, $m = 1, 2, \ldots, N_{k,\lambda}$, and $\ell, \lambda = 0, 1, \ldots, \mathcal{L} - 1$.

In the following, we assume that the US condition in (7.57) is always fulfilled. In this case, the correlation properties of the complex deterministic SOS processes $\tilde{\mu}_\ell(t)$ introduced in (7.56a) can be described by

$$\lim_{T\to\infty} \frac{1}{2T} \int_{-T}^{T} \tilde{\mu}_\ell^*(t)\, \tilde{\mu}_\lambda(t+\tau)dt = \begin{cases} \tilde{r}_{\mu_\ell\mu_\ell}(\tau), & \text{if } \ell = \lambda, \\ 0, & \text{if } \ell \neq \lambda, \end{cases} \tag{7.58}$$

where

$$\tilde{r}_{\mu_\ell\mu_\ell}(\tau) = \sum_{i=1}^{2} \tilde{r}_{\mu_{i,\ell}\mu_{i,\ell}}(\tau), \tag{7.59a}$$

$$\tilde{r}_{\mu_{i,\ell}\mu_{i,\ell}}(\tau) = \sum_{n=1}^{N_{i,\ell}} \frac{c_{i,n,\ell}^2}{2} \cos(2\pi f_{i,n,\ell}\tau) \tag{7.59b}$$

hold for $i = 1, 2$ and $\ell, \lambda = 0, 1, \ldots, \mathcal{L} - 1$.

At this stage, it should be mentioned that all parameters determining the statistical behaviour of the time-variant impulse response $\tilde{h}(\tau', t)$ can be calculated in such a way that the scattering function of the deterministic system approximates a given specified or measured scattering function. A procedure for this will be introduced in Subsection 7.4.4. We may therefore assume that all parameters introduced above are not only known, but also constant quantities, which will not be changed during the simulation run. In this case, the time-variant impulse response $\tilde{h}(\tau', t)$ is a deterministic process (sample function) which will consequently be called the *time-variant deterministic impulse response*. This impulse response defines an important class of simulation models for frequency-selective mobile radio channels. In the following, channel models with an impulse response according to (7.55) will be called *SOSUS*[5] *models*. The SOSUS model can be considered as an extension of the SOS model with respect to frequency selectivity. It can also be interpreted as a deterministic counterpart of Bello's [18] stochastic WSSUS model.

Since the discrete propagation delays $\tilde{\tau}'_\ell$ in (7.55) cannot become negative, $\tilde{h}(\tau', t)$ fulfils the causality condition, i.e., it holds

$$\tilde{h}(\tau', t) = 0 \quad \text{for} \quad \tau' < 0. \tag{7.60}$$

[5] SOSUS is introduced here as an abbreviation for "sum-of-sinusoids uncorrelated scattering".

By analogy to (7.11), we can compute the output signal $y(t)$ for any given input signal $x(t)$ by applying

$$y(t) = \int_0^\infty x(t - \tau')\, \tilde{h}(\tau', t)\, d\tau' . \tag{7.61}$$

If we now employ the expression (7.55) for the time-variant deterministic impulse response $\tilde{h}(\tau', t)$, we obtain

$$y(t) = \sum_{\ell=0}^{\mathcal{L}-1} \tilde{a}_\ell\, \tilde{\mu}_\ell(t)\, x(t - \tilde{\tau}'_\ell) . \tag{7.62}$$

Hence, the output signal $y(t)$ of the channel simulator can be interpreted as a superposition of $\mathcal{L}$ delayed replicas of the input signal $x(t - \tilde{\tau}'_\ell)$, where each of the delayed replicas is weighted by a constant path gain $\tilde{a}_\ell$ and a time-variant complex deterministic SOS process $\tilde{\mu}_\ell(t)$. Without restriction of generality, we may ignore the propagation delay of the line-of-sight component in this model. To simplify matters, we define $\tilde{\tau}'_0 := 0$. This does not cause any problem, because only the propagation delay differences $\Delta\tilde{\tau}'_\ell = \tilde{\tau}'_\ell - \tilde{\tau}'_{\ell-1}$ ($\ell = 1, 2, \ldots, \mathcal{L} - 1$) are relevant for the system behaviour. From (7.62) follows the tapped-delay-line structure shown in Figure 7.11 of a deterministic simulation model for a frequency-selective mobile radio channel in the continuous-time representation.

The discrete-time simulation model, required for computer simulations, can be obtained from the continuous-time structure, e.g., by substituting $\tilde{\tau}'_\ell \to \ell T'_s$, $x(t) \to x(kT'_s)$, $y(t) \to y(kT'_s)$, and $\tilde{\mu}_\ell(t) \to \tilde{\mu}_\ell(kT_s)$, where T_s and T'_s denote sampling intervals, k is an integer, and ℓ refers to the ℓth propagation path ($\ell = 0, 1, \ldots, \mathcal{L} - 1$). For the propagation delay differences

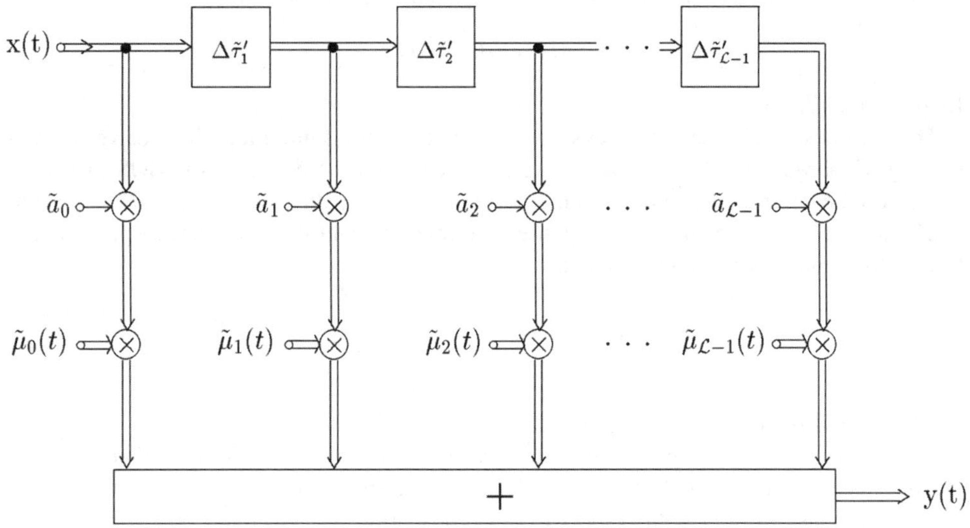

Figure 7.11 Deterministic simulation model for frequency-selective mobile radio channels in the equivalent complex baseband.

$\Delta\tilde{\tau}'_\ell = \tilde{\tau}'_\ell - \tilde{\tau}'_{\ell-1}$, we obtain $\Delta\tilde{\tau}'_\ell \to T'_s$ for all $\ell = 1, 2, \ldots, \mathcal{L}-1$. The sampling intervals T_s and T'_s have to be sufficiently small, but must not necessarily be identical. Between T_s and T'_s, we can establish the general relation $T_s = m'_s T'_s$, where $m'_s \in \mathbb{N}$ is in the following called the *sampling rate ratio*. The larger (smaller) the sampling rate ratio m'_s is chosen, the higher (lower) is the simulation speed of the channel simulator and the larger (smaller) is the error occurring due to the discretization of $\tilde{\mu}_\ell(t)$. The sampling rate ratio m'_s enables the user to find a good compromise between the simulation speed and the precision of the channel model. As a guideline, m'_s should be chosen so that the sampling interval T_s satisfies the condition $T'_s \leq T_s \leq T_{\text{sym}}$ for any given symbol interval T_{sym}. The upper limit $T_s = T_{\text{sym}}$ corresponds to the often made assumption that the impulse response is constant for the duration of one data symbol. However, this quasi-stationary assumption is only justified if the product $f_{\max} T_{\text{sym}}$ is sufficiently small.

From the general relation (7.62), two important special cases can be derived. These are characterized by

$$\text{(i)} \quad \tilde{a}_0 \neq 0, \quad \tilde{a}_1 = \tilde{a}_2 = \cdots = \tilde{a}_{\mathcal{L}-1} = 0 \tag{7.63a}$$

and

$$\text{(ii)} \quad \tilde{\mu}_\ell(t) = \tilde{\mu}_\ell = const., \quad \forall\,\ell = 0, 1, \ldots, \mathcal{L}-1\,. \tag{7.63b}$$

The first special case (i) describes a channel for which all scattering components caused by obstacles located relatively far away from the receiver can be ignored. Using $\tilde{\mu}(t) := \tilde{a}_0\,\tilde{\mu}_0(t)$, we can in this case represent the time-variant deterministic impulse response by

$$\tilde{h}(\tau', t) = \delta(\tau') \cdot \tilde{\mu}(t)\,. \tag{7.64}$$

A comparison with (7.47) shows that we are here dealing with a frequency-nonselective channel model. This also explains the fact that the multiplicative relation

$$y(t) = \tilde{\mu}(t) \cdot x(t) \tag{7.65}$$

follows from (7.61).

The second special case (ii) always occurs if both the transmitter and the receiver are not moving. Consequently, the Doppler effect disappears and the deterministic SOS processes $\tilde{\mu}_\ell(t)$ become complex-valued constants $\tilde{\mu}_\ell$ for all discrete paths $\ell = 0, 1, \ldots, \mathcal{L}-1$. From (7.55), it then follows the impulse response of a time-invariant finite impulse response (FIR) filter with $\mathcal{L}$ complex-valued coefficients

$$\tilde{h}(\tau') = \sum_{\ell=0}^{\mathcal{L}-1} a_\ell\,\delta(\tau' - \tilde{\tau}'_\ell), \tag{7.66}$$

where $a_\ell := \tilde{a}_\ell\,\tilde{\mu}_\ell$ for $\ell = 0, 1, \ldots, \mathcal{L}-1$.

Next, we consider the general case in more detail. By analogy to (7.13), we define the time-variant transfer function $\tilde{H}(f', t)$ by the Fourier transform of the time-variant deterministic impulse response $\tilde{h}(\tau', t)$ with respect to the propagation delay variable τ', which can be expressed symbolically as $\tilde{h}(\tau', t) \overset{\tau'\;\; f'}{\circ\!\!-\!\!\bullet} \tilde{H}(f', t)$. If we replace the impulse response $h(\tau', t)$ by the deterministic impulse response $\tilde{h}(\tau', t)$ in (7.13), and take (7.55) into account, then we can easily find the following closed-form solution for the *time-variant transfer function* $\tilde{H}(f',t)$ *of SOSUS models*

$$\tilde{H}(f',t) = \sum_{\ell=0}^{\mathcal{L}-1} \tilde{a}_\ell \, \tilde{\mu}_\ell(t) \, e^{-j2\pi f' \tilde{\tau}'_\ell} . \tag{7.67}$$

It is obvious that $\tilde{H}(f',t)$ is deterministic, because the Fourier transform of a deterministic function again results in a deterministic function. For the description of the input-output relationship of SOSUS models, we may refer to (7.14), where of course the time-variant transfer function $H(f',t)$ has to be replaced by $\tilde{H}(f',t)$. Notice that the input-output relationship in (7.62) can directly be derived from (7.14) after substituting there $H(f',t)$ by $\tilde{H}(f',t)$, where $\tilde{H}(f',t)$ is given by (7.67).

An insight into the phenomenon of the Doppler effect can be obtained from the Doppler-variant impulse response $\tilde{s}(\tau',f)$. This function is defined by the Fourier transform of $\tilde{h}(\tau',t)$ with respect to the time variable t, i.e., $\tilde{h}(\tau',t) \overset{t \;\; f}{\circ\!\!-\!\!\bullet} \tilde{s}(\tau',f)$. Using the expression (7.55), we obtain the following closed-form solution for the *Doppler-variant impulse response* $\tilde{s}(\tau',f)$ of SOSUS models

$$\tilde{s}(\tau',f) = \sum_{\ell=0}^{\mathcal{L}-1} \tilde{a}_\ell \, \tilde{\Xi}_\ell(f) \, \delta(\tau' - \tilde{\tau}'_\ell), \tag{7.68}$$

where $\tilde{\Xi}_\ell(f)$ denotes the Fourier transform of $\tilde{\mu}_\ell(t)$, which is given by

$$\tilde{\Xi}_\ell(f) = \tilde{\Xi}_{1,\ell}(f) + j\tilde{\Xi}_{2,\ell}(f), \quad \ell = 0, 1, \ldots, \mathcal{L}-1, \tag{7.69a}$$

$$\tilde{\Xi}_{i,\ell}(f) = \sum_{n=1}^{N_{i,\ell}} \frac{c_{i,n,\ell}}{2} \left[\delta(f - f_{i,n,\ell}) \, e^{j\theta_{i,n,\ell}} + \delta(f + f_{i,n,\ell}) \, e^{-j\theta_{i,n,\ell}} \right], \quad i = 1, 2 . \tag{7.69b}$$

Thus, $\tilde{s}(\tau',f)$ is a two-dimensional discrete line spectrum, where the spectral lines are located at the discrete positions $(\tau',f) = (\tilde{\tau}'_\ell, \pm f_{i,n,\ell})$ and weighted by the complex-valued factors $\tilde{a}_\ell (c_{i,n,\ell}/2) \, e^{\pm j\theta_{i,n,\ell}}$. For the description of the input-output behaviour in terms of $\tilde{s}(\tau',f)$, the relation (7.19) is useful, if there the Doppler-variant impulse response $s(\tau',f)$ is substituted by $\tilde{s}(\tau',f)$. It should also be observed that (7.62) follows from (7.19), if in the latter equation $s(\tau',f)$ is replaced by (7.68).

Finally, we consider the *Doppler-variant transfer function* $\tilde{T}(f',f)$ *of SOSUS models* which is defined by the two-dimensional Fourier transform of the time-variant deterministic impulse response $\tilde{h}(\tau',t)$, i.e., $\tilde{h}(\tau',t) \overset{\tau',t \;\; f',f}{\circ\!\!-\!\!\bullet} \tilde{T}(f',f)$. Due to $\tilde{h}(\tau',t) \overset{\tau' \;\; f'}{\circ\!\!-\!\!\bullet} \tilde{H}(f',t)$ and $\tilde{h}(\tau',t) \overset{t \;\; f}{\circ\!\!-\!\!\bullet} \tilde{s}(\tau',f)$, the computation of an expression for $\tilde{T}(f',f)$ can also be carried out via the one-dimensional Fourier transform $\tilde{H}(f',t) \overset{t \;\; f}{\circ\!\!-\!\!\bullet} \tilde{T}(f',f)$ or $\tilde{s}(\tau',f) \overset{\tau' \;\; f'}{\circ\!\!-\!\!\bullet} \tilde{T}(f',f)$. No matter which procedure we decide upon, we obtain in any case the following closed-form expression for the Doppler-variant transfer function $\tilde{T}(f',f)$ of the deterministic system

$$\tilde{T}(f',f) = \sum_{\ell=0}^{\mathcal{L}-1} \tilde{a}_\ell \, \tilde{\Xi}_\ell(f) \, e^{-j2\pi f' \tilde{\tau}'_\ell} . \tag{7.70}$$

In summary, we can say that the four system functions $\tilde{h}(\tau', t)$, $\tilde{H}(f', t)$, $\tilde{s}(\tau', f)$, and $\tilde{T}(f', f)$ can be analyzed explicitly, if the model parameters $\{c_{i,n,\ell}\}$, $\{f_{i,n,\ell}\}$, $\{\theta_{i,n,\ell}\}$, $\{\tilde{a}_\ell\}$, $\{\tilde{\tau}'_\ell\}$, $\{N_{i,\ell}\}$, and $\mathcal{L}$ are known and constant quantities. By analogy to Figure 7.3, the system functions of deterministic channel models are related in pairs by the Fourier transform. The resulting relationships are illustrated in Figure 7.12.

7.4.2 Correlation Functions and Power Spectral Densities of SOSUS Models

With reference to the WSSUS model, analog relations can be established in the general sense for the correlation functions and power spectral densities of the frequency-selective deterministic channel model (SOSUS model). In particular, the correlation functions of the four system functions $\tilde{h}(\tau', t)$, $\tilde{H}(f', t)$, $\tilde{s}(\tau', f)$, and $\tilde{T}(f', f)$ can be represented by the following relations:

$$\tilde{r}_{hh}(\tau_1', \tau_2'; t, t+\tau) = \delta(\tau_2' - \tau_1')\, \tilde{S}_{hh}(\tau_1', \tau), \tag{7.71a}$$

$$\tilde{r}_{HH}(f', f'+\upsilon'; t, t+\tau) = \tilde{r}_{HH}(\upsilon', \tau), \tag{7.71b}$$

$$\tilde{r}_{ss}(\tau_1', \tau_2'; f_1, f_2) = \delta(f_2 - f_1)\, \delta(\tau_2' - \tau_1')\, \tilde{S}(\tau_1', f_1), \tag{7.71c}$$

$$\tilde{r}_{TT}(f', f'+\upsilon'; f_1, f_2) = \delta(f_2 - f_1)\, \tilde{S}_{TT}(\upsilon', f_1)\,. \tag{7.71d}$$

In these equations, $\tilde{S}_{hh}(\tau_1', \tau)$ denotes the delay cross-power spectral density, $\tilde{r}_{HH}(\upsilon', \tau)$ signifies the time-frequency correlation function, $\tilde{S}(\tau_1', f_1)$ is the scattering function, and $\tilde{S}_{TT}(\upsilon', f_1)$ is called the Doppler cross-power spectral density of the deterministic system.

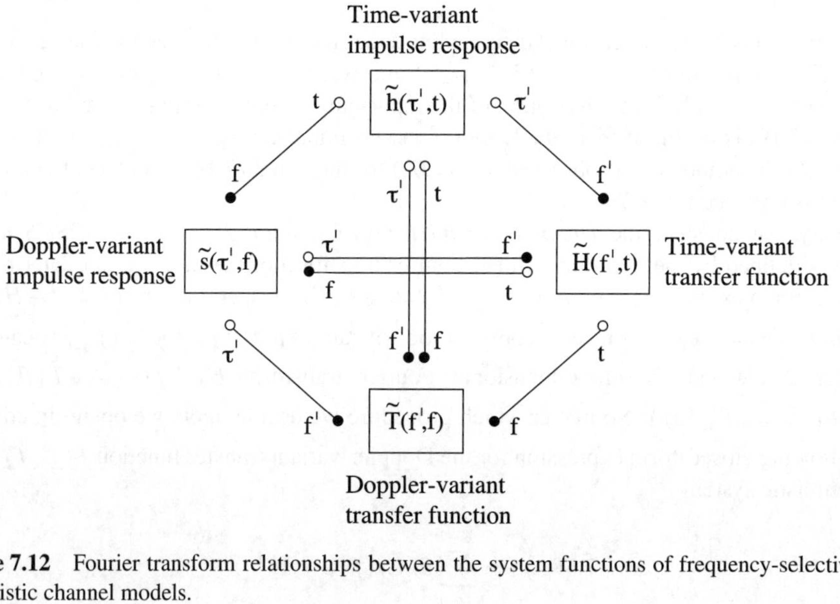

Figure 7.12 Fourier transform relationships between the system functions of frequency-selective deterministic channel models.

Two of these quantities at a time form a Fourier transform pair in the same way as for the WSSUS model. By analogy to Figure 7.5, we obtain the relationships depicted in Figure 7.13 for frequency-selective deterministic channel models. In order to simplify the notation, the variables τ_1' and f_1 have again been replaced by τ' and f, respectively.

The interpretation of $\tilde{h}(\tau', t)$ as a time-variant deterministic process, enables us to derive closed-form solutions for the correlation functions (7.71a)–(7.71d), and thus also for the system functions shown in Figure 7.13. This provides the basis for analyzing the statistical properties of the deterministic channel model analytically. We will deal with this task in the following.

At first, we define the autocorrelation function of the time-variant deterministic impulse response $\tilde{h}(\tau', t)$ by using time averages as such

$$\begin{aligned}\tilde{r}_{hh}(\tau_1', \tau_2'; t, t+\tau) &:= < \tilde{h}^*(\tau_1', t)\, \tilde{h}(\tau_2', t+\tau) > \\ &= \lim_{T\to\infty} \frac{1}{2T} \int_{-T}^{T} \tilde{h}^*(\tau_1', t)\, \tilde{h}(\tau_2', t+\tau)\, dt\,. \end{aligned} \quad (7.72)$$

It should be taken into account that the time average stands in contrast to the definition in (7.23a), where the derivation of the autocorrelation function of the stochastic impulse response $h(\tau', t)$ is achieved by computing statistical averages (ensemble averages). In the

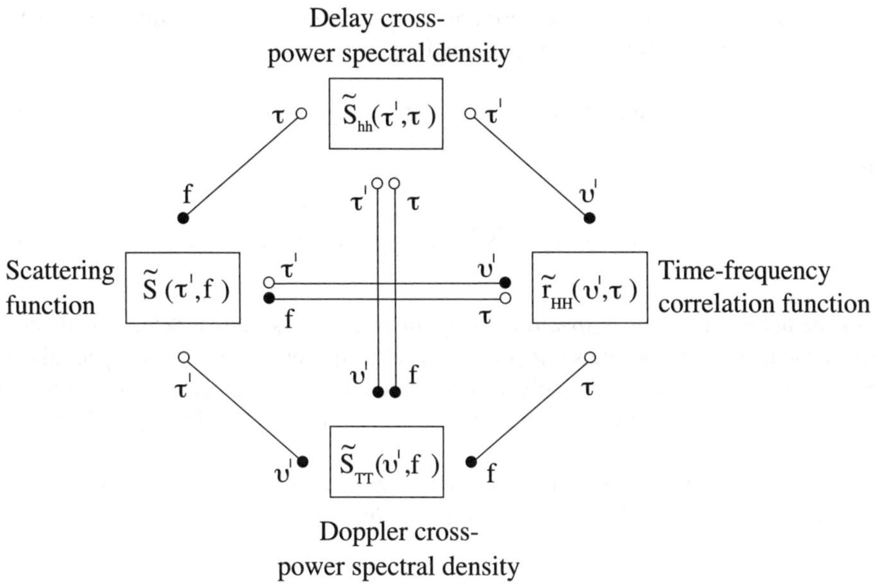

Figure 7.13 Fourier transform relationships between the delay cross-power spectral density $\tilde{S}_{hh}(\tau', \tau)$, the time-frequency correlation function $\tilde{r}_{HH}(\upsilon', \tau)$, the scattering function $\tilde{S}(\tau', f)$, and the Doppler cross-power spectral density $\tilde{S}_{TT}(\upsilon', f)$ of SOSUS models.

equation above, we use the expression (7.55) for $\tilde{h}(\tau', t)$, so that we may write

$$\tilde{r}_{hh}(\tau_1', \tau_2'; t, t+\tau) = \lim_{T\to\infty} \frac{1}{2T} \int_{-T}^{T} \left[\sum_{\ell=0}^{\mathcal{L}-1} \tilde{a}_\ell \, \tilde{\mu}_\ell^*(t) \, \delta(\tau_1' - \tilde{\tau}_\ell') \right] \cdot \left[\sum_{\lambda=0}^{\mathcal{L}-1} \tilde{a}_\lambda \, \tilde{\mu}_\lambda(t+\tau) \, \delta(\tau_2' - \tilde{\tau}_\lambda') \right] dt$$

$$= \lim_{T\to\infty} \sum_{\ell=0}^{\mathcal{L}-1} \sum_{\lambda=0}^{\mathcal{L}-1} \tilde{a}_\ell \, \tilde{a}_\lambda \, \delta(\tau_1' - \tilde{\tau}_\ell') \, \delta(\tau_2' - \tilde{\tau}_\lambda') \cdot \frac{1}{2T} \int_{-T}^{T} \tilde{\mu}_\ell^*(t) \, \tilde{\mu}_\lambda(t+\tau) \, dt \,. \tag{7.73}$$

Using (7.58), it follows

$$\tilde{r}_{hh}(\tau_1', \tau_2'; t, t+\tau) = \sum_{\ell=0}^{\mathcal{L}-1} \tilde{a}_\ell^2 \, \tilde{r}_{\mu_\ell \mu_\ell}(\tau) \, \delta(\tau_1' - \tilde{\tau}_\ell') \, \delta(\tau_2' - \tilde{\tau}_\ell') \,. \tag{7.74}$$

Generally, the product of two Dirac delta functions is not defined. But in the present case, however, the first Dirac delta function depends on the variable τ_1' and the second one on τ_2'. Since τ_1' and τ_2' are independent variables, the product will not cause any problems in the two-dimensional (τ_1', τ_2')-plane. Furthermore, $\delta(\tau_1' - \tilde{\tau}_\ell') \, \delta(\tau_2' - \tilde{\tau}_\ell')$ is equivalent to $\delta(\tau_1' - \tilde{\tau}_\ell') \, \delta(\tau_2' - \tilde{\tau}_1')$, so that (7.74) can be represented by

$$\tilde{r}_{hh}(\tau_1', \tau_2'; t, t+\tau) = \delta(\tau_2' - \tilde{\tau}_1') \, \tilde{S}_{hh}(\tau_1', \tau), \tag{7.75}$$

where

$$\tilde{S}_{hh}(\tau', \tau) = \sum_{\ell=0}^{\mathcal{L}-1} \tilde{a}_\ell^2 \, \tilde{r}_{\mu_\ell \mu_\ell}(\tau) \, \delta(\tau' - \tilde{\tau}_\ell') \tag{7.76}$$

denotes the *delay cross-power spectral density* of frequency-selective deterministic channel models. Note that (7.75) has the same form as (7.33). The delay cross-power spectral density $\tilde{S}_{hh}(\tau', \tau)$ can be computed explicitly in connection with the autocorrelation functions in (7.59a) and (7.59b) if all model parameters $\{c_{i,n,\ell}\}$, $\{f_{i,n,\ell}\}$, $\{\tilde{a}_\ell\}$, $\{\tilde{\tau}_\ell'\}$, $\{N_{i,\ell}\}$, and $\mathcal{L}$ are known.

The Fourier transform of the delay cross-power spectral density $\tilde{S}_{hh}(\tau', \tau)$ with respect to the propagation delay variable τ' results in the *time-frequency correlation function*

$$\tilde{r}_{HH}(\upsilon', \tau) = \sum_{\ell=0}^{\mathcal{L}-1} \tilde{a}_\ell^2 \, \tilde{r}_{\mu_\ell \mu_\ell}(\tau) \, e^{-j2\pi\upsilon'\tilde{\tau}_\ell'} \tag{7.77}$$

of the deterministic system.

Preferably, we also refer to the delay cross-power spectral density $\tilde{S}_{hh}(\tau', \tau)$ in order to calculate an analytical expression for the scattering function. The Fourier transform of (7.76) with respect to τ leads immediately to the expression

$$\tilde{S}(\tau', f) = \sum_{\ell=0}^{\mathcal{L}-1} \tilde{a}_\ell^2 \, \tilde{S}_{\mu_\ell\mu_\ell}(f)\, \delta(\tau' - \tilde{\tau}'_\ell), \tag{7.78}$$

which describes the *scattering function* of frequency-selective deterministic channel models. In this equation,

$$\tilde{S}_{\mu_\ell\mu_\ell}(f) = \sum_{i=1}^{2} \sum_{n=1}^{N_{i,\ell}} \frac{c_{i,n,\ell}^2}{4} \left[\delta(f - f_{i,n,\ell}) + \delta(f + f_{i,n,\ell})\right] \tag{7.79}$$

represents the Doppler power spectral density of the ℓth propagation path ($\ell = 0, 1, \ldots, \mathcal{L} - 1$). Notice that the Doppler power spectral density $\tilde{S}_{\mu_\ell\mu_\ell}(f)$ is defined by the Fourier transform of the autocorrelation function $\tilde{r}_{\mu_\ell\mu_\ell}(\tau)$ given by (7.59a). Now it becomes obvious that the scattering function $\tilde{S}(\tau', f)$ of deterministic channel models can be represented by a finite sum of weighted Dirac delta functions. The Dirac delta functions are located in the two-dimensional (τ', f)-plane at the positions $(\tilde{\tau}'_\ell, \pm f_{i,n,\ell})$ and are weighted by the constants $(\tilde{a}_\ell c_{i,n,\ell})^2/4$. Without restriction of generality, we assume in the following that the scattering function $\tilde{S}(\tau', f)$ is normalized to unity, so that the volume under $\tilde{S}(\tau', f)$ is equal to one, i.e.,

$$\int_{-\infty}^{\infty} \int_{0}^{\infty} \tilde{S}(\tau', f)\, d\tau' df = 1\,. \tag{7.80}$$

To ensure that (7.80) is definitely fulfilled, the gains $c_{i,n,\ell}$ and the path gains $\tilde{a}_\ell$ have to fulfil the boundary conditions

$$\sum_{n=1}^{N_{i,\ell}} c_{i,n,\ell}^2 = 1 \quad \text{and} \quad \sum_{\ell=0}^{\mathcal{L}-1} \tilde{a}_\ell^2 = 1\,. \tag{7.81a,b}$$

The first condition implies that the mean power of $\tilde{\mu}_\ell(t)$ equals unity, i.e., $\tilde{r}_{\mu_\ell\mu_\ell}(0) = 1$. The second condition states that the statistical average of the total path power equals one. Combining the two conditions ensures that the output signal $y(t)$ of the channel has the same mean power as the input signal $x(t)$.

Finally, we determine the Fourier transform of the scattering function $\tilde{S}(\tau', f)$ with respect to τ' in order to obtain the *Doppler cross-power spectral density*

$$\tilde{S}_{TT}(\upsilon', f) = \sum_{\ell=0}^{\mathcal{L}-1} \tilde{a}_\ell^2 \, \tilde{S}_{\mu_\ell\mu_\ell}(f)\, e^{-j2\pi\upsilon'\tilde{\tau}'_\ell} \tag{7.82}$$

of frequency-selective deterministic channel models. We can easily assure ourselves that one also obtains the Doppler cross-power spectral density $\tilde{S}_{TT}(\upsilon', f)$ in the presented form (7.82), if the alternative approach — via the Fourier transform of the time-frequency correlation function $\tilde{r}_{HH}(\upsilon', \tau)$ with respect to τ — is made use of, where in this case the relation (7.77) has to be used for $\tilde{r}_{HH}(\upsilon', \tau)$.

Thus, it has been shown that the four system functions $\tilde{S}_{hh}(\tau', \tau)$, $\tilde{r}_{HH}(\upsilon', \tau)$, $\tilde{S}(\tau, f)$, and $\tilde{S}_{TT}(\upsilon', f)$ can be presented in closed form enabling one to study the deterministic system analytically, if the relevant model parameters $\{c_{i,n,\ell}\}$, $\{f_{i,n,\ell}\}$, $\{\tilde{a}_\ell\}$, $\{\tilde{\tau}'_\ell\}$, $\{N_{i,\ell}\}$, and $\mathcal{L}$ are given. The closed-form solutions of the system functions disclose which model parameters strongly affect the channel characteristics.

7.4.3 Delay Power Spectral Density, Doppler Power Spectral Density, and Characteristic Quantities of SOSUS Models

In this subsection, simple closed-form solutions will be derived for the fundamental characteristic functions and quantities of SOSUS models, such as the delay power spectral density, Doppler power spectral density, and delay spread. For this purpose, we will here discuss the terms introduced for stochastic channel models (WSSUS models) in Subsection 7.3.2.3 for deterministic channel models (SOSUS models).

Delay power spectral density: Let $\tilde{S}(\tau', f)$ be the scattering function of a deterministic channel model, then, by analogy to (7.38), the corresponding *delay power spectral density* $\tilde{S}_{\tau'}(\tau')$ which is also called the *power delay profile* is defined by

$$\tilde{S}_{\tau'}(\tau') := \tilde{S}_{hh}(\tau', 0) = \int_{-\infty}^{\infty} \tilde{S}(\tau', f)\, df\,. \tag{7.83}$$

After employing (7.78) and considering the boundary condition in (7.81a), it follows

$$\tilde{S}_{\tau'}(\tau') = \sum_{\ell=0}^{\mathcal{L}-1} \tilde{a}_\ell^2\, \delta(\tau' - \tilde{\tau}'_\ell)\,. \tag{7.84}$$

Hence, the delay power spectral density $\tilde{S}_{\tau'}(\tau')$ is a discrete line spectrum, where the spectral lines are located at the discrete positions $\tau' = \tilde{\tau}'_\ell$ and weighted by the constants $\tilde{a}_\ell^2$. Consequently, the behaviour of $\tilde{S}_{\tau'}(\tau')$ is completely determined by the model parameters $\tilde{a}_\ell$, $\tilde{\tau}'_\ell$, and $\mathcal{L}$. It should be pointed out that the area under the delay power spectral density $\tilde{S}_{\tau'}(\tau')$ is equal to unity due to (7.81b), i.e., $\int_0^\infty \tilde{S}_{\tau'}(\tau')\, d\tau' = 1$.

Average delay: Let $\tilde{S}_{\tau'}(\tau')$ be the delay power spectral density of a deterministic channel model, then the first moment of $\tilde{S}_{\tau'}(\tau')$ is called the *average delay* $\tilde{B}_{\tau'}^{(1)}$. Thus, by analogy to (7.39), the definition

$$\tilde{B}_{\tau'}^{(1)} := \frac{\int_{-\infty}^{\infty} \tau' \tilde{S}_{\tau'}(\tau')\, d\tau'}{\int_{-\infty}^{\infty} \tilde{S}_{\tau'}(\tau')\, d\tau'} \tag{7.85}$$

holds. After substituting (7.84) in (7.85), we obtain the average delay $\tilde{B}_{\tau'}^{(1)}$ in closed form as

$$\tilde{B}_{\tau'}^{(1)} = \frac{\sum_{\ell=0}^{\mathcal{L}-1} \tilde{\tau}'_\ell \tilde{a}_\ell^2}{\sum_{\ell=0}^{\mathcal{L}-1} \tilde{a}_\ell^2}, \tag{7.86}$$

where the denominator equals one if the condition in (7.81b) is fulfilled.

Delay spread: The square root of the second central moment of $\tilde{S}_{\tau'}(\tau')$ is called the *delay spread* $\tilde{B}_{\tau'}^{(2)}$, which is, by analogy to (7.40), defined by

$$\tilde{B}_{\tau'}^{(2)} := \sqrt{\frac{\int_{-\infty}^{\infty} \left(\tau' - \tilde{B}_{\tau'}^{(1)}\right)^2 \tilde{S}_{\tau'}(\tau')\, d\tau'}{\int_{-\infty}^{\infty} \tilde{S}_{\tau'}(\tau')\, d\tau'}}\,. \tag{7.87}$$

Using (7.84), the closed-form expression

$$\tilde{B}_{\tau'}^{(2)} = \sqrt{\frac{\sum_{\ell=0}^{\mathcal{L}-1} \left(\tilde{\tau}'_\ell \tilde{a}_\ell\right)^2}{\sum_{\ell=0}^{\mathcal{L}-1} \tilde{a}_\ell^2} - \left(\tilde{B}_{\tau'}^{(1)}\right)^2} \tag{7.88}$$

can be derived, where $\tilde{B}_{\tau'}^{(1)}$ is the average delay according to (7.86).

Doppler power spectral density: Let $\tilde{S}(\tau', f)$ be the scattering function of a deterministic channel model, then — by analogy to (7.42) — the corresponding *Doppler power spectral density* $\tilde{S}_{\mu\mu}(f)$ can be determined via the relation

$$\tilde{S}_{\mu\mu}(f) := \tilde{S}_{TT}(0, f) = \int_{-\infty}^{\infty} \tilde{S}(\tau', f)\, d\tau'\,. \tag{7.89}$$

With the scattering function $\tilde{S}(\tau', f)$ given by (7.78), we can now derive a closed-form solution for the Doppler power spectral density $\tilde{S}_{\mu\mu}(f)$ of the deterministic system. As a result, we obtain

$$\tilde{S}_{\mu\mu}(f) = \sum_{\ell=0}^{\mathcal{L}-1} \tilde{a}_\ell^2\, \tilde{S}_{\mu_\ell\mu_\ell}(f), \tag{7.90}$$

where $\tilde{S}_{\mu_\ell\mu_\ell}(f)$ denotes the Doppler power spectral density of the ℓth scattering component determined by (7.79). This equation shows that the Doppler power spectral density $\tilde{S}_{\mu\mu}(f)$ of frequency-selective deterministic channel models is given by the sum of the Doppler power spectral densities $\tilde{S}_{\mu_\ell\mu_\ell}(f)$ of all propagation paths $\ell = 0, 1, \ldots, \mathcal{L}$, where each individual Doppler power spectral density $\tilde{S}_{\mu_\ell\mu_\ell}(f)$ has to be weighted by the square of the corresponding path gain. Here, the square of the path gain $\tilde{a}_\ell^2$ represents the *path power*, that is the mean (average) power of all scattered components associated with the ℓth propagation path $\tilde{\tau}'_\ell$.

With knowledge of the Doppler power spectral density $\tilde{S}_{\mu\mu}(f)$ or $\tilde{S}_{\mu_\ell\mu_\ell}(f)$, the *average Doppler shift* $\tilde{B}_{\mu\mu}^{(1)}$ and the *Doppler spread* $\tilde{B}_{\mu\mu}^{(2)}$ can be computed. The definitions of $\tilde{B}_{\mu\mu}^{(1)}$ and $\tilde{B}_{\mu\mu}^{(2)}$ have already been introduced in Section 4.2, where also the closed-form solution and physical interpretation of these characteristic quantities can be found. We will refrain from a recapitulation of these results at this time.

Frequency correlation function: Let $\tilde{r}_{HH}(\upsilon', \tau)$ be the time-frequency correlation function of a deterministic channel model. Then, by analogy to (7.44), the *frequency correlation function* $\tilde{r}_{\tau'}(\upsilon')$ is defined by the time-frequency correlation function $\tilde{r}_{HH}(\upsilon', \tau)$ at $\tau = t_2 - t_1 = 0$, i.e.,

$$\begin{aligned}\tilde{r}_{\tau'}(\upsilon') &:= \tilde{r}_{HH}(\upsilon', 0)\\ &= \int_{-\infty}^{\infty}\int_{-\infty}^{\infty} \tilde{S}(\tau', f)\, e^{-j2\pi\upsilon'\tau'}\, d\tau'\, df\\ &= \int_{-\infty}^{\infty} \tilde{S}_{\tau'}(\tau')\, e^{-j2\pi\upsilon'\tau'}\, d\tau' .\end{aligned} \tag{7.91}$$

This result identifies the frequency correlation function $\tilde{r}_{\tau'}(\upsilon')$ as the Fourier transform of the delay power spectral density $\tilde{S}_{\tau'}(\tau')$. A closed-form expression for $\tilde{r}_{\tau'}(\upsilon')$ is obtained in a simple way by setting $\tau = 0$ in (7.77). Taking the boundary condition in (7.81a) into consideration, which implies that $\tilde{r}_{\mu_\ell\mu_\ell}(0) = 1$ holds for all $\ell = 0, 1, \ldots, \mathcal{L} - 1$, we then obtain

$$\tilde{r}_{\tau'}(\upsilon') = \sum_{\ell=0}^{\mathcal{L}-1} \tilde{a}_\ell^2\, e^{-j2\pi\upsilon'\tilde{\tau}'_\ell} . \tag{7.92}$$

The same closed-form expression is obtained after substituting (7.84) in (7.91) and solving the integral.

In the following, some properties of the frequency correlation function $\tilde{r}_{\tau'}(\upsilon')$ are highlighted. The function $\tilde{r}_{\tau'}(\upsilon')$ is periodic with period Υ' given by

$$\Upsilon' = 1/\gcd\{\tilde{\tau}'_0, \tilde{\tau}'_1, \ldots, \tilde{\tau}'_{\mathcal{L}-1}\}, \tag{7.93}$$

where $\gcd\{\,\cdot\,\}$ denotes the greatest common divisor[6]. Thus, we can write $\tilde{r}_{\tau'}(\upsilon') = \tilde{r}_{\tau'}(\upsilon' + k \cdot \Upsilon')$, where k is an integer. Note that $\Upsilon' \to \infty$ as $\gcd\{\tilde{\tau}'_0, \tilde{\tau}'_1, \ldots, \tilde{\tau}'_{\mathcal{L}-1}\} \to 0$. Furthermore, it follows from (7.91) that the frequency correlation function exhibits the Hermitian symmetry property, i.e.,

$$\tilde{r}_{\tau'}(\upsilon') = \tilde{r}^*_{\tau'}(-\upsilon') . \tag{7.94}$$

[6] Here, the greatest common divisor $\gcd\{\tilde{\tau}'_0, \tilde{\tau}'_1, \ldots, \tilde{\tau}'_{\mathcal{L}-1}\}$ is defined as follows. Let $\tilde{\tau}'_\ell = q_\ell \cdot \Delta\tau'$, where q_ℓ are integers for $\ell = 0, 1, \ldots, \mathcal{L} - 1$ and $\Delta\tau'$ is a real-valued constant, then $\gcd\{\tilde{\tau}'_0, \tilde{\tau}'_1, \ldots, \tilde{\tau}'_{\mathcal{L}-1}\} = \Delta\tau'$.

From this relation and $\tilde{r}_{\tau'}(\upsilon') = \tilde{r}_{\tau'}(\upsilon' + k \cdot \Upsilon')$, it also follows that $\tilde{r}_{\tau'}(\upsilon') = \tilde{r}^*_{\tau'}(k \cdot \Upsilon' - \upsilon')$ and $\tilde{r}_{\tau'}([2k+1]/2 \cdot \Upsilon' - \upsilon') = \tilde{r}^*_{\tau'}([2m+1]/2 \cdot \Upsilon' + \upsilon')$ where k and m are integers. Thus, the real and imaginary parts of the frequency correlation function are even and odd functions, respectively, and the frequency correlation function is Hermitian symmetric with respect to one half of the period, i.e., to the frequency $\upsilon' = \Upsilon'/2$. Consequently, the complete information on the frequency correlation function is contained in one half of the period of the frequency correlation function. Finally, since $\tilde{S}_{\tau'}(\tau') = 0$ for $\tau' < 0$, the real and imaginary parts of $\tilde{r}_{\tau'}(\upsilon')$ are related to each other by the Hilbert transform

$$\mathrm{Re}\{\tilde{r}_{\tau'}(\upsilon')\} = \mathcal{H}\{\mathrm{Im}\{\tilde{r}_{\tau'}(\upsilon')\}\}, \tag{7.95}$$

where $\mathcal{H}\{\cdot\}$ denotes the Hilbert transform operator. Recall that (7.94) and (7.95) also apply to the frequency correlation function of the WSSUS model.

Coherence bandwidth: Let $\tilde{r}_{\tau'}(\upsilon')$ be the frequency correlation function given by (7.92), then the frequency separation variable $\upsilon' = \tilde{B}_C$ for which

$$|\tilde{r}_{\tau'}(\tilde{B}_C)| = \frac{1}{2}|\tilde{r}_{\tau'}(0)| \tag{7.96}$$

holds, is called the *coherence bandwidth* of deterministic channel models. With (7.92) and taking the boundary condition (7.81b) into consideration, we obtain the transcendental equation

$$\left| \sum_{\ell=0}^{\mathcal{L}-1} \tilde{a}_\ell^2 \, e^{-j2\pi \tilde{B}_C \tau'_\ell} \right| - \frac{1}{2} = 0\,. \tag{7.97}$$

The smallest positive value for $\tilde{B}_C$ which fulfils the equation above defines the coherence bandwidth. Apart from simple special cases, (7.97) has to be solved generally by means of numerical *root-finding techniques*. The *Newton's* (or the *Newton-Raphson) method* is one of the most powerful and well-known numerical methods for solving root-finding problems [213].

Time correlation function: Let $\tilde{r}_{HH}(\upsilon', \tau)$ be the time-frequency correlation function of a deterministic channel model. Then, by analogy to (7.50), the *time correlation function*$\tilde{r}_{\mu\mu}(\tau)$ is defined by the time-frequency correlation function $\tilde{r}_{HH}(\upsilon', \tau)$ at $\upsilon' = f_2' - f_1' = 0$, i.e.,

$$\begin{aligned}\tilde{r}_{\mu\mu}(\tau) &:= \tilde{r}_{HH}(0, \tau) \\ &= \int_{-\infty}^{\infty}\int_{-\infty}^{\infty} \tilde{S}(\tau', f)\, e^{j2\pi f \tau}\, d\tau'\, df \\ &= \int_{-\infty}^{\infty} \tilde{S}_{\mu\mu}(f)\, e^{j2\pi f \tau}\, df\,.\end{aligned} \tag{7.98}$$

This result shows again that the time correlation function and the Doppler power spectral density form a Fourier transform pair. We consider (7.77) at $\upsilon' = 0$, and thus obtain

$$\tilde{r}_{\mu\mu}(\tau) = \sum_{\ell=0}^{\mathcal{L}-1} \tilde{a}_\ell^2 \, \tilde{r}_{\mu_\ell\mu_\ell}(\tau), \tag{7.99}$$

where $\tilde{r}_{\mu_\ell\mu_\ell}(\tau)$ is given by (7.59a).

Coherence time: Let $\tilde{r}_{\mu\mu}(\tau)$ be the time correlation function given by (7.99), then the time interval $\tau = \tilde{T}_C$ for which

$$|\tilde{r}_{\mu\mu}(\tilde{T}_C)| = \frac{1}{2}|\tilde{r}_{\mu\mu}(0)| \tag{7.100}$$

holds, is called the *coherence time* of deterministic channel models. Substituting (7.99) in (7.100), and taking (7.59a) and (7.59b) into account, results in the transcendental equation

$$\left| \sum_{i=1}^{2} \sum_{\ell=0}^{\mathcal{L}-1} \sum_{n=1}^{N_{i,\ell}} \frac{(\tilde{a}_\ell c_{i,n,\ell})^2}{2} \cos(2\pi f_{i,n,\ell} \tilde{T}_C) \right| - \frac{1}{2} = 0, \tag{7.101}$$

from which the coherence time $\tilde{T}_C$ can be computed by applying numerical root-finding techniques. The smallest positive value for $\tilde{T}_C$ which solves (7.101) defines the coherence time.

In order to facilitate an overview, the relationships derived above between the correlation functions and the power spectral densities as well as the characteristic quantities of frequency-selective deterministic channel models are illustrated in Figure 7.14.

7.4.4 Determination of the Model Parameters of SOSUS Models

In this subsection, we are concerned with the determination of the model parameters $\tilde{\tau}'_\ell$, $\tilde{a}_\ell$, $f_{i,n,\ell}$, $c_{i,n,\ell}$, and $\theta_{i,n,\ell}$ of the simulation model shown in Figure 7.11 representing the SOSUS model determined by (7.55). The starting point of the procedure described here is the scattering function $S(\tau', f)$ of a given stochastic channel model. Since the procedure is universally valid, $S(\tau', f)$ can in principle be any specified scattering function. The method may be applied just as well if $S(\tau', f)$ is the result of an evaluation of a single snapshot measurement obtained from a real-world channel.

From the scattering function $S(\tau', f)$, which is assumed to be known henceforth, the corresponding delay power spectral density $S_{\tau'}(\tau')$ and the Doppler power spectral density $S_{\mu\mu}(f)$ are determined first. For this purpose, we use the relations

$$S_{\tau'}(\tau') = \int_{-\infty}^{\infty} S(\tau', f)\, df \quad \text{and} \quad S_{\mu\mu}(f) = \int_{-\infty}^{\infty} S(\tau', f)\, d\tau', \tag{7.102a,b}$$

which have been introduced in (7.38) and (7.42), respectively. The causality condition in (7.12) leads to $S_{\tau'}(\tau') = 0$ if $\tau' < 0$. Furthermore, we assume that all scattering components with propagation delays $\tau' > \tau'_{\max}$ can be ignored. The delay power spectral density is then confined

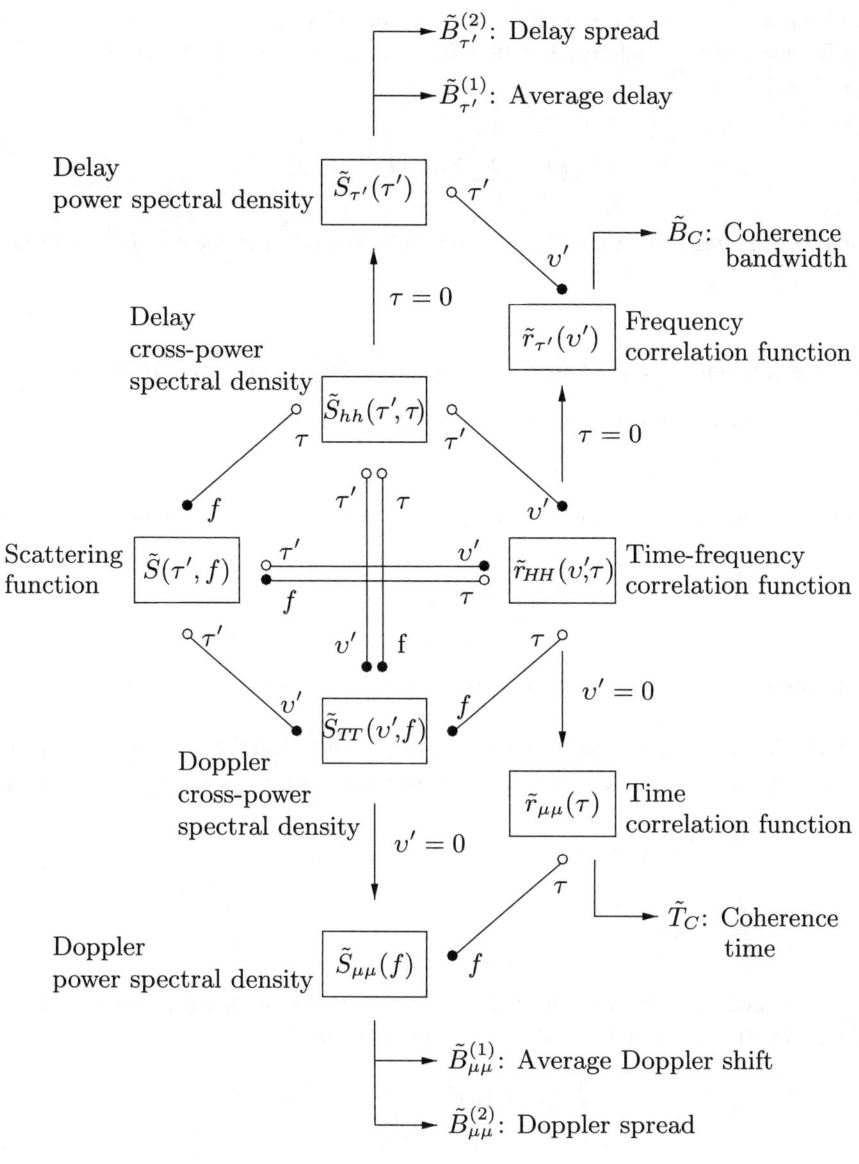

Figure 7.14 Relationships between the correlation functions, the power spectral densities, and the characteristic quantities of SOSUS models.

to the interval $I = [0, \tau'_{\max}]$, and we can write

$$S_{\tau'}(\tau') = 0 \quad \text{for} \quad \tau' \notin I = [0, \tau'_{\max}] . \tag{7.103}$$

Next, we perform a partition of the interval $I = [0, \tau'_{\max}]$ into a number of $\mathcal{L}$ disjointed subintervals I_ℓ such that $I = \bigcup_{\ell=0}^{\mathcal{L}-1} I_\ell$ and $I_\ell \cap I_\lambda = 0$ for $\ell \neq \lambda$. This partition is realized in a way that allows us to consider the delay power spectral density $S_{\tau'}(\tau')$ and the Doppler power spectral density $S_{\mu_\ell\mu_\ell}(f)$ appertaining to I_ℓ as independent within each subinterval I_ℓ. From this, it follows that the scattering function $S(\tau', f)$ can be expressed in terms of $S_{\tau'}(\tau')$ and $S_{\mu_\ell\mu_\ell}(f)$ as

$$S(\tau', f) = \sum_{\ell=0}^{\mathcal{L}-1} S_{\mu_\ell\mu_\ell}(f)\, S_{\tau'}(\tau')\Big|_{\tau' \in I_\ell} \,. \tag{7.104}$$

Continuing from this form, we will now determine the model parameters of the deterministic system.

7.4.4.1 Determination of the Discrete Propagation Delays and the Path Gains

The discrete propagation delays $\tilde{\tau}'_\ell$ are integer multiples of the sampling interval T'_s, i.e.,

$$\tilde{\tau}'_\ell = \ell \cdot T'_s, \quad \ell = 0, 1, \ldots, \mathcal{L} - 1, \tag{7.105}$$

where the number of discrete propagation paths $\mathcal{L}$ with different propagation delays is given by

$$\mathcal{L} = \left\lfloor \frac{\tau'_{\max}}{T'_s} \right\rfloor + 1 \,. \tag{7.106}$$

Thus, the ratio $\tau'_{\max}/T'_s$ determines the number of delay elements shown in Figure 7.11. Note that $\mathcal{L} \to \infty$ as $T'_s \to 0$.

With the discrete propagation delays $\tilde{\tau}'_\ell$ given by (7.105) and the sampling interval T'_s, the subintervals I_ℓ required for the partition of the interval $I = [0, \tau'_{\max}] = \bigcup_{\ell=0}^{\mathcal{L}-1} I_\ell$ can be defined as follows:

$$I_\ell := \begin{cases} [0,\ T'_s/2) & \text{for } \ \ell = 0, \\ \left[\tilde{\tau}'_\ell - T'_s/2,\ \tilde{\tau}'_\ell + T'_s/2\right) & \text{for } \ \ell = 1, 2, \ldots, \mathcal{L} - 2, \\ \left[\tilde{\tau}'_\ell - T'_s/2,\ \tau'_{\max}\right] & \text{for } \ \ell = \mathcal{L} - 1 \,. \end{cases} \tag{7.107}$$

Next, we demand that the areas under the delay power spectral densities $S_{\tau'}(\tau')$ and $\tilde{S}_{\tau'}(\tau')$ are identical within each subinterval I_ℓ, i.e., we require that

$$\int\limits_{\tau' \in I_\ell} S_{\tau'}(\tau')\, d\tau' = \int\limits_{\tau' \in I_\ell} \tilde{S}_{\tau'}(\tau')\, d\tau' \tag{7.108}$$

holds for all $\ell = 0, 1, \ldots, \mathcal{L} - 1$. Substituting $\tilde{S}_{\tau'}(\tau')$ by the expression (7.84) in the right-hand side of the equation above and applying the sifting property of Dirac delta functions leads directly to the following explicit formula for the path gains

$$\tilde{a}_\ell = \sqrt{\int\limits_{\tau' \in I_\ell} S_{\tau'}(\tau')\, d\tau'}, \quad \ell = 0, 1, \ldots, \mathcal{L} - 1, \tag{7.109}$$

where I_ℓ are the subintervals defined in (7.107). This result shows that the path gain $\tilde{a}_\ell$ of the ℓth propagation path is the square root of the average path power within the subinterval I_ℓ.

The method above is known as the method of equal distances (MED), which is described in detail in Subsection 7.5.2. There it is shown that $\tilde{S}_{\tau'}(\tau')$ converges to $S_{\tau'}(\tau')$ as $\mathcal{L} \to \infty$. From the definitions in (7.85) and (7.87), it follows that this property must also hold for the average delay $\tilde{B}^{(1)}_{\tau'}$ and the delay spread $\tilde{B}^{(2)}_{\tau'}$ of the simulation model. For $\mathcal{L} < \infty$ ($T'_s > 0$), however, we generally have to write $\tilde{B}^{(1)}_{\tau'} \approx B^{(1)}_{\tau'}$ and $\tilde{B}^{(2)}_{\tau'} \approx B^{(2)}_{\tau'}$. By way of example for the delay power spectral densities of the channel models specified by COST 207, which are depicted in Figure 7.7 (see also Table 7.1), the quality of the approximation $\tilde{B}^{(i)}_{\tau'} \approx B^{(i)}_{\tau'}$ is shown for $i = 1, 2$ in Figures 7.15(a)–7.15(d) as a function of the number of discrete propagation paths $\mathcal{L}$.

7.4.4.2 Determination of the Discrete Doppler Frequencies and the Gains

The discrete Doppler frequencies $f_{i,n,\ell}$ and the gains $c_{i,n,\ell}$ can be determined by applying the methods described in Section 5.1. Apart from the method of exact Doppler spread (MEDS), we preferably also use the L_p-norm method (LPNM). The first method mentioned is especially

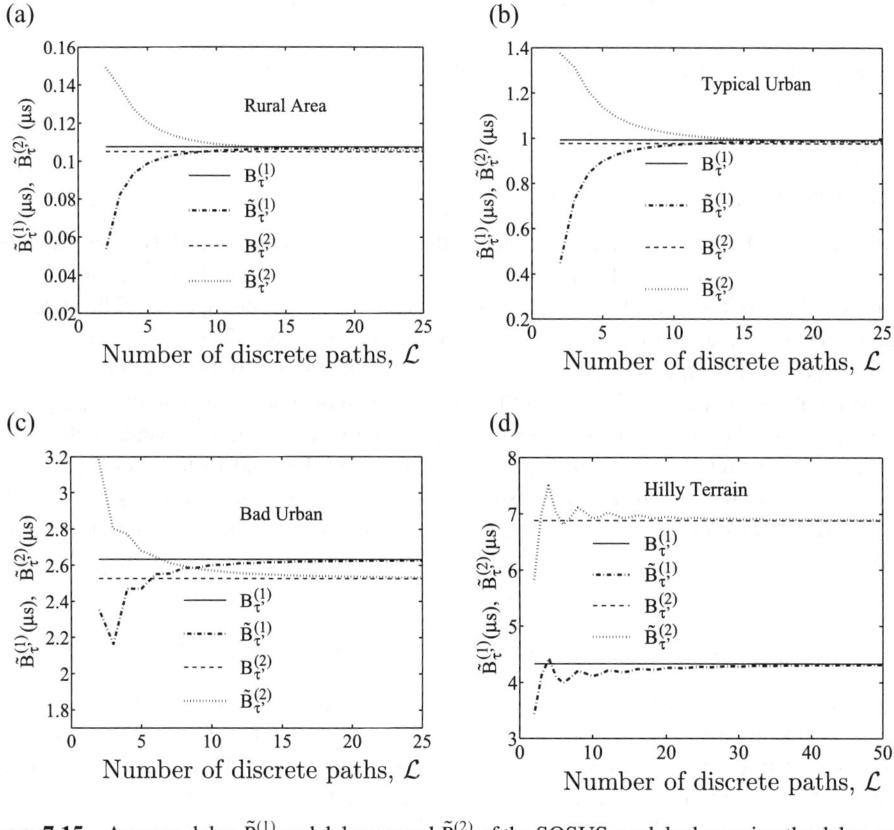

Figure 7.15 Average delay $\tilde{B}^{(1)}_{\tau'}$ and delay spread $\tilde{B}^{(2)}_{\tau'}$ of the SOSUS model when using the delay power spectral densities specified by COST 207 [19]: (a) Rural Area, (b) Typical Urban, (c) Bad Urban, and (d) Hilly Terrain.

recommended for the Jakes power spectral density. However, to fulfil the US condition, it has to be taken into account that the complex deterministic SOS processes $\tilde{\mu}_\ell(t)$ have to be designed such that $\tilde{\mu}_\ell(t)$ and $\tilde{\mu}_\lambda(t)$ are uncorrelated for $\ell \neq \lambda$ $(\ell, \lambda = 0, 1, \ldots, \mathcal{L} - 1)$. Recall that this is always the case if the discrete Doppler frequencies $f_{i,n,\ell}$ fulfil the condition in (7.57b). Using the MEDS, this condition is always fulfilled in case the number of sinusoids $N_{i,\ell}$ are chosen such that the inequality $N_{i,\ell}/N_{k,\lambda} \neq (2n-1)/(2m-1)$ holds for $n = 1, 2, \ldots, N_{i,\ell}$ and $m = 1, 2, \ldots, N_{k,\lambda}$, where $i, k = 1, 2$ and $\ell, \lambda = 0, 1, \ldots, \mathcal{L} - 1$ ($i = k$ and $\ell = \lambda$ do not hold at the same time) [214]. However, when using the LPNM method, we do not have to take this inequality into account, since even for $N_{i,\ell} = N_{k,\lambda}$, disjoint sets $\{f_{i,n,\ell}\}$ and $\{\pm f_{k,m,\lambda}\}$ with $\ell \neq \lambda$ can easily be found, to guarantee that the resulting deterministic processes $\tilde{\mu}_\ell(t)$ and $\tilde{\mu}_\lambda(t)$ are uncorrelated for $\ell \neq \lambda$. For this purpose, it is sufficient to either minimize the L_p-norm in (5.74) by using different values for the parameter p, or by performing the optimization of each set $\{f_{i,n,\ell}\}$ of discrete Doppler frequencies by choosing different values for the quantity $\tau_{\max}$ defining the upper limit of the integral in (5.74). Having this in mind, the numerical optimization of the autocorrelation functions $\tilde{r}_{\mu_{i,\ell}\mu_{i,\ell}}(\tau)$ [see (7.59b)] guarantees that the desired condition

$$\{f_{i,n,\ell}\} \bigcap \{\pm f_{k,m,\lambda}\} = \emptyset \quad \Longleftrightarrow \quad \ell \neq \lambda \tag{7.110}$$

is usually fulfilled for all $i, k = 1, 2$, $n = 1, 2, \ldots, N_{i,\ell}$, $m = 1, 2, \ldots, N_{k,\lambda}$, and $\ell, \lambda = 0, 1, \ldots, \mathcal{L} - 1$.

7.4.4.3 Determination of the Phases

In Subsections 7.4.2 and 7.4.3, it turned out that the phases $\theta_{i,n,\ell}$ have no influence on the system functions shown in Figure 7.14. Hence, we may conclude that the fundamental statistical properties of SOSUS models are independent of the choice of the phases $\theta_{i,n,\ell}$. The statements made in Section 5.2 are still valid for the frequency-selective case. Therefore, we once again may assume that the phases $\theta_{i,n,\ell}$ are realizations of a random variable, which is uniformly distributed in the interval $(0, 2\pi]$. Alternatively, $\theta_{i,n,\ell}$ can also be determined by applying the deterministic procedure described in Section 5.2. In both cases different events (sets) $\{\theta_{i,n,\ell}\}$ always result in different realizations (sample functions) of the time-variant impulse response $\tilde{h}(\tau', t)$ but, nevertheless, all impulse responses have the same statistical properties. In other words: Every realization of the impulse response $\tilde{h}(\tau', t)$ contains the complete statistical information.

7.4.5 Simulation Models for the COST 207 Channel Models

In this subsection, we will once more pick up the channel models according to COST 207 [19] and will show how to develop efficient simulation models for them. For this purpose, we restrict our attention to the 4-path and 6-path channel models (RA, TU, BU, HT) specified in Table 7.3. As these models are already given in a discrete form with respect to τ', the discrete propagation delays $\tilde{\tau}'_\ell$ can be equated directly with the values for τ'_ℓ as listed in Table 7.3, i.e., $\tilde{\tau}'_\ell = \tau'_\ell$. The adaptation of the sampling interval T'_s to the discrete propagation delays $\tilde{\tau}'_\ell$ is achieved by $\tilde{\tau}'_\ell = q_\ell \cdot T'_s$, where q_ℓ denotes an integer and T'_s is the greatest common divisor of $\tau_1', \tau_2', \ldots, \tau_{\mathcal{L}-1}'$, i.e., $T_s' = \gcd\{\tau_\ell'\}_{\ell=1}^{\mathcal{L}-1}$. The corresponding path gains $\tilde{a}_\ell$ are identical to the square root of the path powers as listed in Table 7.3.

The specifications for the Doppler power spectral density can be found in Table 7.2. In the case of the Jakes power spectral density, we determine the model parameters $f_{i,n,\ell}$ and $c_{i,n,\ell}$ by applying the L_p-norm method described in Subsection 5.1.5 by taking into account that (7.110) is fulfilled. The simulation model for the resulting complex deterministic SOS process $\tilde{\mu}_\ell(t)$ by using the Jakes power spectral density equals that in Figure 7.10. For the Gaussian power spectral densities (Gauss I and Gauss II), the third variant of the L_p-norm method (LPNM III) is of advantage. For the solution of the present problem, it is recommended to start with a Gaussian random process $\nu_{i,\ell}(t)$ having a symmetrical Gaussian power spectral density of the form

$$S_{\nu_{i,\ell}\nu_{i,\ell}}(f) = A_{i,\ell}\, e^{-\frac{f^2}{2s_{i,\ell}^2}}, \quad i = 1, 2, \tag{7.111}$$

and then perform a frequency shift of $f_{i,0,\ell}$, which finally results in

$$S_{\mu_\ell\mu_\ell}(f) = \sum_{i=1}^{2} S_{\nu_{i,\ell}\nu_{i,\ell}}(f - f_{i,0,\ell}), \tag{7.112}$$

where $A_{i,\ell}$, $s_{i,\ell}$, and $f_{i,0,\ell}$ denote the parameters specified in Table 7.2. The autocorrelation function required for the minimization of the error function in (5.78) is in the present case given by the inverse Fourier transform of the Gaussian power spectral density in (7.111), i.e.,

$$r_{\nu_{i,\ell}\nu_{i,\ell}}(\tau) = \sigma_{i,\ell}^2\, e^{-2(\pi s_{i,\ell}\tau)^2}, \tag{7.113}$$

where $\sigma_{i,\ell}^2 = \sqrt{2\pi}\, A_{i,\ell}\, s_{i,\ell}$ describes the variance of the Gaussian random process $\nu_{i,\ell}(t)$. For the simulation model, this means that we first have to determine the model parameters $f_{i,n,\ell}$ and $c_{i,n,\ell}$ of the deterministic process

$$\tilde{\nu}_{i,\ell}(t) = \sum_{n=1}^{N_{i,\ell}} c_{i,n,\ell} \cos(2\pi f_{i,n,\ell} + \theta_{i,n,\ell}) \tag{7.114}$$

by using the LPNM III. The application of the frequency shift theorem of the Fourier transform then provides the desired complex deterministic SOS process in the form

$$\begin{aligned}\tilde{\mu}_\ell(t) &= \sum_{i=1}^{2} \tilde{\nu}_{i,\ell}(t)\, e^{-j2\pi f_{i,0,\ell} t} \\ &= \sum_{i=1}^{2} \tilde{\nu}_{i,\ell}(t) \cos(2\pi f_{i,0,\ell} t) - j \sum_{i=1}^{2} \tilde{\nu}_{i,\ell}(t) \sin(2\pi f_{i,0,\ell} t)\,.\end{aligned} \tag{7.115}$$

The resulting simulation model for the complex deterministic process $\tilde{\mu}_\ell(t)$ is shown in Figure 7.16. Finally, the overall simulation model for the COST 207 channel models is obtained by integrating the structures shown in Figures 7.10 and 7.16 in the tapped-delay-line structure presented in Figure 7.11.

Using the L_p-norm method, we have the chance to choose an equal number of sinusoids $N_{i,\ell}$ not only for all propagation paths, but also for the corresponding real and imaginary

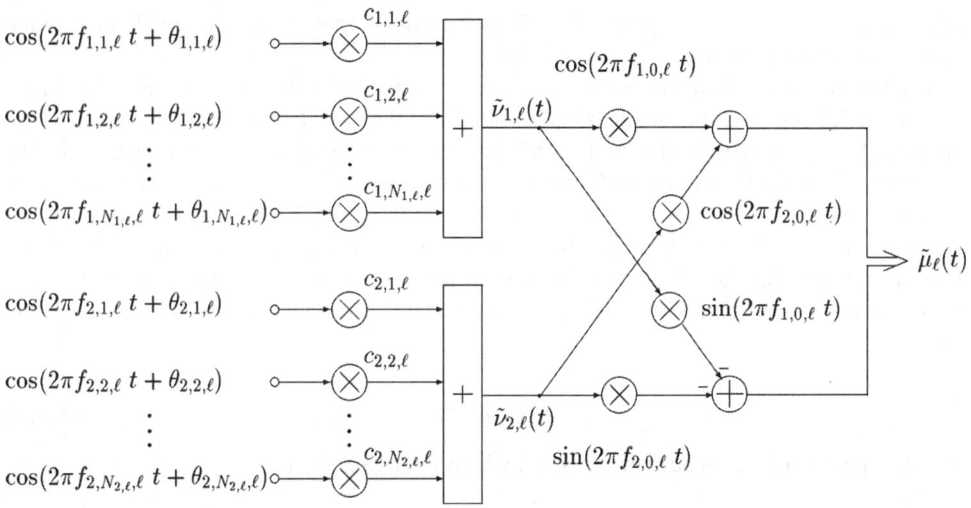

Figure 7.16 Simulation model for complex deterministic SOS processes $\tilde{\mu}_\ell(t)$ by using the frequency-shifted Gaussian power spectral densities specified by COST 207 [see Table 7.2].

parts, without violating the condition in (7.110). As an example, we fix $N_{i,\ell}$ for the $\mathcal{L}$-path channel models according to COST 207 by $N_{i,\ell} = 10$ ($\forall i = 1, 2$, and $\ell = 0, 1, \ldots, \mathcal{L}-1$). For the maximum Doppler frequency $f_{\max}$, we choose the value 91 Hz. The remaining model parameters can now be computed by following the procedure described above. Knowing the model parameters, not only the scattering function $\tilde{S}(\tau', f)$ [see (7.78)], but also all other correlation functions, power spectral densities, and characteristic quantities shown in Figure 7.14 can be determined analytically. For example, the resulting scattering functions $\tilde{S}(\tau', f)$ of the deterministic simulation models are shown in Figures 7.17(a)–7.17(d) for the COST 207 channel models specified in Table 7.3.

We notice that the processing of the discrete input signal $x(kT'_s)$ and the corresponding output signal $y(kT'_s)$ is performed with the sampling rate $f'_s = 1/T'_s$, whereas the sampling of the complex deterministic SOS process $\tilde{\mu}_\ell(t)$ ($\ell = 0, 1, \ldots, \mathcal{L}-1$) takes place at the discrete time instants $t = kT_s = k\,m'_s\,T'_s$. Recall that the statements made in Subsection 7.4.1 have to be taken into account for the choice of the sampling rate ratio $m'_s = T_s/T'_s$.

7.5 Methods for Modelling of Given Power Delay Profiles

In this section, altogether five fundamental methods are proposed for the modelling of the power delay profile of frequency-selective channels. We exploit the duality relation between the Doppler power spectral density of the sum-of-sinusoids model and the power delay profile of the tapped-delay-line model. This feature allows us to profit from the parameter computation methods introduced in Chapter 5. The procedures proposed here for the modelling of given power delay profiles are the method of equal distances (MED), mean-square-error method (MSEM), method of equal areas (MEA), Monte Carlo method (MCM), and L_p-norm method (LPNM). These methods have different degrees of sophistication, where four of them are purely deterministic methods, namely the MED, MSEM, MEA, and LPNM, while the MCM

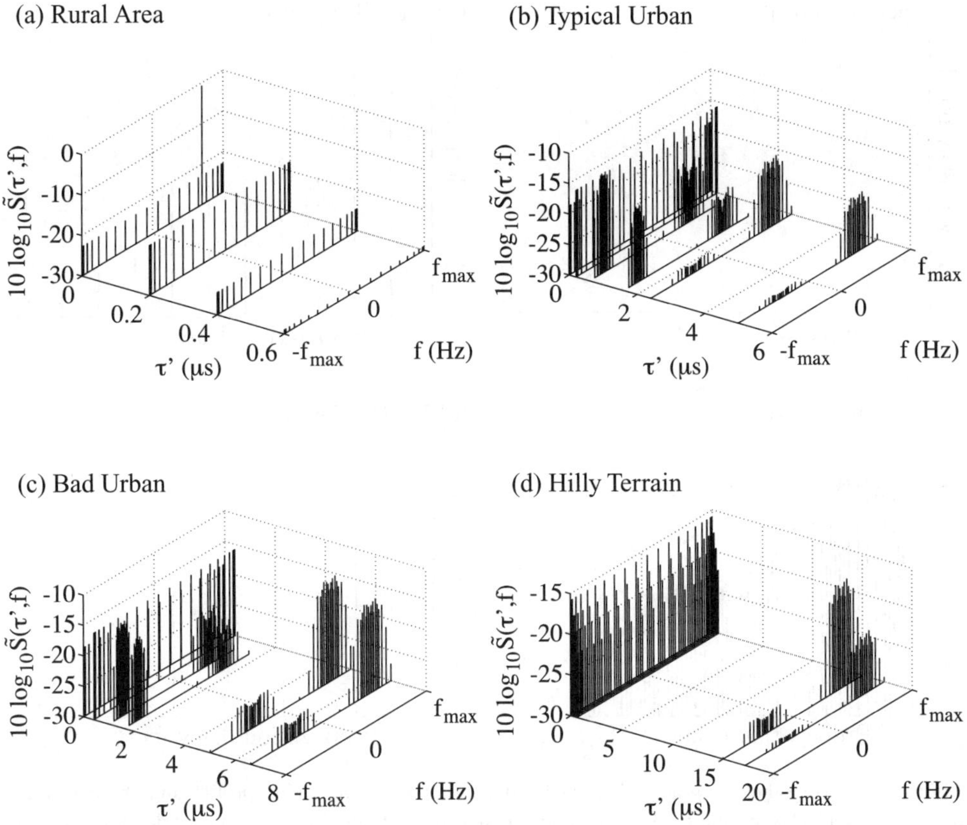

Figure 7.17 Scattering functions $\tilde{S}(\tau', f)$ of deterministic channel models designed on the basis of the $\mathcal{L}$-path channel models specified by COST 207 [19]: (a) Rural Area, (b) Typical Urban, (c) Bad Urban, and (d) Hilly Terrain.

is a statistic procedure. All procedures are universally valid, so that they can be applied to any specified or measured power delay profile. However, to work out the strengths and weaknesses of the proposed methods, we employ each procedure to the negative exponential power delay profile specified for the typical urban (TU) channel in [19]. The performance of each procedure will be evaluated analytically with respect to the frequency correlation function (FCF), the average delay, and the delay spread. The numerical results show that the LPNM has the best performance and the MCM the worst.

7.5.1 Problem Description

The problem we are focussing on can be formulated as follows: Given is a specified or measured power delay profile. How can the given profile efficiently be modelled by using the tapped-delay-line model? This channel modelling problem is illustrated in Figures 7.18(a)–(d) for two different cases. In the first case, a specified power delay profile $S_{\tau'}(\tau')$ is given, where it is supposed that $S_{\tau'}(\tau')$ is a continuous function confined to the interval $[0, \tau'_{\max}]$, as shown in

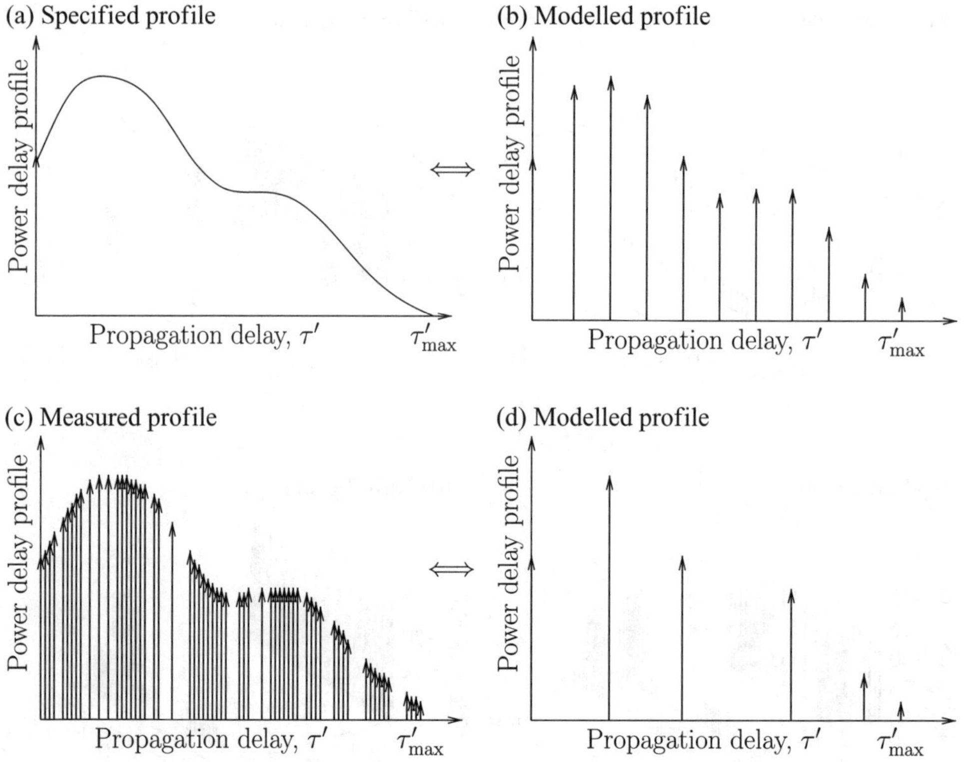

Figure 7.18 Hypothetical graph of a specified continuous power delay profile and (b) equivalent discrete power delay profile of the simulation model. (c) Measured discrete power delay profile and (d) equivalent discrete power delay profile of the simulation model with reduced complexity.

Figure 7.18(a). The power delay profile $\tilde{S}_{\tau'}(\tau')$ of the tapped-delay-line model is — according to (7.84) — discrete [see Figure 7.18(b)]. Our aim is to compute the model parameters $\{\tilde{a}_\ell\}_{\ell=0}^{\mathcal{L}-1}$ and $\{\tilde{\tau}'_\ell\}_{\ell=0}^{\mathcal{L}-1}$ such that the power delay characteristics of the simulation model (tapped-delay-line model) are sufficiently close to those of the reference model described by the specified power delay profile. In the second case, $S_{\tau'}(\tau')$ is obtained from channel measurements, where the measured power delay profile $S_{\tau'}(\tau')$ is discrete and it is supposed that $\mathcal{L}^\star \gg 1$ propagation paths could be resolved. The problem is now to find the model parameters $\{\tilde{a}_\ell\}_{\ell=0}^{\mathcal{L}-1}$ and $\{\tilde{\tau}'_\ell\}_{\ell=0}^{\mathcal{L}-1}$ of a simulation model with reduced complexity, i.e., $\mathcal{L} < \mathcal{L}^\star$, but with similar power delay characteristics as the measured channel. This parameter computation problem is illustrated in Figures 7.18(c) and 7.18(d).

As reference profile, we consider the TU channel specified in [19]. According to Table 7.1, the power delay profile $S_{\tau'}(\tau')$ of the TU channel is given by

$$S_{\tau'}(\tau') = \begin{cases} c_{TU}\, e^{-b\tau'}, & 0 \le \tau' \le \tau'_{\max}, \\ 0, & \text{else}, \end{cases} \tag{7.116}$$

where $b = 1$ MHz and $\tau'_{\max} = 7\,\mu$s. The real-valued constant $c_{TU} = 1/(1 - e^{-7})$ has been introduced here to normalize the power of the profile to unity, i.e., $\int_0^\infty S_{\tau'}(\tau')\, d\tau' = 1$. The average delay $B_{\tau'}^{(1)}$ and the delay spread $B_{\tau'}^{(2)}$ of the TU channel are given by $B_{\tau'}^{(1)} = 0.993\ \mu$s and $B_{\tau'}^{(2)} = 0.977\ \mu$s, respectively. Using the relation (7.44), the frequency correlation function corresponding to the TU channel is obtained as

$$r_{\tau'}(\upsilon') = c_{TU} \frac{1 - e^{-\tau'_{\max}(b + j2\pi\upsilon')}}{b + j2\pi\upsilon'} . \tag{7.117}$$

Finally, it should be mentioned that the expressions (7.116) and (7.117) derived for the TU channel can easily be modified to describe also the RA channel specified in [19] by merely changing the values of the constants c_{TU}, $\tau'_{\max}$, and b.

With reference to (7.84), the power delay profile $\tilde{S}_{\tau'}(\tau')$ of the tapped-delay-line channel simulator can be expressed by

$$\tilde{S}_{\tau'}(\tau') = \sum_{\ell=0}^{\mathcal{L}-1} \tilde{a}_\ell^2 \, \delta(\tau' - \tilde{\tau}'_\ell) . \tag{7.118}$$

Hence, $\tilde{S}_{\tau'}(\tau')$ is a discrete function determined by a finite sum of $\mathcal{L}$ weighted Dirac delta functions, where the delta functions are located at $\tilde{\tau}'_\ell$ and the corresponding weighting factors are given by the squared path gains $\tilde{a}_\ell^2$ ($\ell = 0, 1, \ldots, \mathcal{L} - 1$). Taking the Fourier transform of (7.118) results in the frequency correlation function $\tilde{r}_{\tau'}(\upsilon')$ of the simulation model. Thus,

$$\tilde{r}_{\tau'}(\upsilon') = \sum_{\ell=0}^{\mathcal{L}-1} \tilde{a}_\ell^2 \, e^{-j2\pi\upsilon'\tilde{\tau}'_\ell} . \tag{7.119}$$

Considering the expressions in (7.118) and (7.119), we realize that the power delay characteristics of the simulation model can be investigated analytically provided that the parameter sets $\{\tilde{a}_\ell\}_{\ell=0}^{\mathcal{L}-1}$ and $\{\tilde{\tau}'_\ell\}_{\ell=0}^{\mathcal{L}-1}$ are known.

Comparing (7.118) with the expression in (4.18) that has been obtained for the Doppler power spectral density of the sum-of-sinusoids model, it turns out that the power delay profile and the Doppler power spectral density are dual to each other. Due to this duality relation, most methods introduced in Chapter 5 for the modelling of Doppler power spectral densities are also proper candidates for the modelling of power delay profiles. The methods which cannot be applied are the Jakes method (JM) and the method of exact Doppler spread (MEDS). Both approaches are too restrictive, as they have been developed exclusively for the modelling of the classical Jakes Doppler power spectral density. Among the useful methods are the MED, MSEM, MEA, MCM, LPNM [215]. In the following, it will be shown how these methods can be applied to solve the parameter computation problem described above.

7.5.2 Methods for the Computation of the Discrete Propagation Delays and the Path Gains

It is the task of this section to compute the sets $\{\tilde{a}_\ell\}_{\ell=0}^{\mathcal{L}-1}$ and $\{\tilde{\tau}'_\ell\}_{\ell=0}^{\mathcal{L}-1}$ in such a way that the power delay characteristics of the simulation model are close to those of the reference model. Altogether, five different methods are presented for the computation of the discrete propagation delays $\tilde{\tau}'_\ell$ and the path gains $\tilde{a}_\ell$. Each method is first formulated in universally applicable terms and then exemplified by setting the number of propagation paths $\mathcal{L}$ to 12, which enables a fair comparison with the performance of the 12-path COST 207 models. Nevertheless, it should be stressed that the case $\mathcal{L} = 12$ is just an example. Subsequently, in Subsection 7.5.3, the methods are all compared over a larger range of $\mathcal{L}$ values.

7.5.2.1 Method of Equal Distances (MED)

The MED was originally presented in [216] aiming the modelling of given power delay profiles. Later, in [215], this procedure has been considered as a reference method when discussing the performance of other, more sophisticated procedures. The basic idea of the MED is to compute the model parameters $\tilde{\tau}'_\ell$ by introducing equal distances between adjacent discrete propagation delays $\tilde{\tau}'_\ell$ and $\tilde{\tau}'_{\ell-1}$. Let $S_{\tau'}(\tau')$ be any given specified or measured power delay profile, which is equal to zero outside the interval $I = [0,\, \tau'_{\max}]$. Then, the discrete propagation delays $\tilde{\tau}'_\ell$ are defined by multiples of a constant quantity $\Delta\tau'$ according to

$$\tilde{\tau}'_\ell = \ell \cdot \Delta\tau', \quad \ell = 0, 1, \ldots, \mathcal{L} - 1, \tag{7.120}$$

where $\Delta\tau'$ is related to $\tau'_{\max}$ and $\mathcal{L} > 1$ according to $\Delta\tau' = \tau'_{\max}/(\mathcal{L} - 1)$. From (7.93) and (7.120) it follows that the frequency correlation function $\tilde{r}_{\tau'}(\upsilon')$ is a periodic function in the frequency separation variable υ' with the period $\Upsilon' = 1/\Delta\tau'$.

Next, we partition the interval $I = [0,\, \tau'_{\max}]$ into $\mathcal{L}$ subintervals I_ℓ according to $I = \cup_{\ell=0}^{\mathcal{L}-1} I_\ell$, where $I_0 = [0,\, \Delta\tau'/2)$, $I_\ell = [\tilde{\tau}'_\ell - \Delta\tau'/2,\, \tilde{\tau}'_\ell + \Delta\tau'/2)$ for $\ell = 0, 1, \ldots, \mathcal{L} - 2$, and $I_{\mathcal{L}-1} = [\tilde{\tau}'_{\mathcal{L}-1} - \Delta\tau'/2,\, \tau'_{\max}]$. Furthermore, we impose on our channel model that the integrals of $\tilde{S}_{\tau'}(\tau')$ and $S_{\tau'}(\tau')$ over $\tau' \in I_\ell$ are identical, i.e.,

$$\int\limits_{\tau' \in I_\ell} S_{\tau'}(\tau')\, d\tau' = \int\limits_{\tau' \in I_\ell} \tilde{S}_{\tau'}(\tau')\, d\tau', \quad \ell = 0, 1, \ldots, \mathcal{L} - 1\,. \tag{7.121}$$

Thus, after substituting (7.118) in the right-hand side of (7.121), we obtain the following equation for the path gains

$$\tilde{a}_\ell = \sqrt{\int\limits_{\tau' \in I_\ell} S_{\tau'}(\tau')\, d\tau'}, \quad \ell = 0, 1, \ldots, \mathcal{L} - 1\,. \tag{7.122}$$

Notice that the square of the path gain, $\tilde{a}_\ell^2$, of the ℓth discrete propagation path is a measure of the average path power $S_{\tau'}(\tau')$ within the subinterval I_ℓ.

In this context, it is interesting to study the limit of the power delay profile $\tilde{S}_{\tau'}(\tau')$ as $\mathcal{L} \to \infty$ and $\Delta\tau' \to 0$. For this purpose, we substitute (7.122) into (7.84) and obtain

$$\begin{aligned}\lim_{\substack{\mathcal{L}\to\infty \\ \Delta\tau'\to 0}} \tilde{S}_{\tau'}(\tau') &= \lim_{\substack{\mathcal{L}\to\infty \\ \Delta\tau'\to 0}} \sum_{\ell=0}^{\mathcal{L}-1} \left[\int\limits_{\tau' \in I_\ell} S_{\tau'}(\tau')\, d\tau' \right] \delta(\tau' - \tilde{\tau}'_\ell) \\ &= \lim_{\mathcal{L}\to\infty} \sum_{\ell=0}^{\mathcal{L}-1} S_{\tau'}(\tilde{\tau}'_\ell)\, \delta(\tau' - \tilde{\tau}'_\ell)\, \Delta\tilde{\tau}'_\ell\end{aligned}$$

$$= \int_0^\infty S_{\tau'}(\tilde{\tau}'_\ell)\,\delta(\tau' - \tilde{\tau}'_\ell)\,d\tilde{\tau}'_\ell$$

$$= S_{\tau'}(\tau')\,. \tag{7.123}$$

It becomes obvious that $\tilde{S}_{\tau'}(\tau')$ approaches against $S_{\tau'}(\tau')$ if the number of discrete propagation paths $\mathcal{L}$ tends to infinity. Consequently, this property holds also for the average delay $\tilde{B}^{(1)}_{\tau'}$ and the delay spread $\tilde{B}^{(2)}_{\tau'}$ of the simulation model, i.e., we obtain $\tilde{B}^{(1)}_{\tau'} \to B^{(1)}_{\tau'}$ and $\tilde{B}^{(2)}_{\tau'} \to B^{(2)}_{\tau'}$ as $\mathcal{L} \to \infty$ $(\Delta\tau' \to 0)$. For $\mathcal{L} < \infty$ $(\Delta\tau' > 0)$, however, we generally have to write $\tilde{B}^{(1)}_{\tau'} \approx B^{(1)}_{\tau'}$ and $\tilde{B}^{(2)}_{\tau'} \approx B^{(2)}_{\tau'}$.

Figure 7.19(a) shows the reference power delay profile $S_{\tau'}(\tau')$ according to (7.116) in comparison with the power delay profile $\tilde{S}_{\tau'}(\tau')$ of the simulation model by applying the MED with $\mathcal{L} = 12$. Taking the absolute value of the frequency correlation function $\tilde{r}_{\tau'}(\upsilon')$ given in (7.117) leads to the results shown in Figure 7.19(b). For comparison, we have also presented the corresponding analytical results of the simulation model in the same figure. These results have been obtained by substituting (7.120) and (7.122) in (7.119) and then taking the absolute value of $\tilde{r}_{\tau'}(\upsilon')$. Moreover, Figure 7.19(c) shows the convergence behaviour of the average delay $\tilde{B}^{(1)}_{\tau'}$ [see (7.86)] and the delay spread $\tilde{B}^{(2)}_{\tau'}$ [see (7.88)] in terms of the number of discrete paths $\mathcal{L}$.

7.5.2.2 Mean-Square-Error Method (MSEM)

The basic idea of the MSEM is to determine the sets $\{\tilde{\tau}'_\ell\}_{\ell=0}^{\mathcal{L}-1}$ and $\{\tilde{a}_\ell\}_{\ell=0}^{\mathcal{L}-1}$ such that the simulation model's mean-square error of the frequency correlation function

$$E_{r_{\tau'}} := \frac{1}{\upsilon'_{\max}} \int_0^{\upsilon'_{\max}} |r_{\tau'}(\upsilon') - \tilde{r}_{\tau'}(\upsilon')|^2\,d\upsilon' \tag{7.124}$$

is minimal within a given frequency interval $[0, \upsilon'_{\max}]$. The upper limit of this interval, denoted here by $\upsilon'_{\max}$, will be defined below. A closed-form solution to this problem exists if, and only if, the discrete propagation delays $\tilde{\tau}'_\ell$ are given by (7.120). Hence, the set $\{\tilde{\tau}'_\ell\}$ is obtained as $\{\tilde{\tau}'_\ell \,|\, \tilde{\tau}'_\ell := \ell\Delta\tau',\ \ell = 0, 1, \ldots, \mathcal{L}-1\}$, where $\Delta\tau' = \tau'_{\max}/(\mathcal{L}-1)$. As a consequence of the equally spaced discrete propagation delays, it follows that the frequency correlation function $\tilde{r}'_\tau(\upsilon')$ is a periodic function with the period $\Upsilon' = 1/\Delta\tau'$. One half of this period is therefore a proper value for the upper limit of the integral in (7.124). Thus, we choose $\upsilon'_{\max} = 1/(2\Delta\tau')$. The periodicity of $\tilde{r}_{\tau'}(\upsilon')$ does not cause any reason for concern as long as the period Υ' is larger than or equal to twice the system bandwidth B, i.e., $\Upsilon' = 1/\Delta\tau' \geq 2B$. This condition can easily be satisfied by choosing $\mathcal{L} = \lceil 2B\tau'_{\max}\rceil + 1$, where $\lceil x \rceil$ denotes the smallest integer greater than or equal to x.

Inserting (7.119) into (7.124) and setting the partial derivatives of the mean-square error $E_{r_{\tau'}}$ with respect to the path gains $\tilde{a}_\ell$ equal to zero, i.e., $\partial E_{r_{\tau'}}/\partial \tilde{a}_\ell = 0$, results in the following

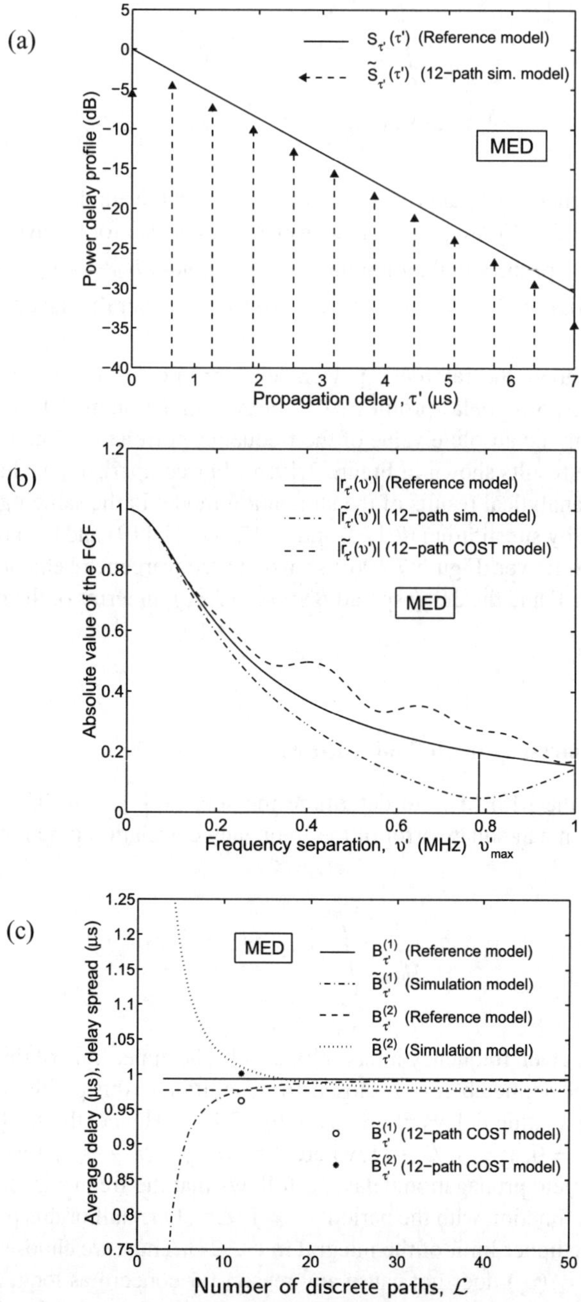

Figure 7.19 (a) Power delay profile $S_{\tau'}(\tau')$ of the TU channel, (b) absolute value of the frequency correlation function $|r_{\tau'}(\upsilon')|$, and (c) average delay $B_{\tau'}^{(1)}$ as well as the delay spread $B_{\tau'}^{(2)}$ in comparison with the corresponding quantities of the simulation model when using the MED.

closed-form expression for the path gains

$$\tilde{a}_\ell = \sqrt{\operatorname{Re}\left\{\frac{1}{\upsilon'_{max}} \int_0^{\upsilon'_{max}} r_{\tau'}(\upsilon')\, e^{j2\pi \upsilon' \tilde{\tau}'_\ell}\, d\upsilon'\right\}}, \quad \ell = 0, 1, \ldots, \mathcal{L}-1, \tag{7.125}$$

where $\upsilon'_{max} = 1/(2\Delta\tau')$. By substituting the expression (7.125) in (7.118), it can be shown that the power delay profile $\tilde{S}_{\tau'}(\tau')$ of the simulation model approaches the desired profile $S_{\tau'}(\tau')$ if the number of discrete paths $\mathcal{L}$ tends to infinity, i.e., $\tilde{S}_{\tau'}(\tau') \to S_{\tau'}(\tau')$ if $\mathcal{L} \to \infty$ $(\Delta\tau' \to 0)$. However, for finite values of $\mathcal{L}$, we have to write $\tilde{r}_{\tau'}(\upsilon') \approx r_{\tau'}(\upsilon')$, where $\upsilon' \in [0, \upsilon'_{max}]$, and $\tilde{B}^{(i)}_{\tau'} \approx B^{(i)}_{\tau'}$ $(i = 1, 2)$.

As an example, we apply the MSEM to the TU power delay profile as defined in (7.116). The results obtained for the power delay profile $\tilde{S}_{\tau'}(\tau')$ of the simulation model and the absolute value of the frequency correlation function $|\tilde{r}_{\tau'}(\upsilon')|$ are shown in Figures 7.20(a) and (b), respectively. Here, the number of discrete paths $\mathcal{L}$ was set to 12. This enables a direct comparison with the results shown in Figures 7.19(a) and (b). The comparison reveals that the quality of the approximation $r_{\tau'}(\upsilon') \approx \tilde{r}_{\tau'}(\upsilon')$ over the interval $[0, \upsilon'_{max}]$ is slightly better for the MSEM than for the MED. The simulation model's average delay $\tilde{B}^{(1)}_{\tau'}$ and delay spread $\tilde{B}^{(2)}_{\tau'}$ are both illustrated in Figure 7.20(c) in terms of the number of discrete paths $\mathcal{L}$.

A closer investigation of the MSEM and the MED indicates that the equal spacing between two adjacent discrete propagation delays limits the performance and the efficiency of these two methods.

7.5.2.3 Method of Equal Areas (MEA)

The MEA was originally derived in [148] aiming to model given Doppler power spectral densities. Due to the duality relation between the Doppler power spectral density of the sum-of-sinusoids model and the power delay profile of the tapped-delay-line model, the MEA is also a suitable candidate for the computation of the path gains and the discrete propagation delays. The basic idea of the MEA is to choose the same constant quantity

$$\tilde{a}_\ell = \frac{1}{\sqrt{\mathcal{L}}}, \quad \ell = 0, 1, \ldots, \mathcal{L}-1, \tag{7.126}$$

for all path gains $\tilde{a}_\ell$, and to determine the discrete propagation delays $\tilde{\tau}'_\ell$ such that the area under the power delay profile $S_{\tau'}(\tau')$ over the interval $[\tilde{\tau}'_\ell, \tilde{\tau}'_{\ell+1})$ equals the squared value of the path gain (average path power) $\tilde{a}^2_\ell$ for all ℓ, i.e.,

$$\int_{\tilde{\tau}'_\ell}^{\tilde{\tau}'_{\ell+1}} S_{\tau'}(\tau')\, d\tau' = \tilde{a}^2_\ell = \frac{1}{\mathcal{L}}, \quad \ell = 0, 1, \ldots, \mathcal{L}-1, \tag{7.127}$$

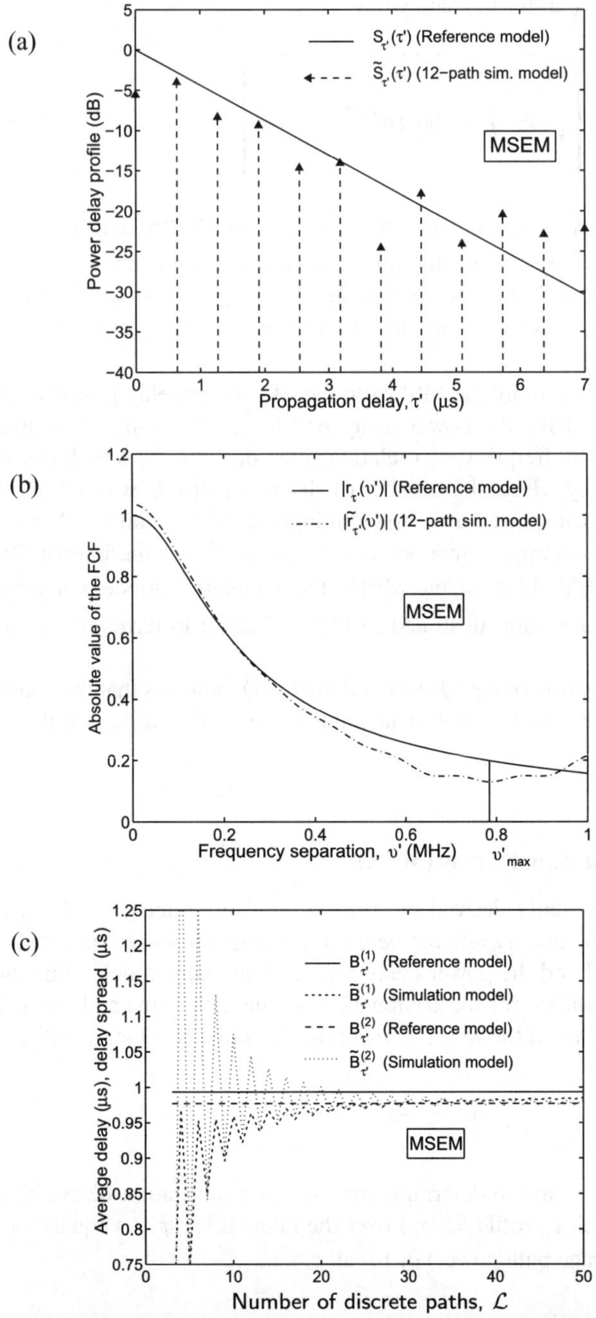

Figure 7.20 (a) Power delay profile $S_{\tau'}(\tau')$ of the TU channel, (b) absolute value of the frequency correlation function $|r_{\tau'}(\upsilon')|$, and (c) average delay $B_{\tau'}^{(1)}$ as well as the delay spread $B_{\tau'}^{(2)}$ in comparison with the corresponding quantities of the simulation model when using the MSEM.

with $\tilde{\tau}'_0 := 0$. To solve this problem, we first introduce an auxiliary function $G(x)$, which is defined as

$$G(x) := \int_0^x S_{\tau'}(\tau')\, d\tau' . \tag{7.128}$$

Using (7.127), we may then write

$$G(\tilde{\tau}'_\ell) = \sum_{\lambda=0}^{\ell-1} \int_{\tilde{\tau}'_\lambda}^{\tilde{\tau}'_{\lambda+1}} S_{\tau'}(\tau')\, d\tau' = \frac{\ell}{\mathcal{L}} . \tag{7.129}$$

If the inverse function G^{-1} of G exists, then a general expression for the discrete propagation delays is obtained as

$$\tilde{\tau}'_\ell = G^{-1}\left(\frac{\ell}{\mathcal{L}}\right) . \tag{7.130}$$

Now, we apply this procedure to our chosen reference model by substituting (7.116) in (7.129). After solving the resulting equation for $\tilde{\tau}'_\ell$, we finally obtain the closed-form solution

$$\tilde{\tau}'_\ell = -\frac{1}{b} \cdot \ln\left(1 - \frac{b \cdot \ell}{c_{TU} \cdot \mathcal{L}}\right) . \tag{7.131}$$

It is important to notice that the implementation of the $\mathcal{L}$-path tapped-delay-line model on a computer requires a quantization of the discrete propagation delays such that the greatest common divisor $\Delta\tau' = \gcd\{\tilde{\tau}'_0, \tilde{\tau}'_1, \ldots, \tilde{\tau}'_{\mathcal{L}-1}\}$ is equal to the sampling interval of the resulting digital channel simulator. However, if the sampling interval is sufficiently small, then the period of the frequency correlation function — given by $\Upsilon' = 1/\Delta\tau'$ — is in general much larger than the period obtained by using the MED or the MSEM. The impact of the quantization of the discrete propagation delays on the performance of the simulation model is discussed in Subsection 7.5.2.5.

Analogously to Figures 7.19 and 7.20, the Figures 7.21(a) and (b) show the results for the power delay profile and the frequency correlation function, respectively. The results presented for the simulation model with non-quantized discrete propagation delays were obtained by substituting the parameters described by (7.126) and (7.131) in the closed-form expressions (7.118) and (7.119). Particularly, as shown in Figure 7.21(c), the average delay and the delay spread converge relatively slowly in comparison with the corresponding results obtained by applying the MED [see Figure 7.19(c)] and the MSEM [see Figure 7.20(c)].

7.5.2.4 Monte Carlo Method (MCM)

The design of frequency-selective mobile fading channels using the Monte Carlo principle was first introduced in [100], refined in [99], and improved in [101]. The basic idea of the MCM is to generate the discrete propagation delays $\tilde{\tau}'_\ell$ according to a given probability density function $p_{\tau'}(\tau')$ that models the distribution of the propagation delays τ'. It can be shown without any

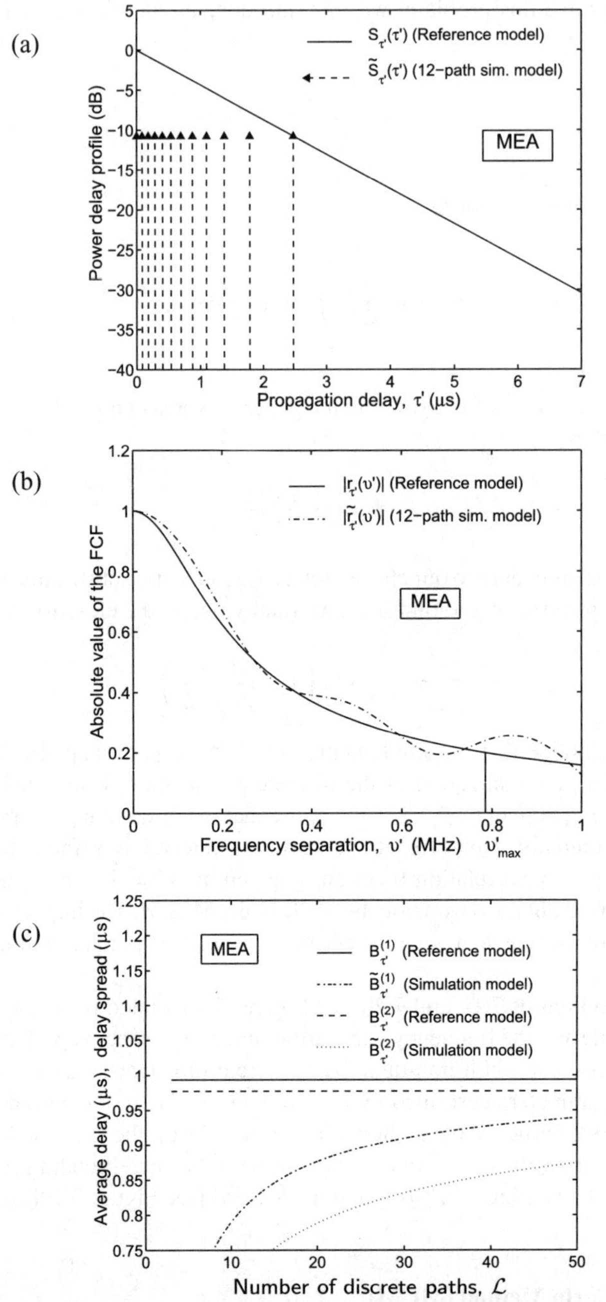

Figure 7.21 (a) Power delay profile $S_{\tau'}(\tau')$ of the TU channel, (b) absolute value of the frequency correlation function $|r_{\tau'}(\upsilon')|$, and (c) average delay $B_{\tau'}^{(1)}$ as well as the delay spread $B_{\tau'}^{(2)}$ in comparison with the corresponding quantities of the simulation model when using the MEA.

difficulty that $p_{\tau'}(\tau')$ is proportional to the power delay profile $S_{\tau'}(\tau')$, i.e.,

$$p_{\tau'}(\tau') = c_{\tau'} \cdot S_{\tau'}(\tau'), \tag{7.132}$$

where $c_{\tau'}$ is a real-valued constant defined by $c_{\tau'} := 1/\int_0^{\infty} S_{\tau'}(\tau')\,d\tau'$. According to [99], it is convenient to start from a random number generator that generates i.i.d. random numbers u_ℓ, each of which is uniformly distributed over the interval [0, 1], and then to perform the mapping

$$g_{\tau'}: \quad [0,1] \to [0, \tau'_{\max}], \quad u_\ell \mapsto \tilde{\tau}'_\ell = g_{\tau'}(u_\ell), \tag{7.133}$$

such that the cumulative distribution function of the discrete propagation delays $\tilde{\tau}'_\ell$ equals the desired cumulative distribution function $F_{\tau'}(\tilde{\tau}'_\ell) = \int_0^{\tilde{\tau}'_\ell} p_{\tau'}(\tau')\,d\tau'$. As it is shown in [41], the function $g_{\tau'}(u_\ell)$ is the inverse of $F_{\tau'}(\tilde{\tau}'_\ell) = u_\ell$, so that the discrete propagation delays $\tilde{\tau}'_\ell$ can be computed as

$$\tilde{\tau}'_\ell = g_{\tau'}(u_\ell) = F_{\tau'}^{-1}(u_\ell), \quad \ell = 0, 1, \ldots, \mathcal{L}-1. \tag{7.134}$$

The application of the MCM to the TU power delay profile [see (7.116)] results in the following formulas for the discrete propagation delays $\tilde{\tau}'_\ell$ and path gains $\tilde{a}_\ell$:

$$\tilde{\tau}'_\ell = -\frac{1}{b}\ln\left(1 - \frac{b \cdot u_\ell}{c_{TU}}\right), \tag{7.135a}$$

$$\tilde{a}_\ell = \frac{1}{\sqrt{\mathcal{L}}}, \tag{7.135b}$$

where $u_\ell \sim U[0, 1]$ for $\ell = 0, 1, \ldots, \mathcal{L}-1$. Computing the sets $\{\tilde{\tau}'_\ell\}_{\ell=0}^{\mathcal{L}-1}$ and $\{\tilde{a}_\ell\}_{\ell=0}^{\mathcal{L}-1}$ for $\mathcal{L} = 12$ and substituting the obtained coefficients in (7.118) and (7.119) leads to the results shown in Figures 7.22(a) and (b), respectively.

Since the elements of the set $\{u_\ell\}_{\ell=0}^{\mathcal{L}-1}$ are random variables, it follows from the mapping $\{u_\ell\}_{\ell=0}^{\mathcal{L}-1} \mapsto \{\tilde{\tau}'_\ell\}_{\ell=0}^{\mathcal{L}-1}$ that the characteristics of $\tilde{S}_{\tau'}(\tau')$ and $\tilde{r}_{\tau'}(\upsilon')$ depend strongly on the actual realization of the elements of the set $\{\tilde{\tau}'_\ell\}_{\ell=0}^{\mathcal{L}-1}$. However, the approximations $\tilde{r}_{\tau'}(\upsilon') \approx r_{\tau'}(\upsilon')$ and $\tilde{B}_{\tau'}^{(i)} \approx B_{\tau'}^{(i)}$ [see Figure 7.22(c)] are poor for $i = 1, 2$, even for large values of $\mathcal{L}$. Nevertheless, one can show that $\tilde{S}_{\tau'}(\tau')$ tends to $S_{\tau'}(\tau')$ in the limit $\mathcal{L} \to \infty$, and thus $\tilde{r}_{\tau'}(\upsilon') \to r_{\tau'}(\upsilon')$ and $\tilde{B}_{\tau'}^{(i)} \to B_{\tau'}^{(i)}$ $(i = 1, 2)$. A further characteristic feature of the MCM is that the statistical average of $\tilde{S}_{\tau'}(\tau')$ with respect to the distribution of the random variables $\tilde{\tau}'_\ell$ leads to the desired result $E\{\tilde{S}_{\tau'}(\tau')\} = S_{\tau'}(\tau')$, no matter which value has been chosen for $\mathcal{L}$, i.e., even if $\mathcal{L}$ equals 1. Consequently, we can write $E\{\tilde{r}_{\tau'}(\upsilon')\} = r_{\tau'}(\upsilon')$ and $E\{\tilde{B}_{\tau'}^{(i)}\} = B_{\tau'}^{(i)}$ $(i = 1, 2)$.

7.5.2.5 L_p-Norm Method (LPNM)

The basic idea of the LPNM [215] is to fit the frequency correlation function $\tilde{r}_{\tau'}(\upsilon')$ of the simulation model as close as possible to the frequency correlation function $r_{\tau'}(\upsilon')$ of the reference model over the frequency range of interest determined by the interval $[0, \upsilon'_{\max}]$. An

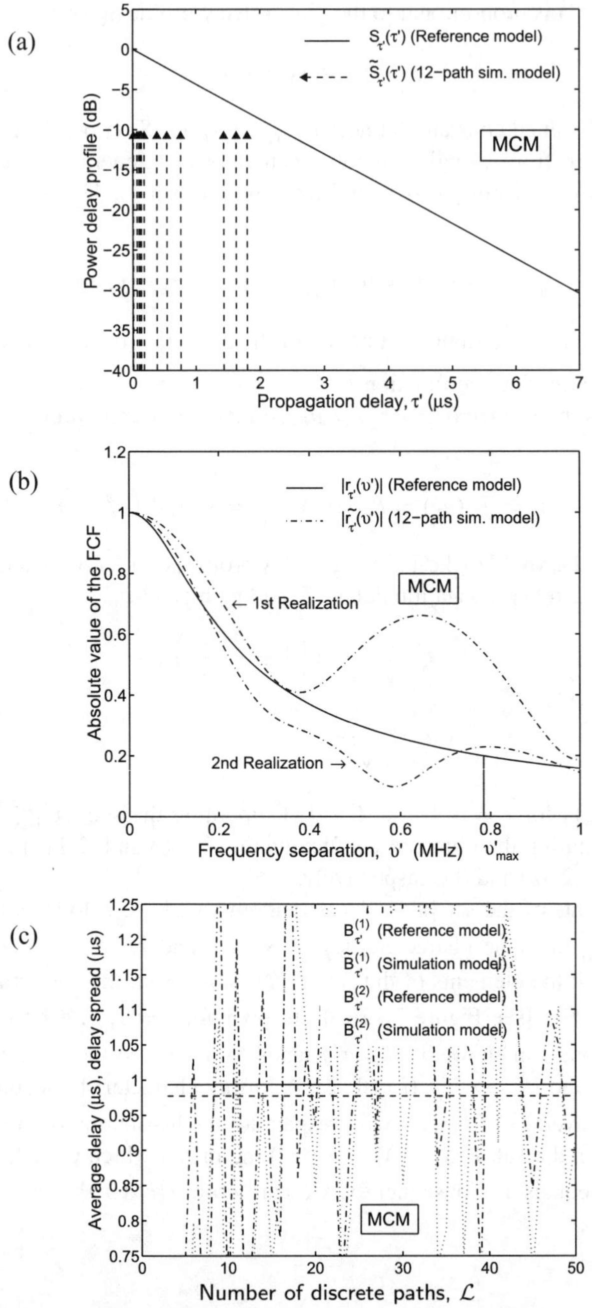

Figure 7.22 (a) Power delay profile $S_{\tau'}(\tau')$ of the TU channel, (b) absolute value of the frequency correlation function $|r_{\tau'}(\upsilon')|$, and (c) average delay $B_{\tau'}^{(1)}$ as well as the delay spread $B_{\tau'}^{(2)}$ in comparison with the corresponding quantities of the simulation model when using the MCM.

appropriate error function for this purpose is the L_p-norm defined by

$$E_{r_{\tau'}}^{(p)} := \left[\int_0^{\upsilon'_{\max}} \left| r_{\tau'}(\upsilon') - \tilde{r}_{\tau'}(\upsilon') \right|^p d\upsilon' \right]^{1/p} , \qquad (7.136)$$

where p denotes a real-valued number $1 \leq p < \infty$ and $\upsilon'_{\max}$ is a sufficiently large quantity that defines the upper bound of the frequency range of interest. Here, we define $\upsilon'_{\max}$ by $\upsilon'_{\max} = (\mathcal{L} - 1)/(2\tau'_{\max})$, so that the upper limits of the integrals appearing in (7.124) and (7.136) are identical. This is considered as a prerequisite for a fair comparison of the performance of LPNM and the MSEM. However, detailed investigations for different types of power delay profiles have revealed that sufficiently good results can generally be obtained if $\upsilon'_{\max}$ is larger than twice this value, i.e., $\upsilon'_{\max} \geq (\mathcal{L} - 1)/\tau'_{\max}$. The task we are confronted with is to optimize the elements of the sets $\{\tilde{\tau}'_\ell\}_{\ell=0}^{\mathcal{L}-1}$ and $\{\tilde{a}_\ell\}_{\ell=0}^{\mathcal{L}-1}$, which determine the shape of the frequency correlation function $\tilde{r}_{\tau'}(\upsilon')$ [see (7.92)] such that the L_p-norm $E_{r_{\tau'}}^{(p)}$ in (7.136) is minimal. A local minimum of $E_{r_{\tau'}}^{(p)}$ can be found numerically, e.g., by using the Fletcher-Powell optimization algorithm [162, 207]. This procedure requires proper initial values for the quantities $\tilde{\tau}'_\ell$ and $\tilde{a}_\ell$, which can be obtained, e.g., by using the MED or the MSEM. In this case, it follows that $\tilde{S}_{\tau'}(\tau')$ converges to $S_{\tau'}(\tau')$ in the limit $\mathcal{L} \to \infty$, and thus $\tilde{r}_{\tau'}(\upsilon')$ tends to $r_{\tau'}(\upsilon')$. Consequently, we can write $\tilde{B}_{\tau'}^{(i)} = B_{\tau'}^{(i)}$ as $\mathcal{L} \to \infty$, where $i = 1, 2$.

It should be noted explicitly that the discrete propagation delays $\tilde{\tau}'_\ell$ are included in the optimization procedure without imposing any boundary condition. In general, the optimization algorithm therefore results in real-valued quantities, which are not equally spaced. However, to enable the implementation of the tapped-delay-line model on a computer, it is necessary to quantize the obtained optimized values $\tilde{\tau}'_\ell$, so that they have a greatest common divisor, denoted by $\Delta\tau'$, which defines the sampling interval of the channel simulator. Of course, the quantization of $\tilde{\tau}'_\ell$ has an impact on the quality of the approximation $\tilde{r}_{\tau'}(\upsilon') \approx r_{\tau'}(\upsilon')$ as well as on the length of the period $\Upsilon' = 1/\Delta\tau'$. However, the influence of the quantization effect can be ignored if $\Delta\tau'$ is sufficiently small. This can be achieved, e.g., by choosing $\Delta\tau' \leq \tau'_{\max}/(50\mathcal{L})$. In such cases, the period of $\tilde{r}_{\tau'}(\upsilon')$ equals $\Upsilon' = 1/\Delta\tau' \geq 50\mathcal{L}/\tau'_{\max}$. Notice that this is more than fifty times the period obtained by using the MED and the MSEM by keeping the same number of discrete paths $\mathcal{L}$. In general, the LPNM results in a much higher period of the frequency correlation function than the MED and the MSEM.

To demonstrate the power of the LPNM, we apply this procedure to the TU power delay profile [see (7.116)] by choosing $p = 2$, $\mathcal{L} = 12$, and $\Delta\tau' = \tau'_{\max}/(50\mathcal{L})$. The obtained results are presented in Figures 7.23(a)–(c). Especially by comparing the frequency correlation function $\tilde{r}_{\tau'}(\upsilon')$ shown in Figure 7.23(b) with those presented in the Figures 7.19(b)–7.22(b), we can realize that the LPNM has a much better performance than the four methods described in the previous subsections. Figure 7.23(c) reveals that the approximation $\tilde{B}_{\tau'}^{(i)} \approx B_{\tau'}^{(i)}$ $(i = 1, 2)$ is nearly perfect if $\mathcal{L} \geq 8$.

7.5.3 Comparison of the Parameter Computation Methods

In this subsection, we compare the performance of the parameter computation methods (MED, MSEM, MEA, MCM, and LPNM) described in the previous section. As performance measure,

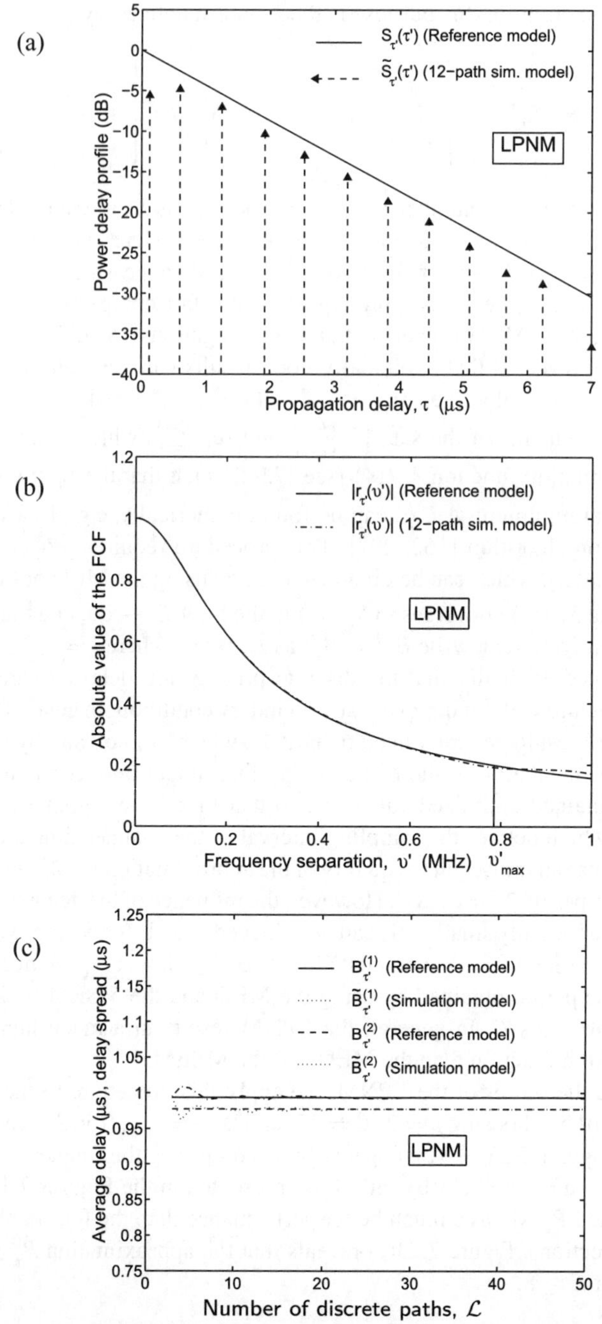

Figure 7.23 (a) Power delay profile $S_{\tau'}(\tau')$ of the TU channel, (b) absolute value of the frequency correlation function $|r_{\tau'}(\upsilon')|$, and (c) average delay $B_{\tau'}^{(1)}$ as well as the delay spread $B_{\tau'}^{(2)}$ in comparison with the corresponding quantities of the simulation model when using the LPNM with $p = 2$.

we study the mean-square error of the frequency correlation function $E_{r_{\tau'}}$ [see (7.124)] in terms of the number of discrete paths $\mathcal{L} \in \{3, 4, \ldots, 50\}$. In addition, we evaluate the average delay as well as the delay spread. Figure 7.24(a) summarizes the results found for the mean-square error of the frequency correlation function evaluated for the five parameter computation methods by choosing $\upsilon'_{\max} = (\mathcal{L} - 1)/(2\tau'_{\max})$. Moreover, this figure includes the results obtained when using the original and the alternative 12-path COST 207 TU models with the model parameters (path gains and discrete propagation delays) as specified in Table 7.A.2. Notice that $\upsilon'_{\max}$ is a function of $\mathcal{L}$. Figures 7.24(b) and (c) illustrate the average delay and the delay spread, respectively. In these figures, the results for the 6-path and the 12-path COST 207 TU models also appear with two points for each $\mathcal{L}$ value, namely one for the original and one for the alternative COST 207 TU model [see Tables 7.3 and 7.A.2].

Summarizing the results, we can say that the power delay profile of a tapped-delay-line model of order $\mathcal{L}$ is completely defined by two different kinds of parameters: the discrete propagation delays and the path gains. We have discussed five fundamental methods (MED, MSEM, MEA, MCM, and LPNM), which enable us to compute these parameters such that the power delay characteristics of the designed channel model are close to those of a given (specified or measured) reference model. The MED, MSEM, MEA, and LPNM are entirely deterministic methods, whereas the MCM is a statistic procedure. Owing to closed-form solutions, the performance of the presented methods could be studied analytically with respect to the frequency correlation function, the average delay, and the delay spread. It turned out that all deterministic procedures outperform the Monte Carlo method with respect to the three considered performance measures. The method with the best performance is definitely the LPNM. The reason why this method works so well — even when the number of paths $\mathcal{L}$ is small — lies in the fact that the LPNM exploits the full flexibility of the tapped-delay-line model by optimizing $2\mathcal{L}$ parameters. This is in contrast to all other methods, where $\mathcal{L}$ parameters are pre-defined and the remaining $\mathcal{L}$ parameters have to be determined according to a method-specific criterion. Concerning the LPNM, it should also be mentioned that the computational complexity of this top-performance procedure is higher than that of the other four methods. However, this disadvantage is of minor importance when using today's high-speed computers.

By way of example, we will apply in the following the outperforming LPNM method to measurement data of power delay profiles collected in different propagation environments. The objective is twofold. First, we will demonstrate the usefulness of the LPNM for the modelling of real-world channels. Second, we intend to show that the realization complexity of tapped-delay-line channel simulators can be reduced considerably by using this method.

7.5.4 Applications to Measured Power Delay Profiles

This subsection deals with the application of the LPNM to two different measured power delay profiles of equivalent satellite channels. The measurement campaign was initiated by the ESA[7] and carried out by the DLR[8] in different propagation environments and under various elevation angles at a carrier frequency of 1.82 GHz [203]. Here, we use two measurements

[7] ESA: European Space Agency

[8] DLR: German Aerospace Center (German: Deutsches Zentrum für Luft- und Raumfahrt e.V.)

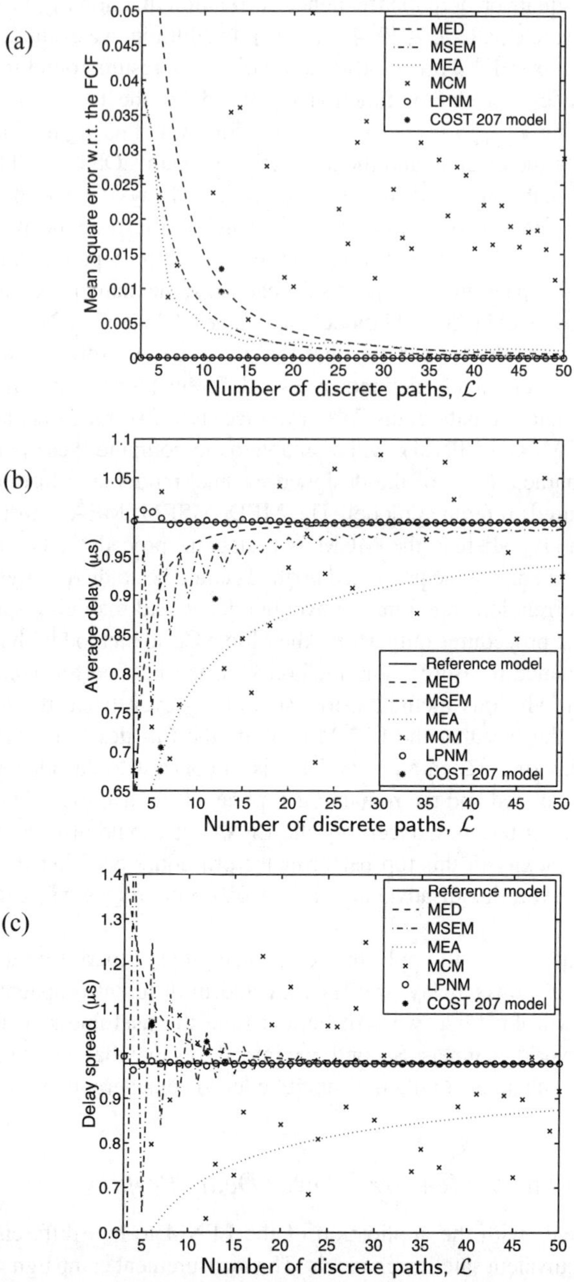

Figure 7.24 (a) Mean-square error of the frequency correlation function, (b) average delay $B_{\tau'}^{(1)}$, and (c) delay spread $B_{\tau'}^{(2)}$ of the continuous TU channel model in comparison with the corresponding quantities of the simulation model when using the MED, MSEM, MEA, MCM, LPNM, as well as the discrete 6- and 12-path TU channel models specified by COST 207.

carried out in an open rural area and in an urban environment with elevation angles in the range from 10° to 20°. The average delay $B^{(1)}_{\tau'}$ and the delay spread $B^{(2)}_{\tau'}$ are given by $B^{(1)}_{\tau'} = 54$ ns and $B^{(2)}_{\tau'} = 17$ ns for the open rural area, and by $B^{(1)}_{\tau'} = 90$ ns and $B^{(2)}_{\tau'} = 106$ ns for the urban environment.

Figure 7.25(a) presents the power delay profile of the measured channel in comparison with the corresponding discrete profile of the simulation model for the open rural area. Figure 7.25(b) shows the absolute values of the respective frequency correlation functions corresponding to the measured channel and the simulation model. For the simulation model, the LPNM has been used with $\mathcal{L} = 20$ and $\upsilon'_{\max} = (\mathcal{L} - 1)/(2\tau'_{\max})$, where $\tau'_{\max}$ is given by $\tau'_{\max} = 412.5$ ns. In Figures 7.26(a) and (b), the corresponding functions for the urban environment are presented. We have again applied the LPNM with $\mathcal{L} = 20$ and $\upsilon'_{\max} = (\mathcal{L} - 1)/(2\tau'_{\max})$, whereas $\tau'_{\max}$ is in this case given by $\tau'_{\max} = 612.5$ ns. Note that in Figures 7.25(b) and 7.26(b), the frequency correlation function is represented up to one half of the period of the measured signal which was equal to 80 MHz. The relative errors of the average delay and the delay spread are smaller than $5 \cdot 10^{-4}$ for both measurements. Thus, one can say that the characteristic quantities of the simulation model and the measured channel are almost the same. Figures 7.25(b) and 7.26(b) exhibit the fact that the frequency correlation functions of the simulation model approximate the corresponding functions of the measured channel very closely over the interval $[0, \upsilon'_{\max}]$.

Here, we have chosen a tapped-delay-line model of order $\mathcal{L} = 20$. According to $\upsilon'_{\max} = (\mathcal{L} - 1)/(2\tau'_{\max})$, this results in $\upsilon'_{\max} = 23.03$ MHz [Figure 7.25(b)] and $\upsilon'_{\max} = 15.51$ MHz [Figure 7.26(b)]. For system studies or practical applications, however, $\upsilon'_{\max}$ should be adapted to the (total) bandwidth B of the transmitted signal of the system being simulated, e.g., by defining $\upsilon'_{\max} := B$. In this case, the required number of taps $\mathcal{L}$ is given by

$$\mathcal{L} = \lceil 2B\tau'_{\max} \rceil + 1 \,. \tag{7.137}$$

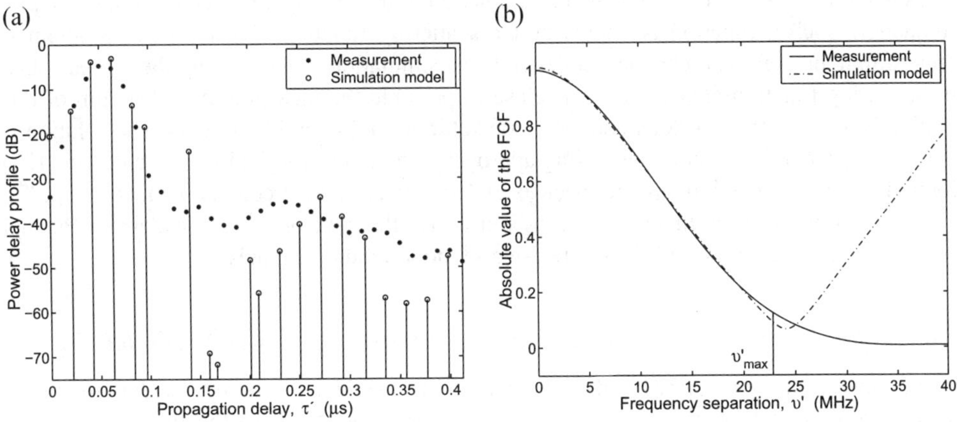

Figure 7.25 (a) Power delay profile $S_{\tau'}(\tau')$ and (b) absolute value of the frequency correlation function $|r_{\tau'}(\upsilon')|$ of a measured satellite channel in an open rural area in comparison with the corresponding system functions of the simulation model when using the LPNM with $\mathcal{L} = 20$.

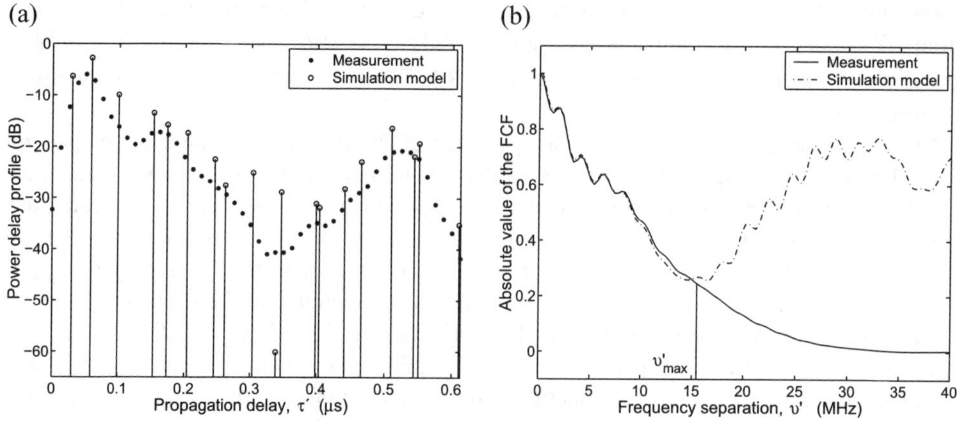

Figure 7.26 (a) Power delay profile $S_{\tau'}(\tau')$ and (b) absolute value of the frequency correlation function $|r_{\tau'}(\upsilon')|$ of a measured satellite channel in an urban area in comparison with the corresponding system functions of the simulation model when using the LPNM with $\mathcal{L} = 20$.

Invoking some of the major cellular standards and assuming that the power delay profile is given by (7.116), then the required number of taps is $\mathcal{L} = 4$ for GSM ($B = 200\,\text{kHz}$), $\mathcal{L} = 19$ for IS-95 ($B = 1.25\,\text{MHz}$), and $\mathcal{L} = 71$ for W-CDMA ($B = 5\,\text{MHz}$).

7.6 Perfect Modelling and Simulation of Measured Wideband Mobile Radio Channels

In recent years, various types of channel sounders have been developed aiming to measure and to investigate the propagation characteristics of wideband mobile radio channels in indoor and outdoor environments. After post-processing the collected data, one often presents the measured output power of the channel as a function of the propagation delay and the Doppler frequency. Such a function is known as the scattering function. An important problem is then to find an analytical channel model and/or a simulation model having the property that its scattering function approximates as close as possible the measured one. The aim of this section is to show that an exact and a general solution to this problem exists. The solution is known as the perfect channel modelling approach, which was originally introduced in [121] for wideband channels and has later been extended to space-time-frequency channels in [122]. The method proposed in this section can be used for the development of wideband channel simulators enabling the simulation of real-world mobile radio channels.

7.6.1 The Sum-of-Cisoids Uncorrelated Scattering (SOCUS) Model

In this subsection, we describe briefly a slightly modified version the SOSUS model. In contrast to the SOSUS model discussed in Section 7.4, we describe the Doppler effect in the present case by a sum-of-cisoids (SOC) instead of a sum-of-sinusoids (SOS). Recall that the SOC model is preferred to the SOS model if the Doppler power spectral density has an asymmetrical shape, which is always the case in real-world (measured) channels. The new class of model processes is called the *sum-of-cisoids uncorrelated scattering (SOCUS)* model. It can be interpreted as an extension of the SOC model with respect to frequency selectivity. The SOCUS model provides the basis for the perfect channel modelling approach. The understanding of this approach requires some basic knowledge about the statistical properties of SOCUS models, which will be provided in the next subsection.

7.6.1.1 System Functions of SOCUS Models

Time-variant impulse response: The time-variant impulse response $\tilde{h}(\tau', t)$ of the SOCUS model is given by a sum of $\mathcal{L}$ discrete paths with different propagation delays according to

$$\tilde{h}(\tau', t) = \sum_{\ell=0}^{\mathcal{L}-1} \tilde{a}_\ell \, \tilde{\mu}_\ell(t) \, \delta(\tau' - \tilde{\tau}'_\ell), \tag{7.138}$$

where the real-valued quantities $\tilde{a}_\ell$ are denoting the *path gains*, $\tilde{\mu}_\ell(t)$ are representing the SOC processes, and the quantities $\tilde{\tau}'_\ell$ are the *discrete propagation delays*, which are supposed to be non-negative real numbers for reasons of causality, i.e., $\tilde{\tau}'_\ell \geq 0$, $\forall\, \ell = 0, 1, \ldots, \mathcal{L} - 1$.

The channel variations caused by the Doppler effect are modelled here by SOC processes $\tilde{\mu}_\ell(t)$ having the form

$$\tilde{\mu}_\ell(t) = \sum_{n=1}^{N_\ell} c_{n,\ell} \, e^{j(2\pi f_{n,\ell} t + \theta_{n,\ell})}, \tag{7.139}$$

where $\ell = 0, 1, \ldots, \mathcal{L} - 1$. Thereby, N_ℓ denotes the number of complex sinusoids (cisoids) assigned to the ℓth propagation path, $c_{n,\ell}$ is the gain of the nth component of the ℓth propagation path, and $f_{n,\ell}$ and $\theta_{n,\ell}$ are the corresponding discrete Doppler frequency and the phase, respectively. Figure 7.27 shows the structure of the SOC process $\tilde{\mu}_\ell(t)$ in the continuous-time representation. The overall structure of the simulation model for frequency-selective mobile radio channels is the same as in Figure 7.11, if we replace there the SOS processes by SOC processes. In Subsection 7.6.2, it will be shown how the model parameters appearing in (7.138) and (7.139) can be determined from measured channels. In the following, we assume that these

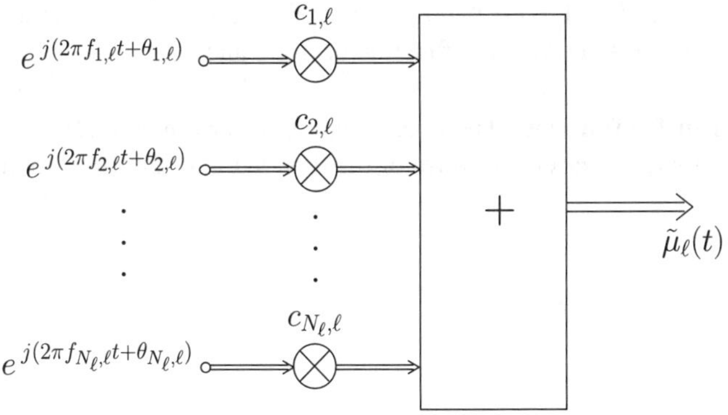

Figure 7.27 Simulation model for deterministic SOC processes $\tilde{\mu}_\ell(t)$.

parameters are known and constant quantities. Under this condition, the time-variant impulse response $\tilde{h}(\tau', t)$ and the SOC processes $\tilde{\mu}_\ell(t)$ are completely deterministic. Therefore, the correlation properties of $\tilde{h}(\tau', t)$ have to be derived by means of time averages instead of statistical averages.

From the assumption that the scattered components with different propagation delays are uncorrelated, it follows that the SOC processes $\tilde{\mu}_\ell(t)$ must be uncorrelated for different discrete propagation delays, i.e., the deterministic processes $\tilde{\mu}_\ell(t)$ and $\tilde{\mu}_\lambda(t)$ have to be designed such that they are uncorrelated for $\ell \neq \lambda$, where $\ell, \lambda = 0, 1, \ldots, \mathcal{L} - 1$. This US condition can readily be satisfied if the discrete Doppler frequencies $f_{n,\ell}$ are chosen such that the sets $\{f_{n,\ell}\}$ and $\{f_{m,\lambda}\}$ are mutually disjoint for different propagation delays. Thus, the US condition can be expressed by the following three equivalent statements:

$$US \Longleftrightarrow \tilde{\mu}_\ell(t) \text{ and } \tilde{\mu}_\lambda(t) \text{ are uncorrelated for } \ell \neq \lambda, \tag{7.140a}$$

$$US \Longleftrightarrow \{f_{n,\ell}\} \cap \{f_{m,\lambda}\} = \emptyset \text{ for } \ell \neq \lambda, \tag{7.140b}$$

$$US \Longleftrightarrow f_{n,\ell} \neq f_{m,\lambda} \text{ for } \ell \neq \lambda, \tag{7.140c}$$

where $n = 1, 2, \ldots, N_\ell$, $m = 1, 2, \ldots, N_\lambda$, and $\ell, \lambda = 0, 1, \ldots, \mathcal{L} - 1$.

Henceforth, we assume that the SOCUS model satisfies the US condition. In this case, the correlation properties of the SOC processes $\tilde{\mu}_\ell(t)$ in (7.139) can be expressed in closed form by using

$$\tilde{r}_{\mu_\ell \mu_\lambda}(\tau) := \lim_{T\to\infty} \frac{1}{2T} \int_{-T}^{T} \tilde{\mu}_\ell^*(t)\, \tilde{\mu}_\lambda(t+\tau)\, dt \tag{7.141}$$

as follows

$$\tilde{r}_{\mu_\ell \mu_\ell}(\tau) = \sum_{n=1}^{N_\ell} c_{n,\ell}^2\, e^{j2\pi f_{n,\ell}\tau}, \text{ if } \ell = \lambda, \tag{7.142a}$$

$$\tilde{r}_{\mu_\ell \mu_\lambda}(\tau) = 0, \text{ if } \ell \neq \lambda, \tag{7.142b}$$

where $\ell, \lambda = 0, 1, \ldots, \mathcal{L} - 1$. Note that $\tilde{r}_{\mu_\ell \mu_\ell}(\tau)$ is the autocorrelation function of $\tilde{\mu}_\ell(t)$, and $\tilde{r}_{\mu_\ell \mu_\lambda}(\tau)$ denotes the cross-correlation function of $\tilde{\mu}_\ell(t)$ and $\tilde{\mu}_\lambda(t)$.

Time-variant transfer function: The time-variant transfer function $\tilde{H}(f', t)$ of the SOCUS model is defined by the Fourier transform of the time-variant impulse response $\tilde{h}(\tau', t)$ with respect to τ', i.e.,

$$\begin{aligned}\tilde{H}(f', t) &:= \int_{-\infty}^{\infty} \tilde{h}(\tau', t)\, e^{-j2\pi f'\tau'} d\tau' \\ &= \sum_{\ell=0}^{\mathcal{L}-1} \tilde{a}_\ell\, \tilde{\mu}_\ell(t)\, e^{-j2\pi f'\tilde{\tau}_\ell'} .\end{aligned} \tag{7.143}$$

Time-frequency correlation function: The autocorrelation function $\tilde{r}_{HH}(f', f' + \upsilon'; t, t + \tau)$ of the time-variant transfer function $\tilde{H}(f', t)$, defined by

$$\tilde{r}_{HH}(f', f' + \upsilon'; t, t + \tau) := \lim_{T\to\infty} \frac{1}{2T} \int_{-T}^{T} \tilde{H}^*(f', t)\, \tilde{H}(f' + \upsilon', t + \tau)\, dt, \tag{7.144}$$

is called the time-frequency correlation function. Putting (7.143) in (7.144) and taking the US condition into account results in

$$\tilde{r}_{HH}(\upsilon', \tau) = \sum_{l=0}^{\mathcal{L}-1} \tilde{a}_\ell^2\, \tilde{r}_{\mu_\ell\mu_\ell}(\tau)\, e^{-j2\pi\upsilon'\tilde{\tau}'_\ell}. \tag{7.145}$$

The result above shows that $\tilde{r}_{HH}(\upsilon', \tau)$ is only a function of the frequency separation variable υ' and the time separation variable τ. Consequently, the time-variant transfer function $\tilde{H}(f', t)$ of SOCUS processes has — with respect to both variables f' and t — similar properties like a wide-sense stationary process.

7.6.1.2 The Scattering Function of SOCUS Models

The scattering function $\tilde{S}(\tau', f)$ of SOCUS models is related to the time-frequency correlation function $\tilde{r}_{HH}(\upsilon', \tau)$ via the two-dimensional Fourier transform

$$\tilde{S}(\tau', f) = \int_{-\infty}^{\infty}\int_{-\infty}^{\infty} \tilde{r}_{HH}(\upsilon', \tau)\, e^{j2\pi(\upsilon'\tau' - \tau f)}\, d\upsilon\, d\tau. \tag{7.146}$$

Substituting (7.145) in (7.146) gives the scattering function $\tilde{S}(\tau', f)$ in closed form as

$$\begin{aligned} \tilde{S}(\tau', f) &= \sum_{\ell=0}^{\mathcal{L}-1} \tilde{a}_\ell^2\, \tilde{S}_{\mu_\ell\mu_\ell}(f)\, \delta(\tau' - \tilde{\tau}'_\ell) \\ &= \sum_{\ell=0}^{\mathcal{L}-1}\sum_{n=1}^{N_\ell} (\tilde{a}_\ell c_{n,\ell})^2\, \delta(f - f_{n,\ell})\, \delta(\tau' - \tilde{\tau}'_\ell), \end{aligned} \tag{7.147}$$

where we have used the function

$$\tilde{S}_{\mu_\ell\mu_\ell}(f) = \sum_{n=1}^{N_\ell} c_{n,\ell}^2\, \delta(f - f_{n,\ell}), \tag{7.148}$$

which can be identified as the Doppler power spectral density of the ℓth propagation path. Note that $\tilde{S}_{\mu_\ell\mu_\ell}(f)$ is the Fourier transform of the autocorrelation function $\tilde{r}_{\mu_\ell\mu_\ell}(\tau)$ [see (7.142a)]. Without loss of generality, it is often assumed that the scattering function $\tilde{S}(\tau', f)$ is normalized such that the area under $\tilde{S}(\tau', f)$ is equal to unity. Such a normalized scattering function is obtained by imposing on the path gains $\tilde{a}_\ell$ and the gains $c_{n,\ell}$ the following conditions

$$\sum_{\ell=0}^{\mathcal{L}-1} \tilde{a}_\ell^2 = 1 \qquad \text{and} \qquad \sum_{n=1}^{N_\ell} c_{n,\ell}^2 = 1, \tag{7.149a,b}$$

where the latter holds for all $\ell = 0, 1, \ldots, \mathcal{L} - 1$.

The closed-form expression in (7.147) reveals that the scattering function of SOCUS models is completely determined by the model parameters of the underlying SOC processes and the parameters determining the delay power spectral density. This property provides the basis for the statement made in Subsection 7.6.2 that all discrete-type scattering functions of given (measured or specified) mobile radio channels can be perfectly modelled by SOCUS processes.

7.6.1.3 Further System Functions and Characteristic Quantities of SOCUS Models

In this section, we describe further important characteristic functions and quantities of SOCUS models such as the delay power spectral density, frequency correlation function, delay spread, etc. These functions and quantities are defined analogously to the corresponding functions and quantities of the stochastic WSSUS model.

Delay power spectral density: The delay power spectral density $\tilde{S}_{\tau'}(\tau')$ is related to the scattering function $\tilde{S}(\tau', f)$ via

$$\tilde{S}_{\tau'}(\tau') = \int_{-\infty}^{\infty} \tilde{S}(\tau', f)\, df\,. \tag{7.150}$$

Using the expression for the scattering function $\tilde{S}(\tau', f)$ [see (7.147)] allows us to present the delay power spectral density $\tilde{S}_{\tau'}(\tau')$ of the SOCUS model in closed form according to

$$\tilde{S}_{\tau'}(\tau') = \sum_{\ell=0}^{\mathcal{L}-1} \tilde{a}_\ell^2\, \delta(\tau' - \tilde{\tau}_\ell')\,. \tag{7.151}$$

Thus, the delay power spectral density $\tilde{S}_{\tau'}(\tau')$ consists of $\mathcal{L}$ Dirac delta functions located at $\tau' = \tilde{\tau}_\ell'$ and weighted by the square of the path gains $\tilde{a}_\ell^2$. Hence, the behaviour of $\tilde{S}_{\tau'}(\tau')$ is fully determined by the model parameters $\mathcal{L}$, $\tilde{a}_\ell$, and $\tilde{\tau}_\ell'$.

Average delay: The average delay $\tilde{D}_{\tau'}^{(1)}$ of the SOCUS model is defined by the first moment of $\tilde{S}_{\tau'}(\tau')$, i.e.,

$$\tilde{D}_{\tau'}^{(1)} := \frac{\int_{-\infty}^{\infty} \tau'\, \tilde{S}_{\tau'}(\tau')\, d\tau'}{\int_{-\infty}^{\infty} \tilde{S}_{\tau'}(\tau')\, d\tau'} = \frac{\sum_{\ell=0}^{\mathcal{L}-1} \tilde{\tau}_\ell'\, \tilde{a}_\ell^2}{\sum_{\ell=0}^{\mathcal{L}-1} \tilde{a}_\ell^2}\,. \tag{7.152}$$

Delay spread: The delay spread $\tilde{D}_{\tau'}^{(2)}$ of the SOCUS model is defined by the square root of the second central moment of $\tilde{S}_{\tau'}(\tau')$, i.e.,

$$\tilde{D}_{\tau'}^{(2)} := \sqrt{\frac{\int_{-\infty}^{\infty} \left(\tau' - \tilde{D}_{\tau'}^{(1)}\right)^2 \tilde{S}_{\tau'}(\tau')\, d\tau'}{\int_{-\infty}^{\infty} \tilde{S}_{\tau'}(\tau')\, d\tau'}}$$

$$= \sqrt{\frac{\sum_{\ell=0}^{\mathcal{L}-1} \left(\tilde{\tau}'_\ell \tilde{a}_\ell\right)^2}{\sum_{\ell=0}^{\mathcal{L}-1} \tilde{a}_\ell^2} - \left(\tilde{D}_{\tau'}^{(1)}\right)^2} . \tag{7.153}$$

Doppler power spectral density: The Doppler power spectral density $\tilde{S}_{\mu\mu}(f)$ of the SOCUS model can be determined from the scattering function $\tilde{S}(\tau', f)$ as follows

$$\tilde{S}_{\mu\mu}(f) = \int_{-\infty}^{\infty} \tilde{S}(\tau', f)\, d\tau'$$

$$= \sum_{\ell=0}^{\mathcal{L}-1} \tilde{a}_\ell^2 \tilde{S}_{\mu_\ell\mu_\ell}(f), \tag{7.154}$$

where $\tilde{S}_{\mu_\ell\mu_\ell}(f)$ is given by (7.148) for all $\ell = 0, 1, \ldots, \mathcal{L} - 1$.

Average Doppler shift: The average Doppler shift $\tilde{D}_{\mu\mu}^{(1)}$ of the SOCUS model is defined by the first moment of $\tilde{S}_{\mu\mu}(f)$, i.e.,

$$\tilde{D}_{\mu\mu}^{(1)} := \frac{\int_{-\infty}^{\infty} f\, \tilde{S}_{\mu\mu}(f)\, df}{\int_{-\infty}^{\infty} \tilde{S}_{\mu\mu}(f)\, df}$$

$$= \frac{\sum_{\ell=0}^{\mathcal{L}-1} \left[\tilde{a}_\ell^2 \sum_{n=1}^{N_\ell} f_{n,l}\, c_{n,l}^2 \right]}{\sum_{\ell=0}^{\mathcal{L}-1} \left[\tilde{a}_\ell^2 \sum_{n=1}^{N_\ell} c_{n,l}^2 \right]} . \tag{7.155}$$

Doppler spread: The Doppler spread $\tilde{D}^{(2)}_{\mu\mu}$ of the SOCUS model is defined by the square root of the second central moment of $\tilde{S}_{\mu\mu}(f)$, i.e.,

$$\tilde{D}^{(2)}_{\mu\mu} := \sqrt{\frac{\int\limits_{-\infty}^{\infty} \left(f - \tilde{D}^{(1)}_{\mu\mu}\right)^2 \tilde{S}_{\mu\mu}(f)\, df}{\int\limits_{-\infty}^{\infty} \tilde{S}_{\mu\mu}(f)\, df}}$$

$$= \sqrt{\frac{\sum\limits_{\ell=0}^{\mathcal{L}-1} \left[\tilde{a}_\ell^2 \sum\limits_{n=1}^{N_\ell} (f_{n,l}\, c_{n,l})^2 \right]}{\sum\limits_{\ell=0}^{\mathcal{L}-1} \left[\tilde{a}_\ell^2 \sum\limits_{n=1}^{N_\ell} c_{n,l}^2 \right]} - \left(\tilde{D}^{(1)}_{\mu\mu}\right)^2} \, . \tag{7.156}$$

Frequency correlation function: The frequency correlation function $\tilde{r}_{\tau'}(\upsilon')$ of the SOCUS model can be obtained from the time-frequency correlation function $\tilde{r}_{HH}(\upsilon', \tau)$ [see (7.145)] by setting $\tau = 0$, i.e.,

$$\tilde{r}_{\tau'}(\upsilon') := \tilde{r}_{HH}(\upsilon', 0) = \sum_{\ell=0}^{\mathcal{L}-1} \tilde{a}_\ell^2\, e^{-j2\pi\upsilon'\tilde{\tau}'_\ell} \, . \tag{7.157}$$

Note that the frequency correlation function $\tilde{r}_{\tau'}(\upsilon')$ is the Fourier transform of the delay power spectral density $\tilde{S}_{\tau'}(\tau')$. It should also be observed that $\tilde{r}_{\tau'}(\upsilon')$ is a continuous function, in contrast to $\tilde{S}_{\tau'}(\tau')$ which is a discrete line spectrum.

Coherence bandwidth: The smallest value of the frequency separation variable $\upsilon' = \tilde{B}_C$ that satisfies the condition

$$|\tilde{r}_{\tau'}(\tilde{B}_C)| = \frac{1}{2}|\tilde{r}_{\tau'}(0)| \tag{7.158}$$

is called the *coherence bandwidth* of the SOCUS model. Using the expression above for $\tilde{r}_{\tau'}(\upsilon')$, we obtain the following transcendent equation

$$\left| \sum_{\ell=0}^{\mathcal{L}-1} \tilde{a}_\ell^2\, e^{-j2\pi\tilde{B}_C\tilde{\tau}'_\ell} \right| - \frac{1}{2} = 0 \tag{7.159}$$

which must be solved for $\tilde{B}_C$, in general, numerically.

Time correlation function: The time correlation function $\tilde{r}_{\mu\mu}(\tau)$ of the SOCUS model may be determined from the time-frequency correlation function $\tilde{r}_{HH}(\upsilon', \tau)$ [see (7.145)] by setting $\upsilon' = 0$, i.e.,

$$\tilde{r}_{\mu\mu}(\tau) := \tilde{r}_{HH}(0, \tau) = \sum_{\ell=0}^{\mathcal{L}-1} \tilde{a}_\ell^2\, \tilde{r}_{\mu_\ell\mu_\ell}(\tau), \tag{7.160}$$

where $\tilde{r}_{\mu_\ell\mu_\ell}(\tau)$ is the autocorrelation function of $\tilde{\mu}_\ell(t)$, as given by (7.142a).

Coherence time: The smallest value of the time separation variable $\tau = \tilde{T}_C$ that fulfills the condition

$$|\tilde{r}_{\mu\mu}(\tilde{T}_C)| = \frac{1}{2}|\tilde{r}_{\mu\mu}(0)| \tag{7.161}$$

is said to be the *coherence time* of the SOCUS model. Using the expression derived above for the autocorrelation function $\tilde{r}_{\mu\mu}(\tau)$ results in the following transcendent equation

$$\left|\sum_{\ell=0}^{\mathcal{L}-1}\sum_{n=1}^{N_\ell}(\tilde{a}_\ell c_{n,\ell})^2 e^{j2\pi f_{n,\ell}\tilde{T}_C}\right| - \frac{1}{2} = 0 \tag{7.162}$$

which has in general to be solved for $\tilde{T}_C$ by employing numerical methods.

7.6.2 *The Principle of Perfect Channel Modelling*

This subsection describes the principle of perfect channel modelling. Our aim is to fit the scattering function $\tilde{S}(\tau', f)$ of the SOCUS model to a given (measured or specified) scattering function $S^\star(\tau', f)$ having an arbitrary shape. The only condition that we impose on the scattering function $S^\star(\tau', f)$ is that this function must be discrete in the variables τ' and f. The basic idea of the procedure is to determine the model parameters of the SOCUS model directly from the given discrete scattering function $S^\star(\tau', f)$ without any approximation or reducing the number of parameters. A channel model, the scattering function $\tilde{S}(\tau', f)$ of which is identical to a measured scattering function $S^\star(\tau', f)$, i.e.,

$$\tilde{S}(\tau', f) = S^\star(\tau', f), \tag{7.163}$$

is called a *perfect channel model*. Its software or hardware realization is said to be a *perfect channel simulator*. A perfect channel simulator enables the simulation of measured wideband mobile radio channels without any model or approximation error.

Next, we describe the procedure for the determination of the model parameters $\{c_{n,\ell}\}$, $\{f_{n,\ell}\}$, $\{\tilde{a}_\ell\}$, $\{\tilde{\tau}'_\ell\}$, $\{N_\ell\}$, and $\mathcal{L}$ of the SOCUS model from a given measured scattering function $S^\star(\tau', f)$. Since the measured scattering function $S^\star(\tau', f)$ is assumed to be discrete, $S^\star(\tau', f)$ can equivalently be represented in form of a so-called *scattering matrix*

$$\mathbf{S}^\star = \begin{pmatrix} s_{1,0} & s_{1,1} & \cdots & s_{1,L-1} \\ s_{2,0} & s_{2,1} & \cdots & s_{2,L-1} \\ \vdots & & \ddots & \vdots \\ s_{N,0} & & \cdots\cdots & s_{N,L-1} \end{pmatrix} \begin{matrix} -f_{\max} \\ \downarrow f \\ \\ f_{\max} \end{matrix} \tag{7.164}$$
$$\qquad 0 \qquad \overrightarrow{\tau'} \qquad \tau'_{\max}$$

of size $N \times L$, where the number of rows N and the number of columns L are directly related to the maximum Doppler frequency $f_{\max}$ and the largest measured propagation delay $\tau'_{\max}$, respectively. The number of discrete propagation paths $\mathcal{L}$ of the SOCUS model is equal to the number of columns L of $\mathbf{S}^\star$, i.e., $\mathcal{L} = L$. Furthermore, the number N_ℓ of complex sinusoids of the sum-of-cisoids processes $\tilde{\mu}_\ell(t)$ [see (7.139)] is determined by the number of rows of

$\boldsymbol{S}^{\star}$, i.e., we choose $N_\ell = N$ for all $\ell = 0, 1, \ldots, \mathcal{L}-1$. The remaining model parameters $\{\tilde{a}_\ell\}$, $\{\tilde{\tau}'_\ell\}$, $\{c_{n,\ell}\}$, and $\{f_{n,\ell}\}$ of the SOCUS model can be calculated readily from the measured scattering matrix (function). By equating $\tilde{S}(\tau', f)$ [see (7.147)] with $S^*(\tau', f)$, as imposed by (7.163), we obtain directly

$$\tilde{a}_\ell = \sqrt{\sum_{n=1}^{N} s_{n,\ell}}, \tag{7.165a}$$

$$\tilde{\tau}'_\ell = \ell \cdot \Delta\tau', \tag{7.165b}$$

$$c_{n,\ell} = \frac{\sqrt{s_{n,\ell}}}{\tilde{a}_\ell}, \tag{7.165c}$$

$$f_{n,\ell} = -f_{\max} + (n-1) \cdot \Delta f, \tag{7.165d}$$

where $n = 1, 2, \ldots, N_\ell$ and $\ell = 0, 1, \ldots, \mathcal{L}-1$. Equation (7.165a) follows from the fact that the mean path power $\tilde{a}_\ell^2$ of the ℓth propagation path of the SOCUS model must be equal to the sum of the entries of the ℓth column of the measured scattering matrix $\boldsymbol{S}^{\star}$. Furthermore, due to (7.163), we can identify directly the weighting factor $(\tilde{a}_\ell c_{n,\ell})^2$ appearing in the scattering function $\tilde{S}(\tau', f)$ of the SOCUS model as the entry $s_{n,\ell}$ of the scattering matrix $\boldsymbol{S}^{\star}$, i.e., $(\tilde{a}_\ell c_{n,\ell})^2 = s_{n,\ell}$, which explains the expression in (7.165c). The symbol $\Delta\tau'$ in (7.165b) denotes the resolution of the channel sounder in τ'-direction, and Δf in (7.165d) is the resolution in f-direction. These parameters are related to the quantities L, N, $f_{\max}$, and $\tau'_{\max}$ as follows:

$$\Delta\tau' = \frac{\tau'_{\max}}{L-1}, \qquad \Delta f = \frac{2f_{\max}}{N-1}. \tag{7.166a,b}$$

The only model parameters which cannot be derived from (7.163) are the phases $\theta_{n,\ell}$, since $\tilde{S}(\tau', f)$ is independent of them [see (7.147)]. Therefore, we assume that the phases $\theta_{n,\ell}$ are realizations of i.i.d. random variables, each of which is uniformly distributed over the interval $[0, 2\pi)$. With the model parameters determined in that way, all correlation functions and characteristic quantities describing the SOCUS model can be evaluated easily by using the equations given in Section 7.6.1. Moreover, after putting the obtained model parameters in (7.138) and (7.139), one can directly simulate sample functions of the time-variant impulse response of the measured wideband mobile radio channel. Notice that different sets of phases $\{\theta_{n,\ell}^{(k)}\}$ result in different sample functions of the time-variant impulse response $\tilde{h}^{(k)}(\tau', t)$ $(k = 1, 2, \ldots)$. The ensemble of all sample functions constitutes the stochastic time-variant impulse response $\hat{h}(\tau', t) = \{\tilde{h}^{(1)}(\tau', t), \tilde{h}^{(2)}(\tau', t), \ldots\}$. It should be obvious that all sample functions $\tilde{h}^{(k)}(\tau', t)$ have the same correlation properties for $k = 1, 2, \ldots,$ as the SOCUS model is correlation ergodic if only the phases $\theta_{n,\ell}$ are random variables and all other model parameters are constant.

7.6.3 Application to a Measured Wideband Indoor Channel

In the sequel, we apply the perfect channel modelling approach to a real-world measured scattering function which has been obtained from an indoor measurement campaign [217,

218] carried out by the Technical University of Ilmenau, Germany. The centre frequency and the bandwidth of the channel sounder was 5.2 GHz and 120 MHz, respectively. To reduce the effect of measurement noise, we neglect the values of the measured scattering function which were smaller than -35 dB with respect to the line-of-sight component. The delay resolution $\Delta\tau'$ and the Doppler frequency resolution Δf were $\Delta\tau' = 6.25\,\text{ns}$ and $\Delta f = 0.945\,\text{Hz}$, respectively. Due to the limited resolution in the Doppler frequency domain, one cannot rule out the possibility that some of the measured Doppler frequencies are identical for different propagation delays. In such situations, the US condition in (7.140a) is violated. To obtain a US model, it is appropriate to substitute $f_{n,\ell}^{us}$ for the original Doppler frequencies $f_{n,\ell}$, where $f_{n,\ell}^{us}$ is defined by the mapping

$$f_{n,\ell} \mapsto f_{n,\ell}^{us} = f_{n,\ell} + \left(u_{n,\ell} - \frac{1}{2}\right)\Delta f\,. \tag{7.167}$$

In the equation above, the symbol $u_{n,\ell}$ designates i.i.d. random variables, each has a uniform distribution over the interval [0, 1). Notice that a large value of the Doppler frequency resolution Δf results in a small period T_ℓ of $\tilde{\mu}_\ell(t)$, because $T_\ell = 1/\text{gcd}\{f_{n,\ell}\} = 1/\Delta f$. Also this problem can be solved by randomizing the discrete Doppler frequencies $f_{n,\ell}$ according to (7.167).

As an example, we consider Figure 7.28, which illustrates a measured scattering function $S^\star(\tau', f)$ of an indoor mobile radio channel. The corresponding scattering matrix $\boldsymbol{S}^\star$ has a size of $N \times L = 128 \times 129$. The number of entries $s_{n,\ell}$ which are different from zero equals $261 \ll 16\,512$ ($= 128 \times 129$). This means that the total number of cisoids $\sum_{\ell=0}^{\mathcal{L}-1} N_\ell$ amounts to 261. Determining from $\boldsymbol{S}^\star$ the model parameters according to the perfect channel modelling approach and substituting the obtained parameters in (7.147) results in the scattering function $\tilde{S}(\tau', f)$ of the SOCUS model shown in Figure 7.29. Using the extracted model parameters, one can easily compute all system and correlation functions as well as the characteristic quantities of the SOCUS model by using the closed-form expressions derived in Subsection 7.6.1. By

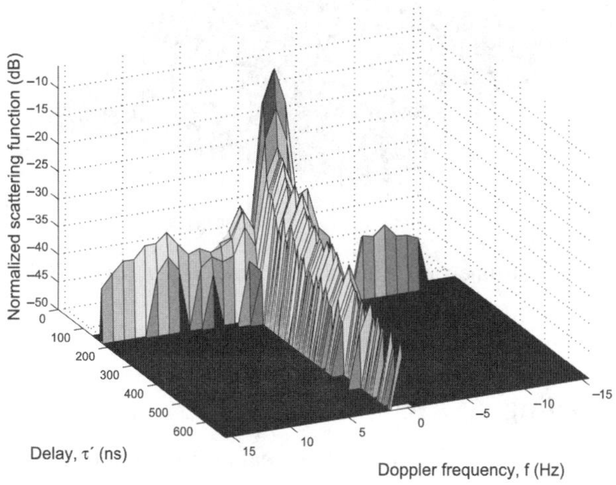

Figure 7.28 Measured scattering function $S^\star(\tau', f)$ of an indoor mobile radio channel.

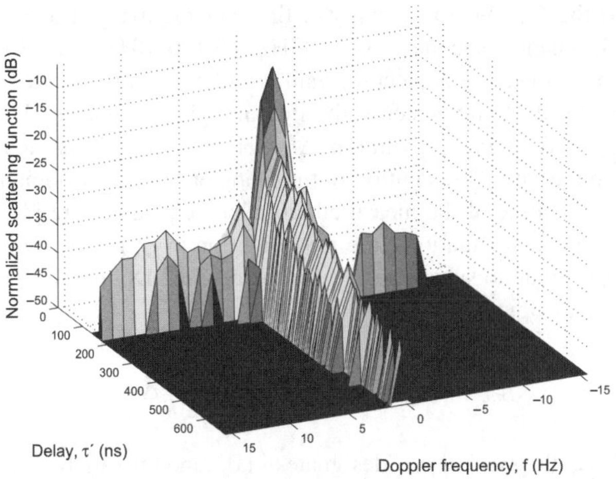

Figure 7.29 Scattering function $\tilde{S}(\tau', f)$ of the channel simulator (SOCUS model).

way of example, Figure 7.30 presents the magnitude of the time-variant impulse response, $|\tilde{h}(\tau', t)|$, which has been obtained by evaluating (7.138) with the parameters obtained from (7.165a)–(7.165d).

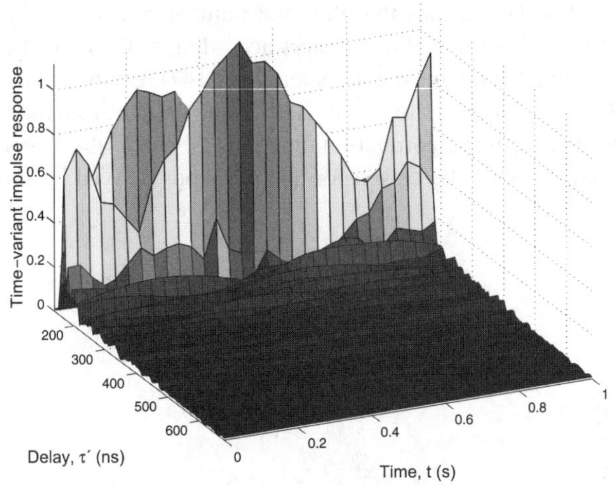

Figure 7.30 Magnitude of the time-variant impulse response $|\tilde{h}(\tau', t)|$ of the SOCUS model (simulation results).

7.7 Further Reading

In the literature, the number of publications that deal merely with frequency-selective mobile radio channels has grown so much that even scientists working at full capacity in this area are running the risk of losing track. Therefore, it is impossible to mention here every author who has made a contribution to this subject, particularly since this survey can only present a small selection of publications. In order to organize this group of themes systematically, it is sensible to make a rough classification of the publications, dividing them into the following categories: theory, simulation, and measurement.

The first category comprises works in which the description and analysis of mobile radio channels are in the main treated theoretically. The most important article in this category is indisputably [18]. In this fundamental work on stochastic time-variant linear systems, Bello introduces the WSSUS model that is employed almost exclusively for the description of frequency-selective mobile radio channels. With this model, the input-output behaviour of mobile radio channels can be described in the equivalent complex baseband in a relatively simple manner, since the channel is assumed to be quasi-stationary during the observation time interval. Empirically grounded statements have shown [208] that the assumption of quasi-stationarity is justified, when the mobile unit moves a distance in the order of less than a few tens of the wavelength of the carrier signal. The articles [64, 208, 219] give a good overview of the most important characteristics of time-variant channels, to which naturally mobile radio channels also belong. Books discussing this subject are, for example, [11, 118, 220, 221] and [222]. A deep insight into the system theoretical description of WSSUS models can be obtained from the study of [211, 223]. The analysis of WSSUS models was also the subject of investigation in [224], where correlation functions as well as scattering functions have been derived on the assumption of non-uniformly distributed angles-of-arrival. A procedure for the modelling of wideband mobile radio channels using orthogonalization techniques has been presented in [225]. The article [226] provides an overview of the state of research in the field of channel modelling, carried out until 1996 by European research projects such as COST 207, RACE CODIT, and RACE ATDMA. Since then, intensive research on spatial-temporal channel models for future mobile radio systems with adaptive antennas has been carried out [227]. Detailed articles giving an overview with many references concerning this subject are [228–231]. The maximum time duration over which the WSSUS assumption is valid is called the stationarity interval. Real-world frequency-selective mobile radio channels do not fulfil the WSSUS condition if the observation time interval is larger than the stationarity interval. The identification of stationarity intervals is thus important for the estimation of the autocorrelation function and other characteristic quantities. In [232], test procedures for the wide-sense stationarity of multipath channels have been developed and applied to measured multiple-input multiple-output (MIMO) mobile radio channels. The objective of this paper was to determine whether a time sequence of a measured channel can be considered to be wide-sense stationary, and if so, what are the typical durations of the stationarity and non-stationarity intervals. An empirical study on the stationarity properties of mobile radio channels in urban environments can be found in [233]. A framework for the statistical characterization of non-WSSUS wireless fading channels has been presented in [234].

The second category includes works stressing the development of simulation models for frequency-selective mobile radio channels. Concerning the method of realization normally used for these kinds of channel simulators, one distinguishes between hardware realizations and software realizations. Hardware realizations can in addition be divided into analog and digital channel simulators. Analog channel simulators (e.g., [235, 236]) model the channel in the high-frequency band or in the intermediate-frequency band, where surface acoustic wave (SAW) filters are applied to realize different multipath propagation delays. Digital channel simulators in general perform all arithmetic operations that become necessary in the complex baseband in real-time using digital signal processors [128] or vector processors (e.g., [237, 238]). In most applications, however, the channel simulation does not take place under real-time conditions, but is performed on a workstation or a personal computer. As proper design methods for the required algorithms, in principle the filter method (e.g., [129, 131]) and the Rice method (e.g., [99–102, 156, 216]) can be used. Incidentally, both of these methods are eligible for the design of channel simulators for mobile communication systems with frequency-hopping capabilities, which has been shown by using the filter method in [239] and by applying the Rice method in [240].

The third and last category covers works reporting on experimental measurement results of mobile radio channels as well as papers describing the equipment enabling these measurements. Certainly, the works by Young [61], Nylund [62], Cox [74, 241], Nielson [242], as well as by Bajwa and Parsons [243] belong to the pioneering works in the field of channel measurements. The subject of measuring the mobile radio channel is treated in a clear and easily understandable manner in the overview article by Andersen et al. [244]. In this publication, mobile radio channels are divided into classes depending on the environment; and typical measured characteristic quantities for different propagation scenarios are provided. In connection with the measurement of system functions of mobile radio channels, the papers [245] and [246] are of particular interest. The article [247] reports on wideband propagation measurements for indoor broadband Wireless Local Area Networks (WLANs) operating in the unlicensed 60 GHz frequency band and providing a data rate higher than 2 Mbit/s up to 150 Mbit/s [248]. For the measurement of the propagation properties of mobile radio channels, special measurement devices, called channel sounders, are required. Review articles introducing the principles of various channel sounding techniques can be found in [249] and [250]. At the Telecommunications Institute of the University of Erlangen–Nuremberg, Germany, the three channel sounders RUSK 400, RUSK 5000, and RUSK X have been developed [251]. Detailed information on the principle of the applied measurement methods of the channel sounders can be found in [130, 251, 252]. The results of measurement campaigns performed with the device type RUSK 5000 are, for example, reported in [253, 254, 255]. The channel sounders RUSK 400 and RUSK 5000 have been produced merely as prototypes, whereas RUSK X and the succeeding models RUSK SX and RUSK WLL have for some time already been sold commercially by MEDAV GmbH. The RUSK ATM is another device that became part of the RUSK channel sounders' family. This device arose from the ATMmobil project, which was supported by the BMFT in cooperation with MEDAV GmbH [217, 256]. With this vector channel sounder, in particular directional resolved measurements of mobile radio channels can be carried out in the 1.8–2.5 GHz (UMTS) and the 5–6 GHz (WLAN/HIPERLAN) frequency band with a measurement bandwidth of up to 240 MHz. Examples of different measurements carried out at 5.2 GHz using the RUSK ATM channel sounder are presented in [257]. This paper also gives an overview of the various applications of channel sounder measurements. Recently, the RUSK channel sounder hardware concept has been extended to enable the measurement of MIMO channels. The successor of RUSK ATM has been introduced as the RUSK MIMO channel sounder. Another commercial system is the Elektrobit Propsound CS^{TM} MIMO channel sounder, which is described in [258]. The channel sounder SIMOCS 2000 has been produced by Siemens in

Munich, Germany. The principle of the measurement setup used in SIMOCS 2000 is described in [259] and [260]. Furthermore, it should be mentioned that Zollinger [261] has developed a channel sounder at the Swiss Federal Institute of Technology Zurich (ETH) in Zurich, Switzerland. The channel sounder ECHO 24 (**E**TH **Ch**annel **So**under operating at **24** GHz) also traces its descent from the ETH. With this channel sounder, complex channel impulse responses can be measured in indoor environments with a temporal resolution of 2 ns [262]. Noncommercial wideband channel sounders have also been developed at Helsinki University of Technology (HUT) [263], Aalborg University (AAU) [264], and Durham University [265], to name but a few.

From the beginning of mobile communications every effort was made to find realistic models for the power delay profile and for the characteristic quantities derivable from it, such as, e.g., the frequency correlation function. The problem of estimating the frequency correlation function from measurements of mobile radio channels is discussed in [266]. Material on measurement results for power delay profiles and delay spreads can be found, e.g., in [243] and [267]. Theoretical investigations on this topic have been done, e.g., in [268] and [224]. Specifications of continuous power delay profiles for GSM channels at 900 MHz are presented in [19]. Discrete power delay profiles and Doppler spectral densities for HIPERLAN/2 channels in typical indoor and outdoor environments have been specified by ETSI BRAN [20]. The HIPERLAN/2 channel models have been employed in [269] for a comparative performance study of HIPERLAN/2 and IEEE 802.11a systems.

Appendix 7.A Specification of the $\mathcal{L}$-Path COST 207 Channel Models

In addition to the 4-path and 6-path channel models presented in Table 7.3, further $\mathcal{L}$-path channel models have been specified by COST 207 [19]. They are quoted in this appendix for completeness.

Table 7.A.1 Specification of the 6-path RA channel model according to COST 207 [19].

Path no. ℓ	Propagation delay τ'_ℓ	Relative path power (linear)	Relative path power (dB)	Doppler PSD $S_{\tau'}(\tau')$	Delay spread $B_{\tau'}^{(2)}$
Rural Area (RA): 6-path channel model (alternative)					
0	$0\ \mu s$	1	0	"Rice"	
1	$0.1\ \mu s$	0.4	−4	"Jakes"	
2	$0.2\ \mu s$	0.16	−8	"Jakes"	
3	$0.3\ \mu s$	0.06	−12	"Jakes"	$0.1\ \mu s$
4	$0.4\ \mu s$	0.03	−16	"Jakes"	
5	$0.5\ \mu s$	0.01	−20	"Jakes"	

Table 7.A.2 Specification of the 12-path and 6-path TU channel models according to COST 207 [19].

Path no. ℓ	Propagation delay τ'_ℓ	Relative path power (linear)	Relative path power (dB)	Doppler PSD $S_{\tau'}(\tau')$	Delay spread $B_{\tau'}^{(2)}$
(i) Typical Urban (TU): 12-path channel model					
0	0.0 μs	0.4	−4	"Jakes"	
1	0.2 μs	0.5	−3	"Jakes"	
2	0.4 μs	1	0	"Jakes"	
3	0.6 μs	0.63	−2	"Gauss I"	
4	0.8 μs	0.5	−3	"Gauss I"	
5	1.2 μs	0.32	−5	"Gauss I"	
6	1.4 μs	0.2	−7	"Gauss I"	1.0 μs
7	1.8 μs	0.32	−5	"Gauss I"	
8	2.4 μs	0.25	−6	"Gauss II"	
9	3.0 μs	0.13	−9	"Gauss II"	
10	3.2 μs	0.08	−11	"Gauss II"	
11	5.0 μs	0.1	−10	"Gauss II"	
(ii) Typical Urban (TU): 12-path channel model (alternative)					
0	0.0 μs	0.4	−4	"Jakes"	
1	0.1 μs	0.5	−3	"Jakes"	
2	0.3 μs	1	0	"Jakes"	
3	0.5 μs	0.55	−2.6	"Jakes"	
4	0.8 μs	0.5	−3	"Gauss I"	
5	1.1 μs	0.32	−5	"Gauss I"	
6	1.3 μs	0.2	−7	"Gauss I"	1.0 μs
7	1.7 μs	0.32	−5	"Gauss I"	
8	2.3 μs	0.22	−6.5	"Gauss II"	
9	3.1 μs	0.14	−8.6	"Gauss II"	
10	3.2 μs	0.08	−11	"Gauss II"	
11	5.0 μs	0.1	−10	"Gauss II"	
(iii) Typical Urban (TU): 6-path channel model (alternative)					
0	0.0 μs	0.5	−3	"Jakes"	
1	0.2 μs	1	0	"Jakes"	
2	0.5 μs	0.63	−2	"Jakes"	
3	1.6 μs	0.25	−6	"Gauss I"	1.0 μs
4	2.3 μs	0.16	−8	"Gauss II"	
5	5.0 μs	0.1	−10	"Gauss II"	

Table 7.A.3 Specification of the 12-path and 6-path BU channel models according to COST 207 [19].

Path no. ℓ	Propagation delay τ'_ℓ	Relative path power (linear)	Relative path power (dB)	Doppler PSD $S_{\tau'}(\tau')$	Delay spread $B_{\tau'}^{(2)}$
(i) Bad Urban (BU): 12-path channel model					
0	0.0 μs	0.2	−7	"Jakes"	2.5 μs
1	0.2 μs	0.5	−3	"Jakes"	
2	0.4 μs	0.79	−1	"Jakes"	
3	0.8 μs	1	0	"Gauss I"	
4	1.6 μs	0.63	−2	"Gauss I"	
5	2.2 μs	0.25	−6	"Gauss II"	
6	3.2 μs	0.2	−7	"Gauss II"	
7	5.0 μs	0.79	−1	"Gauss II"	
8	6.0 μs	0.63	−2	"Gauss II"	
9	7.2 μs	0.2	−7	"Gauss II"	
10	8.2 μs	0.1	−10	"Gauss II"	
11	10.0 μs	0.03	−15	"Gauss II"	
(ii) Bad Urban (BU): 12-path channel model (alternative)					
0	0.0 μs	0.17	−7.7	"Jakes"	2.5 μs
1	0.1 μs	0.46	−3.4	"Jakes"	
2	0.3 μs	0.74	−1.3	"Jakes"	
3	0.7 μs	1	0	"Gauss I"	
4	1.6 μs	0.59	−2.3	"Gauss I"	
5	2.2 μs	0.28	−5.6	"Gauss II"	
6	3.1 μs	0.18	−7.4	"Gauss II"	
7	5.0 μs	0.72	−1.4	"Gauss II"	
8	6.0 μs	0.69	−1.6	"Gauss II"	
9	7.2 μs	0.21	−6.7	"Gauss II"	
10	8.1 μs	0.1	−9.8	"Gauss II"	
11	10.0 μs	0.03	−15.1	"Gauss II"	
(iii) Bad Urban (BU): 6-path channel model (alternative)					
0	0.0 μs	0.56	−2.5	"Jakes"	2.5 μs
1	0.3 μs	1	0	"Jakes"	
2	1.0 μs	0.5	−3	"Gauss I"	
3	1.6 μs	0.32	−5	"Gauss I"	
4	5.0 μs	0.63	−2	"Gauss II"	
5	6.6 μs	0.4	−4	"Gauss II"	

Table 7.A.4 Specification of the 12-path and 6-path HT channel models according to COST 207 [19].

Path no. ℓ	Propagation delay τ'_ℓ	Relative path power (linear)	Relative path power (dB)	Doppler PSD $S_{\tau'}(\tau')$	Delay spread $B_{\tau'}^{(2)}$
(i) Hilly Terrain (HT): 12-path channel model					
0	0.0 μs	0.1	−10	"Jakes"	
1	0.2 μs	0.16	−8	"Jakes"	
2	0.4 μs	0.25	−6	"Jakes"	
3	0.6 μs	0.4	−4	"Gauss I"	
4	0.8 μs	1	0	"Gauss I"	
5	2.0 μs	1	0	"Gauss I"	5.0 μs
6	2.4 μs	0.4	−4	"Gauss II"	
7	15.0 μs	0.16	−8	"Gauss II"	
8	15.2 μs	0.13	−9	"Gauss II"	
9	15.8 μs	0.1	−10	"Gauss II"	
10	17.2 μs	0.06	−12	"Gauss II"	
11	20.0 μs	0.04	−14	"Gauss II"	
(ii) Hilly Terrain (HT): 12-path channel model (alternative)					
0	0.0 μs	0.1	−10	"Jakes"	
1	0.1 μs	0.16	−8	"Jakes"	
2	0.3 μs	0.25	−6	"Jakes"	
3	0.5 μs	0.4	−4	"Jakes"	
4	0.7 μs	1	0	"Gauss I"	
5	1.0 μs	1	0	"Gauss I"	5.0 μs
6	1.3 μs	0.4	−4	"Gauss I"	
7	15.0 μs	0.16	−8	"Gauss II"	
8	15.2 μs	0.13	−9	"Gauss II"	
9	15.7 μs	0.1	−10	"Gauss II"	
10	17.2 μs	0.06	−12	"Gauss II"	
11	20.0 μs	0.04	−14	"Gauss II"	
(iii) Hilly Terrain (HT): 6-path channel model (alternative)					
0	0.0 μs	1	0	"Jakes"	
1	0.1 μs	0.71	−1.5	"Jakes"	
2	0.3 μs	0.35	−4.5	"Jakes"	5.0 μs
3	0.5 μs	0.18	−7.5	"Jakes"	
4	15 μs	0.16	−8.0	"Gauss II"	
5	17.2 μs	0.02	−17.7	"Gauss II"	

Appendix 7.B Specification of the $\mathcal{L}$-Path HIPERLAN/2 Channel Models

This appendix presents the set of five $\mathcal{L}$-path HIPERLAN/2 channel models, which have been specified by ETSI BRAN [20]. The first channel model, called model A, has been developed for typical office environments under the assumption of non-line-of-sight (NLOS) propagation conditions. Model B is recommended for typical large open space and office environments with large delay spreads assuming NLOS propagation conditions. Models C and E assume also NLOS propagation conditions and correspond to typical large open space indoor as well as outdoor environments with large delay spreads. Model D is basically the same as model C apart from a line-of-sight (LOS) component, which contributes by a 10 dB spike to the power delay profile at zero delay. The number of discrete propagation paths $\mathcal{L}$ equals 18 in all HIPERLAN/2 channel models.

Table 7.B.1 Specification of the 18-path HIPERLAN/2 channel model A according to ETSI BRAN [20].

Path no. ℓ	Propagation delay τ'_ℓ	Relative path power (linear)	Relative path power (dB)	Rice factor c_R	Doppler PSD $S_{\tau'}(\tau')$	Delay spread $B^{(2)}_{\tau'}$
Model A: Typical office environments (NLOS)						
1	0 ns	1.0000	0.0	0	"Jakes"	
2	10 ns	0.1259	−0.9	0	"Jakes"	
3	20 ns	0.6761	−1.7	0	"Jakes"	
4	30 ns	0.5495	−2.6	0	"Jakes"	
5	40 ns	0.4467	−3.5	0	"Jakes"	
6	50 ns	0.3715	−4.3	0	"Jakes"	
7	60 ns	0.3020	−5.2	0	"Jakes"	
8	70 ns	0.2455	−6.1	0	"Jakes"	
9	80 ns	0.2042	−6.9	0	"Jakes"	50 ns
10	90 ns	0.1660	−7.8	0	"Jakes"	
11	110 ns	0.3388	−4.7	0	"Jakes"	
12	140 ns	0.1862	−7.3	0	"Jakes"	
13	170 ns	0.1023	−9.9	0	"Jakes"	
14	200 ns	0.0562	−12.5	0	"Jakes"	
15	240 ns	0.0427	−13.7	0	"Jakes"	
16	290 ns	0.0159	−18.0	0	"Jakes"	
17	340 ns	0.0058	−22.4	0	"Jakes"	
18	390 ns	0.0021	−26.7	0	"Jakes"	

Table 7.B.2 Specification of the 18-path HIPERLAN/2 channel models B and C according to ETSI BRAN [20].

Path no. ℓ	Propagation delay τ'_ℓ	Relative path power (linear)	Relative path power (dB)	Rice factor c_R	Doppler PSD $S_{\tau'}(\tau')$	Delay spread $B_{\tau'}^{(2)}$
Model B: Typical large open space and office environments (NLOS)						
1	0 ns	0.5495	−2.6	0	"Jakes"	
2	10 ns	0.5012	−3.0	0	"Jakes"	
3	20 ns	0.4467	−3.5	0	"Jakes"	
4	30 ns	0.4074	−3.9	0	"Jakes"	
5	50 ns	1.0000	0.0	0	"Jakes"	
6	80 ns	0.7413	−1.3	0	"Jakes"	
7	110 ns	0.5495	−2.6	0	"Jakes"	
8	140 ns	0.4074	−3.9	0	"Jakes"	
9	180 ns	0.4571	−3.4	0	"Jakes"	100 ns
10	230 ns	0.2754	−5.6	0	"Jakes"	
11	280 ns	0.1698	−7.7	0	"Jakes"	
12	330 ns	0.1023	−9.9	0	"Jakes"	
13	380 ns	0.0617	−12.1	0	"Jakes"	
14	430 ns	0.0372	−14.3	0	"Jakes"	
15	490 ns	0.0288	−15.4	0	"Jakes"	
16	560 ns	0.0145	−18.4	0	"Jakes"	
17	640 ns	0.0085	−20.7	0	"Jakes"	
18	730 ns	0.0035	−24.6	0	"Jakes"	
Model C: Large open space indoor and outdoor environments (NLOS)						
1	0 ns	0.4677	−3.3	0	"Jakes"	
2	10 ns	0.4365	−3.6	0	"Jakes"	
3	20 ns	0.4074	−3.9	0	"Jakes"	
4	30 ns	0.3802	−4.2	0	"Jakes"	
5	50 ns	1.0000	0.0	0	"Jakes"	
6	80 ns	0.8128	−0.9	0	"Jakes"	
7	110 ns	0.6761	−1.7	0	"Jakes"	
8	140 ns	0.5495	−2.6	0	"Jakes"	
9	180 ns	0.7080	−1.5	0	"Jakes"	150 ns
10	230 ns	0.5012	−3.0	0	"Jakes"	
11	280 ns	0.3631	−4.4	0	"Jakes"	
12	330 ns	0.2570	−5.9	0	"Jakes"	
13	400 ns	0.2951	−5.3	0	"Jakes"	
14	490 ns	0.1622	−7.9	0	"Jakes"	
15	600 ns	0.1148	−9.4	0	"Jakes"	
16	730 ns	0.0479	−13.2	0	"Jakes"	
17	880 ns	0.0234	−16.3	0	"Jakes"	
18	1050 ns	0.0076	−21.2	0	"Jakes"	

Table 7.B.3 Specification of the 18-path HIPERLAN/2 channel models D and E according to ETSI BRAN [20].

Path no. ℓ	Propagation delay τ'_ℓ	Relative path power (linear)	Relative path power (dB)	Rice factor c_R	Doppler PSD $S_{\tau'}(\tau')$	Delay spread $B^{(2)}_{\tau'}$
Model D: Large open space indoor and outdoor environments (LOS)						
1	0 ns	1.0000	0.0	10	"Rice"	
2	10 ns	0.1000	−10.0	0	"Jakes"	
3	20 ns	0.0933	−10.3	0	"Jakes"	
4	30 ns	0.0871	−10.6	0	"Jakes"	
5	50 ns	0.2291	−6.4	0	"Jakes"	
6	80 ns	0.1906	−7.2	0	"Jakes"	
7	110 ns	0.1549	−8.1	0	"Jakes"	
8	140 ns	0.1259	−9.0	0	"Jakes"	
9	180 ns	0.1622	−7.9	0	"Jakes"	140 ns
10	230 ns	0.1148	−9.4	0	"Jakes"	
11	280 ns	0.0832	−10.8	0	"Jakes"	
12	330 ns	0.0589	−12.3	0	"Jakes"	
13	400 ns	0.0676	−11.7	0	"Jakes"	
14	490 ns	0.0372	−14.3	0	"Jakes"	
15	600 ns	0.0263	−15.8	0	"Jakes"	
16	730 ns	0.0110	−19.6	0	"Jakes"	
17	880 ns	0.0054	−22.7	0	"Jakes"	
18	1050 ns	0.0017	−27.6	0	"Jakes"	
Model E: Large open space indoor and outdoor environments (NLOS)						
1	0 ns	0.3236	−4.9	0	"Jakes"	
2	10 ns	0.3090	−5.1	0	"Jakes"	
3	20 ns	0.3020	−5.2	0	"Jakes"	
4	40 ns	0.8318	−0.8	0	"Jakes"	
5	70 ns	0.7413	−1.3	0	"Jakes"	
6	100 ns	0.6457	−1.9	0	"Jakes"	
7	140 ns	0.9333	−0.3	0	"Jakes"	
8	190 ns	0.7586	−1.2	0	"Jakes"	
9	240 ns	0.6166	−2.1	0	"Jakes"	250 ns
10	320 ns	1.0000	0.0	0	"Jakes"	
11	430 ns	0.6457	−1.9	0	"Jakes"	
12	560 ns	0.5248	−2.8	0	"Jakes"	
13	710 ns	0.2884	−5.4	0	"Jakes"	
14	880 ns	0.1862	−7.3	0	"Jakes"	
15	1070 ns	0.0871	−10.6	0	"Jakes"	
16	1280 ns	0.0457	−13.4	0	"Jakes"	
17	1510 ns	0.0182	−17.4	0	"Jakes"	
18	1760 ns	0.0081	−20.9	0	"Jakes"	

8

MIMO Channel Models

The development of multielement antenna systems is driven by the limited available bandwidth and the increasing demand for high data rate transmission systems. As it was shown in the seminal papers [115, 270], the channel capacity can greatly be increased in rich scattering environments by using multielement antenna arrays at both the transmitter side and the receiver side. It has been demonstrated in [271] that the multiple-input multiple-output (MIMO) channel can be considered as a system of a number of parallel spatial subchannels allowing the transmission of parallel symbol data streams. Depending on the propagation environment, the number of subchannels can reach the minimum of the number of receiver and transmitter antenna elements. Supposing rich scattering propagation conditions, the capacity of a MIMO channel increases linearly with the number of spatial subchannels.

In general, MIMO channel models are important for the optimization, test, and performance evaluation of space-time coding techniques [21] as well as space-time processing systems [22]. As the propagation conditions determine the channel characteristics, and thus the channel capacity of a MIMO system, it is therefore also of importance to develop MIMO channel models to enable studying the impact of the channel parameters on the capacity of MIMO channels in different propagation environments.

This chapter deals with the modelling, analysis, and simulation of MIMO fading channels for mobile communication systems employing multiple antennas at the transmitter and the receiver side. Starting from specific geometrical scattering models, we present a universal technique for the derivation of stochastic reference models for MIMO channels under the assumption of isotropic and non-isotropic scattering. By way of example, we apply the technique on the most important geometrical models, which are known as the *one-ring model*, the *two-ring model*, and the *elliptical model*. Regarding the one-ring model, it is supposed that only the transmitter, which is in our case the mobile station, is surrounded by an infinite number of local scatterers. This is in contrast to the two-ring and the elliptical models for which it is assumed that scattering objects are located near both the transmitter and the receiver. For all presented geometry-based MIMO channel models, the complex channel gains of the reference models are derived starting from a wave propagation model. The statistical properties of the derived MIMO channel models are studied in detail. General analytical solutions are provided for the three-dimensional (3D) space-time cross-correlation function (CCF) from which other important correlation functions, such as the 2D space CCF and the time autocorrelation function (ACF), can easily be derived. Especially for the two-ring model, we show that the 3D space-time CCF can be expressed under certain conditions as the product of two 2D space-time correlation functions (CFs), called the transmitter CF and the receiver CF. Furthermore, from the non-realizable reference models, stochastic and deterministic simulation models are derived using a limited number of complex-valued sinusoids (cisoids). It is shown how the parameters of the simulation models can be determined for any given distribution of the angle-of-departure (AOD) and the angle-of-arrival (AOA). In case of isotropic scattering, closed-form solutions are presented for the parameter computation problem. The principal theoretical results for the designed reference and simulation models are illustrated and validated by simulations. The proposed procedure provides an important framework for designers of advanced mobile communication systems to verify new transmission concepts employing MIMO techniques under realistic propagation conditions.

This chapter is divided into five sections. Section 8.1 presents the generalized principle of deterministic channel modelling. The one-ring model is introduced and analyzed in Section 8.2. The derivation, analysis, and simulation of a mobile-to-mobile MIMO channel model based on the geometrical two-ring model is the topic of Section 8.3. Section 8.4 deals with the elliptical scattering model and shows its usefulness for the modelling of wideband MIMO channels in propagation environments characterized by multiple clusters of scatterers. Section 8.5 concludes this chapter with an overview of related work.

8.1 The Generalized Principle of Deterministic Channel Modelling

The concept of deterministic channel modelling has been introduced in Chapter 4 to enable the design of simulation models for single-input single-output (SISO) channels. In a nutshell, the basic idea of this concept was to start from a non-realizable reference model consisting of one or several Gaussian random processes, and then to replace each Gaussian random process by a finite sum-of-sinusoids with fixed gains, fixed frequencies, and random (or fixed) phases. For narrowband SISO channels, the Gaussian processes determining the reference model were correlated in time only, while wideband SISO channels are correlated in time and frequency. For the modelling and simulation of MIMO channels, however, we have to design multiple Gaussian random processes with scenario-specific space-time-frequency correlation characteristics. Another challenge of MIMO channel modelling is that many reference models for MIMO channels in specific areas are still unknown and have thus to be derived from scratch. This calls for a generalization of the principle of deterministic channel modelling to provide a universal tool enabling the derivation of reference and simulation models for MIMO channels for any given distribution of local scatterers. The description of the generalized principle of deterministic channel modelling is the objective for this section.

To introduce the problem of MIMO channel modelling, we consider first the general block diagram of a MIMO system with M_T transmitter and M_R receiver antennas as shown in Figure 8.1. In this figure, the symbols $s_l(t)$ and $r_k(t)$ denote the complex envelope of the transmitted signal and the received signal, respectively, where $l = 1, 2, \ldots, M_T$ and $k = 1, 2, \ldots, M_R$. Each transmitter antenna is connected by M_R transmission links (subchannels) with all receiver antennas. The total number of transmission links amounts thus to $M_T \cdot M_R$. If the

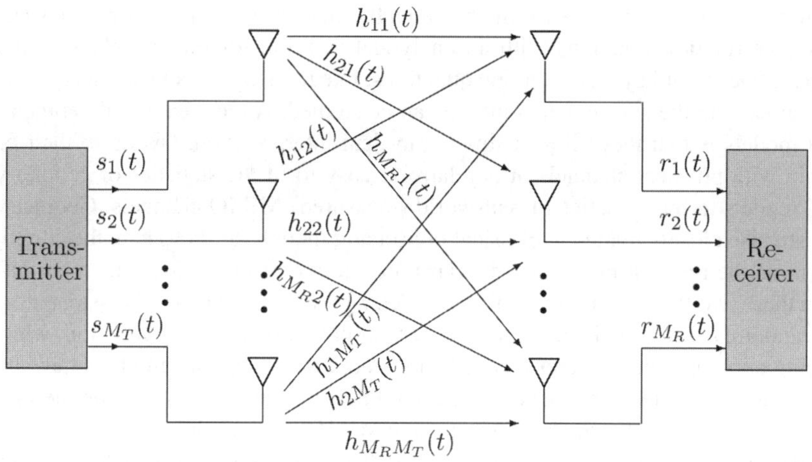

Figure 8.1 General block diagram of an $M_T \times M_R$ MIMO mobile communication system.

MIMO channel is frequency-nonselective, then the transmission link from the lth transmitter antenna to the kth receiver antenna is described by a complex channel gain $h_{kl}(t)$. For such narrowband MIMO channels, the input signal $s_l(t)$ and the output signal $r_k(t)$ are related by $r_k(t) = h_{kl}(t) \cdot s_l(t)$. Otherwise, if the MIMO channel is frequency-selective, then the impulse response $h_{kl}(\tau', t)$ is the appropriate system function for characterizing the transmission link from the lth transmitter antenna to the kth receiver antenna. In the wideband case, the output signal $r_k(t)$ follows from the convolution of the input signal $s_l(t)$ and the impulse response $h_{kl}(\tau', t)$ by using 7.11. This means that the input-output relationship regarding each subchannel of an $M_T \times M_R$ MIMO channel follows the same rules as for SISO channels described in Section 7.2. The main difference here is that two different subchannels $h_{kl}(t)$ and $h_{k'l'}(t)$ are in general not only correlated in time and frequency, but also in space.

The spatial correlation between subchannels can substantially affect the bit error rate performance and the channel capacity of MIMO systems. The amount of spatial correlation depends strongly on the propagation environment (location, shape, and dielectric as well as conductive properties of all relevant objects in the environment). Another important factor is the antenna array configuration (uniform linear array, non-uniform linear array, uniform circular array, hexagonal array) and the distance between the antenna elements. The best performance and the highest capacity of a MIMO system is obtained under isotropic scattering conditions. In real-world channels, however, the scattering conditions are always non-isotropic resulting in a higher bit error rate and a lower channel capacity. The bottom line is that an accurate modelling of the spatial correlation characteristics must be an integral part of each realistic MIMO channel model.

Geometrical models provide a reasonable starting point to describe the random distribution of scattering objects in a variety of propagation environments. Combining geometrical models with plane wave models is an efficient means to derive reference models for MIMO channels with environment-specific space-time-frequency correlation characteristics. These so-called geometry-based MIMO channel models have several advantages. They often result in closed-form solutions or at least mathematically tractable formulas for the space-time-frequency correlation function. With an analytical solution for this correlation function at hand, the influence of key characteristic quantities, like the Doppler spread, delay spread, and angular spread, on the system performance can be studied. A nice feature of geometry-based channel models is that they allow a simple physical interpretation. Owing to their close relationship with physical channels, it is relatively easy to fit the statistics of geometry-based channel models to the statistics of real-world (measured) MIMO channels. Geometry-based models are also closely related to classical ray-tracing models, as they share the common idea to determine the relevant rays from the transmitter to the receiver side. The main difference between these two types of models is that geometry-based channel models assume the scatterers are randomly distributed according to a certain probability density function, while in the ray-tracing approach the scatterer locations are stored in a database, and thus they are deterministic. Last but not least, geometry-based reference channel models provide the fundament for the derivation of stochastic and deterministic MIMO channel simulators.

The idea behind the generalized concept of deterministic channel modelling is to provide universal guidelines for the development chain from a given geometrical model via the reference model to the corresponding stochastic and deterministic simulation models. The starting point of this concept is always a geometrical model that describes the location of the transmitter and the receiver antenna arrays as well as the location of the scatterers. The number of scatterers N is usually infinite and they are randomly distributed in the 2D or 3D plane. The most commonly used geometrical models are known as the one-ring model, the two-ring model, and the elliptical model. These models form the core of the present chapter. From the geometrical model, the reference models can be obtained for the complex channel gains (or the time-variant impulse responses) of all transmission links. For each transmission link, the reference model is derived by using a plane wave model, which can be interpreted as a sum-of-cisoids with constant gains, random Doppler frequencies, and several random phase terms. The reference model is a non-realizable, theoretical, stochastic model from which a stochastic simulation model can be derived by reducing the infinite number of cisoids to a finite number, e.g., by choosing $N \approx 25$. The parameters of the sum-of-cisoids have to be determined such that the statistical properties of the stochastic simulation model are nearly identical to those of the corresponding reference model. In MIMO channel modelling, for example, the time ACF and the 2D space CF function have been identified as proper statistical quantities. In Chapter 5, several parameter computation methods have been proposed. Among them, the extended method of exact Doppler spread (EMEDS), the modified method of equal areas (MMEA), and the L_p-norm method (LPNM) will be used in this chapter. The practical realization of a stochastic simulation model requires theoretically the simulation of an infinite number of sample function. Recall that the sample function of a sum-of-cisoids is always deterministic. If the designed stochastic simulation model is ergodic, then each sample function contains the same statistical information. In this case, we can alternatively fit the statistical properties of the deterministic simulation model to those of the reference model. The difference to the previous parametrization procedure is only that the properties of deterministic simulation models have to be analyzed by using time averages instead of statistical averages. The system that generates sample functions of MIMO channels is called the deterministic MIMO channel simulator. For ergodic channel simulators, it is sufficient to simulate just a single or some few sample functions. The generalized concept of deterministic channel modelling described above can be summarized by the following six design steps.

Step 1: The starting point is a geometrical model with an infinite number of local scatterers.
Step 2: Derive a stochastic reference model from the geometrical model by using a plane wave model.
Step 3: Derive an ergodic stochastic simulation model from the reference model by using only a finite number of N local scatterers.
Step 4: Determine a deterministic simulation model by fixing all model parameters of the stochastic simulation model.
Step 5: Compute the parameters of the simulation model by using a proper parameter computation method, such as the EMEDS, MMEA, or the LPNM.
Step 6: Generate one or some few sample functions by using the deterministic simulation model with fixed parameters.

The six design steps of the generalized principle of deterministic channel modelling are illustrated in Figure 8.2. A comparison with Figure 4.5 reveals that only Step 1 is new, and instead of using sum-of-sinusoids processes, we are now employing sum-of-cisoids processes. The reason is that sum-of-cisoids processes are the better choice for modelling spatially correlated channel models, especially if the propagation environment has strong non-isotropic characteristics. In the following three subsections, we apply this concept to the one-ring model, the two-ring model, and the elliptical model.

8.2 The One-Ring MIMO Channel Model

The one-ring model has originally been proposed in [272] as an appropriate stochastic model for narrowband MIMO Rayleigh fading channels. This model is based on the assumption that the base station is elevated, and thus, not obstructed by local scatterers, whereas the mobile station is surrounded by an infinite number of local scatterers randomly distributed around a ring centred on the mobile station.

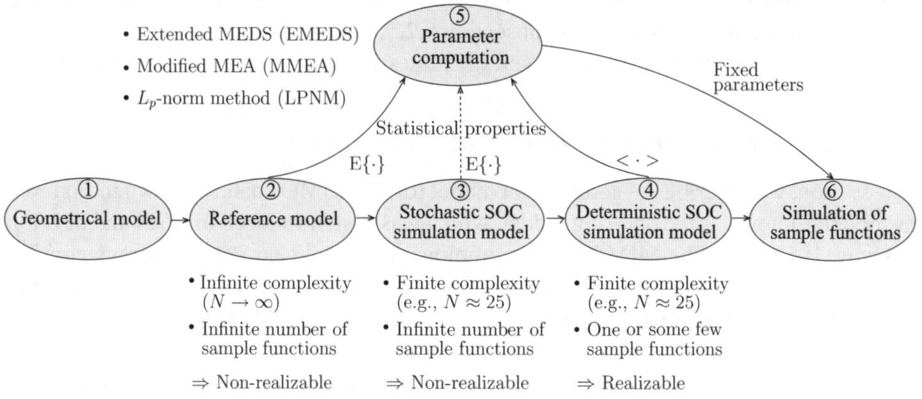

Figure 8.2 Illustration of the design steps from a geometrical model to the simulation of MIMO channels according to the generalized principle of deterministic channel modelling.

This section illustrates how the generalized principle of deterministic channel modelling can be applied to the geometrical one-ring model. Starting from the geometrical one-ring model, the aim is to derive a stochastic and a deterministic simulation model via a non-realizable stochastic reference model. The MIMO channel simulator takes the Doppler effect and the spatial correlation properties into account. The statistical properties of the reference model and the simulation model are studied in detail. For both models, expressions are derived for the 3D space-time CCF from which other important CFs, such as the time ACF and the 2D space CCF, can be obtained immediately. It is shown that the designed stochastic simulation model is always ergodic with respect to the 3D space-time CCF. This is in contrast to Monte-Carlo-method-based channel simulators, which are generally non-ergodic [133, 134, 154]. It will be demonstrated that the parameters of the resulting space-time channel simulator can be determined such that the statistical properties of the simulation model are nearly identical to those of the reference model. More specifically, a performance investigation will show us that the time ACF and the 2D space CCF of the deterministic simulation model fit almost perfectly to the corresponding statistical quantities of the reference model within a certain prescribed domain. This is achieved by employing the EMEDS and the LPNM.

Moreover, it is demonstrated how the channel simulator can be used to analyze the statistics of the capacity of MIMO channels from time-domain simulations. From the simulation of the MIMO channel capacity, one can determine, for example, how often the capacity crosses a given threshold from up to down (or from down to up) within one second, or the expected value for the length of time intervals in which the channel capacity remains below a given threshold. The proposed procedure for the design of deterministic MIMO channel simulators provides a fundamental framework for the test, optimization, design, and analysis of multielement antenna communication systems.

The remainder of this section is organized in six parts exemplifying the six steps of the generalized concept of deterministic channel modelling. Subsection 8.2.1 introduces the geometrical one-ring scattering model (Step 1). Starting from this geometrical model, a stochastic reference model is derived in Subsection 8.2.2 (Step 2). In Subsection 8.2.3, it is demonstrated how a stochastic and a deterministic simulation model can be derived from the reference model (Steps 3 and 4). Subsection 8.2.4 shows how the EMEDS and the LPNM can be applied to compute the model parameters (Step 5). The performance of the developed MIMO narrowband fading channel simulator is studied in Subsection 8.2.5. Finally, Subsection 8.2.6 demonstrates the usefulness of the channel simulator and illustrates some simulation results (Step 6).

8.2.1 The Geometrical One-Ring Scattering Model

In this section, we describe the geometrical model for the MIMO channel shown in Figure 8.3. This model is known as the one-ring model, which was first introduced in [272] and developed further in [105, 273]. The MIMO system employs uniform linear antenna arrays consisting of M_T (M_R) equispaced antenna elements at the transmitter (receiver). The antenna element spacings, also called inter-element spacings, of the transmitter and the receiver antenna array are denoted by δ_T and δ_R, respectively.

For convenience, the base station acts as transmitter, leaving the mobile station the role as receiver. The one-ring model is appropriate for describing environments, in which the base station is elevated and unobstructed, whereas the mobile station is surrounded by a large number of local scatterers S_n ($n = 1, 2, \ldots, N$). Due to high path loss, the contributions of remote scatterers to the total received power are neglected. Hence, the mobile station receives

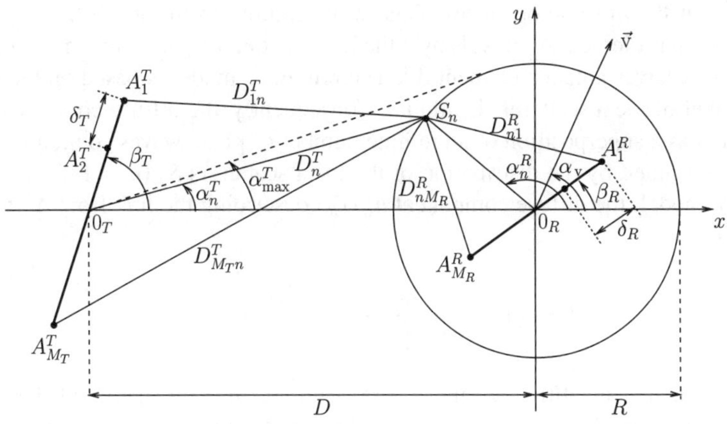

Figure 8.3 Geometrical model (one-ring model) for an $M_T \times M_R$ MIMO channel with local scatterers around the mobile station (receiver).

signals from different directions determined only by the distribution of the local scatterers. In the one-ring model, the local scatterers are scattered around a ring centred on the mobile station. It is usually assumed that the ring radius R is small in comparison with D, denoting the distance between the base station and the mobile station. The AOD of the nth transmitted homogeneous plane wave is denoted by α_n^T, and the corresponding AOA is described by α_n^R $(n = 1, 2, \ldots, N)$. The symbol α_{max}^T in Figure 8.3 designates the maximum AOD seen from the base station. This quantity is related to R and D by $\alpha_{max}^T = \arctan(R/D) \approx R/D$. Furthermore, it is assumed that both R and D are large compared to the geometrical size of the antenna arrays, i.e., $D \gg R \gg \max\{(M_T - 1)\delta_T, (M_R - 1)\,\delta_R\}$.

The remaining parameters of the one-ring model include the tilt angle β_T between the x-axis and the orientation of the base station's antenna array, as well as the tilt angle β_R describing analogously the orientation of the antenna array at the mobile station. As depicted in Figure 8.3, the mobile station moves with speed v in the direction determined by the angle-of-motion α_v. Each transmitted wave is further assumed to be scattered (reflected) only once, and all scattered components that reach the receiving antenna are supposed to be equal in power. For simplicity, it is assumed that the line-of-sight path between the base station and the mobile station is obstructed.

Here, we are focussing on the modelling of fast-term fading effects, where the time-varying properties of the channel are incorporated by allowing the mobile station to move. The validity of the resulting model is limited to relatively short distances over which the local mean of the received envelope is approximately constant. This is usually the case when the mobile station covers a distance in the order of a few tens of wavelengths relative to the centre of the ring [13].

8.2.2 The Reference Model for the One-Ring MIMO Channel Model

8.2.2.1 Derivation of the Reference Model

In this section, we perform Step 2 of the generalized principle of deterministic channel modelling showing how the reference model for a narrowband $M_T \times M_R$ MIMO channel can be derived from the geometrical one-ring model presented in Figure 8.3. From this figure, we observe that the nth homogeneous plane wave emitted from the lth ($l = 1, 2, \ldots, M_T$) transmitter antenna element A_l^T travels over the local scatterer S_n and impinges on the kth ($k = 1, 2, \ldots, M_R$) receiver antenna element A_k^R. The reference model is based on the assumption that the number of local scatterers is infinite. Consequently, the diffuse component at A_k^R can be represented as a superposition of an infinite number of plane waves coming from different directions determined by the distribution of the local scatterers S_n ($n = 1, 2, \ldots, \infty$). Using a plane wave model, the diffuse component $h_{kl}(\vec{r}_R)$ describing the link from A_l^T to A_k^R can be expressed as

$$h_{kl}(\vec{r}_R) = \lim_{N \to \infty} \sum_{n=1}^{N} E_n \, e^{j(\theta_n - \vec{k}_n^R \cdot \vec{r}_R - k_0 D_n)}. \tag{8.1}$$

In this equation, the quantities E_n and θ_n are denoting the path gain and the phase shift, respectively, caused by the interaction of the nth transmitted plane wave with the local scatterer S_n. The symbol $\vec{k}_n^R$ designates the wave vector, which points in the propagation direction of the nth received plane wave, and $\vec{r}_R$ denotes the spatial translation vector pointing in the direction in which the receiver is moving. Furthermore, k_0 is the free-space wave number, which is related to the wavelength λ_0 via $k_0 = 2\pi/\lambda_0$. Finally, D_n denotes the total length which a plane wave travels from A_l^T via S_n to A_k^R.

From (8.1), we notice that the phase of the diffuse component $h_{kl}(\vec{r}_R)$ consists of three terms

$$\arg\{h_{kl}(\vec{r}_R)\} = \theta_n - \vec{k}_n^R \cdot \vec{r}_R - k_0 D_n, \tag{8.2}$$

which have the following physical causes:

(i) θ_n: Phase change due to the interaction with the nth scatterer S_n.
(ii) $\vec{k}_n^R \cdot \vec{r}_R$: Phase change due to the movement of the receiver (mobile station).
(iii) $k_0 D_n$: Phase change due to the total distance travelled.

Each scatterer S_n causes a gain E_n and a phase shift θ_n. The values of E_n and θ_n are dependent on the material of the scatterer S_n and the direction of the nth incoming plane wave. Since $(M_T - 1)\,\delta_T \ll D$, it follows that the plane waves emerging from different transmitter antenna elements arrive at a particular scatterer S_n at approximately the same angle. Similarly, since $(M_R - 1)\,\delta_R \ll R$, we may suppose that plane waves redirected from a particular scatterer S_n arrive at different receiver antenna elements at approximately the same AOA α_n^R. For these reasons, we can assume that for a particular scatterer S_n, the gain E_n and the phase shift θ_n are the same for all plane waves arriving from (or travelling to) different transmitter (or receiver) antenna elements. To simplify the reference model, we assume that each scatterer S_n introduces an infinitesimal constant gain

$$E_n = \frac{1}{\sqrt{N}} \tag{8.3}$$

and a random phase shift θ_n. The phase shifts θ_n are modelled as independent, identically distributed (i.i.d.) random variables, each having a uniform distribution over the interval $[0, 2\pi)$. The above assumptions imply that all plane waves reaching the receiver antenna array from different directions are equal in power and that the average power $E\{|h_{kl}(\vec{r}_k)|^2\}$ of the diffuse component $h_{kl}(\vec{r}_R)$ is normalized to unity.

The second phase term in (8.1), $\vec{k}_n^R \cdot \vec{r}_R$, is caused by the movement of the receiver and accounts for the Doppler effect. This term can be expanded as

$$\vec{k}_n^R \cdot \vec{r}_R = -2\pi f_{\max} \cos\left(\alpha_n^R - \alpha_v\right) t, \tag{8.4}$$

where $f_{\max} = v/\lambda_0$ denotes the maximum Doppler frequency.

The third phase term in (8.1), $k_0 D_n$, is due to the total distance the nth plane wave travelled from A_l^T via S_n to A_k^R. With the help of Figure 8.3, this term can be expressed as

$$k_0 D_n = \frac{2\pi}{\lambda_0}\left(D_{ln}^T + D_{nk}^R\right), \tag{8.5}$$

where D_{ln}^T is the path length from the lth transmitter antenna A_l^T to the scatterer S_n, and D_{nk}^R denotes the path length from the scatterer S_n to the kth receiver antenna A_k^R. Using the law of cosines, the path lengths D_{ln}^T and D_{nk}^R can be written as

$$D_{ln}^T = \sqrt{\left(D_n^T\right)^2 + (M_T - 2l + 1)^2\left(\frac{\delta_T}{2}\right)^2 - (M_T - 2l + 1) D_n^T\,\delta_T \cos\left(\alpha_n^T - \beta_T\right)}, \tag{8.6a}$$

$$D_{nk}^R = \sqrt{R^2 + (M_R - 2k + 1)^2\left(\frac{\delta_R}{2}\right)^2 - (M_R - 2k + 1)\,R\,\delta_R \cos\left(\alpha_n^R - \beta_R\right)}, \tag{8.6b}$$

respectively, where D_n^T is the distance from the centre of the transmitter antenna array to the scatterer S_n. These two path lengths D_{ln}^T and D_{nk}^R can be approximated by using $(M_T - 1)\,\delta_T \ll D_n^T$, $(M_R - 1)\,\delta_R \ll R$, and $\sqrt{1+x} \approx 1 + x/2 \;\; (x \ll 1)$ by

$$D_{ln}^T \approx D_n^T - (M_T - 2\,l + 1)\,\frac{\delta_T}{2}\cos\left(\alpha_n^T - \beta_T\right), \tag{8.7a}$$

$$D_{nk}^R \approx R - (M_R - 2k + 1)\,\frac{\delta_R}{2}\cos\left(\alpha_n^R - \beta_R\right). \tag{8.7b}$$

Applying the law of sines $D/\sin(\alpha_n^R - \alpha_n^T) = R/\sin(\alpha_n^T)$ and noticing that $\alpha_{\max}^T \approx R/D$ and α_n^T are small, one can approximate the AOD α_n^T by $\alpha_n^T \approx \alpha_{\max}^T \sin(\alpha_n^R)$. Now, the path length D_{ln}^T can be expressed in terms of the AOA α_n^R as

$$D_{ln}^T \approx D + R\cos\left(\alpha_n^R\right) - (M_T - 2\,l + 1)\frac{\delta_T}{2}\left[\cos(\beta_T) + \alpha_{\max}^T \sin(\beta_T)\sin\left(\alpha_n^R\right)\right], \tag{8.8}$$

where we have used $D_n^T \approx D + R\cos(\alpha_n^R)$. Finally, after substituting (8.3)–(8.5) in (8.1) and using (8.7b) and (8.8), the complex channel gain $h_{kl}(t)$ of the reference model describing the link from the lth transmitter antenna element A_l^T $(l = 1, 2, \ldots, M_T)$ to the kth receiver antenna element A_k^R $(k = 1, 2, \ldots, M_R)$ can approximately be expressed as

$$h_{kl}(t) = \lim_{N\to\infty} \frac{1}{\sqrt{N}} \sum_{n=1}^{N} g_{kln}\, e^{j(2\pi f_n t + \theta_n + \theta_0)}, \tag{8.9}$$

where

$$g_{kln} = a_{ln}\, b_{kn}\, c_n, \tag{8.10}$$

$$a_{ln} = e^{j\pi(M_T-2l+1)\frac{\delta_T}{\lambda_0}\left[\cos(\beta_T)+\alpha_{\max}^T \sin(\beta_T)\sin(\alpha_n^R)\right]}, \tag{8.11}$$

$$b_{kn} = e^{j\pi(M_R-2k+1)\frac{\delta_R}{\lambda_0}\cos(\alpha_n^R-\beta_R)}, \tag{8.12}$$

$$c_n = e^{-j\frac{2\pi}{\lambda_0}R\cos(\alpha_n^R)}, \tag{8.13}$$

$$f_n = f_{\max}\cos(\alpha_n^R - \alpha_v), \tag{8.14}$$

$$\theta_0 = -\frac{2\pi}{\lambda_0}(D+R). \tag{8.15}$$

Without loss of generality, the constant phase shift θ_0 will henceforth be set to zero and c_n to unity, since it can be shown that theses quantities have neither an influence on the space-time correlation characteristics nor on the distribution of $h_{kl}(t)$. From (8.9) it follows that the mean value and the mean power of the complex channel gain $h_{11}(t)$ are equal to $E\{h_{11}(t)\} = 0$ and $E\{|h_{11}(t)|^2\} = 1$, respectively. Hence, the central limit theorem implies that $h_{11}(t)$ is a zero-mean complex Gaussian process with unit variance. Consequently, the envelope $|h_{11}(t)|$ follows the Rayleigh distribution.

Next, we combine the complex channel gains $h_{kl}(t)$ of the A_l^T-to-A_k^R links to form the so-called *channel matrix*

$$\mathbf{H}(t) = \begin{pmatrix} h_{11}(t) & \cdots & h_{1M_T}(t) \\ \vdots & \ddots & \vdots \\ h_{M_R1}(t) & \cdots & h_{M_RM_T}(t) \end{pmatrix}, \tag{8.16}$$

which describes the reference model of the proposed MIMO frequency-nonselective Rayleigh fading channel. Let $s_l(t)$ $(l = 1, 2, \ldots, M_T)$ represent the complex envelope of the signal transmitted from the antenna element A_l^T and $r_k(t)$ $(k = 1, 2, \ldots, M_R)$ the complex envelope of the received signal at the antenna element A_k^R. By using vector notations $\mathbf{s}(t) = (s_1(t), s_2(t), \ldots, s_{M_T}(t))^T$ and $\mathbf{r}(t) = (r_1(t), r_2(t), \ldots, r_{M_R}(t))^T$, where $(\cdot)^T$ stands for the transpose operator, we can express the input-output relationship of the $M_T \times M_R$ MIMO frequency-nonselective Rayleigh fading channel in the presence of additive white Gaussian noise by

$$\mathbf{r}(t) = \mathbf{H}(t)\cdot\mathbf{s}(t) + \mathbf{n}(t). \tag{8.17}$$

In this equation, the elements of the noise vector $\mathbf{n}(t) = (n_1(t), n_2(t), \ldots, n_{M_R}(t))^T$ are complex uncorrelated additive white Gaussian noise (AWGN) processes, each with zero mean and variance (noise power) N_0. The covariance matrix $\mathbf{R}_{nn}$ of $\mathbf{n}(t)$ equals $\mathbf{R}_{nn} = E\{\mathbf{n}(t)\,\mathbf{n}^H(t)\} = N_0\mathbf{I}_{M_R}$, where $(\cdot)^H$ designates the Hermitian transpose (or conjugate transpose) operator, and $\mathbf{I}_{M_R}$ is the $M_R \times M_R$ identity matrix.

8.2.2.2 Correlation Functions of the Reference Model

A measure of the correlation in space and time between two subchannels is the 3D space-time CCF. To simplify the notation, we will study this function only for a 2×2 MIMO channel. For this case, the 3D space-time CCF of the channel gains $h_{11}(t)$ and $h_{22}(t)$ is defined as [273]

$$\rho_{11,22}(\delta_T, \delta_R, \tau) := E\left\{h_{11}(t)h_{22}^*(t+\tau)\right\}\big|_{\theta_n, \alpha_n^R}, \tag{8.18}$$

where $(\cdot)^*$ denotes the complex conjugation. Using (8.9) and noticing that $M_T = M_R = 2$, we can express $\rho_{11,22}(\delta_T, \delta_R, \tau)$ as

$$\rho_{11,22}(\delta_T, \delta_R, \tau) = \lim_{N\to\infty} \frac{1}{N} \sum_{n=1}^{N} E\left\{a_{1n}^2\, b_{1n}^2\, e^{-j2\pi f_n \tau}\right\}\big|_{\alpha_n^R}. \tag{8.19}$$

Here it is important to recall that the quantities a_{1n}, b_{1n}, and f_n are functions of the AOA α_n^R seen from the receiver (mobile station). Since the number of local scatterers is infinite ($N \to \infty$) in the reference model, it is mathematically convenient to assume that the discrete AOA α_n^R is a continuous random variable α^R with a given distribution $p_{\alpha^R}(\alpha^R)$. The infinitesimal power of the scattered components coming from the differential angle $d\alpha^R$ is proportional to $p_{\alpha^R}(\alpha^R)d\alpha^R$. In the limit as $N \to \infty$, this contribution must be equal to $1/N$, i.e., $1/N = p_{\alpha^R}(\alpha^R)d\alpha^R$. Hence, for any given distribution $p_{\alpha^R}(\alpha^R)$, the 3D space-time CCF in (8.19) can be written as

$$\rho_{11,22}(\delta_T, \delta_R, \tau) = \int_{-\pi}^{\pi} a^2(\delta_T, \alpha^R)\, b^2(\delta_R, \alpha^R)\, e^{-j2\pi f(\alpha^R)\tau}\, p_{\alpha^R}(\alpha^R)\, d\alpha^R, \tag{8.20}$$

where

$$a(\delta_T, \alpha^R) = e^{j\pi \frac{\delta_T}{\lambda_0}[\cos(\beta_T) + \alpha_{\max}^T \sin(\beta_T)\sin(\alpha^R)]}, \tag{8.21}$$

$$b(\delta_R, \alpha^R) = e^{j\pi \frac{\delta_R}{\lambda_0}\cos(\alpha^R - \beta_R)}, \tag{8.22}$$

$$f(\alpha^R) = f_{\max}\cos(\alpha^R - \alpha_v). \tag{8.23}$$

The time ACF of the complex channel gain $h_{kl}(t)$ can be obtained as

$$r_{h_{kl}}(\tau) := E\{h_{kl}(t)h_{kl}^*(t+\tau)\} = \int_{-\pi}^{\pi} e^{-j2\pi f_{\max}\cos(\alpha^R - \alpha_v)\tau}\, p_{\alpha^R}(\alpha^R)\, d\alpha^R \tag{8.24}$$

for $k = 1, 2, \ldots, M_R$ and $l = 1, 2, \ldots, M_T$. The general expression in (8.24) reveals that $r_{h_{kl}}(\tau)$ is independent of k and l, i.e., all complex channel gains $h_{kl}(t)$ modelling the link from A_l^T to A_k^R are characterized by the same time ACF $r_{h_{kl}}(\tau)$ for all $k = 1, 2, \ldots, M_R$ and $l = 1, 2, \ldots, M_T$. Notice also that the time ACF $r_{h_{kl}}(\tau)$ equals the 3D space-time CCF $\rho_{11,22}(\delta_T, \delta_R, \tau)$ at $\delta_T = \delta_R = 0$, i.e., $r_{h_{kl}}(\tau) = \rho_{11,22}(0, 0, \tau)$, as was to be expected.

For our purpose, we also need the 2D space CCF, which is defined as

$$\rho(\delta_T, \delta_R) := E\{h_{11}(t)h_{22}^*(t)\}$$
$$= \int_{-\pi}^{\pi} a^2(\delta_T, \alpha^R)\, b^2(\delta_R, \alpha^R)\, p_{\alpha^R}(\alpha^R)\, d\alpha^R. \tag{8.25}$$

Comparing (8.25) and (8.20) shows that the 2D space CCF $\rho(\delta_T, \delta_R)$ can easily be obtained from the 3D space-time CCF $\rho_{11,22}(\delta_T, \delta_R, \tau)$ by setting τ to zero, i.e., $\rho(\delta_T, \delta_R) = \rho_{11,22}(\delta_T, \delta_R, 0)$.

The expressions for the CFs in (8.20), (8.24), and (8.25) are valid for any given distribution $p_{\alpha^R}(\alpha^R)$ of the AOA α^R. In the literature, several distributions $p_{\alpha^R}(\alpha^R)$ have been proposed, e.g., the uniform distribution [272], the von Mises distribution [176], the Gaussian distribution [274], and the Laplacian distribution [275]. We proceed with the von Mises distribution, which has previously been shown to be successful in describing measured data [176]. This distribution function is given by

$$p_{\alpha^R}(\alpha^R) = \frac{1}{2\pi I_0(\kappa)} e^{\kappa\cos(\alpha^R - m_\alpha)}, \quad \alpha^R \in (0, 2\pi], \tag{8.26}$$

where $I_0(\cdot)$ is the zeroth-order modified Bessel function, the quantity $m_\alpha \in (0, 2\pi]$ accounts for the mean direction of the AOA α^R seen from the mobile station, and $\kappa \geq 0$ is a parameter controlling the angular spread of α^R. For example, if $\kappa = 0$, then isotropic scattering is obtained. In this case, the AOA α^R is uniformly distributed, i.e., $p_{\alpha^R}(\alpha^R) = 1/(2\pi)$, where $\alpha^R \in (0, 2\pi]$. On the other hand, extremely non-isotropic scattering is obtained if $\kappa \to \infty$, and thus, $p_{\alpha^R}(\alpha^R) \to \delta(\alpha^R - m_\alpha)$, where $\delta(\cdot)$ denotes the Dirac delta function. For large values of κ, the angular spread of α^R is approximately equal to $2/\sqrt{\kappa}$ [176].

Substituting (8.26) in (8.20), and solving the integral over α^R by using [23, Equation (3.338–4)], enables us to present the 3D space-time CCF $\rho_{11,22}(\delta_T, \delta_R, \tau)$ in closed form as [273]

$$\rho_{11,22}(\delta_T, \delta_R, \tau) = \frac{e^{j2\pi\frac{\delta_T}{\lambda_0}\cos(\beta_T)}}{I_0(\kappa)} I_0\left(\left\{\kappa^2 - (2\pi)^2\left[\left(\frac{\delta_R}{\lambda_0}\right)^2 - 2\frac{\delta_R}{\lambda_0} f_{\max}\tau\cos(\beta_R - \alpha_v)\right.\right.\right.$$
$$\left. + (f_{\max}\tau)^2 + \left(\frac{\delta_T}{\lambda_0}\alpha_{\max}^T \sin(\beta_T)\right)^2\right] + 2(2\pi)^2\frac{\delta_T}{\lambda_0}\alpha_{\max}^T\sin(\beta_T)$$
$$\cdot\left[f_{\max}\tau\sin(\alpha_v) - \frac{\delta_R}{\lambda_0}\sin(\beta_R)\right] + j4\pi\kappa\left[\frac{\delta_R}{\lambda_0}\cos(m_\alpha - \beta_R) - f_{\max}\tau\right.$$
$$\left.\left.\left.\cdot\cos(m_\alpha - \alpha_v) + \frac{\delta_T}{\lambda_0}\alpha_{\max}^T\sin(\beta_T)\sin(m_\alpha)\right]\right\}^{1/2}\right). \tag{8.27}$$

This 3D space-time CCF includes a number of other well-known correlation models as special cases. For example, the relationship $r_{h_{kl}}(\tau) = \rho_{11,22}(0, 0, \tau)$ can be exploited to find immediately $r_{h_{kl}}(\tau)$ from (8.27), which provides, in case of non-isotropic scattering ($\kappa \neq 0$) around the mobile station with $\alpha_v = \pi$, the equation [176]

$$r_{h_{kl}}(\tau) := \frac{I_0\left(\sqrt{\kappa^2 - (2\pi f_{\max}\tau)^2 + j4\pi\kappa f_{\max}\tau\cos(m_\alpha)}\right)}{I_0(\kappa)}. \tag{8.28}$$

In [176], it has been reported that this correlation model fits very well to measured data. For isotropic scattering ($\kappa = 0$), the expression in (8.28) reduces further to the well-known Jakes (Clarke) ACF $r_{h_{kl}}(\tau) = J_0(2\pi f_{\max}\tau)$ [13], where $J_0(\cdot)$ denotes the zeroth-order Bessel function of the first kind.

Applying the relation $\rho(\delta_T, \delta_R) = \rho_{11,22}(\delta_T, \delta_R, 0)$ to the 3D space-time CCF presented in (8.27) results in case of isotropic scattering ($\kappa = 0$) in [276]

$$\rho(\delta_T, \delta_R) := e^{j2\pi\frac{\delta_T}{\lambda_0}\cos(\beta_T)} J_0\left(2\pi\left\{\left(\frac{\delta_R}{\lambda_0}\right)^2 + \left(\frac{\delta_T}{\lambda_0}\alpha_{\max}^T\sin(\beta_T)\right)^2 + 2\frac{\delta_T\delta_R}{\lambda_0^2}\alpha_{\max}^T\sin(\beta_T)\sin(\beta_R)\right\}^{1/2}\right). \tag{8.29}$$

If the transmitter and receiver antenna arrays are perpendicular to the x-axis, i.e., $\beta_T = \beta_R = \pi/2$, then the expression above reduces to $\rho(\delta_T, \delta_R) = J_0(2\pi(\delta_R/\lambda_0 + \alpha_{\max}^T\delta_T/\lambda_0))$ [276]. On the other hand, if the base station is equipped with a uniform linear array and the mobile station has a single antenna, then the 2D space CCF $\rho(\delta_T, \delta_R)$ in (8.29) simplifies for $\delta_R = 0$ to the spatial correlation model in [277], which can be written as $\rho(\delta_T, 0) = \exp[j2\pi\cos(\beta_T)\delta_T/\lambda_0]\, J_0(2\pi\alpha_{\max}^T\sin(\beta_T)\delta_T/\lambda_0)$.

It is important to note that the reference model described above is non-realizable, since the number of scatterers N is infinite. From the reference model, however, a simulation model can be derived, as we will see in the next section.

8.2.3 Simulation Models for the One-Ring MIMO Channel Model

An efficient simulation model can be obtained from any given reference model derived from the plane wave model by applying the generalized concept of deterministic channel modelling. According to Figure 8.2, this concept comprises in general six steps. In the third step, a stochastic simulation model is derived from the reference model simply by replacing the infinite number of scatterers by a finite number N. In practice, a compromise must be found between the performance and the realization complexity of the resulting channel simulator by choosing a proper value for N. Investigations have shown that values of N around 25 are in the majority of cases sufficient, especially when the design procedures described in Section 8.2.4 are used. The fourth step is the step from the stochastic simulation model to the corresponding deterministic simulation model, which is performed by fixing all model parameters, including the phases. In the following two subsections, we demonstrate the application of Steps 3 and 4 to the one-ring MIMO channel model.

8.2.3.1 The Stochastic Simulation Model

From the reference model described in Section 8.2.2, a stochastic simulation model is obtained by assuming that the number of scatterers N is finite. If N is finite, then the complex channel gain of the link from A_l^T to A_k^R can be modelled by using (8.9) with $\theta_0 = 0$ and $c_n = 1$ as

$$\hat{h}_{kl}(t) = \frac{1}{\sqrt{N}} \sum_{n=1}^{N} a_{ln}\, b_{kn}\, e^{j(2\pi f_n t + \theta_n)}, \tag{8.30}$$

where a_{ln}, b_{kn}, and f_n are given by (8.11), (8.12), and (8.14), respectively. The discrete AOAs α_n^R $(n = 1, 2, \ldots, N)$ are now constants, which will be determined in Section 8.2.4. The phases θ_n $(n = 1, 2, \ldots, N)$ are still i.i.d. random variables, each with a uniform distribution over $(0, 2\pi]$, just as the phases θ_n in (8.1) and (8.9). Hence, $\hat{h}_{kl}(t)$ represents a stochastic process that can be interpreted as a finite sum-of-cisoids with constant gains, constant AOAs, and random phases. It has been shown in [139] that this class of sum-of-cisoids processes is first-order stationary. Subsequently, it will be shown that the stochastic process $\hat{h}_{kl}(t)$ described by (8.30) is also ergodic with respect to the 3D space-time CCF. This is in contrast to sum-of-sinusoids-based channel models designed by employing Monte Carlo techniques, where the AOA and/or the gains are random variables as well. In such cases, the resulting class of stochastic processes is always non-autocorrelation ergodic [278]. The problems of non-autocorrelation ergodic processes have been discussed in [78].

By analogy to (8.16), we combine the complex channel gains $\hat{h}_{kl}(t)$ to form the stochastic channel matrix $\hat{\mathbf{H}}(t) := [\hat{h}_{kl}(t)]$. Its software or hardware realization is called the *stochastic simulation model* for the one-ring MIMO frequency-nonselective Rayleigh fading channel. The resulting structure of this channel simulator is illustrated in Figure 8.4 for $M_T = M_R = 2$, where we have used the notation $a_n = a_{1n} = a_{2n}^*$ and $b_n = b_{1n} = b_{2n}^*$.

The 3D space-time CCF $\hat{\rho}_{11,22}(\delta_T, \delta_R, \tau)$ of the stochastic simulation model can be written as

$$\begin{aligned} \hat{\rho}_{11,22}(\delta_T, \delta_R, \tau) &:= E\left\{\hat{h}_{11}(t)\hat{h}_{22}^*(t+\tau)\right\}\Big|_{\theta_n} \\ &= \frac{1}{N}\sum_{n=1}^{N} a_{1n}^2(\delta_T)\, b_{1n}^2(\delta_R)\, e^{-j2\pi f_n \tau}. \end{aligned} \tag{8.31}$$

Furthermore, the time ACF $\hat{r}_{h_{kl}}(\tau)$ of the complex channel gain $\hat{h}_{kl}(t)$ of the stochastic simulation model can be expressed as

$$\begin{aligned} \hat{r}_{h_{kl}}(\tau) &:= E\left\{\hat{h}_{kl}(t)\hat{h}_{kl}^*(t+\tau)\right\}\Big|_{\theta_n} \\ &= \frac{1}{N}\sum_{n=1}^{N} e^{-j2\pi f_n \tau}, \end{aligned} \tag{8.32}$$

where we observe that the relationship $\hat{r}_{h_{kl}}(\tau) = \hat{\rho}_{11,22}(0, 0, \tau)$ holds for all $k = 1, 2, \ldots, M_R$ and $l = 1, 2, \ldots, M_T$.

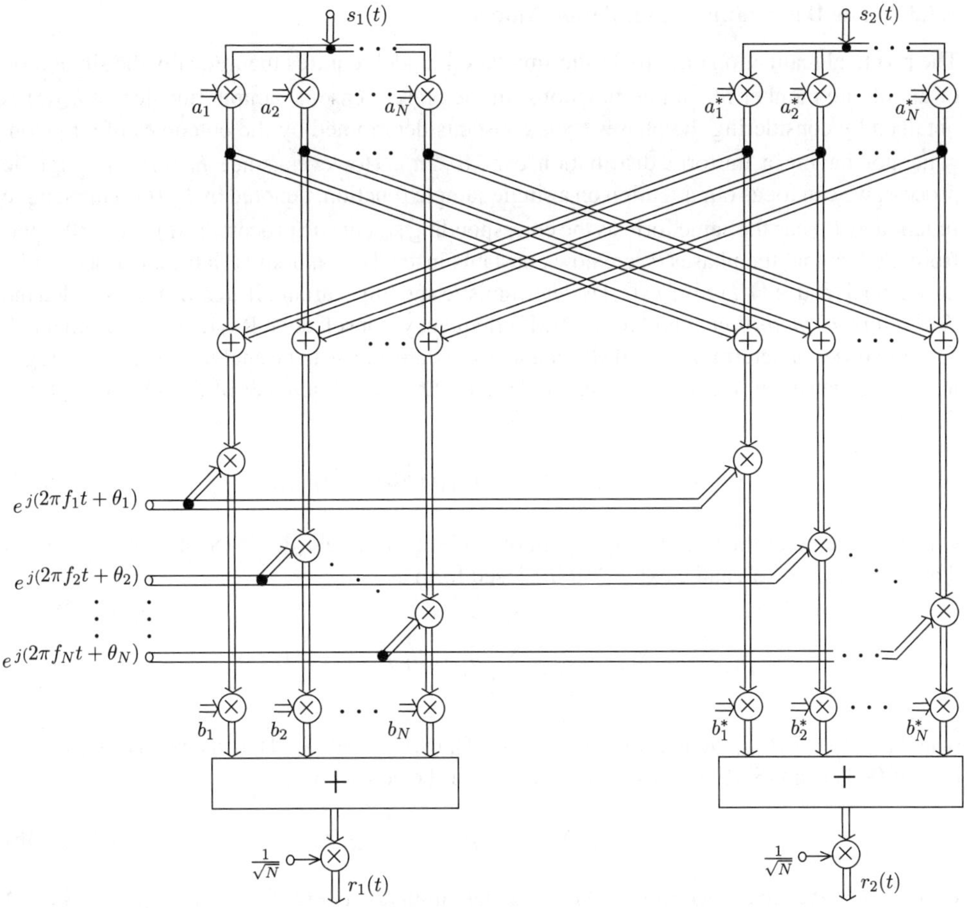

Figure 8.4 Structure of the one-ring 2 × 2 MIMO channel simulator.

Finally, the 2D space CCF $\hat{\rho}(\delta_T, \delta_R)$ of the stochastic simulation model can be derived as follows

$$
\begin{aligned}
\hat{\rho}(\delta_T, \delta_R) &:= E\left\{\hat{h}_{11}(t)\hat{h}_{22}^*(t)\right\}\Big|_{\theta_n} \\
&= \hat{\rho}_{11,22}(\delta_T, \delta_R, 0) \\
&= \frac{1}{N}\sum_{n=1}^{N} a_{1n}^2(\delta_T)\, b_{1n}^2(\delta_R) \\
&= \frac{1}{N}\sum_{n=1}^{N} e^{j2\pi\left\{\frac{\delta_T}{\lambda_0}\left[\cos(\beta_T)+\alpha_{\max}^T \sin(\beta_T)\sin(\alpha_n^R)\right]+\frac{\delta_R}{\lambda_0}\cos(\alpha_n^R-\beta_R)\right\}}.
\end{aligned}
\tag{8.33}
$$

8.2.3.2 The Deterministic Simulation Model

The practical realization of a stochastic simulation model requires theoretically the simulation of an infinite number of sample functions. In the present case, a sample function of $\hat{h}_{kl}(t)$ is obtained by considering the phases θ_n as constants determined by the outcomes of a random generator having a uniform distribution over $[0, 2\pi)$. However, since $\hat{h}_{kl}(t)$ is an ergodic process, we can focus our attention on a single sample function, denoted by $\tilde{h}_{kl}(t)$. This sample function $\tilde{h}_{kl}(t)$ has the same form as the corresponding stochastic process $\hat{h}_{kl}(t)$ in (8.30), apart from the fact that the phases θ_n are now constants instead of random variables. Consequently, the channel matrix $\tilde{\mathbf{H}}(t) := [\tilde{h}_{kl}(t)]$ is deterministic and time variant. Its realization is called the *deterministic simulation model* for a MIMO frequency-nonselective Rayleigh fading channel. The properties of deterministic MIMO channel simulators have to be analyzed by time averages instead of statistical averages. For example, the 3D space-time CCF of $\tilde{h}_{11}(t)$ and $\tilde{h}_{22}(t)$ is defined as

$$\tilde{\rho}_{11,22}(\delta_T, \delta_R, \tau) :=< \tilde{h}_{11}(t)\, \tilde{h}^*_{22}(t+\tau) >, \tag{8.34}$$

where $<\cdot>$ denotes the time average operator.[1] By substituting $\tilde{h}_{kl}(t)$ in (8.34), it can be shown that $\tilde{\rho}_{11,22}(\delta_T, \delta_R, \tau)$ can be expressed in closed form as

$$\tilde{\rho}_{11,22}(\delta_T, \delta_R, \tau) = \frac{1}{N} \sum_{n=1}^{N} a^2_{1n}(\delta_T)\, b^2_{1n}(\delta_R)\, e^{-j2\pi f_n \tau}, \tag{8.35}$$

where $a_{1n}(\delta_T)$, $b_{1n}(\delta_R)$, and f_n are given by (8.11), (8.12), and (8.14), respectively. A comparison of (8.31) and (8.35) proves the correctness of the equation

$$\hat{\rho}_{11,22}(\delta_T, \delta_R, \tau) = \tilde{\rho}_{11,22}(\delta_T, \delta_R, \tau), \tag{8.36}$$

which states that the stochastic MIMO channel simulator is ergodic with respect to the 3D space-time CCF. This statement implies that $\hat{h}_{kl}(t)$ is both autocorrelation ergodic in time and cross-correlation ergodic in space.

By analogy to (8.24), the time ACF of the deterministic process $\tilde{h}_{kl}(t)$ can be obtained as

$$\begin{aligned} \tilde{r}_{h_{kl}}(\tau) &:= < \tilde{h}_{kl}(t)\, \tilde{h}^*_{kl}(t+\tau) > \\ &= \frac{1}{N} \sum_{n=1}^{N} e^{-j2\pi f_{\max} \cos(\alpha_n^R - \alpha_v)\tau} \\ &= \hat{r}_{h_{kl}}(\tau) \end{aligned} \tag{8.37}$$

for $k = 1, 2, \ldots, M_R$ and $l = 1, 2, \ldots, M_T$. Note that the expressions in (8.37) and (8.35) are related by $\tilde{r}_{h_{kl}}(\tau) = \tilde{\rho}_{11,22}(0, 0, \tau)$. Moreover, we will make use of the 2D space CCF of the

[1] Recall that the time average operator $< x(t) >$ is defined as $< x(t) > := \lim_{T\to\infty} \frac{1}{2T} \int_{-T}^{T} x(t)\, dt$.

deterministic simulation system, which can be derived as follows

$$\begin{aligned}\tilde{\rho}(\delta_T, \delta_R) &:= < \tilde{h}_{11}(t)\,\tilde{h}_{22}^*(t) > \\ &= \frac{1}{N}\sum_{n=1}^{N} e^{j2\pi\left\{\frac{\delta_T}{\lambda_0}[\cos(\beta_T)+\alpha_{\max}^T \sin(\beta_T)\sin(\alpha_n^R)]+\frac{\delta_R}{\lambda_0}\cos(\alpha_n^R-\beta_R)\right\}} \\ &= \hat{\rho}(\delta_T, \delta_R). \end{aligned} \tag{8.38}$$

It should be obvious that the 2D space CCF in (8.38) and the 3D space-time CCF in (8.35) are related by $\tilde{\rho}(\delta_T, \delta_R) = \tilde{\rho}_{11,22}(\delta_T, \delta_R, 0)$. In the next subsection, it will be shown how the model parameters of the simulation model can be determined such that the temporal and spatial correlation properties of the simulation model are sufficiently close to those of the reference model.

8.2.4 *Parameter Computation Methods*

The objective of any parameter computation method is to find proper values for model parameters, such that the statistical properties of the deterministic (stochastic) simulation model are sufficiently close to those of the stochastic reference model. The model parameters to be determined are the discrete AOAs α_n^R ($n = 1, 2, \ldots, N$). All other model parameters of the deterministic simulation model are identical to the corresponding parameters of the reference model.

Extended method of exact Doppler spread (EMEDS): For isotropic scattering around the receiver, we employ the EMEDS to determine the set of discrete AOAs $\{\alpha_n^R\}_{n=1}^N$. This method is a special case of the generalized method of exact Doppler spread (GMEDS$_q$), which is described in detail in Subsection 10.1.2. According to the EMEDS, the model parameters α_n^R are determined by

$$\alpha_n^R = \frac{2\pi}{N}\left(n - \frac{1}{2}\right) + \alpha_0^R, \quad n = 1, 2, \ldots, N, \tag{8.39}$$

where α_0^R is called the *angle-of-rotation*. Equation (8.39) implies that the scatterers are located equally spaced on a ring centered on the receiver. The EMEDS reveals its best performance if the angle-of-rotation α_0^R is defined as

$$\alpha_0^R := \frac{\alpha_n^R - \alpha_{n-1}^R}{4} = \frac{\pi}{2N}. \tag{8.40}$$

The performance of the EMEDS will be demonstrated in the next subsection.

L_p-norm method (LPNM): For non-isotropic scattering around the receiver, we apply the LPNM, which is described in detail in Subsection 5.1.6. The application of this method to the present problem requires the minimization of the following two L_p-norms:

$$E_1^{(p)} := \left\{ \frac{1}{\tau_{\max}} \int_0^{\tau_{\max}} |r_{h_{11}}(\tau) - \tilde{r}_{h_{11}}(\tau)|^p \, d\tau \right\}^{1/p}, \tag{8.41}$$

$$E_2^{(p)} := \left\{ \frac{1}{\delta_{\max}^T \, \delta_{\max}^R} \int_0^{\delta_{\max}^T} \int_0^{\delta_{\max}^R} |\rho(\delta_T, \delta_R) - \tilde{\rho}(\delta_T, \delta_R)|^p \, d\delta_R \, d\delta_T \right\}^{1/p}, \tag{8.42}$$

where $p = 1, 2, \ldots$ The quantities $\tau_{\max}$, $\delta_{\max}^T$, and $\delta_{\max}^R$ are real-valued constants defining the upper limits of the domains over which the approximations $\tilde{r}_{h_{11}}(\tau) \approx r_{h_{11}}(\tau)$ and $\tilde{\rho}(\delta_T, \delta_R) \approx \rho(\delta_T, \delta_R)$ are of interest. The optimization of the model parameters α_n^R $(n = 1, 2, \ldots, N)$ can be done in a number of ways. The preferred way is to perform a joint optimization of $E_1^{(p)}$ and $E_2^{(p)}$ by using the Fletcher-Powell algorithm [162]. Another way is to replace α_n^R in (8.42) by $\alpha_n'^R$. This enables an orthogonalization of the optimization problem, meaning that the error norms $E_1^{(p)}$ and $E_2^{(p)}$ can be minimized independently. The resulting performance advantage must be traded off against the model inconsistency. Both the EMEDS and the LPNM will be applied in the next subsection to demonstrate the performance of the one-ring MIMO channel simulator.

8.2.5 *Performance Evaluation*

The performance of the simulation model can best be evaluated by comparing its statistical properties with those of the reference model. In the present case, the time ACF and the 2D space CCF are identified as proper statistical quantities. Concerning the reference model, we have chosen the following parameters: $\beta_T = \beta_R = 90°, \alpha_v = 180°, \alpha_{\max}^T = 2°$, and $f_{\max} =$ 91 Hz. For the distribution $p(\alpha^R)$ of the AOA α^R seen from the receiver (mobile station), we have employed the von Mises density with parameters κ and m_α [see (8.26)].

Let us first study the isotropic scattering case for which $\kappa = 0$ holds. Hence, $p(\alpha^R) = 1/(2\pi)$ and the time ACF $r_{h_{kl}}(\tau)$ in (8.24) results in $r_{h_{kl}}(\tau) = J_0(2\pi f_{\max}\tau)$. A plot of the graph of $r_{h_{kl}}(\tau)$ is shown in Figure 8.5. This figure also shows the resulting time ACF $\tilde{r}_{h_{kl}}(\tau)$ of the simulation model designed by using the EMEDS with $N = 25$ as well as the LPNM with $p = 2$, $\tau_{\max} = N/(4 f_{\max})$, and $N = 25$. As can be seen, the approximation $r_{h_{kl}}(\tau) \approx \tilde{r}_{h_{kl}}(\tau)$ is excellent over the interval from 0 to $\tau_{\max}$. The approximation error outside the interval $[0, \tau_{\max}]$ is not controlled by the L_p-norm in (8.41). This explains the discrepancy between $r_{h_{kl}}(\tau)$ and $\tilde{r}_{h_{kl}}(\tau)$ if $\tau > \tau_{\max}$. Figure 8.6 visualizes the behaviour of the 2D space CCF $\rho(\delta_T, \delta_R)$ of the reference model. The results have been found by evaluating (8.29) with the parameters listed above. The behaviour of the 2D space CCF $\tilde{\rho}(\delta_T, \delta_R)$ of the simulation model can also be studied analytically. Therefore, one has to evaluate (8.38) using the discrete AOAs α_n^R obtained by applying the EMEDS. The corresponding results for the simulation model are illustrated in Figure 8.7. Obviously, there is no visual difference between $\rho(\delta_T, \delta_R)$ and $\tilde{\rho}(\delta_T, \delta_R)$.

Next, we investigate the non-isotropic scattering case by choosing $\kappa = 0$ and $m_\alpha = 0$. Since the EMEDS cannot be used to determine the model parameters α_n^R for non-isotropic

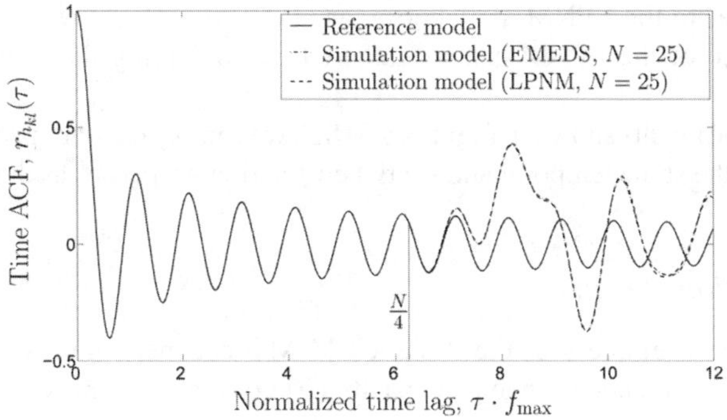

Figure 8.5 Time ACFs $r_{h_{kl}}(\tau)$ (reference model) and $\tilde{r}_{h_{kl}}(\tau)$ (simulation model) for isotropic scattering ($\kappa = 0$).

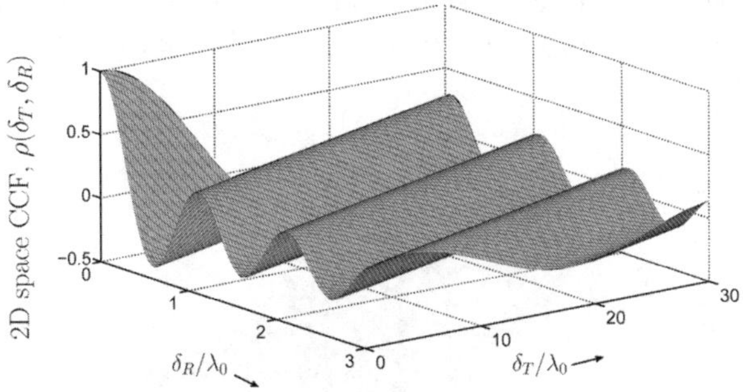

Figure 8.6 The 2D space CCF $\rho(\delta_T, \delta_R)$ of the reference model for isotropic scattering ($\kappa = 0, \beta_T = \beta_R = 90°, \alpha^T_{\max} = 2°$).

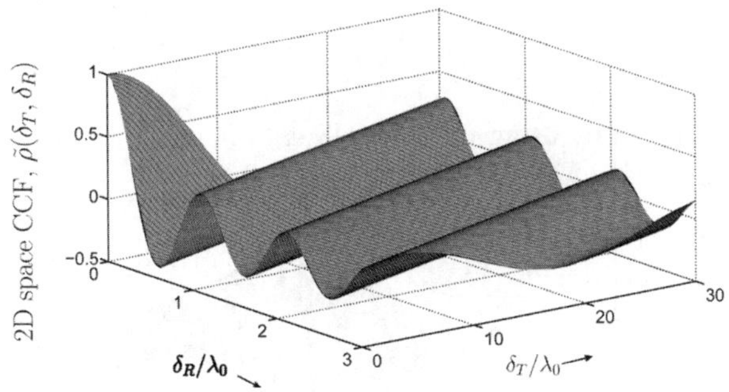

Figure 8.7 The 2D space CCF $\tilde{\rho}(\delta_T, \delta_R)$ of the simulation model for isotropic scattering ($\kappa = 0, \beta_T = \beta_R = 90°, \alpha^T_{\max} = 2°$, EMEDS, $N = 25$).

scattering, we refer to the LPNM and choose, as above, $p = 2$ and $N = 25$. An impression of the performance of the LPNM can be gathered by comparing the results in Figures 8.8 and 8.9.

Summarizing the results shown in Figures 8.5–8.9, we can say that the space-time simulation model has nearly the same temporal and spatial correlation properties as the reference model.

8.2.6 Simulation Results

To demonstrate the usefulness of the designed MIMO channel simulator, we focus on the simulation of the instantaneous channel capacity $\tilde{C}(t)$ of the deterministic simulation model. The instantaneous channel capacity $\tilde{C}(t)$ is defined as

$$\tilde{C}(t) := \log_2 \left[\det \left(\mathbf{I}_{M_R} + \frac{\gamma_c}{M_T} \tilde{\mathbf{H}}(t) \tilde{\mathbf{H}}^H(t) \right) \right] \quad \text{(bits/s/Hz)}, \tag{8.43}$$

where it has been supposed that $M_T \geq M_R$ holds. In (8.43), $\det(\cdot)$ designates the determinant, and γ_c represents the average received signal-to-noise ratio (SNR), which equals the ratio of the total transmitted power $P_{T,\text{total}}$ and the noise power N_0, i.e., $\gamma_c = P_{T,\text{total}}/N_0$. Thereby, it

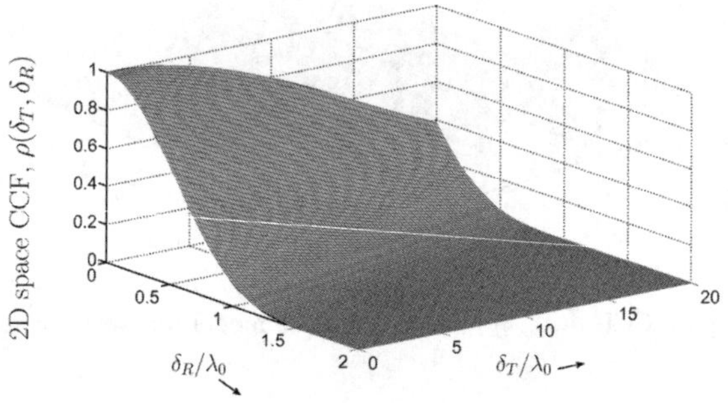

Figure 8.8 The 2D space CCF $\rho(\delta_T, \delta_R)$ of the reference model for non-isotropic scattering ($\kappa = 10$, $m_\alpha = 0, \beta_T = \beta_R = 90°, \alpha^T_{\text{max}} = 2°$).

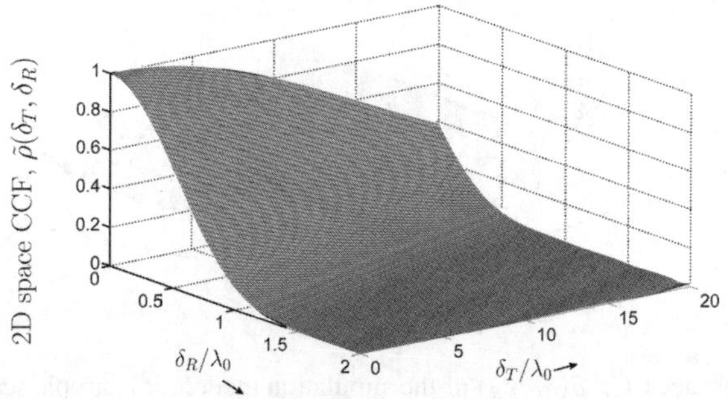

Figure 8.9 The 2D space CCF $\tilde{\rho}(\delta_T, \delta_R)$ of the simulation model for non-isotropic scattering ($\kappa = 10$, $m_\alpha = 0, \beta_T = \beta_R = 90°, \alpha^T_{\text{max}} = 2°$, LPNM, $N = 25$).

has been assumed that the total transmitted power $P_{T,\mathrm{total}}$ has been allocated equally among all of the M_T transmitter antenna elements. Note that the instantaneous capacity $\tilde{C}(t)$ in (8.43) is a real-valued deterministic process, whose temporal behaviour depends on the complex deterministic channel matrix $\tilde{\mathbf{H}}(t) = [\tilde{h}_{kl}(t)]$. An example visualizing how the instantaneous capacity $\tilde{C}(t)$ of the simulation model changes over time is shown in Figure 8.10. This figure also shows the mean capacity C. According to Shannon's coding theorem [279, 280], there exists an error-correcting code, which allows the error-free transmission of an information bit sequence at a rate R (bits/s/Hz) if $R < C$. The converse theorem states that the error probability is always larger than zero if the rate R is higher than the mean capacity C. Unfortunately, there exists no design rule for constructing such error-correcting codes, but after decades of research it is now possible to reach the Shannon limit very closely with turbo codes and low-density parity-check (LDPC) codes. If the instantaneous capacity $\tilde{C}(t)$ stretches above the mean capacity C for a short time interval, as shown in Figure 8.10, then this should not be misinterpreted in the sense that the Shannon limit can be exceeded during this interval. Nevertheless, the information on the temporal behaviour of the channel capacity, especially the level-crossing rate (LCR) and the average duration of the fades (ADF), is very useful for improving the system performance [281].

With the proposed MIMO channel simulator, the LCR $\tilde{N}_C(r)$ and the ADF $\tilde{T}_{C_-}(r)$ of the instantaneous capacity $\tilde{C}(t)$ can be determined by simulation. The LCR $\tilde{N}_C(r)$ describes how often the instantaneous capacity crosses a given threshold from up to down (or from down to up) within one second; and the ADF $\tilde{T}_{C_-}(r)$ is the expected value of the length of time intervals within which the instantaneous capacity $\tilde{C}(t)$ is below a given threshold. Since analytical solutions for the LCR and the ADF of the MIMO channel capacity are difficult to obtain, the proposed channel simulator is quite useful for determining these statistical quantities by

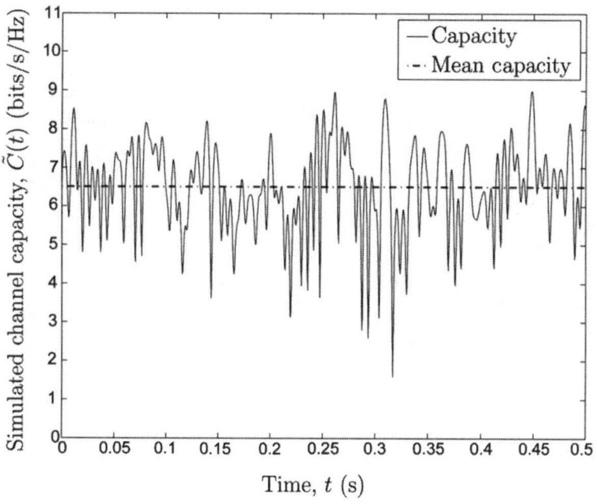

Figure 8.10 Instantaneous channel capacity $\tilde{C}(t)$ of the designed MIMO channel simulator (isotropic scattering, EMEDS, $N = 25$, $M_T = M_R = 2$, $\delta_T = \delta_R = \lambda_0/2$, $\beta_T = \beta_R = 90°$, $\alpha_v = 180°$, $f_{\max} = 91$ Hz, $\alpha^T_{\max} = 2°$, and $\gamma_c = 17$ dB).

simulation. In Figure 8.11, some simulation results for the normalized LCR $\tilde{N}_C(r)/f_{\max}$ are presented. This figure gives insight into the temporal behaviour of the channel capacity. As expected, the parameter κ of the von Mises density [see (8.26)] has a strong influence on the LCR. Figure 8.11 illustrates that the rate of change of the instantaneous channel capacity $\tilde{C}(t)$ is highest in case of isotropic scattering ($\kappa = 0$). With increasing values of κ, the spread of the AOAs becomes smaller, which significantly reduces the LCR $\tilde{N}_C(r)$.

8.3 The Two-Ring MIMO Channel Model

Mobile-to-mobile communications are expected to play an important role in multihop ad-hoc networks and intelligent transportation systems [282–284], where the communication links must be extremely reliable. To cope with problems faced by the development and performance investigation of future mobile-to-mobile radio transmission systems, new adequate multipath fading channel models are required making a detailed knowledge about their statistical properties inevitable.

This section deals with the modelling and simulation of narrowband mobile-to-mobile MIMO multipath fading channels. It is well known from the studies in [115, 270] that MIMO channels generally offer large gains in capacity over SISO channels. Especially this fact, combined with the opportunities opened by large vehicle surfaces on which multielement antennas can be placed, makes MIMO techniques very attractive for mobile-to-mobile communication systems. Although, extensive studies on the simulation of radio channels for classical cellular radio links are available in the literature, their direct applicability for simulating mobile-to-mobile fading channel statistics is not possible. The main reason is that the two channel types are significantly different in terms of the Doppler spectrum characteristics. Therefore, the extension of all these techniques with respect to the design of mobile-to-mobile MIMO channel

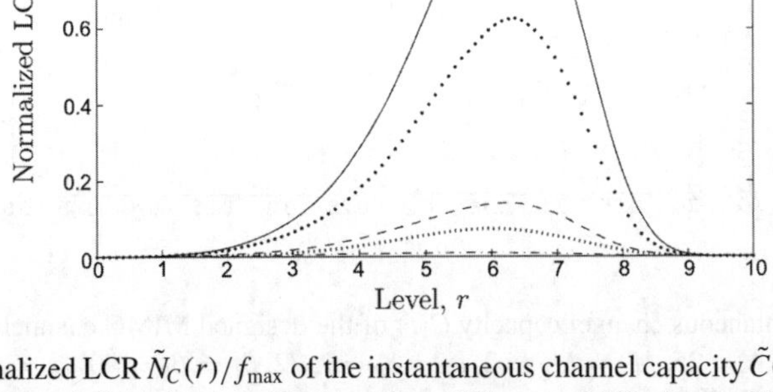

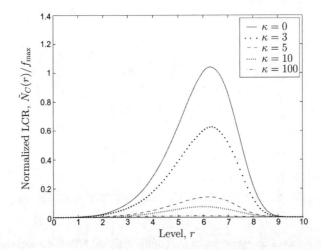

Figure 8.11 Normalized LCR $\tilde{N}_C(r)/f_{\max}$ of the instantaneous channel capacity $\tilde{C}(t)$ (LPNM, $N = 25$, $m_\alpha = 180°$, $M_T = M_R = 2$, $\delta_T = \delta_R = \lambda_0/2$, $\beta_T = \beta_R = 90°$, $\alpha_v = 180°$, $f_{\max} = 91$ Hz, $\alpha^T_{\max} = 2°$, and $\gamma_c = 17$ dB).

simulators is not straightforward. Hence, the lack of concepts for modelling, analyzing, and simulating mobile-to-mobile MIMO channels was the motivating factor in [169, 285].

The main contributions of this section are three-fold. Firstly, we will learn how a reference model for mobile-to-mobile MIMO fading channels can be derived and how its spatial/temporal correlation properties can be analyzed. The scattering environment around the transmitter and receiver is modelled here by invoking the geometrical two-ring scattering model, which is used as a starting point for the derivation of a proper reference model. The geometrical two-ring model for a MIMO channel was originally proposed in [286] for fixed-to-mobile communication links. A study of the two-ring MIMO channel model can also be found in [287] for fixed-to-fixed communication links in indoor environments. This section extends these works by assuming that both terminals are moving. Secondly, it is shown how an ergodic MIMO channel simulator can be derived from the reference channel model and how its model parameters can be computed. The procedure that is utilized to the design of the simulation model is based on a double sum of complex sinusoidal functions. The double sum of complex sinusoidal functions can be interpreted as an extension of Rice's sum-of-sinusoids method [16, 17], which has been used extensively in previous chapters to design accurate and efficient simulation models for fixed-to-mobile SISO fading channels. Thirdly, it is shown how the computation of the model parameters can be performed by applying the EMEDS, MMEA, and LPNM. The analytical and simulation results presented in this section have partially been reported in [169].

The rest of this section is structured as follows. In Subsection 8.3.1, we describe briefly the geometrical two-ring scattering model for a wireless transmission link between two mobile stations. In Subsection 8.3.2, it is then shown how the reference model for the transmission links can be derived from the geometrical model. From the reference model, a stochastic as well as a deterministic simulation model is derived in Subsection 8.3.3. Subsection 8.3.4 discusses isotropic and non-isotropic scattering scenarios and presents closed-form expressions for the most important CFs describing the reference model. Subsection 8.3.5 presents three parameter computation methods and studies the performance of the resulting MIMO channel simulator.

8.3.1 The Geometrical Two-Ring Scattering Model

In this subsection, we describe briefly the geometrical two-ring scattering model for a narrow-band mobile-to-mobile MIMO channel as illustrated in Figure 8.12. For reasons of brevity, we restrict our considerations to a 2×2 antenna configuration, meaning that both the transmitter and the receiver are equipped with only two omnidirectional antennas. Such an elementary antenna configuration can be used to construct many other types of 2D multielement antenna arrays, including uniform linear arrays, rectangular arrays, hexagonal arrays, and circular antenna arrays. For example, a circular antenna array at the transmitter side consisting of M_T (M_T even) antenna elements can be represented by $M_T/2$ antenna pairs with identical antenna element spacings but different tilt angles. For simplicity, it is assumed that there is no line-of-sight path between the two mobile stations.

The propagation scenario shown in Figure 8.12 includes a mobile transmitter and a mobile receiver. In the two-ring model, only local scattering is considered, because it is assumed that the contributions of remote scatterers to the total received power can be neglected due to high path loss. As can be seen in Figure 8.12, the local scatterers around the transmitter, denoted by

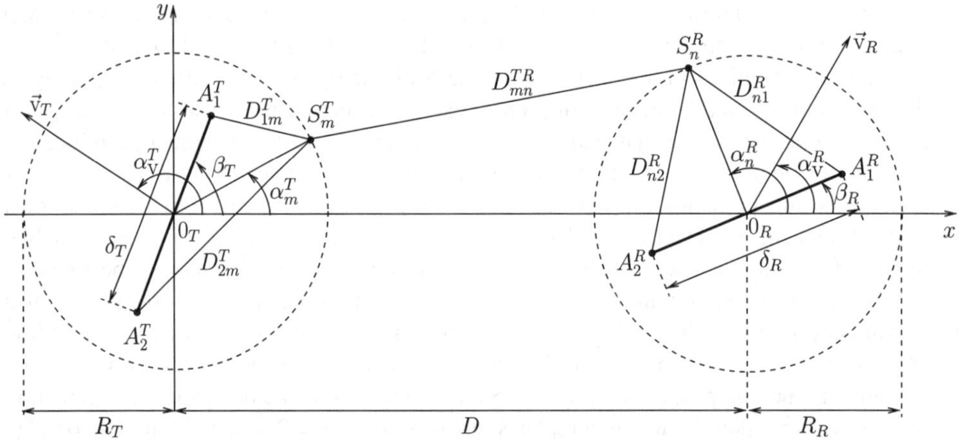

Figure 8.12 The geometrical two-ring model for a 2 × 2 MIMO channel with local scatterers around a mobile transmitter (left) and a mobile receiver (right).

S_m^T $(m = 1, 2, \ldots)$, are located on a ring of radius R_T, and the local scatterers S_n^R $(n = 1, 2, \ldots)$ around the receiver are scattered on a separate ring of radius R_R. The symbols α_m^T and α_n^R denote the AOD of the mth transmitted wave and the AOA of the nth received wave, respectively. It is assumed that the radii R_T and R_R are small in comparison to D, which is the distance between the transmitter and the receiver, i.e., $\max\{R_T, R_R\} \ll D$. The antenna element spacing of the transmitter (receiver) antenna array is denoted by δ_T (δ_R). Since the antenna element spacings δ_T and δ_R are generally small in comparison to the radii R_T and R_R, we will take profit from the inequality $\max\{\delta_T, \delta_R\} \ll \min\{R_T, R_R\}$. The tilt angle between the x-axis and the orientation of the antenna array at the transmitter is denoted by β_T. Analogously, the tilt angle β_R describes the orientation of the antenna array at the receiver. Moreover, it is assumed that the transmitter (receiver) moves with speed v_T (v_R) in the direction determined by the angle-of-motion α_v^T (α_v^R). Finally, the rings of scatterers are assumed to be fixed, so that the mobile environment can be regarded as quasi-stationary for short periods of time $\Delta t \ll R_T/v_T$ $(\Delta t \ll R_R/v_R)$.

The geometrical two-ring model is an appropriate model for describing propagation scenarios in which neither the transmitter nor the receiver is elevated, but both are surrounded by a large number of local scatterers. Such propagation conditions occur often in urban and suburban areas.

8.3.2 *The Reference Model for the Two-Ring MIMO Channel Model*

8.3.2.1 Derivation of the Reference Model

In this subsection, we derive the reference model for the mobile-to-mobile MIMO fading channel based on the propagation scenario described by the geometrical two-ring model shown in Figure 8.12. From this figure, we observe that the mth homogeneous plane wave emitted from the first antenna element A_1^T of the transmitter travels over the local scatterers S_m^T and S_n^R before impinging on the first antenna element A_1^R of the receiver. It is assumed that all waves reaching the receiver antenna array are equal in power. The reference model is based on the assumption that the numbers of local scatterers, M and N, around the transmitter and the receiver are infinite. Hence, the diffuse component at the first receiver antenna A_1^R is composed of an infinite number of homogeneous plane waves, each of which carries a power that is negligible compared to the total mean power of the diffuse component. Under flat fading conditions, the diffuse component of the channel's transmission link from A_1^T to A_1^R can be modelled as

$$h_{11}(\vec{r}_T, \vec{r}_R) = \lim_{\substack{M\to\infty \\ N\to\infty}} \sum_{m=1}^{M} \sum_{n=1}^{N} E_{mn}\, e^{j(\vec{k}_m^T \cdot \vec{r}_T - \vec{k}_n^R \cdot \vec{r}_R + \theta_{mn} - k_0 D_{mn})}. \tag{8.44}$$

In this equation, E_{mn} and θ_{mn} denote the joint gain and joint phase shift, respectively, caused by the interaction of the plane waves with the scatterers S_m^T and S_n^R. The symbol $\vec{k}_m^T$ designates the wave vector pointing in the propagation direction of the mth transmitted plane wave, and $\vec{r}_T$ denotes the spatial translation vector of the transmitter. Analogously, $\vec{k}_n^R$ is the wave vector pointing in the propagation direction of the nth received plane wave, and $\vec{r}_R$ denotes the spatial translation vector of the receiver. Furthermore, the symbol k_0 denotes the free-space wave number, which is related to the wavelength λ_0 via $k_0 = 2\pi/\lambda_0$. Finally, D_{mn} is the length of the total distance which a plane wave travels from A_1^T to A_1^R via the scatterers S_m^T and S_n^R. The plane wave model for the diffuse component $h_{11}(\vec{r}_T, \vec{r}_R)$ in (8.44) provides the starting point for the derivation of the reference model in continuous-time form.

Each local scatterer causes a gain and a phase shift. It is generally assumed that the gain and the phase shift introduced by a particular scatterer is dependent on the direction of the incoming plane waves. Let us denote the gain and the phase shift caused by S_n^R on the condition that a wave is received from S_m^T by $E_{n|m}$ and $\theta_{n|m}$, respectively. Since $D \gg \max\{R_T, R_R\}$, it follows that the gain $E_{n|m}$ and the phase shift $\theta_{n|m}$ are approximately identical to the corresponding quantities obtained on condition that a wave is received from $S_{m'}^T$, where $m \neq m'$. Therefore, we may assume that the gain $E_{n|m}$ and the phase shift $\theta_{n|m}$ caused by S_n^R are independent of the location of S_m^T and vice versa. For this reason, we may write $E_{n|m} = E_n$ ($E_{m|n} = E_m$) and $\theta_{n|m} = \theta_n$ ($\theta_{m|n} = \theta_m$). Furthermore, since $\delta_T \ll R_T$, the waves emerging from different transmitter antenna elements arrive at a particular scatterer S_m^T at approximately the same angle [286]. Similarly, if $\delta_R \ll R_R$, then we may state that waves emerging from a particular scatterer S_n^R arrive at different receiver antennas at approximately the same angle. For these reasons, it has been assumed in [287] that for a particular scatterer S_m^T (or S_n^R), the gain and phase shift is the same for waves arriving from (or travelling to) different antenna elements. In our model, we assume that each scatterer S_m^T on the ring centered on the transmitter introduces an infinitesimal constant gain $E_m = 1/\sqrt{M}$ and a random phase shift θ_m. Analogously, on the receiver side, each scatterer S_n^R introduces a constant gain $E_n = 1/\sqrt{N}$ and a random phase shift θ_n. As a consequence of the assumption $D \gg \max\{R_T, R_R\}$, it is reasonable to assume that the phase shifts θ_m and θ_n are i.i.d. random variables, each having a uniform distribution over the interval $[0, 2\pi)$. Thus, the joint gains E_{mn} and joint phases θ_{mn} in (8.44) can be expressed as

$$E_{mn} = E_m \cdot E_n = 1/\sqrt{MN}, \tag{8.45}$$

$$\theta_{mn} = \theta_m + \theta_n \pmod{2\pi}. \tag{8.46}$$

The phase changes $\vec{k}_m^T \cdot \vec{r}_T$ and $\vec{k}_n^R \cdot \vec{r}_R$ in (8.44) are due to the motion of the transmitter and the receiver, respectively. These two scalar products can be expressed in the form

$$\vec{k}_m^T \cdot \vec{r}_T = 2\pi f_{\max}^T \cos\left(\alpha_m^T - \alpha_v^T\right) t, \tag{8.47}$$

$$\vec{k}_n^R \cdot \vec{r}_R = -2\pi f_{\max}^R \cos\left(\alpha_n^R - \alpha_v^R\right) t, \tag{8.48}$$

where $f_{\max}^T = v_T/\lambda_0$ ($f_{\max}^R = v_R/\lambda_0$) is the maximum Doppler frequency caused by the movement of the transmitter (receiver). In the reference model, the AOD α_m^T and the AOA α_m^R are independent random variables determined by the distribution of the local scatterers.

Furthermore, the phase change $k_0 D_{mn}$ in (8.44) is due to the total distance travelled and can be written as

$$k_0 D_{mn} = \frac{2\pi}{\lambda_0}\left(D_{1m}^T + D_{mn}^{TR} + D_{n1}^R\right), \tag{8.49}$$

where D_{1m}^T, D_{mn}^{TR}, and D_{n1}^R are the path lengths as illustrated in Figure 8.12. These distances can be approximated by using $R_T \gg \delta_T$, $R_R \gg \delta_R$, and $\sqrt{1+x} \approx 1 + x/2$ ($x \ll 1$) as follows:

$$D_{1m}^T \approx R_T - \frac{\delta_T}{2}\cos\left(\alpha_m^T - \beta_T\right), \tag{8.50}$$

$$D_{mn}^{TR} \approx D + R_R \cos\left(\alpha_n^R\right) - R_T \cos\left(\alpha_m^T\right), \tag{8.51}$$

$$D_{n1}^R \approx R_R - \frac{\delta_R}{2}\cos\left(\alpha_n^R - \beta_R\right). \tag{8.52}$$

Finally, after substituting (8.45) and (8.47)–(8.49) in (8.44) and using (8.50)–(8.52), we may express the complex channel gain of the transmission link from A_1^T to A_1^R approximately as

$$h_{11}(t) = \lim_{\substack{M\to\infty \\ N\to\infty}} \frac{1}{\sqrt{MN}} \sum_{m=1}^{M}\sum_{n=1}^{N} g_{mn}\, e^{j[2\pi(f_m^T + f_n^R)t + \theta_{mn} + \theta_0]}, \tag{8.53}$$

where

$$g_{mn} = a_m\, b_n\, c_{mn}, \tag{8.54}$$

$$a_m = e^{j\pi(\delta_T/\lambda_0)\cos(\alpha_m^T - \beta_T)}, \tag{8.55}$$

$$b_n = e^{j\pi(\delta_R/\lambda_0)\cos(\alpha_n^R - \beta_R)}, \tag{8.56}$$

$$c_{mn} = e^{j\frac{2\pi}{\lambda_0}(R_T\cos(\alpha_m^T) - R_R\cos(\alpha_n^R))}, \tag{8.57}$$

$$f_m^T = f_{\max}^T \cos\left(\alpha_m^T - \alpha_v^T\right), \tag{8.58}$$

$$f_n^R = f_{\max}^R \cos\left(\alpha_n^R - \alpha_v^R\right), \tag{8.59}$$

$$\theta_0 = -\frac{2\pi}{\lambda_0}(R_T + D + R_R). \tag{8.60}$$

It should be mentioned that the phase θ_0 can be set to zero in (8.53) without loss of generality, because a constant phase shift does not influence the statistics of the reference model. From the statistical properties of the complex channel gain $h_{11}(t)$ in (8.53) it follows that its mean value and mean power are equal to 0 and 1, respectively.

One can show that the complex channel gain $h_{22}(t)$ of the transmission link from A_2^T to A_2^R can be obtained from (8.53) by replacing δ_T by $-\delta_T$ and δ_R by $-\delta_R$. From (8.55) and (8.56) it follows that this is equivalent to replacing a_m and b_n by their respective complex conjugates a_m^* and b_n^*. Similarly, the complex channel gains $h_{12}(t)$ and $h_{21}(t)$ can directly be obtained from (8.53) by performing the substitutions $a_m \rightarrow a_m^*$ and $b_n \rightarrow b_n^*$, respectively. The four complex channel gains $h_{kl}(t)$ $(k, l = 1, 2)$ of the A_l^T-to-A_k^R transmission links can be combined to the stochastic channel matrix

$$\mathbf{H}(t) = \begin{pmatrix} h_{11}(t) & h_{12}(t) \\ h_{21}(t) & h_{22}(t) \end{pmatrix}, \tag{8.61}$$

which describes completely the reference model of the proposed two-ring frequency-nonselective mobile-to-mobile MIMO fading channel.

If the condition $D \gg \max\{R_T, R_R\}$ is fulfilled, which is usually the case in outdoor environments, then the joint phases θ_{mn} can be written as a sum of two i.i.d. random variables [see (8.46)]. Consequently, the double sum in (8.53) can be expressed as a product of two single sums. Since the central limit theorem states that each of the single sums tends in the limit M, $N \rightarrow \infty$ to a zero-mean complex Gaussian process with unit variance, it follows that the envelope $|h_{kl}(t)|$ has a double-Rayleigh distribution [288].[2] In [289] it has been shown that double-Rayleigh fading channels result in considerably higher system performance degradations compared to classical (single) Rayleigh fading channels. However, if the condition $D \gg \max\{R_T, R_R\}$ is not fulfilled, which is usually the case in indoor environments, then it is reasonable to assume that the joint phases θ_{mn} are i.i.d. random variables. In this case, the double sum in (8.53) cannot be expressed as the product of two single sums, with the consequence that $|h_{kl}(t)|$ follows the classical Rayleigh distribution.

The extension of the reference model to uniform linear antenna arrays consisting of any number of $M_T \geq 2$ $(M_R \geq 2)$ transmitter (receiver) antenna elements can easily be made. Without proof, we mention that the complex channel gain $h_{kl}(t)$ of the transmission link from A_l^T $(l = 1, 2, \ldots, M_T)$ to A_k^R $(k = 1, 2, \ldots, M_R)$ follows from $h_{11}(t)$ [see (8.53)] by replacing δ_T/λ_0 in (8.55) and δ_R/λ_0 in (8.56) by $(M_T - 2l + 1)\,\delta_T/\lambda_0$ and $(M_R - 2k + 1)\,\delta_R/\lambda_0$, respectively.

8.3.2.2 Correlation Functions of the Reference Model

A measure to quantify the spatial and temporal correlation between two different transmission links is the 3D space-time CCF. According to [273], the 3D space-time CCF between the two transmission links from A_1^T to A_1^R and from A_2^T to A_2^R is defined as the correlation between the complex channel gains $h_{11}(t)$ and $h_{22}(t)$, i.e.,

$$\rho_{11,22}(\delta_T, \delta_R, \tau) := E\{h_{11}(t)h_{22}^*(t+\tau)\}. \tag{8.62}$$

Here, the expectation operator applies to all random variables: phases $\{\theta_{mn}\}$, AOD $\{\alpha_m^T\}$, and AOA $\{\alpha_n^R\}$. Starting from (8.53) and making use of the fact that $h_{22}(t)$ can be obtained

[2] A double Rayleigh process $\chi(t)$ is a product process of two independent Rayleigh processes $\zeta^{(1)}(t)$ and $\zeta^{(2)}(t)$, i.e., $\chi(t) = \zeta^{(1)}(t) \cdot \zeta^{(2)}(t)$. The distribution of $\chi(t)$ is called the *double Rayleigh distribution*, which can be derived by using (2.40) and (2.82) as $p_\chi(z) = z/(\sigma_1\sigma_2)^2 K_0(z/(\sigma_1\sigma_2))$. Here, $K_0(\cdot)$ denotes the modified Bessel function [23, Equation (8.432–6)] and σ_k $(k = 1, 2)$ is the standard deviation of the underlying real-valued Gaussian processes $\mu_i^{(k)}(t)$ $(i, k = 1, 2)$ defining the Rayleigh processes $\zeta^{(k)}(t) = |\mu_1^{(k)}(t) + j\mu_2^{(k)}(t)|$ for $k = 1, 2$.

from $h_{11}(t)$ by performing the substitutions $a_m \to a_m^*$ and $b_n \to b_n^*$, we can express the 3D space-time CCF as

$$\rho_{11,22}(\delta_T, \delta_R, \tau) = \lim_{\substack{M\to\infty \\ N\to\infty}} \frac{1}{MN} \sum_{m=1}^{M} \sum_{n=1}^{N} E\{a_m^2 b_n^2 e^{-j2\pi(f_m^T + f_n^R)\tau}\}. \tag{8.63}$$

The expression above has been obtained by averaging over the random phases θ_{mn}. Note that a_m and f_m^T are both functions of the AOD α_m^T, while b_n and f_n^R depend upon the AOA α_n^R. If the number of local scatterers approaches infinity ($M, N \to \infty$), then the discrete random variables α_m^T and α_n^R become continuous random variables α^T and α^R, each of which is characterized by a certain distribution, denoted by $p_{\alpha^T}(\alpha^T)$ and $p_{\alpha^R}(\alpha^R)$, respectively. The infinitesimal power of the complex channel gains corresponding to the differential angles $d\alpha^T$ and $d\alpha^R$ is proportional to $p_{\alpha^T}(\alpha^T)\, p_{\alpha^R}(\alpha^R)\, d\alpha^T\, d\alpha^R$. As $M \to \infty$ and $N \to \infty$, this contribution must be equal to $1/(MN)$, i.e., $1/(MN) = p_{\alpha^T}(\alpha^T)\, p_{\alpha^R}(\alpha^R)\, d\alpha^T\, d\alpha^R$. Hence, it follows from (8.63) that the 3D space-time CCF of the reference model can be expressed as

$$\rho_{11,22}(\delta_T, \delta_R, \tau) = \rho_T(\delta_T, \tau) \cdot \rho_R(\delta_R, \tau), \tag{8.64}$$

where

$$\rho_T(\delta_T, \tau) = \int_{-\pi}^{\pi} a^2(\delta_T, \alpha^T)\, e^{-j2\pi f^T(\alpha^T)\tau}\, p_{\alpha^T}(\alpha^T)\, d\alpha^T \tag{8.65}$$

and

$$\rho_R(\delta_R, \tau) = \int_{-\pi}^{\pi} b^2(\delta_R, \alpha^R)\, e^{-j2\pi f^R(\alpha^R)\tau}\, p_{\alpha^R}(\alpha^R)\, d\alpha^R \tag{8.66}$$

are called the *transmitter CF* and the *receiver CF*, respectively, and

$$a(\delta_T, \alpha^T) = e^{j\pi(\delta_T/\lambda_0)\cos(\alpha^T - \beta_T)}, \tag{8.67}$$

$$b(\delta_R, \alpha^R) = e^{j\pi(\delta_R/\lambda_0)\cos(\alpha^R - \beta_R)}, \tag{8.68}$$

$$f^T(\alpha^T) = f_{\max}^T \cos\left(\alpha^T - \alpha_v^T\right), \tag{8.69}$$

$$f^R(\alpha^R) = f_{\max}^R \cos\left(\alpha^R - \alpha_v^R\right). \tag{8.70}$$

From (8.64), we observe that the 3D space-time CCF $\rho_{11,22}(\delta_T, \delta_R, \tau)$ can be expressed as the product of the transmitter CF $\rho_T(\delta_T, \tau)$ and the receiver CF $\rho_R(\delta_R, \tau)$. This feature can be attributed to the independency of the AOD α^T and the AOA α^R, which was also pointed out for the indoor MIMO channel model proposed in [287]. For the single bounce one-ring model with its strict relationship between α^T and α^R, this property does not hold, as can be seen by viewing the result in (8.20). Furthermore, we also observe that $\rho_{11,22}(\delta_T, \delta_R, \tau)$ in (8.64) is independent of the ring radii (R_T, R_R) and the distance D between the transmitter and the receiver, which is a direct consequence of the assumption $D \gg \max\{R_T, R_R\}$.

The 2D space CCF $\rho(\delta_T, \delta_R)$, defined as $\rho(\delta_T, \delta_R) = E\{h_{11}(t)\, h_{22}^*(t)\}$, equals the 3D space-time CCF $\rho_{11,22}(\delta_T, \delta_R, \tau)$ at $\tau = 0$, i.e.,

$$\begin{aligned}\rho(\delta_T, \delta_R) &= \rho_{11,22}(\delta_T, \delta_R, 0) \\ &= \rho_T(\delta_T, 0) \cdot \rho_R(\delta_R, 0). \end{aligned} \tag{8.71}$$

The time ACF $r_{h_{kl}}(\tau)$ of the complex channel gain $h_{kl}(t)$ of the transmission link from A_l^T to A_k^R $(k, l = 1, 2)$ is defined by

$$r_{h_{kl}}(\tau) := E\{h_{kl}(t)\, h_{kl}^*(t + \tau)\}. \tag{8.72}$$

After applying similar techniques as introduced at the beginning of this subsection, we can show that the equation above can be expressed as

$$r_{h_{kl}}(\tau) = \rho_T(0, \tau) \cdot \rho_R(0, \tau) \tag{8.73}$$

for all $k, l \in \{1, 2\}$, where $\rho_T(0, \tau)$ and $\rho_R(0, \tau)$ can be obtained from (8.65) and (8.66) as

$$\rho_T(0, \tau) = \int_{-\pi}^{\pi} e^{-j2\pi f^T(\alpha^T)\tau}\, p_{\alpha^T}(\alpha^T)\, d\alpha^T, \tag{8.74}$$

$$\rho_R(0, \tau) = \int_{-\pi}^{\pi} e^{-j2\pi f^R(\alpha^R)\tau}\, p_{\alpha^R}(\alpha^R)\, d\alpha^R, \tag{8.75}$$

respectively. Notice that $r_{h_{kl}}(\tau)$ equals $\rho_{11,22}(\delta_T, \delta_R, \tau)$ at $\delta_T = \delta_R = 0$, i.e., $r_{h_{kl}}(\tau) = \rho_{11,22}(0, 0, \tau) = \rho_T(0, \tau) \cdot \rho_R(0, \tau)$ $\forall\, k, l \in \{1, 2\}$. This result shows that the time ACFs of all four complex channel gains are identical. The same statement holds also for the one-ring model.

The reference model described above is basically a theoretical model, which grounds on the assumption that the numbers of scatterers (N, M) are infinite. Although such a model is non-realizable, it is of central importance for the derivation of stochastic and deterministic simulation models. The design of efficient MIMO channel simulators having approximately the same statistical properties as the reference model is the topic of the next subsection.

8.3.3 Simulation Models for the Two-Ring MIMO Channel Model

This subsection presents a stochastic as well as a deterministic simulation model by applying the generalized principle of deterministic channel modelling described in Section 8.1.

8.3.3.1 The Stochastic Simulation Model

From the complex channel gain $h_{11}(t)$ of the reference model described in (8.53), a stochastic simulation model is obtained after performing the following three steps: (i) using only finite numbers of scatterers M and N, (ii) setting the constant phase shift θ_0 in (8.53) equal to zero, and (iii) considering the discrete AODs α_m^T and AOAs α_n^R as constant quantities. Hence, the complex channel gain of the A_1^T-to-A_1^R transmission link can be expressed as

$$\hat{h}_{11}(t) = \frac{1}{\sqrt{MN}} \sum_{m=1}^{M} \sum_{n=1}^{N} g_{mn}\, e^{j\{2\pi [f_m^T(\alpha_m^T) + f_n^R(\alpha_n^R)]t + \theta_{mn}\}}, \tag{8.76}$$

where g_{mn}, $f_m^T(\cdot)$, and $f_n^R(\cdot)$ are given by (8.54), (8.58), and (8.59), respectively. In contrast to the reference model, the discrete AODs α_m^T and AOAs α_n^R are now constants, which will be determined in Subsection 8.3.5 such that the simulation model has nearly the same statistical properties as the reference model. The phases θ_{mn} are still i.i.d. random variables, each with a uniform distribution over the interval $(0, 2\pi]$. Hence, $\hat{h}_{11}(t)$ represents a stochastic process. In analogy to the reference model, the complex channel gain $\hat{h}_{22}(t)$ of the transmission link from A_2^T to A_2^R can be obtained from (8.76) by replacing a_m and b_n by their respective complex conjugates a_m^* and b_n^*. Similarly, $\hat{h}_{12}(t)$ and $\hat{h}_{21}(t)$ can be derived from (8.76) by replacing a_m by a_m^* and b_n by b_n^*, respectively. This allows us to define the channel matrix of the stochastic MIMO channel simulator as $\hat{\mathbf{H}}(t) := [\hat{h}_{kl}(t)]$. Figure 8.13 presents the structure of the resulting stochastic two-ring MIMO channel simulator for the case $M_T = M_R = 2$.

The 3D space-time CCF of $\hat{h}_{11}(t)$ and $\hat{h}_{22}(t)$ is defined by

$$\hat{\rho}_{11,22}(\delta_T, \delta_R, \tau) := E\{\hat{h}_{11}(t)\, \hat{h}_{22}^*(t+\tau)\}, \tag{8.77}$$

where the expectation operator $E\{\,\cdot\,\}$ applies only to the random phases θ_{mn}. After averaging over the uniformly distributed random phases θ_{mn}, the 3D space-time CCF $\hat{\rho}_{11,22}(\delta_T, \delta_R, \tau)$ can be expressed in closed form as

$$\begin{aligned} \hat{\rho}_{11,22}(\delta_T, \delta_R, \tau) &= \frac{1}{MN} \sum_{m=1}^{M} \sum_{n=1}^{N} a_m^2(\delta_T)\, b_n^2(\delta_R)\, e^{-j2\pi (f_m^T + f_n^R)\tau} \\ &= \hat{\rho}_T(\delta_T, \tau) \cdot \hat{\rho}_R(\delta_R, \tau), \end{aligned} \tag{8.78}$$

where

$$\hat{\rho}_T(\delta_T, \tau) = \frac{1}{M} \sum_{m=1}^{M} a_m^2(\delta_T)\, e^{-j2\pi f_m^T \tau} \tag{8.79}$$

and

$$\hat{\rho}_R(\delta_R, \tau) = \frac{1}{N} \sum_{n=1}^{N} b_n^2(\delta_R)\, e^{-j2\pi f_n^R \tau} \tag{8.80}$$

are the transmitter and receiver CF of the stochastic simulation model, respectively. From (8.78), we realize that the property of separable transmitter and receiver CFs also holds for the stochastic simulation model.

The 2D space CCF $\hat{\rho}(\delta_T, \delta_R)$, defined as $\hat{\rho}(\delta_T, \delta_R) := E\{\hat{h}_{11}(t)\, \hat{h}_{22}^*(t)\}$, equals the 3D space-time CCF $\hat{\rho}_{11,22}(\delta_T, \delta_R, \tau)$ at $\tau = 0$, i.e.,

$$\begin{aligned} \hat{\rho}(\delta_T, \delta_R) &= \hat{\rho}_{11,22}(\delta_T, \delta_R, 0) \\ &= \hat{\rho}_T(\delta_T, 0) \cdot \hat{\rho}_R(\delta_R, 0). \end{aligned} \tag{8.81}$$

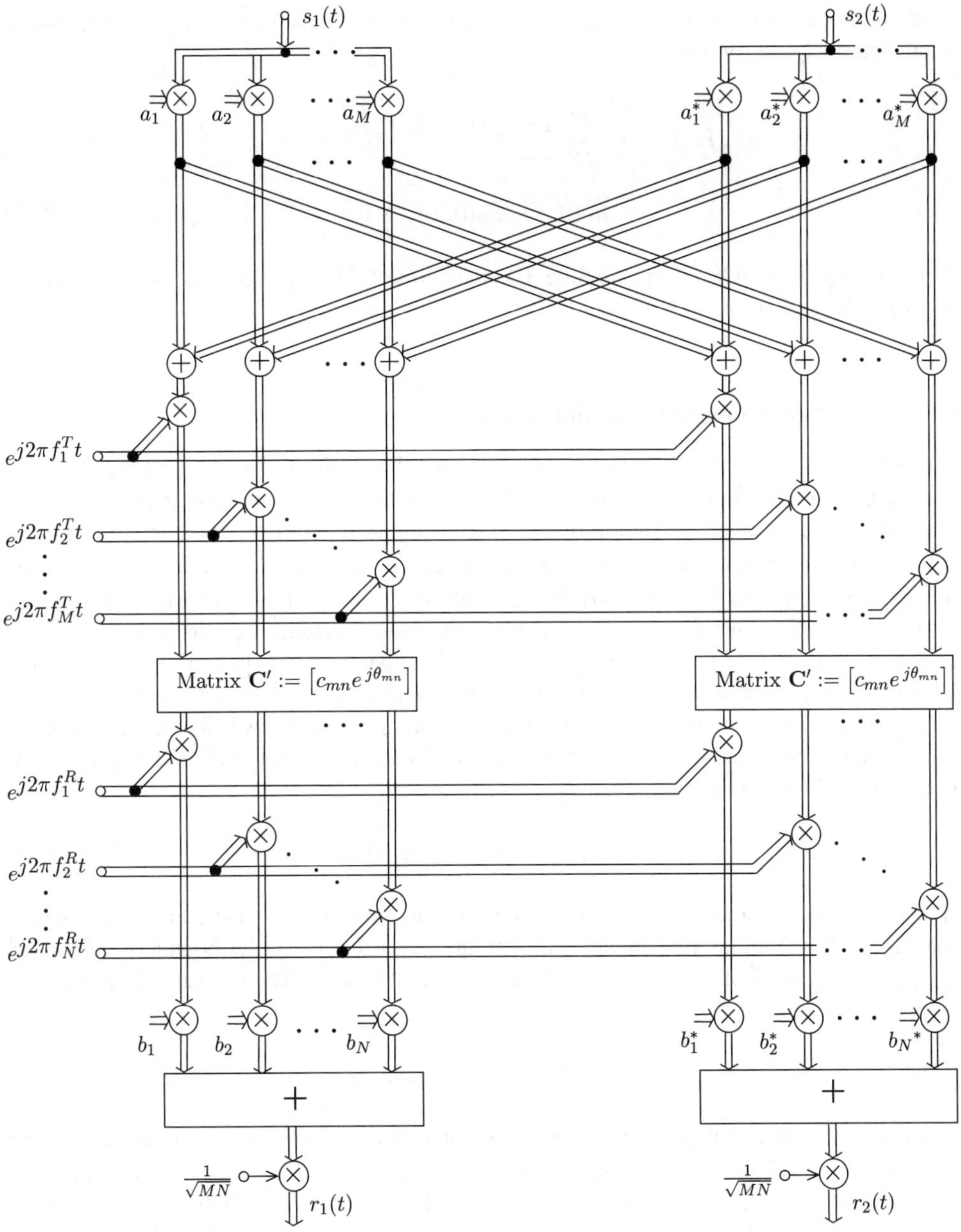

Figure 8.13 Structure of the two-ring 2×2 MIMO channel simulator for mobile-to-mobile channels.

Furthermore, the time ACF $\hat{r}_{h_{kl}}(\tau)$ of $\hat{h}_{kl}(t)$, defined as $\hat{r}_{h_{kl}}(\tau) := E\{\hat{h}_{kl}(t)\,\hat{h}^*_{kl}(t+\tau)\}$, can be expressed in closed form as

$$\hat{r}_{h_{kl}}(\tau) = \frac{1}{MN}\sum_{m=1}^{M}\sum_{n=1}^{N} e^{-j2\pi(f_m^T+f_n^R)\tau}$$
$$= \hat{\rho}_T(0,\tau)\cdot\hat{\rho}_R(0,\tau) \quad \forall k,l \in \{1,2\}. \tag{8.82}$$

Note that the time ACF $\hat{r}_{h_{kl}}(\tau)$ and the 3D space-time CCF $\hat{\rho}_{11,22}(\delta_T,\delta_R,\tau)$ are related by $\hat{r}_{h_{kl}}(\tau) = \hat{\rho}_{11,22}(0,0,\tau)$.

8.3.3.2 The Deterministic Simulation Model

The deterministic simulation model is obtained from the stochastic one by keeping all model parameters fixed, including the phases θ_{mn}. We will see that $\hat{h}_{kl}(t)$ is an ergodic process, which motivates us to focus our attention onto a single sample function of the stochastic process $\hat{h}_{kl}(t)$. In the following, we denote a sample function of $\hat{h}_{kl}(t)$ by $\tilde{h}_{kl}(t)$. A sample function $\tilde{h}_{kl}(t)$ is obtained from $\hat{h}_{kl}(t)$ by considering all phases $\theta_{mn} = \theta_m + \theta_n$ as constants determined by the sum of the outcomes of two independent random generators, each with a uniform distribution over $(0, 2\pi]$. As a consequence of the fact that all model parameters are constant, it follows that the complex channel gains $\tilde{h}_{kl}(t)$ and the associated time-varying channel matrix $\tilde{\mathbf{H}}(t) = [\tilde{h}_{kl}(t)]$ are completely determined for all values of t. The statistical properties of $\tilde{h}_{kl}(t)$ must be analyzed by using time averages rather than statistical averages. For example, the 3D space-time CCF has to be computed by means of

$$\tilde{\rho}_{11,22}(\delta_T,\delta_R,\tau) :=< \tilde{h}_{11}(t)\,\tilde{h}^*_{22}(t+\tau) >, \tag{8.83}$$

where $<\cdot>$ denotes the time average operator. Let us impose the constraint on the channel model that the Doppler frequencies $f_m^T \neq 0$ and $f_n^R \neq 0$ satisfy the boundary condition[3] $f_m^T + f_n^R = f_{m'}^T + f_{n'}^R$ if and only if (iff) $m = m'$ and $n = n'$. Then it can be shown (see Appendix 8.A) that (8.83) results in

$$\tilde{\rho}_{11,22}(\delta_T,\delta_R,\tau) = \hat{\rho}_{11,22}(\delta_T,\delta_R,\tau), \tag{8.84}$$

where $\hat{\rho}_{11,22}(\delta_T,\delta_R,\tau)$ is given by (8.78). Equation (8.84) states that the statistical average of $\hat{h}_{11}(t)\hat{h}^*_{22}(t+\tau)$ equals the time average of $\tilde{h}_{11}(t)\tilde{h}^*_{22}(t+\tau)$. Hence, the proposed stochastic MIMO channel simulator is ergodic with respect to the 3D space-time CCF. This is in contrast to the non-ergodic MIMO channel simulator derived for the two-ring model in [290], where it was assumed that the position of the transmitter is fixed and only the receiver is moving.

[3] It should be mentioned that this condition is sufficient. Necessary conditions are: (i) $f_m^T \neq f_{m'}^T$ if $m \neq m'$, (ii) $f_n^R \neq f_{n'}^R$ if $n \neq n'$, and (iii) $f_m^T \neq f_n^R$ $\forall\, m,n$. These three conditions can be summarized by saying that all Doppler frequencies must be different.

By analogy to (8.81), the 2D space CCF $\tilde{\rho}(\delta_T, \delta_R)$ of the deterministic MIMO channel simulator can be expressed as

$$\begin{aligned}\tilde{\rho}(\delta_T, \delta_R) &:= < \tilde{h}_{11}(t)\, \tilde{h}_{22}^*(t) > \\ &= \tilde{\rho}_{11,22}(\delta_T, \delta_R, 0) \\ &= \tilde{\rho}_T(\delta_T, 0) \cdot \tilde{\rho}_R(\delta_R, 0),\end{aligned} \tag{8.85}$$

where $\tilde{\rho}_T(\delta_T, 0) = \hat{\rho}_T(\delta_T, 0)$ and $\tilde{\rho}_R(\delta_R, 0) = \hat{\rho}_R(\delta_R, 0)$ hold, implying that the stochastic MIMO channel simulator is ergodic with respect to the 2D space CCF.

Similarly, the time ACF $\tilde{r}_{h_{kl}}(\tau)$ of the deterministic process $\tilde{h}_{kl}(t)$ can be obtained as

$$\begin{aligned}\tilde{r}_{h_{kl}}(\tau) &:= < \tilde{h}_{kl}(t)\, \tilde{h}_{kl}^*(t+\tau) > \\ &= \frac{1}{MN} \sum_{m=1}^{M} \sum_{n=1}^{N} e^{-j2\pi (f_m^T + f_n^R)\tau} \\ &= \tilde{\rho}_T(0, \tau) \cdot \tilde{\rho}_R(0, \tau) \quad \forall k, l \in \{1, 2\},\end{aligned} \tag{8.86}$$

where $\tilde{\rho}_T(0, \tau) = \hat{\rho}_T(0, \tau)$ and $\tilde{\rho}_R(0, \tau) = \hat{\rho}_R(0, \tau)$. Hence, the stochastic MIMO channel simulator is ergodic with respect to the time ACF. Note that (8.86) can also be obtained from $\tilde{\rho}_{11,22}(\delta_T, \delta_R, \tau)$ by setting $\delta_T = \delta_R = 0$, i.e., $\tilde{r}_{h_{kl}}(\tau) = \tilde{\rho}_{11,22}(0, 0, \tau)$.

8.3.4 *Isotropic and Non-Isotropic Scattering Scenarios*

In this section, we study the correlation properties of the reference model under both isotropic and non-isotropic scattering conditions. Since the 3D space-time CCF $\rho_{11,22}(\delta_T, \delta_R, \tau)$ of the reference model can be expressed as the product of the transmitter CF $\rho_T(\delta_T, \tau)$ and the receiver CF $\rho_R(\delta_R, \tau)$, it is sufficient to focus our attention on the transmitter CF. The corresponding results for the receiver CF $\rho_R(\delta_R, \tau)$ can directly be obtained from $\rho_T(\delta_T, \tau)$ by replacing the index T by R.

8.3.4.1 Isotropic Scattering Scenarios

Isotropic scattering around the transmitter is characterized by a uniform distribution of the AOD α^T, i.e.,

$$p_{\alpha^T}(\alpha^T) = \frac{1}{2\pi}, \quad \alpha^T \in [0, 2\pi). \tag{8.87}$$

Substituting (8.87) in (8.65) and solving the integral by using [23, Equation (3.338–4)], [23, Equation (3.339)], and [77, Equation (9.6.3)], we obtain the following closed-form solution for the transmitter CF [169]

$$\rho_T(\delta_T, \tau) = J_0\left(2\pi\left[\left(\frac{\delta_T}{\lambda_0}\right)^2 + (f_{\max}^T \tau)^2 - 2\left(\frac{\delta_T}{\lambda_0}\right) f_{\max}^T \tau \cos(\alpha_v^T - \beta_T)\right]^{1/2}\right), \tag{8.88}$$

where $J_0(\cdot)$ denotes the Bessel function of the first kind of order zero. Assuming likewise isotropic scattering around the receiver, then the receiver CF $\rho_R(\delta_R, \tau)$ is obtained from (8.88) simply by replacing the subscript and superscript symbol T by R. By means of (8.64), this allows us to represent the 3D space-time CCF of the reference model as [169]

$$\rho_{11,22}(\delta_T, \delta_R, \tau) = J_0\left(2\pi\left[\left(\frac{\delta_T}{\lambda_0}\right)^2 + (f_{\max}^T\tau)^2 - 2\left(\frac{\delta_T}{\lambda_0}\right) f_{\max}^T \tau \cos(\alpha_v^T - \beta_T)\right]^{1/2}\right)$$
$$\cdot J_0\left(2\pi\left[\left(\frac{\delta_R}{\lambda_0}\right)^2 + (f_{\max}^R\tau)^2 - 2\left(\frac{\delta_R}{\lambda_0}\right) f_{\max}^R \tau \cos(\alpha_v^R - \beta_R)\right]^{1/2}\right). \tag{8.89}$$

From (8.89) and (8.71), it follows that the 2D space CCF $\rho(\delta_T, \delta_R)$ can be expressed as the product of two Bessel functions according to $\rho(\delta_R, \delta_R) = \rho_{11,22}(\delta_T, \delta_R, 0) = J_0(2\pi\delta_T/\lambda_0) \cdot J_0(2\pi\delta_R/\lambda_0)$. Finally, we notice that the time ACF $r_{h_{kl}}(\tau)$ of the reference model results in $r_{h_{kl}}(\tau) = \rho_{11,22}(0, 0, \tau) = J_0(2\pi f_{\max}^T\tau) \cdot J_0(2\pi f_{\max}^R\tau)$ for all $k, l = 1, 2$. The observation that the time ACF can be expressed as the product of two Bessel functions has also been made in [79], where a mobile-to-mobile channel with a single transmitter and a single receiver antenna has been studied.

8.3.4.2 Non-Isotropic Scattering Scenarios

To describe non-isotropic scattering scenarios, we assume that the AOD α^T follows the von Mises distribution, which has been shown to be successful in describing measured data [273]. Another advantage of the von Mises distribution is that the 3D space-time CCF of the reference model can be expressed in closed form. The von Mises distribution is given by

$$p_{\alpha^T}(\alpha^T) = \frac{1}{2\pi I_0(\kappa_T)} e^{\kappa_T \cos(\alpha^T - m_\alpha^T)}, \quad \alpha^T \in [0, 2\pi). \tag{8.90}$$

The parameter $m_\alpha^T \in [0, 2\pi)$ denotes the mean AOD, and $\kappa_T \geq 0$ is a parameter that controls the angular spread around the mean m_α^T. If $\kappa_T = 0$, then the density in (8.90) reduces to the uniform distribution in (8.87). As κ_T increases, the scatterers become more clustered around m_α^T, which is typical for non-isotropic scattering environments. Substituting (8.90) in (8.65) and using [23, Equation (3.338–4)], we obtain the transmitter CF $\rho_T(\delta_T, \tau)$ of the reference model in closed form as [169]

$$\rho_T(\delta_T, \tau) = \frac{1}{I_0(\kappa_T)} I_0\left(\left\{\kappa_T^2 - 4\pi^2\left[\left(\frac{\delta_T}{\lambda_0}\right)^2 + \left(f_{\max}^T\tau\right)^2 - 2\left(\frac{\delta_T}{\lambda_0}\right) f_{\max}^T \tau \cos\left(\alpha_v^T - \beta_T\right)\right]\right.\right.$$
$$\left.\left. + j4\pi\kappa_T\left[\frac{\delta_T}{\lambda_0}\cos\left(\beta_T - m_\alpha^T\right) - f_{\max}^T\tau\cos\left(\alpha_v^T - m_\alpha^T\right)\right]\right\}^{1/2}\right). \tag{8.91}$$

The equation above reduces to (8.88) in case of isotropic scattering ($\kappa_T = 0$). If the AOA α^R is also von Mises distributed, then the receiver CF $\rho_R(\delta_R, \tau)$ is obtained from (8.91) after substituting the index R for T. From the CFs $\rho_T(\delta_T, \tau)$ and $\rho_R(\delta_T, \tau)$, we can then find directly closed-form solutions for the 3D space-time CCF $\rho_{11,22}(\delta_T, \delta_R, \tau)$ [see (8.64)], the 2D space CCF $\rho(\delta_T, \delta_R)$ [see (8.71)], and the time ACF $r_{h_{kl}}(\tau)$ [see (8.73)].

8.3.5 Parameter Computation Methods

In this section, we present three methods for the computation of the parameters determining the statistics of the MIMO channel simulation model. The first method is the EMEDS, which is highly recommended in case of isotropic scattering. The other methods are known as the MMEA [291] and the LPNM. Especially the MMEA and the LPNM can be applied to any given distribution of the AOD (AOA), including the von Mises distribution [176], the Gaussian distribution [274], the truncated Laplacian distribution [275, 292], and the truncated uniform distribution [293].

For the simulation model, the user must first choose adequate values for the numbers of discrete scatterers M and N. Reasonable values for these parameters are in the range from 40 to 50, which might be considered here as a good trade-off between realization complexity and performance if the fitting to the correlation properties of the reference model is taken as a target. The unknown model parameters to be determined are only the discrete AODs α_m^T and AOAs α_n^R, since the remaining parameters of the simulation model are identical to the corresponding parameters of the reference model. Our aim is thus to determine the two sets $\{\alpha_m^T\}_{m=1}^M$ and $\{\alpha_n^R\}_{n=1}^N$ in such a way that the transmitter and receiver CFs of the simulation model are sufficiently close to the corresponding CFs describing the reference model.

8.3.5.1 Extended Method of Exact Doppler Spread (EMEDS)

The original MEDS is described in detail in Subsection 5.1.7. This method was first introduced in [96] as a high-performance parameter computation method for sum-of-sinusoids-based Rayleigh fading channel simulators. When applying the original MEDS, then the AODs (AOAs) are confined to the interval $(0, \pi/2]$; but here we consider the two-ring model under the condition of uniformly distributed local scatterers around both the transmitter and the receiver, where the AODs and AOAs are spanning the range from 0 to 2π. This calls for an extension of the MEDS. By avoiding redundancies and exploiting the symmetry properties of the cosine functions in (8.58) and (8.59), the extension results in the following closed-form solution of the AODs α_m^T and AOAs α_n^R:

$$\alpha_m^T = \frac{2\pi}{M}\left(m - \frac{1}{4}\right) + \alpha_v^T, \qquad m = 1, 2, \ldots, M, \tag{8.92}$$

$$\alpha_n^R = \frac{2\pi}{N}\left(n - \frac{1}{4}\right) + \alpha_v^R, \qquad n = 1, 2, \ldots, N, \tag{8.93}$$

where we have to take into account that the condition $\{f_m^T\} \cap \{f_n^R\} = \emptyset$ must be fulfilled. This is in general the case if $f_{\max}^T \neq f_{\max}^R$ or $\alpha_v^T \neq \alpha_v^R$ and can also most easily be achieved if $f_{\max}^T = f_{\max}^R$ and $\alpha_v^T = \alpha_v^R$, e.g., by defining $M := N + 1$ (N odd) or $M := N + 2$ (N even). When both M and N are even numbers, it can be shown by using (8.92) and (8.93) that the transmitter CF

$\tilde{\rho}_T(\delta_T, \tau)$ and the receiver CF $\tilde{\rho}_R(\delta_R, \tau)$ are given by [169]

$$\tilde{\rho}_T(\delta_T, \tau) = \frac{2}{M}\sum_{m=1}^{M/2} \cos\left\{2\pi(\delta_T/\lambda_0)\cos\left(\frac{2\pi}{M}\left(m-\frac{1}{4}\right)+\alpha_v^T-\beta_T\right)\right.$$
$$\left. -2\pi f_{\max}^T \tau \cos\left(\frac{2\pi}{M}\left(m-\frac{1}{4}\right)\right)\right\}, \tag{8.94}$$

$$\tilde{\rho}_R(\delta_R, \tau) = \frac{2}{N}\sum_{n=1}^{N/2} \cos\left\{2\pi(\delta_R/\lambda_0)\cos\left(\frac{2\pi}{N}\left(n-\frac{1}{4}\right)+\alpha_v^R-\beta_R\right)\right.$$
$$\left. -2\pi f_{\max}^R \tau \cos\left(\frac{2\pi}{N}\left(n-\frac{1}{4}\right)\right)\right\}. \tag{8.95}$$

One can show that $\tilde{\rho}_T(\delta_T, \tau) \to \rho_T(\delta_T, \tau)$ as $M \to \infty$, where $\rho_T(\delta_T, \tau)$ is the transmitter CF of the reference model as given by (8.88). Due to the symmetry of the two-ring model, it is obvious that the same property holds also for the receiver CF $\tilde{\rho}_R(\delta_R, \tau)$, i.e., $\tilde{\rho}_R(\delta_R, \tau) \to \rho_R(\delta_R, \tau)$ as $N \to \infty$.

Unless indicated otherwise, the following parameters have been used in all illustrative examples given below. At the transmitter side, the antenna tilt angle β_T was chosen as $\beta_T = 90°$ and the angle-of-motion α_v^T was set to $0°$. Using these parameters and evaluating (8.88) enables us to visualize graphically the behaviour of the reference model's transmitter CF $\rho_T(\delta_T, \tau)$ as shown in Figure 8.14. A plot of the corresponding transmitter CF $\tilde{\rho}_T(\delta_T, \tau)$ characterizing the simulation model is illustrated in Figure 8.15. These results have been obtained by evaluating (8.94) by choosing $M = 40$. An impression of the performance of the EMEDS can be gathered from comparing the results of the reference model depicted in Figure 8.14 with the results of the simulation model presented in Figure 8.15. A more detailed insight into the performance of the EMEDS requires numerical investigations, which have revealed that an excellent approximation $\rho_T(\delta_T, \tau) \approx \tilde{\rho}_T(\delta_T, \tau)$ can be achieved over the area in the (δ_T, τ)-plane limited by

$$\sqrt{(\delta_T/\lambda_0)^2 + (f_{\max}^T \tau)^2 - 2\left(\frac{\delta_T}{\lambda_0}\right) f_{\max}^T \tau \cos\left(\alpha_v^T - \beta_T\right)} \leq M/8. \tag{8.96}$$

Let $\delta_T = 0$, then the approximation $\rho_T(0, \tau) \approx \tilde{\rho}_T(0, \tau)$ is even excellent if $\tau \cdot f_{\max}^T$ is within the range $[0, M/4]$. Similar observations have been reported in [96], where the original MEDS has been used to design sum-of-sinusoids Rayleigh fading channel simulators for fixed-to-mobile links under isotropic scattering conditions. Analogously, the approximation $\rho_T(\delta_T, 0) \approx \tilde{\rho}_T(\delta_T, 0)$ is very accurate if δ_T/λ_0 is confined to the range $[0, M/4]$ and $\alpha_v^T - \beta_T = \pm 2\pi k/M \; (k = 0, 1, 2, \ldots)$. As an appropriate measure for the quality of the approximation $\rho_T(\delta_T, \tau) \approx \tilde{\rho}_T(\delta_T, \tau)$, we consider the absolute error of the simulation model's transmitter CF

$$e_T(\delta_T, \tau) = |\rho_T(\delta_T, \tau) - \tilde{\rho}_T(\delta_T, \tau)|, \tag{8.97}$$

which is illustrated in Figure 8.16. It goes without saying that similar results hold also for the quality of the approximation $\rho_R(\delta_R, \tau) \approx \tilde{\rho}_R(\delta_R, \tau)$.

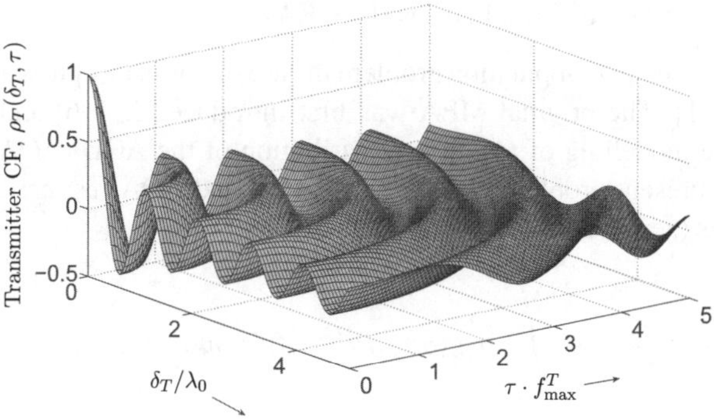

Figure 8.14 The transmitter CF $\rho_T(\delta_T, \tau)$ of the 2×2 MIMO mobile-to-mobile reference channel model (isotropic scattering).

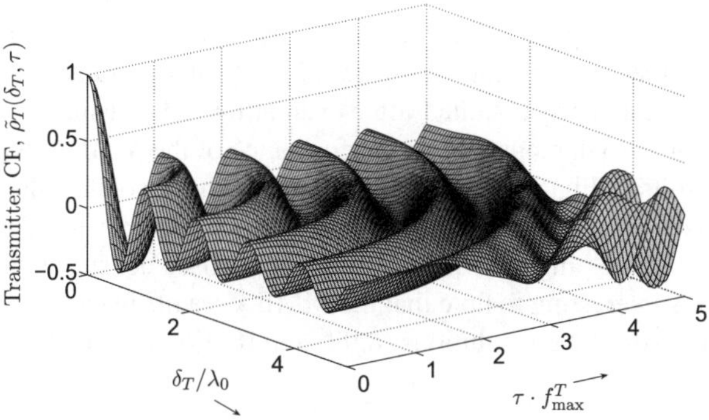

Figure 8.15 The transmitter CF $\tilde{\rho}_T(\delta_T, \tau)$ of the 2×2 MIMO mobile-to-mobile channel simulator designed by applying the EMEDS with $M = 40$ (isotropic scattering).

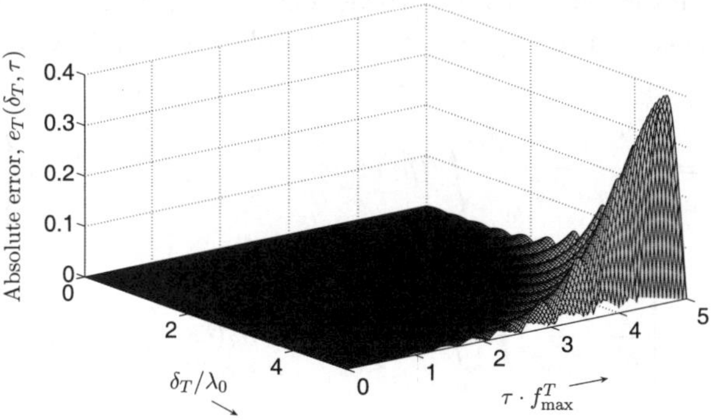

Figure 8.16 Absolute error $e_T(\delta_T, \tau) = |\rho_T(\delta_T, \tau) - \tilde{\rho}_T(\delta_T, \tau)|$ by using the EMEDS with $M = 40$ (isotropic scattering).

8.3.5.2 Modified Method of Equal Areas (MMEA)

A solution to the parameter computation problem in case of non-isotropic scattering is provided by the MMEA [291]. The original MEA was first introduced in [96] and later modified in [291] to enable the modelling of any given distribution of the AOD (AOA). When applying the MMEA to the present problem, then the AODs α_m^T and AOAs α_n^R have to be determined by finding the roots (zeros) of the following sets of equations

$$\frac{m-1/4}{M} - \int_{m_\alpha^T-\pi}^{\alpha_m^T} p_{\alpha^T}(\alpha^T)\, d\alpha^T = 0, \quad m = 1, 2, \ldots, M, \tag{8.98}$$

$$\frac{n-1/4}{N} - \int_{m_\alpha^R-\pi}^{\alpha_n^R} p_{\alpha^R}(\alpha^R)\, d\alpha^R = 0, \quad n = 1, 2, \ldots, N. \tag{8.99}$$

Using numerical root-finding techniques provides us with proper sets $\{\alpha_m^T\}$ and $\{\alpha_n^R\}$. As a reference, we consider the absolute value $|\rho_T(\delta_T, \tau)|$ of the transmitter CF presented in Figure 8.17. The plot has been obtained after evaluating (8.91) using $m_\alpha^T = 60°$, $\kappa_T = 40$, $\alpha_v^T = 0°$, and $\beta_T = 90°$. To demonstrate the performance of the MMEA, we set M equal to 50 and consider the results in Figures 8.18 and 8.19, where the absolute value of the transmitter CF $|\tilde{\rho}_T(\delta_T, \tau)|$ and the absolute error $e_T(\delta_T, \tau)$ are shown, respectively. The presented graphs have been obtained by applying the MMEA to the von Mises density with parameters $m_\alpha^T = 60°$ and $\kappa_T = 40$. It is interesting to note that the MMEA results in case of isotropic scattering ($\kappa_T = \kappa_R = 0$) in the closed-form solutions (8.92) and (8.93) provided by the EMEDS.

8.3.5.3 L_p-Norm Method (LPNM)

If the AODs (AOAs) are non-uniformly distributed on rings centered on the transmitter (receiver), then the LPNM is recommended, which is described in detail in Subsection 5.1.6. The application of the LPNM to the two-ring MIMO channel model requires the computation of the model parameters α_m^T and α_n^R by minimizing the following two L_p-error norms:

$$E_T^{(p)} := \left\{ \frac{1}{\delta_{\max}^T \tau_{\max}^T} \int_0^{\delta_{\max}^T} \int_0^{\tau_{\max}^T} |\rho_T(\delta_T, \tau) - \tilde{\rho}_T(\delta_T, \tau)|^p \, d\delta_T \, d\tau \right\}^{1/p}, \tag{8.100}$$

$$E_R^{(p)} := \left\{ \frac{1}{\delta_{\max}^R \tau_{\max}^R} \int_0^{\delta_{\max}^R} \int_0^{\tau_{\max}^R} |\rho_R(\delta_R, \tau) - \tilde{\rho}_R(\delta_R, \tau)|^p \, d\delta_R \, d\tau \right\}^{1/p}, \tag{8.101}$$

where $p = 1, 2, \ldots$ Here $\rho_T(\delta_T, \tau)$ and $\rho_R(\delta_R, \tau)$ are the transmitter and receiver CFs of the reference model as they are given by (8.65) and (8.66), respectively. The pairs of parameters $(\delta_{\max}^T, \tau_{\max}^T)$ and $(\delta_{\max}^R, \tau_{\max}^R)$ are used to control the domains within which the errors of the approximations $\rho_T(\delta_T, \tau) \approx \tilde{\rho}_T(\delta_T, \tau)$ and $\rho_R(\delta_R, \tau) \approx \tilde{\rho}_R(\delta_R, \tau)$ shall be minimized,

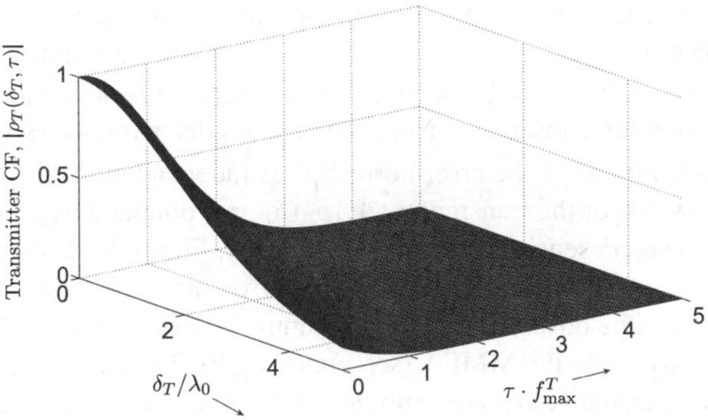

Figure 8.17 Absolute value of the transmitter CF $|\rho_T(\delta_T, \tau)|$ of the 2×2 MIMO mobile-to-mobile reference channel model under non-isotropic scattering conditions (von Mises density with $m_\alpha^T = 60°$ and $\kappa_T = 40$).

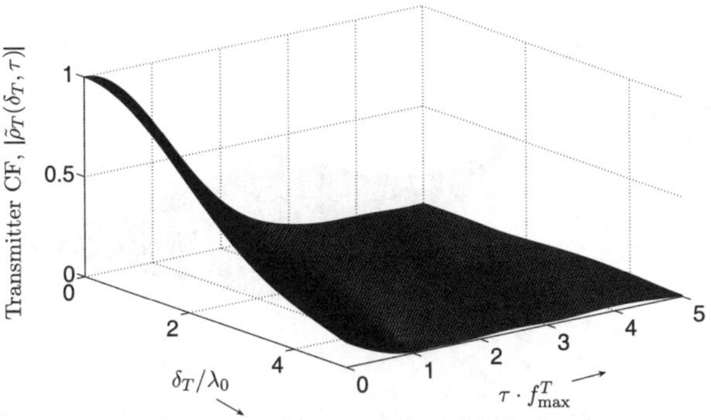

Figure 8.18 Absolute value of the transmitter CF $|\tilde{\rho}_T(\delta_T, \tau)|$ of the 2×2 MIMO mobile-to-mobile channel simulator designed by applying the MMEA with $M = 50$ (non-isotropic scattering, von Mises density with $m_\alpha^T = 60°$ and $\kappa_T = 40$).

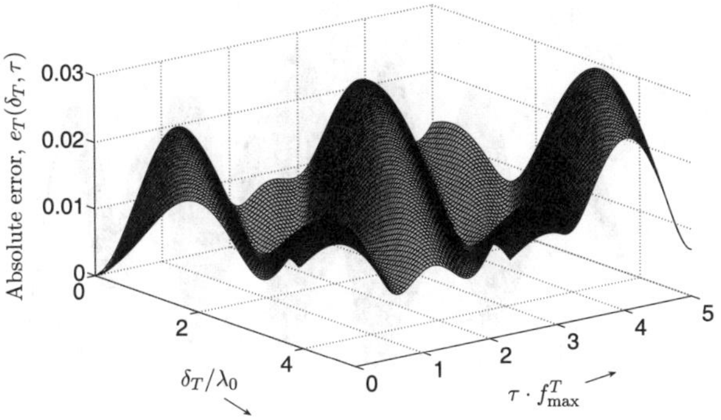

Figure 8.19 Absolute error $e_T(\delta_T, \tau) = |\rho_T(\delta_T, \tau) - \tilde{\rho}_T(\delta_T, \tau)|$ by using the MMEA with $M = 50$ (non-isotropic scattering, von Mises density with $m_\alpha^T = 60°$ and $\kappa_T = 40$).

respectively. The Fletcher-Powell optimization algorithm [162] can be used to perform the numerical minimization of $E_T^{(p)}$ and $E_R^{(p)}$. This optimization algorithm profits from a good choice of the starting values of α_m^T and α_n^R, which can be obtained, e.g., from the MMEA described in the previous subsection. Note that the model parameters α_m^T and α_n^R can be optimized independently, since the error norm $E_T^{(p)}$ is independent of $E_R^{(p)}$ and vice versa. A plot of the absolute value of the transmitter CF $|\tilde{\rho}_T(\delta_T, \tau)|$ obtained by using the LPNM with $p = 100$ and $M = 50$ is presented in Figure 8.20. Finally, Figure 8.21 illustrates the resulting absolute error $e_T(\delta_T, \tau)$ introduced in (8.97). This error function shows a nearly equiripple behaviour, where it can be observed that the maximum values of $e_T(\delta_T, \tau)$ are much smaller than those achieved by using the MMEA (see Figure 8.19). The LPNM has also successfully been applied to the one-ring model (see Subsection 8.2.4) and the two-ring model [290] with a fixed transmitter.

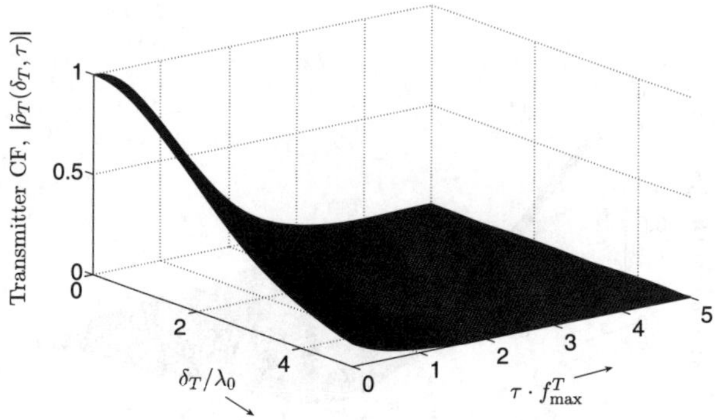

Figure 8.20 Absolute value of the transmitter CF $|\tilde{\rho}_T(\delta_T, \tau)|$ of the 2×2 MIMO mobile-to-mobile channel simulator designed by applying the LPNM with $p = 100$ and $M = 50$ (non-isotropic scattering, von Mises density with $m_\alpha^T = 60°$ and $\kappa_T = 40$).

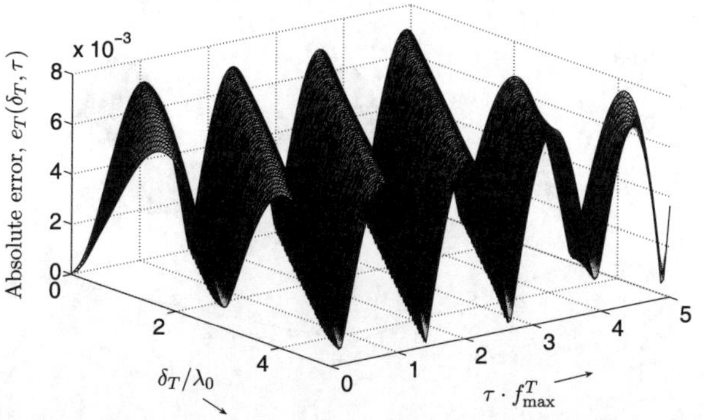

Figure 8.21 Absolute error $e_T(\delta_T, \tau) = |\rho_T(\delta_T, \tau) - \tilde{\rho}_T(\delta_T, \tau)|$ by using the LPNM with $p = 100$ and $M = 50$ (non-isotropic scattering, von Mises density with $m_\alpha^T = 60°$ and $\kappa_T = 40$).

8.4 The Elliptical MIMO Channel Model

For the design and performance evaluation of MIMO wireless communication systems, it is of crucial importance to have accurate and realistic MIMO channel models [228]. Especially for the development of wideband wireless communication systems employing MIMO technologies, such as MIMO orthogonal frequency division multiplexing (OFDM) systems, channel models are required, which take into account the temporal, spatial, and frequency correlation properties. Such space-time-frequency MIMO channel models have been studied, for example, in [294].

This section presents a space-time-frequency MIMO channel model, which is derived from the geometrical elliptical scattering model. Together with the one-ring model [105, 272, 273] and the two-ring model [169, 295], the elliptical model belongs to the most important geometrical models from which spatial channel models have been derived in the past. However, the one-ring model and the two-ring model have been used primarily to model narrowband MIMO channels with specific temporal and spatial correlation properties. This limits the applications of these channel models to performance studies of narrowband mobile communication systems. In contrast to the one-ring and the two-ring models, the elliptical model is predestinated for the modelling of wideband MIMO channels with characteristic temporal, spatial, and frequency correlation properties, as we will see in this section. In particular, the frequency-selectivity feature makes the elliptical scattering model very attractive for developers of future mobile communication systems, since the interest has been turned to the design and performance investigation of high data rate wireless systems employing MIMO-OFDM techniques [296]. Originally, the geometrical elliptical model was proposed in [297] for spatial channels in micro- and picocell environments, where the antenna heights are low, so that multipath scattering is just as likely near the base station as near the mobile station. The geometrically based elliptical model was developed further, e.g., in [106], where a spatial channel model has been proposed for single-input multiple-output (SIMO) channels.

The presentation of this section profits from the results in [298, 299], where the geometrical elliptical scattering SIMO channel model in [106] has been extended to multiple antennas at both the transmitter side and the receiver side. The model takes into account that the AOD and the AOA are dependent. The exact relationship between the AOD and the AOA was first derived in [106] and was later simplified in [300]. Using the relationship in [300], general analytical solutions are derived for the 3D space-time CCF, the time ACF, the 2D space CCF, and the frequency CF. The most important correlation properties will be discussed and visualized assuming isotropic and non-isotropic scattering conditions. Owing to its infinite computational complexity, the proposed MIMO channel model is introduced here as a reference model. The reference model is an important framework for the derivation of an efficient MIMO channel simulator with given space-time-frequency correlation properties. Furthermore, the reference model is also quite useful for studying the capacity of MIMO channels under realistic propagation conditions imposed by the underlying geometrical scattering model.

The organization of this section is as follows. Subsection 8.4.1 reviews the geometrical elliptical scattering model with local scatterers randomly distributed on an ellipse. Starting from the elliptical scattering model, we derive in Subsection 8.4.2 the corresponding narrowband space-time MIMO channel model assuming an infinite number of random scatterers characterized by a given distribution function. The resulting model is considered as a reference model. Its statistical properties will also be analyzed in Subsection 8.4.2 under the assumption of isotropic and non-isotropic scattering conditions. The corresponding stochastic and deterministic simulation models are described in Subsection 8.4.2.3. Subsection 8.4.4 presents two important model extensions. The first incorporates multi-cluster scenarios in the reference model. The second extension introduces several ellipses with multiple clusters of scatterers on each ellipse resulting in a reference model for a large class of wideband space-time MIMO channels.

8.4.1 The Geometrical Elliptical Scattering Model

The starting point for the derivation of an $M_T \times M_R$ MIMO channel model is the geometrical elliptical scattering model shown in Figure 8.22. This figure illustrates that all local scatterers S_n $(n = 1, 2, \ldots, N)$ associated with a certain path length are located on an ellipse, where the base station and the mobile station are located at the focal points. The distance between the two focal points is $2f$. The semi-major axis length and semi-minor axis length of the ellipse are denoted by a and b, respectively.

It is assumed that the base station is the transmitter, whereas the mobile station plays the role of the receiver. Furthermore, we assume that both the transmitter and the receiver are equipped with a uniform linear antenna array consisting of M_T and M_R antenna elements, respectively. The angle β_T (β_R) denotes the tilt angle of the transmitter (receiver) antenna array, and the symbol δ_T (δ_R) describes the antenna element spacing of the transmitter (receiver) antenna array. Since the antenna dimensions are generally small in comparison to the parameters a and f, we will profit from the reasonable assumption that the inequalities $(M_T - 1)\,\delta_T \ll a - f$

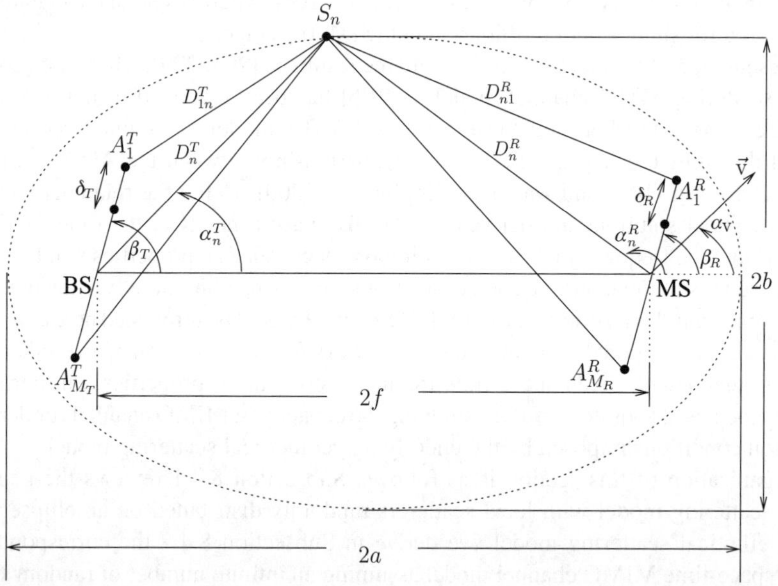

Figure 8.22 Geometrical elliptical scattering model for an $M_T \times M_R$ MIMO channel with local scatterers S_n placed on an ellipse.

and $(M_R - 1)\,\delta_R \ll a - f$ hold. The angle-of-motion α_v describes the angle between the x-axis and the direction of motion. Finally, the AOD of the nth transmitted homogeneous plane wave is denoted by α_n^T, and the corresponding AOA is designated as α_n^R $(n = 1, 2, \ldots, N)$.

8.4.2 The Reference Model for the Elliptical MIMO Channel Model

8.4.2.1 Derivation of the Reference Model

The derivation of the reference model for the elliptical MIMO channel model starts from the geometrical scattering model shown in Figure 8.22. From this figure, we realize that the nth homogeneous plane wave emitted from the lth transmitter antenna element A_l^T $(l = 1, 2, \ldots, M_T)$ travels over the nth local scatterer S_n $(n = 1, 2, \ldots, N)$ before impinging on the kth receiver antenna element A_k^R $(k = 1, 2, \ldots, M_R)$. The reference model is based on the assumption that the number N of local scatterers S_n is infinite. As a consequence, the diffuse component at the kth receiver antenna element A_k^R is composed of an infinite number of homogeneous plane waves. With reference to Figure 8.22, the diffuse component $h_{kl}(\vec{r}_R)$ describing the link from A_l^T $(l = 1, 2, \ldots, M_T)$ to A_k^R $(k = 1, 2, \ldots, M_R)$ can thus be expressed in form of an infinite sum of homogeneous plane waves as follows

$$h_{kl}(\vec{r}_R) = \lim_{N \to \infty} \sum_{n=1}^{N} E_n\, e^{j(\theta_n - \vec{k}_n^R \cdot \vec{r}_R - k_0 D_n)}\,. \tag{8.102}$$

Here, E_n and θ_n are the path gain and phase shift, respectively, caused by the interaction of the nth transmitted plane wave with the local scatterer S_n. The symbol $\vec{k}_n^R$ denotes the wave vector pointing in the propagation direction of the nth received plane wave, and $\vec{r}_R$ is the spatial translation vector of the receiver. Furthermore, k_0 is the free-space wave number, which is defined by $k_0 = 2\pi/\lambda_0$ with λ_0 being the wavelength, and finally D_n denotes the length of the total distance that a plane wave travels from A_l^T via S_n to A_k^R.

It is assumed that the gain E_n and the phase shift θ_n caused by a particular scatterer S_n are generally dependent on the directions of the incoming and emerging waves seen from S_n. Since $(M_T - 1)\,\delta_T \ll a - f$, the waves emerging from different transmitter antennas arrive at a particular scatterer S_n at approximately the same angle. Analogously, since $(M_R - 1)\,\delta_R \ll a - f$ holds, we may conclude that waves redirected from the scatterer S_n arrive at different receiver antennas at approximately the same angle. This allows us to conclude that the gain E_n and phase shift θ_n caused by a particular scatterer S_n are the same for waves arriving from (or travelling to) different transmitter (receiver) antenna elements. Furthermore, it is assumed for reasons of simplicity that each scatterer S_n introduces a constant gain

$$E_n = \frac{1}{\sqrt{N}} \tag{8.103}$$

and a random phase shift θ_n. The phase shifts θ_n are i.i.d. random variables, each having a uniform distribution over the interval $[0, 2\pi)$. Note that (8.103) implies first that all plane waves reaching the receiver antenna array are equal in power, and second that the average power $E\{|h_{kl}(\vec{r}_R)|^2\}$ of the diffuse component $h_{kl}(\vec{r}_R)$ equals unity.

The second phase component in (8.102), $\vec{k}_n^R \cdot \vec{r}_R$, is due to the movement of the receiver and can be written as

$$\vec{k}_n^R \cdot \vec{r}_R = -2\pi f_{\max} \cos\left(\alpha_n^R - \alpha_v\right) t, \tag{8.104}$$

where $f_{\max}$ denotes the maximum Doppler frequency.

Furthermore, the third phase component in (8.102), $k_0 D_n$, is due to the total distance travelled and can be expressed with reference to Figure 8.22 as

$$k_0 D_n = \frac{2\pi}{\lambda_0}\left(D_{ln}^T + D_{nk}^R\right), \tag{8.105}$$

where D_{ln}^T describes the distance from the lth transmitter antenna element A_l^T to the scatterer S_n, and analogously D_{nk}^R denotes the distance from the scatterer S_n to the kth receiver antenna element A_k^R. These two distances can be approximated by using $(M_T - 1)\,\delta_T \ll a - f$, $(M_R - 1)\,\delta_R \ll a - f$, and $\sqrt{1+x} \approx 1 + x/2$ $(x \ll 1)$ as

$$D_{ln}^T \approx D_n^T - (M_T - 2l + 1)\frac{\delta_T}{2}\cos\left(\alpha_n^T - \beta_T\right), \tag{8.106}$$

$$D_{nk}^R \approx D_n^R - (M_R - 2k + 1)\frac{\delta_R}{2}\cos\left(\alpha_n^R - \beta_R\right), \tag{8.107}$$

where D_n^T and D_n^R are the distances from the two focal points to the scatterer S_n as illustrated in Figure 8.22.

Now, after substituting (8.103)–(8.105) in (8.102) and using the approximations in (8.106) and (8.107), the complex channel gain $h_{kl}(t)$ of the proposed reference model describing the link from the lth transmitter antenna element A_l^T $(l = 1, 2, \ldots, M_T)$ to the kth receiver antenna element A_k^R $(k = 1, 2, \ldots, M_R)$ can be expressed as

$$h_{kl}(t) = \lim_{N\to\infty} \frac{1}{\sqrt{N}} \sum_{n=1}^{N} a_{ln}\, b_{kn}\, e^{j(2\pi f_n t + \theta_n + \theta_0)}, \tag{8.108}$$

where

$$a_{ln} = e^{j\pi (M_T - 2l + 1)(\delta_T/\lambda_0)\cos(\alpha_n^T - \beta_T)}, \tag{8.109}$$

$$b_{kn} = e^{j\pi (M_R - 2k + 1)(\delta_R/\lambda_0)\cos(\alpha_n^R - \beta_R)}, \tag{8.110}$$

$$f_n = f_{\max}\cos(\alpha_n^R - \alpha_v), \tag{8.111}$$

$$\theta_0 = -\frac{4\pi a}{\lambda_0}. \tag{8.112}$$

Just as for the one-ring model and the two-ring model, the constant phase shift θ_0 can be set to zero in (8.108), since this parameter has no influence on the statistics of the reference model. Analyzing the statistical properties of the complex channel gain $h_{kl}(t)$ in (8.108) reveals that the mean value and the variance of $h_{kl}(t)$ are equal to 0 and 1, respectively. Thus, by invoking the central limit theorem, it follows that $h_{kl}(t)$ is a complex Gaussian process with zero mean and unit variance. Consequently, we can conclude that the distribution of the envelope $|h_{kl}(t)|$ equals the Rayleigh distribution.

In the reference model, the AOD α_n^T and the AOA α_n^R are dependent. By using trigonometric identities, we can express the AOD α_n^T in terms of the AOA α_n^R as follows [106, 300]

$$\alpha_n^T = \begin{cases} f(\alpha_n^R) & \text{if} \quad 0 < \alpha_n^R \leq \alpha_0, \\ f(\alpha_n^R) + \pi & \text{if} \quad \alpha_0 < \alpha_n^R \leq 2\pi - \alpha_0, \\ f(\alpha_n^R) + 2\pi & \text{if} \quad 2\pi - \alpha_0 < \alpha_n^R \leq 2\pi, \end{cases} \tag{8.113}$$

where

$$f(\alpha_n^R) = \arctan\left[\frac{(\kappa_0^2 - 1)\sin(\alpha_n^R)}{2\kappa_0 + (\kappa_0^2 + 1)\cos(\alpha_n^R)}\right] \tag{8.114}$$

and

$$\alpha_0 = \pi - \arctan\left(\frac{\kappa_0^2 - 1}{2\kappa_0}\right). \tag{8.115}$$

The parameter κ_0 in (8.114) and (8.115) equals the reciprocal value of the eccentricity e of the ellipse, i.e., $\kappa_0 = 1/e = a/f$.

8.4.2.2 Correlation Functions of the Reference Model

According to [273], the 3D space-time CCF of the A_l^T-to-A_k^R and $A_{l'}^T$-to-$A_{k'}^R$ links is defined as the correlation of the complex channel gains $h_{kl}(t)$ and $h_{k'l'}(t)$, i.e.,

$$\rho_{kl,k'l'}(\delta_T, \delta_R, \tau) := E\{h_{kl}(t)\, h_{k'l'}^*(t+\tau)\}. \tag{8.116}$$

Note that the expectation operator has to be applied to all random variables (θ_n and α_n^R), where we recall that the AOD α_n^T is a function of the AOA α_n^R according to (8.113). The solution of (8.116) can be achieved by substituting (8.108) in (8.116) and averaging in the first step over the random phases θ_n. This allows us to express the 3D space-time CCF in the form

$$\rho_{kl,k'l'}(\delta_T, \delta_R, \tau) = \lim_{N\to\infty} \frac{1}{N} \sum_{n=1}^{N} E\left\{a_{ll'n}^2\, b_{kk'n}^2\, e^{-j2\pi f_n \tau}\right\}, \tag{8.117}$$

where

$$a_{ll'n} = e^{-j\pi(l-l')(\delta_T/\lambda_0)\cos(\alpha_n^T - \beta_T)}, \tag{8.118}$$

$$b_{kk'n} = e^{-j\pi(k-k')(\delta_R/\lambda_0)\cos(\alpha_n^R - \beta_R)}. \tag{8.119}$$

In the second step, we compute the statistical average with respect to the distribution of the random variable α_n^R. We notice that if the number of local scatterers N approaches infinity, then the discrete AOD α_n^T and discrete AOA α_n^R become continuous random variables denoted by α^T and α^R, respectively, where we recall that α^T is a function of α^R according to (8.113). The infinitesimal power of the complex channel gain corresponding to the differential angles $d\alpha^R$ is proportional to $p_{\alpha^R}(\alpha^R)\, d\alpha^R$, where $p_{\alpha^R}(\alpha^R)$ denotes the distribution of α^R. As $N \to \infty$, this infinitesimal contribution must be equal to $1/N$, i.e., $1/N = p_{\alpha^R}(\alpha^R)\, d\alpha^R$. From this fact, it follows from (8.117) that the 3D space-time CCF of the reference model can be written as

$$\rho_{kl,k'l'}(\delta_T, \delta_R, \tau) = \int_{-\pi}^{\pi} a_{ll'}^2(\delta_T, \alpha^T)\, b_{kk'}^2(\delta_R, \alpha^R)\, e^{-j2\pi f(\alpha^R)\tau}\, p_{\alpha^R}(\alpha^R)\, d\alpha^R, \quad (8.120)$$

where

$$a_{ll'}(\delta_T, \alpha^T) = e^{-j\pi(l-l')(\delta_T/\lambda_0)\cos(\alpha^T - \beta_T)}, \quad (8.121)$$

$$b_{kk'}(\delta_R, \alpha^R) = e^{-j\pi(k-k')(\delta_R/\lambda_0)\cos(\alpha^R - \beta_R)}, \quad (8.122)$$

$$f(\alpha^R) = f_{\max}\cos(\alpha^R - \alpha_v). \quad (8.123)$$

This result shows that the 3D space-time CCF $\rho_{kl,k'l'}(\delta_T, \delta_R, \tau)$ is independent of the parameters a, b, and f describing the ellipse.

The 2D space CCF $\rho_{kl,k'l'}(\delta_T, \delta_R)$, which is defined as $\rho_{kl,k'l'}(\delta_T, \delta_R) := E\{h_{kl}(t)\, h_{k'l'}^*(t)\}$, is equal to the 3D space-time CCF $\rho_{kl,k'l'}(\delta_T, \delta_R, \tau)$ at $\tau = 0$, i.e., $\rho_{kl,k'l'}(\delta_T, \delta_R) = \rho_{kl,k'l'}(\delta_R, \delta_R, 0)$. Hence,

$$\rho_{kl,k'l'}(\delta_T, \delta_R) = \int_{-\pi}^{\pi} a_{ll'}^2(\delta_T, \alpha^T)\, b_{kk'}^2(\delta_R, \alpha^R)\, p_{\alpha^R}(\alpha^R)\, d\alpha^R. \quad (8.124)$$

The result above shows that the 2D space CCF $\rho_{kl,k'l'}(\delta_T, \delta_R)$ generally does not allow a Kronecker representation, meaning a representation of $\rho_{kl,k'l'}(\delta_T, \delta_R)$ as a product consisting of two terms — the first term being a function of δ_T only and the second one of δ_R. This statement is in agreement with the MIMO channel model based on the geometrical one-ring scattering model, but it is in contrast to the two-ring scattering model in which the AOD α^T and the AOA α^R are independent.

The time ACF $r_{h_{kl}}(\tau)$ of the complex channel gain $h_{kl}(t)$ is defined as $r_{h_{kl}}(\tau) := E\{h_{kl}(t)\, h_{kl}^*(t+\tau)\}$. Alternatively, the time ACF $r_{h_{kl}}(\tau)$ can be obtained from the 3D space-time CCF $\rho_{kl,k'l'}(\delta_T, \delta_R, \tau)$ by setting the antenna element spacings δ_T and δ_R to zero, i.e., $r_{h_{kl}}(\tau) = \rho_{kl,k'l'}(0, 0, \tau)$. In both cases, we obtain

$$r_{h_{kl}}(\tau) = \int_{-\pi}^{\pi} e^{-j2\pi f_{\max}\cos(\alpha^R - \alpha_v)\tau}\, p_{\alpha^R}(\alpha^R)\, d\alpha^R \quad (8.125)$$

for all $k = 1, 2, \ldots, M_R$ and $l = 1, 2, \ldots, M_T$. Notice that the time ACFs $r_{h_{kl}}(\tau)$ of the complex channel gains $h_{kl}(t)$ are identical for all links from $A_l^T\,(l = 1, 2, \ldots, M_T)$ to A_k^R $(k = 1, 2, \ldots, M_R)$. In case of isotropic scattering, characterized by $p_{\alpha^R}(\alpha^R) = 1/(2\pi)$, the above integral can be solved analytically leading to $r_{h_{kl}}(\tau) = J_0(2\pi f_{\max}\tau)$.

8.4.2.3 Illustrative Examples and Numerical Results

In this section, we will present some illustrative examples for the 2D space CCF $\rho_{kl,k'l'}(\delta_T, \delta_R)$ and the time ACF $r_{h_{kl}}(\tau)$ in case of isotropic and non-isotropic scattering. For this purpose, we employ the von Mises density to characterize the distribution of the AOA α^R. The von Mises density with parameters κ and m_α is given by (8.26).

Some numerical results obtained for the 2D space CCF $\rho_{11,22}(\delta_T, \delta_R)$ and the time ACF $r_{h_{kl}}(\tau)$ are presented in Figures 8.23 and 8.24, respectively. These results are valid under the assumption of isotropic scattering ($\kappa = 0$). Figures 8.25 and 8.26 illustrate the behaviour of $|\rho_{11,22}(\delta_T, \delta_R)|$ and $|r_{h_{kl}}(\tau)|$ under non-isotropic scattering conditions assuming that the AOA α^R follows the von Mises distribution with parameters $\kappa = 10$ and $m_\alpha = 0$. In all examples, it has been assumed that the mobile station moves towards the base station, meaning that the angle-of-motion α_v was set to π.

8.4.3 Simulation Models for the Elliptical MIMO Channel Model

8.4.3.1 The Stochastic Simulation Model

According to the generalized concept of deterministic channel modelling, a stochastic simulation model is obtained from the reference model described in Subsection 8.4.2 by using only

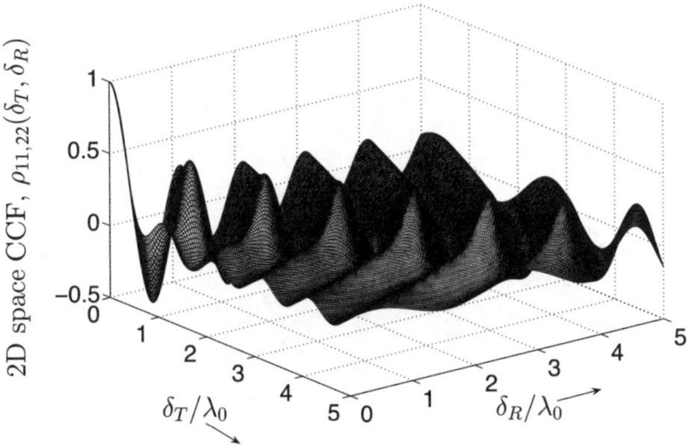

Figure 8.23 The 2D space CCF $\rho_{11,22}(\delta_T, \delta_R)$ of the reference model (isotropic scattering).

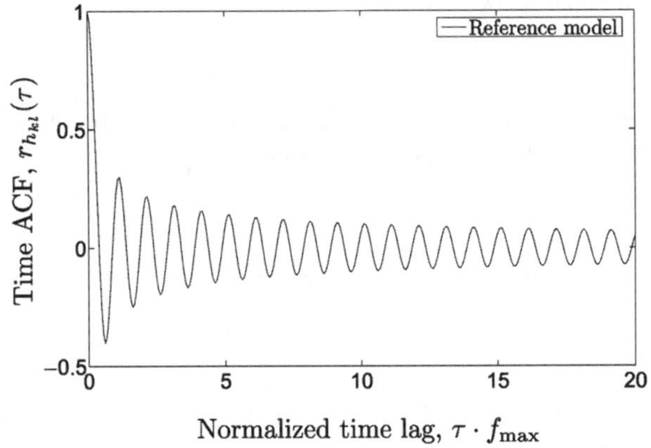

Figure 8.24 The time ACF $r_{h_{kl}}(\tau)$ of the reference model (isotropic scattering).

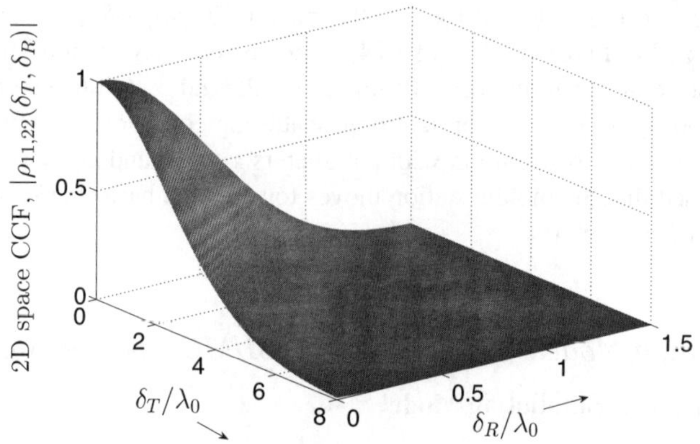

Figure 8.25 Absolute value of the 2D space CCF $|\rho_{11,22}(\delta_T, \delta_R)|$ of the reference model (non-isotropic scattering, von Mises density with parameters $\kappa = 10$ and $m_\alpha = 0$).

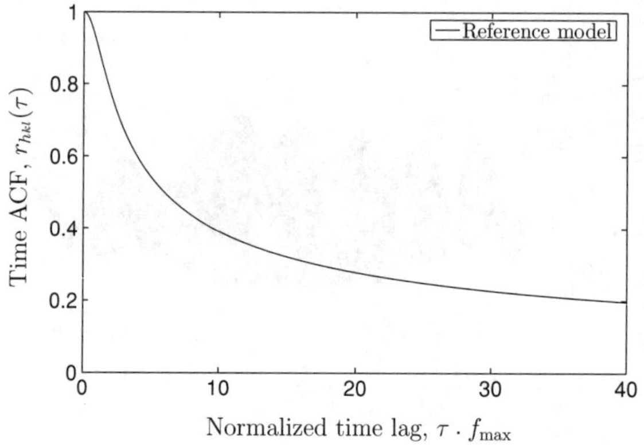

Figure 8.26 Absolute value of the time ACF $|r_{h_{kl}}(\tau)|$ of the reference model (non-isotropic scattering, von Mises density with parameters $\kappa = 10$ and $m_\alpha = 0$).

a finite number of scatterers. Hence, the complex channel gain $\hat{h}_{kl}(t)$ of the link from A_l^T $(l = 1, 2, \ldots, M_T)$ to A_k^R $(k = 1, 2, \ldots, M_R)$ can be written as

$$\hat{h}_{kl}(t) = \frac{1}{\sqrt{N}} \sum_{n=1}^{N} a_{ln}\, b_{kn} e^{j(2\pi f_n t + \theta_n)}, \tag{8.126}$$

where a_{ln}, b_{kn}, and f_n are given by (8.109), (8.110), and (8.111), respectively. In contrast to the reference model, the discrete AODs α_n^T and AOAs α_n^R are now constants, which will be determined in Subsection 8.4.3.3. The phases θ_n are still i.i.d. uniformly distributed random variables. Consequently, $\hat{h}_{kl}(t)$ represents a stochastic process.

The 3D space-time CCF $\hat{\rho}_{kl,k'l'}(\delta_T, \delta_R, \tau)$ of the stochastic simulation model can be expressed in closed form as

$$\begin{aligned}\hat{\rho}_{kl,k'l'}(\delta_T, \delta_R, \tau) &:= E\left\{\hat{h}_{kl}(t)\hat{h}^*_{k'l'}(t+\tau)\right\} \\ &= \frac{1}{N}\sum_{n=1}^{N} a^2_{ll'n}(\delta_T)\, b^2_{kk'n}(\delta_R)\, e^{-j2\pi f_n \tau}, \end{aligned} \tag{8.127}$$

where $a_{ll'n}(\delta_T)$ and $b_{kk'n}(\delta_R)$ are given by (8.118) and (8.119), respectively.

Furthermore, the 2D space CCF $\hat{\rho}_{kl,k'l'}(\delta_T, \delta_R)$ of the stochastic simulation model can be expressed as

$$\begin{aligned}\hat{\rho}_{kl,k'l'}(\delta_T, \delta_R) &:= E\left\{\hat{h}_{kl}(t)\hat{h}^*_{k'l'}(t)\right\} \\ &= \hat{\rho}_{kl,k'l'}(\delta_T, \delta_R, 0) \\ &= \frac{1}{N}\sum_{n=1}^{N} a^2_{ll'n}(\delta_T)\, b^2_{kk'n}(\delta_R). \end{aligned} \tag{8.128}$$

Finally, we notice that the time ACF $\hat{r}_{h_{kl}}(\tau)$ of the stochastic simulation model equals the expression in (8.32). This means that the one-ring model and the elliptical model have the same temporal correlation properties for any given set $\{f_n\}_{n=1}^{N}$.

8.4.3.2 The Deterministic Simulation Model

The deterministic simulation model follows from the stochastic one by fixing all model parameters, including the phases θ_n. Again, a single sample function of the stochastic process $\hat{h}_{kl}(t)$ will be denoted by $\tilde{h}_{kl}(t)$.

The 3D space-time CCF $\tilde{\rho}_{kl,k'l'}(\delta_T, \delta_R, \tau)$ of the deterministic simulation model can be obtained as

$$\begin{aligned}\tilde{\rho}_{kl,k'l'}(\delta_T, \delta_R, \tau) &:= < \tilde{h}_{kl}(t)\, \tilde{h}^*_{k'l'}(t+\tau) > \\ &= \hat{\rho}_{kl,k'l'}(\delta_T, \delta_R, \tau), \end{aligned} \tag{8.129}$$

where $\hat{\rho}_{kl,k'l'}(\delta_T, \delta_R, \tau)$ is given by (8.127). This result proves the ergodicity of the stochastic MIMO channel simulator with respect to the 3D space-time CCF.

Analogously, the 2D space CCF $\tilde{\rho}_{kl,k'l'}(\delta_T, \delta_R)$, defined as $\tilde{\rho}_{kl,k'l'}(\delta_T, \delta_R) := < \tilde{h}_{kl}(t)\, \tilde{h}^*_{k'l'}(t+\tau) >$, can be identified as

$$\tilde{\rho}_{kl,k'l'}(\delta_T, \delta_R) = \hat{\rho}_{kl,k'l}(\delta_T, \delta_R), \tag{8.130}$$

where $\hat{\rho}_{kl,k'l}(\delta_T, \delta_R)$ equals the expression in (8.128).

Finally, we mention that the time ACF $\tilde{r}_{h_{kl}}(\tau) :=< \tilde{h}_{kl}(t)\, \tilde{h}^*_{kl}(t+\tau) >$ of the deterministic simulation model equals that of the stochastic simulation model, i.e., $\tilde{r}_{h_{kl}}(\tau) = \hat{r}_{h_{kl}}(\tau)$, where $\hat{r}_{h_{kl}}(\tau)$ is given by (8.32).

8.4.3.3 Parameter Computation Methods

Since the AODs α_n^T are related to the AOAs α_n^R via (8.113), it follows that the only model parameters to be determined are the elements of the set $\{\alpha_n^R\}_{n=1}^N$. In case of isotropic scattering, the quantities α_n^R should preferably be computed by using the EMEDS, while the MMEA and the LPNM provide high quality solutions when the scattering is non-isotropic. The EMEDS, MMEA, and LPNM are described in detail in Subsection 8.3.5.

8.4.4 Model Extensions

In this section, we will show how the proposed MIMO channel model can be extended to multiple clusters of scatterers as well as to frequency selectivity.

8.4.4.1 Extension to Multiple Clusters of Scatterers

Let us consider Figure 8.27, which illustrates a propagation scenario consisting of $\mathcal{C}$ clusters of scatterers located on a single ellipse. To distinguish between different clusters, we add the subscript $(\cdot)_c$ $(c = 1, 2, \ldots, \mathcal{C})$ to all affected symbols, i.e., we write $h_{kl,c}(t)$, κ_c, $m_{\alpha,c}$, etc. For different values of c, each complex channel gain $h_{kl,c}(t)$ might be characterized by different parameters κ_c and $m_{\alpha,c}$. The resulting complex channel gain $z_{kl}(t)$ describing the link from A_l^T $(l = 1, 2, \ldots, M_T)$ to A_k^R $(k = 1, 2, \ldots, M_R)$ in a multi-cluster $M_T \times M_R$ MIMO channel is then given by the superposition of the received scattered components of all $\mathcal{C}$ clusters, i.e.,

$$z_{kl}(t) = \sum_{c=1}^{\mathcal{C}} w_c \, h_{kl,c}(t), \tag{8.131}$$

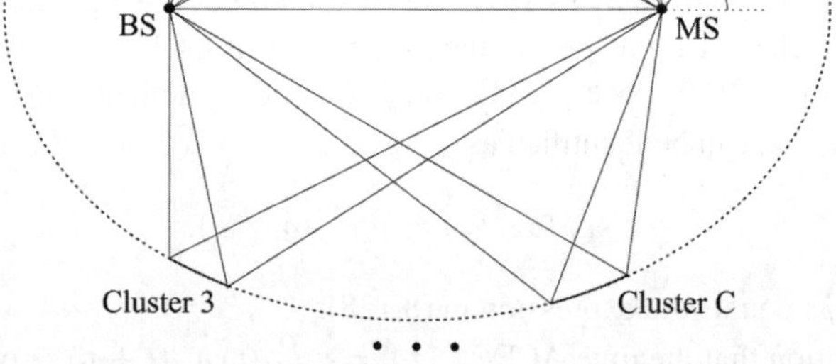

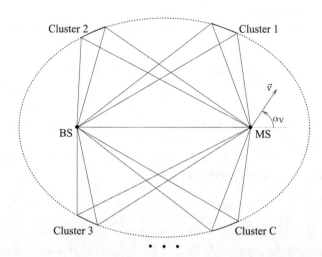

Figure 8.27 Geometrical elliptical scattering model with multiple clusters of scatterers located on an ellipse.

where w_c is a real-valued constant representing the weighting factor of the cth cluster. We impose the boundary condition $\sum_{c=1}^{C} w_c^2 = 1$ on the weighting factors w_c to normalize the mean power of $z_{kl}(t)$ to unity.

8.4.4.2 Extension to Frequency Selectivity

The underlying geometrical elliptical scattering model for the proposed frequency-selective MIMO channel is presented in Figure 8.28. The complex channel gain associated with the lth discrete propagation path τ'_ℓ will be denoted by $z_{kl,\ell}(t)$. This notation allows us to express the time-variant impulse response of the proposed reference model as

$$h_{kl}(\tau', t) = \sum_{\ell=0}^{\mathcal{L}-1} a_\ell \, z_{kl,\ell}(t) \, \delta(\tau' - \tau'_\ell), \tag{8.132}$$

where $\mathcal{L}$ denotes the number of discrete propagation paths with different propagation delays τ'_ℓ, and a_ℓ represents the gain of the lth path.

The Fourier transform of the impulse response $h_{kl}(\tau', t)$ with respect to τ' is known as the time-variant transfer function $H_{kl}(f', t)$. Hence, from (8.132) it follows

$$H_{kl}(f', t) = \sum_{\ell=0}^{\mathcal{L}-1} a_\ell \, z_{kl,\ell}(t) \, e^{-j2\pi f' \tau'_\ell}. \tag{8.133}$$

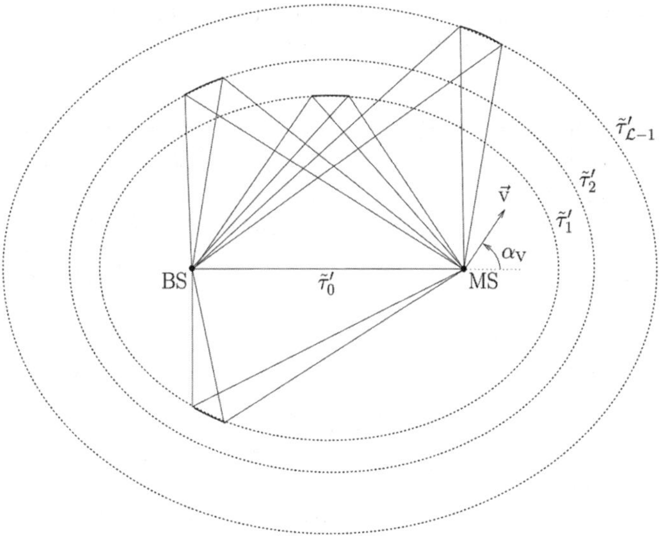

Figure 8.28 Geometrical elliptical scattering model for a frequency-selective MIMO channel with multiple clusters of scatterers located on $\mathcal{L}$ ellipses.

The frequency CF of the reference model, denoted as $r_{\tau'}(\upsilon')$ and defined by $r_{\tau'}(\upsilon') := E\{H_{kl}(f', t)\, H^*_{kl}(f' + \upsilon', t)\}$, can be expressed as

$$r_{\tau'}(\upsilon') = \sum_{\ell=0}^{\mathcal{L}-1} a_\ell^2\, e^{j2\pi\upsilon'\tau'_\ell}. \tag{8.134}$$

The Fourier transform of $r_{\tau'}(\upsilon')$ provides the power delay profile (delay power spectral density)

$$S_{\tau'}(\tau') = \sum_{\ell=0}^{\mathcal{L}-1} a_\ell^2\, \delta(\tau' - \tau'_\ell). \tag{8.135}$$

The results above show that the frequency CF $r_{\tau'}(\upsilon')$ and the power delay profile $S_{\tau'}(\tau')$ are completely determined by the number of propagation paths $\mathcal{L}$, the path gains a_ℓ, and the propagation delays τ'_ℓ. This is an important observation, since it allows us to fit the frequency CF $r_{\tau'}(\upsilon')$ of the proposed channel model to any given specified (or measured) frequency CF characterized by the sets $\{a_\ell\}_{\ell=0}^{\mathcal{L}-1}$ and $\{\tau'_\ell\}_{\ell=0}^{\mathcal{L}-1}$. The same statement holds of course for any given (specified or measured) discrete power delay profile due to the Fourier transform relationship between the frequency CF and the power delay profile. Proper values for the model parameters can be found in many specifications for channel models, such as the COST 207 channel models [19], the SUI channel models [301], and the HIPERLAN/2 channel models [20].

To illustrate the results, we pick up the 18-path HIPERLAN/2 channel model C, which has been specified in [20] for large open areas under the assumption that the line-of-sight path is obstructed. The specified model parameters are listed in Table 7.B.2 of Appendix 7.B. Substituting these parameters in (8.134) and (8.135) enables us to study the frequency CF $r_{\tau'}(\upsilon')$ and the power delay profile $S_{\tau'}(\tau')$ analytically. For illustration purposes, a plot of the absolute value of the frequency CF, $|r_{\tau'}(\upsilon')|$, is shown in Figure 8.29. The corresponding power delay profile $S_{\tau'}(\tau')$ is presented in Figure 8.30.

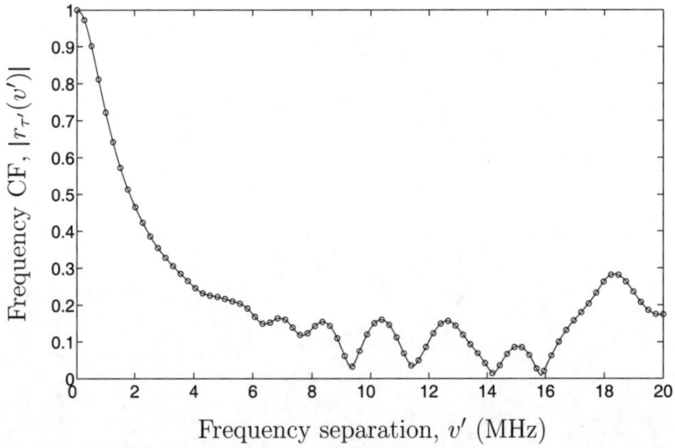

Figure 8.29 Absolute value of the frequency CF $|r_{\tau'}(\upsilon')|$ of the reference model using the 18-path HIPERLAN/2 channel model C [20].

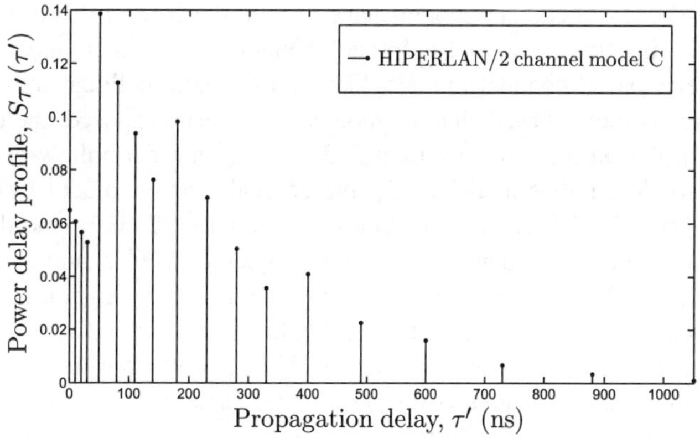

Figure 8.30 Power delay profile $S_{\tau'}(\tau')$ of the 18-path HIPERLAN/2 channel model C [20].

8.5 Further Reading

An informative overview of MIMO wireless systems is provided in [117, 302]. The article [302] reviews some aspects of MIMO channel modelling. The channel models described therein are classified as theoretical models, heuristic models, broadband channel models, and measured channels. The same lead author has also published a tutorial paper [303] on MIMO space-time coded wireless systems. This paper contains a section dealing with MIMO channel modelling, where an introduction to pseudo-static narrowband and time-varying wideband MIMO channel models is given.

An interesting overview of spatial channel models—not necessarily MIMO channel models—is provided in [228]. The paper introduces briefly Lee's model, various types of geometry-based single-bounce statistical channel models, the modified Saleh-Valenzuela model, the extended tap-delay-line model, measurement-based channel models, and ray tracing models. A summary of the introduced spatial channel models is presented in Table 3 of [228], where the main features of each model are briefly described and references for further studies are given.

Another overview of MIMO propagation channel models can be found in [230], where both narrowband and wideband channel models are considered. The paper is a short version of the lead author's PhD thesis [304]. Different MIMO channel models are briefly discussed, including the one-ring and two-ring models, the extended Saleh-Valenzuela model [305], the COST 259 directional channel model [275], the EM scattering model [306], and the virtual channel model [307]. The paper reports also on various measurement results obtained within the IST SATURN project and compares the measurement results with some proposed channel models. The measurement data have been collected by using the Medav RUSK BRI vector sounder. This channel sounder has an eight-element omnidirectional uniform linear array (ULA) at the transmitter side and an eight-element ULA with 120° beamwidth at the receiver side. A thorough survey of channel and radio propagation models for wireless MIMO systems can also be found in [231].

The classification of MIMO channel models can be done in a number of ways. Here, we will classify them into geometry-based MIMO channel models, measurement-based MIMO channel models, analytical MIMO channel models, and standardized MIMO channel models.

Geometry-based MIMO channel models have formed the core of this chapter. Apart from the discussed one-ring, two-ring, and elliptical channel models, there exists a huge variety of further geometry-based channel models. For outdoor radio cellular networks, for example, the following geometry-based channel models have been proposed: the uniform circular model [308–313], the uniform pie-cut model [313], the uniform hollow-disk model [314], the inverted-parabolic circular model [315], the conical circular model [308], the uniform elliptical model [297, 309, 311, 316], the Gaussian model [317–319], and the Rayleigh circular model [320]. The assessment of all these geometry-based channel models by using empirical data sets has been the subject of investigation in [321]. A geometry-based channel model for macrocellular environments with hyperbolically distributed scatterers can be found in [322]. This model has also been examined in [323], where the influence of the spread of hyperbolically distributed scatterers on the channel capacity was studied. Several papers dealing with the modelling of mobile-to-mobile propagation channels have appeared in the literature for SISO systems. Early studies were reported in [79, 80], where the statistics of a mobile-to-mobile fading channel was investigated by focussing on the characterization of the time-varying transfer function. Later, in [324], the temporal correlation properties and the Doppler power spectral density were studied for the case of 3D scattering environments. In [325], an approach based on ray-optical techniques was used to model the wave propagation. The resulting impulse responses incorporate the complete channel information in the form of time series, which can be used directly for system simulations. Experimental measurements of outdoor-to-indoor mobile-to-mobile channels were described in [288]. There, it was suggested that the statistics of realistic mobile-to-mobile fading channels can appropriately be described by a multiple-Rayleigh channel model. Experimental results of the propagation characteristics of radio waves at 60 GHz between moving vehicles have been reported in [326], where also the feasibility of data transmission was demonstrated. The development of simulation models for mobile-to-mobile fading channels has also been addressed in [112], where the authors concentrate on the simulation of narrowband SISO channels. A geometry-based three-ring channel model for mobile-to-mobile MIMO channels in cooperative networks has been introduced in [123]. Moreover, we refer to [327], where a geometry-based stochastic MIMO channel model for vehicle-to-vehicle communications can be found.

Measurement-based channel models aim to fit the statistical properties of the channel model to the statistics of snapshot measurement data. Target functions include the probability density function of the envelope, correlation functions in time, space, and frequency, or alternatively, the corresponding power spectral densities. If the channel model is perfectly fitted against the measured channel, then the channel model is said to be a perfect channel model. The concept of perfect channel modelling has been described for wideband SISO channels in Section 7.6.2. An application of this concept on a measured wideband indoor channel is presented in Section 7.6.3. Extensions of the perfect channel modelling approach to spatio-temporal channels and MIMO channels can be found in [122] and [328], respectively. Fundamental techniques for the development of non-perfect but still highly accurate measurement-based MIMO channel models have been reported in [329, 330]. Various kinds of measurement-based channel models for MIMO and ultra wideband channels are presented in [331]. A measurement-based approach for the modelling of vehicle-to-vehicle MIMO channels is described in [332].

Analytical MIMO channel models characterize the time-variant impulse response (or equivalently the time-variant transfer function) of a channel in an analytical way without explicitly accounting for the physical propagation effects of plane waves. Analytical channel models are very attractive for synthesizing MIMO channel matrices in the context of system performance studies and the evaluation of space-time coding algorithms. The most commonly used analytical MIMO channel models are the Kronecker model [333], the Weichselberger model [334], and the virtual channel representation [335]. The Kronecker model assumes separable transmitter and receiver correlation functions. The Weichselberger model enables to model the MIMO channel in a more general way and includes the Kronecker model as a special case. In contrast to these two models, the virtual channel representation [335] models the MIMO channel in the beamspace instead of the eigenspace. A comparison of these analytical models has shown that the Weichselberger model provides a better match than the Kronecker model and the virtual channel representation in predicting a variety of channel metrics [336]. Another analytical MIMO channel model developed by using orthogonalization techniques, where prolate spheroidal wave functions have been used as a universal basis, is presented in [337].

Standardized MIMO channel models provide an important tool for the development of new wireless MIMO systems. This class of channel models includes the COST 273 MIMO channel model [338], the 3GPP-3GPP2 spatial channel model (SCM) [339], the WINNER II channel models [340], the TGn channel models [341], and the Stanford University Interim (SUI) channel models [342]. The COST 273 MIMO channel model is described in detail in Section 6.8 of [338] and the references therein. The key concept is the geometry-based stochastic modelling philosophy. The COST 273 MIMO channel model is the successor of the COST 259 directional channel model [275], but it differs in several aspects. One is that the modelling of the distribution of the AOA and AOD is different in comparison to the COST 273 model. Another different aspect is that a number of new propagation scenarios have been defined, e.g., peer-to-peer and fixed-wireless-access scenarios. The COST 273 MIMO channel model incorporates several new features, such as double scattering and polarization. The 3GPP-3GPP2 SCM [339] has been developed by 3GPP-3GPP2 for evaluating the performance of MIMO systems in outdoor environments at a carrier frequency of 2 GHz and a system bandwidth of 5 MHz. It uses the geometry-based stochastic channel modelling approach, where a subset of the model parameters is stochastic. The 3GPP-3GPP2 SCM incorporates important parameters, such as propagation delays, Doppler frequency, AOD, AOA, angular spread, and phases. It also takes into account the antenna spacings at the transmitter and the receiver antenna arrays, which makes the investigation of mutual coupling effects feasible. The WINNER II channel models [340] have also been developed by applying the geometry-based stochastic channel modelling approach. This enables creating arbitrary double directional MIMO channel models. The WINNER II channel models are antenna independent, but they allow inclusion of different antenna configurations and different element patterns. The channel parameters are determined statistically according to distributions extracted from measurement data [343]. The distributions are specified, e.g., for the delay spread, propagation delays, angular spread, shadow fading, and the cross-polarization ratio. For each channel snapshot, the channel parameters are calculated from their respective distributions. The channel realizations are generated by summing up the contributions of rays with specific channel parameters like delay, power, AOA, and AOD. Different scenarios can be modelled by using the same approach, but with different parameters. The TGn channel models [341] of IEEE 802.11n have been developed for indoor MIMO Wireless Local Area Networks (WLAN) operating at carrier frequencies of 2 and 5 GHz with bandwidths of up to 100 MHz. The TGn channel models comprise a set of six channel models, labelled A to F, which cover environments like small offices, typical offices, large offices, residential homes, and large spaces (indoor and outdoor). For each of the six environments, and thus for the corresponding TGn channel models, a set of parameters has been specified. Each channel model includes a path loss model, a shadowing model, and a MIMO multipath fading channel model characterized by the power delay profile, the spatial properties, the K-factor distribution, and the Doppler spectrum. The SUI channel models [342] have been developed under the IEEE 802.16 working group for macrocellular fixed wireless access networks operating at a carrier frequency of 2.5 GHz in continental areas that are typical for the United States. They comprise a set of six channel models representing three terrain types, labelled A to C, with a variety of values for the Doppler spreads, delay spreads, and line-of-sight/non-line-of-sight components. The SUI channel models were later enhanced in [344]. These so-called modified SUI channel models account for both omnidirectional antennas and directional antennas having a beamwidth of 30°.

Finally, we mention that the impact of the antenna array configuration and the array orientation on the capacity of MIMO systems has been studied in [345–347]. A comparative analysis using five different array configurations (uniform linear array, non-uniform linear array, uniform circular array, hexagonal array, and star array) can be found in [348].

Appendix 8.A Proof of Ergodicity

A sample function of $\hat{h}_{11}(t)$, denoted by $\tilde{h}_{11}(t)$, can be expressed as

$$\tilde{h}_{11}(t) = \frac{1}{\sqrt{MN}} \sum_{m=1}^{M} \sum_{n=1}^{N} g_{mn}\, e^{j[2\pi(f_m^T + f_n^R)t + \theta_{mn}]}, \tag{8.A.1}$$

where g_{mn}, f_m^T, and f_n^R are given by (8.54), (8.58), and (8.59), respectively, and the phases $\theta_{mn} = \theta_m + \theta_n$ are constant values. Using $g_{mn} = a_m b_n c_{mn}$ and recalling that $\tilde{h}_{22}(t)$ follows from $\tilde{h}_{11}(t)$ by performing the substitutions $a_m \to a_m^*$ and $b_n \to b_n^*$, the 3D space-time CCF of the simulation model can be written as

$$\begin{aligned}
\tilde{\rho}_{11,22}(\delta_T, \delta_R, \tau) &:= < \tilde{h}_{11}(t)\, \tilde{h}_{22}^*(t+\tau) > \\
&= < \frac{1}{\sqrt{MN}} \sum_{m=1}^{M} \sum_{n=1}^{N} a_m\, b_n\, c_{mn}\, e^{j[2\pi(f_m^T + f_n^R)t + \theta_{mn}]} \\
&\quad \cdot \frac{1}{\sqrt{MN}} \sum_{m'=1}^{M} \sum_{n'=1}^{N} a_{m'}\, b_{n'}\, c_{m'n'}^*\, e^{-j[2\pi(f_{m'}^T + f_{n'}^R)(t+\tau) + \theta_{m'n'}]} > \\
&= \frac{1}{MN} \sum_{m=1}^{M} \sum_{m'=1}^{M} \sum_{n=1}^{N} \sum_{n'=1}^{N} a_m\, a_{m'}\, b_n\, b_{n'}\, c_{mn}\, c_{m'n'}^* \\
&\quad \cdot < e^{j2\pi\left(f_m^T + f_n^R - f_{m'}^T - f_{n'}^R\right)t} > e^{-j[2\pi\left(f_{m'}^T + f_{n'}^R\right)\tau - \theta_{mn} + \theta_{m'n'}]}.
\end{aligned} \tag{8.A.2}$$

From the boundary condition: $f_m^T + f_n^R = f_{m'}^T + f_{n'}^R$ iff $m = m'$ and $n = n'$, it follows

$$< e^{j2\pi\left(f_m^T + f_n^R - f_{m'}^T - f_{n'}^R\right)t} > = \begin{cases} 1, & \text{if } m = m',\ n = n', \\ 0, & \text{otherwise.} \end{cases} \tag{8.A.3}$$

Using this relationship in (8.A.2) gives

$$\tilde{\rho}_{11,22}(\delta_T, \delta_R, \tau) = \frac{1}{MN} \sum_{m=1}^{M} \sum_{n=1}^{N} a_m^2(\delta_T)\, b_n^2(\delta_R)\, e^{-j2\pi\,(f_m^T + f_n^R)\tau}\,. \tag{8.A.4}$$

Finally, after comparing (8.A.4) with (8.78), we obtain the identity

$$\tilde{\rho}_{11,22}(\delta_T, \delta_R, \tau) = \hat{\rho}_{11,22}(\delta_T, \delta_R, \tau), \tag{8.A.5}$$

which states that the stochastic MIMO channel simulator is ergodic with respect to the 3D space-time CCF.

9

High-Speed Channel Simulators

The description of the channel simulators considered up to now has always been performed by using continuous-time representation. In Section 4.1, it was stated that a discrete-time simulation model, which is required for computer simulations, can be obtained directly from the continuous-time simulation model by substituting in the latter the time variable t by $t = kT_s$, where T_s denotes the sampling interval. This way of implementation will henceforth be designated as the *direct realization* and the corresponding simulation model will be called the *direct system*. In order to realize a real-valued deterministic Gaussian process by using direct realization, N_i sinusoids as well as several multiplications and additions have to be computed at each time instant k. Since the number of sinusoids N_i is the decisive quantity determining the computation time, the efficiency can only be increased essentially by reducing N_i. On the other hand, we know from our investigations in Chapter 5 that a natural lower limit at $N_i = 7$ exists, and, consequently, choosing $N_i < 7$ will result in heavy losses in quality. Thus, the possibilities for a further increase in the speed of direct systems with $N_i = 7$ are exhausted to a large extent. A speed-up of the simulator without accepting losses in precision can only be attained with indirect realization forms.

In this chapter, several ways of indirect implementation forms will be investigated. The basic idea which enables the derivation of new structures for the simulation of deterministic processes is based on taking advantage of the periodicity of sinusoidal functions. During the set-up phase, each of the N_i sinusoids is sampled only once within its basic period. The samples are then stored in N_i look-up tables. During the simulation run, the entries of each table are read out cyclically and summed up.

In this manner, it is possible to realize simulation models for complex-valued Gaussian random processes by merely using adders, storage elements, and a simple address generator. Time-consuming trigonometric operations as well as the implementation of multiplications are then no longer required. This results in *high-speed channel simulators* or *fast channel simulators* [120, 349], which can be implemented efficiently in software and hardware. The price for the decrease in computational complexity is an increase in the required memory capacity. Guidelines are provided to determine the required number of look-up tables and their respective lengths, which supports the reader to find a trade-off between computational complexity and performance. Since the proposed implementation technique can easily be

Mobile Radio Channels, Second Edition. Matthias Pätzold.

extended to the development of fast simulation models for all kinds of narrowband, wideband, and MIMO channels that can be derived from (complex-valued) Gaussian random processes, we will restrict our attention in this chapter to the design of high-speed channel simulators for Rayleigh channels.

For that purpose, we will employ the discrete-time representation to describe the so-called discrete-time deterministic processes in Section 9.1. This class of processes opens up new possibilities to establish indirect realization forms, where three of the most relevant will be presented in Section 9.2. The elementary and statistical properties of discrete-time deterministic processes will be examined in Section 9.3. Section 9.4 deals with the analysis of the required realization complexity as well as with the measurement of the simulation speed of high-speed channel simulators. A comparison with a filter-method-based simulation model for Rayleigh processes will be carried out in Section 9.5. The chapter closes with Section 9.6, where the reader will find further useful information and pointers to references in which implementation techniques and complexity issues of mobile radio channel simulators are discussed.

9.1 Discrete-Time Deterministic Processes

Our starting point is the deterministic Gaussian process $\tilde{\mu}_i(t)$ introduced in (4.4). Sampling this process at $t = kT_s$ results in a discrete-time signal (sequence) given by

$$\tilde{\mu}_i[k] := \tilde{\mu}_i(kT_s) = \sum_{n=1}^{N_i} c_{i,n} \cos(2\pi f_{i,n} kT_s + \theta_{i,n}). \tag{9.1}$$

With respect to a preferably efficient realization, the range of values has to be limited for the discrete Doppler frequencies $f_{i,n}$ as well as for the phases $\theta_{i,n}$. Thus, for the reciprocal value of the discrete Doppler frequencies $1/f_{i,n}$, for example, only integer multiples of the sampling interval T_s are henceforth permissible. The phases $\theta_{i,n}$ are subject to a similar restriction. According to two mappings, defined below, we obtain from $f_{i,n} \to \bar{f}_{i,n}$ and $\theta_{i,n} \to \bar{\theta}_{i,n}$ *quantized Doppler frequencies* $\bar{f}_{i,n}$ and *quantized phases* $\bar{\theta}_{i,n}$, respectively. Provided that the deviations between $f_{i,n}$ and $\bar{f}_{i,n}$ are sufficiently small, and, consequently, $\bar{f}_{i,n} \approx f_{i,n}$ holds, then

$$\bar{\mu}_i[k] := \bar{\mu}_i(kT_s) = \sum_{n=1}^{N_i} c_{i,n} \cos(2\pi \bar{f}_{i,n} kT_s + \bar{\theta}_{i,n}) \tag{9.2}$$

describes a sequence, which is equivalent to $\tilde{\mu}_i[k]$ in (9.1) (with respect to the relevant statistical properties). In the following, the sequence $\bar{\mu}_i[k]$ is called *discrete-time deterministic Gaussian process*. Thereby, the gains $c_{i,n}$ in (9.2) are identical to those used in (9.1), whereas the quantized Doppler frequencies $\bar{f}_{i,n}$ are related to the quantities $f_{i,n}$ and T_s according to

$$\bar{f}_{i,n} := \frac{1}{T_s \,\text{round}\{1/(f_{i,n}T_s)\}} \tag{9.3}$$

for all $n = 1, 2, \ldots, N_i$.[1] We call

$$L_{i,n} = \frac{1}{\bar{f}_{i,n}T_s} = \text{round}\left\{\frac{1}{f_{i,n}T_s}\right\} \tag{9.4}$$

[1] The operator round$\{x\}$ in (9.3) rounds the real-valued number x to the nearest integer.

the period of the individual elementary discrete-time sinusoidal function $\bar{\mu}_{i,n}[k] = c_{i,n}\cos(2\pi \bar{f}_{i,n}kT_s + \bar{\theta}_{i,n})$, i.e., one has the equation $\bar{\mu}_{i,n}[k] = \bar{\mu}_{i,n}[k + L_{i,n}]$. Note that the rounding operation used in (9.4) always results in a natural number for the period $L_{i,n}$. In the next section, we will see that this will turn out to be a clear advantage for the realization.

The quantized phases $\bar{\theta}_{i,n}$ in (9.2) are calculated from the given quantities $\theta_{i,n}$ according to the expression

$$\bar{\theta}_{i,n} := \frac{2\pi}{L_{i,n}} \operatorname{round}\left\{\frac{L_{i,n}}{2\pi}\,\theta_{i,n}\right\} \tag{9.5}$$

for all $n = 1, 2, \ldots, N_i$. Recall that the phases $\theta_{i,n}$ are real-valued numbers within the interval $(0, 2\pi]$, whereas the quantized values $\bar{\theta}_{i,n}$ according to (9.5) are elements of the set

$$\bar{\Theta}_{i,n} = \left\{2\pi\frac{1}{L_{i,n}},\ 2\pi\frac{2}{L_{i,n}},\ \ldots,\ 2\pi\frac{L_{i,n}-1}{L_{i,n}},\ 2\pi\right\}. \tag{9.6}$$

The mapping $\theta_{i,n} \to \bar{\theta}_{i,n}$ established by (9.5) has been chosen in such a way that $\bar{\theta}_{i,n} \in \bar{\Theta}_{i,n}$ is as close as possible to $\theta_{i,n}$.

By using $x - 1/2 \le \operatorname{round}\{x\} \le x + 1/2$, one can show that in the limit $T_s \to 0$ it follows $\bar{f}_{i,n} = f_{i,n}$ and $\bar{\theta}_{i,n} = \theta_{i,n}$ from (9.3) and (9.5), respectively. However, for sufficiently small sampling intervals T_s, we may write $\bar{f}_{i,n} \approx f_{i,n}$ and $\bar{\theta}_{i,n} \approx \theta_{i,n}$. At this point, we want to note that the quality of the approximation $\bar{\theta}_{i,n} \approx \theta_{i,n}$ under particular conditions, which will be discussed in detail in Section 9.3, does not affect the statistical properties of $\bar{\mu}_i[k]$. On the other hand, the deviations between $\bar{f}_{i,n}$ and $f_{i,n}$ determined by the sampling interval T_s cannot be ignored without hesitation, which will also be substantiated in Section 9.3. As an appropriate measure of the deviation between $\bar{f}_{i,n}$ and $f_{i,n}$, we consider the relative error

$$\varepsilon_{\bar{f}_{i,n}} = \frac{\bar{f}_{i,n} - f_{i,n}}{f_{i,n}} \tag{9.7}$$

illustrated in Figure 9.1. From this figure, it can be realized that the quality of the approximation $\bar{f}_{i,n} \approx f_{i,n}$ decreases if the sampling interval T_s increases. This result indicates that the statistical properties of $\bar{\mu}_i[k]$ depend on the value of the sampling interval T_s. From Figure 9.1, we can also conclude that if $T_s < 1/(10 f_{i,n})$, then the absolute value of the relative error $|\varepsilon_{\bar{f}_{i,n}}|$ is below 5 per cent, which can be tolerated in most practical applications.

Obviously, the discrete-time deterministic Gaussian process $\bar{\mu}_i[k]$ introduced in (9.2) can be derived from the continuous-time deterministic process $\tilde{\mu}_i(t)$ by sampling the latter at the time instants $t = kT_s$ and, furthermore, by replacing the quantities $f_{i,n}$ and $\theta_{i,n}$ by their quantized versions $\bar{f}_{i,n}$ and $\bar{\theta}_{i,n}$, respectively, i.e.,

$$\tilde{\mu}_i(t) \xrightarrow{\ t \to kT_s\ } \tilde{\mu}_i[k] := \tilde{\mu}_i(kT_s) \xrightarrow[\theta_{i,n} \to \bar{\theta}_{i,n}]{f_{i,n} \to \bar{f}_{i,n}} \bar{\mu}_i[k] := \bar{\mu}_i(kT_s). \tag{9.8}$$

From the fact that $\bar{f}_{i,n}$ and $\bar{\theta}_{i,n}$ converge to $f_{i,n}$ and $\theta_{i,n}$, respectively, as T_s decreases, it follows: $\bar{\mu}_i[k] \to \tilde{\mu}_i(t)$ as $T_s \to 0$. With reference to the results of Chapter 4, it becomes obvious that the discrete-time deterministic Gaussian process $\bar{\mu}_i[k]$ tends to a sample function of the Gaussian random process $\mu_i(t)$ as $T_s \to 0$ and $N_i \to \infty$.

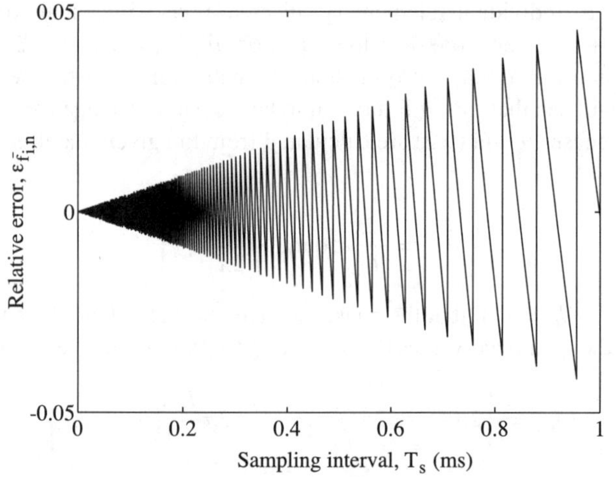

Figure 9.1 Relative error $\varepsilon_{\bar{f}_{i,n}}$ of $\bar{f}_{i,n}$ according to (9.7) for $f_{i,n} = 91$ Hz as a function of the sampling interval T_s.

By analogy to (4.5), we introduce the complex-valued sequence

$$\bar{\mu}[k] = \bar{\mu}_1[k] + j\bar{\mu}_2[k] \tag{9.9}$$

as *complex discrete-time deterministic Gaussian process* and we call its absolute value

$$\bar{\zeta}[k] = |\bar{\mu}[k]| = |\bar{\mu}_1[k] + j\bar{\mu}_2[k]| \tag{9.10}$$

the *discrete-time deterministic Rayleigh process*. Moreover, we will in the following study the phase $\bar{\vartheta}[k] = \arg\{\bar{\mu}[k]\}$ defined by the discrete-time deterministic process

$$\bar{\vartheta}[k] = \arctan\left\{\frac{\bar{\mu}_2[k]}{\bar{\mu}_1[k]}\right\}. \tag{9.11}$$

9.2 Realization of Discrete-Time Deterministic Processes

The discrete-time deterministic processes introduced in the previous section open up new possibilities for the development of high-speed channel simulators. In the following, three procedures will be presented.

9.2.1 Look-Up Table System

The basic idea of the look-up table system is to store the samples of one period of the sequence $\bar{\mu}_{i,n}[k] = c_{i,n}\cos(2\pi \bar{f}_{i,n}kT_s + \bar{\theta}_{i,n})$ into a look-up table and to read out the table entries cyclically during the simulation [120]. For the design of a simulation model for Rayleigh channels, $N_1 + N_2$ look-up tables instead of $N_1 + N_2$ sinusoids are required. By means of an address generator, the values stored in the look-up tables are accessed. At any discrete time instant $k = 0, 1, 2, \ldots$, the discrete sequence $\bar{\mu}[k] = \bar{\mu}_1[k] + j\bar{\mu}_2[k]$ can simply be reconstructed by summing up the selected entries of the look-up table as shown in Figure 9.2.

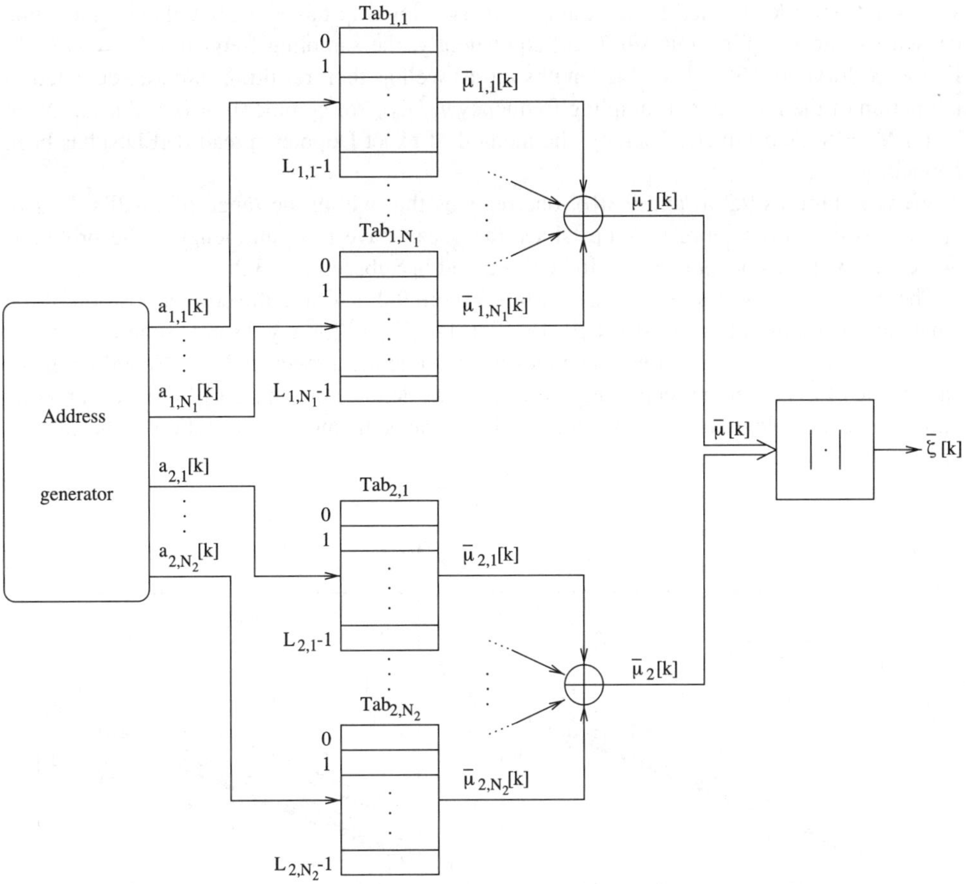

Figure 9.2 Look-up table system for the fast simulation of Rayleigh channels.

After taking the absolute value, the desired discrete-time deterministic Rayleigh process $\bar{\zeta}[k]$ is then available.

The look-up table, in which the information of one period of an elementary discrete-time sinusoidal function $\bar{\mu}_{i,n}[k]$ is stored, will be denoted by $\text{Tab}_{i,n}$. The entry of the look-up table $\text{Tab}_{i,n}$ at position $l \in \{0, 1, \ldots, L_{i,n} - 1\}$ corresponds to the value of $\bar{\mu}_{i,n}[k]$ at $k = l$, i.e., it holds

$$\bar{\mu}_{i,n}[l] = c_{i,n} \cos(2\pi \bar{f}_{i,n} l T_s + \bar{\theta}_{i,n}) \tag{9.12}$$

for all $n = 1, 2, \ldots, N_i$ $(i = 1, 2)$. Now, reading out the entries of the look-up table $\text{Tab}_{i,n}$ cyclically, results in the sequence $\{\bar{\mu}_{i,n}[0],\ \bar{\mu}_{i,n}[1],\ \ldots,\ \bar{\mu}_{i,n}[L_{i,n} - 1],\ \bar{\mu}_{i,n}[L_{i,n}] = \bar{\mu}_{i,n}[0],\ \ldots\}$. Hence, by exploiting the periodicity, one can reconstruct $\bar{\mu}_{i,n}[k]$ completely for all $k = 0, 1, 2, \ldots$ The length of the look-up table $\text{Tab}_{i,n}$ is identical to the period $L_{i,n}$ of $\bar{\mu}_{i,n}[k]$. In consequence, the total amount of storage elements required for the implementation of a discrete-time deterministic process $\bar{\mu}_i[k]$ is given by the sum $\sum_{n=1}^{N_i} L_{i,n}$. Owing to (9.4), the total memory size is not only determined by the number of used look-up tables N_i, but also by the value chosen for the sampling interval T_s or, equivalently, the sampling frequency $f_s = 1/T_s$. In Figures 9.3(a) and 9.3(b), the table lengths $L_{i,n}$ as well as their resulting sums are depicted as a function of the normalized sampling frequency $f_s/f_{\max}$ for commonly used values of $N_1 = 7$ and $N_2 = 8$, respectively. Thereby, the method of exact Doppler spread (MEDS) has been applied.

Viewing Figures 9.3(a) and 9.3(b), one realizes that within the range of small values of $f_s/f_{\max}$, two or even more look-up tables $\text{Tab}_{i,n}$ can have the same length. The problems associated with this phenomenon will be discussed in Subsection 9.3.2.

The task of the address generator shown in Figure 9.2 is to find the correct position of the table entries required to reconstruct $\bar{\mu}[k] = \bar{\mu}_1[k] + j\bar{\mu}_2[k]$ for any discrete time instant $k = 0, 1, 2, \ldots$ Therefore, the address generator has to generate altogether $N_1 + N_2$ addresses for each value of k. As can be seen in Figure 9.2, $a_{i,n}[k]$ denotes the address of the look-up table $\text{Tab}_{i,n}$ at the discrete time instant k. Figure 9.4 illustrates the mode of operation of the address generator.

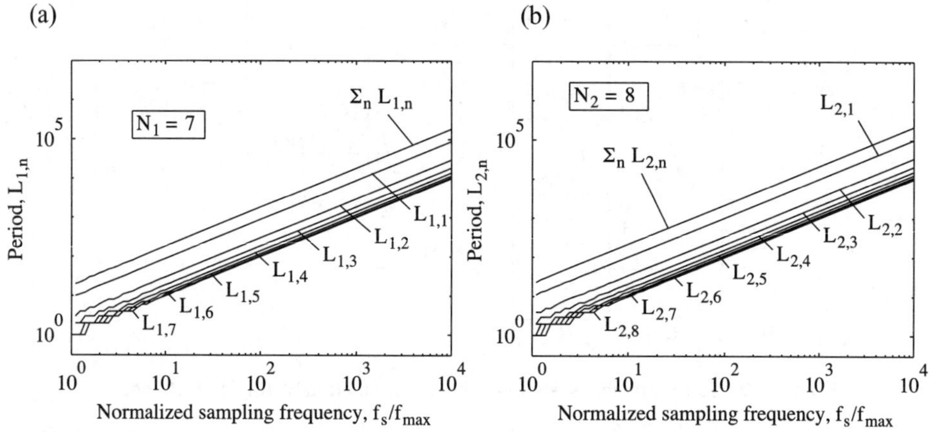

Figure 9.3 Table lengths $L_{i,n}$ as a function of the normalized sampling frequency $f_s/f_{\max}$: (a) $L_{1,n}$ for $N_1 = 7$ and (b) $L_{2,n}$ for $N_2 = 8$ (MEDS, Jakes PSD, $f_{\max} = 91$ Hz, $\sigma_0^2 = 1$).

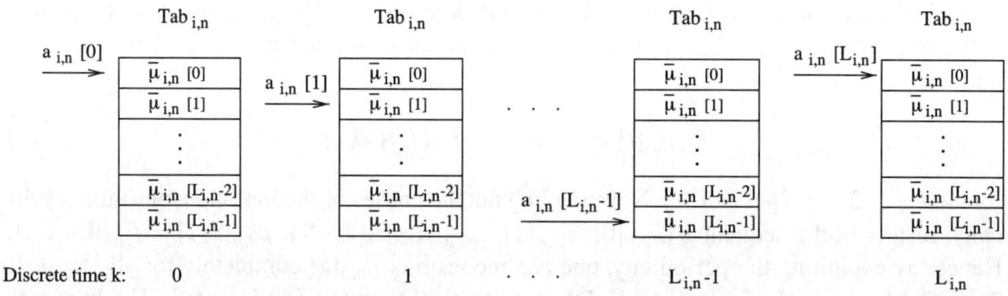

Figure 9.4 The address generator's mode of operation.

At the time instant $k = 0$, the address $a_{i,n}[0]$ points at the register $\bar{\mu}_{i,n}[0]$ of the look-up table $\text{Tab}_{i,n}$. At the next time instant $k = 1$, $a_{i,n}[1]$ refers to $\bar{\mu}_{i,n}[1]$, etc., up to the time instant $k = L_{i,n} - 1$, where the address $a_{i,n}[L_{i,n} - 1]$ points to the last position of the table with the entry $\bar{\mu}_{i,n}[L_{i,n} - 1]$. At the following time instant $k = L_{i,n}$, the address $a_{i,n}[L_{i,n}]$ is reset to $a_{i,n}[0]$, which points again at the initial position $\bar{\mu}_{i,n}[0]$.

Starting with the initial addresses $a_{i,n}[0] = 0$ and applying the modulo operation, all addresses $a_{i,n}[k]$ can be found for any time instant $k > 0$ by using the following recursive algorithm

$$a_{i,n}[k] = (a_{i,n}[k-1] + 1) \bmod L_{i,n}, \tag{9.13}$$

where $n = 1, 2, \ldots, N_i$ $(i = 1, 2)$. It should be mentioned that the modulo operation in (9.13) has been applied here only for mathematical convenience. For the realization of the algorithm on a computer, only one adder and a simple conditional control flow statement (*if-else* statement) have to be implemented for the computation of $a_{i,n}[k]$.

Thus, the entire look-up table system (see Figure 9.2) only consists of adders, storage elements, and simple conditional operators. Multiplications as well as trigonometric operations no longer have to be carried out for the computation of $\bar{\mu}[k] = \bar{\mu}_1[k] + j\bar{\mu}_2[k]$.

9.2.2 *Matrix System*

The matrix system combines the N_i look-up tables to a *channel matrix*, denoted by $\boldsymbol{M}_i$. The number of rows of the channel matrix $\boldsymbol{M}_i$ is identical to the number of look-up tables N_i. Thereby, the nth row of $\boldsymbol{M}_i$ contains the entries of the look-up table $\text{Tab}_{i,n}$. As a result, the length of the largest look-up table, i.e., $L_{i,\max} = \max\,\{L_{i,n}\}_{n=1}^{N_i}$, defines the number of columns of the channel matrix $\boldsymbol{M}_i$. Without loss of generality, we assume in the following that $L_{i,\max} = L_{i,1}$ holds, which is actually always the case if the MEDS is used (see Figure 9.3). The first $L_{i,n}$ entries of the nth row of $\boldsymbol{M}_i$ are exactly identical to the entries of the look-up table $\text{Tab}_{i,n}$, whereas the rest of the row is filled up with zeros. Thus, the channel matrix $\boldsymbol{M}_i \in \mathbb{R}^{N_i \times L_{i,1}}$ can be represented as follows

$$\boldsymbol{M}_i = \begin{pmatrix} \bar{\mu}_{i,1}[0] & \cdots\cdots\cdots\cdots\cdots\cdots & \cdots\cdots\cdots & \cdots & \bar{\mu}_{i,1}[L_{i,1}-1] \\ \bar{\mu}_{i,2}[0] & \cdots\cdots\cdots & \bar{\mu}_{i,2}[L_{i,2}-1] & 0 \;\; \cdots & 0 \\ \vdots & & & \ddots & \vdots \\ \bar{\mu}_{i,N_i}[0] & \cdots \;\; \bar{\mu}_{i,N_i}[L_{i,N_i}-1] & 0 & \cdots & 0 \end{pmatrix}. \tag{9.14}$$

The channel matrix $\boldsymbol{M}_i$ contains the complete information needed for the reconstruction of $\bar{\mu}_i[k]$. In order to guarantee the correct reconstruction of $\bar{\mu}_i[k]$ for all values of $k = 0,\ 1,\ 2, \ldots$, it is necessary to select from each row of $\boldsymbol{M}_i$ one entry at the correct position. This can be achieved by introducing a further matrix $\boldsymbol{S}_i$, which will henceforth be called the *selection matrix*. The entries of the selection matrix $\boldsymbol{S}_i$ are time variant quantities, which can only take the values 0 or 1. There is a close relation between the address generator introduced in the previous subsection and the selection matrix $\boldsymbol{S}_i$. This becomes obvious by noting that the entries of $\boldsymbol{S}_i = (s_{l,n}) \in \{0, 1\}^{L_{i,1} \times N_i}$ can be calculated at any time instant k by using the addresses $a_{i,n}[k]$ in (9.13) according to

$$s_{l,n} = s_{l,n}[k] = \begin{cases} 1 & \text{if} \quad l = a_{i,n}[k] \\ 0 & \text{if} \quad l \neq a_{i,n}[k] \end{cases} \tag{9.15}$$

for all $l = 0, 1, \ldots, L_{i,1} - 1$ and $n = 1, 2, \ldots, N_i$ $(i = 1, 2)$.

The discrete-time deterministic Gaussian process $\bar{\mu}_i[k]$ can now be expressed in terms of the product of the channel matrix $\boldsymbol{M}_i$ and the selection matrix $\boldsymbol{S}_i$ as follows:

$$\bar{\mu}_i[k] = \operatorname{tr}(\boldsymbol{M}_i \cdot \boldsymbol{S}_i), \tag{9.16}$$

where $\operatorname{tr}(\cdot)$ denotes the trace[2] [350, 351].

Using (9.16), we can thus also express the complex discrete-time deterministic Gaussian process $\bar{\mu}[k]$ [see (9.9)] in the following alternative form

$$\bar{\mu}[k] = \operatorname{tr}(\boldsymbol{M}_1 \cdot \boldsymbol{S}_1) + j \operatorname{tr}(\boldsymbol{M}_2 \cdot \boldsymbol{S}_2). \tag{9.17}$$

It is worth mentioning that the number of columns (rows) of the channel matrix $\boldsymbol{M}_i$ (selection matrix $\boldsymbol{S}_i$) tends to infinity as $T_s \to 0$, and thus $\bar{\mu}[k]$ converges to $\tilde{\mu}(t)$. In the limits $T_s \to 0$ and $N_i \to \infty$, the number of columns and the number of rows of both the channel matrix $\boldsymbol{M}_i$ and the selection matrix $\boldsymbol{S}_i$ tend to infinity. In this case, the complex discrete-time deterministic Gaussian process $\bar{\mu}[k]$ converges to a sample function of the complex stochastic Gaussian random process $\mu(t)$, as it was expected.

An equivalent representation of the discrete-time deterministic Rayleigh process $\bar{\zeta}[k]$, introduced in (9.10), can be obtained by taking the absolute value of $\bar{\mu}[k]$ in (9.17), i.e.,

$$\bar{\zeta}[k] = |\bar{\mu}[k]| = |\operatorname{tr}(\boldsymbol{M}_1 \cdot \boldsymbol{S}_1) + j \operatorname{tr}(\boldsymbol{M}_2 \cdot \boldsymbol{S}_2)|. \tag{9.18}$$

For the sake of completeness, we express the phase $\bar{\vartheta}[k]$ of $\bar{\mu}[k] = \bar{\mu}_1[k] + j\bar{\mu}_2[k]$ in the form

$$\bar{\vartheta}[k] = \arctan\left\{\frac{\operatorname{tr}(\boldsymbol{M}_2 \cdot \boldsymbol{S}_2)}{\operatorname{tr}(\boldsymbol{M}_1 \cdot \boldsymbol{S}_1)}\right\}. \tag{9.19}$$

It is evident that in the limit $T_s \to 0$ it follows: $\bar{\zeta}[k] \to \tilde{\zeta}(t)$ and $\bar{\vartheta}[k] \to \tilde{\vartheta}(t)$. Furthermore, the sequences $\bar{\zeta}[k]$ and $\bar{\vartheta}[k]$ are converging to a sample function of the corresponding stochastic processes $\zeta(t)$ and $\vartheta(t)$, respectively, as $T_s \to 0$ and $N_i \to \infty$.

It should be mentioned that the computation of the discrete-time deterministic processes in (9.16)–(9.19) by taking the trace of the product of two matrices is not a very efficient approach due to the large number of multiplications and additions that have to be carried out. However, considerable simplifications are possible if all unnecessary operations such as multiplications with zero and one are avoided at the beginning. In this case, the matrix system reduces to the look-up table system. In other words: The matrix system actually represents no genuine alternative realization form to the look-up table system, but provides some new aspects

[2] The trace of a square matrix $\mathbf{A} = (a_{n,m}) \in \mathbb{R}^{N \times N}$ is defined by the sum of the main diagonal entries $a_{n,m}$, i.e., $\operatorname{tr}(\mathbf{A}) = \sum_{n=1}^{N} a_{n,n}$.

regarding the interpretation and representation of discrete-time deterministic processes and their relationships to stochastic processes.

9.2.3 Shift Register System

From the look-up table system (see Figure 9.2), we can derive the shift register system depicted in Figure 9.5 by replacing the look-up tables $\text{Tab}_{i,n}$ by feedback *shift registers* $\text{Reg}_{i,n}$. Instead of $N_1 + N_2$ look-up tables, now $N_1 + N_2$ shift registers are required for the realization of $\bar{\mu}[k] = \bar{\mu}_1[k] + j\bar{\mu}_2[k]$. The length of the shift register $\text{Reg}_{i,n}$ is thereby identical to the length $L_{i,n}$ of the corresponding look-up table $\text{Tab}_{i,n}$. During the simulation set-up phase, the shift registers $\text{Reg}_{i,n}$ are filled at the positions $l \in \{0, 1, \ldots, L_{i,n} - 1\}$ with the values $\bar{\mu}_{i,n}[l] = c_{i,n} \cos(2\pi \bar{f}_{i,n} l T_s + \bar{\theta}_{i,n})$, where $n = 1, 2, \ldots, N_i$ and $i = 1, 2$. Throughout the simulation run, the contents of the shift registers are shifted by one position to the right at every clock cycle (see Figure 9.5). Owing to the links created between the shift register outputs (positions 0) with their respective inputs (positions $L_{i,n} - 1$), it is ensured that the discrete-time deterministic processes $\bar{\mu}_{i,n}[k]$ and, consequently, also $\bar{\mu}[k] = \bar{\mu}_1[k] + j\bar{\mu}_2[k]$ as well as $\bar{\zeta}[k] = |\bar{\mu}[k]|$ can be reconstructed for all values of $k = 0, 1, 2, \ldots$

Note that in comparison with the look-up table system, an address generator is not needed, but instead of this, $\sum_{i=1}^{2} \sum_{n=1}^{N_i} L_{i,n}$ register entries have to be shifted at every clock cycle, which — especially for software realizations in connection with large register lengths — does

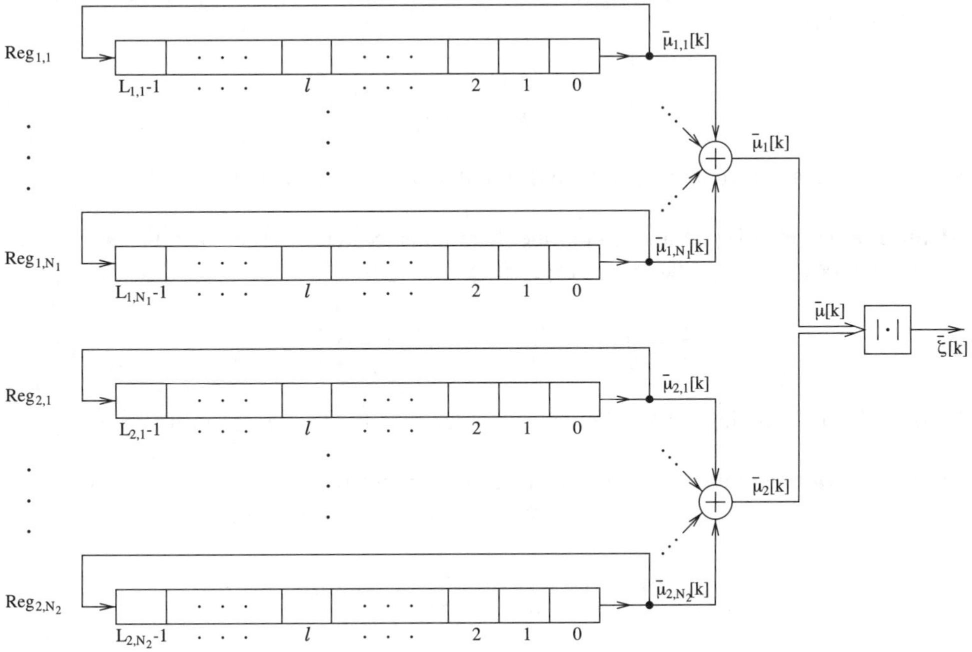

Figure 9.5 Realization of discrete-time deterministic Rayleigh processes $\bar{\zeta}[k]$ by using shift registers.

not lead to a satisfying solution. For this reason, we prefer the look-up table system to the shift register system and turn our attention in the next section to the analysis of the properties of discrete-time deterministic processes.

9.3 Properties of Discrete-Time Deterministic Processes

By analogy to the analysis of continuous-time deterministic processes (Chapter 4), we start in Subsection 9.3.1 with the investigation of the elementary properties of discrete-time deterministic processes, followed by the analysis of their statistical properties in Subsection 9.3.2.

9.3.1 Elementary Properties of Discrete-Time Deterministic Processes

The interpretation of $\bar{\mu}_i[k]$ as a discrete-time deterministic process, i.e., as a mapping of the form

$$\bar{\mu}_i : \mathbb{Z} \to \mathbb{R}, \qquad k \mapsto \bar{\mu}_i[k], \tag{9.20}$$

allows us to establish a close relationship with the investigations performed in Section 4.2. Therefore, we proceed here analogously to Section 4.2 and derive simple closed-form solutions for the fundamental characteristic quantities of $\bar{\mu}_i[k]$, such as the mean value, mean power, autocorrelation sequence, etc.

Mean value: Let $\bar{\mu}_i[k]$ be a discrete-time deterministic process with $\bar{f}_{i,n} \neq 0$ ($n = 1, 2, \ldots, N_i$). Then, it can be shown by using (2.150) in combination with (9.2) that the mean value of $\bar{\mu}_i[k]$ is equal to

$$\bar{m}_{\mu_i} = \lim_{K\to\infty} \frac{1}{2K+1} \sum_{k=-K}^{K} \bar{\mu}_i[k] = 0. \tag{9.21}$$

It is henceforth assumed that $\bar{f}_{i,n} \neq 0$ holds for all $n = 1, 2, \ldots, N_i$ and $i = 1, 2$.

Mean power: Let $\bar{\mu}_i[k]$ be a discrete-time deterministic process. Then, it follows by using (2.151) and (9.2) that its mean power is given by

$$\bar{\sigma}^2_{\mu_i} = \lim_{K\to\infty} \frac{1}{2K+1} \sum_{k=-K}^{K} \bar{\mu}_i^2[k] = \sum_{n=1}^{N_i} \frac{c_{i,n}^2}{2}. \tag{9.22}$$

In particular, by applying the MEDS, we obtain due to (5.86) the desired result $\bar{\sigma}^2_{\mu_i} = \sigma_0^2$.

Autocorrelation sequence: Let $\bar{\mu}_i[k]$ be a discrete-time deterministic process. Then, it follows from (2.152) and (9.2) that the autocorrelation sequence of $\bar{\mu}_i[k]$ can be expressed by

$$\begin{aligned} \bar{r}_{\mu_i\mu_i}[\kappa] &= \lim_{K\to\infty} \frac{1}{2K+1} \sum_{k=-K}^{K} \bar{\mu}_i[k]\, \bar{\mu}_i[k+\kappa] \\ &= \sum_{n=1}^{N_i} \frac{c_{i,n}^2}{2} \cos(2\pi \bar{f}_{i,n} T_s \kappa). \end{aligned} \tag{9.23}$$

A comparison with (4.11) shows that $\bar{r}_{\mu_i\mu_i}[\kappa]$ can be obtained from $\tilde{r}_{\mu_i\mu_i}(\tau)$ if $\tilde{r}_{\mu_i\mu_i}(\tau)$ is sampled at $\tau = \kappa T_s$ and if additionally the quantities $f_{i,n}$ are substituted by $\bar{f}_{i,n}$. Moreover, we realize that also in the discrete-time case, the quantized phases $\bar{\theta}_{i,n}$ have no influence on the behaviour of the autocorrelation sequence $\bar{r}_{\mu_i\mu_i}[\kappa]$. Observe that from (9.22) and (9.23) the relation $\bar{\sigma}^2_{\mu_i} = \bar{r}_{\mu_i\mu_i}[0]$ can directly be obtained.

The deviations between $\tilde{r}_{\mu_i\mu_i}[\kappa] := \tilde{r}_{\mu_i\mu_i}(\kappa T_s)$ and $\bar{r}_{\mu_i\mu_i}[\kappa]$, caused by the quantization of the discrete Doppler frequencies $f_{i,n}$, can be observed in Figure 9.6. Thereby, the MEDS has been applied by using $N_i = 8$ sinusoids (look-up tables). Figure 9.6(a) shows that for sufficiently small sampling intervals ($T_s = 0.1$ ms) no significant differences occur between $\tilde{r}_{\mu_i\mu_i}[\kappa]$ and $\bar{r}_{\mu_i\mu_i}[\kappa]$ if $\tau = \kappa T_s$ is within its range of interest, i.e., $\tau \in [0, N_i/(2f_{\max})]$. However, this does not apply for large values of T_s, as can be seen when considering Figure 9.6(b), where the corresponding results in case of $T_s = 1$ ms are shown.

Cross-correlation sequence: Let $\bar{\mu}_1[k]$ and $\bar{\mu}_2[k]$ be two discrete-time deterministic processes. Then, it follows from (2.153) in connection with (9.2) that the cross-correlation sequence is equal to

$$\bar{r}_{\mu_1\mu_2}[\kappa] = 0, \tag{9.24}$$

if $\bar{f}_{1,n} \neq \pm\bar{f}_{2,m}$ is fulfilled for all $n = 1, 2, \ldots, N_1$ and $m = 1, 2, \ldots, N_2$, or

$$\bar{r}_{\mu_1\mu_2}[\kappa] = \sum_{\substack{n=1 \\ \bar{f}_{1,n}=\pm\bar{f}_{2,m}}}^{\max\{N_1,N_2\}} \frac{c_{1,n}c_{2,m}}{2} \cos(2\pi \bar{f}_{1,n} T_s \kappa - \bar{\theta}_{1,n} \pm \bar{\theta}_{2,m}), \tag{9.25}$$

if $\bar{f}_{1,n} = \pm\bar{f}_{2,m}$ holds for one or several pairs (n,m). Notice that $\bar{r}_{\mu_1\mu_2}[\kappa]$ can immediately be derived from $\tilde{r}_{\mu_1\mu_2}(\tau)$ if the continuous variable τ is replaced in (4.12) and (4.13) by κT_s, and, additionally, the quantities $f_{i,n}$ and $\theta_{i,n}$ are substituted by their quantized quantities $\bar{f}_{i,n}$

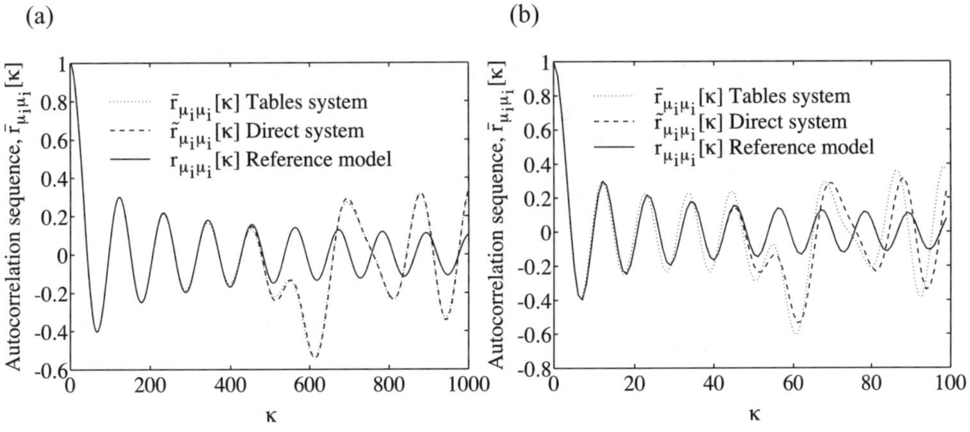

Figure 9.6 Autocorrelation sequence $\bar{r}_{\mu_i\mu_i}[\kappa]$ of discrete-time deterministic Gaussian processes $\bar{\mu}_i[k]$ for (a) $T_s = 0.1$ ms and (b) $T_s = 1$ ms (MEDS, Jakes PSD, $N_i = 8$, $f_{\max} = 91$ Hz, $\sigma_0^2 = 1$).

and $\bar{\theta}_{i,n}$, respectively. The two cross-correlation sequences $\bar{r}_{\mu_1\mu_2}[\kappa]$ and $\bar{r}_{\mu_2\mu_1}[\kappa]$ are related by $\bar{r}_{\mu_2\mu_1}[\kappa] = \bar{r}^*_{\mu_1\mu_2}[-\kappa] = \bar{r}_{\mu_1\mu_2}[-\kappa]$.

Power spectral density: Let $\bar{\mu}_i[k]$ be a discrete-time deterministic process. Then, it follows by applying the discrete Fourier transform (2.154) in connection with (9.23) that the power spectral density $\bar{S}_{\mu_i\mu_i}(f)$ of $\bar{\mu}_i[k]$ can be written as

$$\bar{S}_{\mu_i\mu_i}(f) = \frac{1}{T_s}\sum_{\nu=-\infty}^{\infty}\sum_{n=1}^{N_i}\frac{c_{i,n}^2}{4}\left[\delta(f-\bar{f}_{i,n}-\nu f_s)+\delta(f+\bar{f}_{i,n}-\nu f_s)\right], \tag{9.26}$$

where $f_s = 1/T_s$ denotes the sampling frequency. Thus, the power spectral density $\bar{S}_{\mu_i\mu_i}(f)$ is a symmetrical line spectrum, where the spectral lines are located at $f = \pm\bar{f}_{i,n} + \nu f_s$ and weighted by the factor $c_{i,n}^2/(4T_s)$. Using (2.155) and taking the power spectral density $\tilde{S}_{\mu_i\mu_i}(f)$ of $\tilde{\mu}_i(t)$ according to (4.18) into account, one can express $\bar{S}_{\mu_i\mu_i}(f)$ in terms of $\tilde{S}_{\mu_i\mu_i}(f)$ as follows

$$\bar{S}_{\mu_i\mu_i}(f) = \frac{1}{T_s}\sum_{\nu=-\infty}^{\infty}\tilde{S}_{\mu_i\mu_i}(f-\nu f_s)\bigg|_{f_{i,n}=\bar{f}_{i,n}}. \tag{9.27}$$

The equation above makes clear that the power spectral density $\bar{S}_{\mu_i\mu_i}(f)$ of the discrete-time deterministic process $\bar{\mu}_i[k]$ can be represented by an infinite sum of weighted and frequency-shifted versions of the power spectral density $\tilde{S}_{\mu_i\mu_i}(f)$ of the corresponding continuous-time deterministic process $\tilde{\mu}_i(t)$, where the weighting factor is equal to $1/T_s$ and the frequency shifts are integer multiples of the sampling frequency f_s. In addition, the quantities $f_{i,n}$ have to be replaced by $\bar{f}_{i,n}$.

Cross-power spectral density: Let $\bar{\mu}_1[k]$ and $\bar{\mu}_2[k]$ be two discrete-time deterministic processes. Then, it follows from (2.157) by using (9.24) and (9.25) that the cross-power spectral density of $\bar{\mu}_1[k]$ and $\bar{\mu}_2[k]$ can be expressed by

$$\bar{S}_{\mu_1\mu_2}(f) = 0, \tag{9.28}$$

if $\bar{f}_{1,n} \neq \pm\bar{f}_{2,m}$ holds for all $n = 1, 2, \ldots, N_1$ and $m = 1, 2, \ldots, N_2$, or

$$\bar{S}_{\mu_1\mu_2}(f) = \frac{1}{T_s}\sum_{\nu=-\infty}^{\infty}\sum_{\substack{n=1\\ \bar{f}_{1,n}=\pm\bar{f}_{2,m}}}^{\max\{N_1,N_2\}}\frac{c_{1,n}c_{2,m}}{4}\Big[\delta(f-\bar{f}_{1,n}-\nu f_s)\cdot e^{-j(\bar{\theta}_{1,n}\mp\bar{\theta}_{2,m})}$$
$$+\,\delta(f+\bar{f}_{1,n}-\nu f_s)\cdot e^{j(\bar{\theta}_{1,n}\mp\bar{\theta}_{2,m})}\Big], \tag{9.29}$$

if $\bar{f}_{1,n} = \pm\bar{f}_{2,m}$ is valid for one or several pairs (n,m). Employing (4.19), (4.20), and (2.155), the results in (9.28) and (9.29) can be combined as follows

$$\bar{S}_{\mu_1\mu_2}(f) = \frac{1}{T_s}\sum_{\nu=-\infty}^{\infty}\tilde{S}_{\mu_1\mu_2}(f-\nu f_s)\bigg|_{\substack{f_{i,n}=\bar{f}_{i,n}\\ \theta_{i,n}=\bar{\theta}_{i,n}}}. \tag{9.30}$$

The cross-power spectral densities $\bar{S}_{\mu_1\mu_2}(f)$ and $\bar{S}_{\mu_2\mu_1}(f)$ are related by $\bar{S}_{\mu_2\mu_1}(f) = \bar{S}^*_{\mu_1\mu_2}(f)$.

Average Doppler shift: Let $\bar{\mu}_i[k]$ be a discrete-time deterministic process described by the power spectral density $\bar{S}_{\mu_i\mu_i}(f)$ as given by (9.26). Then, the corresponding average Doppler shift $\bar{B}^{(1)}_{\mu_i\mu_i}$ is defined by

$$\bar{B}^{(1)}_{\mu_i\mu_i} := \frac{\int\limits_{-f_s/2}^{f_s/2} f\,\bar{S}_{\mu_i\mu_i}(f)\,df}{\int\limits_{-f_s/2}^{f_s/2} \bar{S}_{\mu_i\mu_i}(f)\,df} = \frac{1}{2\pi j}\cdot\frac{\dot{\bar{r}}_{\mu_i\mu_i}[0]}{\bar{r}_{\mu_i\mu_i}[0]}. \tag{9.31}$$

In contrast to (3.28a) and (4.24), where the integration is carried out over the entire frequency range, the limits of the integration in (9.31) are restricted to the Nyquist range defined by the frequency interval $[-f_s/2, f_s/2)$. In the special case that the Doppler power spectral density has a symmetrical shape, i.e., $\bar{S}_{\mu_i\mu_i}(f) = \bar{S}_{\mu_i\mu_i}(-f)$, it follows directly

$$\bar{B}^{(1)}_{\mu_i\mu_i} = B^{(1)}_{\mu_i\mu_i} = 0. \tag{9.32}$$

A comparison with (4.25) shows that neither the effect caused by the substitution of the time variable t by $t = kT_s$ nor the quantization of the discrete Doppler frequencies has an influence on the average Doppler shift.

Doppler spread: Let $\bar{\mu}_i[k]$ be a discrete-time deterministic process with the power spectral density $\bar{S}_{\mu_i\mu_i}(f)$ as given by (9.26). Then, the corresponding Doppler spread $\bar{B}^{(2)}_{\mu_i\mu_i}$ is defined by

$$\bar{B}^{(2)}_{\mu_i\mu_i} := \sqrt{\frac{\int\limits_{-f_s/2}^{f_s/2} (f - \bar{B}^{(1)}_{\mu_i\mu_i})^2\,\bar{S}_{\mu_i\mu_i}(f)\,df}{\int\limits_{-f_s/2}^{f_s/2} \bar{S}_{\mu_i\mu_i}(f)\,df}}$$

$$= \frac{1}{2\pi}\sqrt{\left(\frac{\dot{\bar{r}}_{\mu_i\mu_i}[0]}{\bar{r}_{\mu_i\mu_i}[0]}\right)^2 - \frac{\ddot{\bar{r}}_{\mu_i\mu_i}[0]}{\bar{r}_{\mu_i\mu_i}[0]}}. \tag{9.33}$$

Using (9.31), (9.32), and $\bar{\sigma}^2_{\mu_i} = \bar{r}_{\mu_i\mu_i}[0]$, we can express the last equation for the special case of symmetrical Doppler power spectral densities as

$$\bar{B}^{(2)}_{\mu_i\mu_i} = \frac{\sqrt{\bar{\beta}_i}}{2\pi\bar{\sigma}_{\mu_i}}, \tag{9.34}$$

where

$$\bar{\beta}_i = -\ddot{\bar{r}}_{\mu_i\mu_i}[0] = 2\pi^2 \sum_{n=1}^{N_i} (c_{i,n}\bar{f}_{i,n})^2. \tag{9.35}$$

It should be recalled that the MEDS has been developed especially for the Jakes power spectral density. In Subsection 5.1.7, we learned that by using the MEDS, the Doppler spread of the continuous-time simulation model is identical to the Doppler spread of the reference model, i.e., $\tilde{B}^{(2)}_{\mu_i\mu_i} = B^{(2)}_{\mu_i\mu_i}$. This relationship is now only approximately valid. The reason for this is that although $\bar{\sigma}^2_{\mu_i} = \tilde{\sigma}^2_{\mu_i} = \sigma_0^2$ holds, it follows $\bar{\beta}_i \approx \tilde{\beta}_i = \beta_i$ due to $\bar{f}_{i,n} \approx f_{i,n}$, and thus

$$\bar{B}^{(2)}_{\mu_i\mu_i} \approx \tilde{B}^{(2)}_{\mu_i\mu_i} = B^{(2)}_{\mu_i\mu_i}. \tag{9.36}$$

The deviation between $\bar{B}^{(2)}_{\mu_i\mu_i}$ and $B^{(2)}_{\mu_i\mu_i}$ or between $\bar{\beta}_i$ and β_i is basically determined by the chosen value for the sampling interval T_s. We will identify the reason for this by analyzing the model error of discrete-time simulation systems for fading channels.

Model error: Let $\bar{\mu}_i[k]$ be a discrete-time deterministic process introduced by (9.2). Then, the model error $\Delta\bar{\beta}_i$ of the discrete-time simulation system is defined by

$$\Delta\bar{\beta}_i := \bar{\beta}_i - \beta_i. \tag{9.37}$$

Using (3.68) and (9.35), the model error $\Delta\bar{\beta}_i$ can easily be evaluated for all parameter computation methods described in Chapter 5 as a function of N_i and T_s or alternatively $f_s = 1/T_s$.

An example of the behaviour of the relative model error $\Delta\bar{\beta}_i/\beta_i$ of discrete-time systems is shown in Figure 9.7 in terms of the normalized sampling frequency $f_s/f_{\max}$. Here, the MEDS has been applied with $N_i = 7$ (Jakes PSD, $f_{\max} = 91$ Hz, $\sigma_0^2 = 1$). Figure 9.7 illustrates clearly that the relative model error $\Delta\bar{\beta}_i/\beta_i$ decreases if the sampling frequency f_s increases. In the limit $f_s \to \infty$ or $T_s \to 0$, we obtain $\Delta\bar{\beta}_i/\beta_i \to 0$ as expected, since it is well known that the quantized Doppler frequencies $\bar{f}_{i,n}$ are approaching the quantities $f_{i,n}$ as $T_s \to 0$. In case of the MEDS, this directly results in $\bar{\beta}_i \to \tilde{\beta}_i = \beta_i$, and thus $\Delta\bar{\beta}_i \to 0$.

Periodicity: Let $\bar{\mu}_i[k]$ be a discrete-time deterministic process with arbitrary but non-zero parameters $c_{i,n}$, $\bar{f}_{i,n}$ (and $\bar{\theta}_{i,n}$). Then, $\bar{\mu}_i[k]$ is periodic with the least common multiple (lcm) of the set $\{L_{i,n}\}_{n=1}^{N_i}$, i.e., the period L_i of $\bar{\mu}_i[k]$ equals

$$L_i = \operatorname{lcm}\{L_{i,n}\}_{n=1}^{N_i}. \tag{9.38}$$

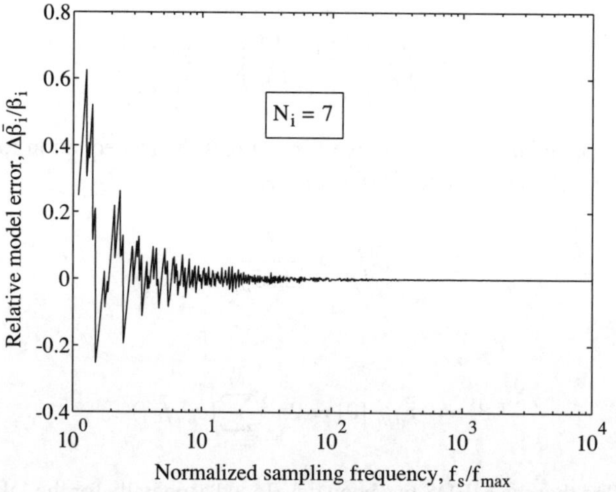

Figure 9.7 Relative model error $\Delta\bar{\beta}_i/\beta_i$ of the discrete-time system (MEDS, Jakes PSD, $N_i = 7$, $f_{\max} = 91$ Hz, $\sigma_0^2 = 1$).

In order to prove this theorem, we have to show that

$$\bar{\mu}_i[k] = \bar{\mu}_i[k + L_i] \tag{9.39}$$

is valid for all $k \in \mathbb{Z}$. Since L_i is the least common multiple of the set $\{L_{i,n}\}_{n=1}^{N_i}$, L_i must be an integer multiple of every table length $L_{i,n}$. Thus, we may write

$$L_{i,n} = \frac{L_i}{q_{i,n}}, \tag{9.40}$$

where $q_{i,n}$ is a natural number, which might be different for every $L_{i,n}$. Since the table length $L_{i,n}$ is identical to the period of $\bar{\mu}_{i,n}[k]$, the product $q_{i,n} \cdot L_{i,n}$ has to fulfil the relation

$$\bar{\mu}_{i,n}[k] = \bar{\mu}_{i,n}[k + q_{i,n}L_{i,n}] \qquad \forall\, k \in \mathbb{Z}. \tag{9.41}$$

Using the last two equations, we can prove the validity of (9.39) in the following way:

$$\begin{aligned} \bar{\mu}_i[k] &= \sum_{n=1}^{N_i} \bar{\mu}_{i,n}[k] \\ &= \sum_{n=1}^{N_i} \bar{\mu}_{i,n}[k + q_{i,n}L_{i,n}] \\ &= \sum_{n=1}^{N_i} \bar{\mu}_{i,n}[k + L_i] \\ &= \bar{\mu}_i[k + L_i] \qquad \forall\, k \in \mathbb{Z}. \end{aligned} \tag{9.42}$$

From L_i being the least common multiple of the set $\{L_{i,n}\}_{n=1}^{N_i}$, it follows that L_i is the smallest (positive) value for which (9.39) is valid. Consequently, L_i is said to be the period of the discrete-time deterministic process $\bar{\mu}_i[k]$.

We will point out here that an upper bound on the period L_i in (9.38) is given by the product of all table lengths $L_{i,n}$, i.e.,

$$\hat{L}_i = \prod_{n=1}^{N_i} L_{i,n}. \tag{9.43}$$

Taking the remarks above into account, it can easily be shown that $\hat{L}_i$ also fulfils (9.39). However, the period L_i and its upper bound $\hat{L}_i$ are related by $\hat{L}_i \geq L_i$.

From the fact that the table length $L_{i,n}$ depends on the sampling frequency f_s, it follows that the period L_i depends on f_s as well. This dependency is illustrated in Figure 9.8, where the period L_i and its upper bound $\hat{L}_i$ are plotted as a function of the normalized sampling frequency $f_s/f_{\max}$. Thereby, the results are presented advisedly for a small, medium, and large number of look-up tables ($N_i = 7$, $N_i = 14$, $N_i = 21$) to point out that both N_i and f_s have a decisive influence on the period L_i. We can also observe that the period L_i is often close to its upper bound $\hat{L}_i$, especially for low values of N_i. The easily computable expression in (9.43) therefore allows in general to estimate the period L_i with sufficient precision. Furthermore, it can be realized by considering Figure 9.8 that the period L_i is very large even for small values of $f_s/f_{\max}$. For this reason, we may say that $\bar{\mu}_i[k]$ is a quasi-nonperiodic discrete-time

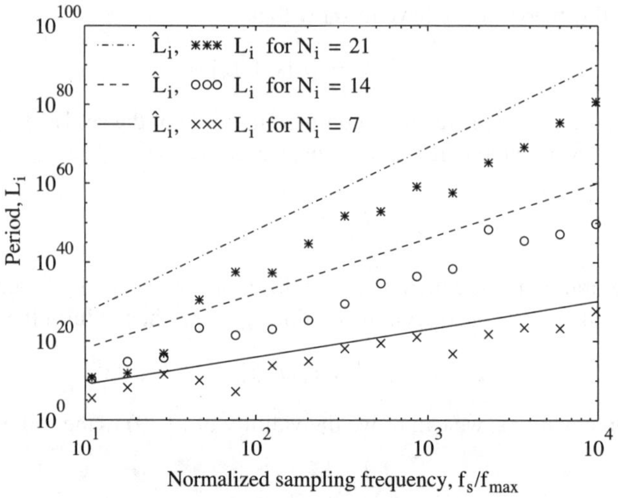

Figure 9.8 Period L_i of $\bar{\mu}_i[k]$ and its upper limit $\hat{L}_i$ as a function of the normalized sampling frequency $f_s/f_{\max}$ (MEDS, Jakes PSD, $f_{\max} = 91$ Hz, $\sigma_0^2 = 1$).

deterministic Gaussian process, provided that the sampling frequency f_s is sufficiently large, i.e., $f_s > 20 f_{\max}$.

Next, we will examine the period of discrete-time deterministic Rayleigh processes $\bar{\zeta}[k]$. Therefore, we consider the following theorem:

Let $\bar{\mu}_1[k]$ and $\bar{\mu}_2[k]$ be two discrete-time deterministic Gaussian processes, which are periodic with periods L_1 and L_2, respectively. Then, the discrete-time deterministic Rayleigh process $\bar{\zeta}[k] = |\bar{\mu}_1[k] + j\bar{\mu}_2[k]|$ is periodic with period

$$L = \operatorname{lcm}\{L_1, L_2\}. \tag{9.44}$$

The proof of this theorem is similar to the proof of (9.39) allowing us this time to present an abridged version. Owing to (9.44), two natural numbers q_1 and q_2 exist, which fulfil the equations $L = q_1 L_1$ and $L = q_2 L_2$, respectively. Thus, it follows

$$\begin{aligned}\bar{\zeta}[k] &= |\bar{\mu}_1[k] + j\bar{\mu}_2[k]| \\ &= |\bar{\mu}_1[k + L_1] + j\bar{\mu}_2[k + L_2]| \\ &= |\bar{\mu}_1[k + q_1 L_1] + j\bar{\mu}_2[k + q_2 L_2]| \\ &= |\bar{\mu}_1[k + L] + j\bar{\mu}_2[k + L]| \\ &= \bar{\zeta}[k + L] \qquad \forall\, k \in \mathbb{Z}.\end{aligned} \tag{9.45}$$

This shows that $\bar{\zeta}[k]$ is periodic with L. Since L is due to (9.44) the smallest integer number which fulfils (9.45), it follows that $L = \operatorname{lcm}\{L_1, L_2\}$ must be the period of the discrete-time deterministic Rayleigh process $\bar{\zeta}[k]$. An upper limit of L is given by

$$\hat{L} = L_1 L_2 \geq L = \operatorname{lcm}\{L_1, L_2\}. \tag{9.46}$$

9.3.2 Statistical Properties of Discrete-Time Deterministic Processes

This subsection begins with the analysis of the probability density function and the cumulative distribution function of the envelope and phase of complex discrete-time deterministic Gaussian processes $\bar{\mu}[k] = \bar{\mu}_1[k] + j\bar{\mu}_2[k]$. Subsequently, it follows the investigation of the level-crossing rate and the average duration of fades of discrete-time deterministic Rayleigh processes $\bar{\zeta}[k]$ introduced by (9.10). When analyzing the statistical properties of discrete-time deterministic processes, we always assume that all model parameters ($c_{i,n}$, $\bar{f}_{i,n}$, and $\bar{\theta}_{i,n}$) are constant quantities. However, we get access to the analysis of the statistical properties by picking up the values (samples) of the discrete-time deterministic Gaussian process $\bar{\mu}_i[k]$ at random time instants k, i.e., we assume in this subsection that k is a random variable, uniformly distributed in the interval $\mathbb{Z}$.

9.3.2.1 Probability Density Function and Cumulative Distribution Function of the Envelope and the Phase

In this subsection, we will derive analytical expressions for the probability density function and cumulative distribution function of the envelope as well as the phase of complex discrete-time deterministic Gaussian processes $\bar{\mu}[k]$. Let us start by considering a single elementary sinusoidal sequence of the form

$$\bar{\mu}_{i,n}[k] = c_{i,n}\cos(2\pi \bar{f}_{i,n}kT_s + \bar{\theta}_{i,n}), \tag{9.47}$$

where the model parameters $c_{i,n}$, $\bar{f}_{i,n}$, and $\bar{\theta}_{i,n}$ are arbitrary but non-zero quantities and k is the uniformly distributed random variable mentioned above. Since $\bar{\mu}_{i,n}[k]$ is periodic with period $L_{i,n}$, we can assume without restriction of generality that the random variable k is limited to the half-open interval $[0, L_{i,n})$. In this case, $\bar{\mu}_{i,n}[k]$ has no longer to be regarded as a deterministic sequence but as a random variable, whose possible elementary events (outcomes or realizations) are the elements of the set $\{\bar{\mu}_{i,n}[0], \bar{\mu}_{i,n}[1], \ldots, \bar{\mu}_{i,n}[L_{i,n}-1]\}$. Thereby, it should be noted that each elementary event occurs with the probability $1/L_{i,n}$. Consequently, the probability density function of $\bar{\mu}_{i,n}[k]$ can be written as

$$\bar{p}_{\mu_{i,n}}(x) = \frac{1}{L_{i,n}}\sum_{l=0}^{L_{i,n}-1}\delta(x - \bar{\mu}_{i,n}[l]), \tag{9.48}$$

where $n = 1, 2, \ldots, N_i$ $(i = 1, 2)$. Since the elementary sinusoidal sequence $\bar{\mu}_{i,n}[k]$ converges to the corresponding elementary sinusoidal function $\tilde{\mu}_{i,n}(t)$ [see (4.34)] as the sampling interval T_s tends to zero, the discrete probability density function $\bar{p}_{\mu_{i,n}}(x)$ converges consequently to the continuous probability density function $\tilde{p}_{\mu_{i,n}}(x)$ introduced in (4.35), i.e., in the limit $T_s \to 0$ it follows $\bar{p}_{\mu_{i,n}}(x) \to \tilde{p}_{\mu_{i,n}}(x)$. An example of the probability density function $\bar{p}_{\mu_{i,n}}(x)$ of $\bar{\mu}_{i,n}[k]$ is shown in Figure 9.9(a) for the case $T_s = 0.1$ ms. In addition to that, Figure 9.9(b) illustrates the results obtained after taking the limit $T_s \to 0$.

Following the approach described above, we proceed with the derivation of the probability density function $\bar{p}_{\mu_i}(x)$ of discrete-time deterministic Gaussian processes $\bar{\mu}_i[k]$. Owing to the periodicity of $\bar{\mu}_i[k]$, we can restrict k to the half-open interval $[0, L_i)$. Therefore, let k be a random variable, uniformly distributed over $[0, L_i)$, then $\bar{\mu}_i[k]$ [see (9.2)] is also a random variable, whose elementary events $\bar{\mu}_i[0], \bar{\mu}_i[1], \ldots, \bar{\mu}_i[L_i - 1]$ occur with the

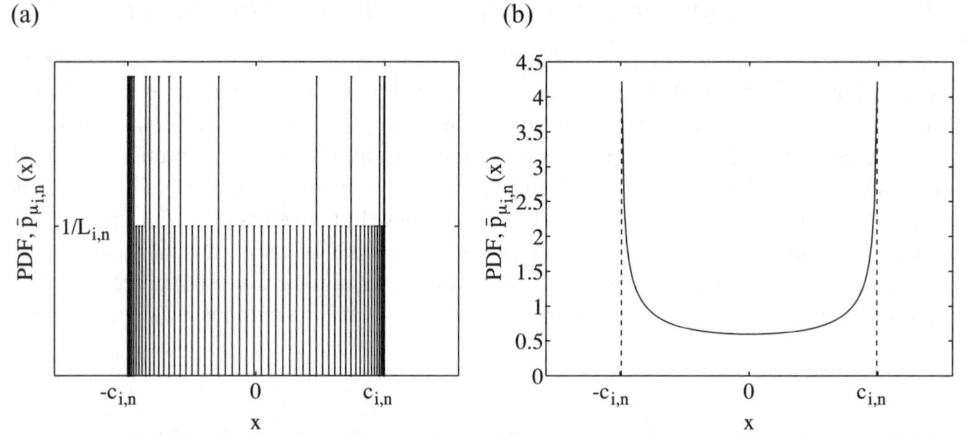

Figure 9.9 Probability density function $\bar{p}_{\mu_{i,n}}(x)$ of $\bar{\mu}_{i,n}(x)$ with (a) $T_s = 0.1$ ms and (b) $T_s \to 0$ (MEDS, Jakes PSD, $N_i = 7$, $n = 7$, $f_{\max} = 91$ Hz, $\sigma_0^2 = 1$).

same probability $1/L_i$. By analogy to (9.48), the probability density function of discrete-time deterministic Gaussian processes $\bar{\mu}_i[k]$ can be expressed by

$$\bar{p}_{\mu_i}(x) = \frac{1}{L_i} \sum_{l=0}^{L_i-1} \delta(x - \bar{\mu}_i[l]). \tag{9.49}$$

This result shows that the density $\bar{p}_{\mu_i}(x)$ of $\bar{\mu}_i[k]$ can be represented as a weighted sum of Dirac delta functions. Thereby, the Dirac delta functions are located at $\bar{\mu}_i[0]$, $\bar{\mu}_i[1], \ldots, \bar{\mu}_i[L_i - 1]$ and weighted by the reciprocal value of the period L_i. Notice that $\bar{p}_{\mu_i}(x)$ does not result from the convolution $\bar{p}_{\mu_{i,1}}(x) * \bar{p}_{\mu_{i,2}}(x) * \cdots * \bar{p}_{\mu_{i,N_i}}(x)$, because the random variables $\bar{\mu}_{i,1}[k]$, $\bar{\mu}_{i,2}[k], \ldots, \bar{\mu}_{i,N_i}[k]$ are, strictly speaking, not statistically independent. Regarding the look-up table system, for example, the statistical dependency finds its manifestation in the fact that the address generator in general does not produce the maximum number of different address combinations (states). This, in the ultimate analysis, is the reason why the actual period L_i and the maximum period $\hat{L}_i$ are related by the inequality $L_i \leq \hat{L}_i$. It should also be noted that in the limit $T_s \to 0$ it follows $\bar{p}_{\mu_i}(x) \to \tilde{p}_{\mu_i}(x)$, where $\tilde{p}_{\mu_i}(x)$ [see (4.41)] is the probability density function of $\tilde{\mu}_i(t)$. Moreover, $\bar{p}_{\mu_i}(x)$ approaches to the Gaussian probability density function $p_{\mu_i}(x)$ defined by (4.43) as $T_s \to 0$ and $N_i \to \infty$. For $T_s > 0$, it is not advisable to analyze the difference between the probability density functions $\bar{p}_{\mu_i}(x)$ and $\tilde{p}_{\mu_i}(x)$ directly, because the former density is a discrete function and the latter is continuous. However, this problem can easily be avoided by considering the cumulative distribution function $\bar{F}_{\mu_i}(r)$ of the discrete-time deterministic Gaussian process $\bar{\mu}_i[k]$. From (9.49), we obtain immediately

$$\bar{F}_{\mu_i}(r) = \frac{1}{L_i} \sum_{l=0}^{L_i-1} \int_0^r \delta(x - \bar{\mu}_i[l])\, dx, \quad r \geq 0. \tag{9.50}$$

A comparison of $\bar{F}_{\mu_i}(r)$ with the cumulative distribution function $\tilde{F}_{\mu_i}(r)$ of the corresponding continuous-time deterministic Gaussian process $\tilde{\mu}_i(t)$

$$\tilde{F}_{\mu_i}(r) = \frac{1}{2} + 2r \int_0^{\infty} \left[\prod_{n=1}^{N_i} J_0(2\pi c_{i,n} \nu) \right] \text{sinc}\,(2\pi \nu r)\, d\nu, \quad r \geq 0, \tag{9.51}$$

is shown in Figure 9.10. The analytical expression for the cumulative distribution function $\tilde{F}_{\mu_i}(r)$ given above can be obtained directly after substituting the probability density function $\tilde{p}_{\mu_i}(x)$ introduced in (4.41) in $\tilde{F}_{\mu_i}(r) = \int_{-\infty}^{r} \tilde{p}_{\mu_i}(x)\, dx$ and then solving the integral with respect to the independent variable x.

In addition, the cumulative distribution function

$$F_{\mu_i}(r) = \frac{1}{2}\left[1 + \text{erf}\left(\frac{r}{\sqrt{2}\sigma_0}\right)\right], \quad r \geq 0, \tag{9.52}$$

of the zero-mean Gaussian random process $\mu_i(t)$ represents in Figure 9.10 the behaviour of the reference model.

For sufficiently small values of the sampling interval T_s, the period L_i becomes very large (see Figure 9.8) and, consequently, the sample space $\{\bar{\mu}_i[l]\}_{l=0}^{L_i-1}$ becomes very large as well. In such cases, it is not possible to evaluate the cumulative distribution function $\bar{F}_{\mu_i}(r)$ according to (9.50) exactly without exceeding any reasonably chosen time-out interval for the computer simulation. Fortunately, this problem can be avoided, because one even obtains excellent results by merely evaluating $K \ll L_i$ elements of the subset $\{\bar{\mu}_i[k]\}_{k=0}^{K-1}$, as demonstrated in Figure 9.10. This figure shows an almost perfect correspondence between $\bar{F}_{\mu_i}(r)$ and $\tilde{F}_{\mu_i}(r)$ or $F_{\mu_i}(x)$, although (9.50) has been evaluated by using only $K = 50 \cdot 10^3 \ll L_i$ samples $\bar{\mu}_i[k]$ $(k = 0, 1, \ldots, K-1)$.

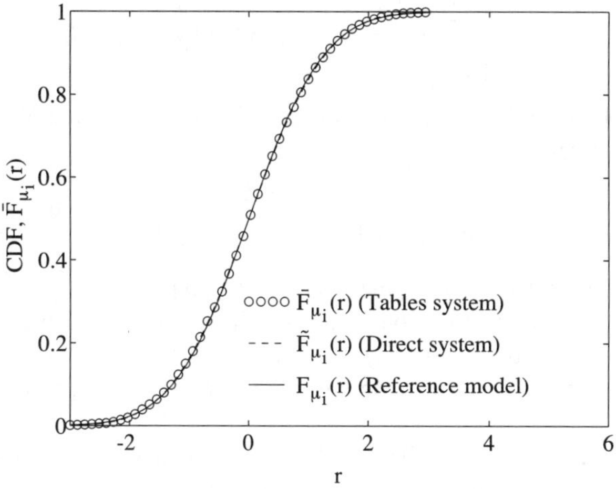

Figure 9.10 Cumulative distribution function $\bar{F}_{\mu_i}(x)$ of discrete-time deterministic Gaussian processes $\bar{\mu}_i[k]$ for $T_s = 0.1$ ms (MEDS, Jakes PSD, $N_i = 7$, $f_{\max} = 91$ Hz, $\sigma_0^2 = 1$).

Next, we will examine the probability density function and the cumulative distribution function of discrete-time deterministic Rayleigh processes $\bar{\zeta}[k]$. Thereby, we take into account that $\bar{\zeta}[k]$ is periodic with period $L = \mathrm{lcm}\{L_1, L_2\}$. Let us assume until further notice that k is a random variable, which is uniformly distributed in the interval $[0, L)$. Then, it follows that $\bar{\zeta}[k]$ defined by (9.10) is also a random variable, where each of the possible outcomes $\bar{\zeta}[0], \bar{\zeta}[1], \ldots, \bar{\zeta}[L-1]$ occurs with the probability $1/L$. By analogy to (9.49), we can thus express the probability density function $\bar{p}_{\zeta}(z)$ of discrete-time deterministic Rayleigh processes $\bar{\zeta}[k]$ as

$$\bar{p}_{\zeta}(z) = \frac{1}{L}\sum_{l=0}^{L-1} \delta(z - \bar{\zeta}[l]), \quad z \geq 0. \tag{9.53}$$

This result enables us to express the cumulative distribution function $\bar{F}_{\zeta_-}(r)$ of $\bar{\zeta}[k]$ as

$$\bar{F}_{\zeta_-}(r) = \frac{1}{L}\sum_{l=0}^{L-1} \int_0^r \delta(z - \bar{\zeta}[l])\, dz, \quad r \geq 0. \tag{9.54}$$

Note that due to $\bar{\zeta}[k] \to \tilde{\zeta}(t)$ as $T_s \to 0$, it follows $\bar{p}_{\zeta}(z) \to \tilde{p}_{\zeta}(z)$ and $\bar{F}_{\zeta_-}(r) \to \tilde{F}_{\zeta_-}(r)$. Thereby, $\tilde{p}_{\zeta}(z)$ is obtained from (4.54a) by setting $\rho = 0$, which allows us to present the cumulative distribution function $\tilde{F}_{\zeta_-}(r)$ of $\tilde{\zeta}(t)$ as

$$\begin{aligned}\tilde{F}_{\zeta_-}(r) &= \int_0^r \tilde{p}_{\zeta}(z)\, dz \\ &= 4r \int_0^{\infty} J_1(2\pi r z) \int_0^{\pi/2} \left[\prod_{n=1}^{N_1} J_0(2\pi c_{1,n} z \cos\theta)\right] \\ &\quad \left[\prod_{n=1}^{N_2} J_0(2\pi c_{2,n} z \sin\theta)\right] d\theta\, dz, \quad r \geq 0. \end{aligned} \tag{9.55}$$

Finally, it should be noted that after performing the limits $T_s \to 0$ and $N_i \to \infty$, the identity $\bar{F}_{\zeta_-}(r) = F_{\zeta_-}(r)$ is obtained, where

$$F_{\zeta_-}(r) = 1 - e^{-\frac{r^2}{2\sigma_0^2}}, \quad r \geq 0, \tag{9.56}$$

describes the cumulative distribution function of Rayleigh processes.

The cumulative distribution functions (9.54)–(9.56) are depicted in Figure 9.11. For the evaluation of $\bar{F}_{\zeta_-}(r)$ according to (9.54), $K = 50 \cdot 10^3 \ll L$ samples $\bar{\zeta}[k]$ $(k = 0, 1, \ldots, K-1)$ have been used, where the sampling interval T_s has been chosen sufficiently small $(T_s = 0.1\,\mathrm{ms})$.

Now, let us analyze in detail the influence of the sampling interval T_s on the statistics of $\bar{\zeta}[k]$. In particular, it is our intention to answer the following question: What is the maximum value of T_s for which $\bar{F}_{\zeta_-}(r)$ does not differ perceptibly from $\tilde{F}_{\zeta_-}(r)$? Up to now, we have in general assumed that T_s is sufficiently small without saying concretely what the phrase 'sufficiently small' really means. In the following, we want to make up for this by deriving a lower limit

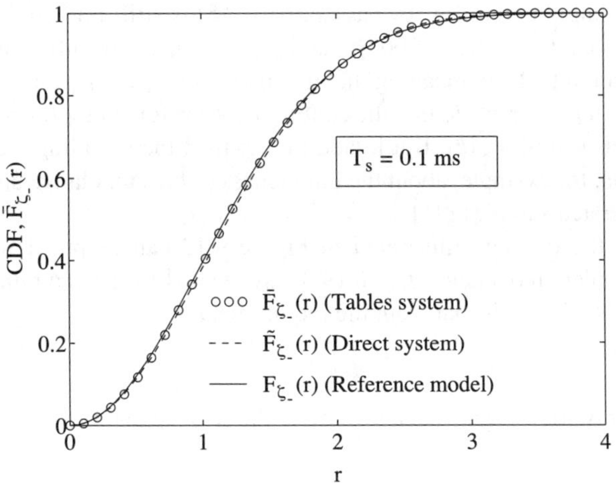

Figure 9.11 Cumulative distribution function $\bar{F}_{\zeta_-}(r)$ of discrete-time deterministic Rayleigh processes $\bar{\zeta}[k]$ for $T_s = 0.1$ ms (MEDS, Jakes PSD, $N_1 = 7$, $N_2 = 8$, $f_{\max} = 91$ Hz, $\sigma_0^2 = 1$).

for T_s. To illustrate the problem that occurs when T_s exceeds a certain critical threshold, we consider the graphs presented in Figure 9.12. In contrast to the cumulative distribution function $\bar{F}_{\zeta_-}(r)$ shown in Figure 9.11, we have used in the present case $K = L = 9240$ samples $\bar{\zeta}[k]$ ($k = 0, 1, \ldots, K - 1$) for the computation of $\bar{F}_{\zeta_-}(r)$ by using (9.54). At the same time, the sampling interval T_s has been increased from $T_s = 0.1$ ms up to $T_s = 5$ ms. Obviously, this seems to be problematical, because different realizations of the quantized phases $\{\bar{\theta}_{i,n}\}_{n=1}^{N_i}$ are now leading to different cumulative distribution functions $\bar{F}_{\zeta_-}(r)$, which may differ considerably from each other, as can be seen in Figure 9.12. It should be observed that in this

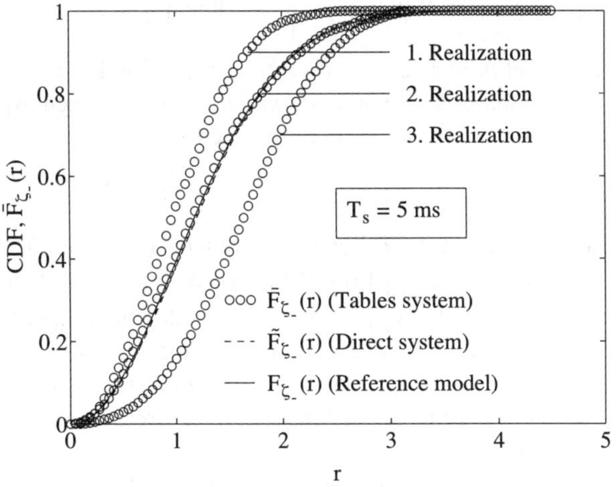

Figure 9.12 Cumulative distribution function $\bar{F}_{\zeta_-}(r)$ of discrete-time deterministic Rayleigh processes $\bar{\zeta}[k]$ for $T_s = 5$ ms and different realizations of the discrete-time phases $\{\bar{\theta}_{i,n}\}_{n=1}^{N_i}$ (MEDS, Jakes PSD, $N_1 = 7$, $N_2 = 8$, $f_{\max} = 91$ Hz, $\sigma_0^2 = 1$).

example the sampling theorem for lowpass signals [45] is still fulfilled, because the chosen values $T_s = 5\,\text{ms}$, i.e., $f_s = 1/T_s = 200\,\text{Hz}$, and $f_{\max} = 91\,\text{Hz}$ are sufficient for the sampling theorem in the form of (2.158), meaning that the inequality $f_s > 2f_{\max}$ holds. By fulfilling the sampling theorem, it is guaranteed that the continuous-time function $\bar{\zeta}(t)$ can be reconstructed completely from its samples $\bar{\zeta}[k]$. But in addition to that, the sampling theorem provides no further information, for example, about the uniqueness of the cumulative distribution function $\bar{F}_{\zeta_-}(r)$ of the generated samples $\bar{\zeta}[k]$.

The reason for the problem illustrated in Figure 9.12 can be put down to the fact that the quantized Doppler frequencies $\bar{f}_{i,n}$ in (9.3) are related to the sampling interval T_s. The consequence of this relation is that from the requirement

$$f_{i,n} \neq f_{j,m} \tag{9.57}$$

it does not inevitably follow that by increasing T_s the two inequalities

$$\bar{f}_{i,n} \neq \bar{f}_{j,m} \iff L_{i,n} \neq L_{j,m} \tag{9.58}$$

are also fulfilled, where $n = 1, 2, \ldots, N_i$ and $m = 1, 2, \ldots, N_j$ $(i, j = 1, 2)$. If T_s exceeds a certain threshold, then one or several pairs (n, m) exist for which $\bar{f}_{1,n} = \bar{f}_{2,m}$ and thus $L_{1,n} = L_{2,m}$ hold. In such a case, the elementary sinusoidal sequences $\bar{\mu}_{1,n}[k]$ and $\bar{\mu}_{2,n}[k]$ are identical apart from a phase shift. Hence, it follows that the discrete-time deterministic Gaussian processes $\bar{\mu}_1[k]$ and $\bar{\mu}_2[k]$ are correlated. Moreover, by increasing T_s it can also be the case that $\bar{f}_{i,n} = \bar{f}_{i,m} \Leftrightarrow L_{i,n} = L_{i,m}$ holds for $i = 1, 2$ and $n \neq m$. This, by the way, becomes obvious by examining the graphs shown in Figures 9.3(a) and 9.3(b) for $f_s/f_{\max} < 10$.

Regarding the derivation of a lower limit on the sampling frequency $f_{s,\min}$, the auxiliary function

$$\Delta_{n,m}^{(i,j)} := L_{i,n} - L_{j,m} \tag{9.59}$$

will turn out to be helpful. Using (9.4), we can also write

$$\Delta_{n,m}^{(i,j)} = \text{round}\left\{\frac{f_s}{f_{i,n}}\right\} - \text{round}\left\{\frac{f_s}{f_{j,m}}\right\}, \tag{9.60}$$

where $n = 1, 2, \ldots, N_i$ and $m = 1, 2, \ldots, N_j$ $(i, j = 1, 2)$. The lower limit on the sampling frequency $f_{s,\min}$ is determined by those pairs (n, m) and (i, j) for which, by decreasing f_s, the auxiliary function in (9.60) is zero for the first time. Hence,

$$f_{s,\min} = \max\left\{f_s \,\middle|\, \Delta_{n,m}^{(i,j)} = 0 \quad \forall\, i, j = 1, 2\right\}_{n,m=1}^{N_i,N_j}. \tag{9.61}$$

This result can be summarized in the following statement: Let us assume that the elements of the two sets $\{f_{1,n}\}_{n=1}^{N_1}$ and $\{f_{2,m}\}_{m=1}^{N_2}$ fulfil the property $f_{i,n} \neq f_{j,m}$, then the corresponding elements of the sets $\{\bar{f}_{1,n}\}_{n=1}^{N_1}$ and $\{\bar{f}_{2,m}\}_{m=1}^{N_2}$ fulfil the analogous property $\bar{f}_{i,n} \neq \bar{f}_{j,m}$ for all $n = 1, 2, \ldots, N_1$ and $m = 1, 2, \ldots, N_2$ $(i, j = 1, 2)$, if the sampling frequency f_s is above the threshold defined by (9.61), meaning that $f_s > f_{s,\min}$. In this case, we may conclude that if the processes $\tilde{\mu}_1(t)$ and $\tilde{\mu}_2(t)$ are uncorrelated, then the corresponding sequences $\bar{\mu}_1[k]$ and $\bar{\mu}_2[k]$ are also uncorrelated.

Two examples showing the results of the evaluation of (9.61) by using the MEDS are presented in Figure 9.13. In particular by deploying the MEDS, the lower limit for the sampling frequency $f_{s,\min}$ is determined by that value for f_s for which the auxiliary function $\Delta_{N_1,N_2}^{(1,2)}$ becomes zero for the first time. The problems caused by the correlation, as illustrated in

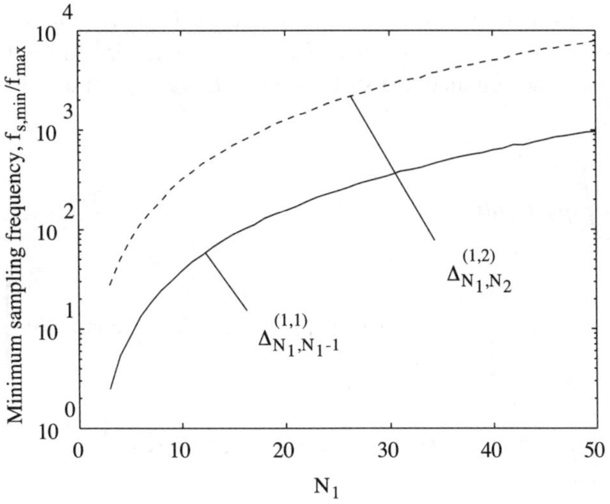

Figure 9.13 Lower limit on the sampling frequency $f_{s,\min}$ in terms of N_1 (MEDS, Jakes PSD, $N_2 = N_1 + 1$, $f_{\max} = 91\,\text{Hz}$, $\sigma_0^2 = 1$).

Figure 9.12, can thus be avoided if the sampling frequency f_s is above the threshold shown in Figure 9.13, which is not the case for the negative examples presented in Figure 9.12.

It should finally be mentioned that the lower limit $f_{s,\min}$ according to (9.61) is sufficient, but not necessary to fulfil the condition in (9.58), i.e., it cannot be excluded that there exist values for f_s which are below $f_{s,\min}$ and even though the two inequalities in (9.58) are fulfilled. Here, it is not our intention to discuss this problem in detail. Instead, we might consider $f_s > 20 f_{\max}$ as a rule of thumb, which has been turned out to be very useful for most practical applications. In conclusion, we can say that the sampling frequency f_s is sufficiently large if f_s is larger than $20 f_{\max}$.

Next, we will proceed with the analysis of the probability density function $\bar{p}_{\vartheta}(\theta)$ of the phase of complex discrete-time deterministic Gaussian processes $\bar{\mu}[k]$. Of course, similar arguments to those used for the derivation of the density $\bar{p}_{\zeta}(z)$ in (9.53) can be applied here to achieve the present goal. However, we prefer a more simple and straightforward approach by substituting the phase $\bar{\vartheta}[l]$ for the envelope $\bar{\zeta}[l]$ in (9.53), which allows us directly to express the probability density function $\bar{p}_{\vartheta}(\theta)$ of the phase as

$$\bar{p}_{\vartheta}(\theta) = \frac{1}{L} \sum_{l=0}^{L-1} \delta(\theta - \bar{\vartheta}[l]), \quad |\theta| \leq \pi, \tag{9.62}$$

where $\bar{\vartheta}[l] = \arctan\{\bar{\mu}_2[l]/\bar{\mu}_1[l]\}$ denotes the phase of the complex deterministic Gaussian process $\bar{\mu}[k] = \bar{\mu}_1[k] + j\bar{\mu}_2[k]$ at time instants $k = l \in \{0, 1, \ldots, L-1\}$. Using (9.62), the corresponding cumulative distribution function $\bar{F}_{\vartheta}(\varphi)$ can be written as

$$\bar{F}_{\vartheta}(\varphi) = \frac{1}{L} \sum_{l=0}^{L-1} \int_{-\pi}^{\varphi} \delta(\theta - \bar{\vartheta}[l])\, d\theta, \quad |\varphi| \leq \pi. \tag{9.63}$$

It should be mentioned that due to $\bar{\vartheta}[k] \to \tilde{\vartheta}(t)$ as $T_s \to 0$, it also follows $\bar{p}_\vartheta(\theta) \to \tilde{p}_\vartheta(\theta)$, and thus $\bar{F}_\vartheta(\varphi) \to \tilde{F}_\vartheta(\varphi)$ as $T_s \to 0$, where $\tilde{p}_\vartheta(\theta)$ can be obtained from (4.54b) for $\rho = 0$, allowing us to express the cumulative distribution function $\tilde{F}_\vartheta(\varphi)$ of the phase $\tilde{\vartheta}(t)$ of $\tilde{\mu}(t) = \tilde{\mu}_1(t) + j\tilde{\mu}_2(t)$ as

$$\begin{aligned}\tilde{F}_\vartheta(\varphi) &= \int_{-\pi}^{\varphi} \tilde{p}_\vartheta(\theta)\,d\theta \\ &= 4\int_{-\pi}^{\varphi}\int_{0}^{\infty} z \left\{\int_0^\infty \left[\prod_{n=1}^{N_1} J_0(2\pi c_{1,n}\nu_1)\right]\cos(2\pi\nu_1 z\cos\theta)\,d\nu_1\right\} \\ &\quad \left\{\int_0^\infty \left[\prod_{m=1}^{N_2} J_0(2\pi c_{2,m}\nu_2)\right]\cos(2\pi\nu_2 z\sin\theta)\,d\nu_2\right\} dz\,d\theta, \quad |\varphi| \le \pi. \end{aligned} \tag{9.64}$$

As $T_s \to 0$ and $N_i \to \infty$, it follows $\bar{F}_\vartheta(\varphi) \to F_\vartheta(\varphi)$, where

$$F_\vartheta(\varphi) = \frac{1}{2}\left(1 + \frac{\varphi}{\pi}\right), \quad |\varphi| \le \pi, \tag{9.65}$$

is the cumulative distribution function of the uniformly distributed phase of zero-mean complex Gaussian random processes $\mu(t) = \mu_1(t) + j\mu_2(t)$.

Figure 9.14 illustrates the cumulative distribution functions in (9.63)–(9.65). The evaluation of (9.63) has been performed by using $K = 50 \cdot 10^3 \ll L$ samples (outcomes) of the sample

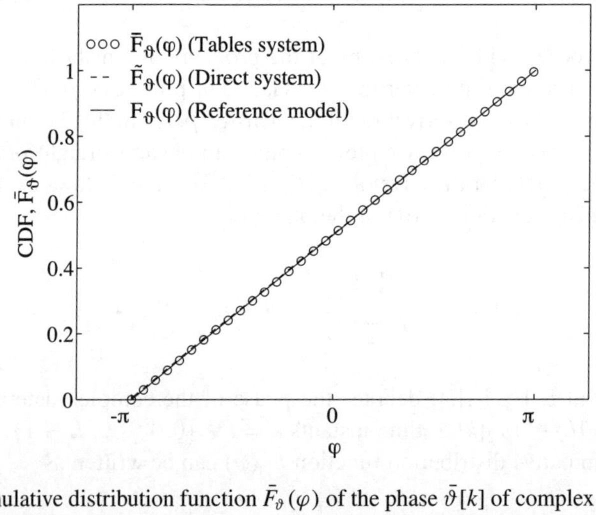

Figure 9.14 Cumulative distribution function $\bar{F}_\vartheta(\varphi)$ of the phase $\bar{\vartheta}[k]$ of complex discrete-time deterministic Gaussian processes $\bar{\mu}[k] = \bar{\mu}_1[k] + j\bar{\mu}_2[k]$ for $T_s = 0.1$ ms (MEDS, Jakes PSD, $N_1 = 7$, $N_2 = 8$, $f_{\max} = 91$ Hz, $\sigma_0^2 = 1$).

space $\{\bar{\vartheta}[l]\}_{l=0}^{L-1}$. For the sampling interval T_s, the value $T_s = 0.1$ ms has been chosen. Thus, by using the MEDS with the parameters specified in the caption of Figure 9.14, the ratio $f_s/f_{\max}$ is close to the threshold $f_{s,\min}/f_{\max}$ (see Figure 9.13).

9.3.2.2 Level-Crossing Rate and Average Duration of Fades

In contrast to continuous-time deterministic Rayleigh processes for which analytical expressions for both the level-crossing rate and the average duration of fades have been derived (see Appendix 4.B), there exist up to now no comparable solutions for discrete-time deterministic Rayleigh processes. In the following, we restrict our investigation to the derivation of approximate formulas by assuming that the normalized sampling frequency $f_s/f_{\max}$ lies above the threshold shown in Figure 9.13. Thereby, the number of sinusoids (look-up tables) N_i is assumed to be sufficiently large, i.e., $N_i \geq 7$. Moreover, we assume that the relative model error $\Delta\bar{\beta}_i/\beta_i$ of the discrete-time system is small, which is in particular the case when the MEDS is applied on condition that $f_s > f_{s,\min}$ is fulfilled (see Figure 9.7). Taking into account that the probability density function $\bar{p}_{\mu_i}(x)$ of discrete-time deterministic processes $\bar{\mu}_i[k]$ is asymptotically equal to the probability density function $\tilde{p}_{\mu_i}(x)$ of continuous-time deterministic processes $\tilde{\mu}_i(t)$, i.e., $\bar{p}_{\mu_i}(x) \sim \tilde{p}_{\mu_i}(x)$, then we can summarize the above mentioned statements and assumptions as follows:

$$\text{(i)} \quad \bar{p}_{\mu_i}(x) \sim \tilde{p}_{\mu_i}(x) \approx p_{\mu_i}(x), \tag{9.66a}$$

$$\text{(ii)} \quad \bar{\beta} = (\bar{\beta}_1 + \bar{\beta}_2)/2 \approx \beta = \beta_1 = \beta_2. \tag{9.66b}$$

Under these assumptions, we may conclude that the level-crossing rate $\bar{N}_\zeta(r)$ and the average duration of fades $\bar{T}_{\zeta_-}(r)$ of discrete-time deterministic Rayleigh processes $\bar{\zeta}[k]$ are in principle still given by the approximations in (4.73) and (4.77), respectively. However, here we have to evaluate these equations for the case $\rho = 0$, and in addition, we must replace the model error $\Delta\beta$ by $\Delta\bar{\beta}$. This results in the approximations

$$\bar{N}_\zeta(r) \approx N_\zeta(r)\left(1 + \frac{\Delta\bar{\beta}}{2\beta}\right), \tag{9.67a}$$

$$\bar{T}_{\zeta_-}(r) \approx T_{\zeta_-}(r)\left(1 - \frac{\Delta\bar{\beta}}{2\beta}\right), \tag{9.67b}$$

where $\Delta\bar{\beta} = \bar{\beta} - \beta$. In (9.67a), $N_\zeta(r)$ denotes the level-crossing rate of Rayleigh processes as defined by (2.117), and in (9.67b), $T_{\zeta_-}(r)$ refers to the average duration of fades introduced in (2.122). As $T_s \to 0$ and $N_i \to \infty$, it follows $\bar{N}_\zeta(r) \to N_\zeta(r)$ and $\bar{T}_{\zeta_-} \to T_{\zeta_-}(r)$.

Figure 9.15(a) shows an example for the normalized level-crossing rate $\bar{N}_\zeta(r)/f_{\max}$ of discrete-time deterministic Rayleigh processes $\bar{\zeta}[k]$. Just as in the previous examples, we computed here the model parameters by using the MEDS with $N_1 = 7$ and $N_2 = 8$. For the sampling interval T_s, again the value $T_s = 0.1$ ms has been chosen. The corresponding normalized average duration of fades $\bar{T}_{\zeta_-}(r) \cdot f_{\max}$ is illustrated in Figure 9.15(b).

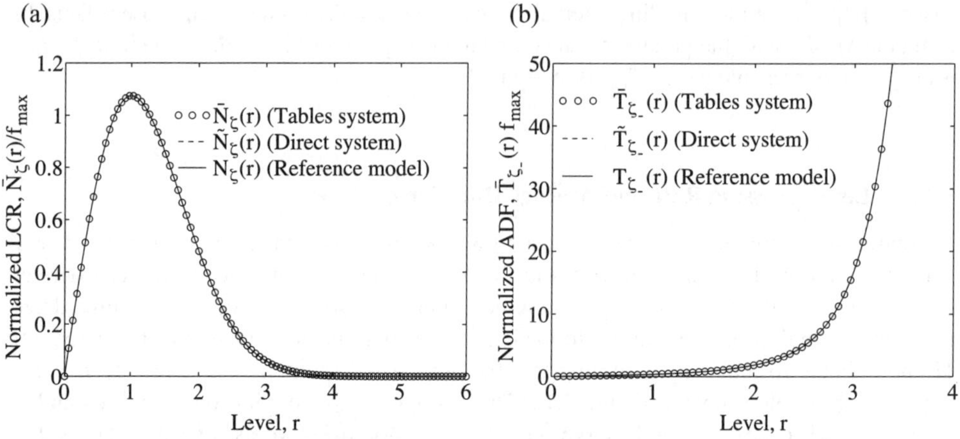

Figure 9.15 (a) Normalized level-crossing rate $\bar{N}_\zeta(r)/f_{\max}$ and (b) normalized average duration of fades $\bar{T}_{\zeta_-}(r)\cdot f_{\max}$ of discrete-time deterministic Rayleigh processes $\bar{\zeta}[k]$ for $T_s = 0.1$ ms (MEDS, Jakes PSD, $N_1 = 7, N_2 = 8, f_{\max} = 91$ Hz, $\sigma_0^2 = 1$).

9.4 Realization Complexity and Simulation Speed

In this section, we will examine in detail the efficiency of the look-up table system presented in Figure 9.2. Moreover, the look-up table system's efficiency will be compared to that of the corresponding discrete-time direct system, which is obtained from the simulation model shown in Figure 4.4 by replacing the continuous-time variable t by kT_s. For convenience, we ignore the influence of the line-of-sight component by choosing $\rho = 0$. Let us assume in the following that the set-up phase has been completed, so that we can restrict our investigations to the computational complexity required for the generation of the respective complex-valued channel output sequence.

It can easily be seen from Figure 9.2 that the operations listed in Table 9.1 have to be carried out at each time instant k in order to compute one sample of the complex discrete-time deterministic Gaussian process $\bar{\mu}[k] = \bar{\mu}_1[k] + j\bar{\mu}_2[k]$. One realizes that merely additions and simple conditional control flow statements (*if-else* statements) are required for generating the output samples $\bar{\mu}[k]$. The additions are needed for the generation of the addresses within the address generator as well as for adding up the look-up tables' outputs, whereas the conditional

Table 9.1 Number of operations required for the computation of $\bar{\mu}[k]$ (look-up table system) and $\tilde{\mu}[k]$ (direct system).

Number of operations	Look-up table system	Direct system
Number of multiplications	0	$2(N_1 + N_2)$
Number of additions	$2(N_1 + N_2) - 2$	$2(N_1 + N_2) - 2$
Number of trig. operations	0	$N_1 + N_2$
Number of if-else operations	$N_1 + N_2$	0

control flow statements are only required for the generation of the addresses within the address generator.

The number of operations required for the generation of the complex-valued sequence $\tilde{\mu}[k] = \tilde{\mu}_1[k] + j\tilde{\mu}_2[k]$ by employing the direct system is also listed in Table 9.1. Thereby, the normalized Doppler frequencies $\Omega_{i,n} = 2\pi f_{i,n} T_s$ have been used in order to avoid unnecessary multiplications within the arguments of the sinusoidal functions.

The results shown in Table 9.1 can be summarized as follows: All multiplications can be avoided, the number of additions remains unchanged, and all trigonometric operations can be replaced by simple if-else statements, when the look-up table system is used instead of the direct system for the generation of complex-valued channel output sequences. It is therefore not surprising if it turns out in the following that the look-up table system has clear advantages in comparison to the direct system with respect to the simulation speed.

As an appropriate measure of the simulation speed of channel simulators, we introduce the *iteration time* defined by

$$\Delta T_{\text{sim}} = \frac{T_{\text{sim}}}{K}, \tag{9.68}$$

where T_{sim} denotes the simulation time required for the computation of K samples of the complex-valued channel output sequence. Thus, the quantity ΔT_{sim} represents the average computation time per complex-valued channel output sample. Figure 9.16 illustrates the iteration time ΔT_{sim} for both the direct system and the look-up table system as a function of the number of sinusoids (look-up tables) N_1. The model parameters $f_{i,n}$ and $c_{i,n}$ have been computed by applying the MEDS with $N_2 = N_1 + 1$ and by using the Jakes method (JM) with $N_2 = N_1$. The algorithms of the channel simulators have been implemented on a computer by using MATLAB and the simulation results for T_{sim} are obtained by running the programs on a workstation. For each run, the number of samples of the complex-valued channel output sequence was equal to $K = 10^4$.

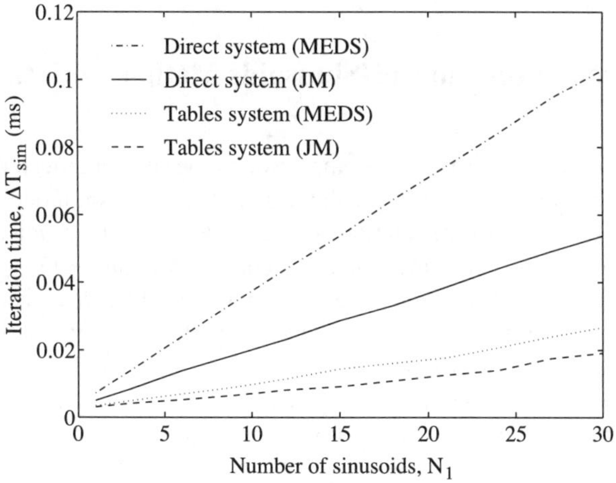

Figure 9.16 Iteration time ΔT_{sim} as a function of the number of sinusoids (look-up tables) N_1 (MEDS with $N_2 = N_1 + 1$, JM with $N_2 = N_1$, $f_{\max} = 91$ Hz, $\sigma_0^2 = 1$, $T_s = 0.1$ ms).

The results illustrated in Figure 9.16 show clearly the difference in speed of the treated channel simulators. When using the MEDS, for example, the simulation speed of the look-up table system is approximately 3.8 times higher than that of the direct system. Applying the JM, we can exploit the fact that the discrete Doppler frequencies $f_{1,n}$ and $f_{2,n}$ are identical, whereas the corresponding phases $\theta_{1,n}$ and $\theta_{2,n}$ are zero for all $n = 1, 2, \ldots, N_1$ ($N_1 = N_2$). This enables a drastic reduction of the complexity of both simulation systems. The consequence for the direct system is that only N_1 instead of $N_1 + N_2$ sinusoidal functions have to be evaluated at each time instant k. The speed of the direct system can thus be increased by approximately a factor of two (see Figure 9.16). The properties of the JM ($f_{1,n} = f_{2,n}, c_{1,n} \neq c_{2,n}, \theta_{1,n} = \theta_{2,n} = 0, N_1 = N_2$) furthermore imply that the look-up tables $\mathrm{Tab}_{1,n}$ and $\mathrm{Tab}_{2,n}$ of the look-up table system have the same length, i.e., it holds $L_{1,n} = L_{2,n}$ for all $n = 1, 2, \ldots, N_1$ ($N_1 = N_2$). Bearing this in mind and noticing that from $\theta_{i,n} = 0$ it follows immediately $\bar{\theta}_{i,n} = 0$, it is seen that the address generator only needs to compute one half of the usually required number of addresses. This is the reason for the observation that the speed of the look-up table system increases approximately by another 40 per cent (see Figure 9.16).

Summing up, we can say that the look-up table system is — by using the MEDS (JM) — approximately four times (three times) faster than the corresponding direct system. The benefit of higher speed is shadowed by the demand for larger storage elements, which is the only disadvantage of the look-up table system worth mentioning. Remember that the total demand for storage elements is proportional to the sampling frequency (see Figure 9.3). By choosing the sampling frequency f_s just above $f_{s,\min}$, then the required number of storage elements is reduced to a minimum without getting appreciable losses in precision. However, a good compromise between the model's precision and complexity is obtained by choosing f_s within the range $20 f_{\max} \leq f_s \leq 30 f_{\max}$. When such a designed channel simulator is used as a link between the transmitter and the receiver of a mobile communication system, then a sampling rate conversion by means of an interpolation (a decimation) filter is in general required to fit the sampling frequency of the channel simulator to the sampling frequency of the receiver's input (transmitter's output).

9.5 Comparison of the Sum-of-Sinusoids Method with the Filter Method

At this point, it is advisable to carry out a comparison of the sum-of-sinusoids method with the filter method, which is also often used for the design of simulation models for mobile radio channels. Here, we restrict our investigations to the modelling of Rayleigh processes. For this purpose, we consider the discrete-time structure depicted in Figure 9.17.

Since white Gaussian noise is, strictly speaking, not realizable, we consider $\tilde{\nu}_1[k]$ and $\tilde{\nu}_2[k]$ as two realizable noise sequences whose statistical properties are sufficiently close to those of ideal white Gaussian random processes. In particular, we demand that these pseudo-random sequences $\tilde{\nu}_i[k]$ ($i = 1, 2$) are uncorrelated. Moreover, it is supposed that $\tilde{\nu}_i[k]$ has a very long period and fulfils the properties $E\{\tilde{\nu}_i[k]\} = 0$ and $\mathrm{Var}\{\tilde{\nu}_i[k]\} = 1$.

In Figure 9.17, the symbol $\tilde{H}(z)$ denotes the transfer function of a digital filter in the z-domain. In practice, recursive digital filters are widely in use for the modelling of narrow-band

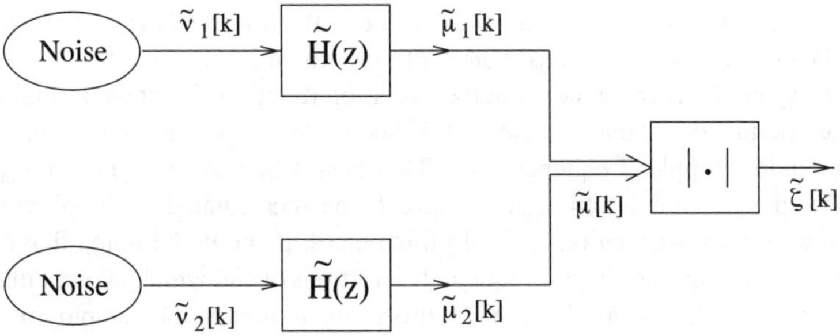

Figure 9.17 A simulation model for Rayleigh processes on basis of the filter method.

random processes. The transfer function of such filters can be represented in the z-domain as follows

$$\tilde{H}(z) = A_0 \frac{\prod_{n=1}^{N_0/2} \left(z - \rho_{0n}\, e^{j\varphi_{0n}}\right)\left(z - \rho_{0n}\, e^{-j\varphi_{0n}}\right)}{\prod_{n=1}^{N_0/2} \left(z - \rho_{\infty n}\, e^{j\varphi_{\infty n}}\right)\left(z - \rho_{\infty n}\, e^{-j\varphi_{\infty n}}\right)}, \tag{9.69}$$

where N_0 denotes the order of the filter and A_0 is a constant, which is determined such that the mean power of the output sequence of the digital filter is equal to σ_0^2. As we already know, the principle of the filter method is to determine the coefficients of the transfer function of the filter such that the deviations between the magnitude of the transfer function $|\tilde{H}(e^{j2\pi f T_s})|$ and the square root of the desired Doppler power spectral density $\sqrt{S_{\mu_i\mu_i}(f)}$ are minimal, or at least as small as possible, with respect to an appropriate error criterion. This problem can in general be solved by applying numerical optimization procedures, such as the Fletcher-Powell algorithm [162] or the Remez exchange procedure. An overview of commonly used optimization procedures can be found in [207, 352, 353].

Particularly for the widely used Jakes power spectral density in (3.23), a recursive digital filter of order eight has been designed in [127], which very closely approximates the desired frequency response. In Table 9.2, the coefficients of the recursive digital filter adopted from [354] are listed for a cut-off frequency f_c that has been normalized to the sampling frequency f_s according to $f_c = f_s/(110.5)$.

Table 9.2 Coefficients of the transfer function of the eighth order recursive filter proposed in [354].

n	ρ_{0n}	φ_{0n}	$\rho_{\infty n}$	$\varphi_{\infty n}$
1	1.0	$5.730778 \cdot 10^{-2}$	0.991177	$4.542547 \cdot 10^{-2}$
2	1.0	$7.151706 \cdot 10^{-2}$	0.980664	$1.912862 \cdot 10^{-2}$
3	1.0	0.105841	0.998042	$5.507401 \cdot 10^{-2}$
4	1.0	0.264175	0.999887	$5.670618 \cdot 10^{-2}$

The resulting graph of the squared absolute value of the transfer function $|\tilde{H}(e^{j2\pi f T_s})|^2$ and the desired Jakes power spectral density are both presented in Figure 9.18(a). The very good conformity between the corresponding autocorrelation functions is shown in Figure 9.18(b).

The cut-off frequency f_c has in case of the Jakes power spectral density been identified with the maximum Doppler frequency $f_{\max}$. This means that by changing of $f_{\max}$ or f_c, all coefficients of the transfer function $\tilde{H}(z)$ have to be recalculated by employing common lowpass-to-lowpass transformations [12]. In this regard, it should be noted that due to these frequency transformations nonlinear frequency distortions occur, which in particular cannot be ignored if the ratio f_c/f_s is small. In practice, this problem is solved by employing a sampling rate converter. Thereby, the digital filter operates with a small sampling rate that has to be converted afterwards by means of an interpolation filter to the mostly much higher sampling rate of the transmission system. We will not go into the details of sampling rate conversion, since our aim here is to compare the computation speed of different channel simulators, which disregards anyway the conversion of the sampling rate for the reason of fairness. Otherwise, an interpolator would also be necessary for both the look-up table system and the direct system. This is in principle always feasible, but with regard to a simple measurement of the computation speed, this will only lead to the fact that the computation speed becomes dependent on the chosen interpolation factor and other system parameters.

In [120], the structure shown in Figure 9.17 of the eighth order recursive digital filter described above has been implemented on a workstation by using MATLAB, and the iteration time ΔT_{sim} has been measured according to the rule in (9.68). The result of this performance test was $\Delta T_{\text{sim}} = 0.02$ ms. It also turned out that approximately 70 per cent of the total computation time is required for the generation of the real-valued random sequences $\tilde{\nu}_1[k]$ and $\tilde{\nu}_2[k]$, whereas the filtering of these sequences only occupies the remaining 30 per cent of the computation time. From this result, we conclude that a reduction of the filter order does not automatically lead to a significant reduction in the iteration time ΔT_{sim}.

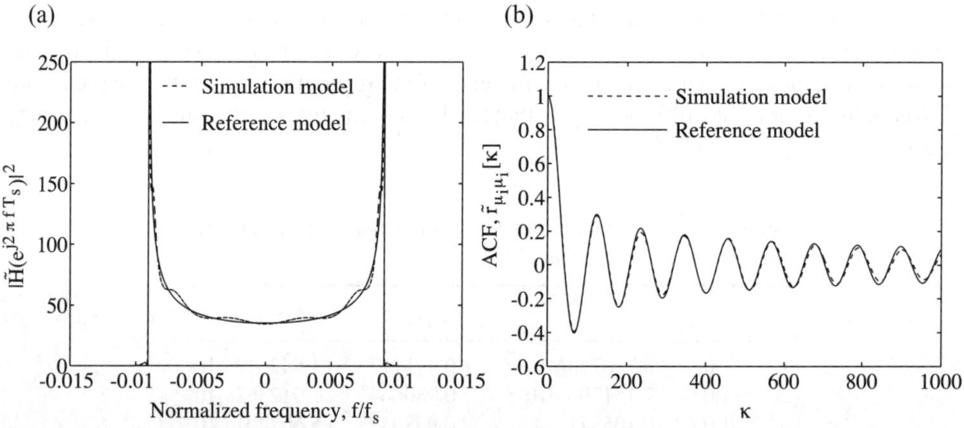

Figure 9.18 (a) Squared absolute value of the transfer function $|\tilde{H}(e^{j2\pi f T_s})|^2$ of the eighth order recursive filter and (b) autocorrelation sequence $\tilde{r}_{\mu_i\mu_i}[\kappa]$ of the filtered random process $\tilde{\mu}_i[k]$ $(i = 1, 2)$ [127].

Relating now the iteration time obtained for the filter system of eighth order to the corresponding iteration time of the direct system and the look-up table system, it becomes apparent that by using the MEDS ($N_1 = 7$, $N_2 = 8$) the direct system is approximately 25 per cent slower than the filter system, whereas the look-up table system outperforms the filter system by approximately 300 per cent with respect to the simulation speed.

9.6 Further Reading

Hardware architectural features and implementation techniques of wideband mobile radio channel simulators have been presented in [355]. A comparison of the computational complexity of the sum-of-cisoids and the filter method, applied to the modelling of MIMO channels, can be found in [356]. The main conclusion of the authors is that the sum-of-cisoids method is recommendable, especially if the number of antennas is large. The study of the computational complexity of narrowband and wideband channel simulators, designed by using the sum-of-sinusoids method and the filter method, was the subject of investigation in [357].

The implementation of sum-of-sinusoids-based fading channel simulators on a field-programmable gate array (FPGA) has been described for Rayleigh channels in [358] and for Rice channels in [359]. Modelling and hardware implementation aspects of fading channel simulators designed by using the filter method and the sum-of-sinusoids method were discussed in turn in [360]. The filter method has also been applied successfully to the development of computationally efficient hardware channel simulators. For example, a compact FPGA channel simulator for Rayleigh fading channels has been presented in [361]. The implementation of wideband mobile radio channel models on a floating point digital signal processor (DSP) is described, e.g., in [362–364]. A fast and accurate software-based Rayleigh fading channel simulator consisting of fixed infinite impulse response (IIR) filters followed by variable polyphase interpolators to accommodate different maximal Doppler frequencies was proposed in [365].

A low complexity algorithm for the implementation of geometry-based MIMO channel simulators on a hardware platform has been proposed in [366]. There it is stated that the proposed algorithm enables a reduction of the computational complexity of a hardware channel simulator with 14-bit precision by more than one order of magnitude in comparison to conventional implementation techniques. In a related paper [367], the method was used to develop a real-time hardware channel emulator for the geometry-based stochastic channel model according to COST 259 [275]. A versatile software-based channel simulator for MIMO wideband channel models has been described in [368] with emphasis on implementation aspects.

10

Selected Topics in Mobile Radio Channel Modelling

10.1 Design of Multiple Uncorrelated Rayleigh Fading Waveforms

This section deals with the design of a set of multiple uncorrelated Rayleigh fading waveforms. The Rayleigh fading waveforms are mutually uncorrelated, but each waveform is correlated in time. The design of a set of multiple uncorrelated Rayleigh fading waveforms is important for the simulation of diversity-combined Rayleigh fading channels, amplify-and-forward fading channels, frequency-selective channels, and multiple-input multiple-output (MIMO) channels with uncorrelated subchannels. The waveforms are generated by using the sum-of-sinusoids (SOS) channel modelling principle.

The development of channel simulators enabling the accurate and efficient generation of multiple uncorrelated Rayleigh fading processes has been a subject of research for many years. The SOS principle, originally introduced in [16, 17], has widely been applied to the development of simulation models for Rayleigh fading channels [13, 96, 97, 99, 100, 111–113, 118, 148, 150, 153, 158, 369, 370]. To generate multiple uncorrelated Rayleigh fading waveforms by using SOS channel simulators, many different parameter computation methods have been proposed [13, 96, 97, 99, 111–114, 118, 158, 214, 369, 370]. Depending on the underlying parameter computation method, SOS channel simulators can generally be classified as deterministic [13, 113, 114, 118, 158, 214, 369, 370], ergodic stochastic, or non-ergodic stochastic [97, 99, 111–113, 153]. A deterministic SOS channel simulator has constant model parameters (gains, frequencies, and phases) for all simulation trials. An ergodic stochastic SOS channel simulator has constant gains and constant frequencies but random phases [133, 134, 153]. Owing to the ergodic property, only a single simulation trial is needed to reveal its complete statistical properties. A sample function, i.e., a single simulation trial of a stochastic SOS channel simulator actually results in a deterministic process (waveform). In this sense, we can also say that a deterministic channel simulator can be used to generate sample functions of a stochastic process. A non-ergodic stochastic SOS channel simulator assumes that the frequencies and/or gains are random variables. The statistical properties of non-ergodic stochastic SOS channel simulators vary for each simulation trial and have to

Mobile Radio Channels, Second Edition. Matthias Pätzold.

be calculated by averaging over a sufficient large number of simulation trials. Both ergodic stochastic (deterministic) and non-ergodic stochastic SOS channel simulators have pros and cons, which have been discussed, e.g., in [113]. Generally one can say that deterministic SOS channel simulators have a higher efficiency compared to non-ergodic stochastic SOS channel simulators [150].

The objective of this section is to present solutions to the problem of designing multiple uncorrelated Rayleigh fading waveforms. We introduce a generalized version of the method of exact Doppler spread (MEDS), which can be interpreted as a class of parameter computation methods that includes many other well-known approaches as special cases. A special case of this class is considered here, which enables the efficient and accurate design of multiple ergodic uncorrelated Rayleigh fading waveforms using deterministic concepts. Analytical and numerical results show that the resulting deterministic SOS-based channel simulator fulfills all main requirements imposed by the reference model with given correlation properties derived under two-dimensional isotropic scattering conditions. The presented method keeps the model complexity low and provides a simple closed-form solution for the computation of the model parameters.

This section is organized as follows. Section 10.1.1 describes the problem and the conditions that must be fulfilled to obtain K mutually uncorrelated Rayleigh fading waveforms using the SOS principle. In Section 10.1.2, the so-called generalized MEDS is introduced as a class of parameter computation methods, which provides a closed-form solution of the problem under isotropic scattering conditions. In Section 10.1.3, it is shown that the class of parameter computation methods includes several other well-known methods as special cases. Section 10.1.4 looks at the cross-correlation properties under practical aspects, where the simulation time is always limited. Finally, a listing of suggested readings on the subject is provided in Section 10.1.5.

10.1.1 Problem Description

We want to simulate K mutually uncorrelated Rayleigh fading waveforms

$$\tilde{\zeta}^{(k)}(t) = |\tilde{\mu}^{(k)}(t)| = |\tilde{\mu}_1^{(k)}(t) + j\tilde{\mu}_2^{(k)}(t)|, \quad k = 1, 2, \ldots, K, \tag{10.1}$$

by using an SOS channel simulator, which generates the waveforms

$$\tilde{\mu}_i^{(k)}(t) = \sqrt{\frac{2}{N_i}} \sum_{n=1}^{N_i} \cos(2\pi f_{i,n}^{(k)} t + \theta_{i,n}^{(k)}), \quad i = 1, 2, \tag{10.2}$$

where N_i denotes the number of sinusoids, $f_{i,n}^{(k)}$ is called the discrete Doppler frequency, and $\theta_{i,n}^{(k)}$ is the phase of the nth sinusoid of the inphase component $\tilde{\mu}_1^{(k)}(t)$ or quadrature component $\tilde{\mu}_2^{(k)}(t)$ of the kth complex waveform $\tilde{\mu}^{(k)}(t)$. The phases $\theta_{i,n}^{(k)}$ are considered as outcomes of independent and identically distributed (i.i.d.) random variables, each having a uniform distribution over the interval $(0, 2\pi]$. For increased clarity, the structure of the SOS channel simulator is shown in Figure 10.1.

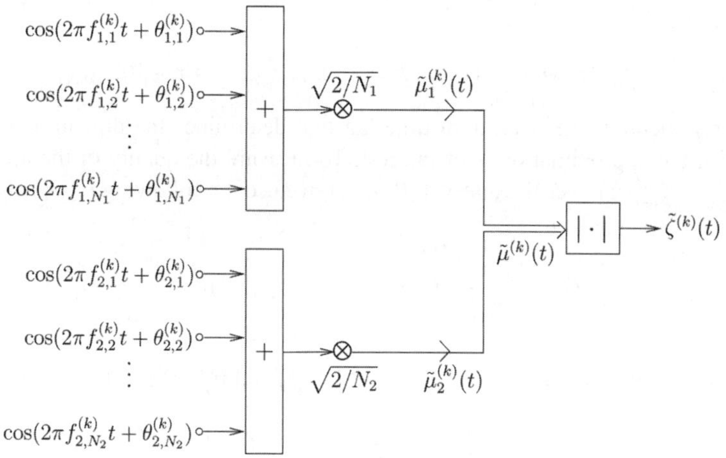

Figure 10.1 Structure of the SOS channel simulator for generating Rayleigh fading waveforms $\tilde{\zeta}^{(k)}(t)$ $(k=1,2,\ldots,K)$.

For given sets of constant model parameters $\{f_{i,n}^{(k)}\}$ and $\{\theta_{i,n}^{(k)}\}$, the time-averaged autocorrelation function $\tilde{r}_{\mu_i\mu_i}^{(k)}(\tau)$ of the kth waveform $\tilde{\mu}_i^{(k)}(t)$ can be expressed as

$$\begin{aligned}\tilde{r}_{\mu_i\mu_i}^{(k)}(\tau) &= \lim_{T\to\infty}\frac{1}{2T}\int_{-T}^{T}\tilde{\mu}_i^{(k)}(t)\tilde{\mu}_i^{(k)}(t+\tau)\,dt\\ &= \frac{1}{N_i}\sum_{n=1}^{N_i}\cos(2\pi f_{i,n}^{(k)}\tau)\end{aligned} \tag{10.3}$$

for $i=1,2$ and $k=1,2,\ldots,K$. Analogously, the time-averaged cross-correlation function $\tilde{r}_{\mu_i\mu_\lambda}^{(k,l)}(\tau)$ of $\tilde{\mu}_i^{(k)}(t)$ and $\tilde{\mu}_\lambda^{(l)}(t)$ can be obtained as

$$\begin{aligned}\tilde{r}_{\mu_i\mu_\lambda}^{(k,l)}(\tau) &= \lim_{T\to\infty}\frac{1}{2T}\int_{-T}^{T}\tilde{\mu}_i^{(k)}(t)\,\tilde{\mu}_\lambda^{(l)}(t+\tau)\,dt\\ &= \begin{cases}\dfrac{1}{N_iN_\lambda}\displaystyle\sum_{n=1}^{N_i}\sum_{m=1}^{N_\lambda}\cos(2\pi f_{i,n}^{(k)}\tau-\theta_{i,n}^{(k)}\pm\theta_{\lambda,m}^{(\ell)}), & \text{if } f_{i,n}^{(k)}=\pm f_{\lambda,m}^{(\ell)},\\ 0, & \text{if } f_{i,n}^{(k)}\neq\pm f_{\lambda,m}^{(\ell)},\end{cases}\end{aligned} \tag{10.4}$$

for $i,\lambda=1,2$ and $k,l=1,2,\ldots,K$. The problem is to find proper values for the discrete Doppler frequencies $f_{i,n}^{(k)}$ such that the following two principal conditions are fulfilled:

(i) The autocorrelation function $\tilde{r}_{\mu_i\mu_i}^{(k)}(\tau)$ of the simulation model must be as close as possible to the autocorrelation function $r_{\mu_i\mu_i}(\tau)$ of a given reference model over a certain domain, i.e.,

$$\tilde{r}^{(k)}_{\mu_i\mu_i}(\tau) \approx r_{\mu_i\mu_i}(\tau), \quad k = 1, 2, \ldots, K, \quad \forall \tau \in [0, \tau_{\max}], \tag{10.5}$$

where $\tau_{\max}$ denotes the maximum time lag that determines the domain over which the quality of the approximation is of interest. To measure the quality of the approximation $\tilde{r}^{(k)}_{\mu_i\mu_i}(\tau) \approx r_{\mu_i\mu_i}(\tau)$ over the interval $[0, \tau_{\max}]$, we use the following L_2-norm

$$E_2 = \left[\frac{1}{\tau_{\max}} \int_0^{\tau_{\max}} |\tilde{r}^{(k)}_{\mu_i\mu_i}(\tau) - r_{\mu_i\mu_i}(\tau)|^2 d\tau \right]^{1/2} . \tag{10.6}$$

(ii) The cross-correlation functions $\tilde{r}^{(k,l)}_{\mu_i\mu_\lambda}(\tau)$ of $\tilde{\mu}_i^{(k)}(t)$ and $\tilde{\mu}_\lambda^{(l)}(t)$ must be equal to zero, i.e.,

$$\tilde{r}^{(k,l)}_{\mu_i\mu_\lambda}(\tau) = 0 \quad \forall \tau, \quad i \neq \lambda, \tag{10.7a}$$

$$\tilde{r}^{(k,l)}_{\mu_i\mu_i}(\tau) = 0 \quad \forall \tau, \quad k \neq l, \tag{10.7b}$$

where $i, \lambda = 1, 2$ and $k, l = 1, 2, \ldots, K$.

It is worth noting that the above conditions should be fulfilled by keeping the complexity of the simulation model to a minimum, which means that N_i must be as small as possible.

By considering the cross-correlation function $\tilde{r}^{(k,l)}_{\mu_i\mu_\lambda}(\tau)$ in (10.4), we may conclude that (10.7a) and (10.7b) can be guaranteed if and only if

$$f^{(k)}_{i,n} \neq \pm f^{(l)}_{\lambda,m}, \quad i \neq \lambda, \tag{10.8a}$$

$$f^{(k)}_{i,n} \neq \pm f^{(l)}_{i,n}, \quad k \neq l, \tag{10.8b}$$

hold, respectively, for all $n = 1, 2, \ldots, N_i$ and $m = 1, 2, \ldots, N_\lambda$. The two equations above state that the sets of the absolute values of the discrete Doppler frequencies of different waveforms $\tilde{\mu}_i^{(k)}(t)$ and $\tilde{\mu}_j^{(l)}(t)$ must be disjoint, i.e., $\{|f^{(k)}_{i,n}|\}_{n=1}^{N_i} \cap \{|f^{(l)}_{\lambda,m}|\}_{m=1}^{N_\lambda} = \emptyset$ if $i \neq \lambda$ $(i, \lambda = 1, 2)$ and $k \neq l$ $(k, l = 1, 2, \ldots, K)$, where $\emptyset$ denotes the empty set.

Since it follows from (10.7a) that the cross-correlation functions $\tilde{r}^{(k,k)}_{\mu_1\mu_2}(\tau)$ and $\tilde{r}^{(k,k)}_{\mu_2\mu_1}(\tau)$ are zero, we can express the autocorrelation function $\tilde{r}^{(k)}_{\mu\mu}(\tau)$ of the kth complex waveform $\tilde{\mu}^{(k)}(t) = \tilde{\mu}_1^{(k)}(t) + j\tilde{\mu}_2^{(k)}(t)$ as

$$\tilde{r}^{(k)}_{\mu\mu}(\tau) = \tilde{r}^{(k)}_{\mu_1\mu_1}(\tau) + \tilde{r}^{(k)}_{\mu_2\mu_2}(\tau). \tag{10.9}$$

In the following, we will restrict our reference model to the Rayleigh model under two-dimensional isotropic scattering conditions. Recall that the autocorrelation function $r_{\mu_i\mu_i}(\tau)$ of the reference model is given by [see (3.25) with $\sigma_0^2 = 1$]

$$r_{\mu_i\mu_i}(\tau) = J_0(2\pi f_{\max}\tau), \quad i = 1, 2, \tag{10.10}$$

where $J_0(\,\cdot\,)$ denotes the zeroth-order Bessel function of the first kind and $f_{\max}$ is the maximum Doppler frequency. In this case, the autocorrelation function of the complex process $\mu(t) = \mu_1(t) + j\mu_2(t)$ equals $r_{\mu\mu}(\tau) = 2r_{\mu_i\mu_i}(\tau) = 2J_0(2\pi f_{\max}\tau)$, as the cross-correlation functions $r_{\mu_1\mu_2}(\tau)$ and $r_{\mu_2\mu_1}(\tau)$ are zero.

10.1.2 Generalized Method of Exact Doppler Spread ($GMEDS_q$)

To solve the problem described in the previous subsection, we consider the generalized MEDS ($GMEDS_q$) [114], which represents a class of parameter computation methods for the design of SOS Rayleigh fading channel simulators. According to this method, the discrete Doppler frequencies $f_{i,n}^{(k)}$ are given by

$$f_{i,n}^{(k)} = f_{\max} \cos(\alpha_{i,n}^{(k)})$$
$$= f_{\max} \cos\left[\frac{q\pi}{2N_i}\left(n - \frac{1}{2}\right) + \alpha_{i,0}^{(k)}\right], \tag{10.11}$$

where $\alpha_{i,0}^{(k)}$ is called the *angle-of-rotation* that will be defined subsequently and $q \in \{0, 1, \ldots, 4\}$. Note that the parameter q mainly determines the range of values for the quantities $\alpha_{i,n}^{(k)}$. Empirical studies have shown that for the $GMEDS_q$, the quantity $\tau_{\max}$ in (10.6) is given by $\tau_{\max} = N_i/(2qf_{\max})$ when $q = 1, 2$. From (10.11), it is obvious that the $GMEDS_q$ reduces to the original MEDS if $q = 1$ and $\alpha_{i,0}^{(k)} = 0$.

10.1.2.1 Problem Solution Using the $GMEDS_1$

In the following, we show how the $GMEDS_1$ can be used to generate any number K of multiple uncorrelated Rayleigh fading waveforms without increasing the model complexity. According to the $GMEDS_1$, the discrete Doppler frequencies $f_{i,n}^{(k)}$ are obtained from (10.11) by setting q to 1 and defining the angle-of-rotation $\alpha_{i,0}^{(k)}$ as

$$\alpha_{i,0}^{(k)} := (-1)^{i-1} \frac{\pi}{4N_i} \cdot \frac{k}{K+2}, \tag{10.12}$$

where $i = 1, 2$ and $k = 1, 2, \ldots, K$.

10.1.2.2 Motivation and Analysis of the $GMEDS_1$

In what follows, we present an analysis of the $GMEDS_1$ including some background information that clarifies the motivation for introducing the angle-of-rotation $\alpha_{i,0}^{(k)}$ according to (10.12). For this purpose, we start from (10.11) with $q = 1$ and substitute (10.3) and (10.10) in (10.6), which allows us to study the influence of the angle-of-rotation $\alpha_{i,0}^{(k)}$ on the quality of the approximation in (10.5). The behaviour of the error function $E_2 = E_2(\alpha_{i,0}^{(k)})$ in terms of $\alpha_{i,0}^{(k)}$ is illustrated in Figure 10.2 for various values of N_i. Clearly, with an increase in N_i, $E_2(\alpha_{i,0}^{(k)})$ can be reduced. It can also be observed that the error function $E_2(\alpha_{i,0}^{(k)})$ is periodic with period $\alpha_{\text{per}} = \pi/2$, i.e., $E_2(\alpha_{i,0}^{(k)}) = E_2(\alpha_{i,0}^{(k)} + p\pi/2)$, where p stands for an integer. Moreover, $E_2(\alpha_{i,0}^{(k)})$ is almost even symmetrical, i.e., $E_2(\alpha_{i,0}^{(k)}) \approx E_2(-\alpha_{i,0}^{(k)})$. The minimum and maximum values of $E_2(\alpha_{i,0}^{(k)})$ are obtained when $\alpha_{i,0}^{(k)}$ equals $p\alpha_{\text{per}}$ and $(2p + 1)\alpha_{\text{per}}/2$, respectively. Hence, the original MEDS ($q = 1$, $\alpha_{i,0}^{(k)} = 0$) results in the best fitting to the quadrature autocorrelation function $r_{\mu_i\mu_i}(\tau)$. The properties of the original MEDS include that $E_2(0) \to 0$ as $N_i \to \infty$. For a limited number of N_i, however, the value of $E_2(0)$ is different from zero. An idea of how fast $E_2(0)$ approaches to zero can be gathered from the plot in Figure 5.54, where the

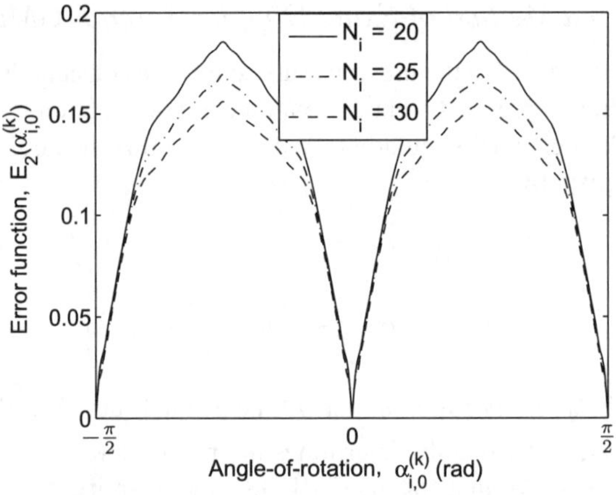

Figure 10.2 The error function $E_2 = E_2(\alpha_{i,0}^{(k)})$ in terms of the angle-of-rotation $\alpha_{i,0}^{(k)}$ by using the GMEDS$_1$ for various values of N_i.

mean-square error of $\tilde{r}_{\mu_i\mu_i}(\tau)$ is shown as a function of N_i. From the inspection of Figure 10.2, we also realize that large approximation errors may occur if $\alpha_{i,0}^{(k)} \neq p\pi/2$. This fact holds even for large values of N_i.

In what follows, we will investigate the conditions we have to impose on $\alpha_{i,0}^{(k)}$, so that (10.8a) and (10.8b) can be satisfied. Since $1 \leq n \leq N_i$, we may conclude that $\frac{\pi}{4N_i} \leq \frac{\pi}{2N_i}(n - \frac{1}{2}) \leq \frac{\pi}{2} - \frac{\pi}{4N_i}$ holds. If we limit the values of $\alpha_{i,0}^{(k)}$ in (10.11) to the interval $[-\frac{\pi}{4N_i}, \frac{\pi}{4N_i}]$ with $N_i < \infty$, we may further write $\alpha_{i,n}^{(k)} \in [0, \pi/2]$. Within this range, it follows from (10.11) that the discrete Doppler frequencies $f_{i,n}^{(k)}$ are monotonically decreasing quantities over the interval $[0, f_{\max}]$ with increasing values of n, i.e., $0 \leq f_{i,n+1}^{(k)} < f_{i,n}^{(k)} \leq f_{\max}$. For simplicity, let $N_1 = N_2$. Then, (10.8a) and (10.8b) can always be fulfilled if and only if

$$\alpha_{i,0}^{(k)} \neq \alpha_{\lambda,0}^{(l)} - \frac{\pi(n-m)}{2N_i}, \quad i \neq \lambda, \tag{10.13a}$$

$$\alpha_{i,0}^{(k)} \neq \alpha_{i,0}^{(l)}, \quad k \neq l, \tag{10.13b}$$

hold, respectively, for all $n, m = 1, 2, \ldots, N_i$, where $k, l = 1, 2, \ldots, K$ and $i, \lambda = 1, 2$. The above conditions are sufficient to guarantee that the cross-correlation functions $\tilde{r}_{\mu_i\mu_\lambda}^{(k,l)}(\tau)$ of $\tilde{\mu}_i^{(k)}(t)$ and $\tilde{\mu}_\lambda^{(l)}(t)$ are equal to zero, i.e., (10.7a) and (10.7b) will follow.

There are a number of ways to satisfy the conditions in (10.13a) and (10.13b) and to design a set of K mutually uncorrelated Rayleigh fading waveforms. Here, we define the angle-of-rotation $\alpha_{i,0}^{(k)}$ as in (10.12). Note that $\alpha_{i,0}^{(k)} \in (0, \pi/(4N_i))$ for $i = 1$ while $\alpha_{i,0}^{(k)} \in (-\pi/(4N_i), 0)$ for $i = 2$. For the GMEDS$_1$, the worst approximation result of the autocorrelation function $\tilde{r}_{\mu_i\mu_i}^{(k)}(\tau)$ is obtained if $k = K$, i.e., $\alpha_{i,0}^{(k)} = \alpha_{i,0}^{(K)} = \pm\frac{\pi}{4N_i}\frac{K}{K+2}$, no matter which value has been chosen for K. In contrast to the MEDS, the introduction of the angle-of-rotation $\alpha_{i,0}^{(k)}$ in (10.11) guarantees that the conditions in (10.8a) and (10.8b) can be satisfied without choosing different values for N_i. In this sense, the GMEDS$_1$ removes the constraint on N_i. Consequently, a very large number (theoretically infinite) K of mutually uncorrelated Rayleigh fading waveforms $\tilde{\zeta}^{(k)}(t)$ $(k = 1, 2, \ldots, K)$ can be designed by using (10.12) without increasing the model complexity determined by N_i.

Following the discussion above, the inequality $\frac{\pi}{2N_i}(n - \frac{1}{2}) + \alpha_{i,0}^{(k)} > \frac{\pi}{2N_i}(n - \frac{1}{2})$ holds if $0 < \alpha_{i,0}^{(k)} < \pi/(4N_i)$. In this case, $f_{i,n}^{(k)}$ according to (10.11) is always smaller for the GMEDS$_1$ than the corresponding $f_{i,n}^{(k)}$ for the MEDS with $\alpha_{i,0}^{(k)} = 0$. From (10.3), it is now clear that when using the GMEDS$_1$, the autocorrelation function $\tilde{r}_{\mu_i\mu_i}^{(k)}(\tau)$ is always larger than or equal to the autocorrelation function $\tilde{r}_{\mu_i\mu_i}(\tau)$, obtained by applying the MEDS, over a certain interval $[0, \tau_{\max}]$. The opposite statement holds if $-\pi/(4N_i) < \alpha_{i,0}^{(k)} < 0$. For the MEDS, it is important to mention that $\tilde{r}_{\mu_i\mu_i}(\tau)$ is very close to $r_{\mu_i\mu_i}(\tau) = J_0(2\pi f_{\max}\tau)$ over the interval $[0, \tau_{\max}]$ with $\tau_{\max} = N_i/(2f_{\max})$, as demonstrated in Figure 10.2. Consequently, we obtain the following important properties for the autocorrelation function $\tilde{r}_{\mu_i\mu_i}^{(k)}(\tau)$ when using the GMEDS$_1$:

$$\tilde{r}_{\mu_i\mu_i}^{(k)}(\tau) \geq r_{\mu_i\mu_i}(\tau), \quad \text{if } 0 < \alpha_{i,0}^{(k)} < \pi/(4N_i), \tag{10.14a}$$

$$\tilde{r}_{\mu_i\mu_i}^{(k)}(\tau) \leq r_{\mu_i\mu_i}(\tau), \quad \text{if } -\pi/(4N_i) < \alpha_{i,0}^{(k)} < 0, \tag{10.14b}$$

where $\tau \in [0, \tau_{\max}]$ and $\tau_{\max} = N_i/(2f_{\max})$. These results can be phrased as follows: A positive (or negative) angle-of-rotation $\alpha_{i,0}^{(k)} \in (0, \pi/(4N_i))$ (or $\alpha_{i,0}^{(k)} \in (-\pi/(4N_i), 0)$) always results in non-negative (or non-positive) errors of the autocorrelation function $\tilde{r}_{\mu_i\mu_i}^{(k)}(\tau)$ of the simulation model with respect to the autocorrelation function $r_{\mu_i\mu_i}(\tau)$ of the reference model within the interval $[0, \tau_{\max}]$.

The property above is confirmed by the results shown in Figure 10.3. This figure presents plots of the autocorrelation functions $\tilde{r}_{\mu_i\mu_i}^{(k)}(\tau)$, which have been obtained by applying the MEDS ($\alpha_{i,0}^{(k)} = 0$) and the GMEDS$_1$ for two selected values of $\alpha_{i,0}^{(k)}$ according to (10.12) with $N_i = 20$ and $k = K = 4$. For comparison, the behaviour of the autocorrelation function $r_{\mu_i\mu_i}(\tau)$ of the reference model is also illustrated. Obviously, the best fitting is obtained when the original MEDS is used. In this case, the approximation $\tilde{r}_{\mu_i\mu_i}^{(k)}(\tau) \approx r_{\mu_i\mu_i}(\tau)$ is excellent

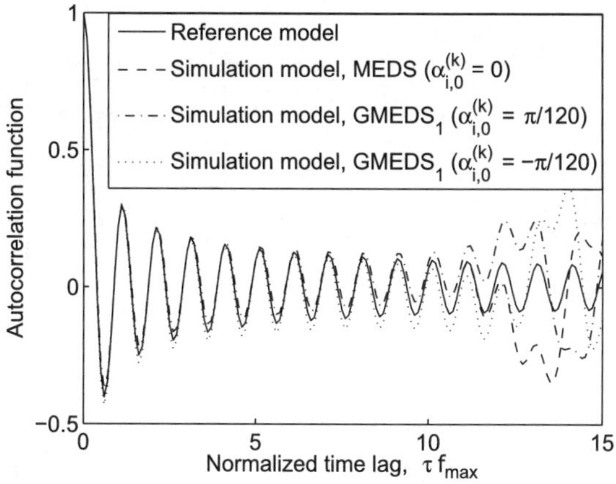

Figure 10.3 The autocorrelation function $r_{\mu_i\mu_i}(\tau)$ of the reference model in comparison to the autocorrelation function $\tilde{r}_{\mu_i\mu_i}^{(k)}(\tau)$ of the simulation model by using the MEDS with $\alpha_{i,0}^{(k)} = 0$ and the GMEDS$_1$ with $\alpha_{i,0}^{(k)} = \pm\pi/120$ ($N_i = 20$, $k = K = 4$).

if $\tau \in [0, N_i/(2f_{\max})]$, i.e., $\tau f_{\max} \in [0, 10]$. It can also be observed that if $\alpha_{i,0}^{(k)} = \pi/120$, then $\tilde{r}_{\mu_i\mu_i}^{(k)}(\tau) \geq r_{\mu_i\mu_i}(\tau)$ holds for $\tau f_{\max} \in [0, 10]$. The opposite conclusion follows if $\alpha_{i,0}^{(k)} = -\pi/120$. Note that Figure 10.3 shows the worst approximation result of the autocorrelation function $\tilde{r}_{\mu_i\mu_i}^{(k)}(\tau)$ for the GMEDS$_1$ when $\alpha_{i,0}^{(k)} = \pm\pi/120$, which follows from (10.12) if $k = K$. For $k < K$, the approximation results are much better, which is also clear from Figure 10.2. A comparative study has revealed that even the worst case of the GMEDS$_1$ provides better approximation results regarding the autocorrelation function $r_{\mu_i\mu_i}(\tau)$ than many other methods, including the Jakes method in [13], the deterministic methods in [158, 369, 370], and the random methods in [97, 99, 111, 113] with respect to single trials.

From (10.12) and (10.14), we can further conclude that the inphase autocorrelation function $\tilde{r}_{\mu_1\mu_1}^{(k)}(\tau)$ with $\alpha_{1,0}^{(k)}$ always has non-negative errors, while the quadrature autocorrelation function $\tilde{r}_{\mu_2\mu_2}^{(k)}(\tau)$ with $\alpha_{2,0}^{(k)} = -\alpha_{1,0}^{(k)}$ always has non-positive errors over the interval $[0, \tau_{\max}]$. Recall from Figure 10.2 that the error function $E_2(\alpha_{i,0}^{(k)})$ is almost even symmetrical. This means that the non-negative and non-positive errors of $\tilde{r}_{\mu_1\mu_1}^{(k)}(\tau)$ and $\tilde{r}_{\mu_2\mu_2}^{(k)}(\tau)$, respectively, compensate each other such that the approximation of $\tilde{r}_{\mu\mu}^{(k)}(\tau) = \tilde{r}_{\mu_1\mu_1}^{(k)}(\tau) + \tilde{r}_{\mu_2\mu_2}^{(k)}(\tau)$ to $r_{\mu\mu}(\tau) = 2J_0(2\pi f_{\max}\tau)$ is excellent at least over the domain $[0, \tau_{\max}]$, where $\tau_{\max} = N_i/(2f_{\max})$. Similar to (10.6), the quality of the approximation $\tilde{r}_{\mu\mu}^{(k)}(\tau) \approx r_{\mu\mu}(\tau)$ over the interval $[0, \tau_{\max}]$ can be measured by using the following L_2-norm

$$E_2'(\alpha_{i,0}^{(k)}) = \left[\frac{1}{\tau_{\max}} \int_0^{\tau_{\max}} |\tilde{r}_{\mu\mu}^{(k)}(\tau) - r_{\mu\mu}(\tau)|^2 d\tau\right]^{1/2} . \tag{10.15}$$

Figure 10.4 shows the behaviour of $E_2'(\alpha_{i,0}^{(k)})$ in terms of $\alpha_{i,0}^{(k)}$ for various values of N_i. From a comparison of Figures 10.2 and 10.4, the above-mentioned compensation effect is obvious, as the error function $E_2'(\alpha_{i,0}^{(k)})$ is overall much smaller than $E_2(\alpha_{i,0}^{(k)})$. Needless to say, $E_2'(\alpha_{i,0}^{(k)})$ is even symmetrical, i.e., $E_2'(\alpha_{i,0}^{(k)}) = E_2'(-\alpha_{i,0}^{(k)})$, and periodic with period $\alpha_{\text{per}} = \pi/(4N_i)$, i.e., $E_2'(\alpha_{i,0}^{(k)}) = E_2'(\alpha_{i,0}^{(k)} + p\alpha_{\text{per}})$, where $p = \pm 1, \pm 2, \ldots$ The maximum and minimum values of

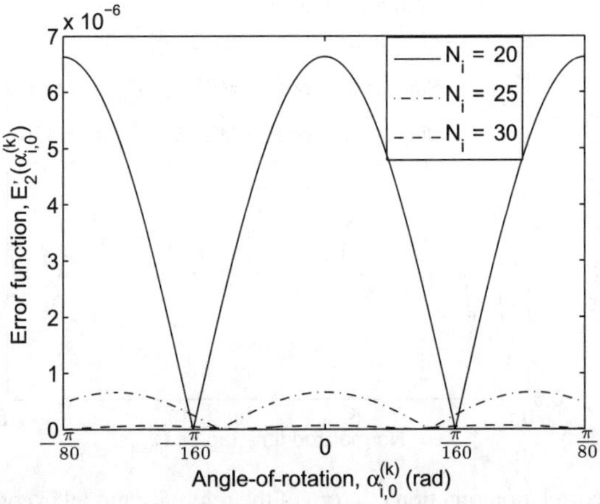

Figure 10.4 The error function $E_2'(\alpha_{i,0}^{(k)})$ in terms of the angle-of-rotation $\alpha_{i,0}^{(k)}$ by using the GMEDS$_1$ for various values of N_i.

$E_2'(\alpha_{i,0}^{(k)})$ are taken at $\alpha_{i,0}^{(k)} = p\alpha_{\text{per}}$ and $\alpha_{i,0}^{(k)} = (2p+1)\alpha_{\text{per}}/2$, respectively. Note that $E_2'(\alpha_{i,0}^{(k)})$ and $E_2(\alpha_{i,0}^{(k)})$ have different periods and that the maximum of $E_2'(\alpha_{i,0}^{(k)})$ at $\alpha_{i,0}^{(k)} = 0$ is coincident with the minimum of $E_2(\alpha_{i,0}^{(k)})$ at the same position $\alpha_{i,0}^{(k)} = 0$. Consequently, the original MEDS with $\alpha_{i,0}^{(k)} = 0$ results in the worst fitting with respect to the autocorrelation function $r_{\mu\mu}(\tau)$ of the complex process $\mu(t)$. This is completely different from the fitting to the autocorrelation function $r_{\mu_i\mu_i}(\tau)$ of the inphase (quadrature) component $\mu_i(t)$, in which case the MEDS provides the best fitting result, as can be seen in Figures 10.2 and 10.3.

In Figure 10.5, plots of the autocorrelation function $\tilde{r}_{\mu\mu}^{(k)}(\tau)$ are shown for four selected values of $\alpha_{i,0}^{(k)}$ computed by using the GMEDS$_1$ with $N_i = 20$ and $K = 4$. Note that $\alpha_{i,0}^{(k)} = \pm\pi/160$ $(k = 3)$ and $\alpha_{i,0}^{(k)} = \pm\pi/480$ $(k = 1)$ represent the best case and the worst case, respectively, according to the results shown in Figure 10.4. The even symmetry property is self-evident. The autocorrelation function $\tilde{r}_{\mu\mu}^{(k)}(\tau)$ of the simulation model obtained by using the original MEDS ($\alpha_{i,0}^{(k)} = 0$) and the autocorrelation function $r_{\mu\mu}(\tau)$ of the reference model are also shown for reasons of comparison. Notice that the performance of the MEDS is worse than the performance of the GMEDS$_1$ in the worst case in terms of the fitting to $r_{\mu\mu}(\tau)$. However, even in the worst case of the GMEDS$_1$, the approximation to $r_{\mu\mu}^{(k)}(\tau)$ is excellent if $\tau \in [0, N_i/(2f_{\max})]$ or, equivalently, $\tau f_{\max} \in [0, 10]$. When $\alpha_{i,0}^{(k)} = \pm\pi/(8N_i) = \pm\pi/160$, the fitting to $r_{\mu\mu}(\tau)$ is excellent even when $\tau_{\max}$ reaches $N_i/f_{\max}$, i.e., $\tau f_{\max} \in [0, 20]$. This is not surprising since this specific constellation corresponds to that of the original MEDS with $2N_i$. In the limited range of $\tau f_{\max} \in [0, 15]$ shown in Figure 10.5, the autocorrelation function $\tilde{r}_{\mu\mu}^{(k)}(\tau)$ obtained by using the GMEDS$_1$ with $\alpha_{i,0}^{(k)} = \pm\pi/160$ and the autocorrelation function $r_{\mu\mu}(\tau)$ of the reference model are indistinguishable. More details on the analysis of the GMEDS$_1$ can be found in [114]. Finally, Figure 10.6 illustrates by way of example for $K = 4$ the temporal behaviour of the resulting four uncorrelated fading envelopes $\tilde{\zeta}^{(k)}(t)$ $(k = 1, 2, \ldots, K)$.

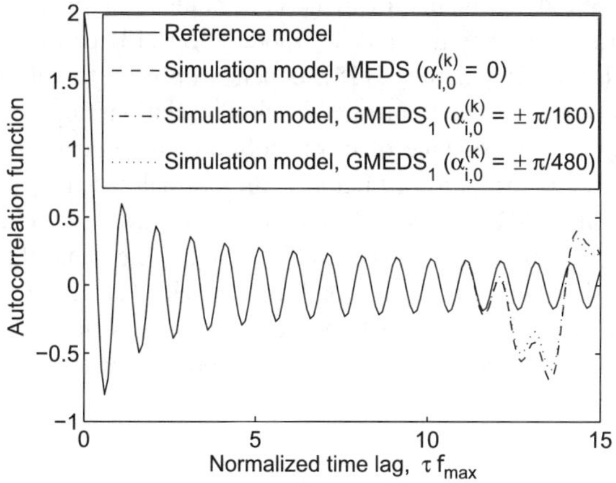

Figure 10.5 The autocorrelation function $r_{\mu\mu}(\tau)$ of the complex Gaussian process of the reference model in comparison with the corresponding autocorrelation function $\tilde{r}_{\mu\mu}^{(k)}(\tau)$ of the simulation model designed by using the MEDS with $\alpha_{i,0}^{(k)} = 0$ and the GMEDS$_1$ with $\alpha_{i,0}^{(k)} = \pm\pi/160$ and $\alpha_{i,0}^{(k)} = \pm\pi/480$ $(N_1 = N_2 = 20, K = 4)$.

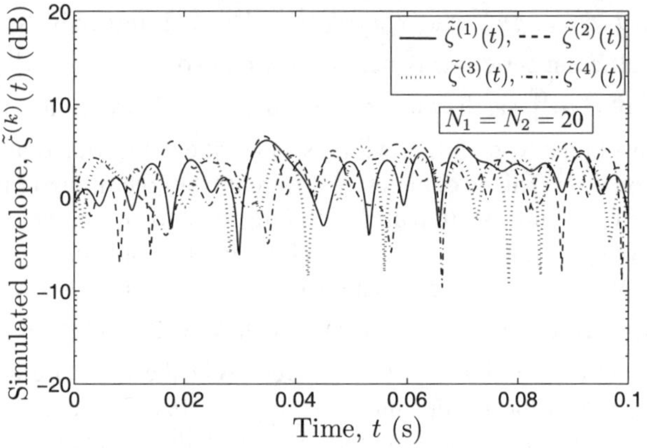

Figure 10.6 Simulated uncorrelated Rayleigh fading waveforms $\tilde{\zeta}^{(k)}(t)$ $(k = 1, 2, \ldots, K)$ by using the GMEDS_1 ($f_{\text{max}} = 91$ Hz, $N_1 = N_2 = 20$, $K = 4$).

10.1.2.3 Discussion of the GMEDS_1

One obvious advantage of the GMEDS_1 over the original MEDS is that the GMEDS_1 can be used to design a very large (theoretically an infinite) number of mutually uncorrelated Rayleigh fading waveforms by keeping the number of sinusoids constant. The complexity of the resulting channel simulator designed using the GMEDS_1 is low and independent of the number K of generated uncorrelated fading waveforms. The drawback of the GMEDS_1 is that small non-negative (non-positive) errors have to be accepted concerning the fitting of the autocorrelation function of the inphase (quadrature) component of the generated complex waveforms. However, these errors compensate each other over the domain of interest when considering the autocorrelation function of the resulting complex waveform. The domain of interest increases linearly with the number of sinusoids and can thus easily be controlled. Note that the performance of many communication systems, e.g., such as those using differential phase shift keying (DPSK) modulation [13] or orthogonal frequency division multiplexing (OFDM) schemes [118], depends only on the autocorrelation function of the complex waveform, rather than on the autocorrelation functions of the quadrature components. In fact, the performance of most mobile communication systems is only sensitive to errors of the autocorrelation function if the normalized time lag is small, meaning $\tau f_{\text{max}} \leq 0.3$ [153]. The GMEDS_1 is well suited for the design of channel simulators for such communication systems. Finally, it should be highlighted that the GMEDS_1 provides a closed-form expression [see (10.11) in combination with (10.12)] for the model parameters, which has been derived on the assumption of isotropic scattering conditions. The extension of the proposed technique to wireless non-isotropic channels might be a topic for future research.

10.1.3 Related Parameter Computation Methods

The GMEDS_q represents a class of parameter computation methods, which includes, among others, the following seven methods as special cases.

Monte Carlo method (MCM): If $q = 0$ and $\alpha_{i,0}^{(k)}$ is replaced by $u_{i,n}^{(k)}$ being i.i.d. random variables uniformly distributed over $(0, \pi/2]$, then we obtain the MCM, which was originally proposed in [100] and further developed in [99]. The MCM and the following randomized MEDS (RMEDS) can be used to generate a large number of uncorrelated Rayleigh fading waveforms without increasing the simulator's complexity. However, due to the non-ergodicity property of the methods, the resulting simulators are not efficient, as discussed in [214].

Randomized MEDS (RMEDS): The RMEDS [111] is obtained if $q = 1$ and $\alpha_{i,0}^{(k)}$ is replaced by $\alpha_{i,0}^{(k)} = u_{i,n}^{(k)}/(4N_i)$, where $u_{i,n}^{(k)}$ are i.i.d. random variables, each having a uniform distribution over $[-\pi, \pi)$.

Original MEDS: The original MEDS [96] results from (10.11) if $q = 1$ and $\alpha_{i,0}^{(k)} = 0$. However, N_i in (10.11) has to be replaced by $N_i^{(k)}$ in order to generate multiple uncorrelated waveforms. The discrete Doppler frequencies are therefore given by $f_{i,n}^{(k)} = f_{\max}\cos[\pi(n - 1/2)/(2N_i^{(k)})]$. It was shown in [214] that the only way to fulfill (10.8a) and (10.8b) with the MEDS is to guarantee that $N_i^{(k)}/N_\lambda^{(l)} \neq (2n-1)/(2m-1)$ for all $n = 1, 2, \ldots, N_i^{(k)}$ and $m = 1, 2, \ldots, N_\lambda^{(l)}$. With an increase in the number of simulated uncorrelated fading waveforms, the number of sinusoids of the deterministic SOS channel simulator increases exponentially if the MEDS is used. For example, to simulate $K = 4$ uncorrelated Rayleigh fading channels by using the MEDS, a possible set of eight values for the number of sinusoids N_i fulfilling (10.8a) and (10.8b) is given by $\{8, 9, 10, 12, 16, 32, 64, 128\}$ [214]. This obvious drawback limits the usefulness of the MEDS if a large number of uncorrelated processes is required.

MEDS with set partitioning (MEDS-SP): Another special case is obtained when $q = 1$ and $\alpha_{i,0}^{(k)} = \pi[k - (K+1)/2]/(2KN_i)$, which results in the so-called MEDS-SP [150]. This method actually belongs to the GMEDS_1, but with a different expression for $\alpha_{i,0}^{(k)}$. The purpose of the MEDS-SP is to take advantage of averaging over multiple sample functions (trials) — a technique that is unavoidable when non-ergodic stochastic methods [97, 99, 111, 113] are used. In [150], it was shown that the MEDS-SP outperforms the non-ergodic method proposed in [111] with respect to both single trials and averaging over multiple trials. The use of the MEDS-SP for the design of multiple uncorrelated Rayleigh fading waveforms is not recommended without additional modifications.

Method of equal areas (MEA): The MEA [148] ensues from the GMEDS_q if $q = 1$ and $\alpha_{i,0}^{(k)} = \pi/(4N_i^{(k)})$. It follows that $f_{i,n}^{(k)} = f_{\max}\cos[n\pi/(2N_i^{(k)})]$. Note that the MEA was originally proposed for the design of a single Rayleigh fading waveform. To generate multiple uncorrelated Rayleigh fading waveforms by using the MEA, N_i in (10.11) has to be replaced by $N_i^{(k)}$. Similar to the MEDS, the only way to fulfill (10.8a) and (10.8b) with the MEA is to change the number of sinusoids for different waveforms, which will greatly increase the simulator's complexity if a large number of uncorrelated waveforms is required.

GMEDS_2: Another solution to the problem of designing multiple uncorrelated Rayleigh fading waveforms follows from the GMEDS_q by setting $q = 2$. With the obtained GMEDS_2, we can design a set of K Rayleigh fading waveforms $\tilde{\zeta}^{(k)}(t)$ if the angle-of-rotation $\alpha_{i,0}^{(k)}$ is given by [114]

$$\alpha_{i,0}^{(k)} = \frac{\pi}{4N_i} \cdot \frac{k-1}{K-1} \tag{10.16}$$

for $k = 1, 2, \ldots, K$ and $i = 1, 2$, where $N_2 := N_1 + 1$. Depending on the chosen values for K and N_1, it was pointed out in [371] that the GMEDS$_2$ can result in waveforms $\tilde{\mu}_i^{(k)}(t)$ and $\tilde{\mu}_\lambda^{(l)}(t)$, which do not fulfill the conditions in (10.8a) and (10.8b) for some few combinations of $i \neq \lambda$ ($i, \lambda = 1, 2$) and $k \neq l$ ($k, l = 1, 2, \ldots, K$). This means that the number of perfectly mutually uncorrelated Rayleigh fading waveforms is $K' \leq K$. To avoid this problem, it is recommended to replace $\alpha_{i,0}^{(k)}$ in (10.16) by [371]

$$\alpha_{i,0}^{(k)} = \frac{\pi}{2N_i} \cdot \frac{k-1/2}{K} \tag{10.17}$$

with $N_2 = N_1 + 2$ (N_1 even), which guarantees that the conditions in (10.8a) and (10.8b) are fulfilled for all combinations of $i \neq \lambda$ ($i, \lambda = 1, 2$) and $k \neq l$ ($k, l = 1, 2, \ldots, K$).

Extended MEDS (EMEDS): Finally, if $q = 4$ and $\alpha_{i,0}^{(k)} = \pi/(2N_i)$, then we obtain the EMEDS, which has been proposed for the design of sum-of-cisoids Rayleigh fading channel simulators under the assumption of isotropic scattering conditions. The EMEDS also plays an important role for the computation of the model parameters of MIMO channel simulators, as demonstrated in Chapter 8. Without further modifications, the EMEDS cannot be used for the design of multiple uncorrelated Rayleigh fading waveforms.

10.1.4 The Effect of Finite Simulation Time on the Cross-Correlation Properties

To gain further insight into the characteristics of multiple uncorrelated Rayleigh fading waveforms, we study the cross-correlation properties under simulation time constraints. That is, we assume that the simulation of $\tilde{\mu}_i^{(k)}(t)$ starts at $t_1 = -T$ and ends at $t_2 = +T$, where T is finite ($T < \infty$). Under this condition, the processes $\tilde{\mu}_i^{(k)}(t)$ and $\tilde{\mu}_\lambda^{(l)}(t)$ are correlated even if the inequalities in (10.8a) and (10.8b) are fulfilled. To study the effect of finite simulation time, we define the following correlation coefficient

$$\tilde{\rho}_{\mu_i\mu_\lambda}^{(k,l)} := \frac{1}{2T} \int_{-T}^{T} \tilde{\mu}_i^{(k)}(t)\, \tilde{\mu}_\lambda^{(l)}(t)\, dt, \quad 0 < T < \infty. \tag{10.18}$$

Notice that the correlation coefficient $\tilde{\rho}_{\mu_i\mu_\lambda}^{(k,l)}$ equals the cross-correlation function $\tilde{r}_{\mu_i\mu_\lambda}^{(k,l)}(\tau)$ at $\tau = 0$ under the condition that T is limited [cf. (10.4)]. Substituting (10.2) in (10.18) gives the closed-form solution

$$\tilde{\rho}_{\mu_i\mu_\lambda}^{(k,l)} = \frac{1}{2\pi T \sqrt{N_i N_\lambda}} \sum_{n=1}^{N_i} \sum_{m=1}^{N_\lambda} \left[\frac{\sin(2\pi(f_{i,n}^{(k)} - f_{\lambda,m}^{(l)})T)\cos(\theta_{i,n}^{(k)} - \theta_{\lambda,m}^{(l)})}{f_{i,n}^{(k)} - f_{\lambda,m}^{(l)}} + \frac{\sin(2\pi(f_{i,n}^{(k)} + f_{\lambda,m}^{(l)})T)\cos(\theta_{i,n}^{(k)} + \theta_{\lambda,m}^{(l)})}{f_{i,n}^{(k)} + f_{\lambda,m}^{(l)}} \right]. \tag{10.19}$$

Obviously, the quantity $\tilde{\rho}^{(k,l)}_{\mu_i\mu_\lambda}$ depends on all model parameters, including the phases $\theta^{(k)}_{i,n}$ and T. For our purpose, it is sufficient to focus on the upper limit of $\tilde{\rho}^{(k,l)}_{\mu_i\mu_\lambda}$. Let us denote the upper limit by $\hat{\rho}^{(k,l)}_{\mu_i\mu_\lambda}$, then it follows from (10.19) by using $\sin(x) \leq 1$ and $\cos(x) \leq 1$ that

$$\hat{\rho}^{(k,l)}_{\mu_i\mu_\lambda} = \frac{1}{\pi T\sqrt{N_i N_\lambda}} \sum_{n=1}^{N_i} \sum_{m=1}^{N_\lambda} \frac{f^{(k)}_{i,n}}{(f^{(k)}_{i,n})^2 - (f^{(l)}_{\lambda,m})^2}. \tag{10.20}$$

It is apparent that $\hat{\rho}^{(k,l)}_{\mu_i\mu_\lambda} \to 0$ as $T \to \infty$. In practise, however, the simulation time $T_{\text{sim}} = 2T$ is limited, and thus the theoretically uncorrelated processes $\tilde{\mu}^{(k)}_i(t)$ and $\tilde{\mu}^{(l)}_\lambda(t)$ remain correlated. For the GMEDS$_1$ [see (10.12)], the GMEDS$_2$ [see (10.16)], and the modified MEDS (MMEDS) introduced in [372], the evaluation results of the upper limit $\hat{\rho}^{(k,l)}_{\mu_i\mu_\lambda}$ are listed in Table 10.1. The results clearly show that the GMEDS$_1$ and the GMEDS$_2$ are superior to the MMEDS under practical considerations, as the latter method results in deterministic processes $\tilde{\mu}^{(k)}_i(t)$ and $\tilde{\mu}^{(l)}_\lambda(t)$ which decorrelate very slowly with increasing values of T. The reason for this drawback is that the MMEDS clusters the discrete Doppler frequencies $f^{(k)}_{i,n}$ tightly around the optimum (original MEDS) for $k = 1, 2, \ldots, K$. This keeps the frequency differences in the denominator of (10.20) very small and results thus in large values for $\hat{\rho}^{(k,l)}_{\mu_i\mu_\lambda}$. This drawback is avoided by the GMEDS$_1$ and the GMEDS$_2$, where the spread of the discrete Doppler frequencies $f^{(k)}_{i,n}$ is much larger.

The results above motivate us to introduce a third important criterion, which states that the K waveforms must decorrelate very fast under simulation time constraints. This condition can only be fulfilled if the spread among the discrete Doppler frequencies used for different waveforms is large. Unfortunately, the third criterion contradicts the first one introduced in Subsection 10.1.1 by (10.5). This means that the design of multiple uncorrelated Rayleigh

Table 10.1 Upper limit $\hat{\rho}^{(k,l)}_{\mu_i\mu_\lambda}$ of the correlation coefficient of $\tilde{\mu}^{(k)}_i(t)$ and $\tilde{\mu}^{(l)}_\lambda(t)$ assuming limited simulation time by using the GMEDS$_1$ ($N_1 = N_2 = 20$), the GMEDS$_2$ ($N_1 = 20$, $N_2 = 21$), and the MMEDS [372] ($N_1 = N_2 = 20$) with $K = 3$ and $f_{\max} = 91$ Hz.

$\hat{\rho}^{(k,l)}_{\mu_i\mu_\lambda}$	GMEDS$_1$	GMEDS$_2$	MMEDS
$\hat{\rho}^{(1,2)}_{\mu_1\mu_1}$	0.6765s/T	$-5.0889 \cdot 10^{-16}$s/T	$-1.5915 \cdot 10^{6}$s/T
$\hat{\rho}^{(1,3)}_{\mu_1\mu_1}$	0.3249s/T	$-1.4843 \cdot 10^{-16}$s/T	$-7.9577 \cdot 10^{5}$s/T
$\hat{\rho}^{(2,3)}_{\mu_1\mu_1}$	0.6391s/T	0.2070s/T	$-1.5915 \cdot 10^{6}$s/T
$\hat{\rho}^{(1,2)}_{\mu_2\mu_2}$	−0.8293s/T	$-5.0889 \cdot 10^{-16}$s/T	$1.5915 \cdot 10^{6}$s/T
$\hat{\rho}^{(1,3)}_{\mu_2\mu_2}$	−0.4103s/T	$-1.4843 \cdot 10^{-16}$s/T	$7.9577 \cdot 10^{5}$s/T
$\hat{\rho}^{(2,3)}_{\mu_2\mu_2}$	−1.0062s/T	0.2070s/T	$1.5915 \cdot 10^{6}$s/T
$\hat{\rho}^{(1,1)}_{\mu_1\mu_2}$	−0.3035s/T	$-1.6554 \cdot 10^{-16}$s/T	$7.9577 \cdot 10^{5}$s/T
$\hat{\rho}^{(2,2)}_{\mu_1\mu_2}$	−0.0817s/T	0.2882s/T	$3.9789 \cdot 10^{5}$s/T
$\hat{\rho}^{(3,3)}_{\mu_1\mu_2}$	0.0386s/T	1.0993s/T	$2.6526 \cdot 10^{5}$s/T

waveforms can only result in a compromise between the first and the third criterion. Such a compromise is provided by the GMEDS_1.

10.1.5 Further Reading

For the simulation of Rayleigh fading, Jakes' deterministic SOS channel simulator [13] is widely in use, although it has some undesirable properties. One of them arises from the non-zero cross-correlation function of the inphase and quadrature components of the generated complex waveforms. In [13], Jakes also proposed an extension of his approach aiming to generate K multiple uncorrelated waveforms, but it was shown in [118, 157] that the maximum (minimum) values of the cross-correlation function of any pair of generated complex waveforms can be quite high. Dent et al. [158] suggested a modification to Jakes' method by using orthogonal Walsh-Hadamard matrices to decorrelate the generated waveforms. This reduces the cross-correlation functions, but they are still not exactly zero. The same problem of non-zero cross-correlation functions between different waveforms is retained for the deterministic method proposed in [369]. Another deterministic method that enables the generation of a set of K mutually uncorrelated Rayleigh fading waveforms has been introduced in [370]. Using this method, the temporal autocorrelation function of each of the K underlying complex waveforms is very close to the specified one. Unfortunately, this is not the case for the autocorrelation functions of the inphase and quadrature parts of the designed complex waveforms. In [113], both a deterministic and a stochastic method have been suggested aiming to tackle the problem of designing multiple uncorrelated Rayleigh fading waveforms. Unfortunately, when applying the deterministic method, the autocorrelation functions of the real and imaginary parts of the generated complex waveforms are quite different from the corresponding autocorrelation functions of the reference model — even if the number of sinusoidal terms tends to infinity — and the proposed stochastic method results in a non-ergodic channel simulator. The L_p-norm method [96] is very powerful and not limited to isotropic channels, but it lacks a simple closed-form solution to the problem of designing multiple uncorrelated Rayleigh fading waveforms and requires professional experience in numerical optimization techniques to achieve the expected results. The usefulness of the method of exact Doppler spread (MEDS) [96] concerning the generation of multiple uncorrelated Rayleigh fading waveforms with a deterministic SOS channel simulator was revisited in [214]. There it was shown that all the main requirements can be fulfilled, but unfortunately the complexity of the resulting channel simulator increases almost exponentially with an increase in the number of uncorrelated waveforms. This makes the original MEDS less efficient if the number of uncorrelated waveforms is large. A detailed comparison of the MEDS with the L_p-norm method and various random methods in [97, 111, 112] can also be found in [214].

Non-ergodic stochastic methods, such as those proposed in [97, 99, 111–113], can be used to guarantee that the cross-correlation functions of different waveforms are zero, but the temporal autocorrelation function of the waveform obtained from a single simulation trial is generally not sufficiently close to the desired autocorrelation function of the reference model. This problem can only be solved by averaging over many simulation trials, which reduces the efficiency of the approach.

Finally, it is mentioned that a general method for the design of ergodic sum-of-cisoids simulators for multiple uncorrelated Rayleigh fading channels has been presented in [170].

This method is not restricted to Jakes' (Clarke's) symmetrical U-shaped Doppler power spectral density. It can be used to generate any number of uncorrelated Rayleigh fading waveforms with arbitrarily (symmetrical and asymmetrical) specified Doppler power spectral characteristics.

10.2 Spatial Channel Models for Shadow Fading

The received envelope of a signal transmitted through a mobile radio channel experiences fast and slow fading. Fast fading results from the superposition of the received scattered multipath components. Slow fading, which is also referred to as *shadow fading* or simply *shadowing*, occurs if propagation paths are blocked by large objects or terrain features, like buildings and hills in macrocells or vehicles in microcells. The effect of shadow fading reveals itself in random fluctuations of the local mean, which can be interpreted as the average of the received envelope over a distance of a few wavelengths. Empirical studies have shown that the signal strength variations of the local mean caused by shadowing can be modelled adequately by a lognormal process [63, 193, 194, 373, 374]. The lognormal process plays an important role in mobile radio channel modelling as well as in performance studies of mobile communication systems. To study the impact of the combined effects caused by fast and slow fading, the lognormal process has become an inherent part of many channel models, such as the Suzuki model [26], the Loo model [202], and the modified Loo model [195]. All these models have been proposed as appropriate stochastic models for land mobile radio channels and satellite channels, which have been described in detail in Chapter 6. Lognormal processes also play an important role in the performance analysis of handover strategies [375, 376], the coverage of multihop cellular systems [377], and the study of the quality of service in mobile ad hoc networks [378]. A good understanding of lognormal processes and their parametrization is therefore an important subject in mobile radio channel modelling.

This section focusses on the modelling, analysis, and simulation of spatial shadowing processes. Such processes are generally assumed to be lognormally distributed. Starting from a non-realizable reference model, we derive a stochastic simulation model for spatial shadowing processes by using the sum-of-sinusoids principle. For both the reference model and the simulation model, analytical expressions are presented for the spatial autocorrelation function, the probability density function, the level-crossing rate, and the average duration of fades. Two user-friendly methods are discussed enabling the fitting of the simulation model to the reference model with respect to the probability density function of the received signal strength as well as to given spatial correlation functions. One of the spatial correlation functions is the well-known Gudmundson correlation model. It is pointed out that Gudmundson's correlation model results in an infinite level-crossing rate. To avoid this problem, we will discuss three alternative correlation models, including one that is based on measurement data. The flexibility and high performance of the shadowing simulator are demonstrated by applying the proposed design methodology on all four types of correlation models. For each of the four correlation models, we will present illustrative examples of the dynamic behaviour of shadow fading. Emphasis will be placed on two realistic propagation scenarios capturing the shadowing effects in suburban and urban areas. The application potential of the spatial shadowing simulator ranges from the simulation of lognormal processes, as they are used in Suzuki processes, over the evaluation of handover algorithms up to studying the quality of service in mobile communication systems.

The remainder of this section is organized as follows. Subsection 10.2.1 introduces a spatial lognormal process, which serves as reference model for shadow fading. The corresponding simulation model is derived in Subsection 10.2.2. This subsection also investigates analytically the statistical properties of the proposed shadowing simulator. Four typical correlation models for shadow fading are presented in Subsection 10.2.3, where it is shown how the proposed channel simulator can be configured to simulate realistic shadowing scenarios in suburban and urban areas. Finally, guides to further reading are provided in Subsection 10.2.4.

10.2.1 The Reference Model for Shadow Fading

The effect of shadowing is generally modelled through a lognormal process $\lambda(t)$, which can be expressed as

$$\lambda(t) = 10^{\,(\sigma_L \nu(t)+m_L)/20}, \tag{10.21}$$

where $\nu(t)$ is a real-valued zero-mean Gaussian process with unit variance. The parameters σ_L and m_L in (10.21) are called the *shadow standard deviation* and the *area mean*, respectively. The area mean m_L is obtained by averaging the received signal strength over an area that is large enough to average over the shadowing effects [118]. The value of m_L is determined by the path loss of the link between the base station and the mobile station. The shadow standard deviation σ_L increases slightly with the carrier frequency and depends upon the antenna heights and the propagation environment. Measurements of σ_L can be found, e.g., in [63, 193, 194, 373, 374]. The shadow standard deviation σ_L is usually in the range from 5 to 12 dB at 900 MHz, where 8 dB is a typical value for macrocellular applications. In the 1800 MHz frequency band, the shadow standard deviation σ_L is about 0.8 dB higher than in the 900 MHz frequency band.

Notice that the lognormal process $\lambda(t)$ in (10.21) is a random function of time t. Let us assume that the mobile station starts at the origin $x_0 = 0$ and moves along the x-axis with velocity v. Then, the time-distance relationship $t = x/\mathrm{v}$ allows us to express the lognormal process $\lambda(t)$ and the Gaussian process $\nu(t)$ as a function of the distance x, indicated by $\lambda(x)$ and $\nu(x)$. Using (10.21), the shadowing effects can thus alternatively be modelled in the spatial domain by the so-called *spatial lognormal process*

$$\lambda(x) = 10^{(\sigma_L \nu(x)+m_L)/20}. \tag{10.22}$$

The distribution of $\lambda(x)$ follows the lognormal distribution

$$p_\lambda(y) = \frac{20}{\sqrt{2\pi}\,\ln(10)\,\sigma_L\, y}\, e^{-\frac{(20\log_{10}(y)-m_L)^2}{2\sigma_L^2}}, \quad y \geq 0. \tag{10.23}$$

In Appendix 10.A [379], it is shown that the spatial autocorrelation function $r_{\lambda\lambda}(\Delta x)$ of the lognormal process $\lambda(x)$ can be expressed in terms of the spatial autocorrelation function $r_{\nu\nu}(\Delta x)$ of the Gaussian process $\nu(x)$ as follows

$$r_{\lambda\lambda}(\Delta x) = e^{\,2m_0+\sigma_0^2(1+r_{\nu\nu}(\Delta x))}, \tag{10.24}$$

where $m_0 = m_L \ln(10)/20$ and $\sigma_0 = \sigma_L \ln(10)/20$. From the equation above, the mean power of the spatial lognormal process $\lambda(x)$ can easily be obtained as $r_{\lambda\lambda}(0) = e^{2(m_0+\sigma_0^2)}$.

It is worth mentioning that the distribution in (10.23) does not provide any information on how fast the spatial lognormal process $\lambda(x)$ changes with distance. However, this information is provided by the level-crossing rate and the average duration of fades. The level-crossing rate of the spatial lognormal process $\lambda(x)$, denoted as $N_\lambda(r)$, is therefore an important characteristic quantity. Here, the level-crossing rate $N_\lambda(r)$ describes how often $\lambda(x)$ crosses on average a given signal level r from up to down (or from down to up) within one unit of length (usually one metre). It is proved in Appendix 10.B [379] that the level-crossing rate $N_\lambda(r)$ of $\lambda(x)$ is given by

$$N_\lambda(r) = \frac{\sqrt{\gamma}}{2\pi} e^{-\frac{(20\log_{10}(r)-m_L)^2}{2\sigma_L^2}}, \quad r \geq 0, \tag{10.25}$$

where $\gamma = -\frac{d^2}{d\Delta x^2} r_{\nu\nu}(\Delta x)|_{\Delta x=0} = -\ddot{r}_{\nu\nu}(0)$.

The average duration of fades, denoted as $T_{\lambda_-}(r)$, is the expected value for the length of the spatial fading intervals in which the lognormal process $\lambda(x)$ is below a given signal level r. If the level-crossing rate $N_\lambda(r)$ is known, then the average duration of fades $T_{\lambda_-}(r)$ can readily be obtained from

$$T_{\lambda_-}(r) = \frac{F_{\lambda_-}(r)}{N_\lambda(r)}, \tag{10.26}$$

where $F_{\lambda_-}(r)$ designates the cumulative distribution function of the spatial lognormal process $\lambda(x)$ being the probability that $\lambda(x)$ is less than or equal to the signal level r, i.e., $F_{\lambda_-}(r) = P\{\lambda(x) \leq r\} = \int_0^r p_\lambda(y)\,dy$.

10.2.2 The Simulation Model for Shadow Fading

10.2.2.1 Derivation of the Shadow Fading Simulator

According to the sum-of-sinusoids principle, a stochastic continuous-time simulation model for a lognormal process $\lambda(t)$ is obtained from the reference model by replacing the Gaussian process $\nu(t)$ in (10.21) by the following sum of N sinusoids

$$\hat{\nu}(t) = \sum_{n=1}^{N} c_n \cos(2\pi f_n t + \theta_n). \tag{10.27}$$

In this equation, the gains c_n and the frequencies f_n are non-zero real-valued constant quantities, and the phases θ_n are independent, identically distributed (i.i.d.) random variables, each having a uniform distribution over the interval $(0, 2\pi]$. Hence, $\hat{\nu}(t)$ represents a stochastic process, which is first-order stationary and ergodic (see Table 4.1). The corresponding stochastic spatial process $\hat{\nu}(x)$ is obtained from (10.27) by making use of the time-distance transformation $t \mapsto x/\mathrm{v} = x/(\lambda_0 f_{\max})$, where λ_0 is the wavelength of the carrier frequency and $f_{\max}$ denotes the maximum Doppler frequency. Thus, the resulting spatial sum-of-sinusoids process $\hat{\nu}(x)$ can be expressed as

$$\hat{\nu}(x) = \sum_{n=1}^{N} c_n \cos(2\pi s_n x + \theta_n), \tag{10.28}$$

where $s_n = f_n/(\lambda_0 f_{max})$ are called the *spatial frequencies*. In Subsection 10.2.3, we will see how the model parameters c_n and s_n can be computed for typical shadow fading correlation models. By analogy to (10.22), a stochastic simulation model for a spatial lognormal process is then described by

$$\hat{\lambda}(x) = 10^{(\sigma_L \hat{\nu}(x)+m_L)/20}. \tag{10.29}$$

The resulting structure of the simulation model for spatial shadowing processes is shown in Figure 10.7. The stochastic spatial process $\hat{\lambda}(x)$ results in a sample function $\tilde{\lambda}(x)$ if the random phases θ_n are replaced by constant quantities, which can be considered as realizations (outcomes) of a random generator with a uniform distribution over $(0, 2\pi]$. Different sample functions $\tilde{\lambda}^{(i)}(x)$ $(i = 1, 2, \ldots)$ with equivalent statistics can be obtained by using different sets of constant phases $\{\theta_1^{(i)}, \theta_2^{(i)}, \ldots, \theta_N^{(i)}\}$. Such kinds of sample functions are called *deterministic spatial lognormal processes*.

10.2.2.2 Statistical Properties of the Shadow Fading Simulator

The following analysis is based on the fact that the gains c_n as well as the spatial frequencies s_n are constant quantities, and the phases θ_n are i.i.d. uniformly distributed random variables. Taking this into account, it follows from (10.28) that the expected value of $\hat{\nu}(x)$ equals $E\{\hat{\nu}(x)\} = 0$, and the variance can be expressed as $\text{Var}\{\hat{\nu}(x)\} = \sum_{n=1}^{N} c_n^2/2$. Let us define the gains c_n as $c_n = \sqrt{2/N}$, then the variance of $\hat{\nu}(x)$ equals unity, and is thus identical with the variance of $\nu(x)$. The spatial autocorrelation function of $\hat{\nu}(x)$, defined as $\hat{r}_{\nu\nu}(\Delta x) := E\{\hat{\nu}(x)\,\hat{\nu}(x+\Delta x)\}$, is obtained as

$$\hat{r}_{\nu\nu}(\Delta x) = \sum_{n=1}^{N} \frac{c_n^2}{2} \cos(2\pi s_n \Delta x). \tag{10.30}$$

Since $\hat{\nu}(x)$ in (10.28) represents a finite sum-of-sinusoids with random phases, we can profit from the result in (4.92) to express the probability density function $\hat{p}_\nu(x)$ of $\hat{\nu}(x)$ as

$$\hat{p}_\nu(x) = 2\int_0^\infty \left[\prod_{n=1}^{N} J_0(2\pi c_n z)\right] \cos(2\pi x z)\, dz, \tag{10.31}$$

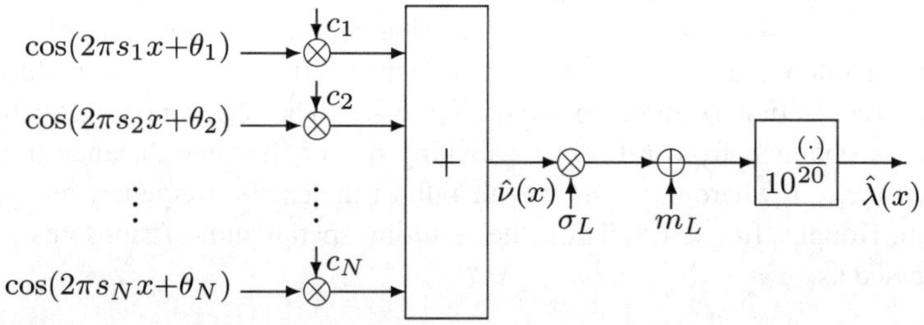

Figure 10.7 Structure of the stochastic spatial shadowing simulator with constant gains c_n, constant spatial frequencies s_n, and random phases θ_n $(n = 1, 2, \ldots, N)$.

where $J_0(\cdot)$ denotes the zeroth-order Bessel function of the first kind. Let $c_n = \sqrt{2/N}$, then the maximum and the minimum value of $\hat{\nu}(x)$ are equal to $\hat{\nu}_{\max} = \max\{\hat{\nu}(x)\} = \sqrt{2N}$ and $\hat{\nu}_{\min} = \min\{\hat{\nu}(x)\} = -\sqrt{2N}$, respectively. Thus, the density $\hat{p}_\nu(x)$ in (10.31) must be zero for all $x \notin [\hat{\nu}_{\min}, \hat{\nu}_{\max}]$.

Applying the concept of transformation of random variables by means of (2.86), we can express the probability density function $\hat{p}_\lambda(y)$ of $\hat{\lambda}(x)$ in terms of the probability density function $\hat{p}_\nu(x)$ of $\hat{\nu}(x)$ as

$$\hat{p}_\lambda(y) = \frac{20\,\hat{p}_\nu\left(\dfrac{20\log_{10}(y) - m_L}{\sigma_L}\right)}{\sigma_L \ln(10) y}. \tag{10.32}$$

Finally, after substituting (10.31) in (10.32), we find the following solution

$$\hat{p}_\lambda(y) = \frac{40}{\sigma_L \ln(10)\, y} \int_0^\infty \left[\prod_{n=1}^{N} J_0(2\pi\, c_n\, z)\right] \cos\left[\frac{2\pi z(20\log_{10}(y) - m_L)}{\sigma_L}\right] dz, \tag{10.33}$$

which allows us to study the distribution of the channel simulator's output process $\hat{\lambda}(x)$ analytically. Figure 10.8 illustrates the density $\hat{p}_\lambda(y)$ in a semi-logarithmic form as a function of the level y for various values of N. In comparison to the lognormal density $p_\lambda(y)$ introduced in (10.23), we can observe noteworthy deviations only at low and high values of y. From this study, we may conclude that the approximation $\hat{p}_\lambda(y) \approx p_\lambda(y)$ is good if the number of sinusoids N is greater than or equal to 25. This is in contrast to classical Rayleigh and Rice densities, which can well be approximated by using two sums-of-sinusoids, each consisting of only seven or eight terms. Furthermore, we notice that if $c_n = \sqrt{2/N}$, then the minimum and the maximum

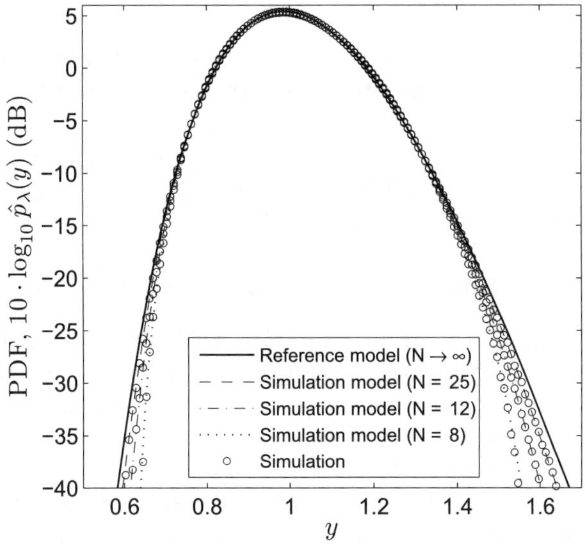

Figure 10.8 The probability density function $\hat{p}_\lambda(y)$ of the spatial shadowing simulator's output process $\hat{\lambda}(x)$ for various values of N ($c_n = \sqrt{2/N}$, $\sigma_L = 1$, and $m_L = 0$).

values of $\hat{\lambda}(x)$ are given by $\hat{\lambda}_{\min(\text{dB})} = -\sigma_L\sqrt{2N} + m_L$ and $\hat{\lambda}_{\max(\text{dB})} = +\sigma_L\sqrt{2N} + m_L$, respectively. Hence, the dynamic range of $\hat{\lambda}(x)$ equals $\Delta\hat{\lambda}_{(\text{dB})} := \hat{\lambda}_{\max(\text{dB})} - \hat{\lambda}_{\min(\text{dB})} = 2\sigma_L\sqrt{2N}$. Finally, it is mentioned that $\hat{p}_\lambda(y)$ tends to $p_\lambda(y)$ as $N \to \infty$. This follows from the fact that the distribution $\hat{p}_\nu(x)$ of $\hat{\nu}(t)$ approaches the standard normal distribution $N(0, 1)$ if $N \to \infty$.

Since the joint distribution $\hat{p}_{\nu\nu'}(x_1, x_2)$ of the spatial random process $\hat{\nu}(\Delta x)$ at two different locations $x_1 = x$ and $x_2 = x + \Delta x$ is unknown, we cannot directly apply the procedure presented in Appendix 10.A for the reference model to find the exact solution to the spatial autocorrelation function $\hat{r}_{\lambda\lambda}(\Delta x)$ of $\hat{\lambda}(x)$. However, if $\hat{p}_\nu(x)$ is close to the standard normal distribution, then we may assume that $\hat{p}_{\nu\nu'}(x_1, x_2)$ is close to the joint Gaussian distribution $p_{\nu\nu'}(x_1, x_2)$ presented in (10.A.2). In this case, a good approximation of $\hat{r}_{\lambda\lambda}(\Delta x)$ is given by (10.24), if we replace there $r_{\nu\nu}(\Delta x)$ by $\hat{r}_{\nu\nu}(\Delta x)$, i.e.,

$$\hat{r}_{\lambda\lambda}(\Delta x) \approx e^{2m_0+\sigma_0^2[1+\hat{r}_{\nu\nu}(\Delta x)]}, \tag{10.34}$$

where $m_0 = m_L \ln(10)/20$, $\sigma_0 = \sigma_L \ln(10)/20$, and $\hat{r}_{\nu\nu}(\Delta x)$ is given by (10.30).

Similar arguments can be used to approximate the level-crossing rate $\hat{N}_\lambda(r)$ of the simulation model by that of the reference model described by (10.25), if we replace there the quantity γ by $\hat{\gamma} = -\ddot{\hat{r}}_{\nu\nu}(0)$. Thus,

$$\hat{N}_\lambda(r) \approx \frac{\sqrt{\hat{\gamma}}}{2\pi} e^{-\frac{(20\log_{10}(r)-m_L)^2}{2\sigma_L^2}}, \quad r \geq 0, \tag{10.35}$$

where

$$\hat{\gamma} = -\frac{d^2}{d\Delta x^2}\hat{r}_{\nu\nu}(\Delta x)\bigg|_{\Delta x=0} = 2\pi^2 \sum_{n=1}^{N} (c_n s_n)^2. \tag{10.36}$$

For reasons of completeness, the derivation of the exact solution of the level-crossing rate $\hat{N}_\lambda(r)$ has been included in Appendix 10.C [380], where the following result can be found

$$\hat{N}_\lambda(r) = \hat{p}_\nu\left(\frac{20\log_{10}(r) - m_L}{\sigma_L}\right) \int_0^\infty \frac{\left[\prod_{n=1}^{N} J_0(4\pi^2 c_n s_n y)\right]}{2(\pi y)^2}$$
$$\cdot [\cos(2\pi y \dot{\hat{\nu}}_{\max}) - 1 + 2\pi y \dot{\hat{\nu}}_{\max} \sin(2\pi y \dot{\hat{\nu}}_{\max})]\, dy, \tag{10.37}$$

with $\dot{\hat{\nu}}_{\max} = 2\pi \sum_{n=1}^{N} |s_n c_n|$. The exact solution gives an insight into the influence of the model parameters on the rate of fading caused by shadowing. In the next subsection, we will see that the model parameters c_n and s_n are always definite. Consequently, the quantity $\hat{\gamma}$ is also definite if $N < \infty$, and thus the level-crossing rate $\hat{N}_\lambda(r)$ according to (10.35) and (10.37) always exists for all finite values of N.

By analogy to (10.26), the average duration of fades $\hat{T}_{\lambda_-}(r)$ of the stochastic simulation model can be obtained from $\hat{T}_{\lambda_-}(r) = \hat{F}_{\lambda_-}(r)/\hat{N}_\lambda(r)$, where $\hat{N}_\lambda(r)$ is given by (10.37) and $\hat{F}_{\lambda_-}(r)$ denotes the cumulative distribution function of $\hat{\lambda}(x)$, which can easily be computed from the probability density function $\hat{p}_\lambda(y)$ of $\hat{\lambda}(x)$ presented in (10.33) via $\hat{F}_{\lambda_-}(r) = \int_0^r \hat{p}_\lambda(y)\, dy$.

10.2.3 *Correlation Models for Shadow Fading*

The spatial autocorrelation function $r_{\nu\nu}(\Delta x)$ of $\nu(x)$ provides an insight into how fast the shadow fading process $\lambda(x)$ changes with distance. The accurate modelling of the spatial autocorrelation function $r_{\nu\nu}(\Delta x)$ is therefore an important subject. In this subsection, altogether four spatial correlation models are presented. For each of them, it will be shown how the parameters of the simulation model can be determined.

10.2.3.1 The Gudmundson Correlation Model

The spatial correlation properties of $\nu(x)$ have been investigated by Gudmundson in [374]. Supported by his empirical studies, Gudmundson has proposed to model the correlation properties of $\nu(x)$ by the following negative exponential function

$$r_{\nu\nu}(\Delta x) = e^{-|\Delta x|/D_c}, \tag{10.38}$$

where the spatial separation $\Delta x := x_2 - x_1$ measures the distance between two locations, and $D_c > 0$ is called the *decorrelation distance*, which is an area-specific real-valued constant. Typical values for D_c and the other parameters describing the reference model are listed in Table 10.2 for suburban and urban areas. The values of the decorrelation distance D_c and the shadow standard deviation σ_L listed in this table are obtained from signal strength measurements [374]. Since the path loss has no influence on the performance of our model, we have assumed without loss of generality that the area mean m_L is zero.

In case of Gudmundson's model, the parameters of the simulation model can be computed by using the method of equal areas (MEA). The MEA has been introduced in Subsection 5.1.3 to model the classical Jakes/Clark Doppler spectrum and the Gaussian Doppler spectrum. In Appendix 10.D [379], it is shown how this procedure can be applied to the spatial autocorrelation function $r_{\nu\nu}(\Delta x)$ of the reference model described by (10.38). There, the following closed-form expressions are derived:

$$c_n = \sqrt{2/N}, \tag{10.39}$$

$$s_n = \frac{1}{2\pi D_c} \tan\left[\frac{\pi(n - 1/2)}{2N}\right], \tag{10.40}$$

where $n = 1, 2, \ldots, N$. A proper value for N can be found, e.g., by substituting (10.39) and (10.40) in (10.30), and then increasing N until the approximation $r_{\nu\nu}(\Delta x) \approx \hat{r}_{\nu\nu}(\Delta x)$ is sufficiently accurate. It should be mentioned that $\hat{r}_{\nu\nu}(\Delta x)$ approaches $r_{\nu\nu}(\Delta x)$ as N tends to infinity, i.e., $\hat{r}_{\nu\nu}(\Delta x) = r_{\nu\nu}(\Delta x)$ as $N \to \infty$. In other words, the simulation model converges to the reference model if the MEA is used and if the number of sinusoids tends to infinity.

Table 10.2 Model parameters of the reference model (from [374]).

Shadowing area	D_c	$\Delta x_{\max}$	σ_L	m_L
Suburban	503.9 m	2500 m	7.5 dB	0
Urban	8.3058 m	40 m	4.3 dB	0

Choosing $N = 25$ and substituting (10.39) and (10.40) in (10.30) results in the spatial autocorrelation functions $\hat{r}_{\nu\nu}(\Delta x)$ illustrated in Figures 10.9(a) and 10.9(b) for suburban and urban areas, respectively. To confirm the correctness of the theory, the corresponding simulation results have also been included in these figures. It can be observed that the Gaussian process $\nu(\Delta x)$ and thus the lognormal process $\lambda(\Delta x)$ decorrelate over a longer distance in suburban areas compared to urban areas. The results of the probability density function $p_\lambda(y)$ [see (10.23)] of the reference model are presented in Figure 10.10. The graphs of the density

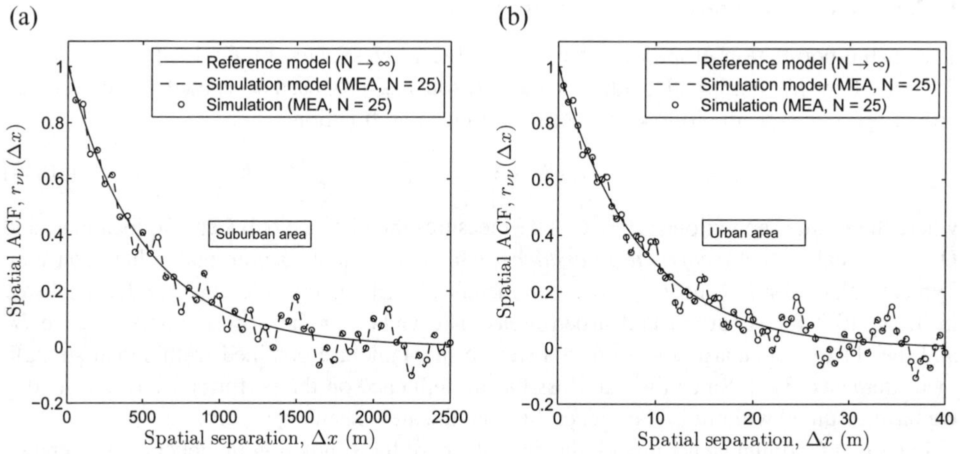

Figure 10.9 Spatial autocorrelation functions $r_{\nu\nu}(\Delta x)$ (reference model) and $\hat{r}_{\nu\nu}(\Delta x)$ (simulation model) for (a) suburban areas and (b) urban areas (Gudmundson's correlation model, MEA with $N = 25$).

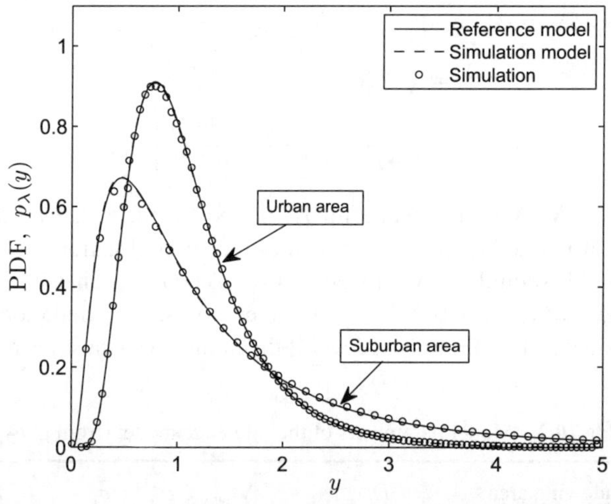

Figure 10.10 The lognormal distribution $p_\lambda(y)$ of the reference model in comparison with the probability density function $\hat{p}_\lambda(y)$ of the simulation model for suburban and urban areas ($c_n = \sqrt{2/N}$ and $N = 25$).

$\hat{p}_\lambda(y)$ describing the simulation model are also shown. These results have been obtained by evaluating the expression in (10.33) by using $c_n = \sqrt{2/N}$ with $N = 25$. The problem with Gudmundson's model is that the level-crossing rate $N_\lambda(r)$ of the reference model is infinite. This is because the second derivative of the spatial autocorrelation function $r_{\nu\nu}(\Delta x)$ in (10.38) is indefinite at the origin, which was first observed in [381]. However, the level-crossing rate $\hat{N}_\lambda(r)$ of the simulation model is limited for finite values of N, but $\hat{N}_\lambda(r)$ approaches infinity as $N \to \infty$. We consider $N = 25$ as an appropriate choice and substitute (10.39) and (10.40) in (10.37) to obtain the exact result of the simulation model's level-crossing rate $\hat{N}_\lambda(r)$ shown in Figure 10.11. This figure also illustrates the simulation results of the level-crossing rate obtained from the simulation of (10.29) and averaging over 100 trials. It can be observed that the simulation results confirm the correctness of the exact solution in (10.37). By comparing the exact solution with the approximate solution presented in (10.35), we may conclude from the graphs that the approximate solution becomes less accurate at high and low signal levels.

10.2.3.2 The Gaussian Correlation Model

The problem with Gudmundson's correlation model is that the level-crossing rate $N_\lambda(r)$ of the lognormal process $\lambda(x)$ becomes infinite. To avoid this problem, a Gaussian correlation model for modelling the shadowing effects has been proposed in [380]. In this case, the spatial correlation properties of $\nu(x)$ are described by

$$r_{\nu\nu}(\Delta x) = e^{-(\Delta x/D_c)^2}, \tag{10.41}$$

where $D_c > 0$. Since the above spatial correlation function has a definite second derivative in the origin, it follows that the level-crossing rate $N_\lambda(r)$ of the reference model exists. The

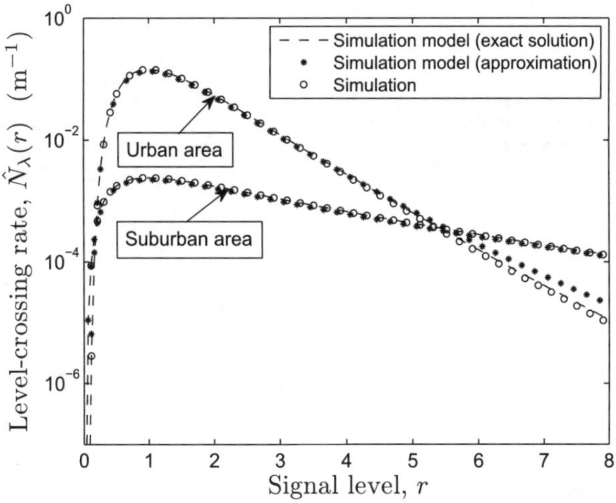

Figure 10.11 Comparison of the exact and approximate solutions for the level-crossing rate $\hat{N}_\lambda(r)$ of the simulation model for shadow fading in suburban and urban areas (Gudmundson's correlation model, MEA with $N = 25$).

quantity γ is given by $\gamma = -\ddot{r}_{\nu\nu}(0) = 2/D_c^2$, so that the behaviour of $N_\lambda(r)$ can easily be analyzed using (10.25).

For the computation of the parameters of the simulation model, we again apply the MEA. This method provides the same expression for the gains c_n as in (10.39), whereas the spatial frequencies s_n are now determined by

$$s_n = \frac{1}{\pi D_c} \operatorname{erf}^{-1}\left(\frac{n - 1/2}{N}\right), \qquad n = 1, 2, \ldots, N, \tag{10.42}$$

where $\operatorname{erf}^{-1}(\cdot)$ denotes the inverse error function.

Choosing again $N = 25$ and substituting (10.39) and (10.42) in (10.30) gives the spatial autocorrelation function $\hat{r}_{\nu\nu}(\Delta x)$ of the simulation model shown in Figures 10.12(a) and 10.12(b). For the probability density function $\hat{p}_\lambda(y)$, we obtain the same graphs as in Figure 10.10, since the gains c_n are the same as in (10.39). Figure 10.13 illustrates both the exact solution (10.37) and the approximate solution (10.35) for the level-crossing rate $\hat{N}_\lambda(r)$ of the simulation model. For comparative purposes, the level-crossing rate $N_\lambda(r)$ of the reference model using (10.25) is also shown.

10.2.3.3 The Butterworth Correlation Model

A second alternative to Gudmundson's model has been proposed in [380]. There, the power spectral density $S_{\nu\nu}(s)$ of the Gaussian process $\nu(x)$ is described by a function having the shape of the kth-order Butterworth filter

$$S_{\nu\nu}(s) = \frac{A_k}{1 + (sD_k)^{2k}}, \tag{10.43}$$

where A_k and D_k are positive constants. The constant A_k has to be chosen such that the mean power of $\nu(x)$ equals unity, i.e., $\int_{-\infty}^{\infty} S_{\nu\nu}(s)\, ds = 1$. Recall that the autocorrelation function

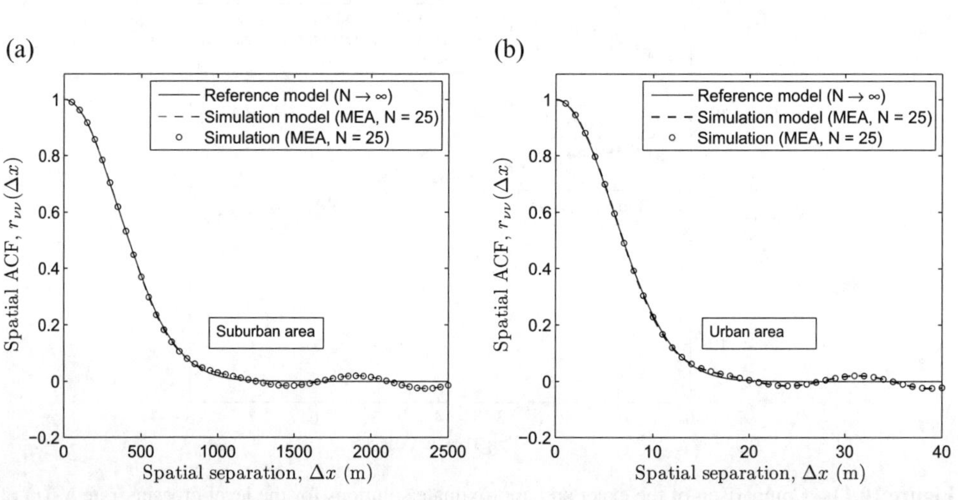

Figure 10.12 Spatial autocorrelation functions $r_{\nu\nu}(\Delta x)$ (reference model) and $\hat{r}_{\nu\nu}(\Delta x)$ (simulation model) for (a) suburban areas and (b) urban areas (Gaussian correlation model, MEA with $N = 25$).

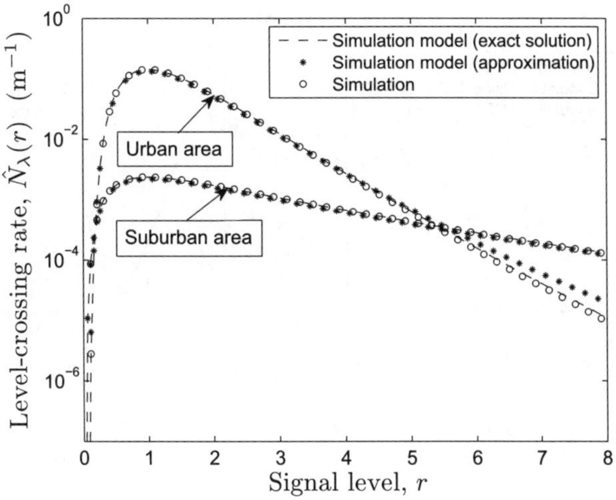

Figure 10.13 Comparison of the exact and approximate solutions for the level-crossing rate $\hat{N}_\lambda(r)$ of the simulation model for shadow fading in suburban and urban areas (Gaussian correlation model, MEA with $N = 25$).

$r_{\nu\nu}(\Delta x)$ is obtained from the inverse Fourier transform of the power spectral density $S_{\nu\nu}(s)$. From the Fourier transform relationship, it follows that the Gudmundson model is included in the Butterworth model as a special case, namely if $k = 1$, $A_1 = 2D_c$, and $D_1 = 2\pi D_c$. Let us restrict our investigations to the 2nd-order Butterworth model, i.e., $k = 2$. In this case, A_2 equals $\sqrt{2}D_2/\pi$ and the spatial correlation properties of $\nu(x)$ are described by

$$r_{\nu\nu}(\Delta x) = \sqrt{2}\, e^{-\pi\sqrt{2}|\Delta x|/D_2} \sin\left(\pi\sqrt{2}\frac{|\Delta x|}{D_2} + \frac{\pi}{4}\right). \tag{10.44}$$

Using the above result, the quantity $\gamma = -\ddot{r}_{\nu\nu}(0)$ can be expressed as $\gamma = (2\pi/D_2)^2$. Notice that the 2nd-order Butterworth model and the Gaussian model result in the same level-crossing rate $N_\lambda(r)$ if we choose $D_2 = \pi\sqrt{2}D_c$. In the following, however, our choice falls on the relation $D_2 = \pi\sqrt{2}D_c/1.2396$, which guarantees that the spatial autocorrelation functions $r_{\nu\nu}(\Delta x)$ in (10.38) and (10.44) are identical at $\Delta x = D_c$.

When using the Butterworth model of order 2, the parameters of the simulation model can also be computed by using the MEA. For the gains c_n, we obtain again the same expression as in (10.39), whereas the spatial frequencies s_n have to be determined for all $n = 1, 2, \ldots, N$ from the following equation by means of numerical integration and root-finding techniques

$$\int_0^{s_n} \frac{D_2}{1 + (sD_2)^4}\, ds - \frac{\left(n - \frac{1}{2}\right)\pi}{2\sqrt{2}N} = 0. \tag{10.45}$$

After computing the parameters c_n and s_n, the autocorrelation function $\hat{r}_{\nu\nu}(\Delta x)$ and the level-crossing rate $\hat{N}_\lambda(r)$ of the simulation model can be analyzed by using the general expressions (10.30) and (10.37), respectively. The results obtained for the spatial autocorrelation function $\hat{r}_{\nu\nu}(\Delta x)$ are illustrated in Figures 10.14(a) and 10.14(b). In these figures, the corresponding

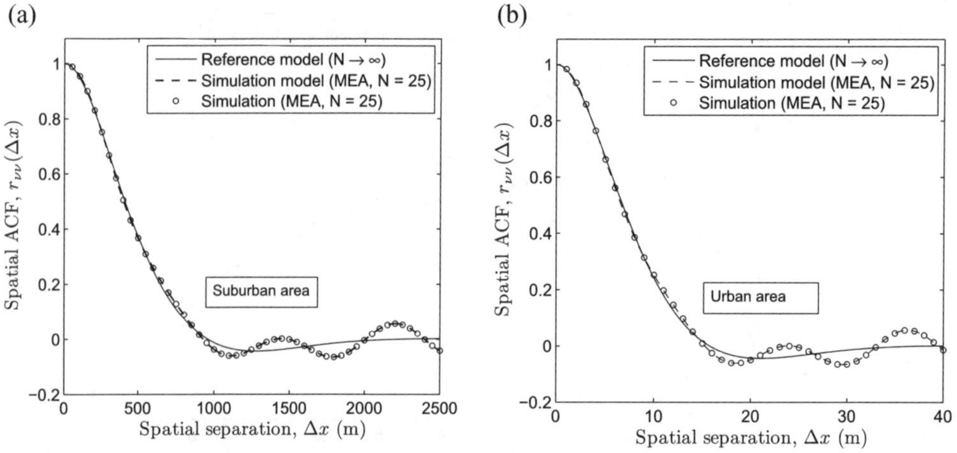

Figure 10.14 Spatial autocorrelation functions $r_{\nu\nu}(\Delta x)$ (reference model) and $\hat{r}_{\nu\nu}(\Delta x)$ (simulation model) for (a) suburban areas and (b) urban areas (2nd-order Butterworth correlation model, MEA with $N = 25$).

graphs found for the reference model are also plotted for comparative purposes. Finally, it is mentioned that the results obtained for the level-crossing rate $\hat{N}_\lambda(r)$ are very similar to those shown in Figure 10.13.

10.2.3.4 Measurement-Based Correlation Model

For the design of measurement-based correlation models, we refer to the L_p-norm method (LPNM). This method is described in detail in Subsection 5.1.6. The advantage of the LPNM versus the MEA is that this powerful procedure enables the fitting of the statistical properties of the channel simulator to real-world channels simply by replacing in the L_p-norm [see (5.74)] the spatial autocorrelation function of the theoretical reference model by the measured one. In this sense, the measured channel plays the role of the reference model. The application of this method on the present problem requires the minimization of the following L_p-norm

$$E^{(p)}_{r_{\nu\nu}} = \left\{ \frac{1}{\Delta x_{\max}} \int_0^{\Delta x_{\max}} |r^\star_{\nu\nu}(\Delta x) - \hat{r}_{\nu\nu}(\Delta x)|^p \, d(\Delta x) \right\}^{1/p}, \quad p = 1, 2, \ldots, \tag{10.46}$$

where $r^\star_{\nu\nu}(\Delta x)$ is the measured spatial autocorrelation function and $\hat{r}_{\nu\nu}(\Delta x)$ is given by (10.30). The quantity $\Delta x_{\max}$ defines the upper limit of the interval $[0, \Delta x_{\max}]$ over which the approximation $r^\star_{\nu\nu}(\Delta x) \approx \hat{r}_{\nu\nu}(\Delta x)$ is of interest. In the proposed method, both the spatial frequencies s_n and the gains c_n are the model parameters which have to be optimized numerically until the L_p-norm $E^{(p)}_{r_{\nu\nu}}$ in (10.46) reaches a local minimum. The numerical optimization can be performed, e.g., by using the Fletcher-Powell algorithm [162]. This optimization procedure requires proper initial values for c_n and s_n, which can be obtained from (10.39) and (10.40), respectively.

Choosing $N = 25$ and applying the LPNM with $p = 2$ on the measured spatial autocorrelation function $r^{\star}_{\nu\nu}(\Delta x)$ published in [374] results in the gains c_n and spatial frequencies s_n listed in Table 10.3. For suburban and urban areas, the resulting spatial autocorrelation functions $\hat{r}_{\nu\nu}(\Delta x)$ of the simulation model are shown in Figures 10.15(a) and 10.15(b), respectively. Both figures also display the graphs of the measured spatial autocorrelation function $r^{\star}_{\nu\nu}(\Delta x)$ and Gudmundson's autocorrelation function $r_{\nu\nu}(\Delta x)$ for comparative purposes. The figures reveal that the measurement-based correlation model allows a much better fitting to the measured data than the Gudmundson correlation model. The graphs of the density $\hat{p}_{\lambda}(y)$ are similar to those in Figure 10.10. The theoretical (exact and approximate) results of the simulation model's level-crossing rate $\hat{N}_{\lambda}(r)$ are shown in Figure 10.16. Finally, Figures 10.17(a) and 10.17(b) illustrate

Table 10.3 Parameters for the measurement-based shadow fading simulator for suburban and urban areas.

Index	Suburban area		Urban area	
	Parameters		Parameters	
n	c_n	s_n	c_n	s_n
1	0.2431	−0.0004	0.3579	0.0200
2	0.2939	−0.0011	0.3374	0.0087
3	0.3583	0.0008	0.3314	0.0710
4	0.2676	−0.0003	0.1461	0.0293
5	0.2023	0.0002	0.3576	0.0252
6	0.3941	0.0002	0.1633	0.0132
7	0.3224	0.0001	0.2347	0.0137
8	0.2704	0.0001	0.3101	−0.0107
9	0.2290	−0.0014	0.4280	0.0099
10	0.2049	0.0000	0.3806	0.0125
11	0.3244	−0.0004	0.1762	0.0095
12	0.2747	0.0007	0.4710	0.0056
13	0.3066	0.0004	0.3361	0.0158
14	0.1513	−0.0186	0.4440	0.0280
15	0.3854	0.0000	0.2701	0.1130
16	0.2880	−0.0005	0.1266	0.1542
17	0.1719	0.0087	0.2201	0.0897
18	0.2911	0.0007	0.3021	0.0490
19	0.3613	−0.0001	0.2036	0.1341
20	0.3067	0.0001	0.1678	0.1815
21	0.3699	0.0001	0.0916	0.1033
22	0.3652	−0.0001	0.1883	0.2008
23	0.0865	0.0050	0.2081	0.2212
24	0.0267	0.0246	0.1718	0.2398
25	0.1966	0.0101	0.1172	0.6227

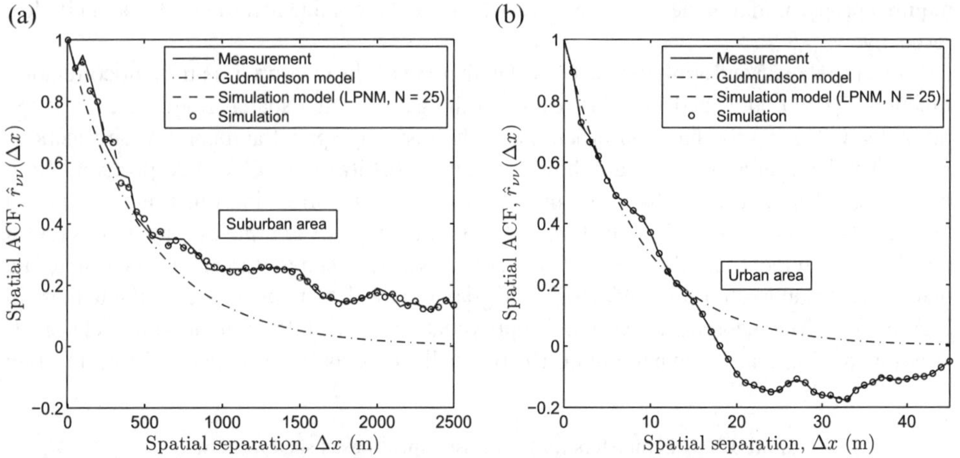

Figure 10.15 Spatial autocorrelation functions $r^{\star}_{\nu\nu}(\Delta x)$ (measured channel [374]) and $\hat{r}_{\nu\nu}(\Delta x)$ (simulation model) for (a) suburban areas and (b) urban areas (measurement-based correlation model, LPNM with $N = 25$).

the fading behaviour of shadowing in suburban and urban areas, respectively. The presented deterministic spatial lognormal processes $\tilde{\lambda}(x)$ are obtained from the respective stochastic process $\hat{\lambda}(x)$ by replacing the random phases θ_n in (10.27) by constant quantities. By using different sets of constant phases $\{\theta_1^{(i)}, \theta_2^{(i)}, \ldots, \theta_N^{(i)}\}$, one can generate different deterministic processes $\tilde{\lambda}^{(i)}(x)$ $(i = 1, 2, \ldots)$ with equivalent statistics.

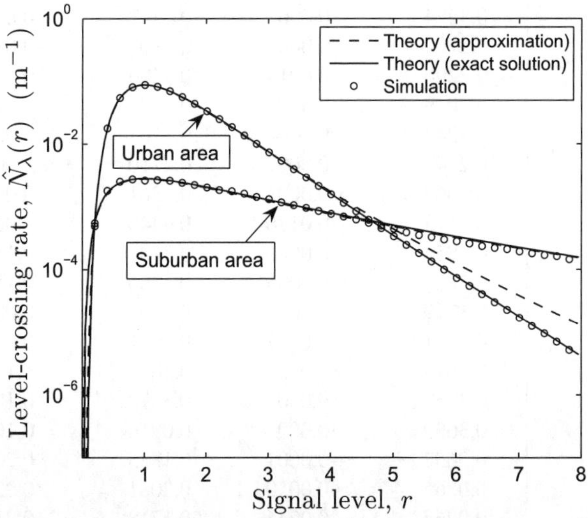

Figure 10.16 Comparison of the exact and approximate solutions for the level-crossing rate $\hat{N}_{\lambda}(r)$ of the simulation model for shadow fading in suburban and urban areas (measurement-based correlation model, LPNM with $N = 25$).

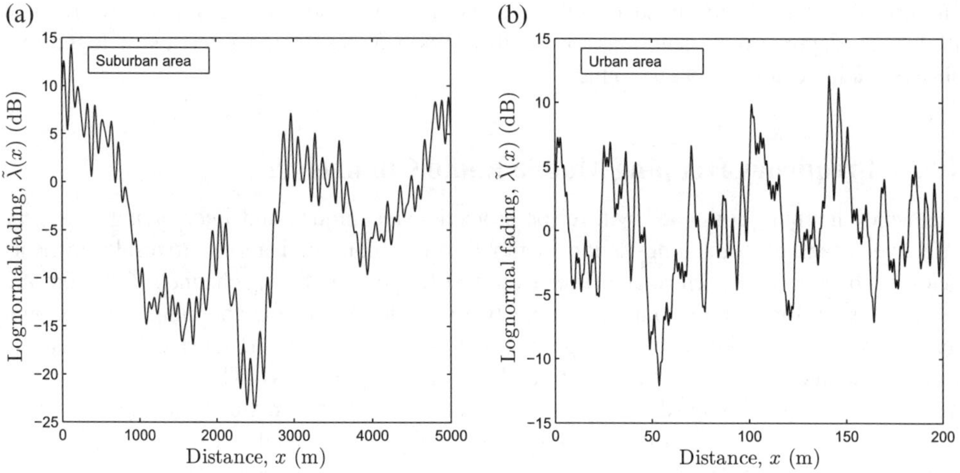

Figure 10.17 Simulation of lognormal fading (shadowing) in (a) suburban areas and (b) urban areas (measurement-based correlation model, LPNM with $N = 25$).

10.2.4 Further Reading

The problem of finding realistic parameters of lognormal processes was the subject of many measurement campaigns conducted in different areas. For example, it has been observed that the shadow standard deviation of the lognormal process varies from 5 to 12 dB in macrocells [63, 193, 194, 373] and from 6.5 to 8.2 dB in urban areas [382]. The spatial correlation properties of shadowing have been studied in a variety of articles, e.g., in [374, 381, 383–385]. The negative exponential correlation model that has been suggested by Gudmundson [374] has been improved later in [381], where it was shown that shadowing in physical channels cannot decorrelate exponentially with distance. To solve this problem, the authors of [381] have proposed a more suitable spatial correlation function resulting from a convolution of the negative exponential function (Gudmundson model) with a Gaussian function. Despite the problems pointed out in [381], Gudmundson's correlation model has been adopted by many researchers and is even part of the evaluation methodology of UMTS [386]. An empirical correlation model based on field measurements at 2.35 GHz in an urban area of a typical medium-sized city in China has been proposed in [387].

A two-dimensional sum-of-sinusoids simulation model for shadowing processes was introduced in [388]. This model has been proposed to study the influence of the shadowing effect on the performance of handover and macro-diversity algorithms assuming that the mobile is moving along a circular road rather than a straight road. The utility of the gamma distribution in modelling shadowing has been examined in [389]. There, it has been shown that the gamma distribution can be used as an alternative to the lognormal distribution for modelling shadow fading. This results from the fact that the gamma and the lognormal distributions can closely approximate each other [390]. By using the gamma distribution instead of the lognormal distribution, one obtains a closed-form expression for the composite distribution, which significantly simplifies the system analysis and design issues. Further improvements have been made in [384, 385], where the individual shadow fading correlation properties between uplink and downlink have been included in the characterization of shadowing. Finally, we refer to [93], where approximate solutions to the distribution of the fading intervals of lognormal shadow fading channels can be found.

10.3 Frequency Hopping Mobile Radio Channels

Frequency hopping combined with error protection techniques and interleaving is a very effective means of combating fading in mobile radio communications. It results in near-Gaussian bit error rate performance even over hostile Rayleigh fading channels [220]. For the Global System for Mobile Communications (GSM), (slow) frequency hopping is an optional feature for each individual cell. The principle of (slow) frequency hopping GSM is that the carrier frequency changes with every time division multiple access (TDMA) frame. By this means, frequency hopping adds frequency diversity to the mobile radio channel [1].

A channel simulator that models accurately the physical channel statistics determined by cyclic or pseudo-random hopping patterns of the carrier frequency is called a *frequency hopping channel simulator*. Such a simulator is important for the design, optimization, and test of frequency hopping mobile communication systems.

In this section, a Rayleigh fading channel simulator with excellent frequency hopping capabilities is described. The channel simulator can easily be designed, so that its statistical properties are in good conformity with those of the underlying physical mobile radio channel. The simulator is based on a finite sum of sinusoidal functions (Rice's sum-of-sinusoids) with determined gains, frequencies, and phases. It is shown, that the occurrence of a frequency hop in the physical channel model corresponds to an exchange of the phases in the simulation model, whereas the other parameters of the simulation model remain unchanged. For the parameters of the simulation model, simple equations in closed form are given. Moreover, the performance of the proposed channel simulator is evaluated with respect to its time-frequency correlation properties.

The organization of this section falls into the following five parts. Subsection 10.3.1 presents a reference model for frequency hopping channels with given time-frequency correlation properties. Starting from the reference model, a deterministic simulation model with accurate frequency hopping capabilities is derived in Subsection 10.3.2, in which also a simple design methodology is presented that results in closed-form solutions for the parametrization of the frequency hopping channel simulator. Subsection 10.3.3 analyzes the channel simulator's performance. It is shown that the time-frequency correlation properties of the simulation model are in good agreement with those of the underlying reference model. Subsection 10.3.4 illustrates by means of simulation the effect of frequency hopping on the received envelope in the GSM system operating in rural areas. This section closes with some guides to further reading, which are provided in Subsection 10.3.5.

10.3.1 The Reference Model for Frequency Hopping Channels

In the following, we derive a reference model for frequency hopping mobile radio channels in the complex baseband. To simplify matters, we restrict our investigations to frequency-nonselective channels, and we assume that no line-of-sight component is present. Under these conditions, the sum of all received scattered and reflected components subjected to two different carrier frequencies, denoted by f_0 and f_0', can be modelled in the complex baseband by the following complex-valued Gaussian random processes:

$$\mu(t) = \lim_{N\to\infty} \sum_{n=1}^{N} c_n \, e^{\,j(2\pi f_n t + \theta_n)}, \qquad \text{at} \quad f_0, \tag{10.47a}$$

$$\mu'(t) = \lim_{N\to\infty} \sum_{n=1}^{N} c_n \, e^{\,j(2\pi f_n t + \theta_n - 2\pi \tau_n' \chi)}, \qquad \text{at} \quad f_0', \tag{10.47b}$$

where the symbol $\chi = f_0' - f_0$ is called the *(carrier) frequency separation variable*, which represents here a measure for the frequency hop from f_0 to f_0'. Recall that the expression in (10.47a) has been derived in Section 3.1 [see (3.15)]. Repeating this derivation under the condition that the carrier frequency f_0 is replaced by $f_0' = f_0 + \chi$ results in (10.47b). In the two models, it is assumed that the gains c_n, Doppler frequencies f_n, phases θ_n, and propagation delays τ_n' are mutually independent zero-mean random variables. The complex Gaussian random processes $\mu(t)$ and $\mu'(t)$ have zero mean and identical variances $\mathrm{Var}\{\mu(t)\} = \mathrm{Var}\{\mu'(t)\} = 2\sigma_0^2 = \lim_{N\to\infty} \sum_{n=1}^{N} E\{c_n^2\}$. By comparing the expressions for $\mu(t)$ and $\mu'(t)$, we realize that if a frequency hop $f_0 \to f_0'$ of size $\chi = f_0' - f_0$ occurs, then the nth scattered component experiences a phase rotation of $-2\pi\tau_n'\chi$.

A suitable reference model for the received envelope at two different carrier frequencies is then given by the following Rayleigh processes:

$$\zeta(t) = |\mu(t)| = |\mu_1(t) + j\mu_2(t)|, \qquad \text{at} \quad f_0, \tag{10.48a}$$

$$\zeta'(t) = |\mu'(t)| = |\mu_1'(t) + j\mu_2'(t)|, \qquad \text{at} \quad f_0', \tag{10.48b}$$

where $\mu_i(t)$ and $\mu_i'(t)$ $(i = 1, 2)$ are both zero-mean real-valued Gaussian random processes, each with variance σ_0^2. In general, the envelopes $\zeta(t)$ and $\zeta'(t)$ are statistically correlated, but they become more and more uncorrelated if the absolute value of the frequency separation variable $|\chi|$ increases. Thus, if $|\chi|$ is sufficiently large, then the Rayleigh processes $\zeta(t)$ and $\zeta'(t)$ can be considered as statistically uncorrelated.

The correlation properties of the Rayleigh processes $\zeta(t)$ and $\zeta'(t)$ are completely determined by the correlation properties of the underlying Gaussian random processes $\mu_i(t)$ and $\mu_\lambda'(t)$ $(i, \lambda = 1, 2)$. Therefore, we can restrict our investigations to the correlation properties of $\mu_i(t)$ and $\mu_\lambda'(t)$. Of specific interest are the following autocorrelation and cross-correlation functions:

$$r_{\mu_i\mu_\lambda}(\tau) := E\{\mu_i(t)\mu_\lambda(t+\tau)\}, \tag{10.49a}$$

$$r_{\mu_i\mu_\lambda'}(\tau, \chi) := E\{\mu_i(t)\mu_\lambda'(t+\tau)\} \tag{10.49b}$$

for $i = 1, 2$ and $\lambda = 1, 2$, where the symbol $E\{\cdot\}$ denotes the statistical (ensemble) average operator. Note that the variance σ_0^2 of $\mu_i(t)$ can easily be obtained by evaluating (10.49a) at $\tau = 0$ for $i = \lambda$, i.e., $r_{\mu_i\mu_i}(0) = \sigma_0^2$. It should also be observed that the cross-correlation function $r_{\mu_i\mu_\lambda'}(\tau, \chi)$ in (10.49b) is a function of both the time separation $\tau = t_2 - t_1$ and the frequency separation $\chi = f_0' - f_0$. Equations (10.49a) and (10.49b) can be solved if the radiation pattern of the receiving antenna, the distribution of the angle-of-arrival of the incident waves, and the distribution of the propagation delays are known. In mobile communications, it is often assumed that omnidirectional receiving antennas are used, the angles-of-arrival are uniformly distributed, and the propagation delays τ' are negative exponentially distributed according to

$$p_{\tau'}(\tau') = \frac{1}{a} e^{-\tau'/a}, \quad \tau' \geq 0, \tag{10.50}$$

where a is a measure of the delay spread. With these assumptions, the autocorrelation and cross-correlation functions in (10.49a) and (10.49b) can be calculated. It is shown in Appendix 10.E that the following closed-form solutions can be obtained:

$$r_{\mu_1\mu_1}(\tau) = r_{\mu_2\mu_2}(\tau) = \sigma_0^2 J_0(2\pi f_{\max}\tau), \tag{10.51a}$$

$$r_{\mu_1\mu_2}(\tau) = r_{\mu_1'\mu_2'}(\tau) = 0, \tag{10.51b}$$

$$r_{\mu_1\mu_1'}(\tau, \chi) = r_{\mu_2\mu_2'}(\tau, \chi) = \frac{\sigma_0^2 J_0(2\pi f_{\max}\tau)}{1 + (2\pi a\chi)^2}, \tag{10.51c}$$

$$r_{\mu_1\mu_2'}(\tau, \chi) = -r_{\mu_2\mu_1'}(\tau, \chi) = -2\pi a\chi r_{\mu_1\mu_1'}(\tau, \chi). \tag{10.51d}$$

The expressions above provide the basis for the performance analysis of the frequency hopping channel simulator presented in the next subsection.

10.3.2 The Simulation Model for Frequency Hopping Channels

In this subsection, an efficient frequency hopping channel simulator is described. The simulation model takes into account that the received envelopes of frequency separated mobile radio channels (e.g., the received envelope before and after the occurrence of a frequency hop) are in general correlated. The proposed frequency hopping channel simulator is based on the concept of deterministic channel modelling introduced in Chapter 4. According to this principle, we approximate the zero-mean real-valued Gaussian random processes $\mu_i(t)$ and $\mu_i'(t)$ [see (10.48a) and (10.48b) for $i = 1, 2$] by the following superpositions of sinusoids (Rice's sum-of-sinusoids)

$$\tilde{\mu}_i(t) = \sum_{n=1}^{N_i} c_{i,n} \cos(2\pi f_{i,n} t + \theta_{i,n}), \quad \text{at} \quad f_0, \tag{10.52a}$$

$$\tilde{\mu}_i'(t) = \sum_{n=1}^{N_i} c_{i,n} \cos(2\pi f_{i,n} t + \theta_{i,n}'), \quad \text{at} \quad f_0', \tag{10.52b}$$

respectively. The number of sinusoids N_i is closely related to the realization complexity and allows to control the performance of the simulation model. The gains $c_{i,n}$ and the discrete Doppler frequencies $f_{i,n}$ are computed by using the method of exact Doppler spread (MEDS), which has been discussed in detail in Subsection 5.1.7. There it is shown that the MEDS results in the following closed-form expressions:

$$c_{i,n} = \sigma_0 \sqrt{\frac{2}{N_i}}, \tag{10.53a}$$

$$f_{i,n} = f_{\max} \sin\left[\frac{\pi}{2N_i}\left(n - \frac{1}{2}\right)\right] \tag{10.53b}$$

for $n = 1, 2, \ldots, N_i$ and $i = 1, 2$. In order to assure that $\tilde{\mu}_1(t)$ $(\tilde{\mu}'_1(t))$ and $\tilde{\mu}_2(t)$ $(\tilde{\mu}'_2(t))$ are uncorrelated, we impose on the number of sinusoids N_1 and N_2 that they are related by $N_2 := N_1 + 1$. A method for the computation of the phases $\theta_{i,n}$ and $\theta'_{i,n}$ is presented in Appendix 10.F, where the following formulas are derived

$$\theta_{i,n} = 2\pi f_0 \varphi_{i,n}, \quad \theta'_{i,n} = 2\pi(f_0 + \chi)\varphi_{i,n}, \tag{10.54a,b}$$

with

$$\varphi_{i,n} = a \ln\left(\frac{1}{1 - \frac{n-1/2}{N_i}}\right) \tag{10.55}$$

for $n = 1, 2, \ldots, N_i$ and $i = 1, 2$. The above expressions show us that a frequency hop $f_0 \to f'_0$ of size $\chi = f'_0 - f_0$ causes phase hops $\theta_{i,n} \to \theta'_{i,n}$ of sizes $2\pi\chi\varphi_{i,n} = \theta'_{i,n} - \theta_{i,n}$ $(n = 1, 2, \ldots, N_i$ and $i = 1, 2)$. Consequently, if a frequency hop $f_0 \to f'_0$ occurs, then the envelope of the channel simulator

$$\tilde{\zeta}(t) = |\tilde{\mu}_1(t) + j\tilde{\mu}_2(t)| \to \tilde{\zeta}'(t) = |\tilde{\mu}'_1(t) + j\tilde{\mu}'_2(t)| \tag{10.56}$$

jumps without any transient behaviour. Figure 10.18 shows on the left-hand side the resulting simulation model for the received envelope $\tilde{\zeta}(t)$ corresponding to the carrier frequency f_0. If we replace in this structure the phases $\theta_{i,n}$ by $\theta'_{i,n}$, then we immediately obtain the model on the right-hand side for the received envelope $\tilde{\zeta}'(t)$ corresponding to f'_0.

It is important to note that the processes $\tilde{\mu}_i(t)$ and $\tilde{\mu}'_i(t)$ are completely deterministic, because the respective model parameters $(c_{i,n}, f_{i,n}, \theta_{i,n})$ and $(c_{i,n}, f_{i,n}, \theta'_{i,n})$ are constant quantities [see (10.53) and (10.54)]. In Chapter 4, it has been shown that deterministic processes having the form in (10.52) follow very closely the statistics of zero-mean real-valued Gaussian random processes if the gains $c_{i,n}$ are given by (10.53a) and if $N_i \geq 7$. Therefore, both $\tilde{\mu}_i(t)$ and $\tilde{\mu}'_i(t)$ are deterministic Gaussian processes, and consequently $\tilde{\zeta}(t)$ and $\tilde{\zeta}'(t)$ are deterministic Rayleigh processes.

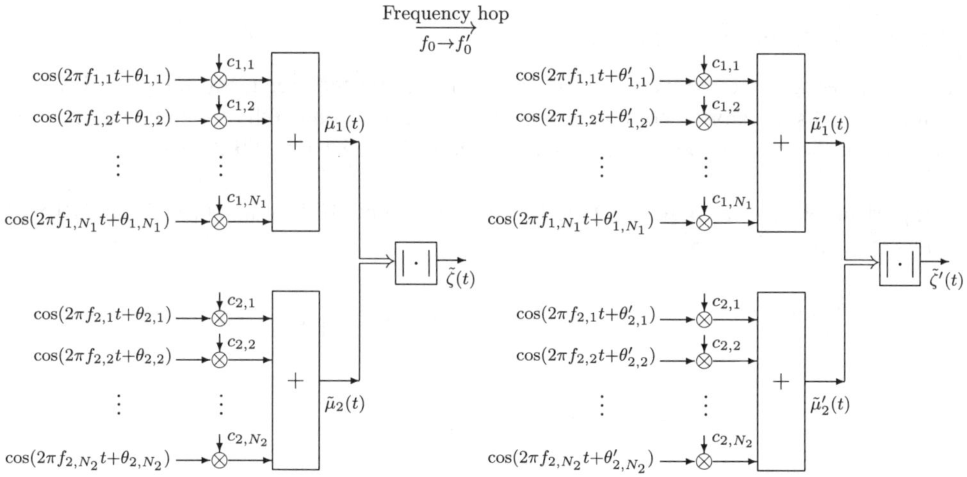

Figure 10.18 Frequency hopping Rayleigh fading channel simulator.

Owing to the deterministic nature of $\tilde{\mu}_i(t)$ and $\tilde{\mu}'_i(t)$, we have to investigate their correlation properties by using time averages instead of statistical averages. Hence, for deterministic processes, the counterparts of (10.49a) and (10.49b) are defined by

$$\tilde{r}_{\mu_i\mu_\lambda}(\tau) := \lim_{T\to\infty} \frac{1}{2T} \int_{-T}^{T} \tilde{\mu}_i(t)\, \tilde{\mu}_\lambda(t+\tau)\, dt, \tag{10.57a}$$

$$\tilde{r}_{\mu_i\mu'_\lambda}(\tau, \chi) := \lim_{T\to\infty} \frac{1}{2T} \int_{-T}^{T} \tilde{\mu}_i(t)\, \tilde{\mu}'_\lambda(t+\tau)\, dt \tag{10.57b}$$

for all $i = 1, 2$ and $\lambda = 1, 2$.

Next, we substitute (10.52a) and (10.52b) in (10.57a) and (10.57b). After solving the integrals for all relevant pairs (i, λ), we finally obtain:

$$\tilde{r}_{\mu_i\mu_i}(\tau) = \sum_{n=1}^{N_i} \frac{c_{i,n}^2}{2} \cos(2\pi f_{i,n}\tau), \tag{10.58a}$$

$$\tilde{r}_{\mu_1\mu_2}(\tau) = \tilde{r}_{\mu'_1\mu'_2}(\tau) = 0, \tag{10.58b}$$

$$\tilde{r}_{\mu_i\mu'_i}(\tau, \chi) = \sum_{n=1}^{N_i} \frac{c_{i,n}^2}{2} \cos(2\pi f_{i,n}\tau + 2\pi\varphi_{i,n}\chi), \tag{10.58c}$$

$$\tilde{r}_{\mu_1\mu'_2}(\tau, \chi) = \tilde{r}_{\mu_2\mu'_1}(\tau, \chi) = 0 \tag{10.58d}$$

for $i = 1, 2$.

Recall that the relation $N_2 = N_1 + 1$ has been imposed on the design of $\tilde{\mu}_i(t)$ and $\tilde{\mu}'_i(t)$. Therefore it follows from (10.53b) that $f_{1,n} \neq f_{2,m}$ holds for all $n = 1, 2, \ldots, N_1$ and $m = 1, 2, \ldots, N_2$. As a further consequence, we obtain the results in (10.58b) and (10.58d), which can be interpreted by saying that the deterministic processes $\tilde{\mu}_i(t)$, $\tilde{\mu}_\lambda(t)$, $\tilde{\mu}'_i(t)$, and $\tilde{\mu}'_\lambda(t)$ are mutually uncorrelated if $i \neq \lambda$.

10.3.3 Performance Analysis

In this subsection, we study the accuracy of the introduced frequency hopping Rayleigh fading channel simulator by comparing the correlation functions of the simulation model [see (10.58a)–(10.58d)] with those of the reference model [see (10.51a)–(10.51d)].

Comparison of $\tilde{r}_{\mu_i\mu_i}(\tau)$ with $r_{\mu_i\mu_i}(\tau)$: The substitution of (10.53a) and (10.53b) in (10.58a) results in the limit $N_i \to \infty$ in

$$\begin{aligned}
\lim_{N_i\to\infty} \tilde{r}_{\mu_i\mu_i}(\tau) &= \lim_{N_i\to\infty} \frac{\sigma_0^2}{N_i} \sum_{n=1}^{N_i} \cos\left\{2\pi f_{\max}\tau \sin\left[\frac{\pi}{2N_i}\left(n - \frac{1}{2}\right)\right]\right\} \\
&= \sigma_0^2 \frac{2}{\pi} \int_0^{\pi/2} \cos(2\pi f_{\max}\tau \sin z)\, dz \\
&= \sigma_0^2 J_0(2\pi f_{\max}\tau) \\
&= r_{\mu_i\mu_i}(\tau).
\end{aligned} \tag{10.59}$$

Thus, $\tilde{r}_{\mu_i\mu_i}(\tau)$ tends to $r_{\mu_i\mu_i}(\tau)$ if $N_i \to \infty$ $(i = 1, 2)$. However, even a limited number of sinusoids N_i gives excellent approximation results for $r_{\mu_i\mu_i}(\tau) \approx \tilde{r}_{\mu_i\mu_i}(\tau)$ in the interval $0 \leq \tau \leq N_i/(2f_{\max})$ (see Subsection 5.1.7), as illustrated in Figure 10.19 for $N_i = 20$.

Comparison of $\tilde{r}_{\mu_1\mu_2}(\tau)$ with $r_{\mu_1\mu_2}(\tau)$: Equation (10.51b) shows that the inphase and quadrature components of the reference model are uncorrelated. From an inspection of (10.58b), we may conclude that the simulation model has exactly the same statistical property if the MEDS is used for the computation of the model parameters $(c_{i,n}, f_{i,n})$ and if N_1 and N_2 are related by $N_2 = N_1 + 1$.

Comparison of $\tilde{r}_{\mu_i\mu_i'}(\tau, \chi)$ with $r_{\mu_i\mu_i'}(\tau, \chi)$: Substituting (10.53a), (10.53b), and (10.55) in (10.58c) leads in the limit $N_i \to \infty$ to

$$
\begin{aligned}
\lim_{N_i\to\infty} \tilde{r}_{\mu_i\mu_i'}(\tau, \chi) &= \lim_{N_i\to\infty} \frac{\sigma_0^2}{N_i} \sum_{n=1}^{N_i} \cos\left\{2\pi f_{\max}\tau \sin\left[\frac{\pi}{2N_i}\left(n - \frac{1}{2}\right)\right] + 2\pi a\chi \ln\left(\frac{1}{1 - \frac{n-1/2}{N_i}}\right)\right\} \\
&= \sigma_0^2 \int_0^1 \cos\left[2\pi f_{\max}\tau \sin\left(\frac{\pi}{2}z\right) + 2\pi a\chi \ln\left(\frac{1}{1-z}\right)\right] dz \\
&= \sigma_0^2 \int_0^\infty e^{-z} \cos\left\{2\pi f_{\max}\tau \sin\left[\frac{\pi}{2}(1 - e^{-z})\right] + 2\pi a\chi z\right\} dz. \qquad (10.60)
\end{aligned}
$$

From a comparison of (10.60) and (10.51c), we realize that $\tilde{r}_{\mu_i\mu_i'}(\tau, \chi)$ does not converge in the two-dimensional (τ, χ)-plane to $r_{\mu_i\mu_i'}(\tau, \chi)$ as $N_i \to \infty$. However, the quality of the approximation $r_{\mu_i\mu_i'}(\tau, \chi) \approx \tilde{r}_{\mu_i\mu_i'}(\tau, \chi)$ is good even for moderate values of N_i. Consider therefore Figures 10.20(a), 10.20(b), and 10.20(c), where the numerical results of (10.51c), (10.60), and (10.58c) are illustrated, respectively, for the COST 207 Rural Area profile [19].

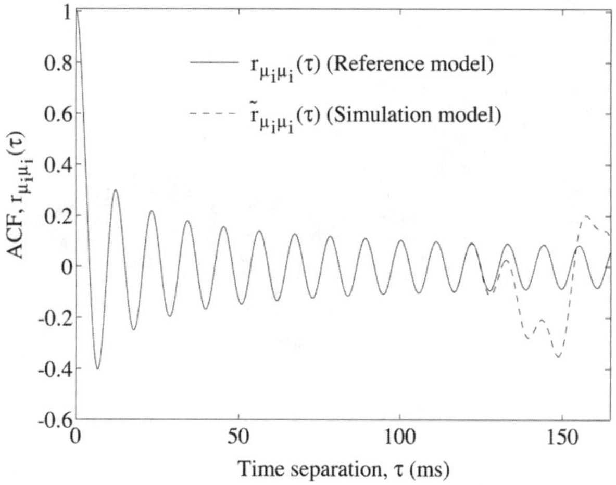

Figure 10.19 Autocorrelation function $r_{\mu_i\mu_i}(\tau)$ of the reference model in comparison with the autocorrelation function $\tilde{r}_{\mu_i\mu_i}(\tau)$ of the simulation model (MEDS with $N_i = 20$, $f_{\max} = 91$ Hz, and $\sigma_0^2 = 1$).

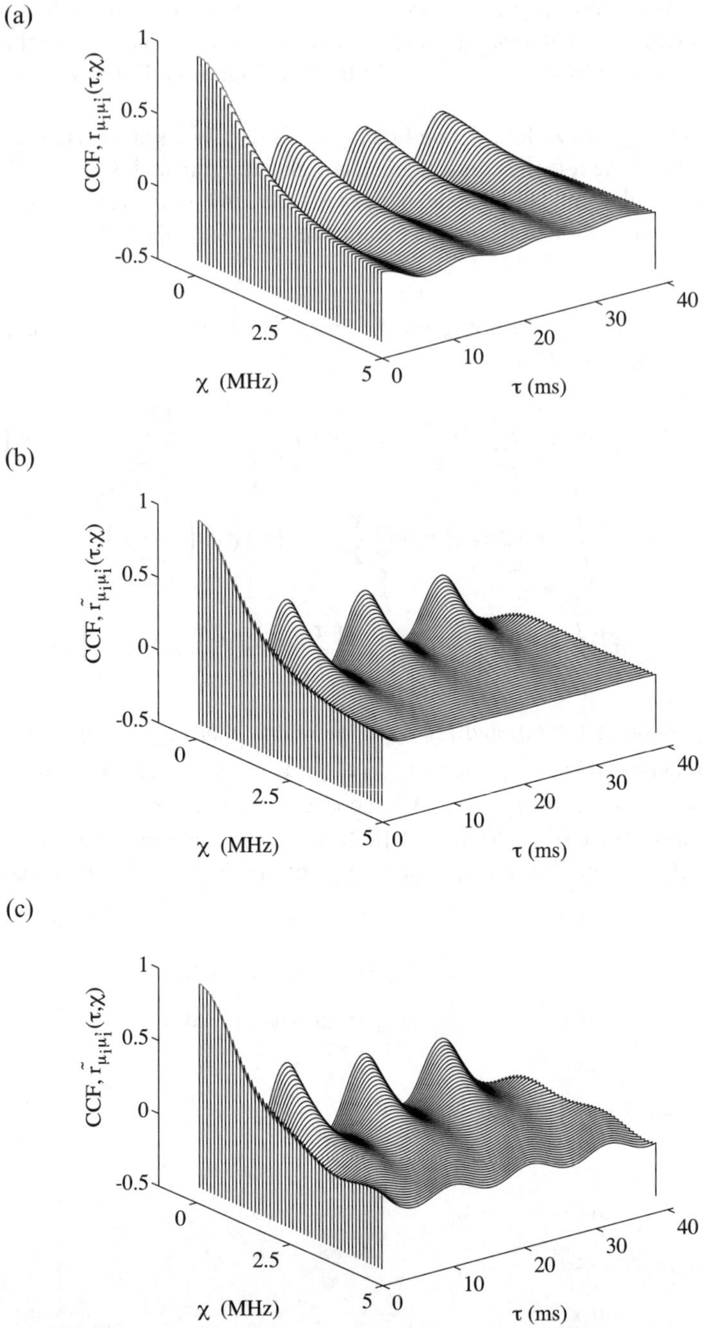

Figure 10.20 Cross-correlation functions $r_{\mu_i\mu_i'}(\tau, \chi)$ and $\tilde{r}_{\mu_i\mu_i'}(\tau, \chi)$ obtained for the COST 207 Rural Area profile ($a = 0.1086\ \mu$s, $f_{\max} = 91$ Hz, $\sigma_0^2 = 1$): (a) $r_{\mu_i\mu_i'}(\tau, \chi)$ (reference model, see (10.51c)), (b) $\tilde{r}_{\mu_i\mu_i'}(\tau, \chi)$ (simulation model for $N_i \to \infty$, see (10.60)), (c) $\tilde{r}_{\mu_i\mu_i'}(\tau, \chi)$ (simulation model for $N_i = 20$, see (10.58c)).

Next, we study the behaviour of the cross-correlation function $\tilde{r}_{\mu_i\mu_i'}(\tau,\chi)$ along the principal axes. Therefore, we evaluate $\tilde{r}_{\mu_i\mu_i'}(\tau,0)$ and $\tilde{r}_{\mu_i\mu_i'}(0,\chi)$ as $N_i \to \infty$ by using (10.60). With regard to (10.51a) and (10.51c), we can then write:

$$\lim_{N_i\to\infty} \tilde{r}_{\mu_i\mu_i'}(\tau,0) = \sigma_0^2 J_0(2\pi f_{\max}\tau) = r_{\mu_i\mu_i}(\tau), \tag{10.61a}$$

$$\lim_{N_i\to\infty} \tilde{r}_{\mu_i\mu_i'}(0,\chi) = \frac{\sigma_0^2}{1+(2\pi a\chi)^2} = r_{\mu_i\mu_i'}(0,\chi). \tag{10.61b}$$

The first result in (10.61a) is not surprising, because $\tilde{r}_{\mu_i\mu_i'}(\tau,0) = \tilde{r}_{\mu_i\mu_i}(\tau)$ holds per definition, and we can thus refer to (10.59) and especially to Figure 10.19, where $\tilde{r}_{\mu_i\mu_i}(\tau) = \tilde{r}_{\mu_i\mu_i'}(\tau,0)$ is shown for finite values of N_i. The second result (10.61b) shows that $\tilde{r}_{\mu_i\mu_i'}(0,\chi)$ tends to $r_{\mu_i\mu_i'}(0,\chi)$ as $N_i \to \infty$ $(i = 1,2)$. For moderate values of N_i, say $N_i = 20$, the cross-correlation function $\tilde{r}_{\mu_i\mu_i'}(0,\chi)$ is close to $r_{\mu_i\mu_i'}(0,\chi)$ (see Figure 10.21).

Comparison of $\tilde{r}_{\mu_1\mu_2'}(\tau,\chi)$ with $r_{\mu_1\mu_2'}(\tau,\chi)$: Due to the fact that N_2 is related to N_1 by $N_2 = N_1 + 1$, not only the cross-correlation of the deterministic processes $\tilde{\mu}_1(t)$ and $\tilde{\mu}_2(t)$ vanishes [see (10.58b)], but also the cross-correlation of $\tilde{\mu}_1(t)$ and $\tilde{\mu}_2'(t)$ [see (10.58d)]. The latter property is not in accordance with the corresponding cross-correlation function $r_{\mu_1\mu_2'}(\tau,\chi)$ of the reference model [see (10.51d)]. From the two-dimensional plot of $r_{\mu_1\mu_2'}(\tau,\chi)$, shown in Figure 10.22, we realize that $r_{\mu_1\mu_2'}(\tau,\chi)$ is small for nearly all pairs (τ,χ). Therefore, we may suppose that the effects caused by (10.58d) do not in general impair the desired statistics.

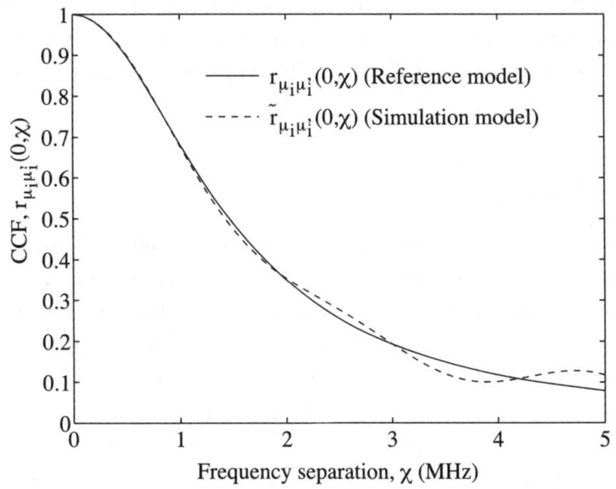

Figure 10.21 Cross-correlation function $r_{\mu_i\mu_i'}(0,\chi)$ (reference model) in comparison with $\tilde{r}_{\mu_i\mu_i'}(0,\chi)$ (simulation model, $N_i = 20$) for the COST 207 Rural Area profile ($a = 0.1086\ \mu$s, $\sigma_0^2 = 1$).

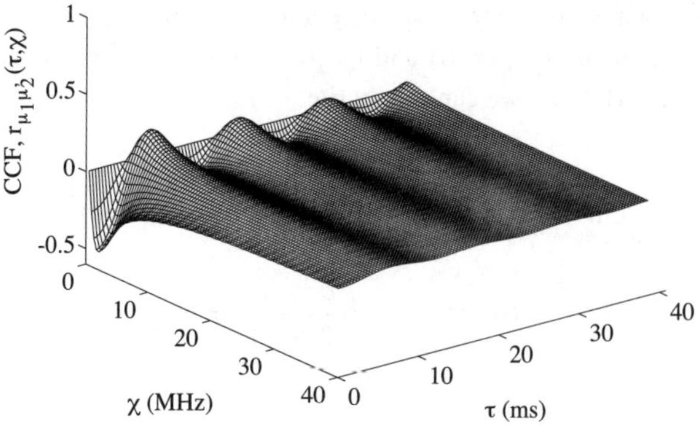

Figure 10.22 Cross-correlation function $r_{\mu_1\mu_2'}(\tau, \chi)$ (reference model) for the COST 207 Rural Area profile ($a = 0.1086\ \mu$s, $f_{max} = 91$ Hz, $\sigma_0^2 = 1$).

10.3.4 Simulation Results

The simulation results described below are based on the Rural Area profile specified for the GSM system by CEPT-COST 207 [19], where the propagation delays τ' are negative exponentially distributed according to (10.50) with parameter $a = 0.1086\ \mu$s. For the carrier frequency f_0 and the maximum Doppler frequency f_{max}, the values $f_0 = 941.2$ MHz and $f_{max} = 91$ Hz have been chosen. The number of sinusoids N_1 and N_2 was fixed to 20 and 21, respectively, and the variance σ_0^2 was normalized to unity. All remaining parameters of the simulation model shown in Figure 10.18 can be computed immediately using (10.53)–(10.55). Without frequency hopping, i.e., for $f_0 = f_0'$, and thus $\chi = f_0' - f_0 = 0$, we obtain the simulated envelope $\tilde{\zeta}(t)$ illustrated in Figure 10.23(a). Now, let us apply the principle of slow frequency hopping GSM, where the carrier frequency $f_0^{(\ell)}$ ($\ell = 1, 2, \ldots$) changes with every TDMA frame of duration 4.615 ms. Under frequency hopping conditions, the resulting output signal of the channel simulator behaves as presented in Figure 10.23(b) for the case that the frequency hops $f_0^{(\ell)} \rightarrow f_0^{(\ell+1)}$ ($\ell = 1, 2, \ldots$) follow a pseudo-random pattern.

10.3.5 Further Reading

Although in the past many channel simulators without frequency hopping capabilities have been developed and implemented in software [103, 127, 129] or hardware [128, 236, 237, 363], little information can be found about the design of frequency hopping channel simulators. This topic is almost untouched in the literature. In [239], a variable data rate frequency hopping channel model has been derived by using frequency transformation techniques and digital filter design methods. The sum-of-sinusoids method was first applied in [240] to the design of frequency hopping Rayleigh fading channel simulator with given correlation properties. A stochastic modelling and simulation approach for frequency hopping wideband fading channels was presented in [391]. This approach was later improved and extended in [392].

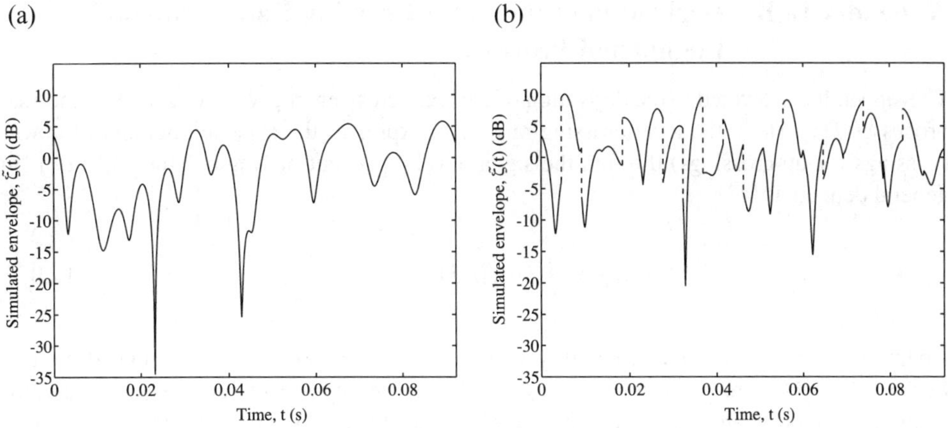

Figure 10.23 Simulation of the received envelope $\tilde{\xi}(t)$ for the rural area environment: (a) without frequency hopping and (b) with frequency hopping.

Appendix 10.A Derivation of the Spatial Autocorrelation Function of Lognormal Processes

In this appendix, we derive the spatial autocorrelation function $r_{\lambda\lambda}(\Delta x)$ of the lognormal process $\lambda(x)$ in terms of the autocorrelation function $r_{\nu\nu}(\Delta x)$ of the spatial Gaussian random process $\nu(x)$. Substituting (10.22) in the definition of the spatial autocorrelation function $r_{\lambda\lambda}(\Delta x) := E\{\lambda(x)\,\lambda(x+\Delta x)\}$ gives

$$\begin{aligned} r_{\lambda\lambda}(\Delta x) &= E\left\{10^{\frac{2m_L+\sigma_L[\nu(x)+\nu(x+\Delta x)]}{20}}\right\} \\ &= E\{e^{2m_0+\sigma_0[\nu(x)+\nu(x+\Delta x)]}\} \\ &= \int_{-\infty}^{\infty}\int_{-\infty}^{\infty} e^{\,2m_0+\sigma_0(x_1+x_2)}\, p_{\nu\nu'}(x_1,x_2)\,dx_1\,dx_2, \end{aligned} \tag{10.A.1}$$

where $m_0 = m_L \ln(10)/20$, $\sigma_0 = \sigma_L \ln(10)/20$, and

$$p_{\nu\nu'}(x_1,x_2) = \frac{e^{-\frac{x_1^2-2r_{\nu\nu}(\Delta x)x_1x_2+x_2^2}{2[1-r_{\nu\nu}^2(\Delta x)]}}}{2\pi\sqrt{1-r_{\nu\nu}^2(\Delta x)}} \tag{10.A.2}$$

denotes the joint probability density function of the spatial Gaussian random process $\nu(x)$ at two different locations $x_1 = x$ and $x_2 = x + \Delta x$. After substituting (10.A.2) in (10.A.1) and solving the double integral by using [77, Equation (7.4.32)], the spatial autocorrelation function $r_{\lambda\lambda}(\Delta x)$ can be brought into the form

$$r_{\lambda\lambda}(\Delta x) = e^{\,2m_0+\sigma_0^2[1+r_{\nu\nu}(\Delta x)]}. \tag{10.A.3}$$

Appendix 10.B Derivation of the Level-Crossing Rate of Spatial Lognormal Processes

This appendix is devoted to the derivation of the level-crossing rate $N_\lambda(r)$ of spatial lognormal processes $\lambda(x)$. Here, the level-crossing rate $N_\lambda(r)$ expresses the expected number of down-crossings (or up-crossings) through the signal level r per unit of length. This quantity is in general defined as [42]

$$N_\lambda(r) = \int_0^\infty \dot{z} p_{\lambda\dot{\lambda}}(r, \dot{z})\, d\dot{z}, \quad r \geq 0, \tag{10.B.1}$$

where $p_{\lambda\dot{\lambda}}(z, \dot{z})$ is the joint probability density function of $\lambda(x)$ and its spatial derivative $\dot{\lambda}(x) = d\lambda(x)/dx$ at the same location x. This joint probability density function can be derived from the joint probability density function of the spatial Gaussian processes $\nu(x)$ and $\dot{\nu}(x)$

$$p_{\nu\dot{\nu}}(y, \dot{y}) = \frac{e^{-y^2/2}}{\sqrt{2\pi}} \cdot \frac{e^{-\dot{y}^2/(2\gamma)}}{\sqrt{2\pi\gamma}}, \tag{10.B.2}$$

where $\gamma = -\ddot{r}_{\nu\nu}(0)$. Applying the concept of transformation of random variables by means of (2.87) and using $z = 10^{(\sigma_L x + m_L)/20}$, we obtain the relation

$$p_{\lambda\dot{\lambda}}(z, \dot{z}) = |J|^{-1} p_{\nu\dot{\nu}} \left(\frac{20 \log_{10}(z) - m_L}{\sigma_L}, \frac{20\dot{z}}{\sigma_L \ln(10)\, z} \right), \tag{10.B.3}$$

where J denotes the Jacobian determinant, which can be expressed as $J = [\sigma_L \ln(10) z/20]^2$. Substituting (10.B.2) in (10.B.3) results in the following joint probability density function

$$p_{\lambda\dot{\lambda}}(z, \dot{z}) = \left(\frac{20}{\sigma_L \ln(10)\, z} \right)^2 \frac{e^{-\frac{(20 \log_{10}(z) - m_L)^2}{2\sigma_L^2}}}{\sqrt{2\pi}} \cdot \frac{e^{-\left(\frac{20\dot{z}/z}{\sigma_L \ln(10)}\right)^2/(2\gamma)}}{\sqrt{2\pi\gamma}}. \tag{10.B.4}$$

Finally, after substituting (10.B.4) in (10.B.1), we can express the level-crossing rate $N_\lambda(r)$ of $\lambda(x)$ in closed form as

$$N_\lambda(r) = \frac{\sqrt{\gamma}}{2\pi} e^{-\frac{(20 \log_{10}(r) - m_L)^2}{2\sigma_L^2}}, \quad r \geq 0. \tag{10.B.5}$$

Appendix 10.C Derivation of the Level-Crossing Rate of Sum-of-Sinusoids Shadowing Simulators

This appendix is concerned with the derivation of the level-crossing rate $\hat{N}_\lambda(r)$ of the stochastic simulation model for spatial lognormal processes $\hat{\lambda}(x)$ defined in (10.29). We recall that the level-crossing rate $\hat{N}_\lambda(r)$ can be computed by using [16, 17]

$$\hat{N}_\lambda(r) = \int_0^\infty \dot{z} \hat{p}_{\lambda\dot{\lambda}}(r, \dot{z})\, d\dot{z}, \quad r \geq 0, \tag{10.C.1}$$

where $\hat{p}_{\lambda\dot{\lambda}}(z, \dot{z})$ denotes the joint probability density function of $\hat{\lambda}(x)$ and its spatial derivative $\dot{\hat{\lambda}}(x) = d\hat{\lambda}(x)/dx$ at the same point on the x-axis. Our starting point for finding the exact solution of $\hat{N}_{\lambda}(r)$ is the stochastic process $\hat{\nu}(x)$ in (10.28) and its spatial derivative $\dot{\hat{\nu}}(x)$, which is given by

$$\dot{\hat{\nu}}(x) = -2\pi \sum_{n=1}^{N} s_n c_n \sin(2\pi s_n x + \theta_n). \tag{10.C.2}$$

The probability density function $\hat{p}_{\dot{\nu}}(\dot{y})$ of $\dot{\hat{\nu}}(x)$ can readily be obtained from (10.31) by replacing c_n by $2\pi s_n c_n$ as

$$\hat{p}_{\dot{\nu}}(\dot{y}) = \begin{cases} 2\left[\displaystyle\int_0^{\infty} \prod_{n=1}^{N} J_0(4\pi^2 s_n c_n z)\right] \cos(2\pi \dot{y} z)\, dz, & |\dot{y}| \leq \dot{\hat{\nu}}_{\max}, \\ 0, & \text{otherwise}, \end{cases} \tag{10.C.3}$$

where $\dot{\hat{\nu}}_{\max} = 2\pi \sum_{n=1}^{N} |s_n c_n|$. The computation of the cross-correlation function of $\hat{\nu}(x)$ and $\dot{\hat{\nu}}(x)$, which is defined by $\hat{r}_{\nu\dot{\nu}}(\Delta x) := E\{\hat{\nu}(x)\, \dot{\hat{\nu}}(x + \Delta x)\}$, gives

$$\hat{r}_{\nu\dot{\nu}}(\Delta x) = -\pi \sum_{n=1}^{N} s_n c_n^2 \sin(2\pi s_n \Delta x). \tag{10.C.4}$$

This result shows that $\hat{\nu}(x)$ and $\dot{\hat{\nu}}(x)$ are in general correlated. However, for the computation of the level-crossing rate, we can restrict our investigations to the behaviour of $\hat{\nu}(x_1)$ and $\dot{\hat{\nu}}(x_2)$ at the same point in space $x = x_1 = x_2$, i.e., $\Delta x = x_2 - x_1 = 0$. From (10.C.4), we observe that $\hat{r}_{\nu\dot{\nu}}(\Delta x) = 0$ holds if $\Delta x = 0$, i.e., $\hat{\nu}(x)$ and $\dot{\hat{\nu}}(x)$ are uncorrelated at the same point in space. In case of Gaussian random processes, it follows that uncorrelatedness is equivalent to independence. Since the probability density functions $\hat{p}_{\nu}(y)$ and $\hat{p}_{\dot{\nu}}(\dot{y})$ in (10.31) and (10.C.3), respectively, are both very close to the Gaussian distribution if $N \geq 7$, we may assume that $\hat{\nu}(x)$ and $\dot{\hat{\nu}}(x)$ are also independent at the same point on the x-axis. This allows us to express the joint probability density function $\hat{p}_{\nu\dot{\nu}}(y, \dot{y})$ of $\hat{\nu}(x)$ and $\dot{\hat{\nu}}(x)$ as

$$\hat{p}_{\nu\dot{\nu}}(y, \dot{y}) = \hat{p}_{\nu}(y) \cdot \hat{p}_{\dot{\nu}}(\dot{y}). \tag{10.C.5}$$

Now, after applying the method of transformation of random variables [see (2.87], we can express the joint probability density function $\hat{p}_{\lambda\dot{\lambda}}(z, \dot{z})$ of $\hat{\lambda}(x)$ and $\dot{\hat{\lambda}}(x)$ as [380]

$$\begin{aligned} \hat{p}_{\lambda\dot{\lambda}}(z, \dot{z}) &= |J|^{-1} \hat{p}_{\nu\dot{\nu}}\left(\frac{20\log_{10}(z) - m_L}{\sigma_L}, \frac{20\dot{z}}{\sigma_L \ln(10) z}\right) \\ &= |J|^{-1} \hat{p}_{\nu}\left(\frac{20\log_{10}(z) - m_L}{\sigma_L}\right) \hat{p}_{\dot{\nu}}\left(\frac{20\dot{z}}{\sigma_L \ln(10) z}\right), \end{aligned} \tag{10.C.6}$$

where $|J|$ is the absolute value of the Jacobian determinant, which is given here by $|J| = (\sigma_L z \ln(10)/20)^2$. Finally, after substituting (10.C.6) in (10.C.1) and using (10.C.3), we find the following exact solution for the level-crossing rate $\hat{N}_\lambda(r)$ of $\hat{\lambda}(x)$

$$\hat{N}_\lambda(r) = \hat{p}_\nu\left(\frac{20\log_{10}(r) - m_L}{\sigma_L}\right)\int_0^\infty \frac{\left[\prod_{n=1}^{N} J_0(4\pi^2 c_n s_n y)\right]}{2(\pi y)^2}$$

$$\cdot\left[\cos(2\pi y\dot{\hat{\nu}}_{\max}) - 1 + 2\pi y\dot{\hat{\nu}}_{\max}\sin(2\pi y\dot{\hat{\nu}}_{\max})\right] dy, \qquad (10.C.7)$$

where $\dot{\hat{\nu}}_{\max} = 2\pi\sum_{n=1}^{N}|s_n c_n|$. We mention without proof that $\hat{N}_\lambda(r) \to N_\lambda(r)$ as $N \to \infty$ if the method of equal areas is chosen for the computation of the model parameters c_n and s_n.

Appendix 10.D Application of the Method of Equal Areas (MEA) on the Gudmundson Correlation Model

In this appendix, we show how the MEA can be applied to the Gudmundson correlation model to find closed-form solutions for the model parameters s_n and c_n. Using the MEA requires the knowledge of the power spectral density of the stochastic process $\nu(x)$. This power spectral density, denoted as $S_{\nu\nu}(s)$, is obtained by computing the Fourier transform of the spatial autocorrelation function $r_{\nu\nu}(\Delta x)$ in (10.38) as

$$S_{\nu\nu}(s) = \frac{2D_c}{1 + (2\pi s D_c)^2}. \qquad (10.D.1)$$

When applying the MEA to the present problem, we have to determine the discrete spatial frequencies s_n such that the area under the power spectral density $S_{\nu\nu}(s)$ within the interval $I_n := (s_{n-1}, s_n]$ equals $1/(2N)$, i.e.,

$$\int_{s\in I_n} S_{\nu\nu}(s)\,ds = \frac{1}{2N}, \qquad (10.D.2)$$

where $s_0 := 0$ and $s_n > 0$ for $n = 1, 2, \ldots, N$. To solve this problem, we introduce the following auxiliary function

$$F_s(s_n) := \int_{-\infty}^{s_n} S_{\nu\nu}(s)\,ds, \qquad (10.D.3)$$

which can be expressed by using (10.D.2) as

$$F_s(s_n) = \frac{1}{2} + \sum_{k=1}^{n}\int_{s_{k-1}}^{s_k} S_{\nu\nu}(s)\,ds$$

$$= \frac{1}{2} + \frac{n}{2N}, \quad s_n > 0. \qquad (10.D.4)$$

On the other hand, substituting (10.D.1) in (10.D.3) gives

$$F_s(s_n) = \frac{1}{2} + \frac{1}{\pi}\arctan(2\pi D_c\, s_n). \qquad (10.D.5)$$

Hence, after equating (10.D.4) with (10.D.5) and solving the resulting expression for s_n, we finally find the following closed-form solution

$$s_n = \frac{1}{2\pi D_c} \tan\left(\frac{\pi n}{2N}\right) \tag{10.D.6}$$

for $n = 1, 2, \ldots, N$. Notice that s_n becomes ∞ if $n = N$. To avoid this problem, we replace n by $n - 1/2$ in (10.D.6), which improves the performance of the MEA and leads to the final solution presented in (10.40). The gains c_n are determined by imposing the following condition on the simulation model

$$\int_{s \in I_n} \hat{S}_{\nu\nu}(s)\, ds = \int_{s \in I_n} S_{\nu\nu}(s)\, ds, \tag{10.D.7}$$

where $\hat{S}_{\nu\nu}(s)$ is the power spectral density of $\hat{\nu}(x)$. According to the Wiener-Khinchin theorem, the power spectral density $\hat{S}_{\nu\nu}(s)$ is related to the autocorrelation function $\hat{r}_{\nu\nu}(\Delta x)$ by the Fourier transform. This relationship enables us to find $\hat{S}_{\nu\nu}(s)$ via the Fourier transform of $\hat{r}_{\nu\nu}(\Delta x)$ [see (10.30)] in the following form

$$\hat{S}_{\nu\nu}(s) = \sum_{n=1}^{N} \frac{c_n^2}{4} \left[\delta(s + s_n) + \delta(s - s_n)\right]. \tag{10.D.8}$$

Substituting (10.D.8) in (10.D.7) and identifying the right-hand side of (10.D.7) with (10.D.2) results in

$$c_n = \sqrt{2/N} \tag{10.D.9}$$

for all $n = 1, 2, \ldots, N$.

Appendix 10.E Derivation of the Time-Frequency Cross-Correlation Function of Frequency Hopping Channels

In this appendix, we derive the time-frequency cross-correlation function of frequency hopping Rayleigh fading channels. From this function, all other relevant autocorrelation and cross-correlation functions can readily be obtained. The time-frequency cross-correlation function $r_{\mu\mu'}(\tau, \chi)$ of the complex Gaussian random processes $\mu(t)$ and $\mu'(t)$ is defined as

$$r_{\mu\mu'}(\tau, \chi) := E\{\mu^*(t)\, \mu'(t + \tau)\}, \tag{10.E.1}$$

where the statistical average has to be taken with respect to the distributions of all random variables c_n, f_n, θ_n, and τ'_n.[1] Substituting (10.47a) and (10.47b) in (10.E.1) gives

$$r_{\mu\mu'}(\tau, \chi) = E\left\{ \lim_{N\to\infty} \sum_{n=1}^{N} \sum_{m=1}^{N} c_n\, c_m\, e^{-j(\theta_n - \theta_m)}\, e^{-j2\pi(f_n - f_m)t}\, e^{j2\pi(f_m\tau - \tau'_m\chi)} \right\}. \tag{10.E.2}$$

[1] Recall that the random variables c_n, f_n, θ_n, and τ'_n are supposed to be statistically independent.

Averaging first over the uniformly distributed phases θ_n (θ_m) $\sim U(0, 2\pi]$ results in

$$r_{\mu\mu'}(\tau, \chi) = E\left\{ \lim_{N\to\infty} \sum_{n=1}^{N} c_n^2 \, e^{\,j2\pi(f_n\tau - \tau_n'\chi)} \right\}\Bigg|_{c_n, f_n, \tau_n'} . \quad (10.E.3)$$

Since the random variables c_n, f_n, and τ_n' are statistically independent, we can write

$$r_{\mu\mu'}(\tau, \chi) = \lim_{N\to\infty} \sum_{n=1}^{N} E\left\{c_n^2\right\} \cdot E\left\{e^{j2\pi f_n\tau}\right\} \cdot E\left\{e^{-j2\pi\tau_n'\chi}\right\} . \quad (10.E.4)$$

With reference to the statements given below (10.47b), the first term can be identified as

$$\lim_{N\to\infty} \sum_{n=1}^{N} E\left\{c_n^2\right\} = 2\sigma_0^2 . \quad (10.E.5)$$

Under the assumption of isotropic scattering, we can profit from the result in (3.A.19), which allows us to express the second term as

$$E\left\{e^{\,j2\pi f_n\tau}\right\} = J_0(2\pi f_{\max}\tau) . \quad (10.E.6)$$

Finally, for negative exponentially distributed propagation delays τ' [see (10.50)], the third term in (10.E.4) leads to

$$E\left\{e^{-j2\pi\tau_n'\chi}\right\} = \frac{1}{1 + j2\pi a\chi} . \quad (10.E.7)$$

Thus, after substituting (10.E.5)–(10.E.7) in (10.E.4), we obtain the following closed-form solution for the time-frequency cross-correlation function

$$r_{\mu\mu'}(\tau, \chi) = \frac{2\sigma_0^2 J_0(2\pi f_{\max}\tau)}{1 + j2\pi a\chi} . \quad (10.E.8)$$

On the analogy of the expression in (2.110), the cross-correlation function $r_{\mu\mu'}(\tau, \chi)$ of the complex Gaussian random processes $\mu(t) = \mu_1(t) + j\mu_2(t)$ and $\mu'(t) = \mu_1'(t) + j\mu_2'(t)$ can be expressed in terms of the cross-correlation functions of their inphase and quadrature components as follows

$$r_{\mu\mu'}(\tau, \chi) = r_{\mu_1\mu_1'}(\tau, \chi) + r_{\mu_2\mu_2'}(\tau, \chi) + j\left(r_{\mu_1\mu_2'}(\tau, \chi) - r_{\mu_2\mu_1'}(\tau, \chi)\right) . \quad (10.E.9)$$

A comparison of (10.E.8) and (10.E.9) provides the following results:

$$r_{\mu_1\mu_1'}(\tau,\chi) = r_{\mu_2\mu_2'}(\tau,\chi) = \frac{\sigma_0^2 J_0(2\pi f_{\max}\tau)}{1+(2\pi a\chi)^2}, \tag{10.E.10a}$$

$$r_{\mu_1\mu_2'}(\tau,\chi) = -r_{\mu_2\mu_1'}(\tau,\chi) = -2\pi a\chi\, r_{\mu_1\mu_1'}(\tau,\chi). \tag{10.E.10b}$$

The autocorrelation function $r_{\mu\mu}(\tau)$ of $\mu(t)$ follows from the time-frequency cross-correlation function $r_{\mu\mu'}(\tau,\chi)$ by setting the frequency separation variable χ to zero, i.e.,

$$\begin{aligned} r_{\mu\mu}(\tau) &:= E\left\{\mu^*(t)\,\mu(t+\tau)\right\} \\ &= r_{\mu\mu'}(\tau,0) \\ &= 2\sigma_0^2 J_0(2\pi f_{\max}\tau). \end{aligned} \tag{10.E.11}$$

Hence, comparing (10.E.11) and (2.110) implies that the following relations hold:

$$r_{\mu_1\mu_1}(\tau) = r_{\mu_2\mu_2}(\tau) = r_{\mu_1\mu_1'}(\tau,0) = \sigma_0^2 J_0(2\pi f_{\max}\tau), \tag{10.E.12a}$$

$$r_{\mu_1\mu_2}(\tau) = -r_{\mu_2\mu_1}(\tau) = r_{\mu_1\mu_2'}(\tau,0) = 0. \tag{10.E.12b}$$

Without any difficulty, it can be shown that the complex Gaussian random processes $\mu(t)$ and $\mu'(t)$ have identical autocorrelation properties, i.e., $r_{\mu\mu}(\tau) = r_{\mu'\mu'}(\tau) = r_{\mu\mu'}(\tau,0)$. Finally, it is mentioned that an alternative proof of the main results in (10.E.10a) and (10.E.10b) can be found in [13, p. 50].

Appendix 10.F Parametrization of Frequency Hopping Channel Simulators

The aim of this appendix is to derive an explicit equation for the quantities $\varphi_{i,n} \geq 0$, so that the approximation error between $\tilde{r}_{\mu_i\mu_i'}(0,\chi)$ [see (10.58c) for $\tau = 0$] and $r_{\mu_i\mu_i'}(0,\chi)$ [see (10.51c) for $\tau = 0$] is sufficiently small.

For the Fourier transforms of $r_{\mu_i\mu_i'}(0,\chi)$ and $\tilde{r}_{\mu_i\mu_i'}(0,\chi)$, we obtain by using (10.51c), (10.53a), and (10.58c) the relations

$$S_{\mu_i\mu_i'}(0,\varphi) = \frac{\sigma_0^2}{2a}\, e^{-\frac{|\varphi|}{a}}, \tag{10.F.1a}$$

$$\tilde{S}_{\mu_i\mu_i'}(0,\varphi) = \frac{\sigma_0^2}{2N_i}\sum_{n=1}^{N_i}\left[\delta(\varphi-\varphi_{i,n}) + \delta(\varphi+\varphi_{i,n})\right], \tag{10.F.1b}$$

respectively. Let us introduce the intervals $I_{i,n} = (\varphi_{i,n-1}, \varphi_{i,n}]$ with $\varphi_{i,0} = 0$ in such a way that (10.F.1a) and (10.F.1b) are related by

$$\int\limits_{\varphi\in I_{i,n}} S_{\mu_i\mu_i'}(0,\varphi)\,d\varphi = \int\limits_{\varphi\in I_{i,n}} \tilde{S}_{\mu_i\mu_i'}(0,\varphi)\,d\varphi, \tag{10.F.2}$$

for all $n = 1, 2, \ldots, N_i$ $(i = 1, 2)$. Furthermore, we define an auxiliary function of the form

$$G(\varphi_{i,n}) := \int_{-\infty}^{\varphi_{i,n}} S_{\mu_i \mu_i'}(0, \varphi)\, d\varphi, \tag{10.F.3}$$

which can be written by using (10.F.1b) and (10.F.2) as

$$\begin{aligned} G(\varphi_{i,n}) &= \frac{\sigma_0^2}{2} + \sum_{l=1}^{n} \int_{\varphi \in I_{i,l}} S_{\mu_i \mu_i'}(0, \varphi)\, d\varphi \\ &= \frac{\sigma_0^2}{2} + \sum_{l=1}^{n} \int_{\varphi \in I_{i,l}} \tilde{S}_{\mu_i \mu_i'}(0, \varphi)\, d\varphi \\ &= \frac{\sigma_0^2}{2} \left(1 + \frac{n}{N_i}\right). \end{aligned} \tag{10.F.4}$$

On the other hand, after substituting (10.F.1a) in (10.F.3), we obtain the following expression for the auxiliary function

$$G(\varphi_{i,n}) = \frac{\sigma_0^2}{2}(2 - e^{-\varphi_{i,n}/a}). \tag{10.F.5}$$

Now, the quantities $\varphi_{i,n}$ can easily be identified from (10.F.4) and (10.F.5) as

$$\varphi_{i,n} = a \ln\left(\frac{1}{1 - \frac{n}{N_i}}\right) \tag{10.F.6}$$

for all $n = 1, 2, \ldots, N_i$ $(i = 1, 2)$. Hence, the closed-form solution in (10.55) follows after replacing on the right-hand side of (10.F.6) the quantity n by $n - 1/2$. We remark (without proof) that this substitution improves the approximation $r_{\mu_i \mu_i'}(0, \chi) \approx \tilde{r}_{\mu_i \mu_i'}(0, \chi)$ considerably.

References

1. S. M. Redl, M. K. Weber, and M. W. Oliphant, *An Introduction to GSM*. Boston, MA: Artech House, 1995.
2. International Telecommunication Union, "The world in 2010: ICT facts and figures," *website*, http://www.itu.int/ITU-D/ict/material/FactsFigures2010.pdf, Oct. 2010.
3. E. Lutz, M. Werner, and A. Jahn, *Satellite Systems for Personal and Broadband Communications*. Berlin: Springer, 2000.
4. P. Chini, G. Giambene, and S. Kota, "A survey on mobile satellite systems," *Int. J. Satell. Commun.*, vol. 28, pp. 29–57, 2010.
5. J. E. Padgett, C. G. Günther, and T. Hattori, "Overview of wireless personal communications," *IEEE Communication Magazine*, vol. 33, no. 1, pp. 28–41, Jan. 1995.
6. Manifest Technology, "Satellite phones: Staying in touch when travelling in remote places," *website*, http://www.manifest-tech.com/ce_wireless/satellite_phones_beacons.htm, Oct. 2010.
7. M. Dinis and J. Fernandes, "Provision of sufficient transmission capacity for broadband mobile multimedia: A step towards 4G," *IEEE Communications Magazine*, vol. 39, no. 8, pp. 46–54, Aug. 2001.
8. ITU Study Group 5 (SG 5) Terrestrial services, *Recommendation ITU-R M.1645, Framework and overall objectives of the future development of IMT-2000 and systems beyond IMT-2000*. Geneva, Switzerland, 2003.
9. S. Sesia, I. Toufik, and M. Baker, Eds., *LTE — The UMTS Long Term Evolution: From Theory to Practice*. Chichester: John Wiley & Sons, 2009.
10. K.-C. Chen and J. R. B. de Marca, Eds., *Mobile WiMAX*. New York: Wiley-IEEE Press, 2008.
11. J. Proakis, *Digital Communications*. New York: McGraw-Hill, 4th ed., 2001.
12. A. V. Oppenheim and R. W. Schafer, *Discrete-Time Signal Processing*. Englewood Cliffs, New Jersey: Prentice-Hall, 2nd ed., 1999.
13. W. C. Jakes, Ed., *Microwave Mobile Communications*. Piscataway, NJ: IEEE Press, 2nd ed., 1994.
14. R. E. Blahut, *Theory and Practice of Error Control Codes*. Reading, Massachusetts: Addison-Wesley, 1984.
15. H. P. Kuchenbecker, "Statistische Eigenschaften von Schwund- und Verbindungsdauer beim Mobilfunk-Kanal," *FREQUENZ*, vol. 36, no. 4/5, pp. 138–144, 1982.
16. S. O. Rice, "Mathematical analysis of random noise," *Bell Syst. Tech. J.*, vol. 23, pp. 282–332, July 1944.
17. S. O. Rice, "Mathematical analysis of random noise," *Bell Syst. Tech. J.*, vol. 24, pp. 46–156, Jan. 1945.
18. P. A. Bello, "Characterization of randomly time-variant linear channels," *IEEE Trans. Commun. Syst.*, vol. 11, no. 4, pp. 360–393, Dec. 1963.
19. M. Failli, Ed., *COST 207: Digital Land Mobile Radio Communications*. Luxembourg City, Luxembourg: Commission of the European Communities, 1989, Final Report.
20. J. Medbo and P. Schramm, "Channel models for HIPERLAN/2 in different indoor scenarios," Technical Report 3ERI085B, ETSI EP, BRAN Meeting #3, March 1998.
21. B. Vucetic and J. Yuan, *Space-Time Coding*. NJ: John Wiley & Sons, 2003.
22. P. van Rooyen, M. Lötter, and D. van Wyk, *Space-Time Processing for CDMA Mobile Communications*. Boston: Kluwer Academic Publishers, 2000.
23. I. S. Gradshteyn and I. M. Ryzhik, *Table of Integrals, Series, and Products*. London, UK: Academic Press, 6th ed., 2000.

Mobile Radio Channels, Second Edition. Matthias Pätzold.

24. J. I. Marcum, "A statistical theory of target detection by pulsed radar," *IEEE Trans. Inform. Theory*, vol. 6, no. 2, pp. 59–267, April 1960.
25. D. Wolf, T. Munakata, and J. Wehhofer, "Die Verteilungsdichte der Pegelunterschreitungszeitintervalle bei Rayleigh-Fadingkanälen," *NTG-Fachberichte 84*, pp. 23–32, 1983.
26. H. Suzuki, "A statistical model for urban radio propagation," *IEEE Trans. Commun.*, vol. 25, no. 7, pp. 673–680, July 1977.
27. M. Nakagami, "The *m*-distribution: A general formula of intensity distribution of rapid fading," in *Statistical Methods in Radio Wave Propagation*, W. G. Hoffman, Ed., Oxford, UK: Pergamon Press, 1960, pp. 3–36.
28. U. Charash, "Reception through Nakagami fading multipath channels with random delays," *IEEE Trans. Commun.*, vol. 27, no. 4, pp. 657–670, April 1979.
29. M. D. Yacoub, J. E. V. Bautista, and L. G. de Rezende Guedes, "On higher order statistics of the Nakagami-*m* distribution," *IEEE Trans. Veh. Technol.*, vol. 48, no. 3, pp. 790–794, May 1999.
30. R. S. Hoyt, "Probability functions for the modulus and angle of the normal complex variate," *Bell Syst. Tech. J.*, vol. 26, pp. 318–359, April 1947.
31. J. F. Paris, "Nakagami-*q* (Hoyt) distribution function with applications," *Electron. Lett.*, vol. 45, no. 4, pp. 210–211, Feb. 2009.
32. G. N. Tavares, "Efficient computation of Hoyt cumulative distribution function," *Electron. Lett.*, vol. 46, no. 7, pp. 537–539, April 2010.
33. N. Youssef, C.-X. Wang, and M. Pätzold, "A study on the second order statistics of Nakagami-Hoyt mobile fading channels," *IEEE Trans. Veh. Technol.*, vol. 54, no. 4, pp. 1259–1265, July 2005.
34. W. Weibull, "A statistical distribution function of wide applicability," *ASME Journal of Applied Mechanics*, vol. 18, no. 3, pp. 293–297, Sep. 1951.
35. R. B. Abernethy, *The New Weibull Handbook*. New York: Barringer & Associates, 2000.
36. H. Hashemi, "The indoor radio propagation channel," *IEEE Proc.*, vol. 81, no. 7, pp. 943–968, July 1993.
37. F. Babich and G. Lombardi, "Statistical analysis and characterization of the indoor propagation channel," *IEEE Trans. Commun.*, vol. 48, no. 3, pp. 455–464, March 2000.
38. N. H. Shepherd, "Radio wave loss deviation and shadow loss at 900 MHz," *IEEE Trans. Veh. Technol.*, vol. 26, no. 4, pp. 309–313, Nov. 1977.
39. N. S. Adawi et al., "Coverage prediction for mobile radio systems operating in the 800/900 MHz frequency range," *IEEE Trans. Veh. Technol.*, vol. 37, no. 1, pp. 3–72, Feb. 1988.
40. M. A. Taneda, J. Takada, and K. Araki, "A new approach to fading: Weibull model," in *Proc. IEEE Int. Symp. Personal, Indoor, Mobile Radio Communications, PIMRC 1999*, Osaka, Japan, Sep. 1999, pp. 711–715.
41. A. Papoulis and S. U. Pillai, *Probability, Random Variables and Stochastic Processes*. New York: McGraw-Hill, 4th ed., 2002.
42. S. O. Rice, "Statistical properties of a sine wave plus random noise," *Bell Syst. Tech. J.*, vol. 27, pp. 109–157, Jan. 1948.
43. L. V. Ahlfors, *Complex Analysis*. New York: McGraw-Hill, 3rd ed., 1979.
44. R. L. Allen and D. W. Mills, *Signal Analysis: Time, Frequency, Scale and Structure*. New York: Wiley-IEEE Press, 2004.
45. A. Fettweis, *Elemente nachrichtentechnischer Systeme*. Stuttgart: Teubner, 2nd ed., 1996, reprinted by Schlembach, Wilburgstetten, Germany, 2004.
46. P. Peebles, *Probability, Random Variables, and Random Signal Principles*. New York: McGraw-Hill Science, 4th ed., 2000.
47. C. W. Therrien, *Discrete Random Signals and Statistical Signal Processing*. Englewood Cliffs, New Jersey: Prentice-Hall, 1992.
48. A. Leon-Garcia, *Probability, Statistics, and Random Processes for Electrical Engineering*. Englewood Cliffs, New Jersey: Prentice-Hall, 3rd ed., 2008.
49. J. A. Gubner, *Probability and Random Processes for Electrical and Computer Engineers*. Cambridge, UK: Cambridge University Press, 2006.
50. J. Dupraz, *Probability, Signals, Noise*. London: North Oxford Academic Publishers, 1986.
51. K. S. Shanmugan and A. M. Breipohl, *Random Signals: Detection, Estimation, and Data Analysis*. New York: John Wiley & Sons, 1988.
52. D. Middleton, *An Introduction to Statistical Communication Theory*. New York: McGraw-Hill, 1960.
53. W. B. Davenport, *Probability and Random Processes*. New York: McGraw-Hill, 1970.

54. W. B. Davenport and W. L. Root, *An Introduction to the Theory of Random Signals and Noise*. Piscataway, NJ: IEEE Press, 1987.
55. V. K. Rohatgi and A. K. Md. E. Saleh, *An Introduction to Probability and Statistics*. New York: Wiley Inter-Science, 2nd ed., 2000.
56. A. Papoulis, *Signal Analysis*. New York: McGraw-Hill, 1977.
57. L. R. Rabiner and B. Gold, *Theory and Applications of Digital Signal Processing*. Englewood Cliffs, New Jersey: Prentice-Hall, 1975.
58. T. Kailath, *Linear Systems*. Englewood Cliffs, New Jersey: Prentice-Hall, 1980.
59. V. K. Ingle and J. G. Proakis, *Digital Signal Processing Using MATLAB*. New York, USA: Cengage-Engineering, 2nd ed., 2007.
60. S. D. Stearns, *Digital Signal Processing with Examples in MATLAB*. Boca Raton, FL, USA: CRC Press, 1st ed., 2002.
61. W. R. Young, "Comparison of mobile radio transmission at 150, 450, 900, and 3700 MHz," *Bell Syst. Tech. J.*, vol. 31, pp. 1068–1085, Nov. 1952.
62. H. W. Nylund, "Characteristics of small-area signal fading on mobile circuits in the 150 MHz band," *IEEE Trans. Veh. Technol.*, vol. 17, pp. 24–30, Oct. 1968.
63. Y. Okumura, E. Ohmori, T. Kawano, and K. Fukuda, "Field strength and its variability in VHF and UHF land mobile radio services," *Rev. Elec. Commun. Lab.*, vol. 16, pp. 825–873, Sep./Oct. 1968.
64. R. W. Lorenz, "Zeit- und Frequenzabhängigkeit der Übertragungsfunktion eines Funkkanals bei Mehrwegeausbreitung mit besonderer Berücksichtigung des Mobilfunkkanals," *Der Fernmelde-Ingenieur*, vol. 39, no. 4, Verlag für Wissenschaft und Leben Georg Heidecker, April 1985.
65. J. D. Parsons, *The Mobile Radio Propagation Channel*. Chichester, England: John Wiley & Sons, 2nd ed., 2001.
66. S.A. Fechtel, *Verfahren und Algorithmen der robusten Synchronisation für die Datenübertragung über dispersive Schwundkanäle*, PhD thesis, RWTH Aachen University, Aachen, Germany, 1993.
67. R. H. Clarke, "A statistical theory of mobile-radio reception," *Bell Syst. Tech. Journal*, vol. 47, pp. 957–1000, July/Aug. 1968.
68. T. Aulin, "A modified model for the fading signal at the mobile radio channel," *IEEE Trans. Veh. Technol.*, vol. 28, no. 3, pp. 182–203, Aug. 1979.
69. M. J. Gans, "A power-spectral theory of propagation in the mobile-radio environment," *IEEE Trans. Veh. Technol.*, vol. 21, no. 1, pp. 27–38, Feb. 1972.
70. A. Krantzik and D. Wolf, "Statistische Eigenschaften von Fadingprozessen zur Beschreibung eines Landmobilfunkkanals," *FREQUENZ*, vol. 44, no. 6, pp. 174–182, June 1990.
71. P. A. Bello, "Aeronautical channel characterization," *IEEE Trans. Commun.*, vol. 21, pp. 548–563, May 1973.
72. A. Neul, J. Hagenauer, W. Papke, F. Dolainsky, and F. Edbauer, "Aeronautical channel characterization based on measurement flights," in *Proc. IEEE Global Communications Conference, IEEE GLOBECOM 1987*, Tokyo, Japan, Nov. 1987, pp. 1654–1659.
73. A. Neul, *Modulation und Codierung im aeronautischen Satellitenkanal*, PhD thesis, University of the Federal Armed Forces Munich, Munich, Germany, Sep. 1989.
74. D. C. Cox, "910 MHz urban mobile radio propagation: Multipath characteristics in New York City," *IEEE Trans. Veh. Technol.*, vol. 22, no. 4, pp. 104–110, Nov. 1973.
75. M. Pätzold, U. Killat, and F. Laue, "An extended Suzuki model for land mobile satellite channels and its statistical properties," *IEEE Trans. Veh. Technol.*, vol. 47, no. 2, pp. 617–630, May 1998.
76. S. O. Rice, "Distribution of the duration of fades in radio transmission: Gaussian noise model," *Bell Syst. Tech. J.*, vol. 37, pp. 581–635, May 1958.
77. M. Abramowitz and I. A. Stegun, *Pocketbook of Mathematical Functions*. Frankfurt/Main: Verlag Harri Deutsch, 1984.
78. M. Pätzold, U. Killat, F. Laue, and Y. Li, "On the problems of Monte Carlo method based simulation models for mobile radio channels," in *Proc. IEEE 4th Int. Symp. on Spread Spectrum Techniques&Applications, ISSSTA'96*, Mayence, Germany, Sep. 1996, pp. 1214–1220.
79. A. S. Akki and F. Haber, "A statistical model of mobile-to-mobile land communication channel," *IEEE Trans. Veh. Technol.*, vol. 35, no. 1, pp. 2–7, Feb. 1986.
80. A. S. Akki, "Statistical properties of mobile-to-mobile land communication channels," *IEEE Trans. Veh. Technol.*, vol. 43, no. 4, pp. 826–831, Nov. 1994.
81. J. A. McFadden, "The axis-crossing intervals of random functions I," *IRE Trans. Inform. Theory*, vol. 2, pp. 146–150, 1956.

82. J. A. McFadden, "The axis-crossing intervals of random functions II," *IRE Trans. Inform. Theory*, vol. 4, pp. 14–24, 1958.
83. M. S. Longuet-Higgins, "The distribution of intervals between zeros of a stationary random function," *Phil. Trans. Royal. Soc.*, vol. A 254, pp. 557–599, 1962.
84. A. J. Rainal, "Axis-crossing intervals of Rayleigh processes," *Bell Syst. Tech. J.*, vol. 44, pp. 1219–1224, 1965.
85. H. Brehm, *Ein- und zweidimensionale Verteilungsdichten von Nulldurchgangsabständen stochastischer Signale*, PhD thesis, University Frankfurt/Main, Frankfurt, Germany, June 1970.
86. H. Brehm, *Sphärisch invariante stochastische Prozesse*, PhD thesis, University Frankfurt/Main, Frankfurt, Germany, 1978.
87. T. Munakata and D. Wolf, "A novel approach to the level-crossing problem of random processes," in *Proc. of the 1982 IEEE Int. Symp. on Inf. Theory*, Les Arcs, France, 1982, vol. IEEE-Cat. 82 CH 1767-3 IT, pp. 149–150.
88. T. Munakata and D. Wolf, "On the distribution of the level-crossing time-intervals of random processes," in *Proc. of the 7th Int. Conf. on Noise in Physical Systems*, Montpelier, USA, M. Savelli, G. Lecoy, and J. P. Nougier, Eds., North-Holland Publ. Co., Amsterdam, The Netherlands, 1983, pp. 49–52.
89. D. Wolf, T. Munakata, and J. Wehhofer, "Statistical properties of Rice fading processes," in *Signal Processing II: Theories and Applications, Proc. EUSIPCO'83 Second European Signal Processing Conference*, H. W. Schüßler, Erlangen, Ed., Elsevier Science Publishers B.V. (North-Holland), 1983, pp. 17–20.
90. T. Munakata, *Mehr-Zustände-Modelle zur Beschreibung des Pegelkreuzungsverhaltens stationärer stochastischer Prozesse*, PhD thesis, University Frankfurt/Main, Frankfurt, Germany, March 1986.
91. R. Tetzlaff, J. Wehhofer, and D. Wolf, "Simulation and analysis of Rayleigh fading processes," in *Proc. of the 9th Int. Conf. on Noise in Physical Systems*, Montreal, Canada, 1987, pp. 113–116.
92. H. Brehm, "Pegelkreuzungen bei verallgemeinerten Gauß-Prozessen," *Archiv Elektr. Übertr.*, vol. 43, no. 5, pp. 271–277, 1989.
93. N. Youssef, M. Pätzold, and K. Yang, "Approximate theoretical results for the distribution of lognormal shadow fading intervals," in *Proc. 4th IEEE International Symposium on Wireless Communication Systems, ISWCS 2007*, Trondheim, Norway, Oct. 2007, pp. 204–208.
94. F. Ramos-Alarcon, V. Kontorovich, and M. Lara, "On the level crossing duration distributions of Nakagami processes," *IEEE Trans. Commun.*, vol. 57, no. 2, pp. 542–552, Feb. 2009.
95. G. E. Uhlenbeck, "Theory of random processes," *Radiation Laboratory*, Technical Report, no. 454, Oct. 1943.
96. M. Pätzold, U. Killat, F. Laue, and Y. Li, "On the statistical properties of deterministic simulation models for mobile fading channels," *IEEE Trans. Veh. Technol.*, vol. 47, no. 1, pp. 254–269, Feb. 1998.
97. Y. R. Zheng and C. Xiao, "Simulation models with correct statistical properties for Rayleigh fading channels," *IEEE Trans. Commun.*, vol. 51, no. 6, pp. 920–928, June 2003.
98. T.-M. Wu and S.-Y. Tzeng, "Sum-of-sinusoids-based simulator for Nakagami-*m* fading channels," in *Proc. 58th IEEE Semiannual Veh. Techn. Conf., VTC'03-Fall*, Orlando, Florida, USA, Oct. 2003, vol. 1, pp. 158–162.
99. P. Höher, "A statistical discrete-time model for the WSSUS multipath channel," *IEEE Trans. Veh. Technol.*, vol. 41, no. 4, pp. 461–468, Nov. 1992.
100. H. Schulze, "Stochastische Modelle und digitale Simulation von Mobilfunkkanälen," in *U.R.S.I/ITG Conf. in Kleinheubach 1988, Germany (FR)*, Proc. Kleinheubacher Reports of the German PTT, Darmstadt, Germany, 1989, vol. 32, pp. 473–483.
101. K.-W. Yip and T.-S. Ng, "Efficient simulation of digital transmission over WSSUS channels," *IEEE Trans. Commun.*, vol. 43, no. 12, pp. 2907–2913, Dec. 1995.
102. P. M. Crespo and J. Jiménez, "Computer simulation of radio channels using a harmonic decomposition technique," *IEEE Trans. Veh. Technol.*, vol. 44, no. 3, pp. 414–419, Aug. 1995.
103. M. Pätzold, U. Killat, Y. Shi, and F. Laue, "A deterministic method for the derivation of a discrete WSSUS multipath fading channel model," *European Trans. Telecommun. and Related Technologies (ETT)*, vol. 7, no. 2, pp. 165–175, March/April 1996.
104. J.-K. Han, J.-G. Yook, and H.-K. Park, "A deterministic channel simulation model for spatially correlated Rayleigh fading," *IEEE Communications Letters*, vol. 6, no. 2, pp. 58–60, Feb. 2002.
105. M. Pätzold and B. O. Hogstad, "A space-time channel simulator for MIMO channels based on the geometrical one-ring scattering model," *Wireless Communications and Mobile Computing, Special Issue on Multiple-Input Multiple-Output (MIMO) Communications*, vol. 4, no. 7, pp. 727–737, Nov. 2004.
106. M. Pätzold and N. Youssef, "Modelling and simulation of direction-selective and frequency-selective mobile radio channels," *International Journal of Electronics and Communications*, vol. AEÜ-55, no. 6, pp. 433–442, Nov. 2001.

107. M. Pätzold, "System function and characteristic quantities of spatial deterministic Gaussian uncorrelated scattering processes," in *Proc. 57th IEEE Semiannual Veh. Technol. Conf., VTC 2003-Spring*, Jeju, Korea, April 2003, pp. 256–261.
108. G. Wu, Y. Tang, and S. Li, "Fading characteristics and capacity of a deterministic downlink MIMO fading channel simulation model with non-isotropic scattering," in *Proc. 14th IEEE Int. Symp. on Personal, Indoor and Mobile Radio Communications, PIMRC'03*, Beijing, China, Sep. 2003, pp. 1531–1535.
109. C. Xiao, J. Wu, S.-Y. Leong, Y. R. Zheng, and K. B. Letaief, "A discrete-time model for spatio-temporally correlated MIMO WSSUS multipath channels," in *Proc. 2003 IEEE Wireless Communications and Networking Conference, WCNC'03*, New Orleans, Louisiana, USA, March 2003, vol. 1, pp. 354–358.
110. C.-X. Wang and M. Pätzold, "Efficient simulation of multiple correlated Rayleigh fading channels," in *Proc. 14th IEEE Int. Symp. on Personal, Indoor and Mobile Radio Communications, PIMRC 2003*, Beijing, China, Sep. 2003, pp. 1526–1530.
111. Y. R. Zheng and C. Xiao, "Improved models for the generation of multiple uncorrelated Rayleigh fading waveforms," *IEEE Communications Letters*, vol. 6, no. 6, pp. 256–258, June 2002.
112. C. S. Patel, G. L. Stüber, and T. G. Pratt, "Comparative analysis of statistical models for the simulation of Rayleigh faded cellular channels," *IEEE Trans. Commun.*, vol. 53, no. 6, pp. 1017–1026, June 2005.
113. A. G. Zajić and G. L. Stüber, "Efficient simulation of Rayleigh fading with enhanced de-correlation properties," *IEEE Trans. Wireless Commun.*, vol. 5, no. 7, pp. 1866–1875, July 2006.
114. M. Pätzold, C.-X. Wang, and B. Hogstad, "Two new sum-of-sinusoids-based methods for the efficient generation of multiple uncorrelated Rayleigh fading waveforms," *IEEE Trans. Wireless Commun.*, vol. 8, no. 6, pp. 3122–3131, June 2009.
115. G. J. Foschini and M. J. Gans, "On limits of wireless communications in a fading environment when using multiple antennas," *Wireless Pers. Commun.*, vol. 6, pp. 311–335, March 1998.
116. A. J. Paulraj and C. B. Papadias, "Space-time processing for wireless communications," *IEEE Signal Processing Magazine*, vol. 14, no. 5, pp. 49–83, Nov. 1997.
117. A. J. Paulraj, D. Gore, R. U. Nabar, and H. Bölcskei, "An overview of MIMO communications — A key to gigabit wireless," *Proc. IEEE*, vol. 92, no. 2, pp. 198–218, Feb. 2004.
118. G. L. Stüber, *Principles of Mobile Communications*. Boston, MA: Kluwer Academic Publishers, 2nd ed., 2001.
119. S. M. Alamouti, "A simple transmit diversity technique for wireless communications," *IEEE J. Select. Areas Commun.*, vol. 16, no. 8, pp. 1451–1458, Oct. 1998.
120. M. Pätzold, R. García, and F. Laue, "Design of high-speed simulation models for mobile fading channels by using table look-up techniques," *IEEE Trans. Veh. Technol.*, vol. 49, no. 4, pp. 1178–1190, July 2000.
121. M. Pätzold, "Perfect channel modeling and simulation of measured wide-band mobile radio channels," in *Proc. 1st International Conference on 3G Mobile Communication Technologies*, IEE 3G2000, London, UK, March 2000, pp. 288–293.
122. M. Pätzold and Q. Yao, "Perfect modeling and simulation of measured spatio-temporal wireless channels," in *Proc. 5th Int. Symp. on Wireless Personal Multimedia Communications, WPMC'02*, Honolulu, Hawaii, Oct. 2002, pp. 563–567.
123. B. Talha and M. Pätzold, "A geometrical three-ring-based model for MIMO mobile-to-mobile fading channels in cooperative networks," *EURASIP Journal on Advances in Signal Processing*, Special Issue on "Cooperative MIMO Multicell Networks," vol. 2011, Article ID 892871, 13 pages, doi:10.1155/2011/892871, 2011.
124. B. Talha, *Mobile-to-Mobile Cooperative Communication Systems: Channel Modeling and System Performance Analysis*, PhD thesis, University of Agder, Kristiansand, 2010.
125. G. Rafiq and M. Pätzold, "The impact of shadowing and the severity of fading on the first and second order statistics of the capacity of OSTBC MIMO Nakagami-lognormal channels," *Wireless Personal Communications*, doi:10.1007/s11277-011-0275-x, 2011.
126. G. Rafiq, *Statistical Analysis of the Capacity of Mobile Radio Channels*, PhD thesis, University of Agder, Kristiansand, 2011.
127. H. Brehm, W. Stammler, and M. Werner, "Design of a highly flexible digital simulator for narrowband fading channels," in *Signal Processing III: Theories and Applications*, I. T. Young, Ed., Amsterdam, The Netherlands: Elsevier Science Publishers B. V. (North-Holland), EURASIP, Sep. 1986, pp. 1113–1116.
128. H. W. Schüßler, J. Thielecke, K. Preuss, W. Edler, and M. Gerken, "A digital frequency-selective fading simulator," *FREQUENZ*, vol. 43, no. 2, pp. 47–55, 1989.
129. S. A. Fechtel, "A novel approach to modeling and efficient simulation of frequency-selective fading radio channels," *IEEE J. Select. Areas Commun.*, vol. 11, no. 3, pp. 422–431, April 1993.

130. U. Martin, "Modeling the mobile radio channel by echo estimation," *FREQUENZ*, vol. 48, no. 9/10, pp. 198–212, Sep./Oct. 1994.
131. D. I. Laurenson and G. J. R. Povey, "Channel modelling for a predictive rake receiver system," in *Proc. 5th IEEE Int. Symp. Personal, Indoor and Mobile Radio Commun., PIMRC'94*, The Hague, The Netherlands, Sep. 1994, pp. 715–719.
132. W. R. Bennett, "Distribution of the sum of randomly phased components," *Quart. Appl. Math.*, vol. 5, pp. 385–393, May 1948.
133. M. Pätzold and B. O. Hogstad, "Classes of sum-of-sinusoids Rayleigh fading channel simulators and their stationary and ergodic properties — Part I," *WSEAS Transactions on Mathematics*, vol. 5, no. 1, pp. 222–230, Feb. 2006.
134. M. Pätzold and B. O. Hogstad, "Classes of sum-of-sinusoids Rayleigh fading channel simulators and their stationary and ergodic properties — Part II," *WSEAS Transactions on Mathematics*, vol. 4, no. 4, pp. 441–449, Oct. 2005.
135. M. Pätzold and B. Talha, "On the statistical properties of sum-of-cisoids-based mobile radio channel simulators," in *Proc. 10th International Symposium on Wireless Personal Multimedia Communications, WPMC 2007*, Jaipur, India, Dec. 2007, pp. 394–400.
136. S. Primak, V. Kontorovich, and V. Lyandres, *Stochastic Methods and their Applications to Communications — Stochastic Differential Equations Approach*. Chichester, England: John Wiley & Sons, 2004.
137. C. A. Gutiérrez, *Channel Simulation Models for Mobile Broadband Communication Systems*, PhD thesis, University of Agder, Norway, 2009.
138. M. Pätzold and C. A. Gutiérrez, "Level-crossing rate and average duration of fades of the envelope of a sum-of-cisoids," in *Proc. IEEE 67th Vehicular Technology Conference, IEEE VTC 2008-Spring*, Marina Bay, Singapore, May 2008, pp. 488–494.
139. B. O. Hogstad and M. Pätzold, "On the stationarity of sum-of-cisoids-based mobile fading channel simulators," in *Proc. IEEE 67th Vehicular Technology Conference, IEEE VTC 2008-Spring*, Marina Bay, Singapore, May 2008, pp. 400–404.
140. C. A. Gutiérrez, A. Meléndez, A. Sandoval, and H. Rodriguez, "On the autocorrelation ergodic properties of sum-of-cisoids Rayleigh fading channel simulators," in *Proc. 17th European Wireless Conference, EW 2011*, Vienna, Austria, 2011, in print.
141. G. Rafiq and M. Pätzold, "Statistical properties of the capacity of multipath fading channels," in *Proc. IEEE International Symposium on Personal, Indoor and Mobile Radio Communications, PIMRC 2009*, Tokyo, Japan, Sep. 2009, pp. 1103–1107.
142. M. Pätzold and D. Kim, "Test procedures and performance assessment of mobile fading channel simulators," in *Proc. 59th IEEE Semiannual Veh. Technol. Conf., VTC 2004-Spring*, Milan, Italy, May 2004, vol. 1, pp. 254–260.
143. M. Pätzold and F. Laue, "The performance of deterministic Rayleigh fading channel simulators with respect to the bit error probability," in *Proc. IEEE 51st Veh. Technol. Conf., VTC2000-Spring*, Tokyo, Japan, May 2000, pp. 1998–2003.
144. Y. Ma and M. Pätzold, "Performance analysis of wideband SOS-based channel simulators with respect to the bit error probability of BPSK OFDM systems with perfect and imperfect CSI," in *Proc. 12th International Symposium on Wireless Personal Multimedia Communications, WPMC 2009*, Sendai, Japan, Sep. 2009.
145. Y. Ma and M. Pätzold, "Performance analysis of wideband sum-of-cisoids-based channel simulators with respect to the bit error probability of DPSK OFDM systems," in *Proc. IEEE 69th Vehicular Technology Conference, IEEE VTC 2009-Spring*, Barcelona, Spain, April 2009.
146. M. Pätzold and F. Laue, "Level-crossing rate and average duration of fades of deterministic simulation models for Rice fading channels," *IEEE Trans. Veh. Technol.*, vol. 48, no. 4, pp. 1121–1129, July 1999.
147. M. Pätzold, U. Killat, and F. Laue, "A deterministic model for a shadowed Rayleigh land mobile radio channel," in *Proc. 5th IEEE Int. Symp. Personal, Indoor and Mobile Radio Commun., PIMRC'94*, The Hague, The Netherlands, Sep. 1994, pp. 1202–1210.
148. M. Pätzold, U. Killat, and F. Laue, "A deterministic digital simulation model for Suzuki processes with application to a shadowed Rayleigh land mobile radio channel," *IEEE Trans. Veh. Technol.*, vol. 45, no. 2, pp. 318–331, May 1996.
149. M. Pätzold, U. Killat, F. Laue, and Y. Li, "A new and optimal method for the derivation of deterministic simulation models for mobile radio channels," in *Proc. IEEE 46th Veh. Technol. Conf., IEEE VTC'96*, Atlanta, Georgia, USA, April/May 1996, pp. 1423–1427.

150. M. Pätzold, B. O. Hogstad, and D. Kim, "A new design concept for high-performance fading channel simulators using set partitioning," *Wireless Personal Communications*, vol. 40, no. 3, pp. 267–279, Feb. 2007.
151. C. A. Gutiérrez and M. Pätzold, "The generalized method of equal areas for the design of sum-of-sinusoids simulators for mobile Rayleigh fading channels with arbitrary Doppler spectra," *Wireless Communications and Mobile Computing*, first published online: July 2011, doi:10.1002/wcm.1154.
152. C. A. Gutiérrez and M. Pätzold, "The Riemann sum method for the design of sum-of-cisoids simulators for Rayleigh fading channels in non-isotropic scattering environments," in *Proc. IEEE Workshop on Mobile Computing and Network Technologies, WMCNT 2009*, St. Petersburg, Russia, Oct. 2009.
153. P. Höher and A. Steingaß, "Modeling and emulation of multipath fading channels using controlled randomness," in *Proc. ITG-Fachtagung Wellenausbreitung bei Funksystemen und Mikrowellensystemen*, Oberpfaffenhofen, Germany, May 1998, pp. 209–220.
154. M. Pätzold, "On the stationarity and ergodicity of fading channel simulators based on Rice's sum-of-sinusoids," *International Journal of Wireless Information Networks (IJWIN)*, vol. 11, no. 2, pp. 63–69, April 2004.
155. G. D. Forney, "Maximum-likelihood sequence estimation for digital sequences in the presence of intersymbol interference," *IEEE Trans. Inform. Theory*, vol. 18, pp. 363–378, May 1972.
156. P. Höher, *Kohärenter Empfang trelliscodierter PSK-Signale auf frequenzselektiven Mobilfunkkanälen — Entzerrung, Decodierung und Kanalparameterschätzung*. Düsseldorf: VDI-Verlag, Fortschritt-Berichte, series 10, no. 147, 1990.
157. M. Pätzold and F. Laue, "Statistical properties of Jakes' fading channel simulator," in *Proc. IEEE 48th Veh. Technol. Conf., IEEE VTC'98*, Ottawa, Ontario, Canada, May 1998, vol. 2, pp. 712–718.
158. P. Dent, G. E. Bottomley, and T. Croft, "Jakes fading model revisited," *Electronics Letters*, vol. 29, no. 13, pp. 1162–1163, June 1993.
159. I. N. Bronstein and K. A. Semendjajew, *Taschenbuch der Mathematik*. Frankfurt/Main: Verlag Harri Deutsch, 25th ed., 1991.
160. E. F. Casas and C. Leung, "A simple digital fading simulator for mobile radio," in *Proc. IEEE Veh. Technol. Conf., IEEE VTC'88*, Sep. 1988, pp. 212–217.
161. E. F. Casas and C. Leung, "A simple digital fading simulator for mobile radio," *IEEE Trans. Veh. Technol.*, vol. 39, no. 3, pp. 205–212, Aug. 1990.
162. R. Fletcher and M. J. D. Powell, "A rapidly convergent descent method for minimization," *Computer Journal*, vol. 6, no. 2, pp. 163–168, 1963.
163. W. R. Braun and U. Dersch, "A physical mobile radio channel model," *IEEE Trans. Veh. Technol.*, vol. 40, no. 2, pp. 472–482, May 1991.
164. U. Dersch and R. J. Rüegg, "Simulations of the time and frequency selective outdoor mobile radio channel," *IEEE Trans. Veh. Technol.*, vol. 42, no. 3, pp. 338–344, Aug. 1993.
165. N. Youssef, T. Munakata, and M. Takeda, "Fade statistics in Nakagami fading environments," in *Proc. IEEE 4th Int. Symp. on Spread Spectrum Techniques&Applications, ISSSTA'96*, Mayence, Germany, Sep. 1996, pp. 1244–1247.
166. G. E. Johnson, "Constructions of particular random processes," *Proc. of the IEEE*, vol. 82, no. 2, pp. 270–285, Feb. 1994.
167. G. Ungerboeck, "Channel coding with multilevel/phase signals," *IEEE Trans. Inform. Theory*, vol. IT-28, pp. 55–67, Jan. 1982.
168. E. Biglieri, D. Divsalar, P. J. McLane, and M. K. Simon, *Introduction to Trellis-Coded Modulation with Applications*. New York: Macmillan Publishing Company, 1991.
169. M. Pätzold, B. O. Hogstad, and N. Youssef, "Modeling, analysis, and simulation of MIMO mobile-to-mobile fading channels," *IEEE Trans. Wireless Commun.*, vol. 7, no. 2, pp. 510–520, Feb. 2008.
170. C. A. Gutiérrez and M. Pätzold, "A generalized method for the design of ergodic sum-of-cisoids simulators for multiple uncorrelated Rayleigh fading channels," in *Proc. 4th International Conference on Signal Processing and Communication Systems, ICSPCS 2010*, Gold Coast, Australia, Dec. 2010.
171. B. O. Hogstad, M. Pätzold, N. Youssef, and D. Kim, "A MIMO mobile-to-mobile channel model: Part II – The simulation model," in *Proc. 16th IEEE Int. Symp. on Personal, Indoor and Mobile Radio Communications, PIMRC 2005*, Berlin, Germany, Sep. 2005, vol. 1, pp. 562–567.
172. H. Zhang, D. Yuan, M. Pätzold, Y. Wu, and V. D. Nguyen, "A novel wideband space-time channel simulator based on the geometrical one-ring model with application in MIMO-OFDM systems," *Wireless Communications and Mobile Computing*, vol. 10, no. 6, pp. 758–771, June 2010.
173. E. O. Brigham, *The Fast Fourier Transform and its Applications*. New Jersey: Prentice-Hall, 1988.

174. N. Blaunstein and Y. Ben-Shimol, "Spectral properties of signal fading and Doppler spectra distribution in urban mobile communication links," *Wireless Communications and Mobile Computing*, vol. 6, no. 1, pp. 113–126, Feb. 2006.
175. R. von Mises, "Über die 'Ganzzahligkeit' der Atomgewichte und verwandte Fragen," *Physikalische Zeitschrift*, vol. 19, pp. 490–500, 1918.
176. A. Abdi, J. A. Barger, and M. Kaveh, "A parametric model for the distribution of the angle of arrival and the associated correlation function and power spectrum at the mobile station," *IEEE Trans. Veh. Technol.*, vol. 51, no. 3, pp. 425–434, May 2002.
177. M. F. Pop and N. C. Beaulieu, "Limitations of sum-of-sinusoids fading channel simulators," *IEEE Trans. on Communications*, vol. 49, no. 4, pp. 699–708, April 2001.
178. C. Loo, "A statistical model for a land mobile satellite link," *IEEE Trans. Veh. Technol.*, vol. 34, no. 3, pp. 122–127, Aug. 1985.
179. C. Loo and N. Secord, "Computer models for fading channels with applications to digital transmission," *IEEE Trans. Veh. Technol.*, vol. 40, no. 4, pp. 700–707, Nov. 1991.
180. E. Lutz, D. Cygan, M. Dippold, F. Dolainsky, and W. Papke, "The land mobile satellite communication channel — Recording, statistics, and channel model," *IEEE Trans. Veh. Technol.*, vol. 40, no. 2, pp. 375–386, May 1991.
181. G. E. Corazza and F. Vatalaro, "A statistical model for land mobile satellite channels and its application to nongeostationary orbit systems," *IEEE Trans. Veh. Technol.*, vol. 43, no. 3, pp. 738–742, Aug. 1994.
182. F. Hansen and F. I. Meno, "Mobile fading — Rayleigh and lognormal superimposed," *IEEE Trans. Veh. Technol.*, vol. 26, no. 4, pp. 332–335, Nov. 1977.
183. A. Krantzik and D. Wolf, "Distribution of the fading-intervals of modified Suzuki processes," in *Signal Processing V: Theories and Applications*, L. Torres, E. Masgrau, and M. A. Lagunas, Eds., Amsterdam, The Netherlands: Elsevier Science Publishers, B.V, 1990, pp. 361–364.
184. M. Pätzold, U. Killat, Y. Li, and F. Laue, "Modeling, analysis, and simulation of nonfrequency-selective mobile radio channels with asymmetrical Doppler power spectral density shapes," *IEEE Trans. Veh. Technol.*, vol. 46, no. 2, pp. 494–507, May 1997.
185. M. Pätzold, U. Killat, F. Laue, and Y. Li, "An efficient deterministic simulation model for land mobile satellite channels," in *Proc. IEEE 46th Veh. Technol. Conf., IEEE VTC'96*, Atlanta, Georgia, USA, April/May 1996, pp. 1028–1032.
186. C. Loo, "Measurements and models of a land mobile satellite channel and their applications to MSK signals," *IEEE Trans. Veh. Technol.*, vol. 35, no. 3, pp. 114–121, Aug. 1987.
187. C. Loo, "Digital transmission through a land mobile satellite channel," *IEEE Trans. Commun.*, vol. 38, no. 5, pp. 693–697, May 1990.
188. B. Vucetic and J. Du, "Channel modeling and simulation in satellite mobile communication systems," *IEEE J. Select. Areas in Commun.*, vol. 10, no. 8, pp. 1209–1218, Oct. 1992.
189. M. J. Miller, B. Vucetic, and L. Berry, Eds., *Satellite Communications: Mobile and Fixed Services*. Boston, MA: Kluwer Academic Publishers, 3rd ed., 1995.
190. B. Vucetic and J. Du, "Channel modeling and simulation in satellite mobile communication systems," in *Proc. Int. Conf. Satell. Mobile Commun.*, Adelaide, Australia, Aug. 1990, pp. 1–6.
191. M. Pätzold, A. Szczepanski, S. Buonomo, and F. Laue, "Modeling and simulation of nonstationary land mobile satellite channels by using extended Suzuki and handover processes," in *Proc. IEEE 51st Veh. Technol. Conf., VTC2000-Spring*, Tokyo, Japan, May 2000, pp. 1787–1792.
192. M. Pätzold, U. Killat, and F. Laue, "Ein erweitertes Suzukimodell für den Satellitenmobilfunkkanal," in *Proc. 40. Internationales Wissenschaftliches Kolloquium*, Technical University Ilmenau, Ilmenau, Germany, Sep. 1995, vol. I, pp. 321–328.
193. D. O. Reudink, "Comparison of radio transmission at X-band frequencies in suburban and urban areas," *IEEE Trans. Ant. Prop.*, vol. 20, no. 4, pp. 470–473, July 1972.
194. D. M. Black and D. O. Reudink, "Some characteristics of mobile radio propagation at 836 MHz in the Philadelphia area," *IEEE Trans. Veh. Technol.*, vol. 21, no. 2, pp. 45–51, May 1972.
195. M. Pätzold, Y. Li, and F. Laue, "A study of a land mobile satellite channel model with asymmetrical Doppler power spectrum and lognormally distributed line-of-sight component," *IEEE Trans. Veh. Technol.*, vol. 47, no. 1, pp. 297–310, Feb. 1998.
196. J. S. Butterworth and E. E. Matt, "The characterization of propagation effects for land mobile satellite services," in *Inter. Conf. on Satellite Systems for Mobile Communications and Navigations*, June 1983, pp. 51–54.

197. R. W. Huck, J. S. Butterworth, and E. E. Matt, "Propagation measurements for land mobile satellite services," in *Proc. IEEE 33rd Veh. Technol. Conf., IEEE VTC'83*, Toronto, Canada, May 1983, vol. 33, pp. 265–268.
198. W. J. Vogel and J. Goldhirsh, "Fade measurements at L-band and UHF in mountainous terrain for land mobile satellite systems," *IEEE Trans. Antennas Propagat.*, vol. 36, no. 1, pp. 104–113, Jan. 1988.
199. W. J. Vogel and J. Goldhirsh, "Mobile satellite system propagation measurements at L-band using MARECS-B2," *IEEE Trans. Antennas Propagat.*, vol. 38, no. 2, pp. 259–264, Feb. 1990.
200. W. J. Vogel and J. Goldhirsh, "Multipath fading at L band for low elevation angle, land mobile satellite scenarios," *IEEE J. Select. Areas Commun.*, vol. 13, no. 2, pp. 197–204, Feb. 1995.
201. C. Loo, "Statistical models for land mobile and fixed satellite communications at Ka band," in *Proc. IEEE 46th Veh. Technol. Conf., IEEE VTC'96*, Atlanta, Georgia, USA, April/May 1996, pp. 1023–1027.
202. C. Loo and J. S. Butterworth, "Land mobile satellite channel measurements and modeling," *Proc. of the IEEE*, vol. 86, no. 7, pp. 1442–1463, July 1998.
203. A. Jahn, "Propagation data and channel model for LMS systems," *ESA Purchase No. 141742*, Final Report, DLR, Institute for Communications Technology, Jan. 1995.
204. S. R. Saunders, C. Tzaras, and B. G. Evans, "Physical-statistical methods for determining state transition probabilities in mobile-satellite channel models," *Int. J. Satell. Commun.*, vol. 19, no. 3, pp. 207–222, 2001.
205. B. Ahmed, S. Buonomo, I. E. Otung, M. H. Aziz, S. Saunders, and B. G. Evans, "Simulation of 20 GHz narrow band mobile propagation data using N-states Markov channel modelling approach," in *Proc. 10th International Conference on Antennas and Propagation*, Edinburgh, UK, April 1997, vol. 2, pp. 48–53.
206. F. Perez-Fontan, M. A. Vazquez-Castro, S. Buonomo, J. P. Poiares-Baptista, and B. Arbesser-Rastburg, "S-band LMS propagation channel behaviour for different environments, degrees of shadowing and elevation angles," *IEEE Trans. Broadcasting*, vol. 44, no. 1, pp. 40–76, March 1998.
207. R. Fletcher, *Practical Methods of Optimization*. New York: Wiley InterScience, 2nd ed., 2000.
208. J. D. Parsons and A. S. Bajwa, "Wideband characterisation of fading mobile radio channels," *Inst. Elec. Eng. Proc.*, vol. 129, no. 2, pp. 95–101, April 1982.
209. J. D. Parsons and J. G. Gardiner, *Mobile Communication Systems*. Glasgow: Blackie & Son, 1989.
210. H. D. Lücke, *Signalübertragung — Grundlagen der digitalen und analogen Nachrichtensysteme*. Berlin: Springer, 1990.
211. B. H. Fleury, *Charakterisierung von Mobil- und Richtfunkkanälen mit schwach stationären Fluktuationen und unkorrelierter Streuung (WSSUS)*, PhD thesis, Swiss Federal Institute of Technology Zurich, Zurich, Switzerland, 1990.
212. ETSI, "Broadband Radio Access Networks (BRAN); HIPERLAN Type 2; System Overview," Technical Report TR 101 683 V1.1.1, ETSI, Feb. 2000.
213. R. L. Burden and J. D. Faires, *Numerical Analysis*. Pacific Grove, CA, USA: Brooks/Cole Publishing, 8th ed., 2004.
214. C.-X. Wang, M. Pätzold, and D. F. Yuan, "Accurate and efficient simulation of multiple uncorrelated Rayleigh fading waveforms," *IEEE Trans. Wireless Commun.*, vol. 6, no. 3, pp. 833–839, March 2007.
215. M. Pätzold, A. Szczepanski, and N. Youssef, "Methods for modelling of specified and measured multipath power delay profiles," *IEEE Trans. Veh. Technol.*, vol. 51, no. 5, pp. 978–988, Sep. 2002.
216. M. Pätzold, U. Killat, and F. Laue, "A new deterministic simulation model for WSSUS multipath fading channels," in *Proc. 2. ITG-Fachtagung Mobile Kommunikation '95*, Neu-Ulm, Germany, Sep. 1995, pp. 301–312.
217. R. S. Thomä and U. Martin, "Richtungsaufgelöste Messung von Mobilfunkkanälen," in *Proc. ITG-Diskussionssitzung Meßverfahren im Mobilfunk*, Günzburg, Germany, March 1999, pp. 34–36.
218. D. Hampicke, A. Richter, A. Schneider, G. Sommerkorn, R. S. Thomä, and U. Trautwein, "Characterization of the directional mobile radio channel in industrial scenarios, based on wide-band propagation measurements," in *Proc. IEEE 50th Veh. Technol. Conf., IEEE VTC'99*, April 1999, pp. 2258–2262.
219. S. Stein, "Fading channel issues in system engineering," *IEEE J. Select. Areas Commun.*, vol. 5, no. 2, pp. 68–89, Feb. 1987.
220. R. Steele and L. Hanzo, Eds., *Mobile Radio Communications*. Chichester, England: John Wiley & Sons, 2nd ed., 1999.
221. S. R. Saunders and A. Aragón-Zavala, *Antennas and Propagation for Wireless Communication Systems*. Chichester, England: John Wiley & Sons, 2nd ed., 2007.
222. A. F. Molisch, *Wireless Communications*. Chichester, England: John Wiley & Sons, 2005.
223. R. Kattenbach, *Charakterisierung zeitvarianter Indoor-Funkkanäle anhand ihrer System- und Korrelationsfunktionen*, PhD thesis, Universität Gesamthochschule Kassel, Kassel, Germany, 1997.

224. J. S. Sadowsky and V. Kafedziski, "On the correlation and scattering functions of the WSSUS channel for mobile communications," *IEEE Trans. Veh. Technol.*, vol. 47, no. 1, pp. 270–282, Feb. 1998.
225. R. Parra-Michel, V. Y. Kontorovitch, and A. G. Orozco-Lugo, "Modelling wide band channels using orthogonalizations," *IEICE Transactions on Electronics*, vol. E85-C, no. 3, pp. 544–551, March 2002.
226. B. H. Fleury and P. E. Leuthold, "Radiowave propagation in mobile communications: An overview of European research," *IEEE Commun. Mag.*, vol. 34, no. 2, pp. 70–81, Feb. 1996.
227. J. C. Liberty and T. S. Rappaport, *Smart Antennas for Wireless Communications: IS-95 and Third Generation CDMA Applications*. Upper Saddle River, New Jersey: Prentice-Hall, 1999.
228. R. B. Ertel, P. Cardieri, K. W. Sowerby, T. S. Rappaport, and J. H. Reed, "Overview of spatial channel models for antenna array communication systems," *IEEE Personal Commun.*, vol. 5, no. 1, pp. 10–22, Feb. 1998.
229. U. Martin, J. Fuhl, I. Gaspard, M. Haardt, A. Kuchar, C. Math, A. F. Molisch, and R. Thomä, "Model scenarios for direction-selective adaptive antennas in cellular mobile communication systems — Scanning the literature," *Wireless Personal Communications, Special Issue on Space Division Multiple Access*, pp. 109–129, Kluwer Academic Publishers, Oct. 1999.
230. K. Yu and B. Ottersten, "Models for MIMO propagation channels: A review," *Wireless Communications and Mobile Computing, Special Issue on Adaptive Antennas and MIMO Systems*, vol. 2, no. 7, pp. 653–666, Nov. 2002.
231. P. Almers, E. Bonnek, A. Burr, N. Czink, M. Debbah, V. Degli-Esposti, H. Hofstetter, P. Kyösti, D. Laurenson, G. Matz, A. F. Molisch, C. Oestges, and H. Özcelik, "Survey of channel and radio propagation models for wireless MIMO systems," *EURASIP Journal on Wireless Communications and Networking*, vol. 2007, doi:10.1155/2007/19070, 2007.
232. T. J. Willink, "Wide-sense stationarity of mobile MIMO radio channels," *IEEE Trans. Veh. Technol.*, vol. 57, no. 2, pp. 704–714, March 2008.
233. A. Gehring, M. Steinbauer, I. Gaspard, and M. Grigat, "Empirical channel stationarity in urban environments," in *Proc. 4th European Personal Mobile Communications Conference, EPMCC 2001*, Feb. 2001.
234. G. Matz, "On non-WSSUS wireless fading channels," *IEEE Trans. Wireless Commun.*, vol. 4, no. 5, pp. 2465–2478, Sep. 2005.
235. E. L. Caples, K. E. Massad, and T. R. Minor, "A UHF channel simulator for digital mobile radio," *IEEE Trans. Veh. Technol.*, vol. 29, no. 2, pp. 281–289, May 1980.
236. D. Berthoumieux and J. M. Pertoldi, "Hardware propagation simulator of the frequency-selective fading channel at 900 MHz," in *Proc. 2nd Nordic Seminar on Land Mobile Radio Communications*, Stockholm, Sweden, 1986, pp. 214–217.
237. L. Ehrman, L. B. Bates, J. F. Eschle, and J. M. Kates, "Real-time software simulation of the HF radio channel," *IEEE Trans. Commun.*, vol. 30, no. 8, pp. 1809–1817, Aug. 1982.
238. R. Schwarze, *Ein Systemvorschlag zur Verkehrsinformationsübertragung mittels Rundfunksatelliten*, PhD thesis, Universität-Gesamthochschule-Paderborn, Paderborn, Germany, 1990.
239. U. Lambrette, S. Fechtel, and H. Meyr, "A frequency domain variable data rate frequency hopping channel model for the mobile radio channel," in *Proc. IEEE 47th Veh. Technol. Conf., IEEE VTC'97*, Phoenix, Arizona, USA, May 1997, vol. 3, pp. 2218–2222.
240. M. Pätzold, F. Laue, and U. Killat, "A frequency hopping Rayleigh fading channel simulator with given correlation properties," in *Proc. IEEE Int. Workshop on Intelligent Signal Processing and Communication Systems, ISPACS'97*, Kuala Lumpur, Malaysia, Nov. 1997, pp. S8.1.1–S8.1.6.
241. D. C. Cox, "Delay Doppler characteristics of multipath propagation at 910 MHz in a suburban mobile radio environment," *IEEE Trans. Antennas Propagat.*, vol. 20, no. 5, pp. 625–635, Sep. 1972.
242. D. Nielson, "Microwave propagation measurements for mobile digital radio application," *IEEE Trans. Veh. Technol.*, vol. 27, no. 3, pp. 117–131, Aug. 1978.
243. A. S. Bajwa and J. D. Parsons, "Small-area characterisation of UHF urban and suburban mobile radio propagation," *Inst. Elec. Eng. Proc.*, vol. 129, no. 2, pp. 102–109, April 1982.
244. J. B. Andersen, T. S. Rappaport, and S. Yoshida, "Propagation measurements and models for wireless communications channels," *IEEE Commun. Mag.*, vol. 33, no. 1, pp. 42–49, Jan. 1995.
245. M. Werner, *Modellierung und Bewertung von Mobilfunkkanälen*, PhD thesis, Technical Faculty of the University Erlangen–Nuremberg, Erlangen, Germany, 1991.
246. R. Kattenbach and H. Früchting, "Calculation of system and correlation functions for WSSUS channels from wideband measurements," *FREQUENZ*, vol. 49, no. 3/4, pp. 42–47, 1995.

247. A. G. Siamarou, "Wideband propagation measurements and channel implications for indoor broadband wireless local area networks at the 60 GHz band," *Wireless Personal Communications*, vol. 27, no. 1, pp. 89–98, Hingham, MA, USA: Kluwer Academic Publishers, Oct. 2003.
248. L. M. Correia and R. Prasad, "An overview of wireless broadband communications," *IEEE Communications Magazine*, vol. 35, no. 1, pp. 28–33, Jan. 1997.
249. J. D. Parsons, D. A. Demery, and A. M. D. Turkmani, "Sounding techniques for wideband mobile radio channels: a review," *Communications, Speech and Vision, IEE Proceedings I*, vol. 138, no. 5, pp. 437–446, Oct. 1991.
250. D. Laurenson and P. Grant, "A review of radio channel sounding techniques," in *Proc. European Signal Processing Conference, EUSIPCO 2006*, Florence, Italy, Sep. 2006.
251. U. Martin, *Ausbreitung in Mobilfunkkanälen: Beiträge zum Entwurf von Meßgeräten und zur Echoschätzung*, PhD thesis, University Erlangen–Nuremberg, Erlangen, Germany, 1994.
252. U. Martin, "Ein System zur Messung der Eigenschaften von Mobilfunkkanälen und ein Verfahren zur Nachverarbeitung der Meßdaten," *FREQUENZ*, vol. 46, no. 7/8, pp. 178–188, 1992.
253. G. Kadel and R. W. Lorenz, "Breitbandige Ausbreitungsmessungen zur Charakterisierung des Funkkanals beim GSM-System," *FREQUENZ*, vol. 45, no. 7/8, pp. 158–163, 1991.
254. M. Göller and K.D. Masur, "Ergebnisse von Funkkanalmessungen im 900 MHz Bereich auf Neubaustrecken der Deutschen Bundesbahn," *Nachrichtentechnik–Elektronik*, vol. 42, no. 4, pp. 146–149, 1992.
255. M. Göller and K. D. Masur, "Ergebnisse von Funkkanalmessungen im 900 MHz Bereich auf Neubaustrecken der Deutschen Bundesbahn," *Nachrichtentechnik–Elektronik*, vol. 42, no. 5, pp. 206–210, 1992.
256. R. Thomä, D. Hampicke, A. Richter, G. Sommerkorn, A. Schneider, U. Trautwein, and W. Wirnitzer, "Identification of time-variant directional mobile radio channels," *IEEE Trans. on Instrumentation and Measurement*, vol. 49, pp. 357–364, April 2000.
257. U. Trautwein, C. Schneider, G. Sommerkorn, D. Hampicke, R. Thomä, and W. Wirnitzer, "Measurement data for propagation modeling and wireless system evaluation," Technical Report COST 273 TD(03)021, EURO-COST, Jan. 2003.
258. L. Hentilä, P. Kyösti, J. Ylitalo, X. Zhao, J. Meinilä, and J.-P. Nuutinen, "Experimental characterization of multidimensional parameters at 2.45 GHz and 5.25 GHz indoor channels," in *Proc. 8th International Symposium on Wireless Personal Multimedia Communications, WPMC 2005*, Aalborg, Denmark, Sep. 2005, pp. 254–258.
259. T. Felhauer, P. W. Baier, W. König, and W. Mohr, "Optimized Wideband System for Unbiasd Mobile Radio Channel Sounding with Periodic Spread Spectrum Signals (Special Issue on Spread Spectrum Techniques and Applications)," *The Institute of Electronics, Information and Communication Engineers (IEICE)*, vol. 76, no. 8, pp. 1016–1029, Aug. 1993.
260. T. Felhauer, *Optimale erwartungstreue Algorithmen zur hochauflösenden Kanalschätzung mit Bandspreizsignalformen*. Düsseldorf: VDI-Verlag, Fortschritt-Berichte, series 10, no. 278, 1994.
261. E. Zollinger, *Eigenschaften von Funkübertragungsstrecken in Gebäuden*, PhD thesis, Swiss Federal Institute of Technology Zurich, Zurich, Switzerland, 1993.
262. R. Heddergott and P. Truffer, "Comparison of high resolution channel parameter measurements with ray tracing simulations in a multipath environment," in *Proc. 3rd European Personal Mobile Communications Conference*, EPMCC'99, Paris, France, March 1999, pp. 167–172.
263. J. Kivinen, T. O. Korhonen, P. Aikio, R. Gruber, P. Vainikainen, and S.-G. Häggman, "Wideband radio channel measurement system at 2 GHz," *IEEE Trans. on Instrumentation and Measurement*, vol. 48, no. 1, pp. 39–44, Feb. 1999.
264. J. Ø. Nielsen, J. B. Andersen, P. C. F. Eggers, G. F. Pedersen, K. Olesen, E. H. Sørensen, and H. Suda, "Measurements of indoor 16×32 wideband MIMO channels at 5.8 GHz," in *Proc. IEEE 8th International Symposium on Spread Spectrum Techniques and Applications, ISSSTA 2004*, Sydney, Australia, Aug./Sep. 2004, pp. 864–868.
265. S. Salous, P. Filippidis, R. Lewenz, I. Hawkins, N. Razavi-Ghods, and M. Abdallah, "Parallel receiver channel sounder for spatial and MIMO characterisation of the mobile radio channel," *IEE Proceedings on Commun.*, vol. 152, no. 6, pp. 912–918, Dec. 2005.
266. R. J. C. Bultitude, "Estimating frequency correlation functions from propagation measurements on fading radio channels: A critical review," *IEEE J. Select. Areas Commun.*, vol. 20, no. 6, pp. 1133–1143, Aug. 2002.
267. V. Erceg, D. G. Michelson, S. S. Ghassemzadeh, L. J. Greenstein, A. J. Rustako Jr., P. B. Guerlain, M. K. Dennison, R. S. Roman, D. J. Barnickel, S.C. Wang, and R.R. Miller, "A model for the multipath delay profile of fixed wireless channels," *IEEE J. Select. Areas Commun.*, vol. SAC-17, no. 3, pp. 399–410, March 1999.

268. L. Yuanqing, "A theoretical formulation for the distribution density of multipath delay spread in a land mobile radio environment," *IEEE Trans. Veh. Technol.*, vol. VT-43, pp. 379–388, 1994.
269. A. Doufexi, S. Armour, P. Karlsson, A. Nix, and D. Bull, "A Comparison of HIPERLAN/2 and IEEE 802.11a," *IEEE Communications Magazine*, vol. 40, no. 5, pp. 172–180, May 2002.
270. I. E. Telatar, "Capacity of multi-antenna Gaussian channels," *European Trans. Telecommun. Related Technol.*, vol. 10, no. 6, pp. 585–595, 1999.
271. G. J. Foschini, "Layered space-time architecture for wireless communication in a fading environment when using multi-element antennas," *Bell Labs Technical Journal 1996*, vol. 1, no. 2, pp. 41–59, 1996.
272. D.-S. Shiu, G. J. Foschini, M. J. Gans, and J. M. Kahn, "Fading correlation and its effect on the capacity of multielement antenna systems," *IEEE Trans. Commun.*, vol. 48, no. 3, pp. 502–513, March 2000.
273. A. Abdi and M. Kaveh, "A space-time correlation model for multielement antenna systems in mobile fading channels," *IEEE J. Select. Areas Commun.*, vol. 20, no. 3, pp. 550–560, April 2002.
274. U. Martin, "A directional radio channel model for densely built-up urban areas," in *Proc. 2nd European Personal Mobile Radio Conference/3rd ITG-Fachtagung Mobile Kommunikation*, Bonn, Germany, 1997, pp. 237–244.
275. L. M. Correia, Ed., *Wireless Flexible Personalized Communications*. Chichester, England: John Wiley & Sons, 2001.
276. M. Pätzold and B. O. Hogstad, "A wideband space-time MIMO channel simulator based on the geometrical one-ring model," in *Proc. 64th IEEE Semiannual Vehicular Technology Conference, IEEE VTC 2006-Fall*, Montreal, Canada, Sep. 2006, pp. 1–6.
277. T. Fulghum and K. Molnar, "The Jakes fading model incorporating angular spread for a disk of scatterers," in *Proc. 48th IEEE Vehicular Technology Conference, IEEE VTC 1998*, Ottawa, Canada, 1998, pp. 489–493.
278. M. Pätzold, "On the stationarity and ergodicity of fading channel simulators basing on Rice's sum-of-sinusoids," in *Proc. 14th IEEE Int. Symp. on Personal, Indoor and Mobile Radio Communications, IEEE PIMRC 2003*, Beijing, China, Sep. 2003, pp. 1521–1525.
279. C. E. Shannon, "A mathematical theory of communication," *Bell Syst. Tech. J.*, vol. 27, pp. 379–423, July 1948.
280. C. E. Shannon, "A mathematical theory of communication," *Bell Syst. Tech. J.*, vol. 27, pp. 623–656, Oct. 1948.
281. M. Luccini, A. Shami, and S. Primak, "Cross-layer optimisation of network performance over multiple-input multiple-output wireless mobile channels," *IET Communications Journal*, vol. 4, no. 6, pp. 683–696, April 2010.
282. H. Hartenstein and K. Laberteaux, Eds., *VANET Vehicular Applications and Inter-Networking Technologies*. Chichester: John Wiley & Sons, 2009.
283. J. Yin, T. ElBatt, G. Yeung, B. Ryu, S. Habermas, H. Krishnan, and T. Talty, "Performance evaluation of safety applications over DSRC vehicular ad hoc networks," in *Proc. 1st ACM Workshop on Vehicular Ad Hoc Networks, VANET'04*, Philadelphia, PA, USA, Oct. 2004, pp. 1–9.
284. F. Kojima, H. Harada, and M. Fujise, "Inter-vehicle communication network with an autonomous relay access scheme," *IEICE Trans. Commun.*, vol. E84-B, no. 3, pp. 566–575, March 2001.
285. A. G. Zajić and G. L. Stüber, "Space-time correlated mobile-to-mobile channels: Modelling and simulation," *IEEE Trans. Veh. Technol.*, vol. 57, no. 2, pp. 715–726, March 2008.
286. G. J. Byers and F. Takawira, "Spatially and temporally correlated MIMO channels: modelling and capacity analysis," *IEEE Trans. Veh. Technol.*, vol. 53, no. 3, pp. 634–643, May 2004.
287. Z. Tang and A. S. Mohan, "A correlated indoor MIMO channel model," in *Canadian Conference on Electrical and Computer Engineering 2003, IEEE CCECE 2003*, Venice, Italy, May 2003, vol. 3, pp. 1889–1892.
288. I. Z. Kovacs, P. C. F. Eggers, K. Olesen, and L. G. Petersen, "Investigations of outdoor-to-indoor mobile-to-mobile radio communication channels," in *Proc. IEEE 56th Veh. Technol. Conf., IEEE VTC 2002-Fall*, Vancouver, BC, Canada, Sep. 2002, vol. 1, pp. 430–434.
289. J. Salo, H. M. El-Sallabi, and P. Vainikainen, "Impact of double-Rayleigh fading on system performance," in *Proc. 1st International Symposium on Wireless Pervasive Computing, ISWPC'06*, Phuket, Thailand, Jan. 2006.
290. M. Pätzold and B. O. Hogstad, "Design and performance of MIMO channel simulators derived from the two-ring scattering model," in *Proc. 14th IST Mobile&Communications Summit, IST 2005*, Dresden, Germany, June 2005, paper no. 121.
291. C. A. Gutiérrez and M. Pätzold, "Sum-of-sinusoids-based simulation of flat fading wireless propagation channels under non-isotropic scattering conditions," in *Proc. 50th IEEE Global Communications Conference, IEEE GLOBECOM 2007*, Washington DC, USA, Nov. 2007, pp. 3842–3846.
292. K. I. Pedersen, P. E. Mogensen, and B. H. Fleury, "Power azimuth spectrum in outdoor environments," *IEEE Electronic Letters*, vol. 33, no. 18, pp. 1583–1584, Aug. 1997.

293. J. Salz and J. H. Winters, "Effect of fading correlation on adaptive arrays in digital mobile radio," *IEEE Trans. Veh. Technol.*, vol. 43, no. 4, pp. 1049–1057, Nov. 1994.
294. H. S. Rad, S. Gazor, and K. Shahtalebi, "Spatial-temporal-frequency decomposition for 3D MIMO microcellular environments," in *Canadian Conference on Electrical and Computer Engineering*, Niagara Falls, Canada, May 2004, vol. 3, pp. 1229–1232.
295. G. J. Byers and F. Takawira, "The influence of spatial and temporal correlation on the capacity of MIMO channels," in *Wireless Communications and Networking Conference, WCNC 2003*, March 2003, pp. 359–364.
296. H. Bölcskei, D. Gesbert, and A. J. Paulraj, "On the capacity of OFDM-based spatial multiplexing systems," *IEEE Trans. Commun.*, vol. 50, no. 2, pp. 225–234, Feb. 2002.
297. J. C. Liberti and T. S. Rappaport, "A geometrically based model for line-of-sight multipath radio channels," in *Proc. IEEE 46th Veh. Technol. Conf., IEEE VTC'96*, Atlanta, GA, USA, May 1996, vol. 2, pp. 844–848.
298. M. Pätzold and B. O. Hogstad, "A wideband MIMO channel model derived from the geometrical elliptical scattering model," *Wireless Communications and Mobile Computing*, vol. 8, pp. 597–605, May 2008.
299. B. O. Hogstad, M. Pätzold, A. Chopra, D. Kim, and K. B. Yeom, "A wideband MIMO channel simulation model based on the geometrical elliptical scattering model," in *Proc. 15th Meeting of the Wireless World Research Forum (WWRF)*, Paris, France, Dec. 2005.
300. B. O. Hogstad, M. Pätzold, and A. Chopra, "A study on the capacity of narrow- and wideband MIMO channel models," in *Proc. 15th IST Mobile&Communications Summit, IST 2006*, Myconos, Greece, June 2006.
301. V. Erceg et al., "Channel models for fixed wireless applications," Technical Report IEEE 802.16a-03/01, IEEE Task Group, June 2003.
302. D. Gesbert and J. Akhtar, "Breaking the barriers of Shannon's capacity: An overview of MIMO wireless systems," *Telektronikk*, vol. 98, no. 1, pp. 53–64, 2002.
303. D. Gesbert, M. Shafi, D.-S. Shiu, P. J. Smith, and A. Naguib, "From theory to practice: an overview of MIMO space-time coded wireless systems," *IEEE J. Select. Areas Commun.*, vol. 21, no. 3, pp. 281–302, 2003.
304. K. Yu, *Modeling of multiple-input multiple-output radio propagation channels*, PhD thesis, School of Electrical Engineering, Royal Institute of Technology, Stockholm, Sweden, 2002.
305. J. W. Wallace and M. A. Jensen, "Modeling the indoor MIMO wireless channel," *IEEE Trans. Antennas Propag.*, vol. 50, no. 5, pp. 591–599, May 2002.
306. T. Svantesson, "A physical MIMO radio channel model for multi-element multi-polarized antenna systems," in *IEEE Vehicular Technology Conference, IEEE VTC Fall*, Atlantic City, NY, USA, Oct. 2001, vol. 2, pp. 1083–1087.
307. M. Stege, J. Jelitto, M. Bronzel, and G. Fettweis, "A multiple-input multiple-output channel model for simulation of Tx- and Rx-diversity wireless systems," in *Proc. IEEE 52nd Veh. Technol. Conf., VTC 2000-Fall*, Boston, MA, USA, Sep. 2000, pp. 833–839.
308. P. C. F. Eggers, "Generation of base station DOA distributions by Jacobi transformation of scattering areas," *Electron. Lett.*, vol. 34, no. 1, pp. 24–26, Jan. 1998.
309. R. J. Piechocki, G. V. Tsoulos, and J. P. McGeehan, "Simple general formula for PDF of angle of arrival in large cell operational environments," *Electron. Lett.*, vol. 34, no. 18, pp. 1784–1785, Sep. 1998.
310. D. R. Van Rheeden and S. C. Gupta, "A geometric model for fading correlation in multipath radio channels," in *Proc. International Conference on Communications, ICC 98*, Atlanta, GA, June 1998, vol. 3, pp. 1655–1659.
311. R. B. Ertel and J. H. Reed, "Angle and time of arrival statistics for circular and elliptical scattering models," *IEEE J. Select. Areas Commun.*, vol. 17, no. 11, pp. 1829–1840, Nov. 1999.
312. P. Petrus, J. H. Reed, and T. S. Rappaport, "Geometrical-based statistical macrocell channel model for mobile environments," *IEEE Trans. Commun.*, vol. 50, no. 3, pp. 495–502, March 2002.
313. L. Jiang and S. Y. Tan, "Simple geometrical-based AOA model for mobile communication systems," *Electron. Lett.*, vol. 40, no. 19, pp. 1203–1205, Sep. 2004.
314. A. Y. Olenko, K. T. Wong, and E. H.-O. Ng, "Analytically derived TOA-DOA statistics of uplink/downlink wireless multipaths arisen from scatterers on a hollow-disc around the mobile," *IEEE Antennas Wireless Propag. Lett.*, vol. 2, pp. 345–348, 2003.
315. A. Y. Olenko, K. T. Wong, and M. Abdulla, "Analytically derived TOA-DOA distributions of uplink/downlink wireless-cellular multipaths arisen from scatterers with an inverted-parabolic spatial distribution around the mobile," *IEEE Signal Processing Lett.*, vol. 12, no. 7, pp. 516–519, July 2005.
316. N. M. Khan, M. T. Simsim, and P. B. Rapajic, "A generalized model for the spatial characteristics of the cellular mobile channel," *IEEE Trans. Veh. Technol.*, vol. 57, no. 1, pp. 22–37, Jan. 2008.

317. R. Janaswamy, "Angle and time of arrival statistics for the Gaussian scatter density model," *IEEE Trans. Wireless Commun.*, vol. 1, no. 3, pp. 488–497, July 2002.
318. A. Andrade and D. Covarrubias, "Radio channel spatial propagation model for mobile 3G in smart antenna systems," *IEICE Trans. Commun.*, vol. E86-B, no. 1, pp. 213–220, Jan. 2003.
319. D. D. Bevan, V. T. Ermolayev, A. G. Flaksman, and I. M. Averin, "Gaussian channel model for mobile multipath environment," *EURASIP Journal on Applied Signal Processing*, vol. 2004, no. 9, pp. 1321–1329, 2004.
320. J. Laurila, A. F. Molisch, and E. Bonek, "Influence of the scatterer distribution on power delay profiles and azimuthal power spectra of mobile radio channels," in *Proc. IEEE 5th International Symposium on Spread Spectrum Techniques and Applications, 1998*, Sun City, South Africa, Sep. 1998, vol. 1, pp. 267–271.
321. K. T. Wong, Y. I. Wu, and M. Abdulla, "Landmobile radiowave multipaths' DOA-distribution: Assessing geometric models by the open literature's empirical datasets," *IEEE Trans. Antennas Propag.*, vol. 58, no. 3, pp. 946–958, March 2010.
322. S. S. Mahmoud, F. S. Al-Qahtani, Z. M. Hussain, and A. Gopalakrishnan, "Spatial and temporal statistics for the geometrical-based hyperbolic macrocell channel model," *Digital Signal Processing, Elsevier*, vol. 18, no. 2, pp. 151–167, 2008.
323. F. S. Al-Qahtani and Z. M. Hussain, "Spatial correlation in wireless space-time MIMO channels," in *Proc. 5th Australasian Telecommunication Networks and Applications Conference, ATNAC 2007*, Christchurch, New Zealand, Dec. 2007, pp. 358–363.
324. F. Vatalaro and A. Forcella, "Doppler spectrum in mobile-to-mobile communications in the presence of three-dimensional multipath scattering," *IEEE Trans. Veh. Technol.*, vol. 46, no. 1, pp. 213–219, Feb. 1997.
325. J. Maurer, T. Fügen, T. Schäfer, and W. Wiesbeck, "A new inter-vehicle communications (IVC) channel model," in *60th IEEE Veh. Technol. Conf., IEEE VTC 2004-Fall*, Los Angeles, California, USA, Sep. 2004, vol. 1, pp. 9–13.
326. A. Kato, K. Sato, M. Fujise, and S. Kawakami, "Propagation characteristics of 60-GHz millimeter waves for ITS inter-vehicle communications," *IEICE Trans. Commun.*, vol. E84-B, no. 9, pp. 2530–2539, Sep. 2001.
327. J. Karedal, F. Tufvesson, N. Czink, A. Paier, C. Dumard, T. Zemen, C. Mecklenbräuker, and A. Molisch, "A geometry-based stochastic MIMO model for vehicle-to-vehicle communications," *IEEE Trans. Wireless Commun.*, vol. 8, no. 7, pp. 3646–3657, July 2009.
328. A. P. G. Ariza, *Modelling and Experimental Analysis of Frequency Dependent MIMO Channels*, PhD thesis, Polytechnical University of Valencia, Valencia, Spain, July 2009.
329. D. Umansky and M. Pätzold, "Design of measurement-based stochastic wideband MIMO channel simulators," in *Proc. IEEE Global Communications Conference, IEEE GLOBECOM 2009*, Honolulu, Hawaii, USA, Nov./Dec. 2009.
330. D. Umansky, *Measurement-Based Channel Simulation Models for Mobile Communication Systems*, PhD thesis, University of Agder, Norway, March 2010.
331. J. Karedal, *Measurement-Based Modeling of Wireless Propagation Channels — MIMO and UWB*, PhD thesis, Lund University, Lund, Sweden, 2009.
332. J. Karedal, F. Tufvesson, N. Czink, A. Paier, C. Dumard, T. Zemen, C. F. Mecklenbräuker, and A. F. Molisch, "Measurement-based modeling of vehicle-to-vehicle MIMO channels," in *Proc. IEEE International Conference on Communications, IEEE ICC'09*, Dresden, Germany, June 2009, pp. 3470–3475.
333. J. P. Kermoal, L. Schumacher, K. I. Pedersen, P. E. Mogensen, and F. Frederiksen, "A stochastic MIMO radio channel model with experimental validation," *IEEE J. Select. Areas Commun.*, vol. 20, no. 6, pp. 1211–1226, Aug. 2002.
334. W. Weichselberger, M. Herdin, H. Özcelik, and E. Bonek, "A stochastic MIMO channel model with joint correlation of both link ends," *IEEE Trans. Wireless Commun.*, vol. 5, no. 1, pp. 90–100, Jan. 2006.
335. A. M. Sayeed, "Deconstructing multiantenna fading channels," *IEEE Trans. Signal Processing*, vol. 50, no. 10, pp. 2563–2579, Oct. 2002.
336. H. Özcelik, N. Czink, and E. Bonek, "What makes a good MIMO channel model?," in *Proc. IEEE 61st Veh. Technol. Conf., IEEE VTC'05-Spring*, Stockholm, Sweden, May/June 2005, pp. 156–160.
337. V. Kontorovich, S. Primak, A. Alcocer-Ochoa, and R. Parra-Michel, "MIMO channel orthogonalisations applying universal eigenbasis," *IET Signal Processing*, vol. 2, no. 2, pp. 87–96, June 2008.
338. L. M. Correia, Ed., *Mobile Broadband Multimedia Networks: Techniques, Models and Tools for 4G*. Amsterdam, The Netherlands: Elsevier Science Publishers, B.V, 2006.
339. 3GPP-3GPP2, "Spatial channel model for multiple input multiple output MIMO simulations," Technical Report 25.996, Version 6.1.0, Release 6, 3rd Generation Partnership Project (3GPP), Sep. 2003.

340. P. Kyösti et al., "WINNER II Channel Models: Part I Channel Models," IST-4-27756 WINNER II Deliverable D1.1.2 V1.1, Sep. 2007.
341. V. Erceg et al., "TGn channel models," Technical Report IEEE P802.11, Wireless LANs, Garden Grove, California, USA, June 2004.
342. V. Erceg et al., "Channel models for fixed wireless applications," Technical Report IEEE 802.16.3c-01/29r1, IEEE Task Group, Feb. 2001.
343. P. Kyösti et al., "WINNER II Channel Models: Part II Radio Channel Measurement and Analysis Results," IST-4-27756 WINNER II Deliverable D1.1.2 V1.0, Sep. 2007.
344. V. Erceg et al., "Channel models for fixed wireless applications," Technical Report IEEE 802.16.3c-01/29r4, IEEE Task Group, July 2001.
345. A. Forenza and R. W. Heath Jr., "Impact of antenna geometry on MIMO communication in indoor clustered channels," in *Proc. IEEE Antennas and Propagation Society International Symposium 2004*, Monterey, California, USA, June 2004, vol. 2, pp. 1700–1703.
346. X. Li and Z.-P. Nie, "Effect of array orientation on performance of MIMO wireless channels," *IEEE Antennas and Wireless Propagation Letters*, vol. 3, pp. 368–371, April 2004.
347. P. Almers, F. Tufvesson, P. Karlsson, and A. F. Molisch, "The effect of horizontal array orientation on MIMO channel capacity," in *Proc. 57th IEEE Semiannual Vehicular Technology Conference, VTC-Spring 2003*, Jeju, Korea, April 2003, vol. 1, pp. 34–38.
348. M. M. Sohul, "Impact of antenna array geometry on the capacity of MIMO communication system," in *Proc. International Conference on Electrical and Computer Engineering, ICECE'06*, Dhaka, Bangladesh, Dec. 2006, pp. 80–83.
349. M. Pätzold and R. García, "Design and performance of fast channel simulators for Rayleigh fading channels," in *Proc. 3rd European Personal Mobile Communications Conference*, EPMCC'99, Paris, France, March 1999, pp. 280–285.
350. R. Zurmühl and S. Falk, *Matrizen und ihre Anwendungen — 1. Grundlagen*. Berlin: Springer, 7th ed., 1997.
351. R. A. Horn and C. R. Johnson, *Matrix Analysis*. New York: Cambridge University Press, reprint ed., 1990.
352. C. Großmann and J. Terno, *Numerik der Optimierung*. Stuttgart, Germany: Teubner, 2nd ed., 1997.
353. S. S. Rao, *Engineering Optimization: Theory and Practice*. New York, Wiley InterScience, 3rd ed., 1996.
354. R. Häb, *Kohärenter Empfang bei Datenübertragung über nichtfrequenzselektive Schwundkanäle*, PhD thesis, RWTH University Aachen, Aachen, Germany, 1988.
355. T. Jämsä, T. Poutanen, and J. Meinilä, "Implementation techniques of broadband radio channel simulators," in *Proc. IEEE 53rd Veh. Technol. Conf., IEEE VTC 2001-Spring*, Rhodes, Greece, May 2001, vol. 1, pp. 433–437.
356. P. Kyösti and T. Jämsä, "Complexity comparison of MIMO channel modelling methods," in *Proc. 4th IEEE International Symposium on Wireless Communication Systems, ISWCS 2007*, Trondheim, Norway, Oct. 2007, pp. 219–223.
357. R. Parra-Michel, V. Y. Kontorovitch, A. G. Orozco-Lugo, and M. Lara, "Computational complexity of narrow band and wide band channel simulators," in *Proc. 58th IEEE Veh. Technol. Conf., IEEE VTC 2003-Fall*, Orlando, FL, USA, Oct. 2003, vol. 1, pp. 143–148.
358. A. Alimohammad, S. F. Fard, B. F. Cockburn, and C. Schlegel, "An improved SOS-based fading channel emulator," in *Proc. 66th IEEE Veh. Technol. Conf., IEEE VTC 2007-Fall*, Baltimore, Maryland, USA, Sep./Oct. 2007, pp. 931–935.
359. A. Alimohammad, S. F. Fard, B. F. Cockburn, and C. Schlegel, "An accurate and compact Rayleigh and Rician fading channel simulator," in *Proc. IEEE 67th Veh. Technol. Conf., IEEE VTC 2008-Spring*, Marina Bay, Singapore, May 2008, pp. 409–413.
360. A. Alimohammad and B. F. Cockburn, "Modeling and hardware implementation aspects of fading channel simulators," *IEEE Trans. Veh. Technol.*, vol. 57, no. 4, pp. 2055–2069, July 2008.
361. A. Alimohammad, S. F. Fard, B. F. Cockburn, and C. Schlegel, "A compact single-FPGA fading-channel simulator," *IEEE Trans. Circ. Syst. — II: Express Briefs*, vol. 55, no. 1, pp. 84–88, Jan. 2008.
362. P. J. Cullen, P. C. Fannin, and A. Garvey, "Real-time simulation of randomly time-variant linear systems: The mobile radio channel," *IEEE Trans. Instrum. Meas.*, vol. 43, no. 4, pp. 583–591, Aug. 1994.
363. A. K. Salkintzis, "Implementation of a digital wide-band mobile channel simulator," *IEEE Trans. Broadcast.*, vol. 45, no. 1, pp. 122–128, March 1999.
364. M. Kahrs and C. Zimmer, "Digital signal processing in a real-time propagation simulator," *IEEE Trans. Instrum. Meas.*, vol. 55, no. 1, pp. 197–205, Feb. 2006.

365. C. Komninakis, "A fast and accurate Rayleigh fading simulator," in *Proc. IEEE Global Communications Conference 2003, IEEE GLOBECOM 2003*, San Francisco, CA, USA, Dec. 2003, pp. 3306–3310.
366. F. Kaltenberger, T. Zemen, and C. W. Ueberhuber, "Low-complexity geometry-based MIMO channel simulation," *EURASIP Journal on Advances in Signal Processing*, vol. 2007, doi:10.1155/2007/95281, 2007.
367. F. Kaltenberger, G. Steinböck, G. Humer, and T. Zemen, "Low-complexity geometry based MIMO channel emulation," in *Proc. First European Conference on Antennas and Propagation, EuCAP 2006*, Nice, France, Nov. 2006, pp. 1–8.
368. J.-M. Conrat and P. Pajusco, "A versatile propagation channel simulator for MIMO link level simulation," *EURASIP Journal on Wireless Communications and Networking*, vol. 2007, doi:10.1155/2007/80194, 2007.
369. Y. B. Li and Y. L. Guan, "Modified Jakes model for simulating multiple uncorrelated fading waveforms," in *Proc. IEEE International Conference on Communications, ICC 2000*, New Orleans, LA, USA, June 2000, vol. 1, pp. 46–49.
370. Y. Li and X. Huang, "The simulation of independent Rayleigh faders," *IEEE Trans. Commun.*, vol. 50, no. 9, pp. 1503–1514, Sep. 2002.
371. Y. Gan and Q. Xu, "An improved SoS method for generating multiple uncorrelated Rayleigh fading waveforms," *IEEE Commun. Lett.*, vol. 14, no. 7, pp. 641–643, July 2010.
372. C.-X. Wang, D. Yuan, H. H. Chen, and W. Xu, "An improved deterministic SoS channel simulator for multiple uncorrelated Rayleigh fading channels," *IEEE Trans. Wireless Commun.*, vol. 7, no. 9, pp. 3307–3311, Sep. 2008.
373. M. F. Ibrahim and J. D. Parsons, "Signal strength prediction in built-up areas. Part 1: Median signal strength," *Proc. IEE*, vol. 130, no. 5, pp. 377–384, Aug. 1983.
374. M. Gudmundson, "Correlation model for shadow fading in mobile radio systems," *Electron. Lett.*, vol. 27, no. 23, pp. 2145–2146, Nov. 1991.
375. G. P. Pollini, "Trends in handover design," *IEEE Commun. Magazine*, vol. 34, no. 3, pp. 82–90, March 1996.
376. N. Zhang and J. M. Holtzman, "Analysis of handoff algorithms using both absolute and relative measurements," *IEEE Trans. Veh. Technol.*, vol. 45, no. 1, pp. 174–179, Feb. 1996.
377. K. Yamamoto, A. Kusuda, and S. Yoshida, "Impact of shadowing correlation on coverage of multihop cellular systems," in *Proc. IEEE International Conference on Communications, ICC'06*, Istanbul, Turkey, June 2006, vol. 10, pp. 4538–4542.
378. L. Hanzo (II) and R. Tafazolli, "The effects of shadow-fading on QoS-aware routing and admission control protocols designed for multi-hop MANETs," *Wireless Communications and Mobile Computing*, vol. 11, no. 1, pp. 1–22, Jan. 2011.
379. M. Pätzold and V. D. Nguyen, "A spatial simulation model for shadow fading processes in mobile radio channels," in *Proc. 15th IEEE Int. Symp. on Personal, Indoor and Mobile Radio Communications, PIMRC 2004*, Barcelona, Spain, Sep. 2004, vol. 3, pp. 1832–1838.
380. M. Pätzold and K. Yang, "An exact solution for the level-crossing rate of shadow fading processes modelled by using the sum-of-sinusoids principle," *Wireless Personal Communications*, vol. 52, no. 1, pp. 57–68, Jan. 2010.
381. D. Giancristofaro, "Correlation model for shadow fading in mobile radio channels," *Electron. Lett.*, vol. 32, no. 11, pp. 958–959, May 1996.
382. P. E. Mogensen, P. Eggers, C. Jensen, and J. B. Andersen, "Urban area radio propagation measurements at 955 and 1845 MHz for small and micro cells," in *IEEE Global Commun. Conf.*, Phoenix, AZ, Dec. 1991, vol. 2, pp. 1297–1302.
383. M. J. Marsan, G. C. Hess, and S. S. Gilbert, "Shadow variability in an urban land mobile radio environment at 900 MHz," *Electron. Lett.*, vol. 26, pp. 646–648, May 1990.
384. E. Perahia and D. C. Cox, "Shadow fading correlation between uplink and downlink," in *IEEE VTC2001-Spring*, May 2001, pp. 308–312.
385. H. Kim and Y. Han, "Enhanced correlated shadowing generation in channel simulation," *IEEE Commun. Letters*, vol. 6, no. 7, pp. 279–281, July 2002.
386. ETSI, "Universal Mobile Telecommunications System (UMTS); Selection procedures for the choice of radio transmission technologies of the UMTS (UMTS 30.03 version 3.2.0)," Technical Report TR 101 112 V3.2.0 (1998-04), ETSI, April 1998.
387. Y. Zhang, J. Zhang, D. Dong, X. Nie, G. Liu, and P. Zhang, "A novel spatial autocorrelation model of shadow fading in urban macro environments," in *Proc. IEEE Global Communications Conference, IEEE GLOBECOM 2008*, New Orleans, LA, USA, Nov./Dec. 2008.

388. X. Cai and G. B. Giannakis, "A two-dimensional channel simulation model for shadowing processes," *IEEE Trans. Veh. Technol.*, vol. 52, no. 6, pp. 1558–1567, Nov. 2003.
389. A. Abdi and M. Kaveh, "On the utility of gamma PDF in modeling shadow fading (slow fading)," in *Proc. IEEE 49th Veh. Technol. Conf., IEEE VTC 1999*, Houston, Texas, USA, May 1999, vol. 3, pp. 2308–2312.
390. N. L. Johnson, S. Kotz, and N. Balakrishnan, *Continuous Univariate Distributions*, vol. 1. New York: John Wiley & Sons, 2nd ed., 1994.
391. C.-X. Wang, M. Pätzold, and Q. Yao, "Stochastic modeling and simulation of frequency hopping wideband fading channels," in *Proc. 57th Int. Semiannual Veh. Technol. Conf., IEEE VTC 2003-Spring*, Jeju, Korea, April 2003, vol. 2, pp. 803–807.
392. C.-X. Wang, M. Pätzold, and Q. Yao, "Stochastic modeling and simulation of frequency-correlated wideband fading channels," *IEEE Trans. Veh. Technol.*, vol. 56, no. 3, pp. 1050–1063, May 2007.

Index

Q

R